Power System Analysis Operation and Control

FOURTH EDITION

Abhijit Chakrabarti
Professor (HAG) Retd.
Department of Electrical Engineering
Indian Institute of Engineering Science & Technology
Shibpur, Howrah, West Bengal
Former Vice Chancellor
Jadavpur University, Kolkata
Former Vice Chairman
West Bengal State Council of Higher Education
Government of West Bengal, Kolkata

Sunita Halder
Professor
Department of Electrical Engineering
Jadavpur University
Kolkata, West Bengal

PHI Learning Private Limited
Delhi-110092
2026

In fond memory of ***Shri Asoke K. Ghosh*** *(October 1942 – February 2024), Founder Chairman and Managing Director of PHI Learning, whose vision endlessly inspires.*

The Legacy Continues....

Published by Pushpita Ghosh, PHI Learning Private Limited, Rimjhim House, 111, Patparganj Industrial Estate, Delhi-110092 and Printed by Syndicate Binders, A-20, Hosiery Complex, Noida, Phase-II Extension, Noida-201305 (N.C.R. Delhi).

₹1125.00

POWER SYSTEM ANALYSIS OPERATION AND CONTROL, 4th ed.
Abhijit Chakrabarti and Sunita Halder

ISBN-978-93-91818-00-5 (Print Book)
ISBN-978-93-91818-01-2 (e-Book)

The export rights of the book are vested solely with the publisher.

To

Our Spouses

Contents

Preface

During the COVID period the civilization suffered enormously across the globe. The education sector has been affected badly and there has been a negative impact on the growth of higher education in most of the countries including our own country. Notably, the learning pattern has been modified among the students in the higher education sector. In the field of Engineering and Technology, there has been a remarkable shift in the process of learning and there has been a change in the attitude of the students. This changing attitude toward the conventional learning process results in preferences for concise text material and syllabus-oriented approaches to textbooks.

With a visible change in the learning process, the authors of this book have revised the text presented here in the fourth edition of the book. The chapters in this book are reorganized, revised, and modified to cope up with the increasing needs of the UG curriculum for a syllabus-oriented textbook. This present edition of the book is suitable for prescribed syllabi of AICTE, GATE, UPSC, Universities and Institutes in our country. Attempts have been made so that this textbook finds its place in the syllabi prescribed by foreign universities across the globe. In suitable places, illustrations and worked-out problems have been included. It is expected that this fourth edition will be more beneficial to UG students in Electrical Engineering.

The authors acknowledge immense assistance from Dipanjan Bose of IIEST Shibpur in revising the book. The authors acknowledge the assistance from their colleagues and family members. Any constructive criticism related to this fourth edition will be appreciated.

Abhijit Chakrabarti
Sunita Halder

Preface to the First Edition

The fundamental aim of this text is to present a number of engineering and economic matters in power system planning operation and control in a comprehensive way. The topics substantiated by a number of illustrations and computer programs describe analytical methods of power system and their operation and control. To understand the text, some acquaintance with the basic concepts in power system as well as advanced calculus methods is needed.

The chapters have been methodically arranged, starting with the basic aspects of power engineering problems. In each chapter, the relevant methods have been dealt with the help of suitable computer-based examples. In a few sections, while dealing with operational problems, optimization methods have been preferred as they can be used without extensive mathematical proofs and are useful in solving practical problems.

The text begins with an introductory discussion on common operating problems and basic aspects of power system operation, including structures of power system, power system representation, and representation of power system elements. Different conventional models are briefly described and analytical treatments are presented to show the modeling concepts of power apparatus like synchronous generator, transformer, transmission lines, motors, etc. Matrix operational methods applicable to power network also get proper attention. Exhaustive analytical treatments are presented for the conventional load flow methods. All the conventional methods of optimization are explained with the help of suitable examples. Some practical and applicational aspects of basic philosophy of ALFC also form part of the discussion. Fundamental aspects of reactive power control and voltage problems in transmission network followed by modern developments in this field including advanced treatments have been detailed. Computerized methods for the analysis of faulted power system have been furnished as well.

The text is self-contained and thorough. It is intended for a one-semester course for postgraduate students as well as a one-year course for senior undergraduate students in electrical engineering. Practising engineers and researchers will also find the book suitable for their use.

The authors acknowledge the constant encouragement they received from the respected Vice-Chancellors Prof. N.R. Banerjea of Bengal Engineering and Science University, Shibpur and Prof. A.N. Basu of Jadavpur University for this project. They also express their gratitude to the respective Deans, Registrars, and Heads of the Departments of both these universities for offering all facilities in course of preparation of the manuscript.

The authors cordially invite any constructive criticism of or comment about the book.

Abhijit Chakrabarti
Sunita Halder

CHAPTER 1

Introduction

1.1 STRUCTURE OF A POWER SYSTEM

Electricity is the only form of energy used in the industrial, domestic, commercial, and transportation sectors. It is a coveted form of energy, since it can be generated in bulk and transmitted economically over long distances. Electric power system deals with the *generation*, *transmission* and *distribution* of electric energy associated with the unique feature of control of the flow or demand of energy at desired nodes throughout the power network. Figure 1.1 represents the fundamental structure of a power network where generators produce electric energy, *transformers* transform this energy into one voltage level from another voltage level, and *transmission lines* wheel the power from the generating stations to the load centres for the final distribution of electrical energy to different loads. *Tie-lines* interconnect one system with the neighbouring electric system belonging to the same grid. The *circuit breakers* isolate a faulty part of the network (the fault being sensed by the relays) while static/rotary *compensators* may be used for voltage control at load or remote buses. Conventionally, loads are represented in a lumped or composite form.

The best location of a generating station being at a place very close to electrical load centre (i.e., the region where the major energy demand exists), the practical location of the primary conventional energy sources does not necessarily coincide with the urban centres. The location of a power plant is frequently governed by its closeness to the energy resource and transportation facility of the fuel as well as availability of nearest load centre. Environmental aspects are also key factors in determining the site of the plant. Mostly, a generating plant consists of generating units complete with necessary accessories. Control elements like different valves, exciters, regulators etc., also step up transformers, and instrument transformers along with breakers are intended in the station switchyard for the transmission of power and protection of the system. Sources of input to the generating system are conventionally *fossil fuels* (e.g., coal, oil and gas), hydro-source and nuclear fuel. However, non-conventional sources such as wind power, solar energy, tidal power, geothermal power etc. are also being used for *stand-alone systems*.

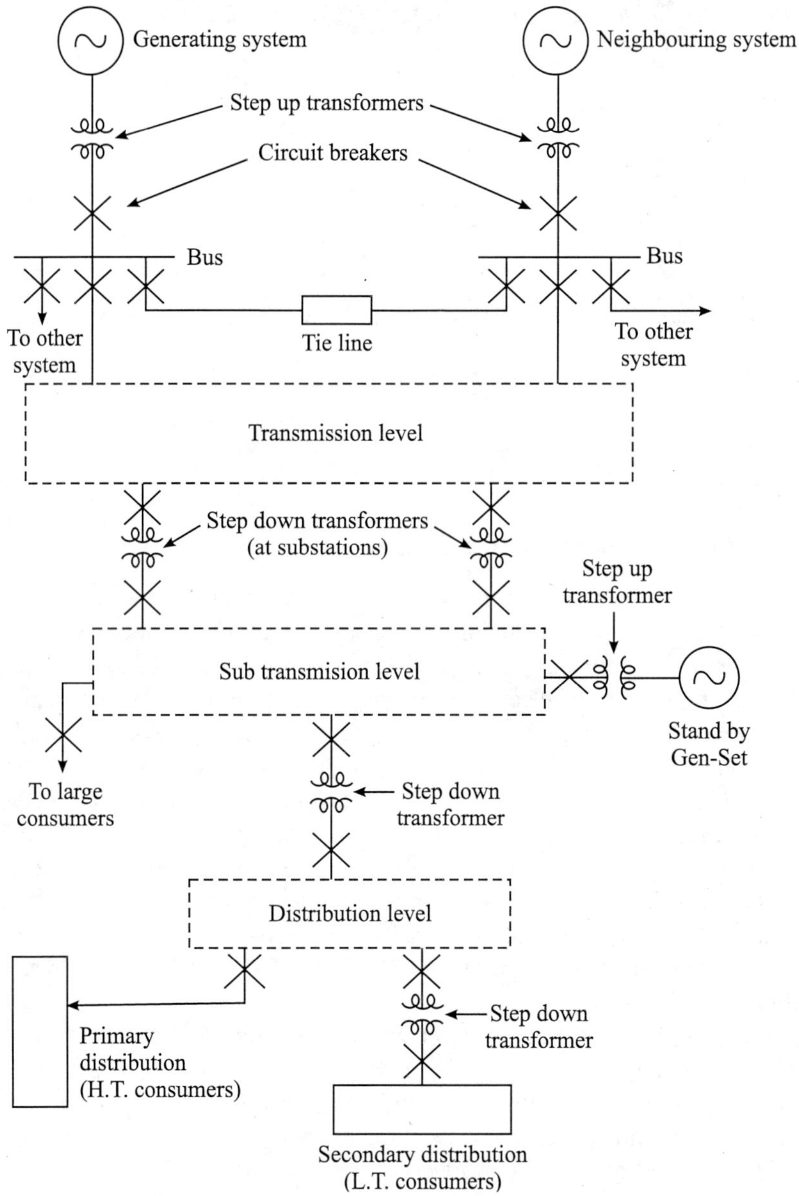

Figure 1.1 Basic structure of a power system.

An electric power system, even a small one, usually constitutes an electric network of vast complexity. The diversity of the system magnitude being great, there is no general rule regarding the structure of the system that applies to any power system. However, any power system could be categorised by a combination of generation, transmission and distribution networks. After generation, transmission plays a vital role in transporting power from the generating station to load centres. Transmission of power is usually done at HV / EHV / UHV range due to the known fact that it reduces the power loss in the line as well as improves stability. The common transmission voltages across the globe are 33 kV/66 kV/114 kV/132 kV/138 kV/161 kV/220 kV/230 kV/345 kV/400 kV/500 kV in the HV and EHV ranges and 765 kV/800 kV/1100 kV/1500 kV in the UHV ranges in most parts of the world while the generation voltages have commonly been 6 kV/11 kV/12.47 kV/13.2 kV/13.8 kV/15 kV/16 kV/22 kV (all are line-to-line voltage).

In sub-transmission level, the circuits distribute electric power to a number of distribution centres in a certain geographical region at a voltage level that typically varies between 23 kV and 138 kV, the most common grades in India being 33 kV/66 kV/110 kV/120 kV/132 kV. The sub-transmission circuits may also receive electric power directly from any generator bus. Larger customers are mostly served by sub-transmission level circuits. In small power systems, the sub-transmission level may coincide with the distribution level.

The distribution level consists of the distribution circuits in the overall region of distribution. The larger consumers, i.e. high tension (H.T.) have been termed as *primary distributors* while low tension (L.T.) consumers are the *secondary distributors*. The consumers consuming energy between 3 kV and 23 kV are H.T. consumers while the consumers in the category of 110 V–400/440 V lie in the class of secondary or L.T. consumers. The increasing demand on the electrical energy has led not only to diversification of the generation, transmission and distribution network but also raised the points of proper utilisation and reliability of the electric power. This, in turn, has necessitated the pooling of larger number of power systems into a common grid and consequently insisting for proper scheduling of generation and demand. It also turned out that the incorporation of a large number of systems into a common grid makes the operation of the entire system very sensitive to the operating conditions. Thus in addition to the study of *power system operation*, the knowledge of *power system control* is very much required in order to run the system economically and to maintain a continuous balance between generation and varying load demand. In one way, the problems of dynamic and transient stability, steady state stability, voltage and frequency regulation, power optimisation need to be properly analysed and on the other hand, a methodology of overall system control is to be devised. Digital computers are the most effective tool for the analysis of a power system.

1.2 POWER SYSTEM REPRESENTATION

As a complete diagram of a practical power system presenting all the three phases (generation, transmission and distribution) is too complicated, it is a normal practice to represent a power system by means of simple systems for each component resulting in single-line diagram, as shown in Figure 1.2.

Any particular component may or may not be shown in the diagram depending on the information required in a system study, e.g. circuit breakers need not be shown in a load flow analysis diagram but are to be shown for a protection study. Different generator and transformer connections are indicated by proper symbols. Equivalent circuits of power circuit components can be represented in the diagrams.

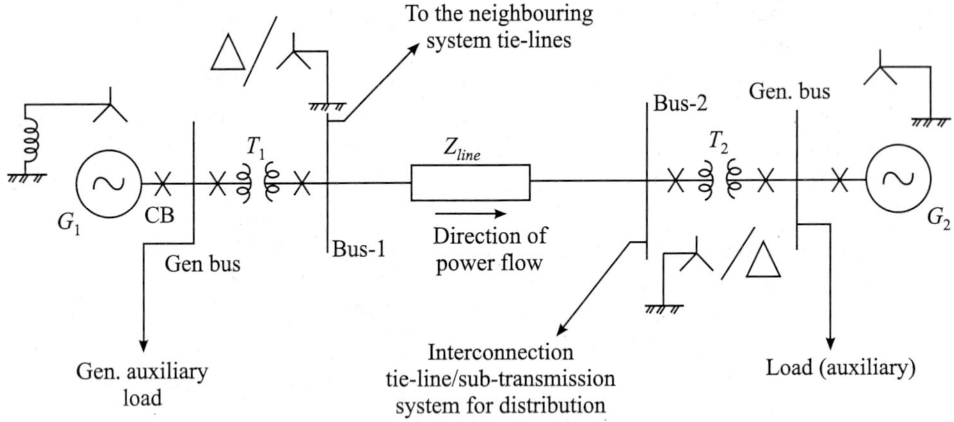

Figure 1.2 Single-line representation of a simple two-bus system [G_1 and G_2 are generators and T_1 and T_2 are transfomers].

1.2.1 Per Unit (P.U.) Representation

The *per unit* (p.u.) representation of an electrical quantity is its ratio to the chosen base quantity of same dimensions. Therefore, an electrical quantity represented in per unit form is basically "normalised" with respect to a selected base value. In single phase circuits the base quantities of current (I_b) and impedance (Z_b) can be represented as

$$I_b = \frac{kVA_b}{kV_b}; Z_b = \frac{V_b}{I_b} = \frac{kV_b^2}{kVA_b} \times 10^3 = \frac{kV_b^2}{MVA_b}$$

For three phase circuits, the expressions of base current and impedance can be modified as

$$I_b = \frac{kVA_b(3\varnothing)}{\sqrt{3}kV_b(L-L)}; Z_b = \frac{V_b\ (L-L)}{\sqrt{3}I_b} = \frac{kV_b^2\ (L-L)}{kVA_b\ (3\varnothing)} \times 10^3 = \frac{kV_b^2\ (L-L)}{MVA_b\ (3\varnothing)}$$

In per unit (p.u.) system, thus the base quantities for single phase circuit can be represented as

$$V_{pu} = \frac{V}{V_b}; I_{pu} = \frac{I}{I_b}; Z_{pu} = \frac{Z}{Z_b} = \left(\frac{MVA_b}{kV_b^2}\right)Z$$

In three phase representation, the p.u. impedance can be represented as

$$Z_{pu} = \frac{Z}{Z_b}$$

$$= \left(\frac{MVA_b(3\varnothing)}{kV_b^2(L-L)}\right)Z \tag{1.1}$$

In case, it is necessary to change p.u. values on one base to p.u. values on different base, then this can be done with the help of following equation

$$Z_{pu,new} = Z_{pu,\ old} \times \frac{MVA_{b,new}}{MVA_{b,\ old}} \times \left[\frac{kV_{b,\ old}}{kV_{b,\ new}}\right]^2 \tag{1.2}$$

1.3 POWER SYSTEM AT NORMAL OPERATING STATE

A power system operates in a normal state if the following conditions are satisfied:

(a) There is a perfect balance between power generation and demand; consequently, the load flow equations are satisfied.

(b) The frequency, f, is constant throughout the system.

(c) The bus voltage magnitude $|V_i|$ is within the prescribed limit, i.e.

$$|V_i|_{min} \le |V_i| \le |V_i|_{max}$$

(This is required as all the power equipment and apparatuses are supposed to be operated at a specified voltage.)

(d) No power system component is to be overloaded.

However, the load is mostly a constantly varying parameter and in order to meet this slow change of the load demand, the normal operating state drifts with time (the load is mostly met by optimal generation scheduling). Change in frequency causes change in the speed of the drives in the consumer's plant. Further, it is necessary to maintain network frequency constant so that the power stations run satisfactorily in parallel, the various motors operating on the system run at the desired speed and other devices function properly. However, the most important reason for keeping frequency of the electrical system constant is that its constancy indicates power balance of the total system.

Overloading of any power system component results in higher temperature of operation and the component is likely to be damaged. System stability, given by the *maximum power* that can be transmitted, also indicates the power system operating at normal state. This *steady state stability limit* (also known as *static transmission capacity*) is given by

$$P_{ij_{max}} = \frac{|V_i||V_j|}{X_{ij}} \tag{1.3}$$

In an attempt to transmit more power than this limit, synchronism is lost and the transmission system collapses. For short lines (less than 100 km), the thermal limit capability fixes the loading of line whereas for medium or long line, the static transmission capacity becomes the limiting factor.

EXERCISES

1. Draw a generalised structure of a power system and discuss it.
2. What do you understand by single line diagram of a power system? Explain with a suitable example.
3. Draw the single line diagram of a two bus power system. What is the usual range of transmission voltage in India?
4. What do you mean by 'normal operating state' of a power system?
5. What is 'steady state stability limit 'of transmission line? Derive an expression for it.
6. Discuss about the ranges operating voltage used in generation, transmission and distribution level of power system.

CHAPTER 2

Electrical Energy Generation

2.1 INTRODUCTION

Electrical energy in bulk quantities is generated mostly using fossil fuels (i.e., coal and oil), hydro power and nuclear energy. Renewable sources of energy like wind, solar radiation, tidal phenomenon, geothermal power etc. can provide localised sources of electrical power generation. Generated electrical energy is mainly of alternating current type employing *synchronous generator* (*alternator*). From the generating station, the electrical energy is transmitted to load centres through transmission lines at very high and extra high voltages. At load centres, the voltage of incoming electrical energy is stepped down and fed to the sub-transmission system and subsequently to the distribution system.

Electrical energy is a very important factor in industrial and economic development of any nation. In the year 2021, the total electrical energy generation in world was close to 29,000 TWh, out of which around 20–22% was the loss in the process of transmission and distribution.

India is considered as a country with emerging economy. Increasing GDP is driving the demand for additional electrical energy. India consumes 4% of global energy and the country's energy demand has grown over average of 9.83% per annum in 2020. In the year 2020, the installed electrical power generation capacity of India is 3,88,134 MW. About 55% of the electrical energy consumed in India comes from thermal power plants while nearly 12% electrical energy comes from hydroelectric power generation. Nearly 2% of the electrical energy comes from the nuclear power plants, 7% of electrical energy from natural gas while availability of around 24% electrical energy is from renewable energy resources. However, to keep pace with growth India needs to increase the electrical energy generation by at least five to six times in coming years. At present the position of India is third in electrical energy generation across the world.

2.1.1 Electrical Energy and Power

The fundamental unit of *electrical energy* is watt-second (W-s) where

$$1 \text{ W-s} = 1 \text{ Joule}$$

However, in commercial use, the electrical energy is measured in Killowatt-hour (kWh).

$$1 \text{ kWh} = 3.6 \times 10^6 \text{ J}$$

Power is defined as the work done per unit time i.e., power is the time rate of change of energy. In any electrical circuit, the power at any instant is given by

$$p = \frac{dU}{dt} = \frac{dW}{dt}$$

where U represents the work and W represents the energy. SI unit of power is watt (W).

Also, $1 \text{ W} = 1 \text{ Joule per second}$.

Horse Power is also a unit of work and capacity. Power ratings of electric motors are commonly expressed in terms of Horse Power (HP).

$$1 \text{ HP} = 746 \text{ W}.$$

2.2 THERMAL POWER GENERATION

An alternator, i.e., the synchronous generator is used for electric power generation from fossil fuel source (i.e., coal) in thermal power generating plants. Modern synchronous generators are efficient electrical machines. Steam turbine acts as a prime mover to rotate the synchronous generator at a speed of 3000 r.p.m. in India and some other countries to produce electricity at a frequency of 50 Hz. Steam boilers generate steam through the process of combustion of coal and this steam is injected into the turbine with a high velocity. It then rotates the turbine which in turn rotates the alternator. Steam turbines have efficiency of around 85% to 90% and can be built as of 1000 MW capacity. The boiler efficiency is to some extent lower and the overall efficiency of the thermal plant is of the order of 20% to 45%.

Generation Process

Figure 2.1 represents a schematic diagram of a simplified thermal power generation plant. In modern thermal power plants the fuel (coal) handling is almost automatic. In the process of generating electric power from coal, the raw pieces of coal is transported to the plant by railway wagon which are tripped into *hopper* in *coal handling plant* (CHP) at the entry of the plant. In the hoper, coal pieces of specified size are allowed to be collected at the base of the hopper by using a *vibrator screen.* A *magnetic separator* is used to segregate the ferrous elements mixed in some of the raw coal pieces collected at the base of the hopper and the raw combustible coal pieces are carried to the *crusher* by employing conveyer belts. In the crusher house, the coal pieces are crushed first to granular form and then to powdered form to produce pulverized coal. The pulverized fuel is carried over to the *bunker* (top) of the boiler and the fuel is channelized to the hearth of the boiler from there. Pulverization helps in complete combustion of the fuel. Preheated air from preheater and traces of liquid fuel (oil) is supplied to the burning chamber (hearth) of the boiler in proper amount for complete burning of coal.

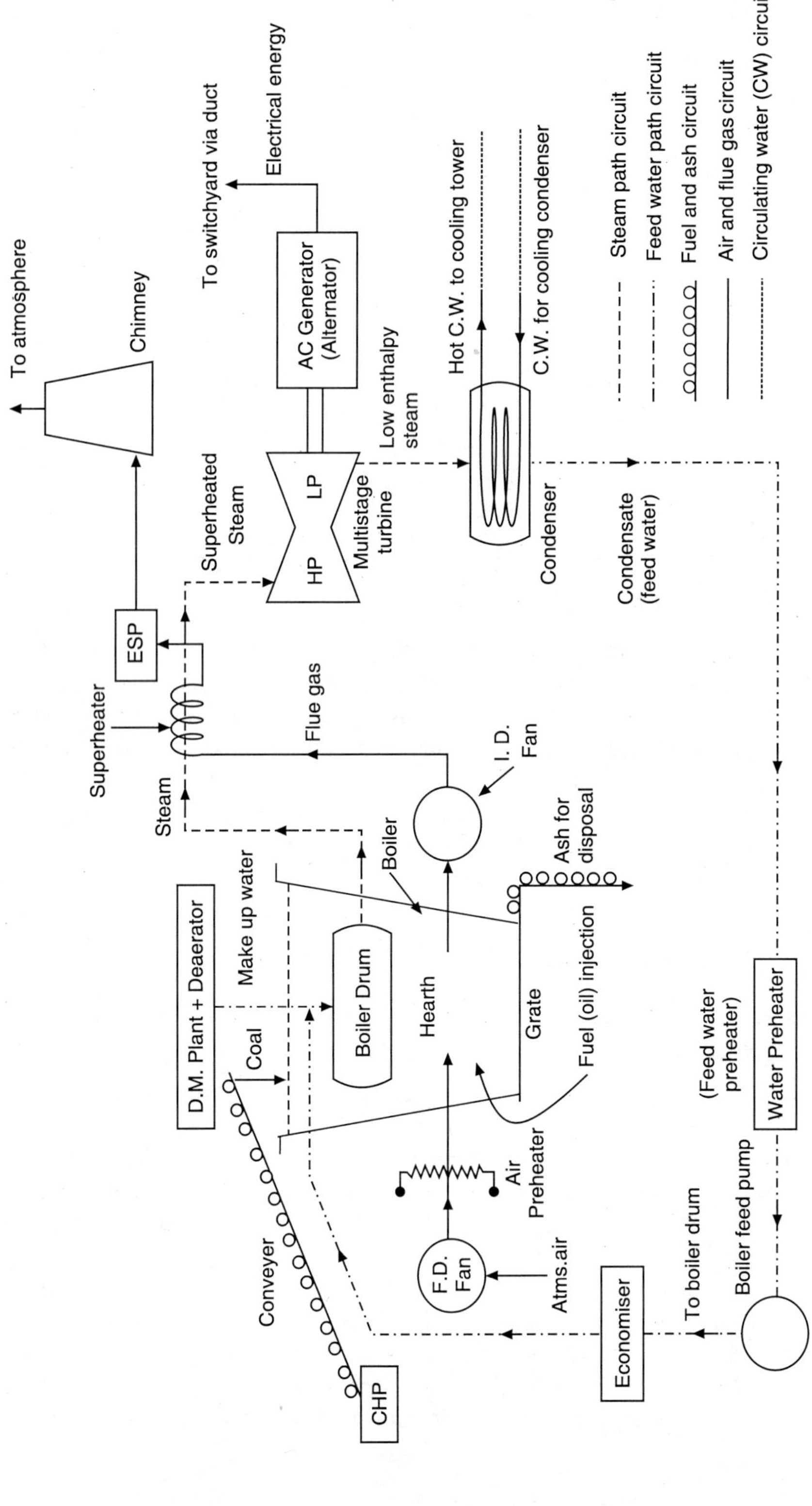

Figure 2.1 Schematic diagram of a thermal power plant.

The entire burning process is automatically regulated to optimize the combustion. The preheated furnace input air is blown using a motorized electric fan, known as *Forced Draft* (FD) *fan* while the flue gas generated after combustion of coal at the hearth of the furnace is induced to pass to the precipitator (for removing pollutants in the flue gas) known as ESP and the *chimney* by another *Induced Draft* (ID) *motorized fan*. This fan is necessary because the air preheater, economizer and precipitators constrict the flow of the exhaust flue gas. Motors as large as 1500–1800 HP, are needed to drive Induced Draught/Draft fans.

The generated heat at around 400°C in the hearth of the furnace is utilized to produce steam from the water kept in boiler drum at the upper section of the boiler. The boiler drum water is a demineralised water (of pH value 7.0) produced by a *demineralisation* (D.M.) *plant*. Raw water from a natural or artificial reservoir is treated in the D.M. plant and once pH value 7.0 is obtained for the treated water, it is injected in the boiler drum as make up or top up water. Such water quality eliminates the risk of scale formation in the water tubes of the steam boiler and subsequent boiler tube leaks are prevented. The boiler drum water is also de-aerated to remove oxygen from the feedwater to reduce corrosion in metal parts of boiler. The generated steam in the boiler is a *high enthalpy* steam and it is passed through the superheater when the heat of the flue gas is exchanged with the steam. *Superheater* is a heat exchanger and the steam, after passing through the superheater, is converted to a higher enthalpy steam at a temperature of 550 to 660°C.

The main purpose of creating high enthalpy steam is to use maximum internal energy of the steam at the *turbine* when the steam gets compressed to develop high pressure jets while passing through suitable nozzles at the entry of the turbine. The high pressure jet strikes the blades of the steam turbine at a high velocity and the turbine rotates at the designated speed. The turbines are usually multistage type and the steam jets go on shedding its enthalpy value once it passes through successive stages (high pressure and low pressure) and transfers the heat energy to kinetic energy at the rotor of the turbine. The electric generator is coupled mechanically with the turbine and it rotates at the same speed of the turbine to produce electricity, provided dc excitation is available at the field of the ac generator, following principle of synchronous generator.

Once the steam losses its enthalpy to a considerable amount, it is passed to the *condenser* where external cooling water (C.W.) helps the low enthalpy steam to be converted into hot water through a process of heat exchange. The cooling water absorbs the heat of the condensate steam and gets heated. This heated external water is treated in a cooling tower so that it gets cooler and is recycled to extract heat from the condenser steam. The condensate, the output hot water of the condenser, is pumped back to the boiler drum through *feed water preheaters* for recycling employing a boiler feed pump (B.F.P.). Use of multiple boiler feed pumps are mandatory as loss of a pump may cause the boiler drum devoid of water and subsequent vacuum in it causing major cracks in the boiler drum. Multiple presence of B.F.P. will not allow the drum to run without water even if the main B.F.P. is at fault.

The electric power generated from the alternator is taken through the bus bar in bus duct to the switchyard generator-transformer for transmission.

2.2.1 Principal Flow Circuits

The flow circuits of a thermal power plant can be classified into four main circuits, such as:

(a) Fuel and ash circuit
(b) Air and the flue gas circuit
(c) Steam and feed water circuit
(d) Cooling water circuit

These sub-circuits are marked in Figure 2.1.

2.2.2 Principal Component of a Thermal Power Plant and their Use

Coal handling plant (CHP): As stated earlier, the coal handling plant uses a wagon tippler, hopper, vibrator, magnetic separator and crusher for collecting the proper coal as fuel and for pulverizing it to gain better combustion at the boiler hearth. Pulverization also improves thermal efficiency of the plant, renders better ash disposal, easy control of fuel supply, no standby loss, provides lower labour cost and increased capacity per unit furnace volume.

Boiler: A thermal power plant uses a steam boiler to produce steam by burning pulverized coal at its hearth with the help of preheated dry air and a little quantity of inflammable oil. Such type of boiler is basically a water tube boiler where the capacity of boiler is expressed in B.T.U/hr. The heat transfer process that takes place in the boiler is a steady flow process for which the heat transferred is equal to the change of enthalpy of the fluid. The overall boiler efficiency at any operating time is the ratio of the transferred heat to the energy supplied by the fuel.

A grate is a sectional metallic structure designed to support the fuel in the bottom part of the boiler furnace and to permit passage of combustion air through openings to the fuel. The fuel is fed using the stoker mechanism. Stoker renders cheaper grade of fuel to have higher efficiency, greater operational flexibility and better maintenance and operation.

Air preheater: Air preheater usually employs either convection or the regeneration principle for heat transmission by using the extracted heat from furnace exhaust in order to preheat the air for successful burning of pulverized fuel.

Draught fans: The draught system consists of two fans, the forced draft (FD) fan and the induced draft (ID) fan. The FD fan supplies preheated air to the boiler hearth. The ID fan creates and induced draft in order to extract the flue gas from the boiler furnace which is produced due to combination of the coal in the hearth of the boiler.

Superheater: Boiler exhaust, the flue gas, is used to raise the enthalpy of steam in the steam circuit in the superheater through the process of heat exchange. Superheated steam is basically the steam at a fixed pressure but with much higher temperature. It reduces the possibility of formation of water droplet in the steam circuit as the droplets may corrode the turbine blades.

Water reheater: A reheater is basically a heat exchanger that injects heat energy to the condensate to raise its temperature in order to increase the thermal efficiency of the plant.

Economiser: The economiser is used to extract heat from the flue gas for heating the feed water of the boiler. The physical location of the economiser is at the last stage of the flue gas path. Economiser increases thermal efficiency of the boiler.

Turbine: Turbine is used to rotate the alternator and conversion of heat energy to mechanical energy take place within this device. Normally high pressure (H.P.) and low pressure (L.P.)

stages of turbine are used. In the H.P. stage the high enthalpy steam expands in the stationary nozzles and attains a higher velocity. There are rows of fixed blades and moving blades in the turbine. The steam pressure in gradually reduced while the steam jet passes through these blades in H.P. and L.P. stages thus transferring the heat energy to mechanical energy and as an effect to this the turbine rotates at a designated speed. Commercial turbines use series combination of impulse and reaction type for better efficiency of the turbine. Most high pressure turbines use input steam at about 560–600°C. The low enthalpy steam goes to condenser after its passage through the L.P. stage of the turbine.

Condenser: Condenser is an important part of thermal power plant. It reduces the enthalpy of the steam coming out of the L.P. stage of the turbine and form warm water (condensate) and this warm water is pumped back to the boiler drum after reheating in order to increase the thermal efficiency of the plant. The condensate is formed by using external cooling water (C.W.) The cooling water extracts heat for the steam and gets reheated. The heated C.W. is taken to the cooling tower (C.T.) for its cooling down by using the process of evaporation and the cold circulating water is pumped to the external surface of the condenser again.

Demineralizing (D.M.) plant and deaerator: The makeup water to the boiler drum is fed from the demineralizing plant where water is demineralized so that the pH value of 7 is maintained. This water is then made free of oxygen in the deaerator and fed to the boiler drum. Demineralized water does not corrode boiler tubes as it cannot form scales in the inner surface of the water tubes of the boiler.

Alternator: In steam power plants, several generating units are used to enhance the total capacity of the plant. A synchronous generator is coupled to the turbine mechanically in order to produce the electrical energy from the mechanical energy. Normally these alternators have dc excitation at the field winding and the ac generated voltage in 3 phase, 50 Hz having the magnitude of the line voltage being in the range of 3 kV to 25 kV depending upon the manufacturer. The capacity is normally expressed in terms of MW and in some present day installations, even 1000 MW generators are in use.

Ash disposal: The ash generated at the grate (chained moving base) of the boiler is collected from the bottom of the boiler for physical disposal.

Electrostatic precipitator (E.S.P.) and chimney: The electrostatic precipitator is used to ionise and remove the pollutants from the flues gas due to environmental restrictions before the flue gas enters the chimney. Draft is created in the chimney for exit of the flue gas.

Bus duct and switchyard: The electric power generated from the alternator is taken to the generator transformer at the switchyard through a set of 3 phase and neutral bus system, kept within aluminium enclosure. This enclosure is non-magnetic so as to eliminate any possibility of induction effect in this bus enclosure. This bus enclosure is known as bus duct.

The bus duct is properly ventilated and the bus bars are insulated. A three phase generator transformer is used to wheel the generated power through transmission lines at higher voltage levels. The overall efficiency of a thermal power plant can be expressed as

$$\eta_{\%\text{plant}} = \text{generator efficiency} \times \text{turbine and condenser efficiency} \times \text{boiler efficiency}$$

The generator efficiency is in the order of 98%–99% while that of turbine and condenser is around 85–90%. The boiler efficiency is around 25%–30% and the overall efficiency of the plant is in the range of 20%–45%.

2.3 ADVANTAGES AND DISADVANTAGES OF THERMAL POWER PLANT

2.3.1 Advantages

1. Since coal is available in mass form in coal mines hence coal fired power plants can provide steady and bulk electric power supply.
2. Expansion and addition of more units in an existing thermal plant is not a difficult job provided land is available.
3. Cost of power produced in coal plants is not too expensive.
4. Thermal power plant acts as a base load plant.

2.3.2 Disadvantages

1. Life of thermal power plant is normally within 25 years.
2. Power production cost may increase due to depletion in coal reserve following mining of coal mines.
3. Maintenance of thermal power plant is comparatively higher.
4. Manpower requirement in thermal power plant is more.
5. Starting and stopping of steam turbine in a thermal power plant is time consuming.
6. Thermal power plant needs extensive arrangements for shipment of coal and disposal of ash.
7. Coal fired thermal plant is not environment friendly and creates noise pollution within the plant building. It has adverse effect in atmospheric pollution too.

2.4 SELECTION OF SITE FOR THERMAL POWER PLANTS

Following factors needed to be considered for selecting the site of a steam power plant.

1. Availability of land for the proposed plant and for further expansion.
2. Availability of water for steam generation and for cooling of condenser.
3. Availability of fuel from nearby mines.
4. Accessibility of road and rail transport.
5. Distance to nearest load centre.
6. Availability of other power sources.
7. Evacuation facility to be created for the generated power.
8. The magnitude and nature of load to be handled.

EXAMPLE 2.1 A thermal power plant consumes daily 3500 tonnes of coal. If coal used in that plant has an average energy content of 10^4 Btu/lb, what is the power output of the plant? Assume the plant efficiency to be 20%.

Solution: Given, Plant consumption = 3000 tonnes/day

$$= \frac{3600 \times 2204.62}{24} \text{ lb/hr} \quad [\because 1 \text{ tonne} = 2204.62 \text{ lb}]$$

$$\therefore \text{ Power available from the coal } = \frac{3600 \times 2204.62}{24} \times 10^4 = 3.306 \times 10^9 \text{ Btu/hr}$$

In terms of electrical energy,

$$\text{The available power } = \frac{3.306 \times 10^9 \times 1.055 \times 10^3}{60 \times 60 \times 10^6} [\because 1 \text{ Btu} = 1.055 \times 10^3 \text{ J}]$$

$$= 969 \text{ MW}$$

At 20% efficiency, the power output (P) is given by

$$P = 969 \times 0.2 = 193.8 \cong 194 \text{ MW} \quad \text{(Ans.)}$$

EXAMPLE 2.2 A coal reserve is estimated to have 2000×10^{15} Btu energy. The energy content of the coal is 10000 Btu/lb. Determine the approximate weight of the coal reserve.

Solution: The approximate weight (Wt) is given by

$$Wt = \frac{2000 \times 10^{15}}{10000 \times 2204.62} \text{ tonnes } [\because 1 \text{ tonne} = 2204.62 \text{ lb}] = 9.07 \times 10^{10} \text{ tonnes} \quad \text{(Ans.)}$$

EXAMPLE 2.3 In a thermal power station, a coal stock can generate 2×10^{12} Btu of energy. The average daily load on this station is 60 MW. Assuming the efficiency of the power station to be 20%, find the period (in days) that the fuel can be used before the stock is finished.

Solution: The energy available from the fuel

$$= 2 \times 10^{12} \text{ Btu} = 2 \times 10^{12} \times 1.055 \times 10^3 \text{ Joules } [\because 1 \text{ Btu} = 1.055 \times 10^3 \text{ Joules}]$$

$$= 2 \times 10^{12} \times 1.055 \times 10^3 \text{ W-s } [\because \text{Joules} = \text{W-s}]$$

$$\text{The energy available in MWh } = E = \frac{2 \times 10^{12} \times 1.055 \times 10^3}{60 \times 60 \times 10^6} = 5.86 \times 10^5 \text{ MWh}$$

The average daily load is 60 MW.

$\therefore$ During 24 hours period, this thermal station supplies $60 \times 24 = 1440$ MWh of energy. Since the efficiency is 20%, hence it requires a daily energy input of $1440/0.2 = 7200$ MWh

$$\text{Hence, the fuel will be consumed in } \frac{5.86 \times 10^5}{7200} \text{ days} = 81.4 \text{ days} \quad \text{(Ans.)}$$

2.5 HYDROELECTRIC POWER PLANTS

Utilization of power of water, falling from a height to rotate a prime mover is the main aim in harnessing the hydro energy and employing it to obtain electrical energy from an alternator by coupling the alternator with the prime mover shaft mechanically. The power of water jet from a height, when falling on proper blades (buckets) of the prime mover, the prime mover rotates to make the coupled alternator rotating and producing alternating current at the output of the alternator.

In practice, water source available in nature (in lakes and rivers) as well as artificially created water reservoirs or dams are used to generate electricity using water power. Water flows down to the prime mover from the reservoir to rotate it. Till 19th century mostly waterwheels were used and the kinetic energy of the water current was employed for rotation of the wheel. Hydroelectric turbines were developed in 19th century itself and these turbines were pressed to service across the world to have more efficient hydroelectric power plant.

The first major hydroelectric station in India started operation in 1902. Subsequently there are a number of hydroelectric stations constructed in India and in recent times much emphasis has been paid to harness the tremendous hydropower potential of India.

The mechanical work that the hydropower can do depends on the head and rate of flow. *Head* of plant is basically the vertical distance through which the water falls from the reservoir to the turbine. Most of the hydroelectric plants are having *high head* (in the range of 70 m to 250 m) while some plants are of *medium head* ((in the range of 20 m to 70 m) and *small head* (upto 20 m). In high head plants large blocks of power is produced using high energy water jet and bigger generating units. Medium and smaller head plants tap smaller sources of water and may be used to meet regional demands for electricity. Pumped water storage plants refer to a recyclable hydro plant when water, after coming out of the water turbine is collected and during night time pumped back to the reservoir. During daytime the reservoir water is utilized to rotate the water turbine and to generate electricity. During night time, when the demand is low, the downstream water is pumped back to the reservoir utilizing same water turbine now operating as water pump. The synchronous machine coupled with the turbine works as a generator during daytime and at night-time when electricity is available at cheaper rate, the same synchronous machine acts as a synchronous motor and employs the coupled turbine to work as pump for the purpose of pumping back the water to the reservoir. This water is then used to power the hydroelectric generator during next demand peak.

Figure 2.2 represents a simple schematic diagram to explain the operation of a conventional hydroelectric power plant. The water from the reservoir (a natural source or a dam) runs to *penstock* (a channel or a pipe to allow the downward flow of water from the reservoir to the turbine bay). Using proper automatic water valves the rate of flow of water is controlled while water enters the turbine bay. The water jet is created using proper nozzles and this water passes to the turbine and impacts with the buckets of the rotor of the turbine in such a way so that the turbine rotor shaft starts rotating. Electricity is produced from the alternator coupled with the shaft of the water turbine.

2.5.1 Main Components of Hydroelectric Power Plant

Dam: A dam (or barrage) is an artificial storage of water to provide a proper head for water turbines. Dams are built on top of hills or in the river. The reservoir is the integral part of the dam for storage of dam water. The capacity of the reservoir is decided by the water requirement for power production. The dam has sluice gates for control of water flow in the penstock.

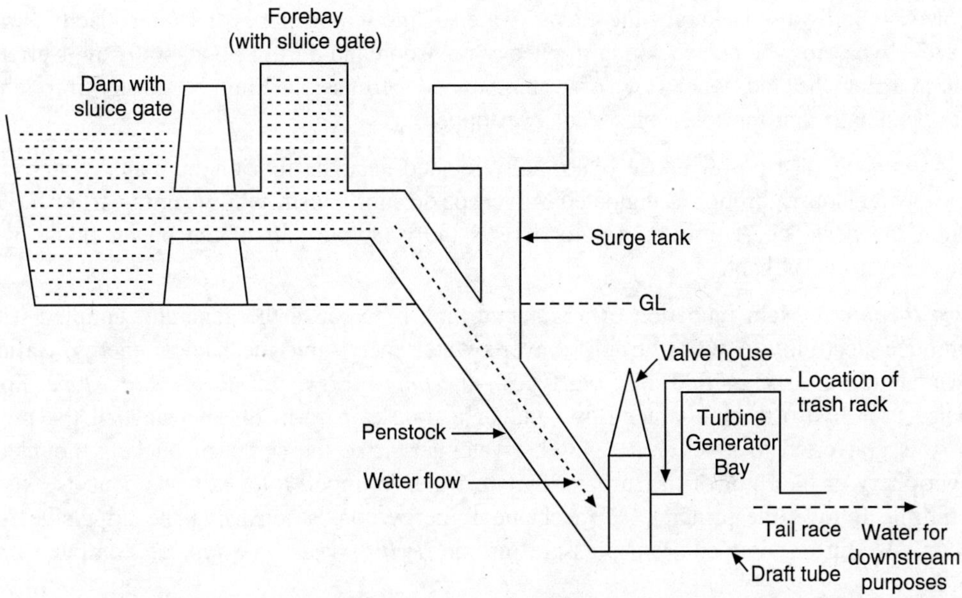

Figure 2.2 Schematic diagram of a hydroelectric power plant.

Forebay: Forebay is the regulating reservoir for temporary water storage when the load on the turbine is reduced and provides water when the load in the turbine is increased. A pond behind the dam on a canal may be used as a forebay.

Penstock: It is basically a water conduit system that carries water from the reservoir down to the turbine through pressure tunnel. Penstock may be constructed above the ground or even under the ground level.

Surge tank: It is a vertical temporary storage facility of water and is connected with the penstock. It is provided to act as pressure release of the water conduit from the effect of water hammer when the valves at the entry of the turbine-generator bay are suddenly shut and water being a non-compressible fluid, tremendous thrust is exerted on the penstock causing water hammer effect. Surge tank allows the water to release its pressure and the water level gets elongated vertically to accommodate the process of release of water pressure created due to sudden stoppage of water flow in the penstock.

Trash rack: It is provided to stop the entry of debris that might damage the valves in the valve house and choke the nozzle of the turbine water inlet. Trash rack is placed across the intake of the water flow at entry of valve house.

Valve house: Valve house has water flow control valves in order to control the mechanical energy input to the water turbine.

Tail race: After passing through the turbine, water is discharged into the tail race and the tail race carries the downstream water to the river. It is either an open channel on a tunnel depending upon the power house location.

Spillway: Spillway discharges the excess water of the reservoir beyond the capacity limit of reservoir bypassing the penstock. During heavy monsoon and during flood, spillway is effective in discharging the additional flow of water beyond permissible limit of storage through the bypass canal so that the reservoir is not overtopped.

Power house: The power house is normally located near the foot of the plant. Water enters to the power house through a regulated valve nozzle mechanism while coming from the dam through the penstock. Turbine-generator bay is the principle part of the power house where the water turbine is kept.

Hydro turbine: Main fuunction of a water turbine is to rotate the generator coupled with it to produce electricity. A water turbine converts water energy into mechanical energy. Basically the water turbine is classified into two types—the *impulse type* and the *reaction type*. For an impulse type hydro turbine water flows out of a nozzle in form of a jet and all the presure energy is converted to kinetic energy. The water jet strike the series of buckets mounted on the periphery of the rotor of the turbine. Due to water jet input, rotor rotates about its axis and the turbine follows the principle of mechanical energy conversion using the impulsive force. The impulse turbine is a constant pressure turbine. *Pelton wheel* is a typical example of such a turbine.

For a reaction turbine, water enters the turbine at high pressure and loaw velocity. Flowing through a runner the water jet creates a reaction on the runner vane and the runner is rotated. In the reaction turbine, water is under pressure and turbine speeds up with water during operation. Reaction turbines are thus encapsulated so that water cannot spalsh out. Reaction turbines are normally of vertical arrangement. *Propeller*, *Francis* and *Kaplan* are in the category of reaction turbine. The use of different types of water turbines depends on available head, directon of water flow and specific speed.

In hydro turbines, the *runaway speed* is defined as the speed that the turbine would reach when the full load on the turbine is suddenly removed and the governor mechanism fails to close the valves. This increase in speed may increase voltage and may cause insulation failure of the generator winding. Also due to high speed of rotation the centrifugal force would be high and may damage the turbine.

Hydro turbine generator: Hydro turbine rotates the rotor of the generator which is coupled to the turbine. In the generator, the field winding is placed on the rotor while the stator contains the armature windings. Hydroelectric genrators are of low speed type machines compared to steam driven generators.

The speed of the hydro generator being less, for same operating frequency the number of poles is high following the equation, $N = \dfrac{120f}{P}$, where N is the rpm of the generator rotor, f is the frequency and P is the number of poles. In order to accmodate the large number of poles, the diameter of the rotor is made larger while the axial length of the poles may be low. A flywheel may attached to the rotor in order to avoid *hunting* and ensure the stabilty of the machine.

2.6 ADVANTAGE AND DISADVANTAGES OF HYDROELECTRIC PLANTS

2.6.1 Advantages

1. Fuel (water) cost is lesser in comparison to the fossil fuel.
2. Water as a fuel is non-wasting and self replenishing. No mining and transport operation is needed to get fuel at the plant.
3. Efficiency of the hydropower plant is nearly 80%–87%, much higher in comparison to fossil fuel plants.
4. Hydropower plants are quick in start and shut down.
5. Downstream water, i.e., the water after use at the power house of hydro-plants can be used in agriculture and other purposes or may be stored and recycled back to the reservoir for pumped storage plants.
6. Maintenance is easy and less expensive in comparison to thermal power plants.
7. Barrage and dams in hydropower plants help to control floods.
8. Hydropower plants have long life, around 60–70 years while that of thermal plants is around 25 years.
9. Hydro-plant offer production of electricity at low cost.
10. Hydro-plant operation is silent and environment friendly.
11. Hydro-plants offer more flexibility in operation and control. Outages are much lesser in comparison to thermal power plants.

2.6.2 Disadvantages

1. The main disadvantage of a hydroelectric power plant is its capital cost and elongated time of construction period.
2. During installation period of dams and reservoirs, a vast portion of area may remain submerged under water and the natural flow path and the characteristics of river may be changed affecting population habitat in that area.
3. Production of power from hydro-plant is dependent on annual rainfall. In dry seasons and during draught, the plant produces power which is much lesser than its capacity.
4. Most of the hydro-plants use perennial rivers and hence they are situated at a hilly terrain and in remote areas. Thus power transmission from such remote place becomes a challenging task.

2.7 SELECTION OF SITE FOR HYDROELECTRIC PLANTS

Following factors need to be considered while the site of a hydro-plant is selected.

1. Availabilty of water.
2. Storage facility of water.
3. Head of water.
4. Accessibilty of site of the plant.
5. Distance of nearest load centre.
6. Social impacts and ecology.

EXAMPLE 2.4 In a hydroelectric power station the head is 50 m. The reservoir of the plant contains 10^6 cubic meters of water to feed the turbine of the plant. Assuming the density of water to be 10^3 kg/m^3, find how much electrical energy this plant will produce neglecting all losses.

Solution: Weight of water $= \text{volume} \times \text{density} = 10^6 \times 10^3 \text{ kg} = 10^9 \text{ kg}$
Potential energy of this water mass would be,

$$\text{PE} = \text{mass of water volume} \times \text{g} \times \text{head of water} = \text{mgh} = 10^9 \times 9.81 \times 50 \text{ J}$$

$$\text{Electrical energy ouput } = \frac{10^9 \times 9.81 \times 50}{60 \times 60 \times 10^6} \text{MWh} = 136.25 \text{ MWh} \quad \text{(Ans.)}$$

2.8 NUCLEAR POWER GENERATION

Nuclear fission process is utilised in electric power geenration from radioactive isotopes. It is a well known fact that uranium isotope is an easily and readily fissionable material. If a neutron bombards an uranium isotope atom, there is splitting of the nucleus of this atom and tremendous amount of energy is released. This will generate heat from the mass of disintegrated uranium atoms. Each fissioned nucleus ejects two or three three neutrons (depeneding upon isotope used) which can again hit the uranium isotope nucleai and accelerate the splitting process. This is known as *chain reaction*. We thus can express this reaction as

$$U\text{(Isotope)} + n \text{ (Bombarding particle)} = \text{Energy} + \text{Radioactive by-product}$$

This radioactive byproduct is a radioactive waste in such a reaction.

Fission process is successfully utilized in nuclear power reactors for heat generation. A nuclear power plant is mostly similar, in operation, to a thermal power plant power genration process except the steam generation part. The conventional steam generation plant of a thermal power plant, i.e., the boiler, is replaced by the reactor and a heat exchanger in a nuclear power plant. The heat produced due to the fission reaction of radioactive isotopes in the reactor chamber is employed for steam generation. The reactor is equipped also with some safety devices. The simple schematic of a liquid coolant nuclear power plant is shown in Figure 2.3. A coolant is used to extract heat from the reactor and this coolant supplies heat to generate steam in the heat exchanger to run the turbine and in order to produce electricity from the coupled generator.

2.8.1 Principal Components of a Reactor

The nuclear reactor is called simply the *reactor*. A mass of radioactive isotope is inserted into the hearth of the reactor in form of fuel rods where the isotope is bombarded by heavy particle. Fission reaction takes place producing another radioactive transformed material as by-product and in this process tremendous heat is generated. This heat is utilised to make the coolant inside the reactor very hot. The hot coolant is taken to heat exchanger where water comes in heat contact with the *coolant* and is converted to superheated steam. It is then used to rotate the turbine.

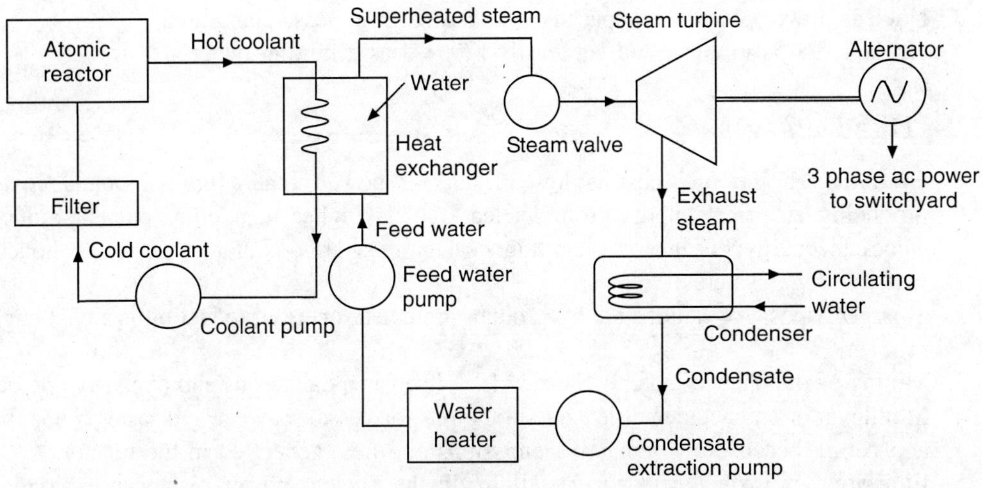

Figure 2.3 Schematic of a nuclear power plant.

Fuel rods: It is typical tube made of an alloy and filled with pellets of radioactive isotope. The fuel rods can be inserted or taken out mechanically.

Moderator: A moderator is used to slow down the neutron released during fission reaction. Normally graphite, beryllium and heavy water (D_2O) are used as moderator for natural uranium isotope while ordinary water can also be used for enriched uranium.

Control rods: Control rods ensure proper control of the chain reaction so that the atomic energy generated remains within control and can be used for power generation. The chain reaction within the nuclear reactor is retarded by removing the fuel rods and by inserting control rods in the reactor. Basically control rod is made of a neutron abosrbing material and is made of either boron carbaide or cadmium or hafnium.

Shielding: Shielding is used to give protection against α, β, and γ radiations produced during the fission process within the reactor so that there is no exposure of such radiations outside the reactor vessel.

Coolant: Coolants are used for extracting heat from the reactor core and tranferring it to the water in the heat exchanger. Normally, heavy water (D_2O), carbon dioxide gas, hydrogen gas, hellium gas, liquid metals and some other organic fluids are used as coolants.

Nuclear fuel: The fuels manily used in nuclear power generation are isotopes of natural uranium, enriched uranium, pluotonium and thorium.

2.9 ADVANTAGES AND DISADVANTAGES OF NUCLEAR POWER PLANT

2.9.1 Advantages

1. Nuclear fuel is an inexpensive source of electrical energy.
2. Quantity of fuel required per MWh is much lesser than other fossil fuels.

3. Nuclear power plant is compact is size and requires lesser land area.
4. Nuclear fuel is in earth and can easily be used as a mining material.

2.9.2 Disadvantages

1. Even though the fuel cost is low in nuclear power generation but actual cost of producing electrical energy from nuclear fuel is higher than other plants producing electrical energy because of very high safety measures, containment and radioactive waste storage systems.
2. Storage/disposal of radioactive by-product in nuclear energy production is a challenging problem.
3. Mining of radioactive isotope, the raw fuel, itself is a dangerous and expensive process.
4. Meltdown or any accidental leakage in the reactor vessel may cause disaster of mankind and nature because of radioactive emissions and heat generated in the reactor.
5. Emissions of toxic level of radioactivity in the nuclear power plant cause biological harm to the population and vegetation in nearby areas.
6. Social, health and environmental restriction make the nuclear energy based electricity production too much expensive and unwelcome in human habitat.

2.10 SELECTION OF SITE FOR NUCLEAR POWER PLANTS

Following factors needs to be considered while selecting the site of a nuclear power plant.

1. Water availability.
2. Plant location must be away from populated areas.
3. Radioactive waste disposal facility.
4. Power evacuation and distribution facility for the nuclear power plant.
5. Accessibility of road and rail transport.
6. Location of fuel mines.

2.11 GAS TURBINE PLANT

Compressed natural gas (CNG) and *gasoline* are two most efficient fossil fuels in addition to coal. Gas turbine based technology of producing electrical power is of great interest in developing countries due to its high degree of efficiency and relatively lower greenhouse effect. Gas turbine plants are used as captive power to places where CNG is not available.

In a gas turbine plant (Figure 2.4), compressed air is delivered to the combustion chamber from an air compressor. The compressor is coupled to the gas turbine so that a part of gas turbine shaft power can be used to operate the air compressor. The gaseous fuel is injected in the combustion chamber and following the principle of internal combustion engine, a high velocity air-fuel mixture is developed after combustion and this mixture forces the gas turbine to rotate. The pressure drops as it flows through the turbine and the gas finally released as exhaust.

Gas turbine plants offer a viable solution to supply power deficit at load centres during the period of load shedding or during peak load periods as the gas turbine plant is easy to control and quick in starting and stopping. It offers inherent advantages of simplicity, compactness, frequent switching and lower gestation period. However, the fuel cost is higher and it leads to increased rate of fuel consumption for electricity produced from gas turbine plants. Also the exhaust needs to be treated before releasing it to the atmosphere to comply with environmental regulation. The maintenance is also easier and requires lesser manpower than thermal power plant of comparable capacity. The capital cost of gas turbine plant is lower than a steam plant and it has low weight to power ratio. The control of such a plant is easier and economical to operate below a given load factor and for a limited period of time.

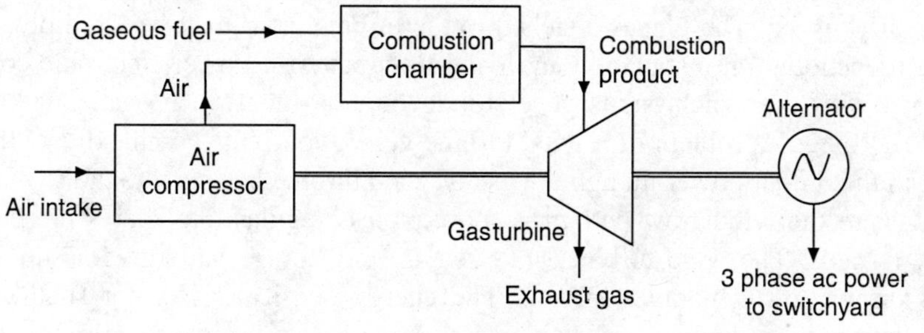

Figure 2.4 Schematic of a gas turbine plant.

Micro-turbines are of same variety like gas turbine plants. Micro-turbines are operated following the same principle as traditional gas turbines. Air is injected into the compressor where it is pressurized and forced into cold side of the *recuperator*. Here, the exhaust heat is utilized to preheat the air before it enters the combustion chamber. In the combustion chamber the air mixes with the gaseous fuel and ignition takes place. The mixture expands through the turbine that drives both the air compressor and generator at very high speeds. The mixture is then taken into the exit path (as exhaust) through the precipitator before being discharged into the atmosphere.

The generator runs at a high speed and hence the frequncy of the generated ac voltage is very high. The frequency of the ac voltage is kep at 50 Hz by using power electronic devices. Micro-turbines have usual power rating of 25 to 500 kW. Some micro-turbines need natural gas as fuel while some use gasoline as fuel.

2.12 RENEWABLE ENERGY SOURCES FOR ELECTRICITY GENERATION

Fossil fuel resources are limited with respect to their availability in nature though the fossil fuel sources are able to provide steady and bulk power generation capabilities. They are categorized as conventional energy sources. Conventional energy sources are coal, gasoline, natural gas, water and nuclear while the electric power generation from wind power, tidal power, solar power, biogas and fuel cells are termed as non-conventional energy sources. Magneto hydrodynamic generation and geo-thermal generation are also termed as non-conventional sources of energy. The fuel sources in the category of non-conventional energy resources are pollution free, inexhaustible but not suitable for bulk power generation as some climatic effects are visible on such resources. These resources are also termed as *renewable sources* of energy.

2.12.1 Wind Energy

Wind power has been in use for sailing boats and ships, in agriculture and atmospheric propulsion since long. Wind energy is plentiful, inexhaustible and pollution free. But the main drawback is its dependence on climatic conditions and is intermittent from the view point of steady flow rate and velocity.

Most of the wind turbines used for electricity generation consists of a two or three bladed propeller rotor rotating around a horizontal axis. Typical system ranges from 25 kW to 2 MW in the commercial sector and may have multiple units. Synchronous generator is coupled with the rotor of the turbine. The blades attached to the turbine rotor may be fixed or with variable blade angle (depending on installation and wind flow pattern). Power electronic converters are also equipped with the synchronous generator in order to control the generated voltage and frequency. The biggest problem of a wind turbine generation is its complexity of the rotor and irregularity in the output power though the use of wind turbine has been found to be economical and viable where the wind flow pattern is more or less regular like shores close to sea beds and at hilly terrains. The wind turbines may require the storage batteries for supplying power to the grid during low generation period. The energy density is low for such wind turbine power installations.

Turbine type wind energy generation transforms the kinetic energy of the wind into rotation of the turbine shaft which in turn rotates the ac generator to produce electricity. The power that can be extracted from wind in such an arrangement is approximately given by

$$P_w = (2.46\times10^{-3})D^2U^3 \text{ Watts.}$$

where D is the wind turbine diameter in feet and U is the wind velocity in miles/hr.

2.12.2 Solar Energy

Solar power has been in use since long in different applications starting from water heating to water purification but its use in electric power generation is most promising. However the availability of solar generation is not uniform even in all parts of the world and during daytime.

India is placed in a geographical region where plenty of sunshine is available during daytime. The zone is suitable for harnessing solar energy. Solar based thermal energy is used for heating water for both industrial and domestic applications.

Solar energy is used for electricity generation using *solar photovoltaic cells*. These cells convert solar radiation to electricity (dc) directly. This electricity can either be used immediately feeding local loads or stored in the batteries. Electricity produced from solar energy can be used in low power and communication applications. However in recent times, grid of solar cells are used to feed generated dc electricity to the grid through converting devices which converts dc generated electricity to ac power.

2.12.3 Tidal Energy

Tides offer a source of energy due to the potential energy of raised tide water column or the kinetic energy of the stream. Tidal energy is originated from the kinetic energy of orbiting and rotating earth, moon and sun. In order to harness tidal energy, a water wheel is placed across the stream. However because of practical installation problems the application of tidal energy in electricity generation has limited scope.

2.12.4 Geothermal Energy

In the geothermal power generation, the heat energy within the interior of the earth is utilised. The heat is trapped through the wells drilled as much as possible inside into the earth. In this process steam is generated at the bottom of the well utilising geo thermal energy and this steam is used to drive the mini steam turbines. The alternators coupled with the turbines produce electricity. The main advantage of geothermalbased energy generation is that it is clean and environment friendly. The main drawback of geothermal generation is the expensive infrastructure and cost of maintenance. Also bulk power generation is not possible through this process.

2.12.5 Biomass Energy

Biomass energy is a form of energy generated from chemical process and is treated as a renewable source of energy. It is derived from the carbonaceous waste of various human and natural resources. Using proper chemical reactions the biomass is treated to produce inflammable methane gas. Methane gas acts as a fuel to drive mini gas turbines and thus ac generators coupled with these gas turbines produces electricity.

2.12.6 Magneto Hydrodynamic Electricity Generation (MHD)

In MHD power generation, an ionised gas or plasma is passed through a strong magnetic field so that the ion transfers its charge which develops the electric potential. Both positive and negative ions are collected by metallic collection plates. Normally superheated steam in plasma state enters the strong magnetic field where the ions in the fuel transfer charge to the collector and become neutral particles. The hot gas coming out of the MHD plant is used to drive steam generator too.

2.13 DISTRIBUTED GENERATION

Distributed Generation (DG) is basically the generation available from small captive or renewable sources strategically located at buses close to load centres or close to bulk consumers and provide economic operation for the interconnected power network. Distributed generations involve gas turbine or diesel generator sets, microturbines, solar photovoltaic panels, wind turbines and fuel cells.

Distributed Generation can be applied in following applications:

1. As standby power source in case of outage of conventional generation.
2. Peak power support in case of abrupt increase of demand on the system.
3. To provide grid support and to defer new instruments for grid expansion.
4. As a service provider to local consumers utilising grid network so that the overall economy is improved.

2.14 COGENERATION

It is a process of capturing the waste heat from manufacturing or industrial process so as to provide heat energy to generate electric power. Basically *cogeneration* is the process of producing two forms of energy from one fuel, like heat and electricity. The exhaust steam from the turbine in a thermal power plant is used for heating purpose. Cogeneration is economical in cold countries as the exhaust steam is utilised in different applications starting from pipeline heating to domestic heating.

EXAMPLE 2.5 A wind generator is rated for 40 kW generation for a wind velocity of 10 miles/hr. what is the approximated blade diameter?

Solution: We know that $P_w = (2.46 \times 10^{-3}) D^2 U^3$ watts.

Given, $P_w = 40$ MW and $U = 10$ miles/hr

Therefore, we have, $40 \times 10^3 = 2.46 \times 10^{-3} \times D^2 \times 10^3$

i.e.,
$$D = \sqrt{\frac{4 \times 10^4}{2.46}} = 40.22 \text{ ft.} \quad \text{(Ans.)}$$

EXAMPLE 2.6 What is the average power output of a wind turbine having a blade diameter of 20 ft. while the average wind velocity is 15 miles/hr?

Solution: We know that $P_w = (2.46 \times 10^{-3}) D^2 U^3$ watts.

Given, $D = 20$ ft and $U = 15$ miles/hr

Therefore, we have, $P_w = 2.46 \times 10^{-3} \times 20^2 \times 15^3 = 3.321$ kW

i.e.,
$$D = \sqrt{\frac{4 \times 10^4}{2.46}} = 40.22 \text{ feet} \quad \text{(Ans.)}$$

[Conversion of answer in Examples 2.5 and 2.6 in MKS system can be done by the reader easily.]

2.15 COMMONLY USED TERMS

Electrical load have diversified characteristics and thus they are termed following their applications. Loads may be domestic type, agriculture type, industrial type and commercial type. Different consumers may draw load at both medium and low voltage (i.e., at 6 kV or 11 kV and at low voltage, i.e., at 400–415 Volts). Excerpt for domestic and a few commercial loads, almost all loads are of 3 phase type. The loads can be classified as follows:

Connected load: It is the summation of the continuous rating of entire load connected with the supply system. It is normally expressed in kW.

Maximum demand: It is the maximum power that the consumer is likely to draw at any time. It is the greatest demand that the consumer may draw during the specific period of time (normally on daily or monthly basis). Maximum demand becomes equal to the connected load provided all equipments at the consumer's premises are operated at rated power simultaneously to full extent. In practice consumers do not use all equipments at full extent simultaneously and hence this maximum demand is normally less than the connected load. Maximum demand is usually expressed in kW or kVA.

Demand factor: It is the ratio of maximum demand to the connected load and usually less than 1.

$$\text{Demand Factor} = \frac{\text{Maximum demand}}{\text{Connected load}}$$

Usually the value of demand factor varies in between 0.5 to 0.8.

Diversity factor: Since the maximum demands of all consumers do no occur simultaneously hence it may not be economical to select the rating of a supply equipment (say, generator or a transformer) equal to the sum of all maximum demands. Diversity factor is thus used to justify this and is defined as

$$\text{Diversity Factor} = \frac{\text{Sum of the individual maximum demand}}{\text{Maximum demand on the system as a whole}}$$

Diversity factor has value always greater than unity.

Coincidence factor: It is the reciprocal of diversity factor.

Load factor: Load factor of a plant is the ratio of average load to peak load for a particular period of time. Frequently this period is 24 hrs.

$$\therefore \quad \text{Load Factor} = \frac{\text{Average demand}}{\text{Maximum demand}}$$

$$= \frac{\text{Energy generated in a particular period of time}}{\text{Maximum demand} \times \text{Hours of operation in the given period of time}}$$

Plant factor: Plant factor is the ratio of average annual load demand on the plant to the capacity of the plant during a given interval of time.

$$\text{Plant Factor} = \frac{\text{Average annual load on the plant}}{\text{Rated capacity of the plant}}$$

$$= \frac{\text{Energy generated by the plant in one year}}{\text{Rated capacity of the plant} \times 8760}$$

The maximum value of plant factor can be one when the plant operates at its rated capacity throughout the year. It is also known as Plant Load Factor (PLF).

Capacity factor: It indicates the extent of use of the generating plant. It is defined as

$$\text{Capacity Factor (CF)} = \frac{\text{Maximum demand on the plant}}{\text{Plant capacity}}$$

It is also known as Plant Capacity Factor.

Plant use factor: It is defined as

$$\text{Plant Use Factor} = \frac{\text{Actual energy produced in a year}}{\text{Plant capacity} \times \text{Plant operating time (hours) in a year}}$$

Plant utilization factor: It is defined as the ratio of maximum demand on the plant to the rated capacity of the plant.

$$\text{Utilization Factor (UF)} = \frac{\text{Maximum load on plant}}{\text{Rated capacity of plant}}$$

Normally the higher value of the utilization factor is desired. The value of UF may be more than unity due to overloading of the plant.

Also, $\quad$ Plant Load Factor (PLF) = Load Factor $\times$ Utilization Factor

Spinning reserve: It is a portion of generating capacity of a plant or the system that is ready to take load. It assists the system during outage of part of generation and starts supplying to the system.

Firm power: It is the power intended to be used under emergency condition. It is normally available in the systems.

Reserve margin: It is the difference between the rated capacity of the plant and the actual loading of on the plant.

Cold reserve: It is that amount of reserve generating capacity which is available for service but is not in operation.

Hot reserve: It is that amount of reserve generating capacity which is in operation but not in service.

EXAMPLE 2.7 A generating station has load factor of 70%. The plant load factor is 50% and plant use factor is 80%, the maximum demand of this generating station is 200 MW. Find (i) Average demand (ii) Electric energy produced daily (iii) Rated capacity of the plant (iv) Reserve capacity of the plant (v) Maximum electric energy that the plant can produce daily for continuous run (vi) Utilization factor.

Solution: (i) Average demand = Maximum demand $\times$ Load Factor

$$= 200 \times 0.7 = 140 \text{ MW} \quad \text{(Ans.)}$$

(ii) Electric energy produced daily = Average demand $\times 24 = 140 \times 24 = 3360$ MWh

(iii) Rated capacity of the plant $= \dfrac{\text{Average demand}}{\text{Plant Load Factor}} = \dfrac{140}{0.5} = 280$ MW (Ans.)

(iv) Reserve capacity of the plant = Installed capacity – Maximum demand

$$= 280 - 200 = 80 \text{ MW} \quad \text{(Ans.)}$$

(v) Maximum electric energy that the plant can produce daily for continuous run

$$= \text{Istalled capacity} \times 24 = 280 \times 24 = 6720 \text{ MWh}$$

(vi) Utilization factor $= \dfrac{\text{Maximum demand}}{\text{Rated capacity}} = \dfrac{200}{280} = 0.714$ (Ans.)

EXERCISES

1. Draw a schematic diagram of a thermal power plant and explain the process of electric power generation utilising fossil fuels. Identify basic circuits in this process.
2. Name the principal components of a thermal power plant and briefly state their uses.
3. List the advantages and disadvantages of a thermal power plant.
4. What are the factors to be considered for selection of site for a thermal power plant set up?
5. Briefly describe the process of hydroelectric power generation.
6. Name the principal components of a hydroelectric power plant and state their uses.
7. What are the advantages of hydroelectric power plants?
8. What are the factors on which the selection of site for a hydroelectric power plant depends?
9. With a schematic diagram, briefly describe the process of generation of electrical energy from nuclear fuel. What are the components of the reactor vessel and what are the functions of these components?
10. State relative advantages and disadvantages of a nuclear power plant.
11. State the nonconventional or renewable energy resources for generation of electrical energies.
12. Briefly state the process of electric power generation from wind energy and solar energy.
13. What do you mean by distributed generation and cogeneration?
14. Define: Maximum demand, demand factor, diversity factor, load factor and plant factors.

CHAPTER 3

Transmission Line Parameters

3.1 INTRODUCTION

Overhead bare conductor lines are extensively used as a medium of electric power transmission in transmission lines. Usually the conductor has an acceptable level of *tensile stress* (i.e. high-breaking load) and low *resistivity* (to make lower I^2R loss) and marginal *impedance drop*. The line conductors are usually made of electrically pure aluminium and are invariably *stranded* to make them more robust, flexible and better inductance property. The stranded conductors have a central steel core conductor to make the overall conductor rigid in terms of mechanical strength. This is then called ACSR (*aluminium conductor steel reinforced*) conductor. Almost all of the transmission lines and most of the distribution lines use such type of conductors. In a few distribution applications AA (*all aluminium*) or even copper conductors are also used. In the following sections the basic aspects of electrical design of overhead lines are discussed.

3.2 CONDUCTORS

For overhead transmission lines, aluminium conductors are normally used. Although copper has a much higher conductivity, an aluminium conductor is lighter, cheaper and has a larger diameter than a copper conductor of the same resistance. Enlarging diameter of conductor results in a lower voltage gradient at the conductor surface and thus reduces the *corona effect* (which will be discussed in another chapter). Moreover, the supply of aluminium is abundant where as that of copper is limited.

Stranded conductors are assembled by winding wires of small diameter in layer, with alternate layers wound in the opposite direction to prevent unwinding and to coincide the radius of outer layer with the inner radius of the next layer. Hence, each strand in the conductor provides a parallel path for the flow of current. Thus it is possible to build flexible conductors with a large cross sectional area.

Aluminium conductor steel reinforced (ACSR) lines consist of stranded steel wires, surrounded by current-carrying layers of aluminium strands. Stranded steel wires are used to increase the mechanical strength of the conductor. Figure 3.1 shows the cross section of an ACSR line with central core of seven steel strands and 24 Al-strands in the two outer layers.

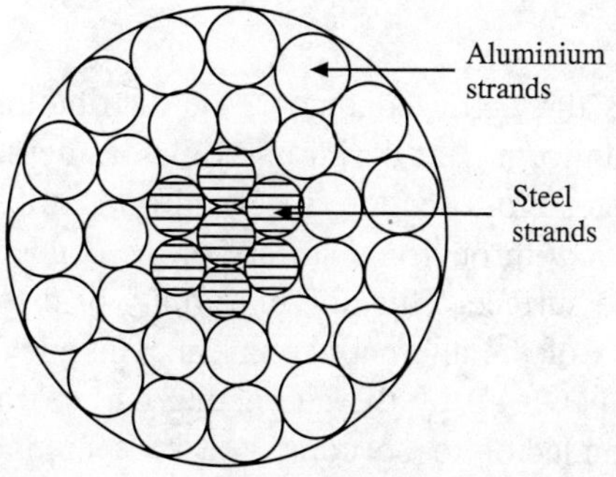

Figure 3.1 Cross section of a 24/7 ACSR conductor.

Other configurations of aluminium conductors that are used in transmission lines are *all-aluminium conductors* (AAC), *all-aluminium-alloy conductors* (AAAC), *aluminium-conductor-alloy-reinforced* (ACAR), expanded ACSR. In order to prevent corona in extra high voltage lines, the diameter of the conductor should be increased, thereby reducing electrical stress, by inserting a filler of paper or *hessian* between steel and aluminium layers. This configuration is known as expanded ACSR conductors.

Conductor manufacturers provide the characteristics of the stranded conductors such as number of strands, material of strands, diameter of strands in mm, diameter of the whole conductor in mm, resistance of the conductor at 20°C in ohms per km and weight of the conductor in kg per km.

3.3 LINE RESISTANCE

Direct current resistance R_{dc} can be obtained from the following formula

$$R_{dc} = \frac{\rho l}{A} \Omega \tag{3.1}$$

where ρ is the resistivity of the conductor material in Ω m, l is the length of the conductor in m and A is the cross sectional area of the conductor in m².

Resistance causes power loss in a transmission line. This power loss causes a rise in temperature of the line conductor. Resistance of the conductor will also vary with the temperature of the conductor. Resistance (R_1) of a condcutor at a temperature t_1°C can be obtained by the following relation.

$$R_1 = R_0 (1 + \alpha_0 t_1) \tag{3.2}$$

where R_0 is the resistance of the conductor at 0°C and α_0 is the temperature co-efficient of the conductor.

The resistance (R_2) of the conductor at t_2°C can be found from the expression

$$\frac{R_2}{R_1} = \frac{(1/\alpha_0) + t_2}{(1/\alpha_0) + t_1} \tag{3.3}$$

3.4 SKIN EFFECT

When a direct current flows through a conductor, the distribution of current across the cross section of the conductor is uniform. But during the flow of alternating current, the distribution of current acorss the conductor cross section is not uniform. In case of circular conductor, the alternating current tends to flow through the surface of the conductor. Thus the current density increases from the interior to surface. This non-uniformity of distribution of current is directly proportional to the frequency of the alternating current. This phenomenon is called skin effect and is due to a magnetic flux set up by the alternating current inside the conductor.

We can imagine the conductor to be composed of a number of annular filaments. The filaments on the surface of the conductor are not linked by the internal flux. Thus the flux linking the filaments decreases progressively from centre to surface. Hence, the inner filament has higher inductance than the filament in the surface. Therefore, the outer filaments carry more current than the inner filaments, which reduce the effective area of cross section of the conductor and consequently the resistance is increased. This causes the ac resistance of a conductor to be higher than the dc resistance. This effect is called *skin effect*. This effective ac resistance is referred as *effective resistance* of the conductor. At 50 Hz, the increase in resistance due to skin effect is around 3–5 percent.

3.5 LINE INDUCTANCE

The inductance of a transmission line is due to the effect of magnetic fields produced by the flow of current in the line conductors. If an electric current (I) flows through the number of turns N of a conductor, then the magnetomotive force (mmf) is $F = IN$ ampere-turns. By Ampere's Circuital Law, the mmf is equal to the line integral of the magnetic field intensity H in AT/metre around a closed magnetic path.

$$\therefore \quad IN = F = \oint H dl = Hl \tag{3.4}$$

where l is the mean flux path in metres.

The magnetic flux density B in Wb/m² in terms of the field intensity H is given by

$$B = \mu H \tag{3.5}$$

where μ is the *permeability constant* of the medium. The total flux (ϕ) through a cross section area of A m² is expressed as

$$\phi = BA \text{ Wb} \tag{3.6}$$

and flux linkages (ψ) in Wb-turns of N turns of the coil is given by

$$\psi = N\phi \tag{3.7}$$

Inductance (L) of a circuit can be defined as the flux linkage (in WbT) per unit current.

$$\therefore \quad L = \frac{\psi}{I} \text{ henry} \tag{3.8}$$

The inductance of a transmission line depends on the size, type and arrangement of the conductors. The arrangement for the phase conductors becomes symmetrical if the conductors of different phases maintain the same distance from each other. Again each phase can have one conductor or a number of conductors bundled together. The transmission line inductance becomes different for different arrangement of conductors.

3.6 INDUCTANCE OF A CONDUCTOR

In order to compute an accurate value for the inductance, the flux inside the conductor as well as the flux external to the conductor must be considered.

3.6.1 Inductance of a Conductor Due to Internal Flux

Cross section of a long cylindrical conductor with radius r and carrying a current (I) amperes (shown in Figure 3.2) is considered for inductance calculation. It is assumed that the return path for the current in the conductor is far away and does not affect the magnetic field of the conductor. So, the magnetic flux lines can be assumed to be concentric with the conductor.

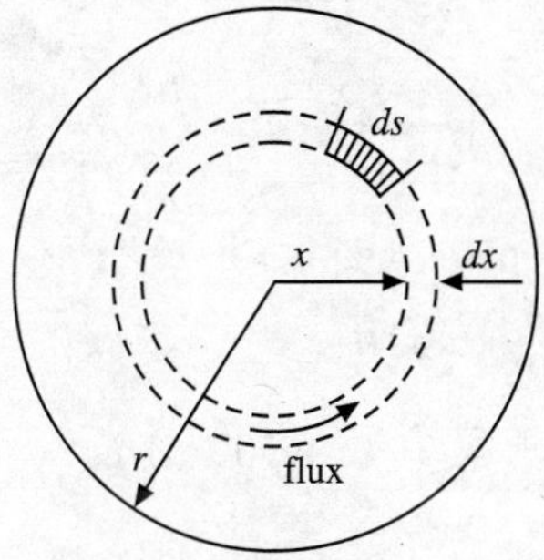

Figure 3.2 Cross section of a current carrying cylindrical conductor.

The field intensity H_x, at a distance x from the centre of the conductor is constant over the circular path.

$\oint H_x\, ds = I_x$, where $\oint$ indicates the integration around a closed path and I_x is current enclosed. However, integration of ds round the closed circular path is $2\pi x$.

Hence,
$$2\pi x H_x = I_x \tag{3.9}$$

If uniform current density across the conductor cross section is considered, I_x can be written as

$$I_x = \frac{\pi x^2}{\pi r^2} I \tag{3.10}$$

Substituting equation (3.10) into equation (3.9),

$$H_x = \frac{x}{2\pi r^2} I \text{ AT/metre} \tag{3.11}$$

Therefore, the flux density B_x, at x metre from the centre of conductor, is expressed as

$$B_x = \mu H_x = \frac{\mu x}{2\pi r^2} I \text{ Wb/m}^2 \tag{3.12}$$

The flux $(d\phi)$ from equation (3.6) in a tubular element of thickness dx per metre of axial length, at x metres from the centre of the conductor is given by

$$d\phi = \frac{\mu x I}{2\pi r^2} dx \text{ Wb/m} \tag{3.13}$$

The flux linkage $d\psi$ per metre length of the conductor is given by

$$d\psi = \frac{\pi x^2}{\pi r^2} d\phi$$

$$= \frac{\mu x^3 I}{2\pi r^4} dx \text{ WbT/m} \tag{3.14}$$

$\therefore$ The total internal flux linkage ψ_{int} is obtained as

$$\psi_{\text{int}} = \int_0^r \frac{\mu I x^3}{2\pi r^4} dx = \frac{\mu I}{8\pi} \text{ WbT/m} \tag{3.15}$$

If relative permeability $\mu_r = 1$, then $(\mu = \mu_0 \mu_r)$ is given by

$$\mu = \mu_0$$

$$= 4\pi \times 10^{-7} \text{ H/m} \tag{3.16}$$

Hence

$$\psi_{\text{int}} = \frac{I}{2} \times 10^{-7} \text{ WbT/m} \tag{3.17}$$

$$L_{\text{int}} = \frac{1}{2} \times 10^{-7} \text{ H/m} = 0.5 \text{ mH/km} \tag{3.18}$$

From equation (3.18), it is clear that inductance due to the internal flux is independent of conductor dimensions (that is, its radius).

3.6.2 Inductance of a Conductor Due to External Flux

The cross section of a conductor of radius 'r' metres and carrying a current I amperes is shown in Figure 3.3.

Here, it is assumed that the return path of the current is far removed from the conductor and does not affect the flux produced by the current flowing in the conductor. The magnetic flux paths are concentric circles around the conductor. Two points A_1 and A_2 at a distance D_1 and D_2 from the conductor surface are considered.

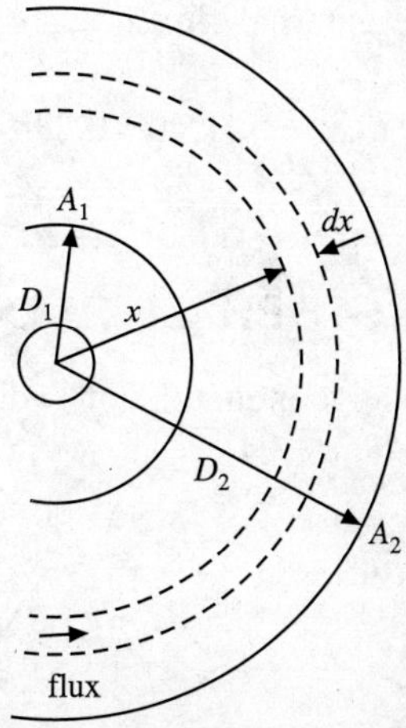

Figure 3.3 External flux linkage due to a current carrying conductor.

The magnetic field intensity at a distance x metres from the centre of the conductor is

$$H_x = \frac{I}{2\pi x} \text{ AT/m} \tag{3.19}$$

The flux density B_x is given by

$$B_x = \frac{\mu I}{2\pi x} \text{ Wb/m}^2 \tag{3.20}$$

The flux $d\phi$ in the tubular element of the thickness dx metres is

$$d\phi = \frac{\mu I}{2\pi x} dx \text{ Wb/m} \tag{3.21}$$

Since the flux external to the conductor links the entire current only, the flux linkages $d\psi$ are given by

$$d\psi = \frac{\mu I}{2\pi x} dx \text{ Wb T/m} \tag{3.22}$$

The total flux between A_1 and A_2 is

$$\psi_{A_1-A_2} = \int_{D_1}^{D_2} \frac{\mu I}{2\pi x} dx$$

$$= \frac{\mu I}{2\pi} \ln \frac{D_2}{D_1} \text{ Wb T/m} \tag{3.23}$$

Since $\mu = \mu_r \cdot \mu_0 = 1 \times 4\pi \times 10^{-7}$, with $\mu_r = 1$,

$$\psi_{A_1-A_2} = 2 \times 10^{-7}\, I \ln \frac{D_2}{D_1} \text{ WbT/m} \tag{3.24}$$

Therefore, the inductance due to the external flux linkage between A_1 and A_2 is

$$L_{1-2} = 2 \times 10^{-7} \ln \frac{D_2}{D_1} \text{ H/m} = 0.2 \ln \frac{D_2}{D_1} \text{ mH/km} \tag{3.25}$$

3.7 INDUCTANCE OF A SINGLE PHASE TWO-WIRE LINE

A single phase circuit of two parallel conductors of radii r_1 and r_2 metres, separated by a distance of D metres, is shown in Figure 3.4.

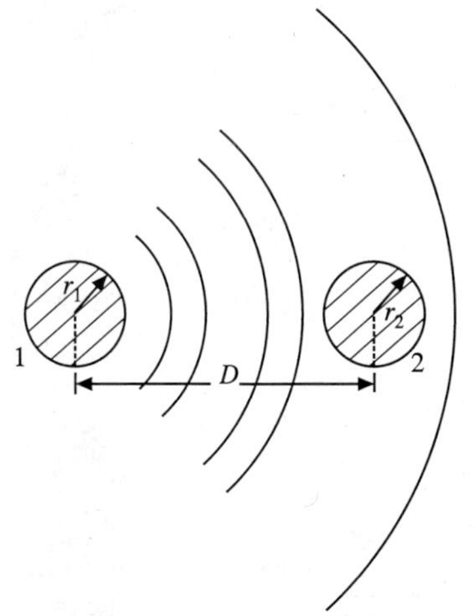

Figure 3.4 Single phase circuit.

It is observed from Figure 3.4 that the flux set up by the current in conductor 1 links it in the following ways:

(i) The external flux between r_1 and $(D - r_2)$, that is, between conductors 1 and 2, links all the current I in conductor 1.

(ii) The external flux set up by the current I in conductor 1 at a distance equal to or greater than $(D + r_2)$.

(iii) The external flux over the surface of conductor 2, that is, between $(D - r_2)$ and $(D + r_2)$ links a fraction of current varying from 1 to zero.

The distance (D) between the conductors, is much greater than r_1 and r_2, hence it can be assumed that D can be used instead of $(D - r_2)$ or $(D + r_2)$. The inductance L_1 of the circuit due to current I in the conductor 1 can be obtained by addition of inductance due to internal flux linkages determined by equation (3.18) and the inductance due to external flux linkages determined by equation (3.25) with r_1 and r_2 replacing D_1 and D_2 to obtain

$$L_1 = 2 \times 10^{-7} \left[\frac{1}{4} + \ln \frac{D}{r_1} \right] \text{H/m} \tag{3.26}$$

In equation (3.26) the term $\frac{1}{4}$ may be written as (ln $e^{1/4}$).

Thus, $$L_1 = 2\times10^{-7}\ \ln\frac{D}{r_1\, e^{-1/4}} = 2\times10^{-7}\ \ln\frac{D}{r_1'}\ \text{H/m} = 0.2\ \ln\frac{D}{r_1'}\ \text{mH/km} \tag{3.27}$$

where $r_1' = r_1\, e^{-1/4} = 0.7788\, r_1$, r_1' is the *geometric mean radius* (GMR) or *self-geometric mean distance* (self-GMD) of a solid round conductor. GMR represents an infinitesimally thin-walled conductor of radius r_1'. This method of replacing a conductor of radius r_1 by an equivalent thin-walled conductor of radius r_1' enables the designers to develop equations for line inductances without considering the field produced due to internal flux.

The inductance due to current in the conductor 2 is

$$L_2 = 2\times10^{-7}\ \ln\frac{D}{r_2'}\ \text{H/ m} = 0.2\ \ln\frac{D}{r_2'}\ \text{mH/km} \tag{3.28}$$

Hence, the total inductance of the circuit taking $r_1' = r_2' = r'$ can be written as

$$L = L_1 + L_2$$

$$L = 4\times10^{-7}\ \ln\frac{D}{r'}\ \text{H/ m}\quad (\text{assuming } r_1' = r_2' = r') \tag{3.29}$$

or, $$L = 0.921\ \log\frac{D}{r'}\ \text{mH/km} \tag{3.30}$$

3.8 INDUCTANCE OF COMPOSITE CONDUCTOR LINES

A single phase line is composed of conductor *A*, having *n* identical parallel filaments and conductor *B*, which is the return circuit for the current (*I*) in conductor *A*, has *m* identical parallel filaments. Each filament (or subconductor) of conductor *A* carries a current $\frac{I}{n}$ and each filament of conductor *B* carries the return current $\left(-\frac{I}{m}\right)$ [negative sign indicates the direction of $\frac{I}{m}$ which is opposite to that of $\frac{I}{n}$].

The flux linkage of the filament (or subconductor) *i* of conductor *A* is obtained by the following equation.

$$\psi_i = 2\times10^{-7}\ \frac{I}{n}\left[\ln\frac{1}{D_{i1}} + \ln\frac{1}{D_{i2}} + \cdots + \ln\frac{1}{r_i'} + \cdots + \ln\frac{1}{D_{in}}\right]$$

$$-\,2\times10^{-7}\ \frac{I}{m}\left[\ln\frac{1}{D_{i1'}} + \ln\frac{1}{D_{i2'}} + \cdots + \ln\frac{1}{D_{im'}}\right]$$

$$\psi_i = 2\times10^{-7}\ I\ \ln\frac{(D_{i1'}\, D_{i2'}\cdots D_{im'})^{1/m}}{(D_{i1}\, D_{i2}\cdots r_i' D_{in})^{1/n}}\ \text{WbT/ m} \tag{3.31}$$

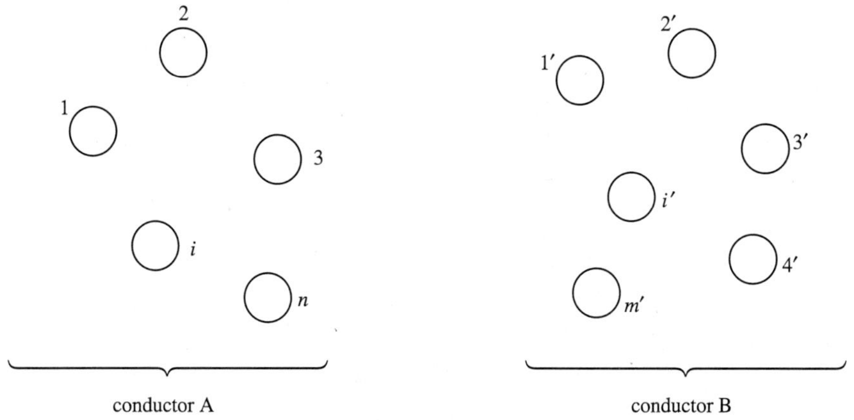

Figure 3.5 Two composite conductors in a single phase line.

If equation (3.31) is divided by the current (I/n), the inductance of filament (or subconductor) i is obtained as

$$L_i = \frac{\psi_i}{(I/n)} = 2n \times 10^{-7} \ln \frac{(D_{i1'}\, D_{i2'} \cdots D_{im'})^{1/m}}{(D_{i1}\, D_{i2} \cdots r_i' D_{in})^{1/n}} \text{ H/m} \tag{3.32}$$

Hence, the average inductance of a subconductor of A is obtained as

$$L_{av} = \frac{L_1 + L_2 + \cdots + L_i + \cdots + L_n}{n} \tag{3.33}$$

The inductance of conductor A then can be written as

$$L_A = \frac{L_{av}}{n} = \frac{L_1 + L_2 + \cdots L_i + \cdots L_n}{n^2} \tag{3.34}$$

The conductor A is composed of n identical electrically parallel elements.

Now substituting expression of each filament from equation (3.32) in equation (3.34) it yields

$$L_A = 2 \times 10^{-7} \ln \frac{\begin{bmatrix} (D_{11'}\, D_{12'} \cdots D_{1m'})\,(D_{21'}\, D_{22'} \cdots D_{2m'}) \\ D_{i1'}\, D_{i2'} \cdots D_{im'})\,(D_{n1'}\, D_{n2'} \cdots D_{nm'}) \end{bmatrix}^{1/mn}}{\begin{bmatrix} (D_{11}\, D_{12} \cdots D_{1n})\,(D_{21}\, D_{22} \cdots D_{2n}) \\ (D_{i1}\, D_{i2} \cdots D_{in})\,(D_{n1}\, D_{n2} \cdots D_{nn}) \end{bmatrix}^{1/n^2}} \text{ H/m} \tag{3.35}$$

where $r_1, r_2, \ldots, r_n$ have been replaced by $D_{11}, D_{22}, \ldots, D_{nn}$ respectively ($D_{11} = D_{22} = D_{33} = D_{nn} = 0.7788$ cm).

It may be observed that the numerator of the argument of the logarithm in equation (3.33) is the mnth root of mn terms, which are the products of the distances from all the n filaments (subconductors) of the conductor A to the all filaments (subconductors) of conductor B. The mnth root of the product of the mn distances is called the *geometric mean distance* (GMD) between the conductor A and B and is denoted by D_m or mutual GMD. The denominator in equation (3.33) is the n^2 root of n^2 terms, which are the products of the distances from every filament in conductor A to itself. The n^2 root of these terms is called the *self GMD* (or GMR), denoted by D_{SA}.

In terms of D_m and D_{SA}, equation (3.33) becomes

$$L_A = 2\times 10^{-7} \ln \frac{D_m}{D_{SA}} \text{ H/m} = 0.2 \ln \frac{D_m}{D_{SA}} \text{ mH/km} \tag{3.36}$$

Similarly, the inductance of conductor B can be determined as

$$L_B = 2\times 10^{-7} \ln \frac{D_m}{D_{SB}} \text{ H/m} = 0.2 \ln \frac{D_m}{D_{SB}} \text{ mH/km} \tag{3.37}$$

Thus the total inductance of the line is obtained as

$$L = L_A + L_B \tag{3.38}$$

3.9 INDUCTANCE OF THREE PHASE LINES

The circuits for transmission of bulk power are three phase circuits. However, the methodology described in the previous sections for calculating the inductance of single phase circuits can be conveniently applied to compute the inductance of three phase transmission lines.

3.9.1 Symmetrical Spacing

Figure 3.6 shows the conductors of three phase lines with equilateral spacing. The conductors are placed at the corners of an equilateral triangle. Each conductor has radius r metres and the spacing between the conductors is D metres. Here, it is assumed that the neutral wire is not present.

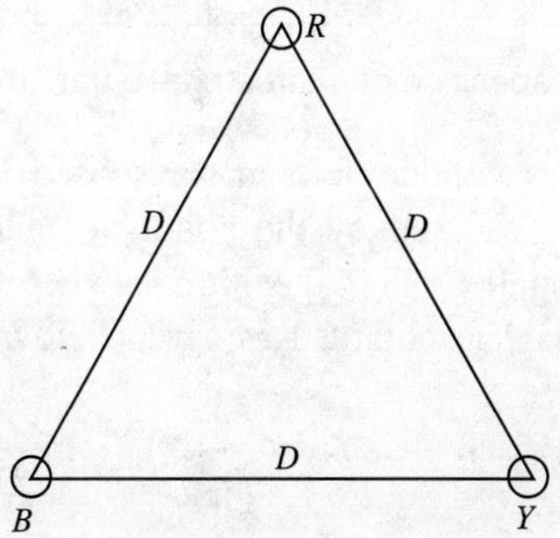

Figure 3.6 Arrangement of conductors in a three phase circuit.

i.e., $$I_R + I_Y + I_B = 0$$

The flux linkages of the conductor R can be written as

$$\psi_R = 2\times 10^{-7}\left[I_R \ln \frac{1}{r'} + I_Y \frac{1}{D} + I_B \frac{1}{D}\right] \text{ WbT/ m} \tag{3.39}$$

Since, $I_R = -(I_Y + I_B)$, the above equation becomes

$$\psi_R = 2\times 10^{-7}\left[I_R \ln \frac{1}{r'} - I_R \frac{1}{D}\right] = 2\times 10^{-7}\, I_R \ln \frac{D}{r_1'} \text{ WbT/ m} \tag{3.40}$$

Hence,

$$L_R = \frac{\psi_R}{I_R} = 2 \times 10^{-7} \ln \frac{D}{r'} \text{ H/m} = 0.2 \ln \frac{D}{r'} \text{ mH/km} \tag{3.41}$$

The inductances of the conductors *Y* and *B* are the same as the inductance of the conductor *R*.

3.9.2 Unsymmetrical Spacing and Transposition

In actual practice, the conductors of a three phase line are not at the corners of an equilateral triangle. The flux linkages and inductance of each phase are not the same. But different inductance in each phase results in an unbalanced circuit. If the line is sufficiently long, the unbalance in the inductance is quite large. To overcome this problem, the lines are *transposed.*

In a transposed line, the phase conductors have their positions interchanged so that the conductor of each phase occupies positions 1, 2 and 3 for about one-third of its total length. Due to the position interchanging methodology, each phase conductor has the same average inductance. Modern transmission lines may not be transposed at regular intervals although the interchange in the positions of the conductors is made at the switching stations to balance the inductance.

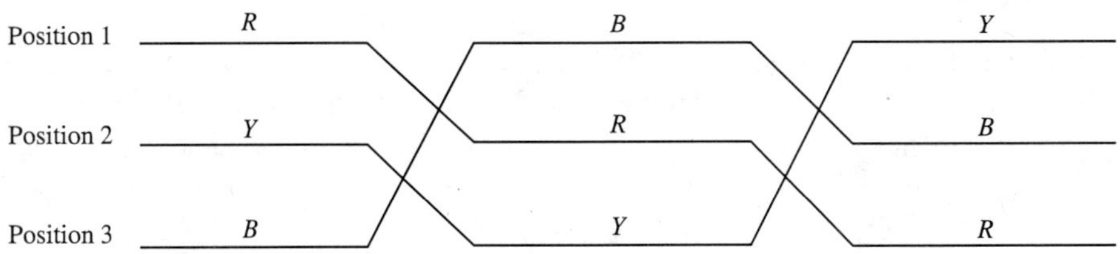

Figure 3.7 Transposition of a three phase transmission line circuit.

The average inductance of a conductor of a transposed line is found by calculating the flux linkages for each position occupied by the conductor and then finding the average flux linkage. The flux linkages of conductor *R* in position 1, when conductor *Y* is in position 2 and conductor *B* is in position 3, may be obtained as given below.

$$\psi_{R1} = 2 \times 10^{-7} \left[I_R \ln \frac{1}{r'_R} + I_Y \ln \frac{1}{D_{RY}} + I_B \ln \frac{1}{D_{BR}} \right] \text{WbT/m} \tag{3.42}$$

Again, with conductor *R* in position 2, when conductor *Y* is in position 3 and conductor *B* is in position 1, the flux linkages are

$$\psi_{R2} = 2 \times 10^{-7} \left[I_R \ln \frac{1}{r'_R} + I_Y \ln \frac{1}{D_{YB}} + I_B \ln \frac{1}{D_{RY}} \right] \text{WbT/m} \tag{3.43}$$

Finally, with conductor *R* in position 3, when conductor *Y* is in position 1 and conductor *B* is in position 2, the flux linkages are

$$\psi_{R3} = 2 \times 10^{-7} \left[I_R \ln \frac{1}{r'_R} + I_Y \ln \frac{1}{D_{BR}} + I_B \ln \frac{1}{D_{YB}} \right] \text{WbT/m} \tag{3.44}$$

Therefore, the average flux linkages of the conductor R is

$$\psi_R = \frac{\psi_{R1} + \psi_{R2} + \psi_{R3}}{3}$$

$$= \frac{2 \times 10^{-7}}{3} \left[3I_R \ln \frac{1}{r'_R} + 3I_Y \ln \frac{1}{D_{RY}\, D_{YB}\, D_{BR}} + 3I_B \ln \frac{1}{D_{RY}\, D_{YB}\, D_{BR}} \right] \tag{3.45}$$

Since,

$$\psi_R = 2 \times 10^{-7}\, I_R \ln \frac{D_{RY}\, D_{YB}\, D_{BR}}{r'_R} \text{ WbT/m} \tag{3.46}$$

The average inductance per phase,

$$L_R = 2 \times 10^{-7} \ln \frac{D_{eq}}{D_S} \text{ H/m} = 0.2 \ln \frac{D_{eq}}{D_S} \text{ mH/km} \tag{3.47}$$

where $$D_{eq} = \sqrt[3]{D_{RY}\, D_{YB}\, D_{BR}} \text{ and } D_S = r'_R \tag{3.48}$$

3.10 INDUCTANCE OF DOUBLE-CIRCUIT THREE PHASE LINES

It is a common practice to run two circuits of three phase transmission lines on the same transmission towers so as to increase the reliability of power transmission. It is also desirable to have low value of mutual GMD, D_m, and high value of self-GMD, D_s, so that the inductance of the parallel lines has a low value and power transfer over the lines increases. Thus, the individual conductors of a phase should be spaced as widely as possible while the distances between phases are kept as low as permissible.

The three sections of a commonly used transposition cycle of a double circuit, three phase transmission line with vertical spacing is shown in Figure 3.8.

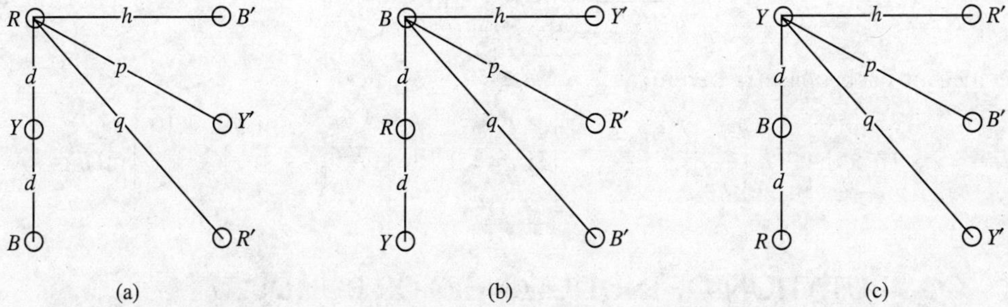

Figure 3.8 Arrangement of conductors in a double circuit three phase transposed line (a) Phase *R* in position 1 (b) Phase *R* in position 2 (c) Phase *R* in position 3.

The conductors *R, Y, B* belong to one circuit while conductors *R′, Y′, B′* belong to the other circuit. The various distances between conductors are shown in Figure 3.8. The flux linkages ψ_{R1}, ψ_{R2}, ψ_{R3} of phase *R* in position-1, position-2, and position-3 respectively are

$$\psi_{R1} = 2\times10^{-7}\left[I_R\left(\ln\frac{1}{r'}+\ln\frac{1}{q}\right)+I_Y\left(\ln\frac{1}{D}+\ln\frac{1}{p}\right)+I_B\left(\ln\frac{1}{2D}+\ln\frac{1}{h}\right)\right] \quad (3.49)$$

$$\psi_{R2} = 2\times10^{-7}\left[I_R\left(\ln\frac{1}{r'}+\ln\frac{1}{h}\right)+I_Y\left(\ln\frac{1}{D}+\ln\frac{1}{p}\right)+I_B\left(\ln\frac{1}{D}+\ln\frac{1}{p}\right)\right] \quad (3.50)$$

$$\psi_{R3} = 2\times10^{-7}\left[I_R\left(\ln\frac{1}{r'}+\ln\frac{1}{q}\right)+I_Y\left(\ln\frac{1}{2D}+\ln\frac{1}{h}\right)+I_B\left(\ln\frac{1}{D}+\ln\frac{1}{p}\right)\right] \quad (3.51)$$

Hence,

$$\psi_R = \frac{\psi_{R1}+\psi_{R2}+\psi_{R3}}{3}$$

$$= \frac{2\times10^{-7}}{3}\left[3I_R\ln\frac{1}{r'}+I_R\ln\frac{1}{q^2h}+(I_Y+I_B)\ln\frac{1}{2D^3}+(I_Y+I_B)\ln\frac{1}{p^2h}\right] \quad (3.52)$$

Since $I_R + I_Y + I_B = 0$, therefore

$$\psi_R = \frac{2\times10^{-7}}{3}\left[3I_R\ln\frac{1}{r'}+I_R\ln\frac{1}{q^2h}-I_R\ln\frac{1}{2D^3}-I_R\ln\frac{1}{p^2h}\right]$$

$$= \frac{2\times10^{-7}}{3}\times 3I_R\left[\ln\left(\frac{2^{1/3}\,Dp^{2/3}\,h^{1/3}}{r'q^{2/3}\,h^{1/3}}\right)\right]$$

$$= 2\times10^{-7}\times I_R\ln\left(\frac{2^{1/3}\,D}{r'}\times\frac{p^{2/3}}{q^{2/3}}\right)$$

Thus, $$L_R = \frac{\psi_R}{I_R} = 2\times10^{-7}\ln\left[\frac{2^{1/3}\,D}{r'}\left(\frac{p}{q}\right)^{2/3}\right]\text{ H/m} \quad (3.53)$$

Inductance of each phase is given by

$$L = \frac{L_R}{2} = 2\times10^{-7}\ln\left[2^{1/6}\left(\frac{D}{r'}\right)^{1/2}\left(\frac{p}{q}\right)^{2/3}\right]\text{ H/m} = 0.2\ln\left[2^{1/6}\left(\frac{D}{r'}\right)^{1/2}\left(\frac{p}{q}\right)^{2/3}\right]\text{ mH/km} \quad (3.54)$$

3.11 COMPUTATION OF INDUCTANCE FOR BUNDLED CONDUCTORS

For EHV and UHV lines, bundled conductors having two or more conductors per phase in close proximity are used. The use of bundled conductors reduces corona loss and decreases the reactance of the transmission line. The reduction of reactance results from increased GMR of the bundle. The arrangements of bundles with two, three, and four conductors are shown in Figure 3.9.

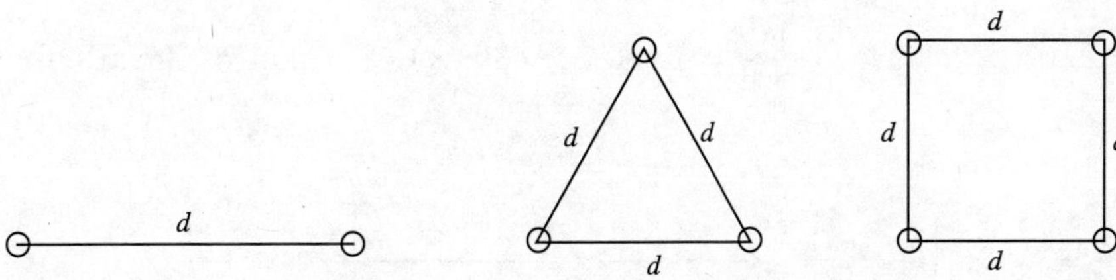

Figure 3.9 Arrangement of bundled conductors.

Let D_s^b indicates GMR of a bundled conductor, D_s, the GMR of the individual conductors, and d, the distance between adjacent conductors of a bundled conductor (Figure 3.9).

For a two-conductor bundle,

$$D_s^b = [(D_s \times d)^2]^{1/4} = (D_s \times d)^{1/2} \tag{3.55}$$

For a three-conductor bundle,

$$D_s^b = [(D_s \times d \times d)^3]^{1/9} = (D_s \times d \times d)^{1/3} \tag{3.56}$$

For a four-conductor bundle,

$$D_s^b = [(D_s \times d \times d \times \sqrt{2}d)^4]^{1/16} = 1.09\,(D_s \times d^3)^{1/4} \tag{3.57}$$

For computing inductance using equation (3.53), D_s^b replaces D_s of a single conductor. To compute D_{eq}, the distance from the centre of one bundle to the centre of another bundle is sufficiently accurate for D_{RY}, D_{YB}, and D_{BR}.

EXAMPLE 3.1 In a single phase line as shown in Figure E3.1, sub-conductors a and a' in parallel form one conductor while sub-conductors b and b' in parallel form the return path. Calculate the total inductance of the line per km assuming that current is equally shared by the two parallel conductors. The diameter of the conductor is 2.0 cm.

Solution: Loop inductance per km

$$= 4 \times 10^{-4}\ \text{In}\ \frac{D_m}{D_s}\ \text{H/km}$$

where D_m = mutual GMD

and D_s = self GMD (GMR)

Here $D_m = \sqrt[4]{D_{ab}D_{ab'}D_{a'b}D_{a'b'}}$

$$= \sqrt[4]{120 \times 140 \times 100 \times 120} = 119\ \text{cm}$$

and $D_s = \sqrt[4]{D_{aa}D_{aa'}D_{a'a'}D_{a'a}}$

where $D_{aa} = D_{a'a'} = 0.7788$ cm and $D_{aa'} = D_{a'a} = 20$ cm

So that $D_s = \sqrt[4]{(0.7788 \times 20)^2} = 3.94$ cm

$$\therefore \quad L = 4 \times 10^{-4} \times 2.303 \log_{10} \frac{119}{3.94} = 4 \times 10^{-4} \times 2.303 \times 1.481 \text{ H/km} = 1.36 \text{ mH/km} \quad \text{(Ans.)}$$

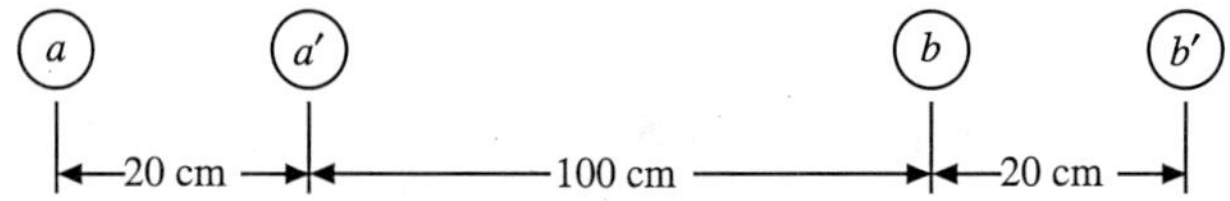

Figure E3.1

EXAMPLE 3.2 Figure E3.2 shows the arrangement of a double circuit single phase line. Conductors a, a', form one connection and conductors b, b' form return connection. The distance between a and a' is 90 cm and between a and b is 175 cm as shown. Calculate the inductance per km of the line (go and return). The diameter of each conductor is 2.5 cm.

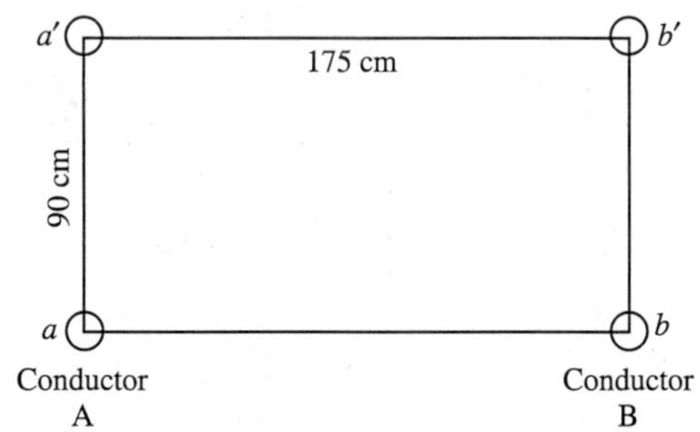

Figure E3.2

Solution: r = 1.25 cm; GMR = 0.7788 × 1.25 = 0.975 cm

Self GMD of conductor A = self GMD (GMR) of conductor B

$$= \sqrt[4]{D_{aa}\, D_{aa'}\, D_{a'a'}\, D_{a'a}} = \sqrt[4]{(0.975 \times 90)^2} = 9.35 \text{ cm}$$

$$D_{ab'} = \sqrt{90^2 + 175^2} = 196.5 \text{ cm}$$

$$\text{Mutual GMD} = \sqrt[4]{(D_{ab}\, D_{ab'})^2} = \sqrt[4]{(175 \times 196.5)^2} = 185 \text{ cm}$$

$$\therefore \quad L = 2 \times 2 \times 10^{-4} \text{ In} \frac{D_m}{D_s} = 4 \times 10^{-4} \times 2.303 \log_{10} \frac{185}{9.35}$$

$$= 4 \times 10^{-4} \times 2.303 \times 1.2967 = 1.2 \times 10^{-3} \text{ H/km} \quad \text{(Ans.)}$$

EXAMPLE 3.3 A three phase 80 km long transmissioin line has its conductors of 1.0 cm diameter spaced at the corners of the equilateral triangle of 100 cm side. Find the inductance per phase of the system.

Solution: Radius of the conductor, $r = \dfrac{1.0}{2}$ = 0.5 cm.

$$\text{GMR} = 0.7788 \times r = 0.7788 \times 0.5 = 0.39 \text{ cm}$$

Spacing of conductors, D = 100 cm.

Inductance per phase, $L = 2 \times 10^{-7} \ln \dfrac{D}{r'}$ H/m

$\therefore$ For the entire length, $L = 2 \times 10^{-7} \ln \left(\dfrac{100}{0.39}\right) \times 80 \times 1000 \text{ H} = 0.8875 \text{ H}$. (Ans.)

EXAMPLE 3.4 Two conductors of a single phase line, each of 3 cm diameter are arranged in a vertical plane with one conductor mounted 1 m above the other. A second identical line is mounted at the same height as the first and spaced horizontally 0.25 m apart from it. The two upper and lower conductors are connected in parallel. Determine the inductance per km of the resulting double circuit line.

Solution: Figure E3.3 shows the arrangement of double circuit single phase line. Conductors a, a' form one connection and conductors b, b' form the return connection.

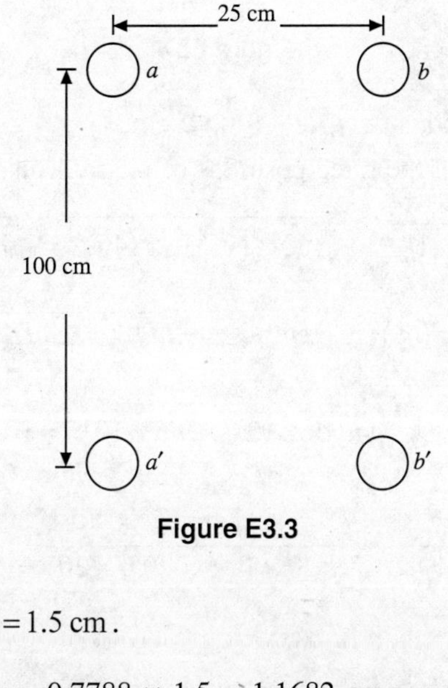

Figure E3.3

The conductor radius, $r = \dfrac{3}{2} = 1.5 \text{ cm}$.

GMR of conductor = 0.7788r = 0.7788 × 1.5 = 1.1682 cm
Self GMD (GMR) of aa' combination.

$$D_S = \sqrt[4]{D_{aa} \times D_{aa'} \times D_{a'a} \times D_{a'a'}} = \sqrt[4]{(1.1682)^2 \times (100)^2} = 10.80 \text{ cm}$$

Mutual GMD between a and b

$$D_m = \sqrt[4]{D_{ab} \times D_{ab'} \times D_{a'b} \times D_{a'b'}} = \sqrt[4]{(25)^2 \times (103)^2} = 50.75 \text{ cm}$$

$$[\because D_{aa} = D_{a'a'} = 0.7788 \times 1.5 = 1.1682 \text{ cm}, D_{a'a} = D_{aa'} = 100 \text{ cm},$$

$$D_{a'b'} = D_{ab} = 25 \text{ cm}, D_{ab'} = D_{a'b} = \sqrt{25^2 + 100^2} = 103 \text{ cm}]$$

∴ Inductance/conductor/mt.

$$= 2 \times 10^{-7} \ln \frac{D_m}{D_S} = 2 \times 10^{-7} \ln \frac{50.75}{10.80} = 3.094 \times 10^{-7} \text{ H}$$

Loop inductance/km of the line = 2 × 3.094 × 10^{-7} × 1000

= 6.18 × 10^{-4} H = 0.618 mH (Ans.)

EXAMPLE 3.5 Find the inductance per phase per km length of the system of conductors shown in Figure E3.4. Self GMD of one conductor is 0.90 cm. Assume transposition.

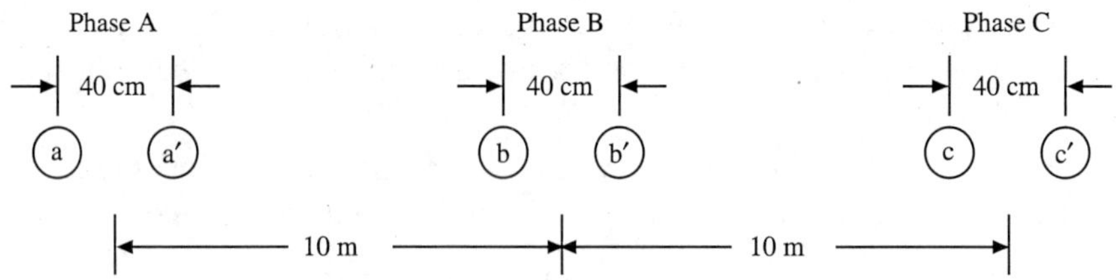

Figure E3.4

Solution: Self GMD of conductor a, D_{aa} = 0.90 cm.

$D_{s1} = D_{s2} = D_{s3}$ because in all the three positions of phase A distance between aa' is the same.

$$D_{s1} = \sqrt[4]{D_{aa}\, D_{aa'}\, D_{a'a'}\, D_{a'a}} = \sqrt[4]{(90 \times 10^{-2})^2 \times (40)^2} = \sqrt{90 \times 10^{-2} \times 40} = 6 \text{ cm}$$

$$\therefore \qquad D_s = \sqrt[3]{D_{s1}\, D_{s2}\, D_{s3}} = (6^3)^{\frac{1}{3}} = 6 \text{ cm}; \; D_m = \sqrt[3]{D_{AB}\, D_{BC}\, D_{CA}}$$

where $D_{AB} = \sqrt[4]{D_{ab}\, D_{ab'}\, D_{a'b}\, D_{a'b'}} = (10 \times 10.4 \times 9.6 \times 10)^{\frac{1}{4}} = 10 \text{ m}$

$$D_{BC} = D_{AB} = 10 \text{ m}$$

$$D_{CA} = \sqrt[4]{D_{ca}\, D_{ca'}\, D_{c'a}\, D_{c'a'}} = \sqrt[4]{(20 \times 19.6 \times 20.4 \times 20)} = 20 \text{ m}$$

$$\therefore \qquad D_m = (10 \times 10 \times 20)^{\frac{1}{3}} = 12.6 \text{ m}$$

Inductance per phase $= 2 \times 10^{-4} \ln (D_m / D_S) = 2 \times 10^{-4} \ln \dfrac{12.6 \times 100}{6} = 1.068 \text{ mH/km}$ (Ans.)

EXAMPLE 3.6 Figure E3.5 shows a stranded conductor having 7 identical strands each of radius r. Find the geometric mean radius of the conductor and the ratio of GMR to overall conductor radius. Comment on the results.

Solution: The distances from strand 1 to the other strands are as follows (ref. Figure E3.5):

$$D_{12} = D_{16} = D_{17} = 2r; \; D_{14} = 4r$$

$$D_{13} = \sqrt{D_{14}^2 - D_{34}^2} = \sqrt{(4r)^2 - (2r)^2}$$

$$D_{13} = 2r\sqrt{3}, D_{15} = D_{13}$$

The geometric mean radius of the seven strands conductor is the 49th root of the product of 49 distances

$$D_s = \sqrt[49]{(r')^7 \, (D_{12}^2 \, D_{13}^2 \, D_{14}^2 \, D_{17}^2)^6 \, (2r)^6}$$

where r' is the GMR of each strand. It is raised to the seventh power to account for seven strands. The term $(D_{12}^2 \, D_{13}^2 \, D_{14} \, D_{17})$ represents the product of distance from one outside strand to every other strand. This term is raised to the sixth power to account for six outside strands. The term $(2r)^6$ accounts for the product of the distances from the central strand to every outside strand

$$D_s = \sqrt[49]{(0.7788r)^7 \, [(2r)^2 \, (2\sqrt{3r})^2 \cdot (4r) \, (2r)] \, (2r)^6} = 2.177r \quad \text{(Ans.)}$$

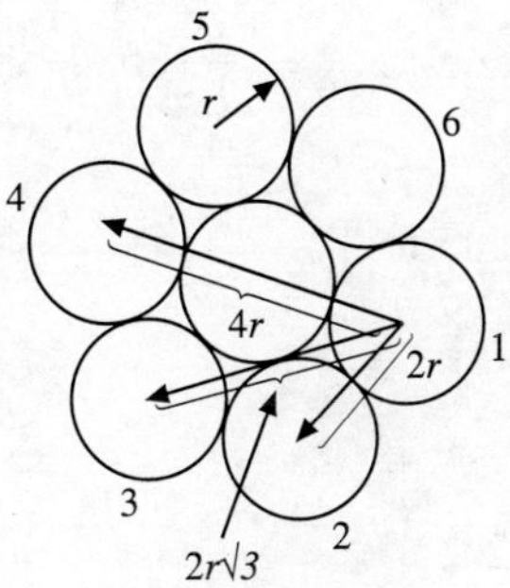

Figure E3.5

The overall conductor radius is $3r$. Therefore, the ratio of GMR to overall condcutor radius is $2.177\ r/3r = 0.7257$. As the number of strands increases, this ratio approaches to 0.7788 which is that of a solid conductor.

3.12 POTENTIAL DIFFERENCE BETWEEN TWO POINTS DUE TO A CHARGE

We know from *Gauss's law* that the electric flux density (D) at a point distance x (metres) from a conductor having a charge q (coulomb/metre length) is given by $D = \dfrac{q}{2\pi x.1}$ coulomb/m^2.

Thus electric field intensity at the point considered $E = D = \dfrac{q}{2\pi x.1}$ volts/metre

If we want to find the potential difference (p.d.) between two points P_1 and P_2 situated at distances D_1 and D_2 from the charge, we can do so by computing the voltage between the equipotential surfaces passing through P_1 and P_2 by integrating the electric field intensity over a radial path between the equipotential surfaces.

Instantaneous voltage drop (potential difference) between P_1 and P_2 is

$$v_{12} = \int_{D_1}^{D_2} E dx = \int_{D_1}^{D_2} \frac{q}{2\pi\varepsilon x} dx = \frac{q}{2\pi\varepsilon} \ln \frac{D_2}{D_1} \text{ volts}$$

where q is the instantaneous charge (coulomb/metre length) of the wire.

3.13 CAPACITANCE OF A TWO WIRE LINE (LINE CAPACITANCE)

Figure 3.10 shows the cross-section of such a line.

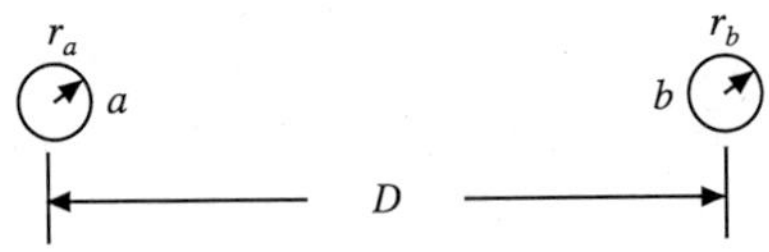

Figure 3.10 A two wire line.

$$v_{ab} \text{ due to } q_a = \frac{q_a}{2\pi\varepsilon} \ln \frac{D}{r_a} \text{ volts}$$

$$v_{ab} \text{ due to } q_b = \frac{q_b}{2\pi\varepsilon} \ln \frac{r_b}{D} \text{ volts}$$

In phasor notation, therefore,

$$V_{ab} = \frac{1}{2\pi\varepsilon} \left\{ q_a \ln \frac{D}{r_a} + q_b \ln \frac{r_b}{D} \right\} \text{ volts}$$

where q_a and q_b are both complex quantities.

We also note that for a two wire line $q_a = - q_b$ so that

$$V_{ab} = \frac{1}{2\pi\varepsilon} q_a \ln \left(\frac{D^2}{r_a r_b} \right) = \frac{1}{\pi\varepsilon} q_a \ln \frac{D}{\sqrt{r_a r_b}} \text{ volts}$$

Therefore

$$C_{ab} = \frac{q_a}{V_{ab}} = \frac{\pi\varepsilon}{\ln (D / \sqrt{r_a r_b})} \text{ Farads/metre (F/m)} \quad (3.58a)$$

If

$$r_a = r_b = r, C_{ab} = \frac{\pi\varepsilon}{\ln (D / r)} \text{ Farads/metre (F/m)} \quad (3.58b)$$

In this expression C_{ab} denotes the capacitance between lines and ε (the permittivity of the medium) = $\varepsilon_0 \varepsilon_r$.

If we assume $\varepsilon_r = 1$, then $\varepsilon = \varepsilon_0 = \frac{1}{4\pi \times 9 \times 10^9}$ Farads/metre (F/m)

Substituting this value of ε, we have

$$C_{ab} = \frac{10^{-9}}{36 \ln \frac{D}{r}} \text{ F/m} = \frac{1}{36 \ln \frac{D}{r}} \mu\text{F/km} \tag{3.59}$$

As capacitance to neutral (C_n) is twice the line-to-line capacitance

$$\therefore \quad C_n = C_{an} = C_{bn} = \frac{1}{18 \ln \frac{D}{r}} = \frac{0.0555}{\ln \frac{D}{r}} \mu\text{F/km} \tag{3.60}$$

The reader will also note that in capacitance formulae the radius is the outside radius of the conductor and not the self GMD of the conductor as was the case with inductance formulae.

3.14 POTENTIAL DIFFERENCE BETWEEN TWO CONDUCTOR OF A GROUP OF CHARGED CONDUCTOR

If a number of conductors are arranged so that they are parallel to each other, the voltage between any two of them can be found by applying equation $v_{12} = \frac{q}{2\pi\varepsilon} \ln \frac{D_2}{D_1}$ volts repeatedly to determine the voltage between the two conductors in question due to the charge on each conductor in the group independently. The voltage drop between the two conductors is the sum of the voltage drops due to each charged conductor. This method is a fundamental method in deriving formulae for capacitance.

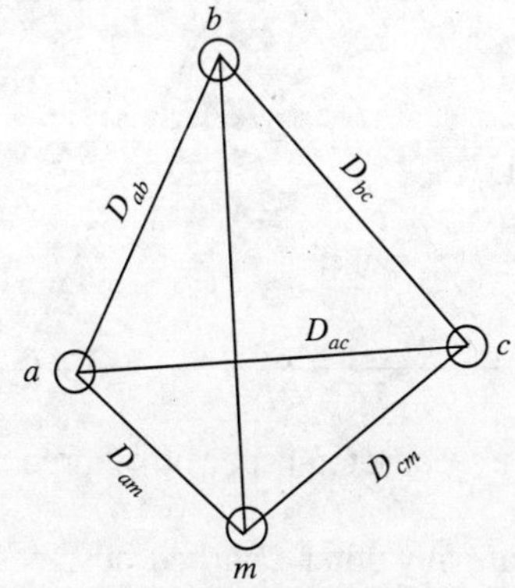

Figure 3.11 Cross-section of a group of parallel conductors.

A group of parallel conductors is shown in Figure 3.11. If we assume that there be no other charged surfaces in the vicinity, the sum of the charges on the conductors is zero. If the ground is far enough away to have negligible effect and if we assume further that the spacing between conductors be large compared with the radius of any one, so that the charge distribution over the surface of a conductor becomes uniform, repeated application of $v_{12} = \frac{q}{2\pi\varepsilon} \ln \frac{D_2}{D_1}$ volts will yield accurate results.

Hence from conductors a to b the voltage drop in phasor notation is

$$V_{ab} = \frac{1}{2\pi\varepsilon}\left(q_a \ln\frac{D_{ab}}{r_a} + q_b \ln\frac{r_b}{D_{ba}} + q_c \ln\frac{D_{cb}}{D_{ca}} + \cdots + q_m \ln\frac{D_{mb}}{D_{ma}}\right) \tag{3.61a}$$

Each term in the above equation is the potential drop from a to b due to the charge on one of the conductors in the group, e.g. first term in the bracket represents the p.d. from a to b due to charge on a, second term represents p.d. from a to b due to charge on b and so on.

$$V_{ac} = \frac{1}{2\pi\varepsilon}\left(q_a \ln\frac{D_{ac}}{r_a} + q_b \ln\frac{D_{bc}}{D_{ba}} + q_c \ln\frac{r_c}{D_{ca}} + \cdots + q_m \ln\frac{D_{mc}}{D_{ma}}\right) \tag{3.61b}$$

and

$$V_{am} = \frac{1}{2\pi\varepsilon}\left(q_a \ln\frac{D_{am}}{r_a} + q_b \ln\frac{D_{bm}}{D_{ba}} + q_c \ln\frac{D_{ca}}{D_{ca}} + \cdots + q_m \ln\frac{r_m}{D_{ma}}\right) \tag{3.61c}$$

If the voltages between conductors a and other conductors are known the group of simultaneous equations expressing the voltage drops in terms of charges may be solved for the charges.

3.15 CAPACITANCE OF THREE PHASE LINES

(*a*) *Equilateral Spacing:* Figure 3.12 shows a three phase line with conductors a, b and c spaced at the equilateral triangle, each side being D. Each conductor is having radius r.

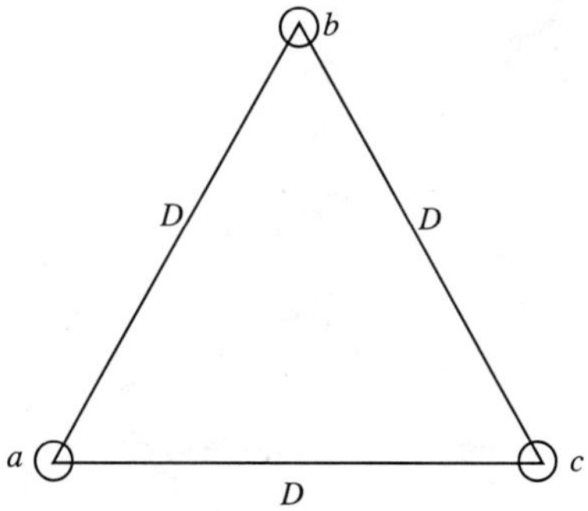

Figure 3.12 Equilateral spacing of three phase line.

We will assume that there are no other charged surfaces in the vicinity so that sum of charges q_a, q_b and q_c is zero (in other words, we are neglecting the effect of ground); also we will assume that the spacing between conductors is usually large compared with the radius of any one conductor and the charges are uniformly distributed over the surfaces of the conductors.

The voltage drop between any two conductors is the sum of the voltage drops due to each charged conductor. So we can write phasor voltages from a to b and a to c as

$$V_{ab} = \frac{1}{2\pi\varepsilon}\left\{q_a \ln\frac{D}{r} + q_b \ln\frac{r}{D} + q_c \ln\frac{D}{D}\right\}$$

and $$V_{ac} = \frac{1}{2\pi\varepsilon}\left\{q_a \ln\frac{D}{r} + q_b \ln\frac{D}{D} + q_c \ln\frac{r}{D}\right\} \text{ volts}$$

Adding V_{ab} and V_{ac} and remembering that $q_b + q_c = -q_a$, we get

$$V_{ab} + V_{ac} = \frac{1}{2\pi\varepsilon}\left\{2q_a \ln\frac{D}{r} - q_a \ln\frac{r}{D}\right\} = \frac{3q_a}{2\pi\varepsilon}\ln\frac{D}{r} \text{ volts}$$

Phasor diagram of voltages V_{ab}, V_{bc} and V_{ca} is shown in Figure 3.13. Phasor addition of V_{ab}, V_{bc} and V_{ca} should give zero; relationship of these voltages with V_{an}, V_{bx} and V_{cn} is also shown.

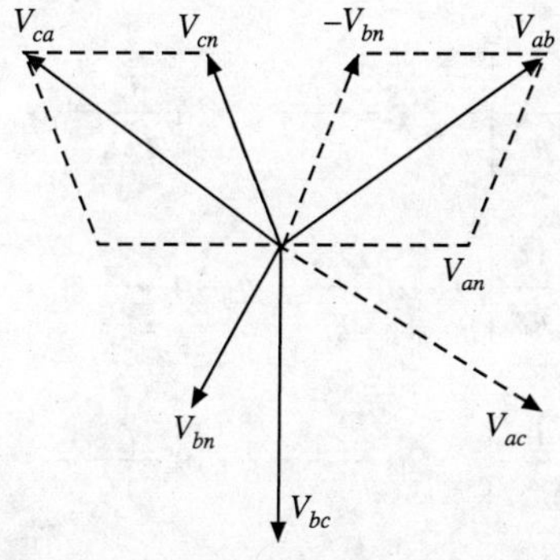

Figure 3.13 Phasor diagram of three phase system.

It is clear from this diagram that $V_{ab} + V_{ac} = 3V_{an}$

So that $V_{an} = \dfrac{q_a}{2\pi\varepsilon}\ln\dfrac{D}{r}$ volts

and $$C_n = \frac{q_a}{V_{an}} = \frac{2\pi\varepsilon}{\ln(D/r)} \text{ F/m} \tag{3.62}$$

As in the case of one phase lines, assuming $\varepsilon_r = 1$; $\varepsilon_0 = [1/4\pi \times 9 \times 10^9]$,

$$C_n = \frac{1}{18\ln(D/r)} = \frac{0.0555}{\ln(D/r)}\,\mu\text{F/km} \tag{3.63}$$

(The reader will recall that this expression is identical with that for one phase lines.)

(b) Unsymmetrical Spacing but Transposed: Such a line presents difficulty in treatment on account of the fact that capacitances from each phase to neutral are unequal. Transposition results in the same average capacitance to neutral of various phases over the entire length of the line.

We may calculate the capacitance of each phase of an untransposed line as the average value of one phase of the same line correctly transposed with negligible error on account of slight dissymmetry in actual practice.

Figure 3.14 shows the arrangement. For a transposed line, a conductor (say, phase *a*) will occupy positions 1, 2 and 3 each for one-third of its length and so on. Denoting the voltage drop (say, V_{ab}) in three positions as V_{ab1}, V_{ab2} and V_{ab3} respectively, we have

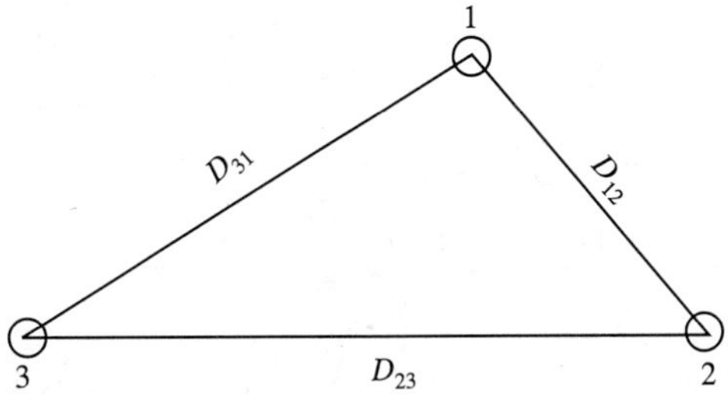

Figure 3.14 Unsymmetrical but transposed spacing of three phase conductors.

$$V_{ab1} = \frac{1}{2\pi\varepsilon}\left[q_a \ln\frac{D_{12}}{r} + q_b \ln\frac{r}{D_{12}} + q_c \ln\frac{D_{23}}{D_{31}}\right]$$

$$V_{ab2} = \frac{1}{2\pi\varepsilon}\left[q_a \ln\frac{D_{23}}{r} + q_b \ln\frac{r}{D_{23}} + q_c \ln\frac{D_{31}}{D_{12}}\right]$$

$$V_{ab3} = \frac{1}{2\pi\varepsilon}\left[q_a \ln\frac{D_{31}}{r} + q_b \ln\frac{r}{D_{31}} + q_c \ln\frac{D_{12}}{D_{23}}\right]$$

Therefore, average voltage between conductors a and b may be written as

$$V_{ab} = \frac{1}{6\pi\varepsilon}\left\{q_a \ln\frac{D_{12}\,D_{23}\,D_{31}}{r^3} + q_b \ln\frac{r^3}{D_{12}\,D_{23}\,D_{31}} + q_c \ln\frac{D_{12}\,D_{23}\,D_{31}}{D_{12}\,D_{23}\,D_{31}}\right\}$$

$$= \frac{1}{2\pi\varepsilon}\left\{q_a \ln\frac{\sqrt[3]{D_{12}D_{23}D_{31}}}{r} + q_b \ln\frac{r}{\sqrt[3]{D_{12}D_{23}D_{31}}}\right\}$$

Denoting $\sqrt[3]{D_{12}D_{23}D_{31}}$ as D_{eq}, $V_{ab} = \frac{1}{2\pi\varepsilon}\left\{q_a \ln\frac{D_{eq}}{r} + q_b \ln\frac{r}{D_{eq}}\right\}$

Similarly, $$V_{ac} = \frac{1}{2\pi\varepsilon}\left\{q_a \ln\frac{D_{cq}}{r} + q_e \ln\frac{r}{D_{eq}}\right\}$$

As $V_{ab} + V_{ac} = 3\,V_{an}$ (See Figure 3.13) and $q_b + q_c = -q_a$

$\therefore$ $$V_{an} = \frac{1}{3\times 2\pi\varepsilon}\left\{2q_a \ln\frac{D_{eq}}{r} - q_c \ln\frac{r}{D_{eq}}\right\} = \frac{1}{2\pi\varepsilon} q_a \ln\frac{D_{eq}}{r} \text{ volts}$$

and $$C_n = \frac{q_a}{V_{an}} = \frac{2\pi\varepsilon}{\ln(D_{eq}/r)} \text{ F/m}$$

Assuming $\varepsilon_r = 1$, the expression yields

$$C_n = \frac{1}{18 \ln\left(\frac{D_{eq}}{r}\right)} = \frac{0.0555}{\ln\left(\frac{D_{eq}}{r}\right)} \ \mu\text{F/km} \qquad (3.64)$$

Thus in the case of unsymmetrical spacing of conductors we have to use equivalent spacing (which is the geometric mean of three distances of the unsymmetrical line) instead of equilateral spacing.

3.16 CHARGING CURRENT DUE TO CAPACITANCE

For a single phase line *charging current* $(I_{chg}) = \dfrac{V}{-j/\omega C} = j\omega CV$ amps (3.65)

where $\omega = 2\pi f$ (f in Hz), C is capacitance between lines (farads) and V = line-to-line voltage (volts).

For a three phase line charging current per phase $I_{chg} = \dfrac{V}{-j/\omega C} = j\omega CV$ amps (3.66)

where C is capacitance to neutral (farads) and V = voltage to neutral (volts), $\omega = 2\pi f$, f being the frequency in Hz.

3.17 EFFECT OF EARTH ON THE CAPACITANCE OF LINE

The presence of earth changes the electric field of the line and so it affects its capacitance. Earth may be assumed to be a perfect conductor in the form of horizontal plane of infinite extent. Therefore, the electric field of charged conductors is forced to conform to a presence of this equipotential surface.

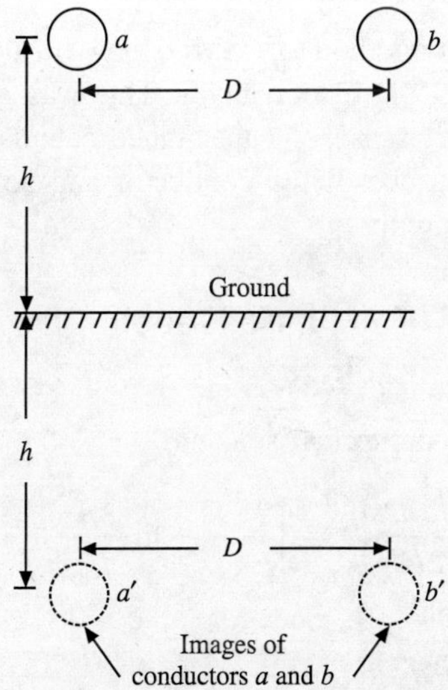

Figure 3.15 Charges and their images of a two wire line.

Figure 3.15 shows a two wire line having both conductors a and b (radius of each conductor r) at height h above ground. In this case for finding out V_{ab} we will have to take into account charges q_a, q_b due to conductors a and b and $-q_a$, $-q_b$ due to their images.

Proceeding as in the case of capacitance of a two wire line we can write,

$$V_{ab} = \frac{1}{2\pi\varepsilon}\left\{q_a \ln\frac{D}{r} + q_b \ln\frac{r}{D}\right\} + \frac{1}{2\pi\varepsilon}\left\{-q_a \ln\frac{\sqrt{D^2+4h^2}}{2h} - q_b \ln\frac{2h}{\sqrt{D^2+4h^2}}\right\}$$

$$= \frac{1}{2\pi\varepsilon}\left\{q_a \ln\frac{2hD}{r\sqrt{D^2+4h^2}} + q_b \ln\frac{r\sqrt{D^2+4h^2}}{2hD}\right\} \tag{3.67}$$

Since $q_a = -q_b$

$$\therefore \quad V_{ab} = \frac{1}{2\pi\varepsilon}\left\{q_a \ln\frac{4h^2D^2}{r^2(D^2+4h^2)}\right\} = \frac{1}{2\pi\varepsilon}\cdot 2q_a \ln\frac{D}{r\sqrt{1+D^2/4h^2}} \text{ volts} \tag{3.68}$$

giving

$$C_{ab} = \frac{q_a}{V_{ab}} = -\frac{\pi\varepsilon}{\ln\left(\dfrac{D}{r\sqrt{1+D^2/4_h^{\,2}}}\right)} \text{ F/m} \tag{3.69}$$

This equation shows that earth effect is to increase the capacitance by a small amount. However, the effect is almost negligible if the conductors are high above ground compared with the distance between them (as can be seen from the above equation).

3.18 CAPACITANCE OF THREE PHASE LINES WITH MORE THAN ONE CIRCUIT

For multi-circuit lines modified method of GMD may be used. In the case of inductance calculation, we had found self GMD of conductor. This was necessary on account of internal flux linkages of conductor. In the case of capacitance calculations we have to bear in mind that all charges reside on the surface of the conductor and so we will have to use the actual radius of the conductor of the circuit.

3.18.1 Capacitance of a Three Phase Double Circuit Line with Symmetrical Spacing

Conductors are arranged with symmetrical spacing.

$$V_{ab} = \frac{1}{2\pi\varepsilon}\left\{q_a\left(\ln\frac{D}{r} + \ln\frac{\sqrt{3}\,D}{2D}\right) + q_b\left(\ln\frac{r}{D} + \ln\frac{2D}{\sqrt{3}\,D}\right) + q_c\left(\ln\frac{D}{\sqrt{3}\,D} + \ln\frac{\sqrt{3}\,D}{D}\right)\right\}$$

$$= \frac{1}{2\pi\varepsilon}\left\{q_a \ln\left(\frac{\sqrt{3}\,D}{2r}\right) + q_b \ln\left(\frac{2r}{\sqrt{3}\,D}\right) + q_c \ln\left(\frac{\sqrt{3}\,D^2}{\sqrt{3}\,D^2}\right)\right\}$$

$$= \frac{1}{2\pi\varepsilon}\left\{q_a \ln \frac{\sqrt{3}\,D}{2r} + q_b \ln \frac{2r}{\sqrt{3}\,D}\right\}$$

Similarly
$$V_{ac} = \frac{1}{2\pi\varepsilon}\left\{q_a \ln \frac{\sqrt{3}\,D}{2r} + q_c \ln \frac{2r}{\sqrt{3}\,D}\right\}$$

$\therefore$
$$V_{ab} + V_{ac} = 3V_{an} = \frac{1}{2\pi\varepsilon}\left\{2q_a \ln \frac{\sqrt{3}\,D}{2r} + (q_b + q_c) \ln \frac{2r}{\sqrt{3}\,D}\right\}$$

Since $q_a + q_b + q_c = 0$

$\therefore$
$$3V_{an} = \frac{3q_a}{2\pi\varepsilon} \ln \frac{\sqrt{3}D}{2r} \quad \text{or} \quad V_{an} = \frac{q_a}{2\pi\varepsilon} \ln \frac{\sqrt{3}D}{2r}$$

and
$$C_{an} = \frac{q_a}{V_{an}} = \frac{2\pi\varepsilon}{\ln \dfrac{\sqrt{3}D}{2r}} \text{ F/m} = \frac{0.0555}{\ln \dfrac{\sqrt{3}D}{2r}}\ \mu\text{F/km} \tag{3.70}$$

[Capacitance per phase (C_n) will be $2C_{an}$.]

3.18.2 Capacitance of a Three Phase Double Circuit Line with Unsymmetrical Spacing (Thoroughly Transposed)

Conductors are arranged corresponding to different positions in the transposition cycle.

$$V_{ab1} = \frac{1}{2\pi\varepsilon}\left\{q_a\left(\ln\frac{D}{r} + \ln\frac{m}{n}\right) + q_b\left(\ln\frac{r}{D} + \ln\frac{h}{m}\right) + q_c\left(\ln\frac{D}{2D} + \ln\frac{m}{h}\right)\right\} \text{ volts}$$

$$V_{ab2} = \frac{1}{2\pi\varepsilon}\left\{q_a\left(\ln\frac{D}{r} + \ln\frac{m}{h}\right) + q_b\left(\ln\frac{r}{D} + \ln\frac{n}{m}\right) + q_c\left(\ln\frac{2D}{D} + \ln\frac{h}{m}\right)\right\} \text{ volts}$$

$$V_{ab3} = \frac{1}{2\pi\varepsilon}\left\{q_a\left(\ln\frac{2D}{r} + \ln\frac{h}{n}\right) + q_b\left(\ln\frac{r}{2D} + \ln\frac{n}{h}\right) + q_c\left(\ln\frac{D}{D} + \ln\frac{m}{m}\right)\right\} \text{ Volts}$$

$\therefore$
$$V_{ab} = \frac{1}{3}(V_{ab1} + V_{ab2} + V_{ab3})$$

$$= \frac{1}{6\pi\varepsilon}\left\{q_a\left(\ln\frac{2D^3}{r^3} + \ln\frac{m^2h}{n^2h}\right) + q_b\left(\ln\frac{r^3}{2D^3} + \ln\frac{n^2h}{m^2h}\right) + q_c\left(\ln\frac{2D^3}{2D^3} + \ln\frac{m^2h}{m^2h}\right)\right\}$$

$\therefore$
$$V_{ab} = \frac{1}{6\pi\varepsilon}\left\{q_a \ln\frac{2D^3m^2}{r^3n^2} + q_b \ln\frac{r^3n^2}{2D^3m^2}\right\} \text{ volts}$$

Similarly

$$V_{ac} = \frac{1}{6\pi\varepsilon}\left\{q_a \ln \frac{2D^3m^2}{r^3n^2} + q_c \ln \frac{r^3n^2}{2D^3m^2}\right\} \text{ volts}$$

$$\therefore \quad V_{ab} + V_{ac} = 3V_{an} = \frac{1}{6\pi\varepsilon}\left\{2q_a \ln \frac{2D^3m^2}{r^3n^2} + (q_b + q_c) \ln \frac{r^3n^2}{2D^3m^2}\right\}$$

Since $q_a + q_b + q_c = 0$

$$3V_{an} = \frac{1}{6\pi\varepsilon}\left\{3q_a \ln \frac{2D^3m^2}{r^3n^2}\right\} \text{ volts}$$

$$\therefore \quad C_{an} = \frac{q_a}{V_{an}} = \frac{6\pi\varepsilon}{\ln \dfrac{2D^3m^2}{r^3n^2}} = \frac{2\pi\varepsilon}{\ln\left\{2^{1/3}\left(\dfrac{D}{r}\right)\left(\dfrac{m}{n}\right)^{2/3}\right\}} \text{ F/m} \tag{3.71}$$

$$\therefore \quad C_{an} = \frac{0.0555}{\ln\left\{2^{1/3}\dfrac{D}{r}\left(\dfrac{m}{n}\right)^{2/3}\right\}} \ \mu\text{F/km}$$

[Capacitance per phase will be $2C_{an}$]

EXAMPLE 3.7 Find the capacitance of a single phase line of 40 km long consisting of 2 parallel wires, each 5 mm in diameter and 1.5 m apart (Fig. E3.6).

Determine the capacitance of the same line taking into account the effect of ground. The height of the conductors above ground is 7 m.

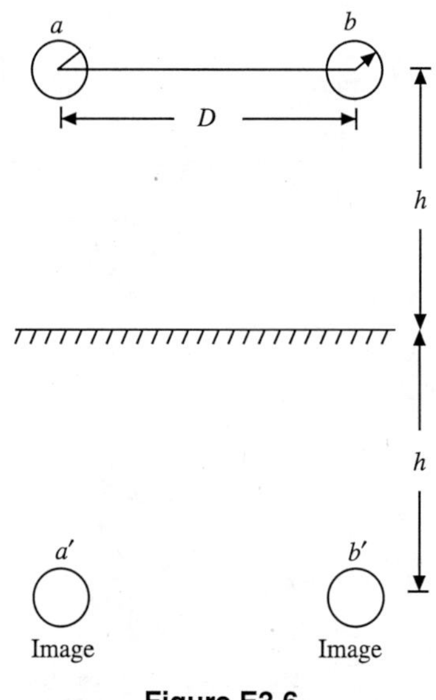

Figure E3.6

Solution: (i) Neglecting the presence of ground

$$C_{ab} = \frac{\pi \in}{\ln (D/r)} \text{ F/m} = \frac{\pi \times 10^{-9}}{36\pi \ln\left(\dfrac{1.5}{2.5 \times 10^{-3}}\right)}$$

$$= \frac{10^{-9}}{36 \ln (600)} = 4.34 \times 10^{-10} \text{ F/m}$$

Capacitance for 40 km = $4.34 \times 10^{-12} \times 40 \times 10^3 \text{ F} = 0.174\ \mu\text{F}$ (Ans.)

(ii) Taking the presence of ground into consideration

$$C_{ab} = \frac{\pi \in}{\ln\left(\dfrac{D}{r\sqrt{1 + D^2/4^2}}\right)} \text{ F/m} = \frac{10^{-9}}{36 \ln \dfrac{1.5}{2.5 \times 10^{-3} \sqrt{1 + 2.25/196}}}$$

$$= \frac{10^{-9}}{36 \ln \dfrac{1.5 \times 10^3}{2.5\sqrt{198.25/196}}} = \frac{10^{-9}}{36 \ln (596)} \text{ F/m} = 4.4 \times 10^{-12} \text{ F/m}$$

Capacitance for 40 km = $4.4 \times 10^{-12} \times 40 \times 10^3 \text{ F} = 0.176\ \mu\text{F}$ (Ans.)

EXAMPLE 3.8 The six conductors of a double circuit three phase line are arranged as shown in Figure E3.7. The diameter of each conductor is 2.5 cm. Find the capacitive reactance to neutral and the charging current per km per phase at 132 kV and 50 Hz assuming that the line is transposed.

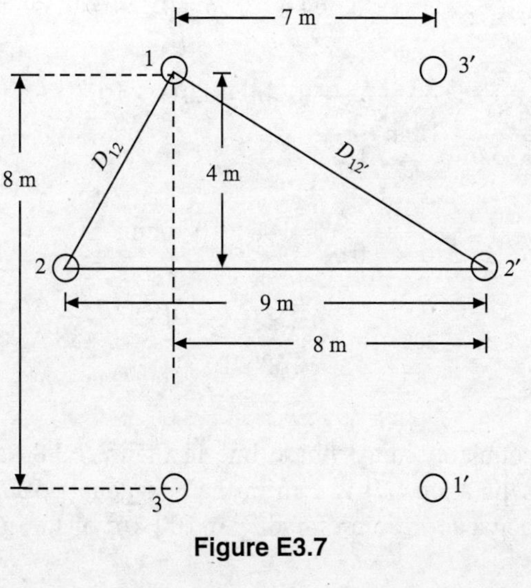

Figure E3.7

Solution:

$$D_{12} = \sqrt{16+1} = 4.12 \text{ m}$$

$$D_{12'} = \sqrt{16+64} = 8.95 \text{ m}$$

$$D_{11'} = \sqrt{64+49} = 10.62 \text{ m}$$

Mutual GMD

$$D_{12} = \sqrt[4]{(4.12\times 8.95)^2} = \sqrt{4.12\times 8.95} = 6.06 \text{ m}$$

$$D_{23} = \sqrt[4]{(4.12\times 8.95)^2} = 6.06 \text{ m}$$

$$D_{31} = \sqrt[4]{(7\times 8)^2} = \sqrt{5612} = 7.49 \text{ m}$$

$$D_{eq} = \sqrt[4]{D_{12}\, D_{23}\, D_{31}} = \sqrt[4]{6.06\times 6.06\times 7.49} = 6.51 \text{ m}$$

Self GMD

$$D_{s1} = \sqrt[4]{\left(\frac{1.25}{100}\times 10.62\right)^2} = \sqrt{\frac{1.25\times 10.62}{100}} = 0.364 \text{ m}$$

$$D_{s2} = \sqrt[4]{\left(\frac{1.25}{100}\times 8.95\right)^2} = \sqrt{\frac{1.25\times 8.95}{100}} = 0.356 \text{ m}$$

$$D_{s3} = \sqrt[4]{\left(\frac{1.25}{100}\times 10.62\right)^2} = 0.364 \text{ m}$$

$$D_s = \sqrt[3]{D_{s1}\, D_{s2}\, D_{s3}} = \sqrt[3]{0.364\times 0.356\times 0.364} = 0.354 \text{ m}$$

$$C_n = \frac{2\pi \in}{\ln \dfrac{6.51}{0.354}} = \frac{10^{-9}}{18 \ln \dfrac{6.5}{0.354}} \text{ F/m} = 0.019\times 10^{-9} \text{ F/m (phase to neutral) (Ans.)}$$

$$X_{cn} = \frac{1}{2\pi f C_n} = \frac{10^9}{314\times 0.019} = 168\times 10^6 \text{ ohms/m}$$

$\therefore$

$$I_{\text{charg}} \text{ (per phase)} = \omega C V = \frac{76.2\times 10^3\times 10^3}{168\times 10^6} = 0.455 \text{ A/km (Ans.)}$$

EXAMPLE 3.9 If the double circuit 3 phase line in Figure E3.8 has conductors of diameter 2.5 cm and distance of separation (D) is 2 m in the hexagonal spacing arrangement, calculate the phase-to-neutral capacitance in microfarads per 100 km of the line.

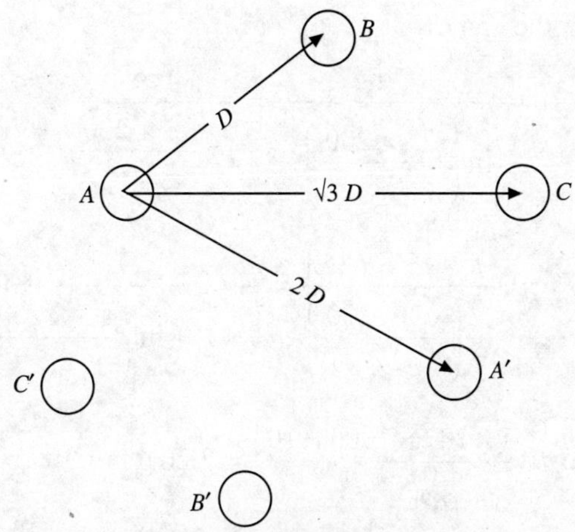

Figure E3.8

Solution: Phase a is composed of conductors A and A', phase b of B and B', phase c of C and C'. The two conductors of each phase are electrically parallel and have the same charge. Because of the symmetrical arrangement, the phases are balanced and the conductors of each individual phase are also balanced if the effect of ground is neglected. Therefore, transposition of the conductors is not necessary to balance the phases.

As shown in text capacitance per phase (to neutral) is given by

$$C_n = 2C_{an} = \frac{2 \times 2\pi\varepsilon}{\left(\ln \dfrac{\sqrt{3}D}{2r}\right)} = \frac{2\pi\varepsilon}{\ln\left(\dfrac{\sqrt{3}D}{2r}\right)^{1/2}} \text{ F/m} \tag{i}$$

Capacitance of one phase to neutral may be expressed as

$$C_n = \frac{2\pi\varepsilon}{\ln \dfrac{D_m}{D_s}}$$

where D_m is the mutual GMD between the phases and D_s is the self GMD between the conductors belonging to the same phase.

It may be seen that $D_m = \sqrt[4]{D \times \sqrt{3}D \times \sqrt{3}D \times D} = \sqrt[4]{3}\, D$ and $D_s = \sqrt{2\, Dr}$

$$\therefore \qquad \frac{D_m}{D_s} = \frac{\sqrt[4]{3}D}{\sqrt{2\, Dr}} = \frac{\sqrt[4]{3}\,\sqrt{D}}{\sqrt{2r}} = \left(\frac{\sqrt{3}D}{2r}\right)^{1/2}$$

$$\therefore \qquad C_n = \frac{2\pi\varepsilon}{\ln\left(\dfrac{\sqrt{3}D}{2r}\right)^{1/2}} \text{ F/m} \tag{ii}$$

(Equations (i) and (ii) are the same)

$$\therefore \quad C_n \text{ (phase to neutral)} = \frac{2\pi\varepsilon}{\ln\left(\dfrac{\sqrt{3}\times 200}{2.5}\right)^{1/2}} = \frac{4\pi\varepsilon}{\ln\dfrac{\sqrt{3}\times 200}{2.5}}$$

$$= \frac{4\pi\times 10^{-9}/36\pi}{\ln\left(\dfrac{\sqrt{3}\times 200}{2.5}\right)} = \frac{1}{9\times 10^{9}\ln\left(\dfrac{\sqrt{3}\times 200}{2.5}\right)} \text{ F/m}$$

$$\text{and } C_n \text{ for 100 km length} = \frac{1\times 100\times 1000\times 10^{6}}{9\times 10^{9}\ln\dfrac{\sqrt{3}\times 200}{2.5}} = 2.26\,\mu\text{F (Ans.)}$$

3.19 CAPACITANCE OF BUNDLED CONDUCTORS

The arrangement of a three phsae line employing bundled conductors is shown in Figure 3.16.

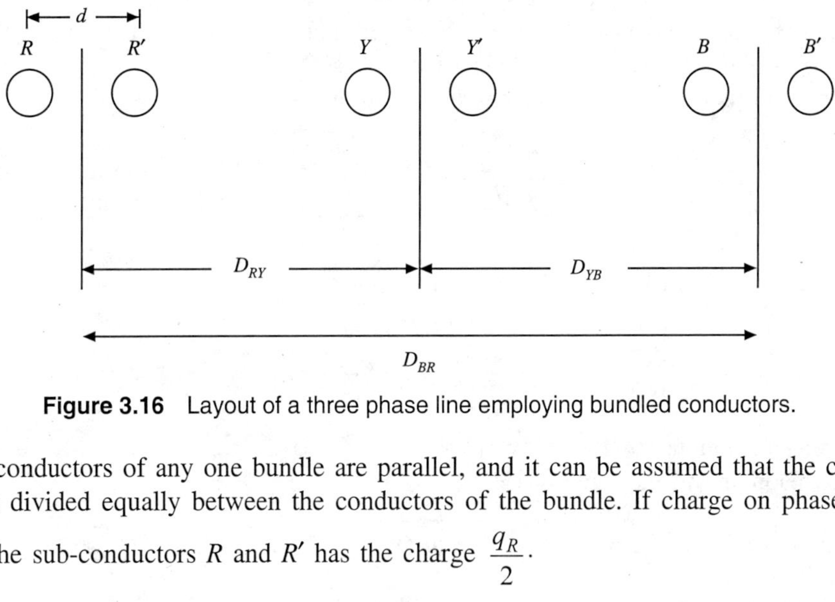

Figure 3.16 Layout of a three phase line employing bundled conductors.

The conductors of any one bundle are parallel, and it can be assumed that the charge per bundle is divided equally between the conductors of the bundle. If charge on phase R is q_R, each of the sub-conductors R and R' has the charge $\dfrac{q_R}{2}$.

$$\text{Then} \quad V_{RY} = \frac{1}{2\pi\varepsilon}\left[\frac{q_R}{2}\left(\ln\frac{D_{RY}}{r}+\ln\frac{D_{RY}}{d}\right)+\frac{q_Y}{2}\left(\ln\frac{r}{D_{RY}}+\ln\frac{d}{D_{RY}}\right)+\frac{q_B}{2}\left(\ln\frac{D_{YB}}{D_{BR}}+\ln\frac{D_{YB}}{D_{BR}}\right)\right]$$

$$= \frac{1}{2\pi\varepsilon}\left[q_R\left(\ln\frac{D_{RY}}{\sqrt{rd}}\right)+q_Y\left(\ln\frac{\sqrt{rd}}{D_{RY}}\right)+q_B\left(\ln\frac{D_{YB}}{D_{BR}}\right)\right] \qquad (3.72)$$

Depending on the line to be transposed, the capacitance may be computed in the usual manner, as described in the earlier section and given by

$$C_n = \frac{2\pi E}{\ln\left(\frac{D_{eq}}{\sqrt{rd}}\right)} \text{ F/m to neutral} \tag{3.73}$$

where

$$D_{eq} = \sqrt[3]{D_{RY}\, D_{YB}\, D_{BR}} \tag{3.74}$$

EXERCISES

1. Derive an expression for the inductance of an overhead line conductor due to (i) internal flux and (ii) external flux.
2. Find the inductance of a single phase two wire line.
3. Derive the expressions for inductance of composite conductor lines.
4. How would you analytically obtain the inductance of a three phase line with (i) symmetrical spacing and (ii) unsymmetrical spacing?
5. Derive analytical expressions for inductance of double circuit three phase lines. How would you find the inductance of bundled conductors?
6. Obtain the expressions for capacitance of a two wire line.
7. Find the expressions for capacitances of three phase lines with (i) symmetrical equilateral and (ii) unsymmetrical spacings.
8. How would you analytically find the charging current due to capacitance of a line?
9. What is the effect of earth on the calculation of capacitance of line?
10. Obtain an expression for three phase lines with more than one circuit.

CHAPTER 4

Steady State Performance and Operation of Transmission Lines

4.1 INTRODUCTION

A *transmission* line is an electrically conductive medium that transmits electrical energy among various circuits in an electric power system. For the purpose of analysis, a transmission line is said to have a sending end and a receiving end. Transmission lines can be used both for transmission of electrical energy at power frequency and for sending communication signals at high frequency. The power frequency transmission lines usually carry higher voltages and currents at 50 Hz (or 60 Hz) frequency while communication lines are employed to transmit low power signals at high frequency. Transmission lines can also be utilised as circuit elements like inductor, transformer, filter, etc. in high frequency circuits. They can also be used as a device that is employed to "match" impedances between the source and load in signal transmission. Popularly, open wire lines are referred to as *transmission lines* but in true sense cables (including co-axial cables) and wave guides can also be termed *transmission lines*. Here, we discuss the performance and operational aspects of power frequency lines only.

4.2 CHARACTERISATION OF A TRANSMISSION LINE

The transmission line is generally characterised by a few parameters like R (resistance), L (inductance), G (shunt conductance) and C (shunt capacitance), Z_0 (characteristic or surge impedance) and γ (propagation constant). Conventionally, R, L, G and C are bunched to be named as *primary constants* while Z_0 and γ are classified as *secondary constants*. When represented by lower case letters (r, l, g and c), then we mean them for per unit length; when expressed by upper case letters (R, L, G and C), they represent the respective values for the entire line.

A transmission line may electrically be represented as a circuit consisting of series resistance R and series inductance L alongwith a shunt capacitance (C) and leakage conductance G (Figure 4.1). When the circuit elements are assumed to be lumped with impedance Z in series and admittance Y in shunt having lumped primary parameters like a single resistance, inductance, capacitance, etc., then this equivalent representation is called *lumped parameter concept* and is generally modelled in terms of T or pi network (two port four terminal network) for simplicity of analysis and assessment of its performance. The parameters may also be assumed to be distributed throughout the entire line and this concept is then known as *distributed parameter concept.*

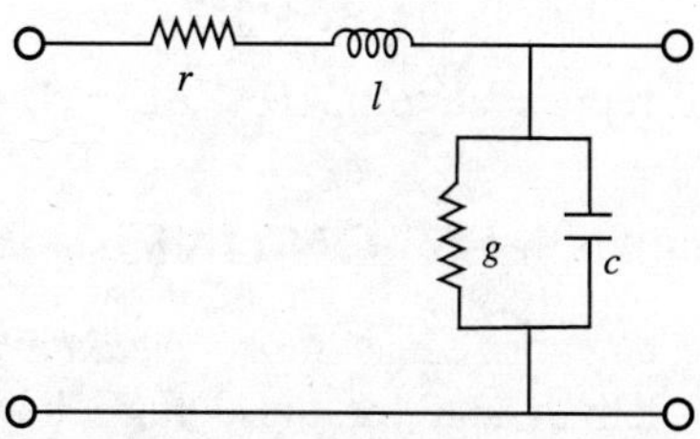

Figure 4.1 Circuit representation of a transmission line section.

In Figure 4.1,

$$z = r + jx_l = r + j\omega\ell$$

$$y = g + jb = b + j\omega c,$$

where x_L is inductive reactance and b is capacitance susceptance.

Also, $Z = z \times \text{line length } (D) = R + jX_L$ and $Y = y \times \text{line length } (D) = G + jB$

Surge impedance (or, *Characteristics impedance*), Z_0 and propagation constant γ, are secondary constants of a transmission line.

The relationships between the primary and secondary constants of the transmission line are as follows

$$z_0 = \sqrt{\frac{z}{y}} = \sqrt{\frac{r + j\omega\ell}{g + j\omega c}} = \sqrt{\frac{R + j\omega L}{G + j\omega C}} = \sqrt{\frac{Z}{Y}}\ \Omega \tag{4.1}$$

and

$$\gamma = \sqrt{z \times y} = \sqrt{(r + j\omega\ell)(g + j\omega c)} \tag{4.2}$$

In analysis, frequently lines are treated as lossless (i.e. $r = 0$; $g = 0$) making

$$z_0 = \sqrt{\frac{\ell}{c}}\,\Omega\ ; \quad \gamma = j\omega\sqrt{\ell c} \tag{4.3}$$

It may be noted here that Z_0 is independent of line length as for the entire line length D, we can write Z_0 as

$$Z_0 = \sqrt{\frac{z \times \text{line length } (D)}{y \times \text{line length } (D)}} = \sqrt{\frac{Z}{Y}} \tag{4.4}$$

Also, for the lossless line, Z_0 is a real quantity and is often called *natural impedance* of the line in addition to its name as characteristic impedance or surge impedance. Conventionally, we use small letters for parameters of line per unit length while capital letters represent the parameter for full line length.

Power frequency transmission lines transmit bulk electrical power with a balanced three phase load. The analysis can conveniently be made on per phase basis and the line on per phase basis can be regarded as a two port network, wherein the sending end voltage V_S and current I_S are related to the receiving end voltage V_R and current I_R through *ABCD* constants as

$$\begin{bmatrix} V_S \\ I_S \end{bmatrix} = \begin{bmatrix} A & B \\ C & D \end{bmatrix} \begin{bmatrix} V_R \\ I_R \end{bmatrix} \tag{4.5}$$

where, for the transmission line representation, $AD - BC = 1$.

4.3 TYPES OF POWER TRANSMISSION LINES

In general, the power lines are categorized as *short*, *medium* and *long* lines. Short lines are usually the lines upto 100 km length and for these lines it is reasonable to neglect Y, the shunt admittance, with assumption that shunt capacitance is low and shunt conductance (g) is zero. Hence short lines can be easily analysed with these approximations. The medium line is generally of length beyond 100 km upto 200 km while the long line is assumed to be of length beyond 200 km. For the medium line it is customary to assume shunt elements, unlike a short line, but it is assumed that these shunt elements (and also series elements) are in lumped form. Thus, medium lines have lumped parameter model with both z and y are present. In long lines, z and y are also assumed to be present but in distributed parameter form, i.e. the primary parameters are distributed althrough the line uniformly.

4.4 NETWORK MODEL FOR TRANSMISSION LINE WITH LUMPED PARAMETER CONCEPT

Conventionally transmission lines are modelled with lumped primary parameters in order to represent the model as a T or π section. Established concepts in circuit theory are applied to study the performance of such a model.

For T section model (Figure 4.2),

$$I_S = I_R + Y(V_R + I_R Z_2) = YV_R + (1 + YZ_2)\, I_R \tag{4.6a}$$

and

$$V_S = V_R + I_R Z_2 + I_S Z_1 = V_R + I_R Z_2 + Z_1 Y V_R + I_R Z_1 + I_R Y Z_1 Z_2$$

$$= (1 + YZ_1)V_R + (Z_1 + Z_2 + YZ_1 Z_2)\, I_R \tag{4.6b}$$

The *ABCD* parameter quantities are given as

$$I_S = CV_R + DI_R \text{ and } V_S = AV_R + BI_R \tag{4.6c}$$

Comparing these two sets of equations 4.6(a) and 4.6(b) with 4.6(c), we have for the T section network,

$$\left.\begin{aligned} &A = (1 + YZ_1);\ B = Z_1 + Z_2 + YZ_1 Z_2; \\ &C = Y;\ D = 1 + YZ_2 \end{aligned}\right\} \tag{4.6d}$$

Also, $AD - BC = 1$ provided $Z_1 = Z_2$, the network being symmetrical. This gives $A = D = 1 + \dfrac{YZ}{2}$ with $Z_1 = Z_2 = \dfrac{Z}{2}$ and $B = Z\left(1 + \dfrac{YZ}{4}\right)$, while $C = Y$.

For π section network representations (Figure 4.3), we have

$$V_S = (I_R + Y_RV_R)\,Z + V_R = V_R(1 + Y_RZ) + ZI_R$$

$$I_S = I_R + Y_RV_R + V_SY_S = I_R + Y_RV_R + Y_S\left[V_R + V_RY_RZ + ZI_R\right]$$

$$= \left[Y_S + Y_R + ZY_SY_R\right]V_R + \left[1 + Y_SZ\right]I_R$$

Equivalence with *ABCD* parameter network holds provided for the π network,

$$A = 1 + Y_SZ; \quad B = Z; \quad C = Y_S + Y_R + ZY_SY_R; \quad D = 1 + Y_SZ$$

With $Y_S = Y_R = \dfrac{Y}{2}$, we get,

$$A = 1 + \frac{YZ}{2};\ B = Z,\ C = Y\left(1 + \frac{YZ}{4}\right);\ D = 1 + \frac{YZ}{2} \tag{4.7}$$

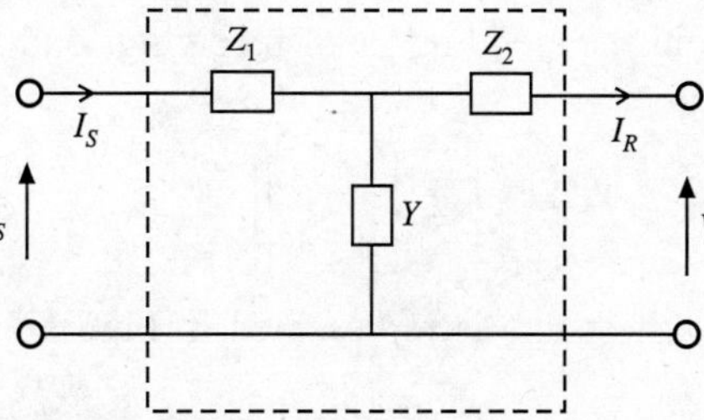

Figure 4.2 *T*-model of a short power line where the series impedance *Z* is segregated in two impedances Z_1 and Z_2 at two sides of the line.

It may be noted here that *A* and *D* do not have any unit while *B* is expressed in ohms; *C* is expressed in mho (or Siemens).

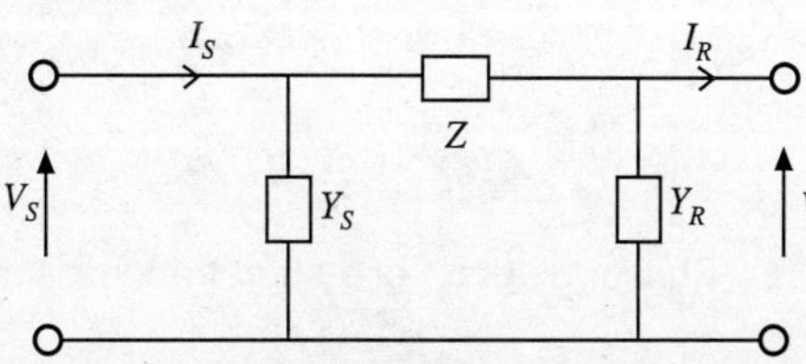

Figure 4.3 π-model of a short line where the shunt admittances are represented as Y_S and Y_R at two sides of the line.

For a short line, since we assume $Y = 0$, we have for the *T* model,

$$A = D = 1$$

$$B = 2Z_1$$

$$C = 0$$

It is customary to assume $Z_1 = Z_2 = Z_l = \dfrac{Z}{2}$ for a power line. Hence, for the short line in *T* form

$$A = D = 1,\ B = Z \text{ and } C = 0.$$

Also, with $Y = 0$ for the π model, we get the same result.

4.5 SHORT TRANSMISSION LINE MODEL

Figure 4.4 represents a short transmission line model.

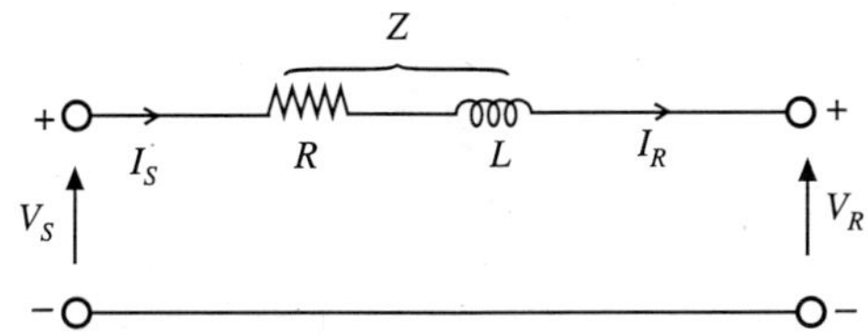

Figure 4.4 Short transmission line model.

The relationship between the sending and receiving end voltages and currents can conveniently be represented as

$$\begin{bmatrix} V_S \\ I_S \end{bmatrix} = \begin{bmatrix} 1 & Z \\ 0 & 1 \end{bmatrix} \begin{bmatrix} V_R \\ I_R \end{bmatrix} \tag{4.8}$$

The phasor diagram of a short line is represented in Figure 4.5.

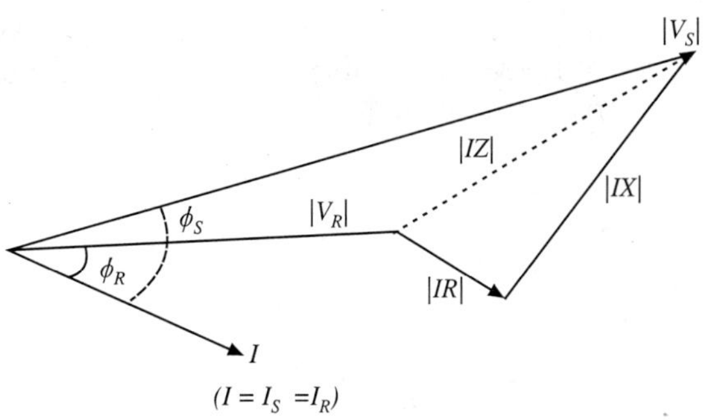

Figure 4.5 Phasor diagram of a short transmission line.

4.6 VOLTAGE REGULATION OF A SHORT LINE

Voltage regulation of a short line is defined as the percentage increase in voltage at the receiving end expressed in terms of full load voltage when the full load at a specified power factor is thrown off and the sending end voltage is kept unchanged.

$$\text{Regulation (\%)} = \frac{|V_{R(NL)}| - |V_{R(FL)}|}{|V_{R(FL)}|} \times 100 \tag{4.9}$$

where, $|V_{R(NL)}|$ = magnitude of no load receiving end voltage and $|V_{R(FL)}|$ = magnitude of full load receiving end voltage at a specified power factor (p.f.).

For short line, $\left|V_{R(NL)}\right| \equiv \left|V_S\right|$ and $\left|V_{R(FL)}\right| \equiv \left|V_R\right|$

$$\text{Regulation (\%)} = \frac{|V_S| - |V_R|}{|V_R|} \times 100 \qquad (4.10)$$

where $|V_S| - |V_R| = |IR| \cos \phi_R + |IX| \sin \phi_R$, for any lagging p.f. load.

For a leading p.f. load, $|V_S| - |V_R| = |IR| \cos \phi_R - |IX| \sin \phi_R$.

Voltage regulation is negative $\left(\text{i.e., } |V_R| > |V_S|\right)$ when $X \sin \phi_R > R \cos \phi_R$ i.e., $\tan \phi_R > (R / X)$.

For zero voltage regulation (i.e. when $|V_S| - |V_R| = 0$), we have $\tan \phi_R = (R / X) = \cot \theta$, θ being the line angle $\left(\tan^{-1} \dfrac{X}{R}\right)$ so that $\phi_R \text{ (leading)} = \left(\dfrac{\pi}{2} - \theta\right)$.

Voltage regulation of a line is dependent on load power factor. Voltage regulation improves $\left(\text{i.e. } |V_S| - |V_R| = \Delta V \text{ decreases}\right)$ as the power factor of a lagging load is increased and it may become zero at a leading power factor. The inclusion of shunt capacitance at the receiving end for a heavily loaded line improves the receiving end voltage magnitude as well as receiving end p.f. and hence improves voltage regulation.

4.7 EFFICIENCY OF TRANSMISSION FOR A SHORT LINE

A certain amount of active power is dissipated as I^2R loss in line conductors. We term this as *line loss*. For three phase power lines, line loss is $3I^2R$. The efficiency of transmission is expressed as

$$\eta_{tr}(\%) = \frac{\text{Power at receiving end}}{\text{Power at sending end}} \times 100 = \frac{P_R}{P_S} \times 100 \qquad (4.11a)$$

Obviously, $P_S = P_R$ + line loss.

$$\therefore \quad \eta_{tr}(\%) = \frac{P_R}{P_R + \text{Line loss}} \times 100$$

$$= \left(1 - \frac{\text{Line loss}}{P_S}\right) \times 100 \qquad (4.11b)$$

[In transmission lines, due to presence of series inductance there is series inductive 'loss', which is I^2X per phase, i.e. series reactive 'loss' $(Q_{SL}) = I^2X$.]

EXAMPLE 4.1 A 3 phase, 50 Hz, 20 km long power transmission line operates with the following data:

Sending end voltage = 11 kV (L-L)

Load delivered at receiving end = 1000 kW at 0.8 p.f. (leg.)

Resistance of each conductor = 0.4 Ω/km

Reactance of each conductor = 0.5 Ω/km

Calculate:

(a) Line current
(b) Receiving end voltage
(c) Efficiency of transmission
(d) Voltage regulation of the line.

Solution: From the given data, we have for the entire line,

Resistance/phase (R) = 0.4 × 20 = 8 Ω

Reactance/phase (X) = 0.5 × 20 = 10 Ω

$$\text{Sending end voltage } (V_S) = \frac{11}{\sqrt{3}} = 6.352 \text{ kV (L-N)}$$

We may now write the expression of sending end voltage as

$$|V_S| = |V_R| + |IR| \cos \phi_R + |IX| \sin \phi_R$$

Here, $|V_S| = 6352$ V/ph; $|I_R| = \dfrac{1000 \times 10^3}{\sqrt{3} \times 11 \times 10^3 \times 0.8} = 65.6 \text{ A } (= I)$

∴ We can write

$$6352 = |V_R| + 65.6 \times 8 \times 0.8 + 65.6 \times 10 \times 0.6$$

$$[\because I = 65.6 \text{ A; p.f.} (= \cos\phi_R) = 0.8; \sin \phi = 0.6]$$

$$\therefore \; |V_R| = 5538.56 \text{ V/ph}$$

as $$|V_R| = \sqrt{3} \times 5538.56 = 9593.06 \text{ V} = 9.59 \text{ kV (L-L)}$$

Again, I^2R loss in transmission line is obtained as $65.6^2 \times 8 \times 10^{-3}$ or 34.427 kW/ph.

∴ Input power to the line is (Power output (P_R) + Losses)

i.e. 100 kW + (3 × 34.427) kW or, 1103.28 kW (= P_S)

Hence, efficiency of transmission η in % is obtained as

$$\eta = \frac{P_R}{P_S} \times 100 = \frac{1000}{1103.28} \times 100 = 90.64\%$$

Voltage regulation of the line is obtained as

$$\Delta V\% = \frac{|V_S| - |V_R|}{|V_R|} \times 100 = \frac{11 - 9.59}{9.54} \times 100 = 14.7\%$$

Hence, we obtained the following results:

Line current = 65.6 A
Receiving end voltage = 9.59 kV (L-L)
Efficiency of transmission = 90.64%
Voltage regulation = 14.7%.

} Ans.

EXAMPLE 4.2 A 6.6 kV (L-L), 3 phase, 50 Hz short power transmission line is 16 km long. The line has a per phase series impedance of (0.125 + *j*0.4375) ohm/km. Determine the sending end voltage, voltage regulation, sending end power and the transmission efficiency when the line delivers at the receiving end

(a) 0.7 MVA, 0.8 p.f. (lag) load at 6.6 kV

(b) 1.2 MW, unity p.f. load at 6.6 kV.

Solution: (a) Receiving end voltage (V_R) = $\dfrac{6.6}{\sqrt{3}} = 3.81\text{ kV (ph-N)}$

Also, at 0.7 MVA receiving end power loading,

$$|I_R| = \frac{0.7\times 10^6}{\sqrt{3}\times 6.6\times 10^3} = 61.240\text{ A}$$

$$\left[\because I_R = \frac{P_R\times 10^6}{\sqrt{3}\times V_R(\text{L-L})\times 10^3}\text{ when } P_R \text{ is expressed in MVA, } V_R \text{ in kV}\right]$$

Assuming V_R as reference phasor,

$$V_R = |V_R|\angle 0^\circ\text{ kV} = 3.81\angle 0^\circ\text{ kV (ph-N)}$$

$\therefore$ I_R at 0.8 p.f. lag can be expressed as

$$I_R = |I_R|\angle -\phi_R \text{ [–ve sign is due to lagging load]}$$

or,

$$I_R = 61.24\angle -36.86^\circ\text{ A } \left[\because \phi_R = \cos^{-1}(0.8)\right]$$

Also, the total series impedance per phase is obtained as

$$Z = (0.125 + j0.4375)\times 16 = (0.455\angle 74.05^\circ\times 16)$$
$$= 7.28\angle 74.05^\circ\ \Omega.$$

Since $V_S = V_R + I_R \times Z$, we can write

$$V_S = 3.81 + 0.06124\angle -36.86^\circ \times 7.28\angle 74.05^\circ$$
$$= 3.81 + 0.4458\angle 37.19^\circ$$
$$= 3.81 + 0.355 + j0.2695$$
$$= (4.165 + j0.2695)\text{ kV} = 4.174\angle 3.7^\circ\text{ kV (L-N)}$$

i.e.

$$|V_S|(\text{L-L}) = \sqrt{3}\times 4.174 = 7.23\text{ kV (Ans.)}$$

$$\text{Voltage regulation} = \frac{7.23-6.6}{6.6}\times 100 = 9.545\%\text{ (Ans)}$$

$$\text{Sending end power } (P_S) = \sqrt{3}\times|V_S|(\text{L-L})\times|I_S|\times\cos\phi_S$$
$$= \sqrt{3}\times 7.23\times 0.06124\times\cos(3.7^\circ + 36.86^\circ)$$
$$= 0.5826\text{ MW (Ans.)}$$

$$|V_R| = 3.81\text{ kV (ph-N)}$$

(b) $$I_R \text{ (when load is 1.2 MW at UPF)} = \frac{1.2 \times 10^6}{\sqrt{3} \times 6.6 \times 10^3 \times 1}$$

$$\left[\because I_R = \frac{P_R \times 10^6}{\sqrt{3} \times V_R(\text{L-L}) \times 10^3 \times \cos \phi_R} \right]$$

$$= 104.98 \text{ A } (\simeq 105 \text{ A})$$

Since V_R is reference phasor and I_R is in same phase, in this case

$$V_S = 3.81 + 0.105\angle 0° \times 7.28\angle 74.05° \qquad \begin{bmatrix} V_R = 3.81\angle 0° \text{ kV (ph-N)} \\ I_R = 0.10498 \simeq 0.105 \text{ kA at angle } 0° \end{bmatrix}$$

$$= 3.81 + 0.7644\angle 74.05°$$

$$= 3.81 + 0.21 + j0.735 = (4.02 + j0.735) \text{ kV}$$

$$= 4.09 \text{ kV(ph-N)}$$

$$\therefore \quad |V_S| \text{ (L-L)} = \sqrt{3} \times 4.09 = 7.08 \text{ kV (Ans.)}$$

Voltage regulation in this case is obtained as

$$\frac{7.08 - 6.6}{6.6} \times 100 = 7.33\% \text{ (Ans.)}$$

It may be observed from above that even if the load increases, the regulation is improved. This is due to the fact that in the second case, the p.f. is unity unlike the previous case when p.f. was 0.8 (lag).

$$\text{Sending end power } (P_S) = \sqrt{3} \times 7.08 \times 0.105 \times \cos(10.36° + 0°)$$

$$= 1.266 \text{ MW (Ans.)}$$

We can now compare the transmission efficiency in both the cases. In the first case,

$$\eta_{\%} = \frac{P_0}{P_S} \times 100 = \frac{0.7 \times 10^6 \times 0.8}{0.5826 \times 10^6} \times 100 = 96.12\% \text{ (Ans.)}$$

In the second case,

$$\eta_{\%} = \frac{1.2 \times 10^6}{1.266 \times 10^6} \times 100 = 94.78\% \text{ (Ans.)}$$

It may be observed that though the regulation is improved in the second case due to p.f. improvement from 0.8 (lag) to unity, the efficiency decreases. The decrease in efficiency is due to increase in loss in the line in the second case due to higher loading. [In the first case, line loss/ph is $(0.06124)^2 \times (0.125 \times 16) = 0.0075$ kW while that in the second case is $(0.105)^2 \times (0.125 \times 16) = 0.022$ kW per phase.]

EXAMPLE 4.3 A 30 km long single phase power transmission line delivers power of 10 MW at 0.8 p.f. (lag) at its receiving end. If the operation frequency is 50 Hz and the receiving end voltage is to be kept at 20 kV, find

(a) The sending end voltage and voltage regulation of the line.
(b) The value of the shunt capacitance to be placed in parallel with the load such that regulation is reduced to 50% of that obtained in (a).
(c) Compare the transmission efficiencies obtained in parts (a) and (b).

Assume resistance and inductance of each conductor is 0.05 Ω/km and 0.7 mH/km, respectively.

Solution: (a) $R = 0.05 \times 30 = 1.5\ \Omega$

$$X = \omega l \times \text{length} = 2 \times \pi \times 50 \times 0.7 \times 10^{-3} \times 30$$
$$= 0.594\ \Omega$$

For the given short line,

$$|I_R|\left(=|I_S|\right) = \frac{P_R}{V_R \times \text{p.f.}} = \frac{10 \times 10^6}{20 \times 10^3 \times 0.8}$$
$$= 625\ \text{A.}$$

The sending end voltage $|V_S| = |V_R| + |I|\,(R\cos\phi_r + X\sin\phi_r)$

$$= 20 + 625\ (1.5 \times 0.8 + 6.594 \times 0.6)10^{-3}$$
$$= 23.22\ \text{kV (Ans.)}$$

$$\text{Voltage regulation} = \frac{23.22 - 20}{20} \times 100\%$$
$$= 16.1\%\ (= \Delta V)\ \text{(Ans.)}$$

(b) As the regulation is desired to be 50% to that obtained in (a), so the desired regulation is $\frac{16.1}{2}\%$ or 8.05%. (= 0.0805 p.u.)

Now, $|V_R| = 20$ kV (fixed)

$$\therefore \quad 0.0805 = \frac{|V_S| - 20}{20} \times 100\%$$
$$\underset{\text{(new value)}}{|V_S|} = (20 \times 0.0805 + 20)\ \text{kV}$$
$$= 21.61\ \text{kV}$$

Figure E4.1 represents the equivalent circuit of the line with capacitor,

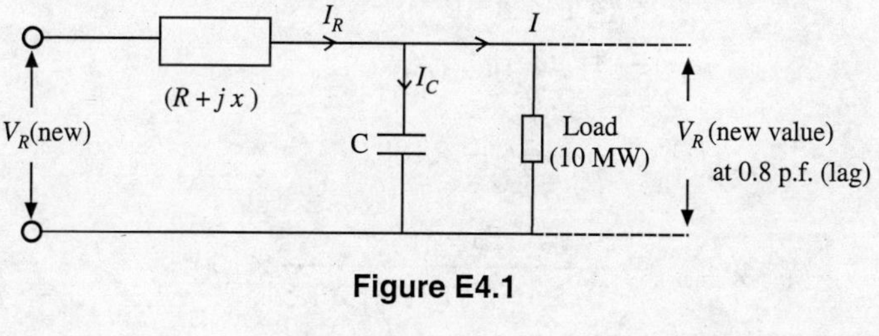

Figure E4.1

Let ϕ_{R_1} be the new p.f. at the receiving end side where the load is in parallel to the shunt capacitor installed.

$$\therefore \qquad |I_{R(\text{new})}| = \frac{10 \times 10^6}{20 \times 10^3 \times \cos \phi_{r_1}}$$

$$= \left(\frac{500}{\cos \phi_{r_1}}\right) \text{A}$$

∴ From the relation $|(V_S - V_r)| = |I_{R(\text{new})}|(R \cos \phi_{R_1} + X \sin \phi_{R_1})$, we get

$$21.61 - 20 = \frac{500}{\cos \phi_{R_1}}(1.5 \cos \phi_{R_1} + 6.594 \sin \phi_{R_1})$$

or, $$1.61 = 750 + 3297 \tan \phi_{R_1}$$

or, $$\tan \phi_{R_1} = -0.226$$

∴ $$\phi_{R_1} = -12.73°$$

$$\cos \phi_{R_1} = 0.975 \text{ (lag)}$$

∴ $$I_{R(\text{new})} = \frac{500}{0.975} = 512.82\angle -\cos^{-1}(0.975) \text{ A}$$

Now, $$I_C = (I_R - I) = 512.82\ (0.975 - j0.22) - 625(0.8 - j0.6)$$

$$= 499.99 - j112.82 - 500 + j375$$

$$= (-0.01 + j262.18) \text{ A}$$

i.e. $$I_C \cong j262.18 \text{ A}$$

We know, $$|X_C| = \frac{1}{2\pi f C} = \left|\frac{V_R}{I_C}\right|$$

Here, $$|X_C| = \frac{20 \times 10^3}{262.18} = 76.283\ \Omega$$

∴ $$76.283 = \frac{1}{2 \times 3.14 \times 50 \times C}$$

i.e. $$C = \frac{1}{2 \times 3.14 \times 50 \times 76.283}$$

$$= (4.174 \times 10^{-5}) \text{ Fss}$$

$$= 41.74\ \mu\text{F} \quad \text{(Ans.)}$$

(c) $$\eta_{\text{transmission}}(\%) = \frac{P_R}{P_R + \text{Loss}} \times 100$$

For case (a), $$\eta_{\text{transmission}} = \frac{10 \times 10^6}{10 \times 10^6 + (625)^2 \times 1.5} \times 100\%$$

$$= 94.46\% \quad \text{(Ans.)}$$

For case (b), $\eta_{\text{transmission}} = \dfrac{10\times10^6}{10\times10^6 + (512.82)^2\times0.4}\times100\%$

$= 97.44\%$. (Ans.)

It may be observed that transmission efficiency improves with installation of shunt capacitor at load bus.

EXAMPLE 4.4 The generalised *ABCD* constants of a three phase power transmission line are as follows:

$$A = D = (0.936 + j0.016)$$
$$B = (33.5 + j138) \text{ ohm}$$
$$C = (-5.18 + j914)\ 10^{-6} \text{ mho.}$$

The load on the receiving end bus is 50 kW at 220 kV with a lagging p.f. of 0.9. Find the magnitude of the sending end voltage and the voltage regulation. Assume that the magnitude of the sending end voltage remains constant.

Solution: We know, for the medium transmission line, the *ABCD* constants are related to the following equations:

$$V_S = AV_R + BI_R \qquad \text{(i)}$$
$$I_S = CV_R + DI_R \qquad \text{(ii)}$$

The receiving end line current (I_R) is obtained as

$$|I_R| = \frac{50\times10^6}{220\times10^3\times\sqrt{3}\times0.9} = 145.78 \text{ A}$$

Also, $$|V_R| = \frac{220\times10^3}{\sqrt{3}} = 127.02\times10^3 \text{ V (L-N)}$$

Taking V_R as reference, we can write

$$V_R \text{ (L-N)} = 127.02\ (1 + j0) \text{ kV}$$
$$I_R = 145.78\angle-\cos^{-1}0.9 = 145.78\angle-25.84^\circ$$
$$= (131.204 - j63.54) \text{ A.}$$

∴ From equation (i), we can find the sending end voltage as

$$V_S = \left\{(0.936 + j0.016)\,127.02\times10^3 + (33.5 + j138)(131.204 - j63.54)\right\}$$
$$= (132054.574 + j18009.882)$$
$$= (132.05 + j18) \text{ kV (app.)}$$
$$= 133.27\angle7.76^\circ \text{ kV/ph}$$

∴ $|V_S| = 230.82$ kV (L-L).

Voltage regulation (ΔV) $= \dfrac{133.27 - 127.02}{127.02} \times 100\%$

or, $\Delta V = 4.92\%$

Thus, we have obtained

$$|V_S| = 230.82 \text{ kV (L-L)}$$

$$|\Delta V| = 4.92\% \qquad \text{(Ans.)}$$

4.8 REPRESENTATION OF MEDIUM LINE

If all the shunt elements are assumed to be lumped at the middle of the line, it leads to the nominal T circuit as shown in Figure 4.6.

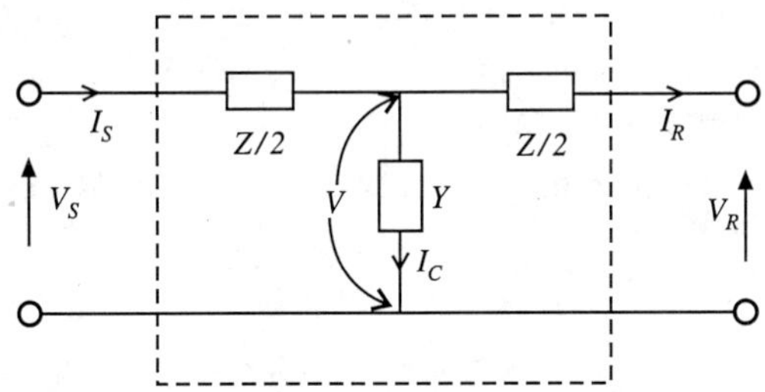

Figure 4.6 Nominal-*T* representation of the medium line.

We obtain the network model representation for the nominal-T section following the concepts presented in article 4.4:

$$\begin{bmatrix} V_S \\ I_S \end{bmatrix} = \begin{bmatrix} \left(1+\dfrac{YZ}{2}\right) & Z\left(1+\dfrac{YZ}{4}\right) \\ Y & \left(1+\dfrac{YZ}{2}\right) \end{bmatrix} \begin{bmatrix} V_R \\ I_R \end{bmatrix}$$

Here,
$$A = D = 1 + \frac{YZ}{2}; \qquad B = Z\left(1+\frac{YZ}{4}\right); \qquad C = Y. \tag{4.12}$$

The phasor diagram of such a circuit is shown in Figure 4.7. We assume the shunt capacitance only to form the shunt admittance. Here,

$$\frac{Z}{2} = \left(\frac{R}{2} + j\frac{X}{2}\right); \qquad Y = j\omega C$$

$$V = V_R + I_R\frac{Z}{2}; \qquad V_S = V + I_S\frac{Z}{2}$$

Also, I_C is leading V by 90° and

$$I_S = I_C + I_R$$

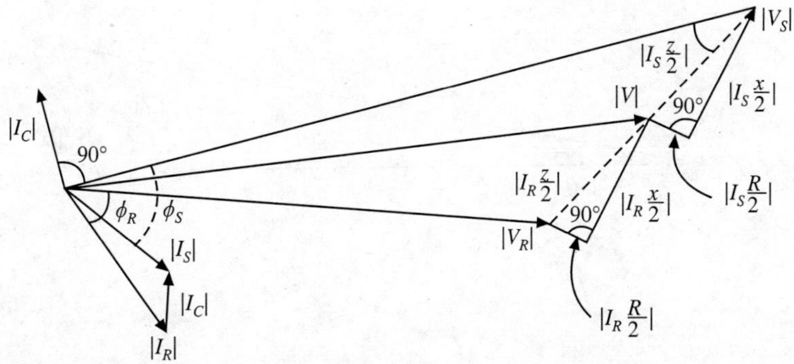

Figure 4.7 Phasor diagram of per phase representation of the medium line as nominal-*T* circuit.

If the total line shunt element (capacitance) is divided into two equal parts and are lumped at the sending and receiving ends, it results in the nominal-π network model (Figure 4.8). From Figure 4.8, we have

$$\begin{bmatrix} V_S \\ I_S \end{bmatrix} = \begin{bmatrix} \left(1+\frac{1}{2}YZ\right) & Z \\ Y\left(1+\frac{1}{4}YZ\right) & \left(1+\frac{1}{2}YZ\right) \end{bmatrix} \begin{bmatrix} V_R \\ I_R \end{bmatrix}$$

Here, $$A = D = 1+\frac{1}{2}YZ;\, B = Z;\, C = Y\left(1+\frac{1}{4}YZ\right) \tag{4.13}$$

It may be noted that in both the nominal-*T* and π representations, *A* and *D* have no units while *B* is expressed in ohm and *C* is expressed in mho. Figure 4.9 represents the phasor diagram of the medium line with nominal-π representation. Here,

$$I = I_{C_2} + I_R; \qquad I_S = I_{C_1} + I$$

$$V_S = V_R + IZ$$

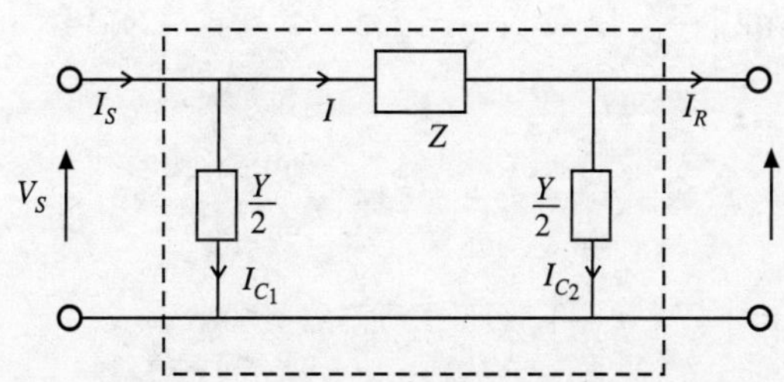

Figure 4.8 Nominal-π representation of a medium transmission line.

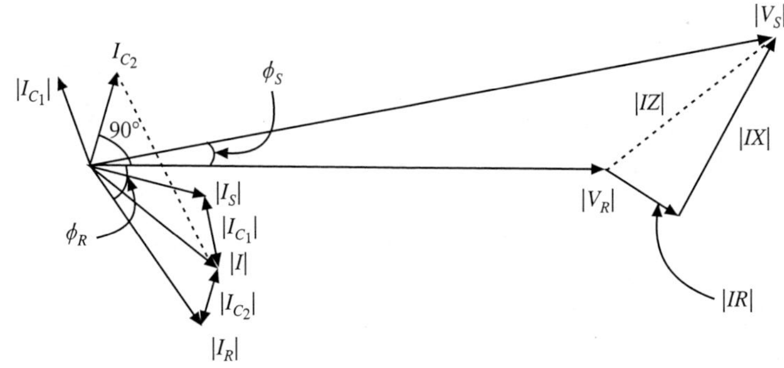

Figure 4.9 Phasor diagram of line operation in nominal-π representation.

EXAMPLE 4.5 A 3 phase, 132 kV, 50 Hz power transmission line has line resistance of 30 ohms and inductive reactance of 70 ohms per phase. The capacitive susceptance is 4 × 10^{-4} mho per phase.

If the load at the receiving end is 50 MVA at 0.8 p.f. lagging with 132 kV line voltage, calculate (a) voltage, (b) current, (c) power factor at the sending end, (d) regulation and (e) efficiency of the line for this load. (f) What is the line loss?

Use nominal-T method and nominal-π method for your calculations.

Solution: *For nominal-T method*

Let us first draw the equivalent circuit of the line (Figure E4.2) on per phase basis.

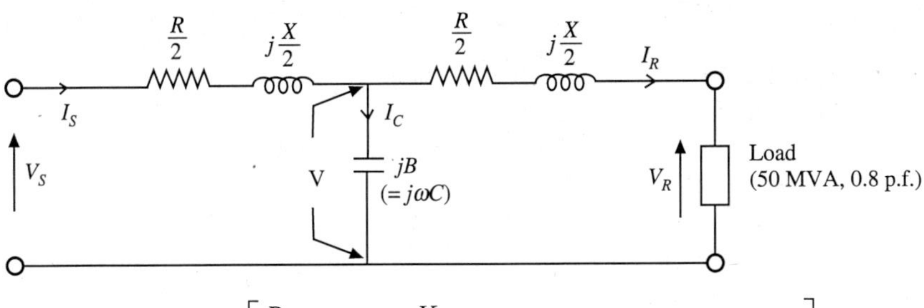

Figure E4.2 $\left[\frac{R}{2} = 15\ \Omega\ ; j\frac{X}{2} = j35\ \Omega\ ;\ Y = j4 \times 10^{-4}\ \text{mho}\right]$.

Since $\quad \frac{R}{2} = 15\Omega; \qquad j\frac{X}{2} = j35\ \Omega$

$\therefore \quad \frac{Z}{2} = \left(\frac{R}{2} + j\frac{X}{2}\right) = (15 + j35)\ \Omega$

Also, $\quad Y = jB = j4 \times 10^{-4}$ mho

$$|V_R| = 132 \text{ kV (L-L)}$$

i.e., $\quad |V_R| \text{ (ph-N)} = \frac{132}{\sqrt{3}} = 76.21 \text{ kV}$

$$V_R \text{ (ph-N)} = 76.21\angle 0° \text{ kV}$$

$$I_R = \frac{3 \text{ ph. power (MVA)} \times 10^6}{\sqrt{3} \times |V_R(\text{L-L})| \times 10^3} \angle -\cos^{-1}(0.8)$$

$$= \frac{50 \times 10^6}{\sqrt{3} \times 132 \times 10^3} \angle -36.86°$$

$$= 218.7 \angle -36.86° \text{ A}$$

From Figure E4.2

$$V = V_R + I_R \frac{Z}{2} = 76210 + 218.7\angle -36.86° \times (15 + j35)$$

$$= 76210 + (174.96 - j131.22)(15 + j35)$$

$$= (83427.1 + j4155.3) \text{ V/ph} = (83.427 + j4.155) \text{ kV/ph}$$

$$I_C = V \times jB = (83427.1 + j4155.3) \times j4 \times 10^{-4}$$

$$= (j33.371 - 1.662) \text{ A}$$

Thus, the sending end current is given by

$$I_S = I_R + I_C$$

$$= (174.96 - j131.22) + (j33.371 - 1.662)$$

$$= 173.30 - j97.85 = 199.02\angle -29.45° \text{ A (Ans. (b))}$$

But $\quad V_S = V + I_S \cdot \dfrac{Z}{2}$

Here, $\quad V_S = (83427.1 + j4155.3) + (173.3 - j97.85)(15 + j35)$

$$= (83427.1 + j4155.3) + 2599.5 + j6065.5 - 1467.75 + 3424.75$$

$$= (89451.35 + j8753.05) \text{ V/ph}$$

$$= 89.88 \angle 5.59° \text{ kV/ph} \quad \therefore V_S(\text{line}) = 155.67 \angle 5.59° \text{ kV(L-L)/ph (Ans. (a))}$$

Power factor at the sending end

$$= \cos(5.59° + 29.45°) = 0.819 \text{ (lag)}$$

$$P_S = \sqrt{3}|V_S||I_S|\cos\phi_S = \sqrt{3} \times 155.67 \times 199.02 \times 0.819$$

$$= 43.947 \text{ MW}$$

$$P_0 = 50 \text{ MVA at 0.8 p.f. (lag)} = 40 \text{ MW (given)}$$

Regulation (%) $\quad = \dfrac{|V_S| - |V_R|}{|V_R|} \times 100 = \dfrac{155.67 - 132}{132} \times 100$

$$= 17.93\% \text{ (Ans. (c))}$$

Transmission efficiency $\quad (\eta_{tr})\% = \dfrac{P_R}{P_S} \times 100$

$$= \frac{40}{43.947} \times 100 = 91\% \text{ (Ans. (d))}$$

Line loss $= P_S - P_0 = 3.947$ MW (for three phases)

The equivalent circuit for nominal-π is shown in Figure E4.3.

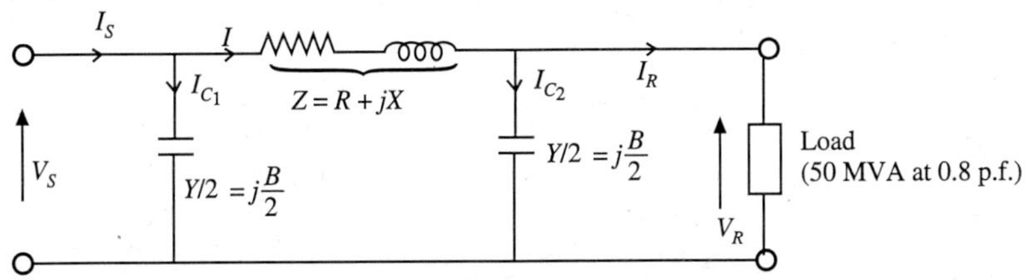

Figure E4.3

$$j\frac{B}{2} = 2 \times 10^{-4} \text{ mho/ph} \left(= j\frac{Y}{2}\right)$$

$$Z = (R + jX) = (30 + j70)\ \Omega$$

$$|V_R| = 76.21 \text{ kV (ph-N)}; \quad I_R = 218.7 \angle -36.86° \text{ A}$$

$$\therefore \quad I_{C_2} = V_r\, j\frac{Y}{2} = (76.21 \angle 0° \times j2 \times 10^{-4}) \text{ kA}$$

$$= j15.24 \text{ A}$$

$$I = I_{C_2} + I_R = j15.24 + 218.7 \angle -36.86°$$

$$= (174.96 - j115.98) \text{ A} = 209.91 \angle -33.54° \text{ A}$$

$$\therefore \quad V_S = V_R + IZ = 76210 + 209.91 \angle -33.54° \times (30 + j70)$$

$$= 76210 + (174.96 - j115.98)\ (30 + j70)$$

$$= (89577.4 + j8767.8) \text{ V/ph}$$

$$= 9005.47 \angle 5.59° \text{ kV/ph}; \; V_S(\text{line}) = 155.89 \angle 5.59° \text{ kV (L-L)} \quad \text{(Ans. (a))}$$

$$I_S = I + I_{C_1} = (174.96 - j115.98) + (j17.915 - 1.75)$$

$$\left[\because I_{C_1} = V_S \times j\frac{B}{2} = (89577.4 + j8767.8) + j2 \times 10^{-4} = j17.915 - 1.75 = 18\angle -5.58° \text{ A}\right]$$

$$= (173.21 - j98.065) \text{ A} = 199.04 \angle -29.52° \text{ A} \qquad \text{(Ans. (b))}$$

$$\text{Regulation} = \frac{155.89 - 132}{132} \times 100 = 18.1\% \qquad \text{(Ans. (c))}$$

$$P_S = \sqrt{3}\,|V_S|\,|I_S| \cos\phi_S = \sqrt{3} \times 155.89 \times 199.04 \cos(29.52 + 5.59°)$$

$$= 43.96 \text{ MW}$$

$$P_0 = 40 \text{ MW (given)}$$

$$\therefore \qquad \eta_{tr}\,\% = \frac{P_R}{P_S} \times 100 = \frac{40}{43.96} \times 90.99\% \qquad \text{(Ans. (d))}$$

$$\therefore \qquad \text{Line loss} = (P_S - P_0) = 3.96 \text{ MW (for three phases)}$$

It is then clear that nominal-T and nominal-π methods give substantially the same results.

EXAMPLE 4.6 A 40 MVA generating station is connected to a 3 phase line having per phase impedance and admittance as follows: $Z = 300\angle 75°\ \Omega$; $Y = 0.0025\ \angle 90°\ \mho$.

The power at the generating station is 40 MVA at unity power factor at a voltage of 120 kV (L-L). There is a load of 10 MW at unity power factor at the midpoint of the line. Calculate the voltage and load at the distant end of the line. Use nominal-T circuit for modelling the line.

Solution:

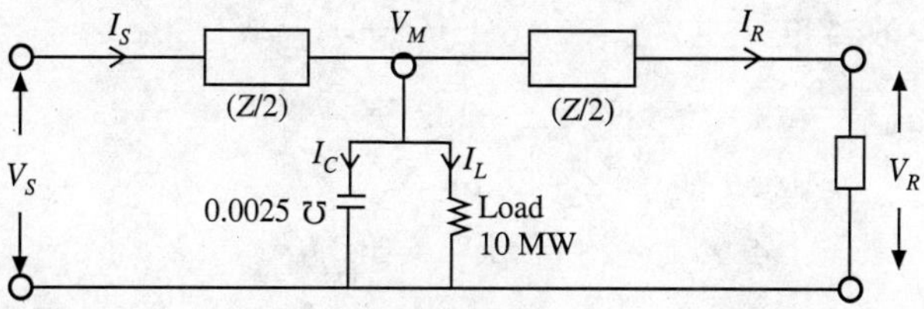

Figure E4.4

In Figure E4.4 (T representation of the line model given in the question),

$$V_{S\text{ (line to neutral)}} = \frac{120}{\sqrt{3}}\angle 0°\text{ kV} \qquad (\because\ V_S \text{ is said to be the reference phasor})$$

$$= 69.282\ \angle 0°\text{ kV.}$$

$$I_S = \frac{40 \times 10^6}{120 \times 10^3 \times \sqrt{3}}\angle 0° \qquad [\because \text{ Generating side p.f. is unity}]$$

$$= 192.45\ \angle 0°\text{ A}$$

$$V_{M\text{ (midpoint voltage)}} = V_S - \left(I_S \cdot \frac{Z}{2}\right) = 69282\angle 0° - 192.45\angle 0° \times 150\angle 75°$$

$$= 69282\angle 0° - 28867.5\angle 75°$$

$$= (69282 + j0) - 7471.458 - j27883.86$$

$$= 61810.54 - j27883.86$$

$$= 67809.42\ \angle{-24.28°}\text{ V (phase value)}$$

$$\therefore \qquad I_C = V_M \times Y = (67808.94\angle -24.28° \times 0.0025\angle 90°)$$

$$= 169.522\angle 65.72°\text{ A}$$

$$I_L = \frac{10 \times 10^6}{3 \times 67808.94\angle -24.28°} \qquad \text{[for load at midpoint of the line]}$$

$$= 49.157\angle 24.28°\text{ A}$$

$$\therefore \quad I_R = I_S - I_C - I_L = 192.45\angle 0° - 169.522\angle 65.72° - 49.157\angle 24.28°$$

$$= (192.45 + j0) - (69.706 + j154.52) - (44.808 + j20.21)$$

$$= (77.936 - j174.73)$$

$$= 191.323\angle 65.96° \text{ A}$$

Also,
$$V_R = V_M - I_R \times \frac{Z}{2}$$

$$= (61810.54 - j27883.86) - 191.323\angle 65.96° \times 150\angle 75°$$

$$= 61810.54 - j27883.86 - 28698.45\angle 140.96°$$

$$= 61810.54 - j27883.86 + 22290.27 - j18076.08$$

$$= 84100.81 - j45959.94$$

$$= 95839.77\angle -28.65° \text{ V (ph-N)}$$

$$\therefore \quad V_{R\text{ (line)}} = \sqrt{3} \times 95839.77\angle -28.65°$$

$$= 165999.351\angle -28.65° \text{ V}$$

$$= 165.99\angle -28.65° \text{ kV (Ans.)}$$

Thus, p.f. at the receiving end is angle between V_R and I_R,

i.e.
$$\cos \phi_R = \cos (28.65° + 65.96°)$$

$$= -0.08 \text{ (lagging)}$$

Load at the receiving end $= \sqrt{3} \times V_R \times I_R \times \cos \phi_R$

$$= \left(\sqrt{3} \times 165.99 \times 0.1913 \times 0.08\right) \text{MW}$$

$$= 4.39 \text{ MW (Ans.)}$$

EXAMPLE 4.7 A 220 kV, 3 phase overload transmission line has an impedance per phase of $(10 + j50)\ \Omega$ and admittance of $j0.002$ ℧. Using π-model, determine the sending end voltage and current when the current at the receiving end is 350 A at 0.7 p.f. lagging.

Solution:

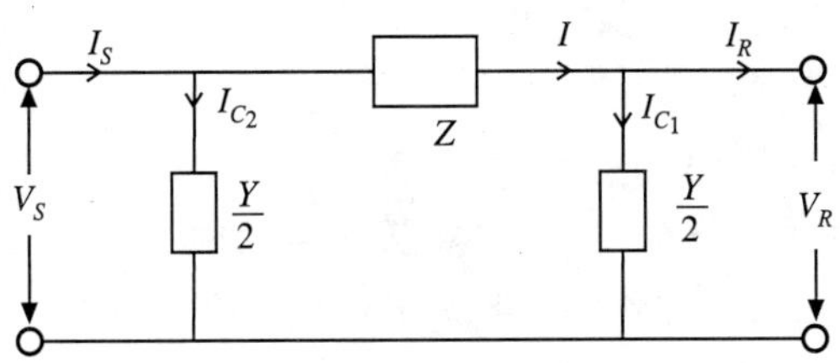

Figure E4.5

$$|V_R| = 220 \text{ kV} \qquad \text{(line to line)}$$

$$|V_R| = \frac{220}{\sqrt{3}} \text{ kV} \qquad \text{(line to neutral)}$$

$$= 127.02 \text{ kV (L–N)}$$

$$Z = (10 + j50)\ \Omega / \text{ph}$$

$$Y = j0.002\ \mho / \text{ph}$$

$$\therefore \quad \frac{Y}{2} = j0.001\ \mho / \text{ph}$$

$$I_R = 350 \text{ A}$$

$$\cos\phi_r = 0.7 \text{ lagging}$$

In the phasor diagram (Fig. E4.5), we can write,

$$V_R = 127.02\angle 0° \text{ kV} \quad (V_R \text{ is reference})$$

$$I_R = 350\,(\cos\phi_r - j\sin\phi_r) \text{ A} \quad [\because\ I_R \text{ lags } V_R \text{ by p.f. angle of } \cos^{-1}(0.7).]$$

$$= 350\ (0.7 - j0.714)$$

$$= 346.5\angle -45° \text{ A}$$

$$I_{C_1} = \left(V_r \times \frac{Y}{2}\right)$$

$$= (127.02\angle 0°) \times 10^3 \times j0.01 \text{ A}$$

$$= \left[(127.02 + j0) \times 10^3 \times j0.001\right] \text{A} = j127.02 \text{ A}$$

$$\therefore \quad I = I_R + I_{C_1}$$

$$= 346.5\angle -45° + j127.02$$

$$= 245.01 - j245.01 + j0127.02$$

$$= (245.01 - j117.99) \text{ A}$$

$$\therefore \quad V_S = V_R + IZ$$

$$= (127020 + j0) + (245.01 - j117.99)\ (10 + j50)$$

$$= 127020 + j0 + 2450.1 - j1179.9 + j12250.5 + 5899.5$$

$$= 135369.6 + j11070.6$$

$$= (135.3 + j11.07) \text{ kV}$$

$$= 135.75\angle -4.67° \text{ V (ph-N)}$$

Also,

$$I_{C_2} = V_S \times \frac{Y}{2}$$

$$= (135369.6 + j11070.6)\ j0.02$$

$$= j270.74 - 22.14$$

$$= (-22.14 + j270.74) \text{ A}$$

$$\therefore \quad I_S = I + I_{C_2}$$
$$= 245.01 - j117.99 + 270.74j - 22.14$$
$$= (222.87 + 152.75j) \text{ A}$$
$$= 270.19\angle 34.42° \text{ A}$$

The sending end voltage and current are thus obtained as

$$\left.\begin{aligned} V_S &= 135.75\angle 4.67° \text{ kV (Line-Neutral)} \\ &= 235.12\angle 4.67° \text{ kV (L-L)} \end{aligned}\right\} \text{(Ans.)}$$

and $I_S = 270.19\angle 34.42°$ A. (Ans.)

4.9 LONG TRANSMISSION LINE MODEL AND GOVERNING EQUATION

The primary constants of a long line are considered to be distributed over the entire length of the line. Figure 4.10 represents one phase of such a line with return through neutral (having zero impedance). Let the line length be D and dx be an elemental section of the line at a distance x from the receiving end having a series impedance $z dx$ and a shunt admittance ydx. The voltage V at any point along this line is given by

$$\frac{d^2V}{dx^2} = \gamma^2 V$$

where $\gamma = \sqrt{yz}$; γ is a propagation constant and is related to the *attenuation constant* α and *phase constant* β by the relation

$$\gamma = \alpha + j\beta$$

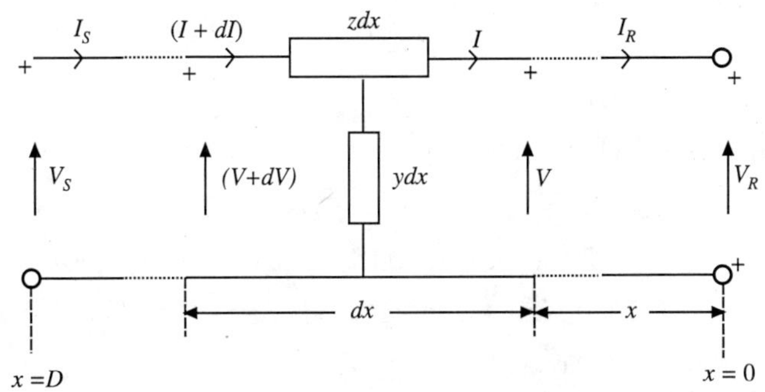

Figure 4.10 Schematic diagram of a longline.

The solution of the equation yields

$$V = \frac{1}{2}V_R(e^{\gamma x} + e^{-\gamma x}) + \frac{1}{2}I_R Z_0(e^{\gamma x} - e^{-\gamma x}) \tag{4.14}$$

where Z_0 is the surge impedance or characteristic impedance or natural impedance of the line $\left(Z_0 = \sqrt{z/y}\right)$ The current I at any point along the line is given by

$$I = \frac{1}{2}\frac{V_R}{Z_0}(e^{\gamma x} - e^{-\gamma x}) + \frac{1}{2}I_R(e^{\gamma x} + e^{-\gamma x}) \tag{4.15}$$

Equations (4.14) and (4.15) may be expressed in terms of hyperbolic function as

$$V = V_R \cosh \gamma x + I_R Z_0 \sinh \gamma x \tag{4.15(a)}$$

$$I = \frac{1}{Z_0} \sinh \gamma x + I_R \cosh \gamma x$$

$$= \frac{1}{Z_0} \sinh \gamma x + \frac{1}{Z_0} V_R \sinh \gamma x \tag{4.15(b)}$$

Equations 4.15(a) and 4.15(b) represent the governing equations for voltage and current at any point 'x' in the line.

With $x = D$, $V = V_S$, $I = I_S$. Hence from the above, we have

$$V_S = V_R \cosh \gamma D + I_R Z_0 \sinh \gamma D$$

$$I_S = \frac{1}{Z_0} V_R \sinh \gamma D + I_R \cosh \gamma D$$

or,

$$\begin{bmatrix} V_S \\ I_S \end{bmatrix} = \begin{bmatrix} \cosh \gamma D & Z_0 \sinh \gamma D \\ \frac{1}{Z_0} \sinh \gamma D & \cosh \gamma D \end{bmatrix} \begin{bmatrix} V_R \\ I_R \end{bmatrix} \tag{4.16}$$

Here,

$$\left.\begin{aligned} A = D &= \cosh \gamma D \\ B &= Z_0 \sinh \gamma D \\ C &= \frac{1}{Z_0} \sinh \gamma D \end{aligned}\right\} \tag{4.17}$$

In case V_S and I_S are known, V_R and I_R can be found just by inverting equation (4.16). Thus,

$$\begin{bmatrix} V_R \\ I_R \end{bmatrix} = \begin{bmatrix} D & -B \\ -C & A \end{bmatrix} \begin{bmatrix} V_S \\ I_S \end{bmatrix} \tag{4.18}$$

For lossless lines, $\alpha = 0$, hence $\gamma = j\beta$. The voltage and current equation are then modified as

$$V_S = V_R \cos \beta D + jZ_0 I_R \sin \beta D \tag{4.19(a)}$$

$$I_S = j\frac{V_R}{Z_0} \sin \beta D + I_R \cos \beta D \tag{4.19(b)}$$

and at any point x, the equations of voltage and current for a lossless line are

$$V(x) = V_R \cos \beta(D - x) + jZ_0 I_R \sin \beta(D - x) \tag{4.19(c)}$$

$$I(x) = j\frac{V_R}{Z_0} \sin \beta(D - x) + I_R \cos \beta(D - x) \tag{4.19(d)}$$

Since the transmission line, in general, may be viewed as a four terminal network, the terminal voltages and currents are related by

$$V_S = AV_R + BI_R$$

$$I_S = CV_R + DI_R$$

The A, B, C and D constants are conventionally known as *transmission parameters* or *generalised circuit constants* or simply *ABCD* parameters. The transmission line being a symmetrical two port network, we have

$$A = D$$

Also, being a reciprocal network, we have

$$AD - BC = 1$$

When two lines having *ABCD* parameters A_1, B_1, C_1, D_1 and A_2, B_2, C_2 and D_2 are in cascade, we can then write the total *ABCD* parameter equation as

$$\begin{bmatrix} A & B \\ C & D \end{bmatrix} = \begin{bmatrix} A_1 & B_1 \\ C_1 & D_1 \end{bmatrix} \begin{bmatrix} A_2 & B_2 \\ C_2 & D_2 \end{bmatrix}$$

Table 4.1 represents the *ABCD* constants for transmission lines on per phase basis.

Table 4.1 ABCD Constants for Lines

Type of Line	Circuit Model	A (no unit)	B (ohm)	C (mho)	D (no unit)
Short	Series R-L Circuit	1	Z	0	1
Medium	Nominal-T	$1+\frac{1}{4}YZ$	$Z\left(1+\frac{1}{4}YZ\right)$	Y	$1+\frac{1}{2}YZ$
	Nominal-π	$1+\frac{1}{2}YZ$	Z	$Y\left(1+\frac{1}{4}YZ\right)$	$1+\frac{1}{2}YZ$
Long	Distributed parameter model	Cosh γD	Z_0 sinh γD	$\frac{\sinh \gamma D}{Z_0}$	cosh γD

4.10 EVALUATION OF *ABCD* CONSTANTS

The hyperbolic function of complex numbers involved in evaluating the *ABCD* constants can be computed by any of the two methods described as follows.

Method 1 (Convergent series (real angle) method)

We know $\gamma = \alpha + j\beta$.

$$\therefore \qquad \cosh(\alpha + j\beta)D = \cosh \alpha D \cos \beta D + j \sinh \alpha D \sin \beta D$$

$$\sinh(\alpha + j\beta)D = \sinh \alpha D \cos \beta D + j \cosh \alpha D \sin \beta D$$

Hence A, B, C and D can be calculated easily.

Method 2 (Convergent series (complex angle) method)
Using Maclaurin's series method where

$$\cosh\theta = 1+\frac{\theta^2}{\lfloor 2}+\frac{\theta^4}{\lfloor 4}+\frac{\theta^6}{\lfloor 6}+\cdots$$

and

$$\sinh\theta = \theta+\frac{\theta^3}{\lfloor 3}+\frac{\theta^5}{\lfloor 5}+\frac{\theta^7}{\lfloor 7}+\cdots$$

We can have with $\gamma=\sqrt{yz}$,

$$\cosh\gamma D = 1+\frac{YZ}{2}+\frac{Y^2Z^2}{24}+\frac{Y^3Z^3}{720}+\cdots$$

and

$$\sinh\gamma D = \sqrt{YZ}\left[1+\frac{YZ}{6}+\frac{Y^2Z^2}{120}+\frac{Y^3Z^3}{5040}+\cdots\right]$$

Thus,

$$A = D = \cosh\gamma D = 1+\frac{YZ}{2}+\frac{Y^2Z^2}{24}+\cdots$$

$$1+\frac{YZ}{2} = \text{(neglecting higher order terms)}$$

and

$$B = Z_0\sinh\gamma D = \sqrt{\frac{Y}{Z}}\sinh\gamma D$$

$$= Z\left[1+\frac{YZ}{6}+\frac{Y^2Z^2}{120}+\cdots\right]$$

$$= Z\left[1+\frac{YZ}{6}\right] \text{ (neglecting higher order terms)}$$

$$C = \frac{1}{Z_0}\sinh\gamma D = \sqrt{\frac{Y}{Z}}\sinh\gamma D$$

$$= Y\left[1+\frac{YZ}{6}\right] \text{ (neglecting higher order terms)}$$

4.11 EQUIVALENT CIRCUIT OF A LONG LINE

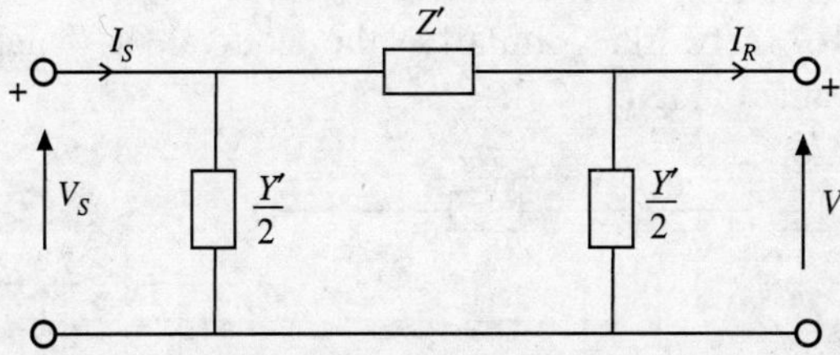

Figure 4.11 Equivalent π representation of a long line.

For a π-network (Figure 4.11), we can write

$$\begin{bmatrix} V_S \\ I_S \end{bmatrix} = \begin{bmatrix} \left(1+\dfrac{Y'Z'}{2}\right) & Z' \\ Y'\left(1+\dfrac{Y'Z'}{4}\right) & \left(1+\dfrac{Y'Z'}{2}\right) \end{bmatrix} \begin{bmatrix} V_R \\ I_R \end{bmatrix}$$

For exact solution of the long line, we also have

$$\begin{bmatrix} V_S \\ I_S \end{bmatrix} = \begin{bmatrix} \cosh \gamma D & Z_0 \sinh \gamma D \\ \dfrac{1}{Z_0} \sinh \gamma D & \cosh \gamma D \end{bmatrix} \begin{bmatrix} V_R \\ I_R \end{bmatrix}$$

For exact equivalence of these two forms, we can write

$$Z' = Z_0 \sinh \gamma D$$

$$1+\frac{Y'Z'}{2} = \cosh \gamma D$$

i.e.

$$Z' = \sqrt{\frac{z}{y}} \sinh \gamma D = ZD \frac{\sinh \gamma D}{D\sqrt{yz}}$$

$$= Z\left(\frac{\sinh \gamma D}{\gamma D}\right)$$

Thus, we observe that $\dfrac{\sinh \gamma D}{\gamma D}$ is the factor by which the series impedance of the nominal-π network must be multiplied to obtain Z' parameter of equivalent π.

Since substituting for $\left(1+\dfrac{1}{2}Y'Z'\right) = 1 + \cosh \gamma D$, $Z' = Z_0 \sinh \gamma D$, we get

$$1+\frac{1}{2}Y'Z_0 \sinh \gamma D = \cosh \gamma D$$

i.e.

$$\frac{1}{2}Y' = \frac{Y}{2}\left[\frac{(\tanh \gamma D)/2}{(\gamma D/2)}\right]$$

Thus, $\left[\dfrac{(\tanh \gamma D)/2}{(\gamma D/2)}\right]$ is the factor by which the shunt admittance arm of nominal-π must be multiplied to obtain the shunt parameter $(Y'/2)$ of the equivalent-π.

For a medium line, $\left[\dfrac{(\tanh \gamma D)/2}{(\gamma D/2)}\right] \simeq 1$ and $\dfrac{\sinh \gamma D}{\gamma D} \simeq 1$. Then the equivalent-$\pi$ network reduces to the nominal-π network. Similarly, the equivalent-T network parameters of a transmission line are obtained (Figure 4.12).

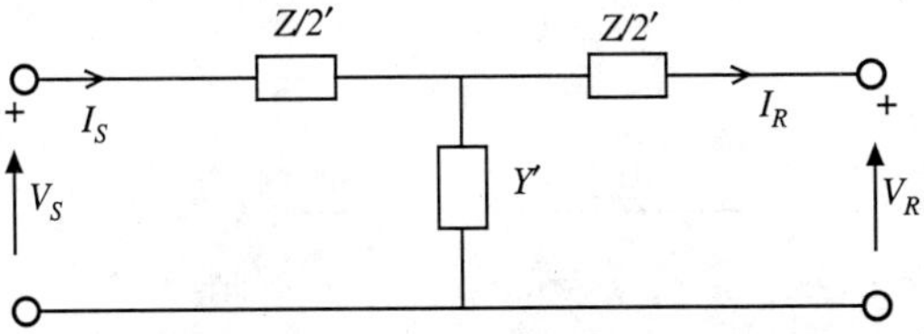

$$\begin{bmatrix} \dfrac{Z'}{2} = \dfrac{Z}{2}\left(\dfrac{\tanh \gamma D/2}{\gamma D/2}\right) \\ Y' = Y(\sinh \gamma D/\gamma D) \end{bmatrix}$$

Figure 4.12 Equivalent-T representation of a long line.

4.12 FERRANTI EFFECT

A long transmission line has considerable shunt capacitance effect. During lightly loaded or unloaded condition, this capacitance effect raises the receiving end voltage beyond the sending end voltage and the ratio ($|V_R| > |V_S|$) increases with line length (Figure 4.13(a)). This is known as *Ferranti effect*.

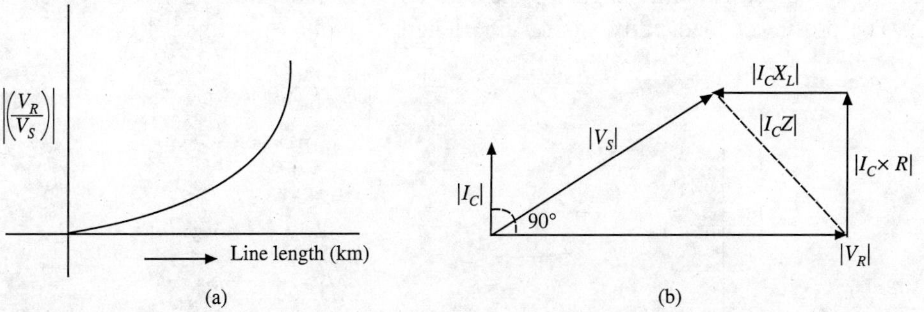

Figure 4.13 (a) Profile of $|V_R|/|V_S|$ with line length, (b) Phasor diagram of Ferranti effect.

Figure 4.13(b) represents the phasor diagram of the line during unloaded condition. It is evident from the diagram that $|V_R| > |V_S|$.

With $I_R = 0$, from the expression of *ABCD* parameter equation,

$$V_S = AV_R + BI_R = AV_R$$

$$\therefore \qquad \frac{V_R}{V_S} = \frac{1}{A}$$

and

$$I_S = CV_R + DI_R = CV_R$$

$$\therefore \qquad \frac{I_S}{V_R} = C$$

4.13 SURGE IMPEDANCE LOADING (SIL)

Surge impedance loading of a power transmission line is the power delivered by a line to a purely resistive load equal to its surge impedance. The line current at SIL is thus given by

$$|I_{\text{line}}| = \frac{|V_{\text{line}}|}{\sqrt{3}\, Z_0}$$

The SIL, P_0, is obtained as

$$P_0 = \left(\sqrt{3}\,|V_L|\right)\left(\frac{|V_L|}{\sqrt{3}\, Z_0}\right) = \frac{V_L^2}{Z_0} \qquad (4.20(a))$$

As P_0 is a real quantity and at SIL the line operates at unity power factor we can express P_0 in either MVA or MW.

Usually, the value of Z_0 for a power transmission line ranges from 400–500 Ω while that of a power cable is between 40 and 60 Ω.

The *loadability* (P_l) of a line is the optimum power transferred through the line at specified operating criterion. Usually, loadability is expressed in terms of SIL.

$$\frac{P_l}{P_0} = \frac{\sin\delta}{\sin\theta} \tag{4.20(b)}$$

where θ is the electrical line length of the line (βD) in radian and δ is the power angle in radian. The profile of loadability against line length is plotted in Figure 4.14.

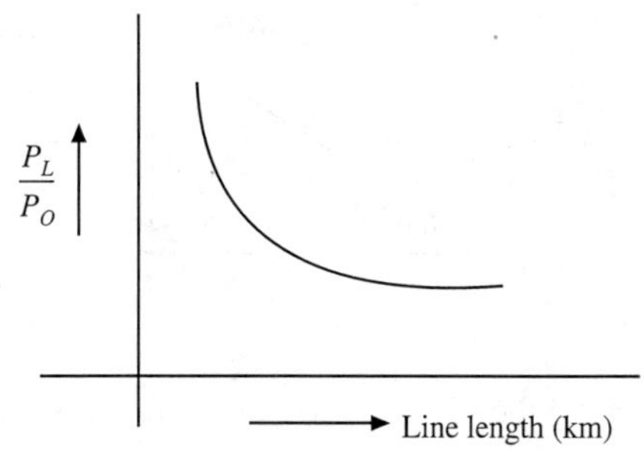

$$\left[P_l = V^2/X \sin\delta, \text{ for lossless live} \right.$$

$$X = \omega l D = \omega D\sqrt{\frac{l}{c}} = \sqrt{lc}$$

$$= Z_0 \beta D = Z_0 \theta$$

$$= Z_0 \sin\theta,\ \theta \text{ being small.}$$

$$\therefore\ P_l = \frac{V^2}{Z_0 \sin\theta} \sin\delta$$

$$\left. = \text{SIL} \cdot \frac{\sin\delta}{\sin\theta} \right]$$

Figure 4.14 Loadability profile of power line.

4.14 POWER FLOW IN TRANSMISSION LINE

It is customary to represent the power expressions of transmission lines in terms of *ABCD* parameters. The real and reactive powers P_R and Q_R at the receiving end of a transmission line are expressed as

$$P_R = \frac{|V_S||V_R|}{|B|}\cos(\beta - \delta) - \frac{|A||V_R|^2}{|B|}\cos(\beta - \alpha) \tag{4.21(a))*}$$

and

$$Q_R = \frac{|V_S||V_R|}{|B|}\sin(\beta - \delta) - \frac{|A||V_R|^2}{|B|}\sin(\beta - \alpha) \tag{4.21(b))*}$$

where $A = |A|\angle\alpha$ and $B = |B|\angle\beta$.

$$I_R = \frac{V_S - AV_R}{B} = \frac{V_S}{B} - \frac{A}{B}V_R = \frac{|V_S|}{|B|}\angle(\delta - \beta) - \frac{|A|}{|B|}\angle(\alpha - \beta)\cdot|V_R|$$

$$S_R = V_R I_R^* = P_R + jQ_R = V_R\left[\frac{|V_S|}{|B|}\angle(\delta - \beta) - \frac{|A|}{|B|}\angle(\alpha - \beta)\cdot|V_R|\right]^*$$

or, $(P_R + jQ_R) = \dfrac{|V_S||V_R|}{|B|}\angle(\beta - \delta) - \dfrac{|V_R|^2|A|}{|B|}\angle(\beta - \alpha)$; V_R being reference phasor.

Separating the real and imaginary parts we get P_R and Q_R as shown in expressions 4.21(a) and 4.21(b).

For a short line, we have $A = 1+\dfrac{ZY}{2} = 1\angle 0$; Y being zero. Thus, $A = 1$; $\alpha = 0°$ and $\beta = |Z|\angle\beta = |X|\angle 90°$ as we can neglect R for transmission line with high X/R ratio. Thus, $|Z| = |X|$ and $\beta = 90°$.

This gives
$$P_R = \frac{|V_S||V_R|}{|X|}\sin\delta \tag{4.22a}$$

and
$$Q_R = \frac{|V_S||V_R|}{|X|}\cos\delta - \frac{|V_R|^2}{|X|} \tag{4.22b}$$

Obviously, $P_{R_{max}}$ is $\dfrac{|V_S||V_R|}{|X|}$ for a lossless line and $P_R = P_{R_{max}}$ at $\delta = 90°$.

On the other hand, if R is not neglected,

$$P_{R_{max}} = \frac{|V_S||V_R|}{|B|} - \frac{1}{|B|}|V_R|^2 \cos\beta = \frac{|V_S||V_R|}{|Z|} - \frac{|V_R|^2}{|Z|}\cos\beta \tag{4.23}$$

Since $\cos\beta = \dfrac{R}{Z}$, we have

$$P_{R_{max}} = \frac{|V_S||V_R|}{|Z|} - \frac{|V_R|^2}{|Z|} \times \frac{R}{|Z|}$$

$$= \frac{|V_S||V_R|}{|Z|} - \frac{|V_R|^2}{Z^2} \times R \tag{4.24}$$

[In the case of a two machine system supplying a series load (Figure 4.15), the expressions of maximum power can be written as

$$P_{1_{max}} = \frac{|V_1||V_2|}{|Z_{12}|} - \frac{|V_2|^2}{Z_{12}^2} \times R_{12}$$

$$P_{2_{max}} = -\frac{|V_1||V_2|}{|Z_{12}|} + \frac{|V_1|^2}{|Z_{12}^2|} \times R_{12}$$

where Z_{12} and R_{12} represent the *transfer impedance* and *resistance* of the two machines and V_1 and V_2 are their respective terminal voltages.]

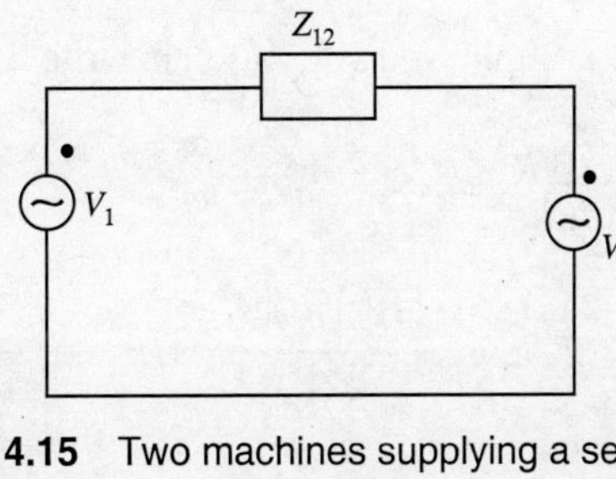

Figure 4.15 Two machines supplying a series load.

4.15 POWER CIRCLE DIAGRAM

The variation of complex power transferred with variation of the power angle δ can be plotted on a complex plane. This plot is known as *power circle diagram* and it provides useful information about the performance of a transmission line. In Figure 4.16, the complex power flow at the sending end can be expressed as

$$S_S = V_S I_S^* = |V_S|\angle\delta_S \left[\frac{|V_S|\angle\delta_S - |V_R|\angle\delta_R}{|Z|\angle\theta}\right]^*$$

$$= |V_S|\angle\delta_S \left[\frac{|V_S|\angle-\delta_S - |V_R|\angle-\delta_R}{|Z|\angle-\theta}\right]$$

$$= \frac{|V_S|^2}{|Z|\angle-\theta} - \frac{|V_S||V_R|\angle(\delta_S - \delta_R)}{|Z|\angle-\theta} \tag{4.25}$$

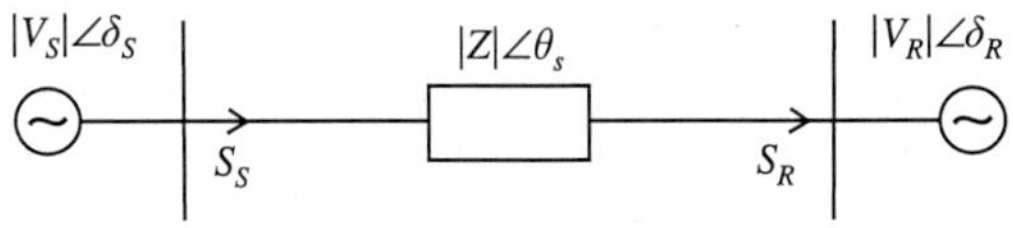

Figure 4.16 A two bus system.

Substituting δ for $(\delta_S - \delta_R)$ in equation (4.25), we get

$$S_S = \frac{|V_S|^2}{|Z|} e^{j\theta} - \frac{|V_S||V_R|}{|Z|} e^{j\theta} . e^{j\delta} \tag{4.26}$$

$$= \frac{|V_S|^2}{|Z|}(\cos\theta + j\sin\theta) - \frac{|V_S||V_R|}{|Z|}\{\cos(\theta+\delta) + j\sin(\theta+\delta)\}$$

$$= P_S + jQ_S$$

Equating the real and imaginary parts, we have

$$P_S = \frac{|V_S|^2}{|Z|}\cos\theta - \frac{|V_S||V_R|}{|Z|}\cos(\theta+\delta)$$

and

$$Q_S = \frac{|V_S|^2}{|Z|}\sin\theta - \frac{|V_S||V_R|}{|Z|}\sin(\theta+\delta)$$

Similarly, for the receiving end complex power, we can write an analogous expression.

$$S_R = -\frac{|V_R|^2}{|Z|} e^{j\theta} + \frac{|V_R||V_S|}{|Z|} e^{j\theta} e^{-j\delta} \tag{4.27}$$

$$= -\frac{|V_R|^2}{|Z|}(\cos\theta + j\sin\theta) + \frac{|V_R||V_S|}{|Z|}\{\cos(\theta+\delta) + j\sin(\theta-\delta)\}$$

$$= P_R + jQ_R$$

$\therefore$
$$P_R = -\frac{|V_R|^2}{|Z|}\cos\theta + \frac{|V_S||V_R|}{|Z|}\cos(\theta-\delta) \quad (4.28a)$$

and
$$Q_R = -\frac{|V_R|^2}{|Z|}\sin\theta + \frac{|V_S||V_R|}{|Z|}\sin(\theta-\delta) \quad (4.28b)$$

[It may be observed that for lossless lines, with $R = 0$, $Z = jX_l$ and $\theta = \pi/2$; the power equations for sending and receiving end get simplified as

$$P_S = P_R = \frac{|V_S||V_R|}{|X_l|}\sin\delta$$

$$Q_S = \frac{|V_S|^2}{|X_l|} - \frac{|V_S||V_R|}{|X_l|}\cos\delta$$

and
$$Q_R = \frac{|V_S||V_R|}{|X_l|}\cos\delta - \frac{|V_R|^2}{|X_l|}\Bigg]$$

Equations (4.26) and (4.27) can be written as

$$S_S = K_S - Me^{j\delta}$$

and
$$S_R = K_R - Me^{-j\delta}$$

where
$$K_S = \frac{|V_S|^2}{|Z|}e^{j\theta}, K_R = -\frac{|V_R|}{|Z|}e^{j\theta}$$

and
$$M = \frac{|V_S||V_R|}{|Z|}e^{j\theta}$$

On a complex plane, if the real power is expressed along the *x*-axis and the reactive power along the *y*-axis, it may be observed that with variation of δ, the loci of both S_S and S_R are circles on complex plane. The radius of each is determined by M. The circle for S_S is called the *sending end power circle diagram* while the circle for S_R is called the *receiving end power circle diagram* (K_S and K_R being the respective centres of the two circles). Figure 4.17 shows the power circle diagrams.

The following aspects of power circle diagrams may be noted:

- The maximum power which can be transmitted is limited by the angle $(\theta + \delta)$ while the maximum power received depends on $(\theta - \delta)$. If $\theta - \delta = 0$, i.e., $\theta = \delta$, the power received is maximum.
- If $\delta \to 0$, K_S, K_R and M lie on the same straight line.
- The real power flow is controlled by adjusting δ while the reactive power flow can be adjusted by controlling the magnitudes of $|V_S|$ and $|V_R|$.

A more convenient form of a circle diagram is described below.

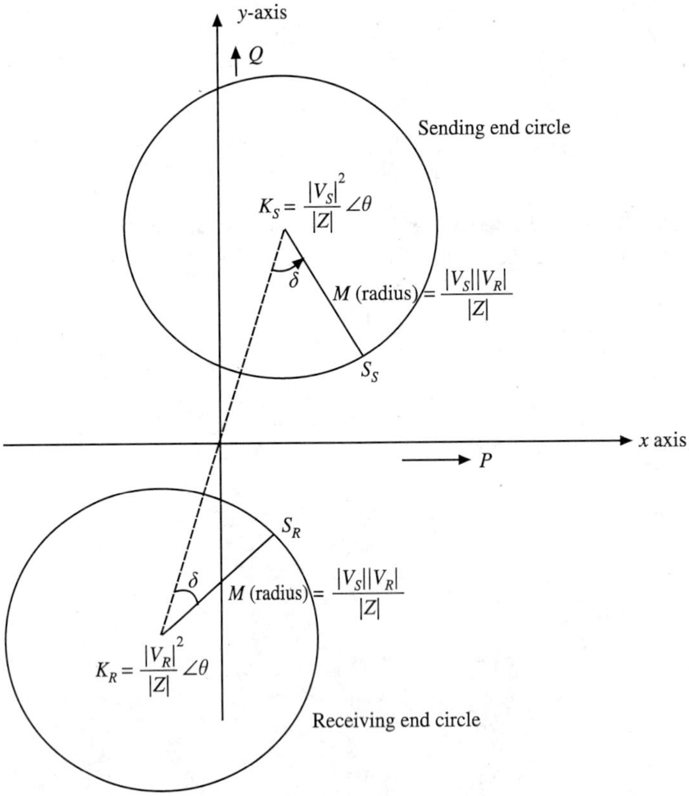

Figure 4.17 Power circle diagrams.

The expressions for complex power at the receiving and sending ends can be written as

$$S_R = -\left|\frac{A}{B}\right| |V_R|^2 \angle(\beta - \alpha) + \frac{|V_S||V_R|}{|B|} \angle(\beta - \delta)$$

and

$$S_S = \left|\frac{D}{B}\right| |V_R|^2 \angle(\beta - \alpha) - \frac{|V_S||V_R|}{|B|} \angle(\beta + \delta)$$

The loci of S_S and S_R are circles with the following parameters:

Receiving end power circle

$$\left.\begin{aligned} &\text{Radius: } \frac{|V_S||V_R|}{|B|} \text{ MVA} \\ &\text{Centre: } -\left|\frac{A}{B}\right| |V_R|^2 \angle(\beta - \alpha) \end{aligned}\right\} \quad (4.29)$$

$$\left[\text{Horizontal co-ordinate: } -\left|\frac{A}{B}\right| |V_R|^2 \cos(\beta - \alpha) \text{ MW}\right.$$

$$\left.\text{Vertical co-ordinate: } -\left|\frac{A}{B}\right| |V_R|^2 \sin(\beta - \alpha) \text{ MVAR}\right]$$

Sending end power circle

$$\left.\begin{aligned} &\text{Radius: } \frac{|V_S||V_R|}{|B|} \text{ MVA} \\ &\text{Centre: } \left|\frac{D}{B}\right| |V_S|^2 \angle(\beta - \alpha) \end{aligned}\right\} \qquad (4.30)$$

$$\left[\text{Horizontal coordinate: } \left|\frac{D}{B}\right| |V_S|^2 \cos(\beta - \alpha) \text{ MW}\right.$$

$$\left.\text{Vertical coordinate: } \left|\frac{D}{B}\right| |V_S|^2 \sin(\beta - \alpha) \text{ MVAR}\right]$$

Figures 4.18 and 4.19 represent the two modified power circle diagrams.

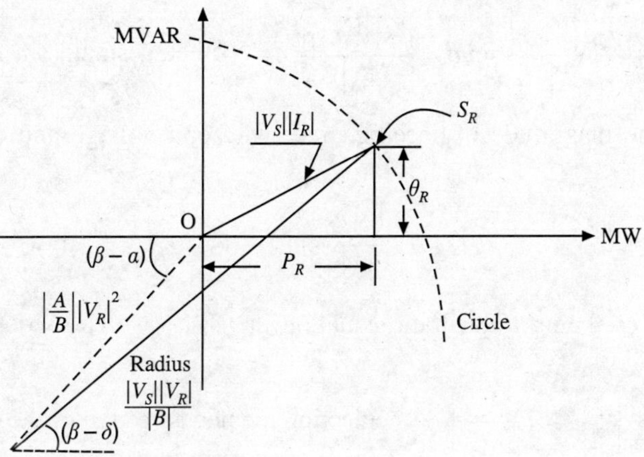

Figure 4.18 Receiving end power circle diagram.

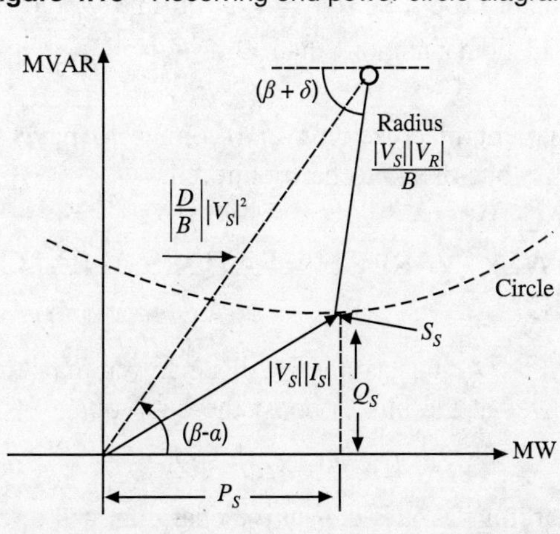

Figure 4.19 Sending end power circle diagram.

4.16 REACTIVE POWER AND VOLTAGE CONTROL IN A TRANSMISSION LINE

Transmission lines have both inductance and capacitance where the inductor absorbs the reactive power and the capacitor generates the reactive power. The net reactive power of the line depends on the loading of the lines. If the line is lossless, the reactive power absorption will be $I^2\omega L$ and the reactive power generation is $V^2\omega C$.

For flat voltage profile along the line, $I^2\omega L = V^2\omega C$, [i.e., when line operation with $\left(Z_0 = \sqrt{\dfrac{L}{C}}\right)$ loading i.e., at SIL value].

If the voltage profile is drooping, $I^2\omega L > V^2\omega C$ and if there is steadystate over voltage, $I^2\omega L < V^2\omega C$.

In a simple two bus line, we can write

$$P_R = \frac{|V_S||V_{R(S)}|}{|X|}\sin\delta \simeq \frac{|V_S||V_{R(S)}|}{|X|}\cdot\delta \quad [\delta \text{ being small, } \sin\delta \simeq \delta] \tag{4.31}$$

where $|V_{R(S)}|$ is the magnitude of specified receiving end voltage and δ is the torque/power angle.

Also,

$$Q_R = \frac{|V_{R(S)}|}{|X|}\left(|V_S| - |V_{R(S)}|\right) \tag{4.32}$$

The expression of Q_R being a quadratic equation in $|V_{R(S)}|$, we can write

$$|V_{R(S)}| = \frac{|V_S|}{2} + \frac{|V_S|}{2}\left(1 - 4\times\frac{Q_R}{|V_S|^2}\right)^{1/2} \quad \left[\text{neglecting the negative sign of the root of } |V_{R(S)}|\right] \tag{4.33}$$

If the reactive power demand be Q_D i.e. $Q_D = Q_R$ the line would operate with $|V_R| = |V_{R(S)}|$. For any other reactive load demand more than Q_R, $|V_R| < |V_{R(S)}|$ and for lesser reactive load demand, $|V_R| > |V_{R(S)}|$.

Thus, it is evident that for the receiving end voltage to have specified bus voltage, the reactive power status of the bus needs to be monitored.

4.17 REACTIVE POWER (VAR) GENERATORS AND LOAD BUS VOLTAGE COMPENSATION

It is now clear that if $Q_D > Q_R$, bus voltage will be lower than the specified value and we need to inject reactive power at the bus to boost the bus voltage.

i.e.

$$Q_D + (-\,Q_C) = Q_R$$

where Q_C is the shunt charging (as it is capacitive charging at the receiving end bus, we have included a negative sign with Q_C).

On the other hand, if the reactive power demand Q is less than Q_R, the receiving end bus voltage is likely to increase and we need to drain this additional reactive power by installing a VAR absorbing device at the receiving end,

i.e. $$Q_D + Q_L = Q_R$$

where Q_L is the inductive VAR added at the receiving end. It has a positive sign as per sign convention.

There are three types of VAR generators. The first one is rotating type (the most rugged and the oldest one, i.e. the *synchronous compensator*) while the second one is static type passive device (a *capacitor* or an *inductor* connected in shunt at the receiving end bus). The third one is a power electronic device known as *static VAR compensator* (*SVC*). The advanced version of power electronic VAR compensation include FACTs (Flexible AC Transmission system) devices.

The rotating VAR generator is an unloaded synchronous motor connected at the receiving end. Since the motor draws negligible real power from the bus, $|E|$ and $|V_R|$ are almost in the same phase. If $|X_S|$ be the synchronous reactance (the resistance of the armature being negligible), we have

$$I_C = \frac{(|V_R| - |E|)\angle 0°}{\sqrt{3} \times jX_S} \text{ A}$$

and $$jQ_C = 3 \times \frac{|V_R|\angle 0°}{\sqrt{3}}(-I_C^*)$$

$$= 3\frac{|V_R|}{\sqrt{3}}\left[-\frac{|V_R| - |E|}{-jX_S \times \sqrt{3}}\right]$$

$$= j|V_R|(|E| - |V_R|)/X_S \times 10^{-6} \text{ VAR} \tag{4.34}$$

if all the voltages are expressed in volts and X_S in ohm.

Therefore, for the rotating VAR generator we have

$$Q_C = |V_R|(|E| - |V_R|)/X_S \times 10^{-6} \text{ MVAR} \tag{4.35}$$

If $|E| > |V_R|$, i.e. the motor is overexcited, the rotating VAR generator will inject Q_C into the line and if $|E| < |V_R|$, i.e. the machine is underexcited, it will then absorb reactive power. Q_C is easily and continuously adjustable by adjusting $|E|$, the excitation voltage.

A passive capacitive VAR generator is a three phase capacitor bank connected at receiving end bus in parallel to the load. If $|V_R|$ be the receiving end line voltage, X_C the per phase capacitive reactance of the bank (or equivalent star basis), we have

I_C (capacitive phase current injected in the line)

$$= j\frac{|V_R|/\sqrt{3}}{X_C} \text{ A} \quad [|V_R| \text{ is expressed in volts while } |X_C| \text{ in ohm.}]$$

$$\therefore \qquad jQ_C = 3\times\frac{|V_R|}{\sqrt{3}}(-I_C^*), \text{ for three phase}$$

$$= j3\times\frac{|V_R|}{\sqrt{3}}\times\frac{|V_R|}{\sqrt{3}X_C}\times 10^{-6} \text{ MVAR}$$

i.e.

$$Q_C = \frac{|V_R|^2}{X_C}10^{-6} \text{ MVAR} \tag{4.36}$$

A three phase static inductor can be employed instead of the capacitor bank in the load bus in shunt with the load in order to control the overvoltage at this bus when the reactive load demand is lesser. The inductive VAR taken from the bus will then be given by

$$Q_L \text{ (3 phase)} = \frac{|V_R|^2}{X_L}\times 10^{-6} \text{ MVAR}$$

The static VAR compensator (the power electronic compensator) is having a controlled inductor in parallel to a fixed capacitor (Figure 4.20) in each phase. It is popularly known as *fixed capacitor thyristor controlled reactor* (FCTCR) and can generate as well as absorb reactive power.

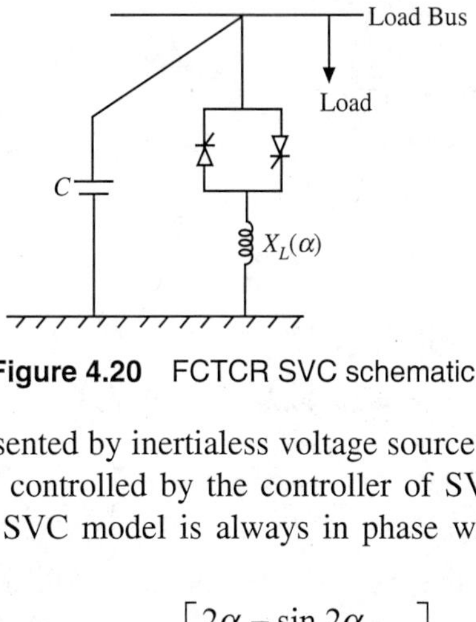

Figure 4.20 FCTCR SVC schematic.

The FCTCR SVC is represented by inertialess voltage source V_{SVC} behind a fixed reactance X_L. The magnitude of V_{SVC} is controlled by the controller of SVC. Since it is a pure reactive power generator, V_{SVC} of the SVC model is always in phase with load bus voltage V_R. V_{SVC} is given by

$$V_{SVC} = V_R\left[\frac{2\alpha - \sin 2\alpha}{\pi} - 1\right] \tag{4.37}$$

α being the firing angle of anti-parallel thyristor. The magnitude of the SVC admittance B_{SVC} is a function of the firing angle α and is obtained as

$$B_{SVC} = \frac{2\pi - 2\alpha + \sin 2\alpha}{\pi X_S} \tag{4.38}$$

for

$$\frac{\pi}{2} \le \alpha \le \pi$$

Thus, B_{SVC} is controllable using SVC.

Here, $$Q_{\text{SVC}} = |V_{\text{SBL}}|^2 B_{\text{SVC}}$$

and $$Q_{\text{SVC}} = Q_C \sim Q_{L(\alpha)}.$$

When the load bus voltage is depressed, α is increased and Q_{SVC} becomes capacitive. It is then possible to bring back the load bus voltage at nominal values. On the other hand, if there is over voltage at load bus, the firing angle α is reduced such that Q_{SVC} shifts to inductive operation thus draining the surplus VAR for the load bus and keeping the bus voltage at nominal value.

4.18 LINE COMPENSATION

Compensation for the line can be done by:

(i) Changing the characteristic impedance (Z_0) of the line and is termed Z_0 compensation (or surge impedance compensation). Ignoring the losses, Z_0 is equal to $\sqrt{z/y}$ (i.e. $\sqrt{x_l \times x_c}$). By changing x_l or x_c, Z_0 can be changed.

(ii) Changing electrical line length of the line. It is termed θ-compensation or line-length compensation. Since $\theta = \beta D = \sqrt{x_l / x_c}$, by reducing x_l, the value of θ can be reduced. Normally, a series capacitor is used for this purpose.

(iii) Sectioning the line. Here, a line is divided into several sections and voltages at the terminals of the sections are maintained using continuous voltage control devices.

Compensation can also be achieved using active devices like *FACTs* (Flexible AC Transmission system devices).

4.19 SHUNT COMPENSATION

If a uniformly distributed shunt compensating inductance be L_c H/m, the effective value of the shunt capacitance C_{sh} can be related as

$$j\omega C_{\text{sh}} = B_{C(\text{sh})} = B_C + B_{\text{sh}} = j\omega C + \frac{1}{j\omega L_c}$$

$$= j\omega C\left(1 - \frac{1}{K_{\text{sh}}}\right) \tag{4.39}$$

where K_{sh} (*degree of shunt compensation*) is defined as

$$K_{\text{sh}} = \frac{x_c}{x_{\text{sh}}} = \frac{B_{\text{sh}}}{B_C} = \frac{1}{\omega^2 C L_c} \tag{4.40}$$

[B_C is the shunt capacitive susceptance per unit length and B_{sh} is the shunt compensating susceptance per unit length.]

K_{sh} would be positive for inductive shunt compensation and is negative for capacitive shunt compensation. The effective values of surge impedance ($Z_{0\,(\text{new})}$) and phase constant ($\beta_{(\text{new})}$) with shunt compensation are given by

$$Z_{0\,(\text{new})} = \sqrt{\frac{x_l}{B_{C(\text{sh})}}} = \frac{Z_0}{\sqrt{1-K_{\text{sh}}}} \tag{4.41}$$

and
$$\beta_{(\text{new})} = \sqrt{x_l . B_{C(\text{sh})}} = \beta\sqrt{1-K_{\text{sh}}} \tag{4.42}$$

Thus, shunt inductive compensation increases Z_0 and decreases β while the shunt capacitive compensation decreases Z_0 and increases β.

4.20 SERIES COMPENSATION

If we now assume uniformly distributed series capacitive compensating capacitance as C_c F/m, the effective value of series reactance x_l can be related as

$$x_l = x_l - x_{\text{se}} = x_l(1 - K_{\text{se}}) \tag{4.43}$$

where K_{se} is the *degree of series compensation.*

Obviously,
$$K_{\text{se}} = \frac{x_{\text{se}}}{x_l} \tag{4.44}$$

K_{se} is positive for capacitive series compensation and is negative for inductive series compensation. The effective value of surge impedance ($Z_{0\,(\text{new})}$) and phase constant ($\beta_{(\text{new})}$) are obtained as

$$Z_{0\,(\text{new})} = \sqrt{x_{l(\text{se})} \times x_c} = Z_0\sqrt{1-K_{\text{se}}} \tag{4.45}$$

$$\beta_{(\text{new})} = \sqrt{\frac{x_{l(\text{se})}}{x_c}} = \beta\sqrt{1-K_{\text{se}}} \tag{4.46}$$

Series capacitive compensation reduces the series impedance (Z) of the line. Generalised constants A, C and D are functions of Z and they are also affected by change in the value of Z, but these changes are much less than change in B (as $B = Z$ in nominal-π and (Z sinh $\gamma D/\gamma D$)) for equivalent-π. The voltage drop ΔV due to series compensation is given by

$$\Delta V = I_R \cos\phi_R + I(X_L - X_C)\sin\phi_R \tag{4.47}$$

where X_C is the capacitive reactance of the series capacitor bank/phase and X_L is the total inductive reactance/phase.

If $(X_L - X_C)\sin\phi_R = R\cos\phi_R$, then $\Delta V = 0$.

4.21 SHUNT AND SERIES COMPENSATION COMBINED

With both shunt and series compensation being applied, the modified values of Z_0 and β are obtained as

$$Z_{0\,(\text{sh + se})} = Z_0\sqrt{\frac{1-K_{\text{se}}}{1-K_{\text{sh}}}} \tag{4.48}$$

$$\beta_{(\text{sh + se})} = \beta\sqrt{(1-K_{\text{sh}})(1-K_{\text{se}})} \tag{4.49}$$

The effective line length $\theta_{(sh + se)}$ is given by

$$\theta_{(sh + se)} = \theta\sqrt{(1 - K_{sh})(1 - K_{se})} \tag{4.50}$$

While the surge impedance loading (SIL) is obtained as

$$P_n = P_0\sqrt{\frac{1 - K_{se}}{1 - K_{sh}}} \tag{4.51}$$

where P_0 is the SIL with no compensation.

EXAMPLE 4.8 A 200 km long three phase, 50 Hz power transmission line has the following per phase distributed parameters:

$$l = 1.3 \times 10^{-3} \text{ H/km}$$
$$c = 10 \times 10^{-9} \text{ F/km}$$
$$r = 0.2 \ \Omega\text{/km}$$
$$g = 0$$

Obtain the *ABCD* constants for this long line.

Solution: Total impedance of the power line is obtained as

$$Z = Z \times \text{Length} = (r + j\omega l) \times \text{Length}$$
$$= (0.2 + j0.408) \times 200 \quad \left[\because \omega l = 2\pi \times 50 \times 1.3 \times 10^{-3} = 0.408\ \Omega\right]$$
$$= (40 + j81.60)\ \Omega$$
$$= 90.87\angle 63.88°\ \Omega.$$

The total admittance is given by

$$Y = y \times \text{length} = (g + j\omega C) \times 200$$
$$= j \times 2 \times \pi \times 50 \times 10 \times 10^{-9} \times 200$$
$$= j0.000629 = 6.29 \times 10^{-4}\angle 90° \text{ mho}$$
$$\gamma D = \sqrt{ZY} = \sqrt{90.87\angle 63.88° \times 6.29 \times 10^{-4}\angle 90°}$$
$$= 0.239\angle 76.94° = (0.054 + j0.232)$$
$$\cosh \gamma D = \cosh\ (0.054 + j0.232)$$
$$= \cosh 0.054 \times \cos 0.232 + j \sinh 0.054 \times \sin 0.232$$
$$= 0.9985 \times 0.9732 + j0.054 \times 0.23$$
$$= 0.9718 + j0.0124 = 0.972\angle 0.73°$$

and

$$\sinh \gamma D = \sinh\ (0.054 + j0.232)$$
$$= \sinh 0.054 \times \cos 0.232 + j\cosh 0.054 \times \sin 0.232$$
$$= 0.054 \times 0.9732 + j1.0015 \times 0.23$$
$$= 0.0526 + j0.2303 = 0.2362\angle 77.13°$$

Also, $$Z_0 = \sqrt{\frac{Z}{Y}} = \sqrt{\frac{90.87\angle 63.88}{0.000629\angle 90°}}$$

$$= 380.08\angle -13.06° \ \Omega$$

∴ From text, we have

$$A = D = \cosh \gamma D = 0.972\angle 0.73°$$

$$B = Z_0 \sinh \gamma D$$

$$= 380.08\angle -13.06° \times 0.2362\angle 77.13°$$

$$= 89.775\angle 64.07° \ \Omega$$

$$C = \frac{1}{Z_0} \times \sinh \gamma D$$

$$= \frac{1}{380.08\angle -13.06°} \times 0.2362\angle 77.13°$$

$$= 0.62 \times 10^{-3} \angle 90.19° \ ℧$$

Thus, we have obtained the generalised $ABCD$ constants as $A = D = 0.972\angle 0.73°$; $B = 89.775\angle 64.07° \ \Omega$ and $C = 0.62 \times 10^{-3} \angle 90.19° \ ℧$. (Ans.)

EXAMPLE 4.9 A 132 kV (L-L), 3 phase, 50 Hz power line has the following distributed parameters per phase per km:

$$l = 1 \times 10^{-3} \text{ H}$$

$$c = 1 \times 10^{-9} \text{ F}$$

$$r = 0.1 \ \Omega$$

$$g = 0$$

If the line be 200 km long and delivers a load of 60 MVA at 0.7 p.f. (lag) at the receiving end, find the sending end voltage, current, power factor and efficiency.

Solution: $$Z = Z \times \text{Length} = (r + j\omega l) \times \text{Length}$$

$$= (0.1 + j \times 2 \times \pi \times 50 \times 1 \times 10^{-3}) \times 200$$

$$= (0.1 + j0.314)\, 200 = (20 + j62.8)\ \Omega$$

$$= 65.9\angle 72.33° \ \Omega$$

$$Y = j\omega c \times \text{length} = j \times 2 \times \pi \times 50 \times 10 \times 10^{-9} \times 200$$

$$= 6.28 \times 10^{-4} \angle 90° \text{ mho.}$$

$$\gamma D = \sqrt{ZY} = \sqrt{65.9\angle 72.33° \times 6.28 \times 10^{-4} \angle 90°}$$

$$= 0.2034\angle 81.165° = (0.031 + j0.2)$$

Thus, $$\cosh \gamma D = \cosh\,(0.031 + j0.2)$$

$$= \cosh 0.031 \times \cos 0.2 + j \sinh 0.031 \times \sin 0.2$$

$$= 1.00048 \times 0.980067 + j0.031005 \times 0.1987$$
$$= (0.98 + j0.0062) = 0.98\angle 0.36°$$
$$\sinh \gamma D = \sinh\,(0.031 + j0.2)$$
$$= \sinh 0.031 \times \cos 0.2 + j \cosh 0.031 \times \sin 0.2$$
$$= 0.031 \times 0.98 + j1.00048 \times 0.1987$$
$$= 0.0304 + j0.1979 = 0.201\angle 81.27°$$

Also,
$$V_R = \frac{132}{\sqrt{3}}\angle 0° = 76.2\angle 0° \text{ kV (ph-N)}$$
$$|I_R| = \frac{3 \text{ phase Power (in MVA)} \times 10^6}{3 \times V_R \text{ (ph-N) kV} \times 10^3}$$
$$= \frac{60 \times 10^6}{3 \times 76.2 \times 10^3} = 262.46 \text{ A}$$

With V_R as reference,
$$I_R = 262.46\angle -\cos^{-1} 0.7 \text{ A}$$
$$= 262.46\angle -45.57° \text{ A}$$

From text, we have
$$V_S = V_R \cosh \gamma D + I_R Z_0 \sinh \gamma D$$
$$= 76.2\angle 0° \times 0.98\angle 0.36° + 262.46\angle -45.57° \times 10^{-3} \times \sqrt{\frac{65.9\angle 72.33°}{6.28 \times 10^{-4}\angle 90°}} \times 0.201\angle 81.27°$$
$$= 74.676\angle 0.36° + 0.26246\angle -45.57° \times 323.94\angle -8.835° \times 0.201\angle 81.27°$$
$$= 74.676\angle 0.36° + 17.09\angle 26.865°$$
$$= (74.674 + j0.4692) + (15.246 + j7.723)$$
$$= (89.92 + j8.192) = 90.29\angle 5.21° \text{ kV (ph-N)}$$
$$I_S = \frac{V_R}{Z_0} \sinh \gamma D + I_R \cosh \gamma D$$
$$= \frac{76.2\angle 0°}{323.94\angle -8.835°} \times 0.201\angle 81.27° + 0.26246\angle -45.57°$$
$$= 0.0473\angle 90.105° + 0.2572\angle -45.21°$$
$$= -0.000087 + j0.0473 + 0.1812 - j0.1825$$
$$= (0.1811 - j0.1352) \text{ kA} = 0226\angle -36.74° \text{ kA}$$

Angle between V_S and I_S is (5.21° + 36.74°), i.e. 41.95° (I_S lags V_S).
The p.f. at the sending end is cos 41.95° = 0.744 (lag)
Next, we find P_S (sending end power).

$$P_S = 3\times|V_S|\text{(ph-N)}\times|I_S|\times\cos 41.95^\circ$$
$$= 3 \times 90.29 \times 0.226 \times 0.744$$
$$= 45.545 \text{ MW}$$

P_R (receiving end power) = 60 MVA × 0.7 = 42 MW

$$\text{transmission efficiency} = \frac{P_R}{P_S}\times 100$$

i.e. $$= \frac{42}{45.545}\times 100 = 92.216\%$$

Hence, we obtain the results as summarized below

$$V_S = 90.29\angle 5.21^\circ \text{ kV (ph-N)}$$
$$= 156.38\angle 5.21^\circ \text{ kV (L-L)} \quad \text{(Ans.)}$$
$$I_S = 226\angle -36.74^\circ \text{ A} \quad \text{(Ans.)}$$

Sending end p.f. = 0.744 (lag). (Ans.)
Transmission efficiency = 92.216%. (Ans.)

EXAMPLE 4.10 A three phase, 50 Hz, 300 km long transmission line has series impedance of $0.087\angle 68^\circ\Omega$/km and shunt admittance of $0.0005\times 10^{-3}\angle 90^\circ$ ℧/km. The load on the line is 100 MW at 200 kV with 0.7 p.f. lag. Find out the voltage, current and power at the sending end and voltage regulation of the line. Also, obtain the per unit values of sending end voltage and current as well as the wavelength and velocity of propagation of wave in the line.

Solution: $$Z = 0.087\angle 68^\circ\times 300\ \Omega \text{ for the entire line.}$$

i.e. $$Z = 26.1\angle 68^\circ\ \Omega = (9.77 + j24.19)\ \Omega$$

$$Y = 10^{-3}\times 0.0005\angle 90^\circ\times 300, \text{ mho for the entire line.}$$

i.e. $$Y = 10^{-3}\times 0.15\angle 90^\circ\ ℧ = j10^{-3}\times 0.15\ ℧.$$

$$\gamma D = \sqrt{YZ} = \sqrt{0.15\times 10^{-3}\times 26.1\angle 90^\circ + 68^\circ}$$
$$= 0.0625\angle 79^\circ = (0.012 + j0.0614)$$

$$Z_0 = \sqrt{\frac{Y}{Z}} = \sqrt{\frac{26.1\angle 68^\circ}{0.15\times 10^{-3}\angle 90^\circ}} = 417.13\angle -11^\circ\ \Omega$$

$$\cosh\gamma D = \cosh(0.012 + j0.0614)$$
$$= \cosh 0.012\times\cos 0.0614 + j\sinh 0.012\times\sin 0.0614$$
$$= 1.00007\times 0.998 + j0.012\times 0.0614$$
$$= 0.998 + j0.00074 = 0.998\angle 3.53^\circ$$

$$\sinh\gamma D = \sinh(0.012 + j0.0614)$$
$$= \sinh 0.012\times\cos 0.0614 + j\cosh 0.012\times\sin 0.0614$$

$$= 0.012 \times 0.998 + j1.00007 \times 0.0614$$

$$= 0.01198 + j0.0614 = 0.0622\angle 26.98°$$

Also, $$V_R = \frac{220 \times 10^3}{\sqrt{3}} \angle 0° \text{ (ph-N)}$$

$$= 127 \times 10^3 \angle 0° \text{ V}$$

$$|I_R| = \frac{100 \times 10^6}{3 \times 127 \times 10^3 \times 0.7} = 374.9 \text{ A}$$

$$I_R = 374.9\angle -\cos^{-1} 0.7 = 374.9\angle -45.57° \text{ A}$$

$$V_S = V_R \cosh \gamma D + I_R Z_0 \sinh \gamma D$$

$$= 127\angle 0° \times 0.998\angle 3.53° + 0.375\angle -45.57° \times 417.13\angle -11° \times 0.0622\angle 26.98°$$

$$= 126.496 + j7.81 + 8.465 - j4.81$$

$$= 134.96 + j3.0 = 135\angle 1.273° \text{ kV (ph-N)}$$

$$= 233.82\angle 1.273° \text{ kV (L-L)} \quad \text{(Ans.)}$$

$$I_S = \frac{V_R}{Z_0} \sinh \gamma D + I_R \cosh \gamma D$$

$$= \frac{127 \times 10^3 \angle 0°}{417.13\angle -11°} \times 0.0622\angle 26.98° + 374.9\angle -45.57° + 0.998\angle 3.53°$$

$$= 304.46\angle 11° \times 0.0622\angle 26.98° + 374.15\angle -42.04°$$

$$= 14.93 + j11.66 + 277.87 - j250.55$$

$$= 292.8 - j238.89 = 377.89\angle -39.21° \text{ A} \quad \text{(Ans.)}$$

$$\text{Voltage regulation} = \frac{135 - 127}{127} \times 100$$

$$= 6.299\% \quad \text{(Ans.)}$$

The angle between V_S and I_S is (1.273° + 39.21°), i.e. 40.48°

$\therefore$ P_S (sending end power)

$$= 3|V_S| \text{(ph-N)} \times |I_S| \times \cos 40.48°$$

$$= 3 \times 135 \times 0.378 \times \cos 40.48°$$

$$= 116.44 \text{ MW} \quad \text{(Ans.)}$$

Next, it is required to obtain p.u. values of sending end voltage and current. We take 220 kV as the base voltage and 374.9 A as the base current.

$$|V_R| = 100 \text{ p.u.} = 220 \text{ kV (L-L)}$$

$$|I_R| = 1.00 \text{ p.u.} = 374.9 \text{ A.}$$

Also, the base impedance is

$$\frac{|V_R|}{|I_R|} = \frac{220 \times 10^3/\sqrt{3}}{374.9} = 338.812\ \Omega$$

Thus, we are now able to represent the values of $|V_S|, |I_S|$ and $|Z_0|$ in terms of base values.

$$|V_S| = 233.82\ \text{kV} = \frac{233.82}{220} = 1.0628\ \text{p.u.} \quad \text{(Ans.)}$$

$$|I_S| = 377.89\ \text{A} = \frac{377.89}{374.9} = 1.008\ \text{p.u.} \quad \text{(Ans.)}$$

$$|Z_0| = 417.13\ \Omega = \frac{417.13}{338.812} = 1.231\ \text{p.u.}$$

We will now find wavelength and velocity of propagation of the wave in the line.

We have $\gamma D = \alpha D + j\beta D$

$= 0.012 + j0.0614.$

i.e. $\beta = \frac{0.0614}{D} = \frac{0.0614}{300} = 2.05 \times 10^{-4}$ rad./km

Wavelength (λ) is given by $\lambda = 2\pi/\beta$

Or, $\lambda = 2 \times 3.14/2.05 \times 10^{-4}$

$= 3.06 \times 104$ km (Ans.)

Velocity $= \lambda f = 3.06 \times 10^4 \times 50$

$= 153 \times 104$ km/sec. (Ans.)

EXAMPLE 4.11 In the previous problem (Ex. 4.10), find the equivalent-π circuit for the line from its nominal-π circuit obtained.

Solution: Let us first draw the nominal-π circuit of the given problem (Figure E4.6).

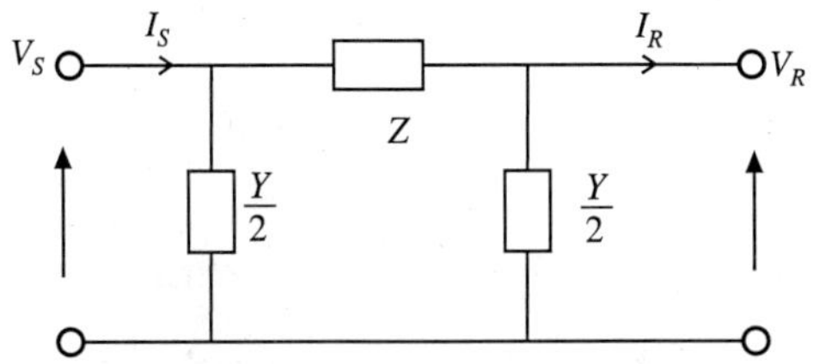

Figure E4.6

As per the given data,

$$Z = 26.1\angle 68^\circ\ \Omega$$

and

$$\frac{Y}{2} = 10^{-3} \times 0.075\angle 90^\circ\ ℧$$

We now draw the equivalent-π network of the nominal-π (Figure E4.7).

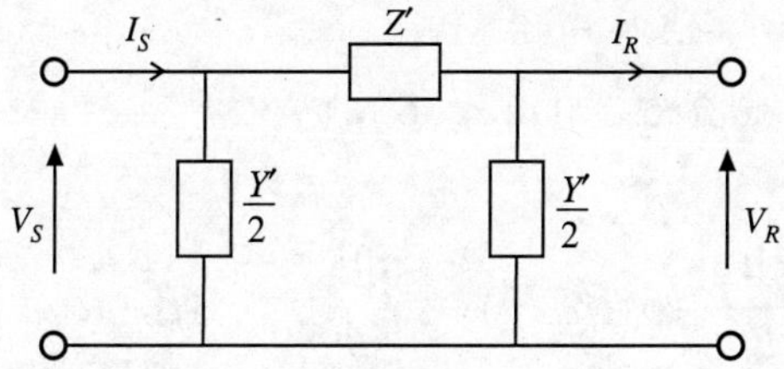

Figure E4.7

We can write now

$$Z' = Z\left(\frac{\sinh \gamma D}{\gamma D}\right)$$

and

$$\frac{Y'}{2} = \frac{Y}{2}\frac{(\tanh \gamma D)/2}{\gamma D/2}.$$

Here

$$Z' = 26.1\angle 68° \times \frac{0.0622\angle 26.98°}{0.0625\angle 79°}$$

$$= 25.97\angle 15.98°\ \Omega$$

and

$$\frac{Y'}{2} = 10^{-3} \times 0.075\angle 90° \times \frac{\left(\frac{\sinh \gamma D}{\cosh \gamma D}\right)\Big/2}{\gamma D/2}$$

$$= 10^{-3} \times 0.075\angle 90° \times \frac{\frac{0.0622\angle 26.98°}{0.998\angle 3.53°}\Big/2}{0.0625\angle 79°}$$

$$= 7.48 \times 10^{-5}\angle 34.45°\ \mho.$$

Thus, for the equivalent π,

$$Z' = 25.97\angle 15.98°\ \Omega$$

and

$$\frac{Y'}{2} = 7.48 \times 10^{-5}\angle 34.45°\ \mho. \quad \text{(Ans.)}$$

Similar exercise can be done for equivalent T circuit.

EXAMPLE 4.12 The generalized circuit constants of a power frequency transmission line are

$$A = (0.95 + j0.015)$$

$$B = (25 + j145)\ \Omega$$

The load at the receiving end is 60 MVA, 50 Hz, 0.8 p.f. lagging. If the voltage at the supply end is 220 kV, calculate the load voltage.

Solution: $A = (0.95 + j0.015)$

$$= 0.95\angle 0.9°$$

$$B = (25 + j145) = 147.13\angle 80.2^\circ\ \Omega$$

$$P_R = 60 \times 0.8 \text{ MW} = 48 \text{ MW}$$

$$Q_R = 60 \times \sin(\cos^{-1} 0.8) = 36 \text{ MVAR.}$$

We know from text,

$$P_R = \frac{|V_S||V_R|}{|B|}\cos(\beta - \delta) - \frac{|A||V_R|^2}{|B|}\cos(\beta - \alpha)$$

$$Q_R = \frac{|V_S||V_R|}{|B|}\sin(\beta - \delta) - \frac{|A||V_R|^2}{|B|}\sin(\beta - \alpha)$$

Here,
$$48 = \frac{220\,|V_R|}{147.13}\cos(80.2^\circ - \delta) - \frac{0.95 \times V_R^2}{147.13}\cos(80.2^\circ - 0.9^\circ) \quad \text{(i)}$$

and
$$36 = \frac{220\,|V_R|}{147.13}\sin(80.2^\circ - \delta) - \frac{0.95 \times V_R^2}{147.13}\sin(80.2^\circ - 0.9^\circ) \quad \text{(ii)}$$

From equation (i), we get

$$1.49\, V_R \cos(80.2^\circ - \delta) = 48 + 1.17 \times 10^{-3}\, V_R^2$$

or,
$$\cos(80.2^\circ - \delta) = \frac{32.21}{V_R} + 0.00078\, V_R \quad \text{(iii)}$$

From equation (ii), we get

$$1.49\, V_R \sin(80.2^\circ - \delta) = 36 + 6.34 \times 10^{-3}\, V_R^2$$

or,
$$\sin(80.2^\circ - \delta) = \frac{24.16}{V_R} + 4.25 \times 10^{-3}\, V_R \quad \text{(iv)}$$

Squaring and adding terms in equations (iii) and (iv), we have

$$1 = \left(\frac{32.21}{V_R} + 0.00078\, V_R\right)^2 + \left(\frac{24.16}{V_R} + 4.25 \times 10^{-3}\, V_R\right)^2$$

or,
$$= (32.21 + 0.00078\, V_R^2)^2 + (24.16 + 0.00425\, V_R^2)^2$$

$$= 1621.18 + 0.255\, V_R^2 + 0.0000186\, V_R^4.$$

or,
$$0.0000186\, V_R^4 - 0.745\, V_R^2 + 1621.18 = 0$$

$$|V_R|^2 = \frac{0.745 \pm \sqrt{(0.745)^2 - 4 \times 0.0000186 \times 1621.18}}{2 \times 0.0000186}$$

$$= \frac{0.745 \pm 0659}{3.72 \times 10^{-5}}$$

$$= \text{either } 37741 \text{ or, } 2311.82$$

$$\therefore \qquad |V_R| = 194.27 \quad \text{or} \quad 48.08 \text{ kV}$$

Taking the higher term, $|V_R| = 194.27$ kV (Ans.)

EXAMPLE 4.13 A 220 kV, 3 phase, 50 Hz power transmission line delivers a load of 80 MW at 0.8 p.f. (lag) at its receiving end. If the series impedance of the entire line is $200\angle 80°\ \Omega$ per phase and the shunt admittance is $0.0013\angle 90°$ mho per phase, find (i) the surge impedance of the line, (ii) SIL, (iii) the *ABCD* constant, (iv) the sending end voltage, current and p.f. of the line, and (v) the transmission efficiency.

Solution:

$$\text{(i) } Z_0 = \sqrt{\frac{Z}{Y}} = \sqrt{\frac{200\angle 80°}{0.0013\angle 90°}} = 392.23\angle -5°\ \Omega \quad \text{(Ans.)}$$

$$\text{(ii) SIL} = \frac{V^2}{Z_0} = \frac{220 \times 220}{392.23} = 123.397 \text{ MW (3 phase)} \quad \text{(Ans.)}$$

$$\text{(iii) } V_R = \frac{220}{\sqrt{3}} = 127 \text{ kV (ph-N)}; \quad I_R = \frac{80 \times 10^6}{\sqrt{3} \times 220 \times 10^3 \times 0.8} = 262.44 \text{ A}$$

Let V_R be the reference phasor.

$$V_R = 127\angle 0° \text{ kV (ph-N)} = 127000\,(1 + j0) \text{ V}$$

and

$$I_R = 262.44\angle \cos^{-1} 0.8 = 262.44\,(0.8 - j0.6)$$

$$= (210 - j157.46) \text{ A} = 262.43\angle -36.87° \text{ A}$$

We have, from nominal-*T* method,

$$A = D = 1 + \frac{YZ}{2} = 1 + 0.0013\angle 90° \times 220\angle 80°$$

$$= 1 + 0.13\angle 170° = 1 - 0.128 + j0.0226$$

$$= (0.872 + j0.0226) = 0.872\angle 1.5°$$

$$B = Z\left(1 + \frac{1}{4}YZ\right) = 200\angle 80°\left[1 + \frac{1}{4} \times 0.0013\angle 90° \times 200\angle 80°\right]$$

$$= 200\angle 80°\,[1 - 0.064 + j0.0113]$$

$$= 200\angle 80°\,(0.936 + j0.0113)$$

$$= 200\angle 80° \times 0.936\angle 0.69° = 187.2\angle 80.69°\ \Omega$$

$$\left.\begin{aligned} C &= Y = 0.0013\angle 90° \text{ mho} \\ A &= D = 0.872\angle 1.5° \\ B &= 187.2\angle 80.69°\ \Omega. \end{aligned}\right\} \text{(Ans.)}$$

(iv) $$V_S = AV_R + BI_R$$

$$= 0.872\angle 1.5° \times 127000\angle 0° + 187.2\angle 80.69° \times 262.43\angle -36.87°$$

$$= 146146 + j3.6915 = 150735\angle 14.18° \text{ V}$$

$$V_S \text{ (ph-N)} = 150.735 \text{ kV}\angle 14.18°$$

$$V_S \text{ (L-L)} = \sqrt{3} \times 150.735\angle 14.18°$$

$$= 261.073\angle 14.18° \text{ kV} \quad \text{(Ans.)}$$

$$I_S = CV_R + DI_R$$

$$= 0.0013\angle 90° \times 127000\angle 0° + 0.872\angle 1.5° \times 262.43\angle -36.87°$$

$$= 165.1\angle 90° + 228.84\angle -35.37°$$

$$= (186.57 + j32.66) \text{ A} = 189.34\angle 9.93° \text{ A}$$

Sending end p.f. angle = 14.18 – 9.93° = 4.25°

∴ cos ϕ_S (sending end p.f.) = cos 4.25° = 0.997 lag.

Thus, $$V_S = 261.073\angle 14.18° \text{ kV (L-L)}$$

$$I_S = 189.34\angle 9.93° \text{ A}$$

$$\cos \phi_S = 0.997 \text{ (lag)}$$

(v) P_S (sending end power)

$$= \sqrt{3} \times |V_S| \times |I_S| \times \cos \phi_S \times 10^{-3} \text{ MW}$$

[where V_S is expressed in kV, and I_S in A]

$$= \sqrt{3} \times 261.073 \times 189.34 \times 0.997 \times 10^{-3}$$

$$= 85.359 \text{ MW}$$

Given P_R (receiving end power) = 80 MW

$\eta_{\%}$ (transmission efficiency)

$$= \frac{P_R}{P_S} \times 100 = \frac{80}{85.359} \times 100 = 93.72\%. \quad \text{(Ans.)}$$

EXAMPLE 4.14 A three phase, 33 kV overhead power line has effective resistance and reactance of 7 Ω and 20 Ω per phase, respectively. The three phase load at the receiving end is 30 MW, operating with 0.8 power factor (lagging). Find the voltage at the sending end. What will be the kVAR rating of the compensating equipment inserted at the receiving end so as to maintain a voltage of 33 kV at each end? Find also the maximum load that can be transmitted.

Solution: From the given data,

$$|I_R| = \frac{\text{Power}}{\sqrt{3} \times V_L \times \cos \phi} = \frac{30 \times 10^6}{\sqrt{3} \times 33 \times 10^3 \times 0.8} \angle -\cos^{-1} 0.8$$

[∵ receiving end p.f. is 0.8 lag]

$$= 524\angle -36.86° \text{ A}.$$

Given $Z = (7 + j20)\,\Omega = 21.18\angle 70.7°\,\Omega$

$$V_S = V_R + I_R Z = \frac{33\times 10^3}{\sqrt{3}} + 524\angle -36.86° \times 21.18\angle 70.7°$$

$$= 19.05\times 10^3 + 11098.32\angle 33.84°$$

$$= (28268.2 + j6180.38)\text{ V} = 28.93\angle 12.33°\text{ kV (ph-N)}$$

or, $$V_S = 28.93\sqrt{3}\angle 12.33°\text{ kV (L-L)} = 50.1\angle 12.33°\text{ kV (L-L)}.$$

According to the given problem,

$$P_R + jQ_R = 30 + j30\tan(\cos^{-1} 0.8)$$

$$= (30 + j22.5)\text{ kVA}$$

We know that

$$P_R = \frac{|V_S||V_R|}{|Z|}\cos(\theta - \delta) - \frac{|V_R|^2}{|Z|}\cos\theta$$

where θ is the line angle (= β).

Here, $|Z| = 21.18\,\Omega, \theta = 70.7°, |V_S| = 33\text{ kV (L-L)} = |V_R|$, with receiving end voltage maintained at the level of sending end voltage.

We have

$$30 = \frac{(33)^2}{21.18}\cos(70.7° - \delta) - \frac{(33)^2}{21.18}\cos 70.7°$$

or, $$30 = 51.41\cos(70.7° - \delta) - 16.99$$

or, $$\cos(70.7° - \delta) = 0.914$$

or, $$\cos(70.7° - \delta) = \cos 23.93°$$

$$\therefore\ \delta = 70.7° - 23.93° = 46.77°.$$

Now, $$Q_R = \frac{|V_S||V_R|}{|Z|}\sin(\theta - \delta) - \frac{|V_R|^2}{|Z|}\sin\theta$$

$$= \frac{(33)^2}{21.18}\sin(70.7° - 46.77°) - \frac{(33)^2}{21.18}\sin 70.7°$$

$$= 20.85 - 48.52 = -27.67\text{ MVAR}$$

where $|V_S| = |V_R|\cdot$ Q is negative, indicating an inductive demand of reactive power at receiving end bus.

But Q_D (reactive power demand at load bus)

$$= P_R\tan\phi = P_R\tan(\cos^{-1} 0.8)$$

$$= 22.5\text{ MVAR}$$

Hence to compensate the total inductive demand of (27.67 + 22.5) MVAR, we need to supply the same amount of reactive power at the load bus and this reactive power injection must be capacitive.

$$Q_C \text{ (capacitive)} = 27.67 + 22.5$$

$$= 50.17 \text{ MVAR (ratting of the capacitive compensation) (Ans.)}$$

for transfer of maximum power. $\theta = \delta$.

This gives
$$P_{max} = \frac{(33)^2}{21.18}(1 - \cos 70.7°)$$

$$= 34.42 \text{ MW (Ans.)}$$

EXAMPLE 4.15 A three phase, 50 Hz, 400 km long transmission line has resistance and inductance (per phase) of 0.035 Ω/km and 1 mH/km, respectively. It has line capacitance (per phase) of 0.01 μF/km. The sending end voltage is 275 kV. Find the value of the shunt reactor having negligible losses inserted at the receiving end so that it is possible to maintain 275 kV at both the ends when the line is at no load.

Solution:

$$R = 0.035 \times 400 = 14\ \Omega \quad \text{(for entire line)}$$

$$X_l = 400 \times 2 \times \pi \times 50 \times 1 \times 10^{-3}$$

$$= 125.6\ \Omega \quad \text{(for entire line)}$$

$$Z = (14 + j125.6)\ \Omega$$

$$= 126.37\angle 83.6°\ \Omega$$

$$Y = j314 \times 0.01 \times 10^{-6} \times 400$$

$$= j1.256 \times 10^{-3}\ ℧$$

$$= 1.256 \times 10^{-3} \angle 90°\ ℧$$

Using nominal π method,

$$A = 1 + \frac{YZ}{2}$$

$$= 1 + \frac{1}{2} \times 1.256 \times 10^{-3} \angle 90° \times 126.37\angle 83.6°$$

$$= 1 + 0.079\angle 173.6°$$

$$= 1 + (-0.078 + j0.008)$$

$$= 0.922 + j0.008$$

$$= 0.922\angle 0.49°$$

$$B = Z = 126.37\angle 83.6°\ \Omega.$$

We know that

$$P_R = \frac{|V_S||V_r|}{|B|}\cos(\beta - \delta) - \left|\frac{A}{B}\right||V_r|^2 \cos(\beta - \alpha)$$

Here

$$|V_S| = |V_r| = 275 \text{ kV}$$

$$|B| = 126.37\ \Omega,\ |A| = 6.922$$

$$\beta = 83.6°$$

$$\alpha = 0.49°$$

$$P_R = 0 \text{ (since line is at no load)}$$

$$0 = \frac{(275)^2}{126.37}\cos(83.6° - \delta) - \left|\frac{0.922}{126.37}\right|(2.75)^2 \cos(83.6° - 0.49°)$$

or,

$$0 = 598.44 \cos(83.6° - \delta) - 66.19$$

$$\cos(83.6 - \delta) = \frac{66.19}{598.44}$$

or,

$$83.6 - \delta = 83.64$$

or,

$$\delta = 0.04°$$

Also,

$$Q_R = \frac{|V_S||V_R|}{|B|}\sin(\beta - \delta) - \frac{|A||V_R|^2}{|B|}\cos(\beta - \alpha)$$

So,

$$Q_R = \frac{(275)^2}{126.37}\sin(83.6° - 0.04°) - \left|\frac{0.922}{126.37}\right|(275)^2 \sin(83.6° - 0.49°)$$

$$= 594.66 - 547.77$$

$$= 46.89 \text{ MVAR}$$

Q is positive, indicating capacitive reactive power at load bus.

Thus, to have $|V_S| = |V_R|$ at light or no load condition, we need to compensate the capacitive Q_R whose value is just obtained. Hence, the rating of the inductive compensator should be 46.89 MVAR so as to keep the bus voltage at the receiving end at nominal value. (Ans.)

EXAMPLE 4.16 The shunt admittance of a 200 km transmission line is $Y = j7 \times 10^{-6}$ ℧/km. Determine the *ABCD* constants of a shunt reactor that will compensate for 60% of the total shunt admittance.

Solution:

$$\text{Total } Y = j7 \times 10^{-6} \times 200\ ℧$$

$$= j1400 \times 10^{-6}\ ℧$$

60% compensation means

i.e.

$$B_L = 0.6 \times 1400 \times 10^{-6}$$

$$= 840 \times 10^{-6}\ ℧$$

We know, for inductor the generalised constants are

$$A = D = 1$$

$$B = 0$$

$$C = -jB_L = -j840 \times 10^{-6}\ ℧$$

[Note: The shunt inductor can be represented by an equivalent T network (Figure E4.8) when we have $A = D = 1$, $B = 0$ (no series impedance), $C = Y_L$ (the admittance of the shunt inductor).]

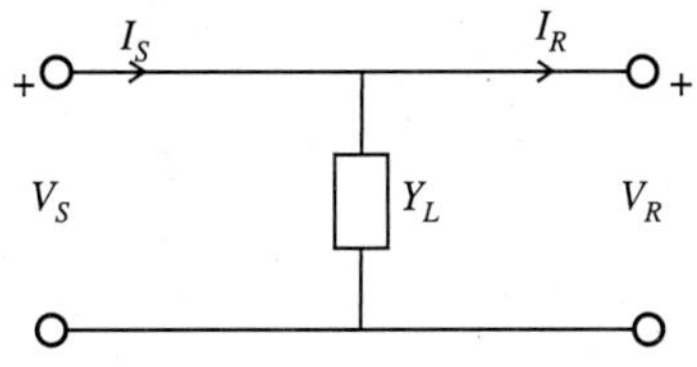

Figure E4.8

EXAMPLE 4.17 For a radial power transmission line delivering complex power S_D at its load bus, find an expression for the receiving end bus voltage. How the expression gets modified if $|V_S| = 1.00$ p.u. and load bus p.f. is unity?

Solution: Let SD represent the demand of complex power at load bus. Obviously,

$$S_D = P_D + jQ_D = P_D + jP_D \tan\phi$$

where ϕ is the p.f. angle at load bus.

Also,
$$\tan\phi = \left(\frac{\sin\phi}{\cos\phi}\right) = \left(\frac{\text{Reactive factor}}{\text{Power factor}}\right)$$
$$= m \text{ (say)}.$$

At load bus, we can write,

$$P_D = \frac{|V_S||V_R|}{|X_{SR}|}\sin\delta_{SR} \qquad \text{(i)}$$

where suffix 'S' stands for sending end and 'R' for receiving end. Also, $|X_{SR}| = |X_{RS}|$.

$$Q_D = -Q_{RS} = \frac{|V_S||V_R|}{|X_{SR}|}\cos\delta_{SR} - \frac{|V_R|^2}{|X_{SR}|} \qquad \text{(ii)}$$

Eliminating δ_{SR} from equations (i) and (ii) and using $\cos^2\delta_{SR} + \sin^2\delta_{SR} = 1$, we get

$$\left(mP_D + \frac{|V_R|^2}{|X_{SR}|}\right)^2 = \left(\frac{|V_S||V_R|}{|X_{SR}|}\right)^2 - P_D^2$$

Rearranging, we get a quadratic equation in $|V_R|^2$.

$$|V_R|^4 + \left(2mP_DX_{SR} - |V_S|^2\right)|V_R|^2 + (1 - m^2)P_D^2X_{SR}^2 = 0$$

$$|V_R|^2 = \left[\frac{|V_S|^2}{2} - mP_DX_{SR}\right] \pm \sqrt{\frac{|V_S|^4}{4} - P_DX_{SR}\left(P_DX + m|V_S|^2\right)}$$

or, $$|V_R| = \left[\left\{\frac{|V_S|^2}{2} - mP_D X_{SR}\right\} \pm \sqrt{\frac{|V_S|^4}{4} - P_D X_{SR}\left(P_D X + m|V_S|^2\right)}\right]^{1/2}$$

With $|V_S| = 1.0$ and $\cos \phi = 1$ (i.e. for $m = 0$), we get

$$|V_R|^2 = \left[\frac{1}{2} \pm \sqrt{\frac{1}{4} - P_D^2 X_{SR}^2}\right]$$

or, $$|V_R| = \left[\frac{1}{2} \pm \sqrt{\frac{1}{4} - P_D^2 X_{SR}^2}\right]^{1/2}$$ (Ans.) [the negative root of V_R^2 is neglected].

This is the required expression.

EXAMPLE 4.18 3 phase power is transferred through a 300 km power line having surge impedance of 350 Ω. If the power angle be 36° and wavelength of the voltage wave be 4 × 10^3 km, find the surge impedance loading (P_n) for this line. Assume the sending end and the receiving voltages to be at 400 kV (L-L) at SIL. Also, find the line loadability in terms of SIL.

Solution: We know from text that for a power line,

$$\text{SIL} = \frac{V^2}{Z_0}$$

Here, $$P_n = \frac{400 \times 400}{350} = 457.14 \text{ MW} \quad \text{(Ans.)}$$

Line loadability in terms of SIL is represented as

$$\frac{P}{\text{SIL}} = \frac{\sin \delta}{\sin \theta}$$

Here $$P = \text{SIL} \times \frac{\sin 36°}{\sin 27°} \left[\begin{aligned} \because \theta &= \beta D = \frac{2\pi}{\lambda} \times D \\ &= \frac{360}{4000} \times 300 \\ &= 27° \end{aligned}\right]$$

or, $$P = 457.14 \times \frac{0.588}{0.454} = 592.067 \text{ MW}$$

Hence the line loadability is 592.067 MW while the SIL value is 457.14 MW. (Ans.)

$$\therefore \quad \frac{\text{Line loadability}}{\text{SIL}} = 1.295.$$

EXAMPLE 4.19 Figure E4.9 represents power flow through an interconnector having reactance of 5 Ω/ph (resistance is negligible) linking two generating stations. If the voltage profile in the interconnector is 33 kV (L-L) throughout the line, estimate the power angle and the station power factors.

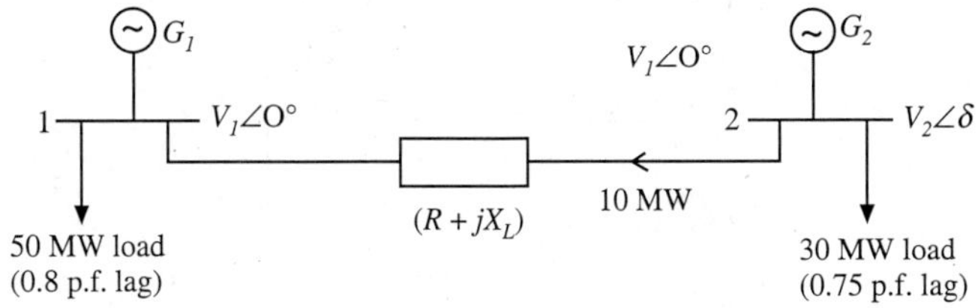

Figure E4.9

Solution: At bus-1,

$$\text{Load} = P_{l_1} + jQ_{l_1} = 50 + j50 \tan(\cos^{-1} 0.8)$$

$$= (50 + j37.5) \text{ MVA}$$

At bus-2,

$$\text{Load} = P_{l_2} + jQ_{l_2} = 30 + j30 \tan(\cos^{-1} 0.75)$$

$$= (30 + j26.46) \text{ MVA}$$

Also,
$$|V_S| = |V_R| = 33 \text{ kV (L-L)}$$

and
$$P_S = P_R = 10 \text{ MW (given)}$$

Also,
$$P_S = P_R = \frac{|V_S||V_R|}{|X|} \sin\delta$$

Here,
$$10 = \frac{33 \times 33}{5} \sin\delta$$

∴
$$\delta = 2.63°.$$

Hence, the operating power angle is 2.63°, i.e. $V_2\angle\delta = 33\,\text{kV}\angle 2.63°$ while $V_1\angle\delta = V_1\angle 0°$ (V_2 leads V_1 to send real power from bus-2 to bus-1).
Next, we compute the reactive power.

$$Q_S = -Q_R = \frac{|V|^2}{|X|} - \frac{|V|^2}{|X|}\cos\delta, \qquad \text{when } |V_S| = |V_R| = |V|$$

Here
$$Q_S = -Q_R = \frac{33^2}{5} - \frac{33^2}{5}\cos 2.63°$$

$$= 8.07 \text{ MVAR}$$

∴ Q_1 (reactive power at station-1) = $Q_{l_1} - 8.07 = 37.5 - 8.07 = 29.43$ MVAR
Thus, station-1 or bus-1 has reactive power of 29.43 MVAR (net).

$$\cos\phi_1 \text{(p.f. at station-1)} = \cos\left(\tan^{-1}\frac{Q_1}{P_{l_1}}\right)$$

$$= \cos\left(\tan^{-1}\frac{29.43}{50}\right)$$

$$= 0.862 \text{ (lag)}$$

Q_2 (reactive power at station-2)

$$= Q_{l_2} + Q_S = 26.46 + 8.07 = 34.53 \text{ MVAR}$$

Station-2 or bus-2 has then reactive power of 34.53 MVAR (net).

$$\therefore \quad \cos\phi_2 \text{ (p.f. at station-2)} = \cos\left(\tan^{-1}\frac{Q_2}{P_{l_2}}\right)$$

$$= \cos\left(\tan^{-1}\frac{34.53}{30}\right)$$

$$= 0.655 \text{ lag.}$$

Thus, the p.f. at the receiving end station (bus-1) is 0.862 lag while that at the sending end station (bus-2) is 0.655 lag. The power angle is 2.63° (Ans.)

EXAMPLE 4.20 A radial power transmission line is terminated by its characteristic impedance Z_0. Find an expression for driving point impedance (Z_{11}) voltage gain $\frac{|V_2|}{|V_1|}$ and current gain $\frac{|I_2|}{|I_1|}$ in terms of secondary constants of the line. What is the expression for transmission efficiency?

Solution: We know that the sending end voltage (V_1) and current (I_1) can be represented in terms of the receiving end voltage (V_2) and current (I_2) using the following relations:

$$V_1 = V_2 \cosh \gamma D + I_2 Z_0 \sinh \gamma D \quad \text{(i)}$$

$$I_1 = \frac{1}{Z_0} V_2 \sinh \gamma D + I_2 \cosh \gamma D \quad \text{(ii)}$$

If the line is terminated in Z_0, then $V_2 = Z_0 I_2$ and hence

$$V_1 = V_2 (\cosh \gamma D + \sinh \gamma D) = V_2 e^{rD} = V_2 e^{\alpha D} . e^{j\beta D}$$

and

$$I_1 = I_2 (\cosh \gamma D + \sinh \gamma D) = I_2 e^{rD} = I_2 e^{\alpha D} . e^{j\beta D}$$

$$[\because \alpha + j\beta = \gamma]$$

$$Z_{11} = \frac{V_1}{I_1} = \frac{V_2}{I_2} = Z_0$$

Thus, the driving point impedance (Z_{11}) becomes Z_0, the surge impedance of the line.

Next, we find from the first equation (equation (i)),

$$|V_1| = |V_2| e^{\alpha D}$$

or,

$$\frac{|V_2|}{|V_1|} = e^{-\alpha D}$$

and from the second equation, we have

$$\frac{|I_2|}{|I_1|} = e^{-\alpha D}.$$

Therefore, the voltage and current gain both are equal and of value $e^{-\alpha D}$, D being the length of the line.

Let us assume that the positive flow of current is that value when it flows from the sending end (end-1) to the receiving end (end-2). Let this direction of receiving end current (I_2) be the reference direction and hence we can represent the complex power at the receiving end as ($-S_{21}$).

Also,

$$-S_{21} = V_2 I_2^* = V_1 e^{-\alpha D}.e^{-j\beta D}.I_1^* e^{-\alpha D}.e^{+j\beta D}$$

$$= S_{12} e^{-2\alpha D}$$

$$\frac{-S_{21}}{S_{12}} = e^{-2\alpha D}$$

Since α is real, transmission efficiency can be calculated as,

$$\eta_{tr} = \frac{-P_{21}}{P_{12}} = e^{-2\alpha D} \text{ in p.u. (Ans.)}$$

[The reader may note here that if the line is lossless, i.e. $R = 0$, $G = 0$, $Z_0 = \sqrt{\frac{L}{C}}$ and $\gamma = j\omega\sqrt{lc}$.]

Hence for a lossless line,

$$\frac{|V_2|}{|V_1|} = \frac{|I_2|}{|I_1|} = -\frac{S_{21}}{S_{12}} = \frac{-P_{21}}{P_{12}} = 1\,(=\eta_{tr})$$

The line becomes an ideal line with 100% transmission efficiency (η_{tr}).

Also,

$$P_{12} = \text{Re}\left[V_1 I_1^*\right] = \text{Re}\left[V_1\left(\frac{V_1}{Z_0}\right)^*\right]$$

$$P_{12} = \text{Re}\left[\frac{V_1 \times V_1^*}{Z_0^*}\right]$$

Since V_1 is the reference and Z_0 is real, we have $P_{12} = \frac{|V_1|^2}{Z_0}$, the SIL value of the line.

Also, $$P_{12} = \text{Re}\left[V_1 I_1^*\right] = \text{Re}\left[Z_0 |I_1|^2\right]$$

$$= \left[Z_0 |I_1|^2\right]$$

EXAMPLE 4.21 A 50 Hz, 3 phase, 220 kV (L-L), 300 km line has the following parameters:

$$R = 0.04\ \Omega/\text{km/phase}$$

$$L = 1\ \text{mH/km/phase}$$

$$C = 0.015\ \mu\text{F/km/phase}$$

If the line is required to maintain 220 kV (L-L) voltage at no load, find the shunt inductive reactor rating at the receiving end using circle diagram.

Solution: $R = 0.04 \times 300 = 12\ \Omega$

$$X = 314 \times 1 \times 10^{-3} \times 300 = 94.2\ \Omega$$

$$Z = (12 + j94.2)\ \Omega = 94.96\angle 82.74°\ \Omega$$

$$Y = 314 \times 0.015 \times 10^{-6} \times 300\angle 90° = 1.41 \times 10^{-3}\angle 90°\ \text{mho}$$

$$A = \left(1 + \frac{YZ}{2}\right) = \left(1 + \frac{1}{2} \times 1.41 \times 10^{-3}\angle 90° \times 94.96\angle 82.74°\right)$$

$$= 1 + 66.95 \times 10^{-3}\angle 172.74°$$

$$= 1 + 0.067\ (\cos 172.74° + j \sin 172.74°)$$

$$= 1 - 0.0665 + j0.0085 = 0.9335 + j00085$$

$$= 0.9335\angle 0.522°$$

$$B = Z = 94.96\angle 82.74°\ \Omega$$

$$|V_S| = |V_R| = 220\ \text{kV}$$

Radius of receiving end circle $= \dfrac{|V_S||V_R|}{|B|} = \dfrac{220^2}{94.96} = 509.69$ MVA

The location of the centre of the receiving end circle is

$$\left|\frac{A}{B}\right| |V_R|^2 = \frac{220 \times 220 \times 0.9335}{94.96} = 475.79\ \text{MAV}$$

$$\angle(\beta - \alpha) = (82.74 - 0.522) = 82.22°$$

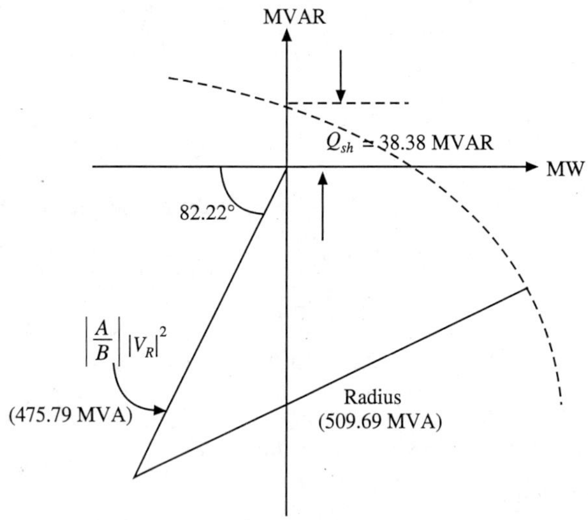

Figure E4.10

By measurement, we find from Figure E4.10 that MVAR (*Q*sh) is 38.38 MVAR. Thus, the shunt reactor should be rated as 38.38 MVAR to maintain 220 kV at the receiving side of the line. (Ans.)

EXAMPLE 4.22 A 3 phase, 50 Hz, 400 kV power transmission line has a series reactance of 81.6 ohm with negligible series resistance and shunt susceptance of $6.29 \times 10^{-4} \angle 90°$ mho. The line length is 200 km.

(i) Determine the reactance and rating of an inductive shunt reactor to be connected at the receiving end if the receiving end voltage is to be maintained at 400 kV or no load.

(ii) If the line delivers 650 MVA at 0.8 p.f. lag at the receiving end, compute the capacitance and MVAR of the shunt capacitor connected at the receiving end to maintain 400 kV load bus voltage.

Solution: (i) Let us assume that the receiving end bus is unloaded. From equation $V_S = V_R \cosh \gamma D + I_R Z_0 \sinh \gamma D$, we get for $I_R = 0$

$$V_S = V_R \cosh \gamma D \quad \text{i.e. } V_R = \frac{V_S}{\cosh \beta D} \qquad [\because r = \alpha + j\beta \text{ and } \alpha = 0]$$

In this problem,

$$Z_0 = \sqrt{\frac{Z}{Y}} = \sqrt{\frac{L}{C}} \qquad [\because r \text{ and } g \text{ both equal to zero}]$$

$$Z = jX_l = j81.6\ \Omega \text{ and } Y = j0.000629 = j\omega C$$

Hence

$$L \text{ (line inductance)} = \frac{X_L}{\omega} = \frac{81.6}{2\pi \times 50} = 0.26 \text{ H}$$

and

$$C = \frac{Y}{\omega} = \frac{0.000629}{2\pi \times 50} = 20 \times 10^{-7} \text{ F}$$

$$|Z_0| = \sqrt{\frac{0.26}{20\times10^{-7}}} = 360\ \Omega$$

$$\gamma D = \sqrt{ZY} = \sqrt{j81.6\times j0.000629}$$

$$= j\times0.227 = j\beta D$$

Thus, $\cosh\gamma D = \cosh(0 + j0.227)$

$= \cosh 0 \times \cos 0.227 + j \sinh 0 \times \sin 0.227$

$= 0.974 + j.0 \times 0225 = 0.974$

$$V_R = \frac{V_S}{\cosh\beta D} = \frac{400/\sqrt{3}}{0.974} = 237.11\text{ kV (ph-N)}$$

or, $V_R = 410.68$ kV (ph-ph)

We need to bring down this higher value of V_R to 400 kV usisng shunt reactor at load/receiving end bus (Figure E4.11).

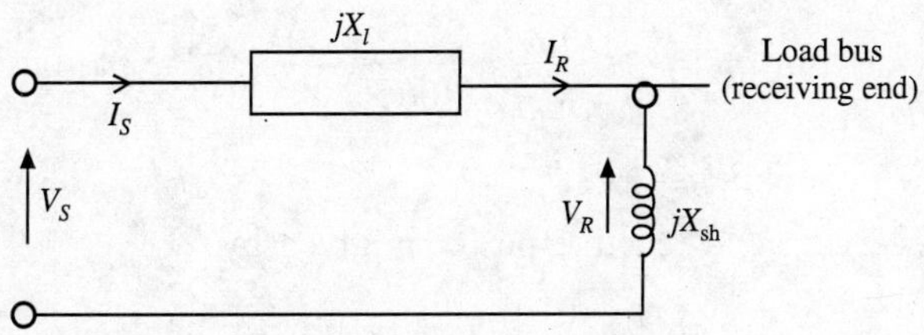

Figure E4.11

Here, the receiving end current is given by

$$I_R = \frac{V_R}{jX_{sh}}$$

By substituting the value of I_R in the expression of V_S, we get

$$V_S = V_R\cos(\beta D) + jZ_0\sin\beta D\times\frac{V_R}{jX_{sh}}$$

[$\therefore \cosh(j\beta D) = \cosh(0 + j\beta D)$

$= \cosh 0.\cos(\beta D) + j\sinh 0.\sin(\beta D)$

$= \cos(\beta D)$]

or, $$V_S = \left[\cos(\beta D) + \frac{Z_0\sin(\beta D)}{X_{sh}}\right]V_R$$

Simplification yields

$$X_{sh} = \frac{Z_0\sin(\beta D)}{\frac{V_S}{V_R} - \cos(\beta D)}$$

X_{sh} represents the value of required inductive reactance to compensate for the voltage rise at no load for the line.

In this problem

$$X_{sh} = \frac{360 \times \sin(0.227)}{1 - \cos(0.227)} = 3158.18\ \Omega/\text{ph} \quad \text{(Ans.)}$$

Shunt reactor rating Q_{sh} is obtained as

$$Q_{sh} = \frac{|V_R|^2}{X_{sh}}\ \text{MVAR/ph.}$$

Since we require $|V_R| = |V_S|$,

$$Q_{sh} = \frac{(237.11)^2}{3158.18} = 17.8\ \text{MVAR/ph.} \quad \text{(Ans.)}$$

(ii) Real power delivered at the receiving end is 650 × 0.8 = 520 MW.

$$P_{12} = \frac{|V_S||V_R|}{X_{12}} \sin \delta_{12}$$ [suffix 1 stands for sending end and suffix 2 stands for receiving end]

Here,

$$\sin \delta_{12} = \frac{P_{12}X_{12}}{|V_S||V_R|} = \frac{520 \times 81.6}{400 \times 400} = 0.2652$$

$$\delta_{12} = 15.38°.$$

The expression for receiving end reactive power is given by

$$Q_R = \frac{|V_S||V_R|}{|B|} \sin(\beta - \delta) - \frac{|A||V_R|^2}{|B|} \sin(\beta - \alpha)$$

Here,

$$|V_S| = |V_R| = 237.11\ \text{kV}$$

$$|B| = Z_0 \sinh \gamma D$$

$$= 360 \times \sinh(0 + j0.0227)$$

$$= 360\left[\sinh 0 . \cos 0.227 + j \cosh 0 . \sin 0.227\right]$$

$$= 360\left[j0.225\right]$$

$$= j81.02\ \Omega = 81.02\angle 90°\ \Omega$$

and

$$A = \cosh \gamma D = 0.974\angle 0°.$$

$$Q_R = \frac{237.11 \times 237.11}{81.02} \sin(90° - 15.38°) - \frac{0.974 \times 237.11^2}{81.02} \sin(90° - 0°)$$

$$= 669.07 - 675.875 = -6.805\ \text{MVAR/ph}$$

Also, the reactive load on the receiving end bus is $\frac{650}{3}$ MVA × 0.6, i.e. 130 MVAR/ph.

Thus, in order to maintain the rated voltage at the receiving end, Q_R = 6.805 MVAR/ph is to be drawn alongwith the reactive load demand. The reactive power status at the load bus is diagrammatically represented in Figure E4.12.

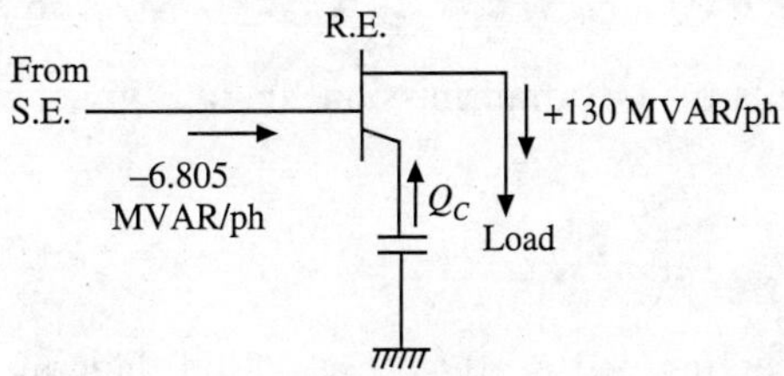

Figure E4.12

We now can write

$$-6.805 + QC = 130$$

or, $$Q_C = 130 + 6.805 = 136.805 \text{ MVAR/ph}$$

The compensating shunt capacitor should have rating of 136.805 MVAR/ph to keep load bus at rated voltage at the specified load level. (Ans.)

Also, $$Q_C = \frac{|V_R|^2}{|X_C|}$$

$$|X_C| = \frac{|V_R|^2}{Q_C} = \frac{237.11^2}{136.805} = 410.95\,\Omega$$

or, $$\frac{1}{\omega C} = 410.95$$

$$C = \left[\frac{1}{2\pi \times 50 \times 410.95}\right]$$

$$= 7.75\,\mu\text{F/ph.} \quad \text{(Ans.)}$$

EXERCISES

1. A 3 phase, 50 Hz, 20 km long transmission line has the following data:
   Receiving end voltage 33 kV
   Load delivered at receiving end 20 MW at 0.8 p.f. lagging
   Impedance per phase per km (0.8 + *j*0.5) Ω.
   Find out the sending end voltage, line current and voltage regulation of the line.
   [**Ans.** V_S(L-L) = $47.21\angle -1.44°$ kV; $I_R = 4.437\angle -36.86°$kA; Reg ($V$) = 43.06%]

2. A single phase 50 Hz, 30 km long transmission line supplies a load of 15 MW of 0.85 p.f. lagging. The line resistance and inductance are 0.06 Ω and 0.4 mH per km.

The voltage at the receiving end is kept constant at 25 kV. Find out the regulation of the line and if a 80 μF capacitor is placed parallel with the load, then what will be the new regulation value of the line.

$\left[\textbf{Ans.}\ \text{Reg}\,(V) = 9.96\%; \text{Reg}\,(V)_{\text{new}} = 1.006\%\right]$

3. The *ABCD* constants of a 3ϕ transmission line are given below:

$A = D = 0.8\angle 5°$

$B = 150\angle 80°\ \Omega$

$C = 0.0004\angle 10°$ ℧

The load at the receiving end is 70 MW at 0.8 p.f. lagging. The receiving end voltage is fixed at 230 kV. Find out the sending end voltage and current of the line.

$\left[\textbf{Ans.}\ V_S = 231.4\angle 13.72°\ \text{kV(L-L)}; I_S = 0.223\angle 26.56°\ \text{kA}\right]$

4. Using nominal '*T*' method, find out the sending end current of a 3ϕ 50 Hz medium length transmission line, where the load is 50 MW at 0.75 p.f. lagging, impedance and admittance are respectively $(10 + j20)$ Ω/phase and $j0.0002$ ℧/phase and receiving end voltage is 220 kV constant.

$\left[\textbf{Ans.}\ I_S = 0.158\angle -34.91°\ \text{kA}\right]$

5. Calculate the *ABCD* constants for a three phase 50 Hz transmission line (160 km long) and having the following distributed constants per phase per km:

$r = 0.15\ \Omega; l = 1.20\ \text{mH}; c = 0.008\ \mu\text{F}; g = 0.$

$\left[\textbf{Ans.}\ A = D = 0.988\angle 0.3°;\quad B = 64\angle 63.3°\ \Omega;\quad C = 0.0004\angle 90.1°\ \text{mho.}\right]$

6. A 350 Hz phase overhead transmission line is 300 km long and it has a total series impedance of $(40 + j125)\ \Omega$ and a total shunt admittance of 10^{-3} mho. The receiving end power demand is 50 MW at 220 kV, 0.8 p.f. lagging. Find the sending end voltage, current, power and power factor using the concept of long line.

$\left[\textbf{Ans.}\ V_S = 237.23\ \text{kV (L-L)};\quad I_S = 0.1286\angle 15.3°\ \text{KA};\right.$

P.F. (sending end) = 0.987 (lead); Power (sending end) = 52.15 MW]

7. A 3-phase 300 km long transmission line supplies a load of 750 MW at 0.8 p.f. lagging. Determine the MVAR and the capacitance of the shunt capacitors value which are to be installed at the receiving end to keep receiving end and sending end voltage constant. The necessary data are given below:

$V_s = V_r = 500$ kV; $Z_c = 290\ \Omega$; $\beta l = 21°$

$\left[\textbf{Ans.}\ \text{MVAR} = 40.49\ \text{MVAR}; C = 6.6\ \mu\text{F}\right]$

8. An interconnnection cable having a reactance of 4 Ω links generating stations 1 and 2 as shown below (Figure P1). The desired voltage profile is $|V_r| = |V_s| = 40$ kV. The loads at the two bus bars 50 MW at 0.85 lagging p.f. and 30 MW at 0.65 lagging p.f. The station loads are equalized by the flow of power in the cable. Find out the station p.f. and the torque angle.

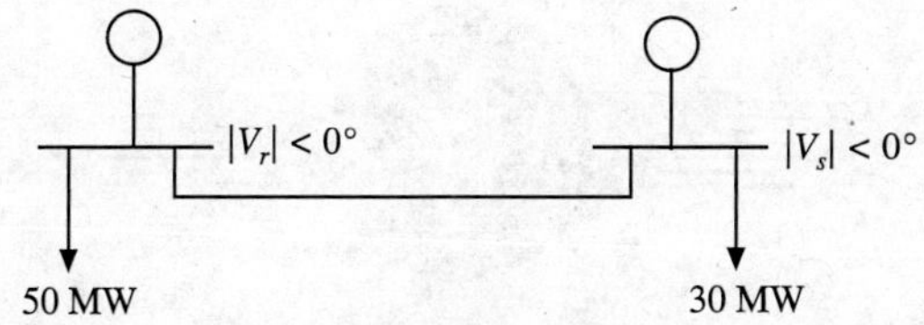

Figure P1

[**Ans.** Q (stn-1) = 31.104 MVAR; P.F. (stn-1) = 0.78 lag;
Q (stn-2) = 35.204 MVAR; P.F. (stn-2) = 0.75 lag]

9. Show that the line regulation is given by

$$|\Delta V| = |V_S| - |V_R| = \text{IR} \cos \phi + \text{IX} \sin \phi, \text{ for a short line}$$

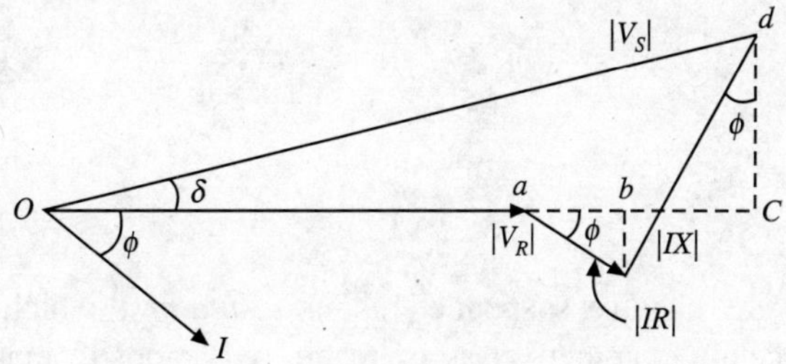

Figure P2

[**Hints:** In Figure P2, we have $|V_R| + ac + dc = od \equiv |V_S|$

Here, $ac = ab + bc$

$= IR \cos \phi + IX \sin \phi$

and $dc = (IX \sin \phi - IR \cos \phi)$

Since ac gives summation of two quantities while dc gives difference of two quantities, we can neglect dc part.

$\therefore$ We have $|V_R| + ac = |V_S|$

i.e. $|V_S| - |V_R| = \Delta V = ac = IR \cos \phi + IX \sin \phi$]

CHAPTER 5

Line Insulators

5.1 INTRODUCTION

Overhead power lines are usually suspended using *insulators*, which are usually designed elements of glazed porcelain, toughed glass or made of composite materials. Insulators have arrangements to support the bare line conductor(s) from arm member(s) of towers or poles. They are designed and manufactured for a certain voltage range (usually the maximum voltage per insulator is 33 kV). In the following sections, we describe the types and characteristics of line insulators.

5.2 INSULATOR MATERIALS

The materials most commonly used for insulators of an overhead transmission line are porcelain, glass, steatite and special composition materials. Porcelain being the most common other materials are used to a limited extent.

5.2.1 Porcelain

Porcelain is produced by firing a mixture of 20% silica, 30% feldspar and 50% china clay at high temperature. It is mechanically stronger (but costlier than glass) and is used as a material to manufacture different types of insulators. Dust deposits and temperature changes normally do not affect much on its surface. The dielectric strength of porcelain is 60 kV per cm of thickness. A single porcelain unit can be used up to 33 kV (r.m.s.). When used at low temperatures, the mechanical properties of the porcelain insulator are better but, at the same time, the material remains porous and may subsequently lead to dielectric failure.

5.2.2 Glass

Glass insulators are extensively used due to their lower costs, high dielectric strength

(140 kV per cm of thickness) and simple design. It is very easy to detect any fault within them because of their optical transparency. When compared to porcelain, glass withstands higher mechanical stresses, has low thermal expansion and it may develop high resistivity after proper annealing. The major disadvantage of glass is that moisture condenses more readily on its surface and facilitates the accumulation of dirt deposits, thus giving a high chance of surface leakage. Hence, its use is limited to a working voltage of about 33 kV.

5.2.3 Steatite

Steatite is actually a magnesium silicate, i.e. a combination of magnesium oxide and silica in various proportions. It has a higher tensile strength and bending stress than that of porcelain. It is used to manufacture insulators at tension towers or where the transmission line takes a sharp turn.

5.2.4 Composite Insulator

Composite insulators are basically non-ceramic insulators (NCI) or polymer insulators and are made of two components of different materials. These insulators are made of lighter weight polymeric materials such as ethylene propylene rubber (EPR), polytetrofluoroethylene (PTFE), silicon rubber, or other similar materials and often used in transmission systems. Composite insulators generally consist of a fibreglass control rod having a number of concentric weather sheds made of highly insulating polymeric material and attached to the rod along its length. They have lightweight structure and hydrophobic property. The insulators employed outdoors are, therefore, highly water-resistant and conducive to repelling dirt. They have low-leakage current losses.

Composite insulators for high voltages should satisfy certain specific electrical requirements as listed below:

(i) The control rod must be insulated electrically in its axial direction and there should be no electrical conduction between the insulating cover and the interior rod.
(ii) The insulating cover should be resistant to the weather and different radiations.
(iii) Composite insulators should have enough mechanical resistance to cold weather conditions.
(iv) These insulators should have enough electrical-tracking resistance.
(v) Composite insulators should be flexible, halogen-free and flame retardant.

5.3 TYPES OF INSULATORS

The porcelain insulators normally used for overhead transmission lines are:

1. Pin type
2. Suspension type
3. Strain type
4. Shackle type

5.3.1 Pin Type Insulators

In the case of *pin type insulator,* the conductor is supported on the top of the insulator. A typical pin-type insulator is shown in Figure 5.1. Maximum working voltage for this type of insulator is 50 kV only as it becomes uneconomical for higher voltages. Upto 25 kV system voltage, it is possible to use pin type insulators as one piece but beyond that voltage range, they are made with more than one piece and practically not very much feasible.

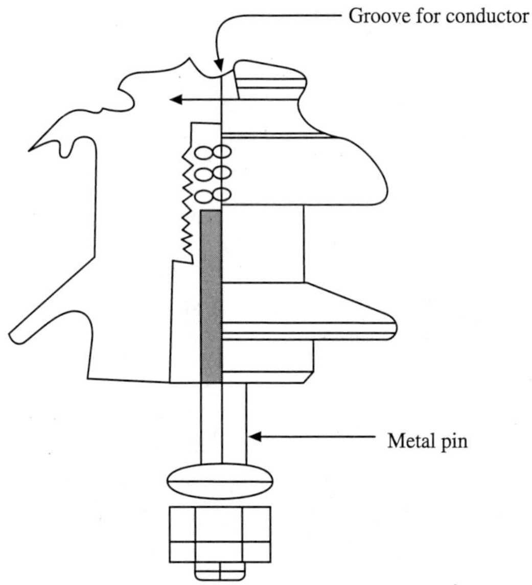

Figure 5.1 Pin type insulator.

5.3.2 Suspension Type Insulators

In the case of *suspension type insulator*, the conductor is supported at the lower end as shown in Figure 5.2. A number of insulators in the form of discs are used together in a flexible string to attain the desired voltage range. The distribution of voltage along the insulator string is not uniform due to the capacitances formed between the discs, conductor, and tower and also due to the proximity of supporting structure. Therefore, the disc (line unit) nearest the conductor has highest voltage stress. It is possible to increase the working voltage by adding insulator units to form a *string* of suspension insulator. Hence, it is advantageous to use suspension type insulators on high voltage transmission lines.

5.3.3 Strain Type Insulators

Suspension type insulators, when used in horizontal position, are known as *strain insulators.* They are used to handle the tension off the conductors at sharp curve/corner or at the end of transmission line or at the anchor towers where lines are dead ended at road or river crossings.

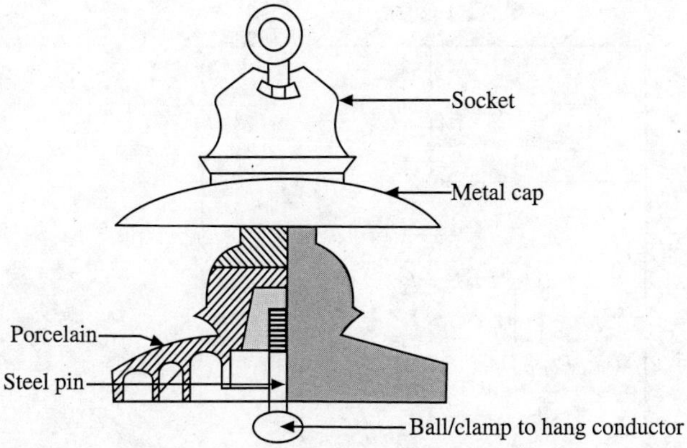

Figure 5.2 Suspension type insulator.

5.3.4 Shackle Type Insulators

Shackle type insulators are frequently used as a strain insulator for low voltage distribution lines. Both the low voltage conductors and the house service wires are attached to the shackle insulator. These insulators are also used in section poles, end poles, on sharp curves and for service connections. These are bolted to the pole or with the cross-arm.

5.4 POTENTIAL DISTRIBUTION OVER A STRING OF SUSPENSION TYPE INSULATORS

Potential distribution among the various units of a string insulator is not uniform due to the capacitance formed between the metal parts of the insulators and the tower. It is possible to decrease these capacitances only by increasing the distance between the insulators and the tower, thereby lengthening the cross arms which are not economical. The equivalent circuit of a string of four identical insulator units is shown in Figure 5.3. The capacitance of each insulator unit is known as *self capacitance* and the capacitance between each insulator unit and earth (i.e. support) is known as *mutual capacitance.*

Let the ratio of mutual to self capacitance of each insulator unit (i.e. disc) be α.

$$\alpha = \frac{\text{Capactiance between insulator disc and ground}}{\text{Capactiance of insulator disc}}$$

It may be noted that α is always less than one.

Also, $$C_1 = C_2 = C_3 = \alpha C \quad (5.1)$$

Let the voltage between the line conductor and earth be V and the voltage drops across the insulator units be V_1, V_2, V_3 and V_4 respectively (Figure 5.3).

i.e. $$V = V_1 + V_2 + V_3 + V_4 \quad (5.2)$$

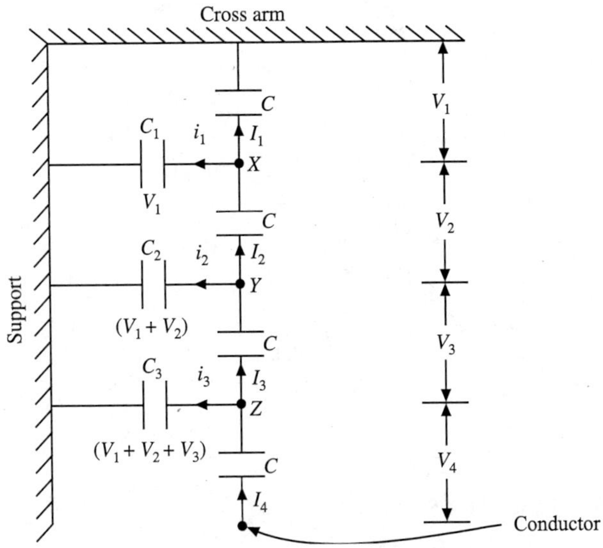

Figure 5.3 Potential distribution over a string of suspension type insulator. (The capacitances between line conductor and metal work of insulators are negligible and therefore not considered here.)

By applying *KCL* at point *X*, we get

$$I_2 = I_1 + i_1$$

or,
$$V_2 \omega C = V_1 \omega C + V_1 \omega \alpha C$$

$\therefore$
$$V_2 = V_1 (1 + \alpha) \tag{5.3}$$

Again by applying *KCL* at point *Y*, we get

$$I_3 = I_2 + i_2$$

or,
$$V_3 \omega C = V_2 \omega C + (V_1 + V_2) \omega \alpha C$$

or,
$$V_3 = \alpha V_1 + V_2 (1 + \alpha)$$

or,
$$V_3 = \alpha V_1 + V_1 (1 + \alpha)^2 \quad [\because V_2 = V_1 (1 + \alpha)]$$

$\therefore$
$$V_3 = V_1 (1 + 3\alpha + \alpha^2) \tag{5.4}$$

Similarly, by applying *KCL* at point *Z*, we get

$$I_4 = I_3 + i_3$$

or,
$$V_4 \omega C = V_3 \omega C + (V_1 + V_2 + V_3) \omega \alpha C$$

or,
$$V_4 = \alpha V_1 + \alpha V_2 + V_3 (1 + \alpha)$$

or,
$$V_4 = \alpha V_1 + \alpha (1 + \alpha) V_1 + (1 + \alpha)(1 + 3\alpha + \alpha^2) V_1$$

$$\left[\because V_2 = (1 + \alpha) V_1, V_3 = (1 + 3\alpha + \alpha^2) V_1\right]$$

$$\therefore \qquad V_4 = (1 + 6\alpha + 5\alpha^2 + \alpha^3)V_1 \tag{5.5}$$

V_1, V_2, V_3 and V_4 can be found out by solving equations (5.2)–(5.5). Normally, the value of α being less than 1, with $\alpha = 0.2$, we have

$$V_2 = 1.2V_1,\quad V_3 = 1.64V_1,\quad V_4 = 2.408V_1$$

Also, $\quad V = V_1 + V_2 + V_3 + V_4 = V_1 + 1.2V_1 + 1.64V_1 + 2.408V_1 = 6.248V_1$ (if $\alpha = 0.2$)

This clearly shows that $V_1 < V_2 < V_3 < V_4$, which indicates that the lowermost disc unit is stressed maximum. This potential distribution can be made uniform by decreasing the value of α.

5.5 STRING EFFICIENCY (η)

The voltage across the unit nearer to the conductor is more than the voltage in the unit nearer to the support. This unequal potential distribution over the string is expressed in terms of string efficiency. *String efficiency* is a measure of uniformity of potential distribution among the disc in a string insulator. The higher is the string efficiency, the more uniform is the potential distribution. The efficiency of a string is defined as the ratio of voltage across the whole string to the product of the number of discs and the voltage across the disc nearest to the line conductor. Mathematically, it can be expressed as,

$$\text{String efficiency}\,(\%) = \frac{\text{Voltage across the string}}{n \times \text{Voltage across the unit nearest to power conductor}} \times 100 \tag{5.6}$$

where n is the number of units in the string.

String efficiency may also be defined as

$$\text{String efficiency}\,(\%) = \frac{\text{Spark over voltage of the string}}{n \times \text{Spark over voltage of lowermost insulator disc}} \times 100 \tag{5.7}$$

So to obtain 100% string efficiency, voltage across each disc is to be exactly same, which is very difficult to achieve. However, efforts may be made to approach this value.

EXAMPLE 5.1 Consider a string of suspension insulator consisting of four units as shown in Figure E5.1. The capacitance between each pin and earth is $\frac{1}{5}$th of the self-capacitance of the unit. The voltage between the line conductor and earth is 76 kV. Find (i) the voltage distribution across each unit, and (ii) the string efficiency.

Solution: Given: Number of units, $n = 4$, $\alpha = \frac{1}{5} = 0.2$, the total voltage, $V = 76$ kV.

Let us assume that the voltage across the first, second, third and fourth units are V_1, V_2, V_3 and V_4 kV, respectively.

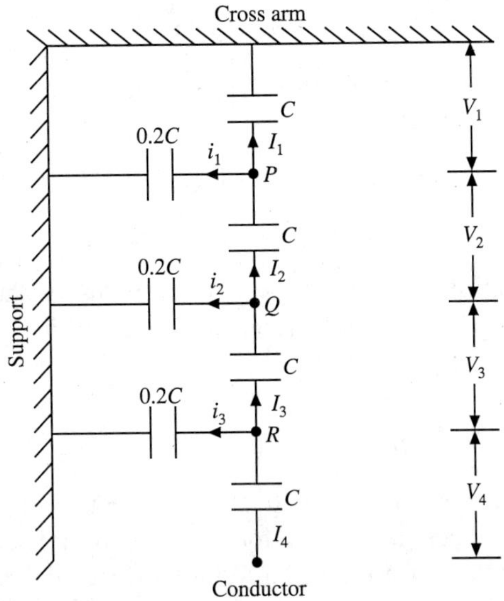

Figure E5.1 String insulator for Example 5.1.

By applying *KCL* at point *P*, we get

$$I_2 = I_1 + i_1$$

or, $$V_2\omega C = V_1\omega C + V_1\omega\alpha C$$

or, $$V_2 = V_1(1+\alpha) = V_1(1+0.2)$$

$\therefore$ $$V_2 = 1.2V_1$$

By applying *KCL* at point *Q*, we get

$$I_3 = I_2 + i_2$$

or, $$V_3\omega C = V_2\omega C + (V_1 + V_2)\omega\alpha C$$

or, $$V_3 = \alpha V_1 + V_2(1+\alpha)$$

or, $$V_3 = 0.2V_1 + 1.2V_1(1+0.2) \quad [\because V_2 = 1.2V_1]$$

$\therefore$ $$V_3 = 1.64V_1$$

By applying *KCL* at point *R*, we get

$$I_4 = I_3 + i_3$$

or, $$V_4\omega C = V_3\omega C + (V_1 + V_2 + V_3)\omega\alpha C$$

or, $$V_4 = \alpha V_1 + \alpha V_2 + V_3(1+\alpha)$$

or, $$V_4 = 0.2V_1 + 0.2\times 1.2V_1 + 1.64V_1(1+0.2) \quad [\because V_2 = 1.2V_1,\ V_3 = 1.64V_1]$$

$\therefore$ $$V_4 = 2.408V_1$$

The total voltage,

$$V = V_4 + V_3 + V_2 + V_1$$

or, $$76 = 2.408V_1 + 1.64V_1 + 1.2V_1 + V_1$$

or,
$$6.248V_1 = 76$$
$\therefore$
$$V_1 = 12.1639 \text{ kV}$$
Then,
$$V_2 = 1.2 \times V_1 = 14.5967 \text{ kV}$$
$$V_3 = 1.64V_1 = 19.9488 \text{ kV}$$
$$V_4 = 2.408V_1 = 29.2907 \text{ kV}$$

Using equation (5.6), the string efficiency is
$$\eta = \frac{76}{4 \times 29.2907} \times 100\% = 64.8670\% \text{ (Ans.)}$$

EXAMPLE 5.2 A string of suspension insulators consists of three units and having capacitance between each pin and earth 10% of the self-capacitance of the unit (Figure E5.2). Determine the greatest working voltage and the string efficiency if the maximum peak voltage per unit is not to exceed 40 kV.

Solution: Given: Number of units, $n = 3$, $\alpha = 0.1$.
Maximum voltage across one unit of string is 40 kV.
i.e. the voltage across the lower disc is 40 kV.

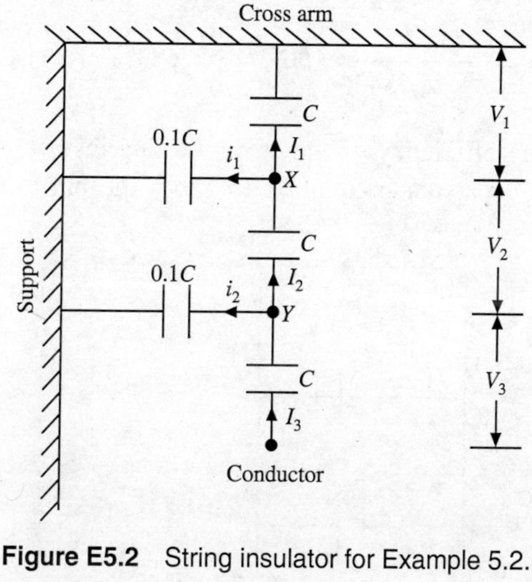

Figure E5.2 String insulator for Example 5.2.

$$V_3 = 40 \text{ kV (peak)} = \frac{40}{\sqrt{2}} = 28.2843 \text{ kV (rms)}$$

The voltage across the string (from Figure E5.2) is given by
$$V = V_3 + V_2 + V_1$$
By applying *KCL* at point *X*, we get
$$I_2 = I_1 + i_1$$
or,
$$V_2\omega C = V_1\omega C + V_1\omega\alpha C$$

or, $$V_2 = V_1(1+\alpha) = V_1(1+0.1)$$

∴ $$V_2 = 1.1V_1$$

Similarly, applying *KCL* at point *Y*, we obtain

$$I_3 = I_2 + i_2$$

or, $$V_3\omega C = V_2\omega C + (V_1 + V_2)\omega\alpha C$$

or, $$V_3 = \alpha V_1 + V_2(1+\alpha)$$

∴ $$V_3 = 1.31V_1 \quad [\text{using } V_2 = 1.1V_1]$$

But $$V_3 = 28.2843 \text{ kV}$$

∴ The voltage across the topmost disc, $V_1 = \dfrac{V_3}{1.31} = \dfrac{28.2843}{1.31} = 21.5910 \text{ kV}$

The voltage across the second disc, $V_2 = 1.1V_1 = 1.1 \times 21.5910 = 23.7501 \text{ kV}$

Maximum working voltage (rms value)

$$V = V_3 + V_2 + V_1 = 28.2843 + 23.7501 + 21.5910 = 73.6254 \text{ kV}$$

From equation (5.6), the string efficiency is

$$\eta = \frac{73.6254}{3 \times 28.2843} \times 100\% = 86.7683\% \quad \text{(Ans.)}$$

EXAMPLE 5.3 Find string efficiency of the string of five insulator units as shown in Figure E5.3. Each unit has a self-capacitance equal to 15 times the pin to earth capacitance.

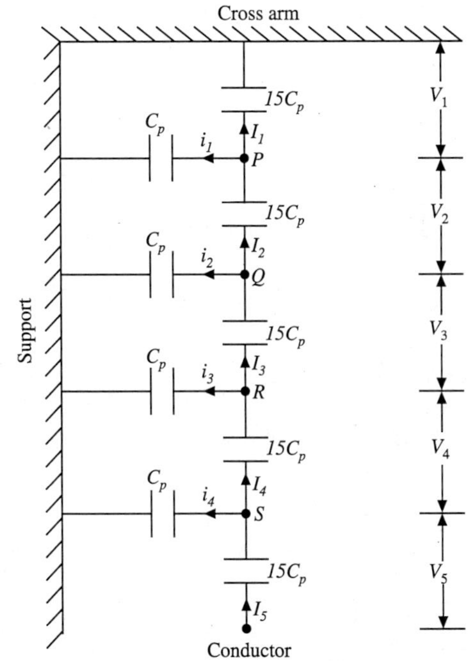

Figure E5.3 String insulator for Example 5.3.

Solution: Given: No of insulators, $n = 5$; C_p = pin to earth capacitance; Self-capacitance, $C = 15C_p$;

$$\therefore \qquad \alpha = \frac{\text{Pin to earth capacitance}}{\text{Self-capacitance}} = \frac{C_p}{C} = \frac{1}{15} = 0.0667$$

By applying *KCL* at point *P*, we get

$$I_2 = I_1 + i_1$$

or, $$V_2 \omega C = V_1 \omega C + V_1 \omega \alpha C$$

or, $$V_2 = V_1(1+\alpha) = V_1(1+0.0667)$$

$\therefore$ Voltage across second unit from cross arm, $V_2 = 1.0667V_1$

By applying *KCL* at point *Q*, we get

$$I_3 = I_2 + i_2$$

or, $$V_3 \omega C = V_2 \omega C + (V_1 + V_2)\omega \alpha C$$

or, $$V_3 = \alpha V_1 + V_2(1+\alpha) = \alpha V_1 + V_1(1+\alpha)^2 \quad [\because V_2 = V_1(1+\alpha)]$$

or, $$V_3 = V_1(1+3\alpha+\alpha^2) = V_1(1+3\times 0.0667 + 0.0667^2)$$

$\therefore$ Voltage across third unit, $V_3 = 1.2045V_1$

By applying *KCL* at point *R*, we obtain

$$I_4 = I_3 + i_3$$

or, $$V_4 \omega C = V_3 \omega C + (V_1 + V_2 + V_3)\omega \alpha C$$

or, $$V_4 = \alpha V_1 + \alpha V_2 + V_3(1+\alpha)$$

or, $$V_4 = \alpha V_1 + \alpha(1+\alpha)V_1 + (1+\alpha)(1+3\alpha+\alpha^2)V_1$$

$$\left[\because \; V_2 = (1+\alpha)V_1 \text{ and } V_3 = (1+3\alpha+\alpha^2)V_1\right]$$

or, $$V_4 = (1+6\alpha+5\alpha^2+\alpha^3)V_1 = (1+6\times 0.0667 + 5\times 0.0667^2 + 0.0667^3)V_1$$

$\therefore$ Voltage across fourth unit, $V_4 = 1.4227V_1$

By applying *KCL* at point *S*, we obtain

$$I_5 = I_4 + i_4$$

or, $$V_5 \omega C = V_4 \omega C + (V_1 + V_2 + V_3 + V_4)\omega \alpha C$$

or, $$V_5 = \alpha V_1 + \alpha V_2 + \alpha V_3 + V_4(1+\alpha)$$

or, $$V_5 = \alpha V_1 + \alpha(1+\alpha)V_1 + \alpha(1+3\alpha+\alpha^2)V_1 + (1+\alpha)(1+6\alpha+5\alpha^2+\alpha^3)V_1$$

$$\left[\because \; V_2 = (1+\alpha)V_1, V_3 = (1+3\alpha+\alpha^2)V_1 \text{ and } V_4 = (1+6\alpha+5\alpha^2+\alpha^3)V_1\right]$$

or, $$V_5 = (1+10\alpha+15\alpha^2+7\alpha^3+\alpha^4)V_1$$

or, $$V_5 = (1 + 10 \times 0.0667 + 15 \times 0.0667^2 + 7 \times 0.0667^3 + 0.0667^4)V_1$$

∴ Voltage across fifth unit, $V_5 = 1.7358V_1$

However, the total string voltage being V, we can write

$$V = V_5 + V_4 + V_3 + V_2 + V_1$$
$$= 1.7358V_1 + 1.4227V_1 + 1.2045V_1 + 1.0667V_1 + V_1$$

∴ $$V = 6.4297V_1$$

Using equation (5.6), the string efficiency, η is

$$\eta = \frac{\text{Voltage across the string}}{n \times \text{Voltage across the unit nearest to power conductor}}$$

$$= \frac{6.4297V_1}{5 \times 1.7358V_1} \times 100\% = 74.0834\% \text{ (Ans.)}$$

EXAMPLE 5.4 Find the maximum voltage that each arrangement shown in Figure E5.4a and Figure E5.4b can withstand so that the peak voltage across any unit is not to exceed 20 kV. Assume the ratio of pin to earth capacitance and self-capacitance to be 0.15 for each unit.

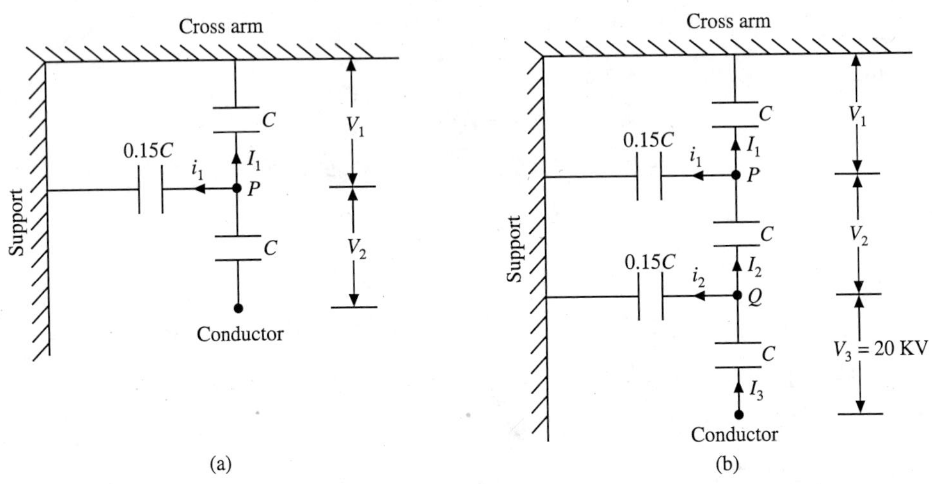

Figure E5.4 String insulator for Example 5.4.

Solution: Given:

$$\alpha = \frac{C_p}{C} = 0.15$$

(i) For Figure E5.4a

The maximum voltage across every unit is not to exceed 20 kV.

∴ The voltage across the unit near to the conductor, $V_2 = V_1(1 + \alpha)$

From equation (5.3), we have,

$$\therefore \qquad V_1 = \frac{V_2}{(1+\alpha)} = \frac{20}{(1+0.15)} = 17.3913 \text{ kV}$$

∴ The maximum voltage across the strings,

$$V = V_1 + V_2 = 17.3913 + 20 = 37.3913 \text{ kV} \quad \text{(Ans.)}$$

(ii) For Figure E5.4b
The maximum voltage across every unit is not to exceed 20 kV.
∴ The voltage across the unit near to the conductor, $V_3 = 20$ kV.
From equation (5.4), we have, $V_3 = (1+3\alpha+\alpha^2)V_1$.

$$\therefore \qquad V_1 = \frac{V_3}{(1+3\alpha+\alpha^2)} = \frac{20}{1+3\times0.15+0.15^2} = \frac{20}{1.4725} = 13.5823 \text{ kV}$$

Again $$V_2 = (1+\alpha)V_1 = (1+0.15)\times 13.5823 \text{ kV}$$

$$\therefore \qquad V_2 = 15.6196 \text{ kV}$$

∴ The maximum voltage across the strings,

$$V = V_1 + V_2 + V_3 = 13.5823 + 15.6196 + 20 = 49.2019 \text{ kV}. \quad \text{(Ans.)}$$

5.6 METHODS OF IMPROVING STRING EFFICIENCY

5.6.1 Selection of α

It is now clear that performance of the string is dependent on the value of α, and as α is decreased (by increasing the length of cross arm which reduces C_p), the potential distribution becomes more uniform. But it is uneconomical beyond a certain value of α, say $\alpha = 0.1$.

5.6.2 Grading of Units

The leakage of current from the insulator pin to the tower structure is the main reason behind unequal distribution of voltage across each disc of string insulator. It is not possible to eliminate this leakage current, but it can be reduced by equalizing the voltage across the disc using different capacities of discs. Thus, to equalize the voltage across each unit, the topmost unit of the string (the unit nearest to the cross arm) must have minimum capacitance and the bottom-most unit (the unit nearest to the power conductor) must have maximum capacitance. Here, the capacitance between the metal pin of the disc and the power conductor is neglected because of small value.

Let C be the capacitance of the topmost unit and the capacitances of other units are $C_2, C_3, \ldots, C_n$ respectively and voltages across each unit be V as shown in Figure 5.4. There are n numbers of units in a string. Here, we assume that the pin to earth capacitance of each unit is same.

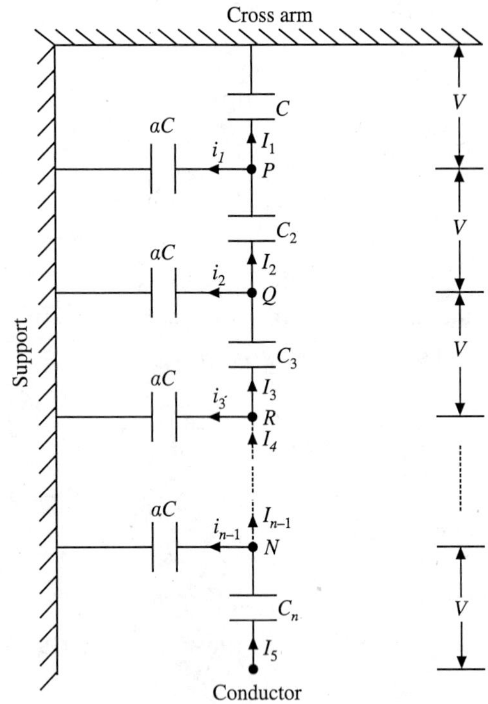

Figure 5.4 String insulator with grading of units.

By applying *KCL* at point *P*, we get

$$I_2 = I_1 + i_1$$

or,
$$V\omega C_2 = V\omega C + V\omega\alpha C$$

or,
$$C_2 = (1+\alpha)C$$

$\therefore$ Capacitance of the second unit is, $C_2 = (1+\alpha)C$

By applying *KCL* at point *Q*, we get

$$I_3 = I_2 + i_2$$

or,
$$V\omega C_3 = V\omega C_2 + 2V\omega\alpha C$$

or,
$$C_3 = C_2 + 2\alpha C$$

or,
$$C_3 = (1+\alpha)C + 2\alpha C \quad [\because C_2 = (1+\alpha)C]$$

$\therefore$ Capacitance of the third unit, $C_3 = (1+3\alpha)C$

By applying KCL at point *R*, we get

$$I_4 = I_3 + i_3$$

or,
$$V\omega C_4 = V\omega C_3 + 3V\omega\alpha C$$

or,
$$C_4 = C_3 + 3\alpha C$$

or, $$C_4 = (1+3\alpha)C + 3\alpha C \quad [\because C_3 = (1+3\alpha)C]$$

$\therefore$ Capacitance of the fourth unit, $C_4 = (1+6\alpha)C$

Similarly, by applying *KCL* at point *N*, we get

$$I_n = I_{n-1} + i_{n-1}$$

or, $$V\omega C_n = V\omega C_{n-1} + (n-1)\, V\omega\alpha C$$

or, $$C_n = C_{n-1} + (n-1)\,\alpha C$$

$\therefore$ Capacitance of *n*-the unit, $C_n = C_{n-1} + (n-1)\,\alpha C$

Hence, it is possible to select capacitance values of insulator discs and to have equal voltage distribution.

The main disadvantages of this method are:

- Difficulty in manufacturing different capacities of discs.
- Difficulty in inventory and handling the discs.

To overcome the above difficulties, *grading of insulators* or *guard ring* is used.

EXAMPLE 5.5 A string of five suspension insulators is to be graded to obtain uniform distribution of voltage across the string shown in Figure E5.5. If the capacitance to ground of each unit is 12% of the capacitance of the top unit, determine the capacitance of the remaining units.

Solution: Let *C* be the capacitance of the topmost unit and the capacitances of other units are C_2, C_3, C_4 and C_5 respectively. The voltage across each unit is *V* (as shown in Figure E5.5).

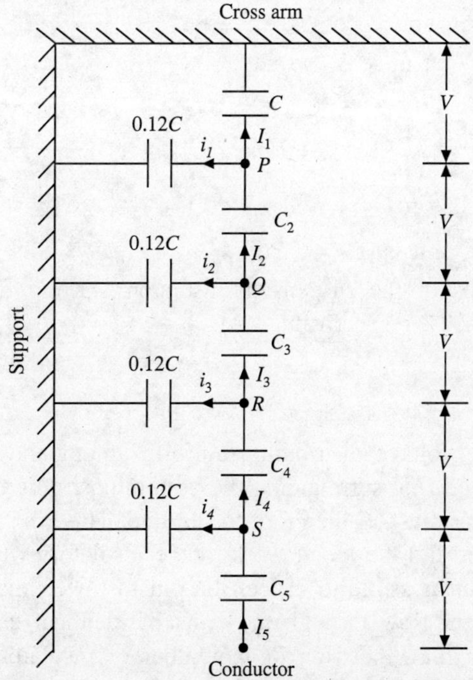

Figure E5.5 String insulator for Example 5.5.

Here, $$\alpha = \frac{\text{Pin to earth capacitance}}{\text{Self-capacitance}} = \frac{C_p}{C} = 12\% = 0.12$$

By applying *KCL* at node *P*, we get

$$I_2 = I_1 + i_1$$

or, $$V\omega C_2 = V\omega C + V\omega\alpha C$$

or, $$C_2 = (1+\alpha)\,C = (1+0.12)\,C$$

$\therefore$ Capacitance of the second unit is, $C_2 = 1.12C$

By applying *KCL* at node *Q*, we get

$$I_3 = I_2 + i_2$$

or, $$V\omega C_3 = V\omega C_2 + 2V\omega\alpha C$$

or, $$C_3 = C_2 + 2\alpha C$$

or, $$C_3 = 1.12C + 2\times 0.12C \quad [\because C_2 = 1.12C]$$

$\therefore$ Capacitance of the third unit becomes, $C_3 = 1.36C$

By applying *KCL* at node *R*, we get

$$I_4 = I_3 + i_3$$

or, $$V\omega C_4 = V\omega C_3 + 3V\omega\alpha C$$

or, $$C_4 = C_3 + 3\alpha C$$

or, $$C_4 = 1.36C + 3\times 0.12C \quad [\because C_3 = 1.36C]$$

$\therefore$ Capacitance of the fourth unit becomes, $C_4 = 1.72C$

Similarly, by applying *KCL* at node *S*, we get

$$I_5 = I_4 + i_4$$

or, $$V\omega C_5 = V\omega C_4' + 4V\omega\alpha C$$

or, $$C_5 = C_4 + 4\alpha C = 1.72C + 4\times 0.12C \quad [\because C_4 = 1.72C]$$

$\therefore$ Capacitance of the fifth unit is, $C_5 = 2.2C$

Therefore, capacitances of the suspension insulator, starting from topmost unit, are *C*, 1.12*C*, 1.36*C*, 1.72*C* and 2.2*C* respectively. (Ans.)

5.6.3 Guard Ring

It is possible to make the voltage distribution uniform among the discs by providing *guard ring*, which is a large metal ring surrounding the bottommost unit and connected to the metal work at the bottom of this unit, and therefore to the line. This ring increases the capacitances between the line conductor and the metal work of the insulators which were neglected in the previous analysis. It is then possible to cancel the pin to tower charging currents by making the same amount of current flow through the units of identical capacities to produce equal voltage drops across each unit. A string of *n* number of insulator discs with guard ring is shown in Figure 5.5.

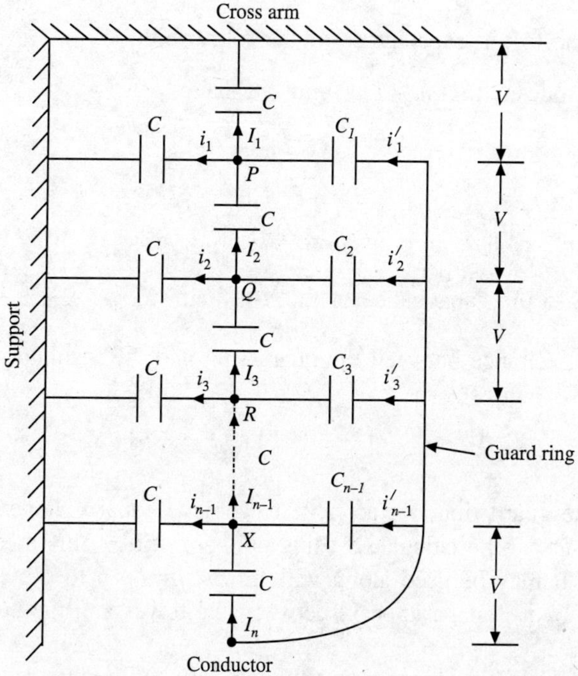

Figure 5.5 Use of guard ring in string insulator.

Let us assume the capacitances formed between the guard ring and the insulator pins (from top to bottom) are $C_1, C_2, C_3, \cdots, C_{n-1}$, respectively. Here voltage across each unit is V and all the insulator units are identical. This makes all charging currents same.

$$I_1 = I_2 = I_3 = \cdots = I_n$$

Applying KCL at node P, we get

$$I_2 + i'_1 = I_1 + i_1$$

or, $$i'_1 = i_1 \quad [\because I_1 = I_2]$$

or, $$(n-1)\,V\omega C_1 = V\omega C$$

$\therefore$ Line to pin capacitance of the second unit is, $C_1 = \dfrac{C}{n-1}$

By applying KCL at node Q, we obtain

$$I_3 + i'_2 = I_2 + i_2$$

or, $$i'_2 = i_2 \quad [\because I_2 = I_3]$$

or, $$(n-2)\,V\omega C_2 = 2V\omega C$$

$\therefore$ Line to pin capacitance of the third unit is, $C_2 = \dfrac{2C}{n-2}$

By applying KCL at node R, we get

$$I_4 + i'_3 = I_3 + i_3$$

or, $$(n-3)\,V\omega C_3 = 3V\omega C \quad [\because I_3 = I_4]$$

∴ Line to pin capacitance of the fourth unit, $C_3 = \dfrac{3C}{n-3}$

By applying *KCL* at node *X*, for $(n - 1)$th unit, we have

$$I_n + i'_{n-1} = I_{n-1} + i_{n-1}$$

or,
$$i'_{n-1} = i_{n-1}$$

or,
$$V\omega C_{n-1} = nV\omega C$$

∴ Line to pin capacitance of the *n*-th unit, $C_{n-1} = \dfrac{nC}{n-(n-1)}$

In general, the capacitance between the guard ring and the insulator pins of *p*-th insulator from the top can be written as

$$C_p = \frac{pC}{n-p} \tag{5.8}$$

Then by using the guard ring, we are able to achieve equal voltage distribution while the capacitance of any unit can be calculated using equation (5.8). This method is normally used due to its advantage. It may be used along with an *arcing horn* to protect the insulator string from flashover when over-voltage appears between the tower and the conductor.

EXAMPLE 5.6 A string of five suspension insulators is to be fitted with a guard ring as shown in Figure E5.6. The pin-to-earth capacitance and pin-to-pin capacitance are all equal to *C*. Find the values of the line to pin capacitances to have uniform voltage distribution over the string.

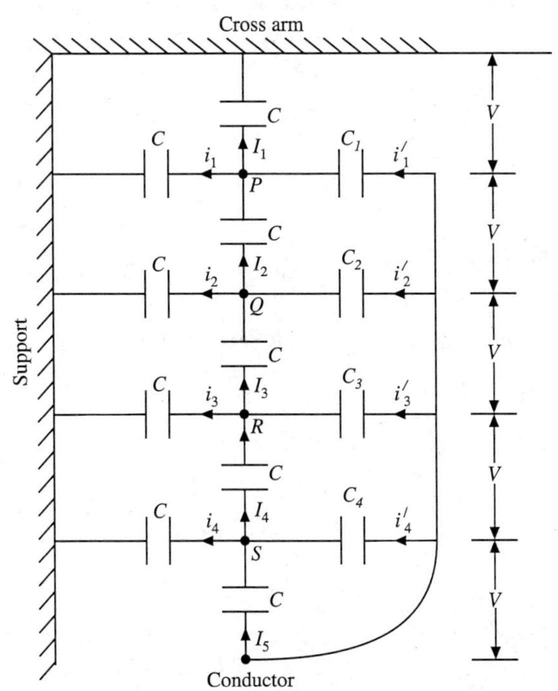

Figure E5.6 String insulator for Example 5.6.

Solution: Voltage across each unit being V, the charging currents of all the units are same.

$$\therefore \quad I_1 = I_2 = I_3 = I_4 = I_5$$

Applying KCL at node P, we get

$$I_2 + i_1' = I_1 + i_1$$

or,
$$i_1' = i_1 \quad [\because I_1 = I_2]$$

or,
$$4V\omega C_1 = V\omega C$$

$\therefore$ Line to pin capacitance of the second unit is, $C_1 = \dfrac{C}{4}$

By applying KCL at node Q, we get

$$I_3 + i_2' = I_2 + i_2$$

or,
$$i_2' = i_2 \quad [\because I_2 = I_3]$$

or,
$$3V\omega C_2 = 2V\omega C$$

$\therefore$ Line to pin capacitance of the third unit, $C_2 = \dfrac{2C}{3}$

By applying KCL at node R, we get

$$I_4 + i_3' = I_3 + i_3$$

or,
$$2V\omega C_3 = 3V\omega C \quad [\because I_3 = I_4]$$

$\therefore$ Line to pin capacitance of the fourth unit, $C_3 = \dfrac{3C}{2}$

By applying KCL at node S, we get

$$I_5 + i_4' = I_4 + i_4$$

or,
$$i_4' = i_4 \quad [\because I_4 = I_5]$$

or,
$$V\omega C_4 = 4V\omega C$$

$\therefore$ Line to pin capacitance of the fifth unit, $C_4 = 4C$.

$\therefore$ Line to pin capacitance of the units become $C_1 = \dfrac{C}{4}$, $C_2 = \dfrac{2C}{3}$, $C_3 = \dfrac{3C}{2}$ and $C_4 = 4C$.

5.7 INSULATION FAILURE

The reason behind the failure of any insulation unit may be flashover, mechanical stresses, crack, porosity and puncture.

When an insulator encounters an overvoltage developed across it due to switching or lighting, and the overvoltage exceeds the flashover voltage of the insulator, an arc appears between the line conductor and the insulator base. This may be a potential cause of insulation failure. The flashover of the insulator is the main reason of its failure in most of the cases. In order to prevent insulator failure due to arc, insulators are often fitted with arcing horns to keep this arc away from the insulator. Usually, after a flashover, an insulator is likely to continue to function properly unless it is damaged by the heat of the arc. Deposition of dirt, salt laden coastal spray, frost and fog over the upper surface of an insulator reduce its flashover voltage and may lead to its failure.

Failure due to mechanical stresses is rare. Alternate and sometimes rapidly changing heat and cold, dryness and dampness may produce high thermal stresses on the body of the insulator which, in turn, develop crack within the insulation causing insulation failure. A porous insulator may absorb water from atmosphere which decreases its electrical resistance and it may lead to insulation failure. The failure of an insulator by puncture is quite rare.

5.8 BUSHING

Bushings are necessary to take out the live conductor through earthed tanks or metallic parts. They are also necessary for transformers, reactors, bulk-oil circuit breakers, metal-clad switchgear, generators, etc. The central conductor of a bushing is in the form of a copper rod (or stranded conductor), passing through the brass tube or aluminium tube supporting the bushing core. The bushing assembly is held in position by lock nuts. Bushings are filled with electric insulating oil. They are designed for vertical upright mounting or horizontal mounting.

The *creepage distance* (i.e. the distance along the surface of the solid insulator across two conduction parts separated by the solid insulator) on the outer surface is decided by the amount of pollution in the surroundings.

5.9 TESTING OF INSULATORS

An insulator should have good mechanical strength to withstand the mechanical load conditions. It should not have any pores or voids which may lead to its breakdown. Its dielectric strength should be large enough to withstand the normal voltage and usual overvoltages. The insulator should face flashover at such voltages which are likely to cause damage to other equipment. The following *type tests* are performed on the insulators:

(i) Mechanical strength, electrical insulation tests, environmental tests
(ii) Porosity, galvanizing test
(iii) Power frequency dry flashover voltage
(iv) Power frequency wet flashover voltage
(v) Impulse voltage withstand flashover test
(vi) Puncture voltage test
(vii) Corona and radio interference test.

Routine tests are performed at site on each insulator unit and it mainly includes physical inspection and high voltage test.

5.10 ARCING HORN

Arcing horns are installed on a few overhead line insulators as a precaution of flashover due to which the porcelain is often cracked or broken up by the power. One of the horns is earthed and the others are electrically connected with the line conductor through the insulators as shown in

Figure 5.6. They are arranged in such a way that the overvoltage in the line will create an arc which is taken by the electrodes and held at sufficient distance from the porcelain to prevent its damage by the heat of the arc.

In addition to this, the horns at the earth-end are shaped in such a way so as to transfer the arc upwards and outwards from the insulator string. This increases the arc length quickly and thereby allowing it to self extinguish. The application of these devices lowers the flashover voltage of the insulator. Arcing horns are often employed across line insulators and bushing in the medium voltage range in subtransmission.

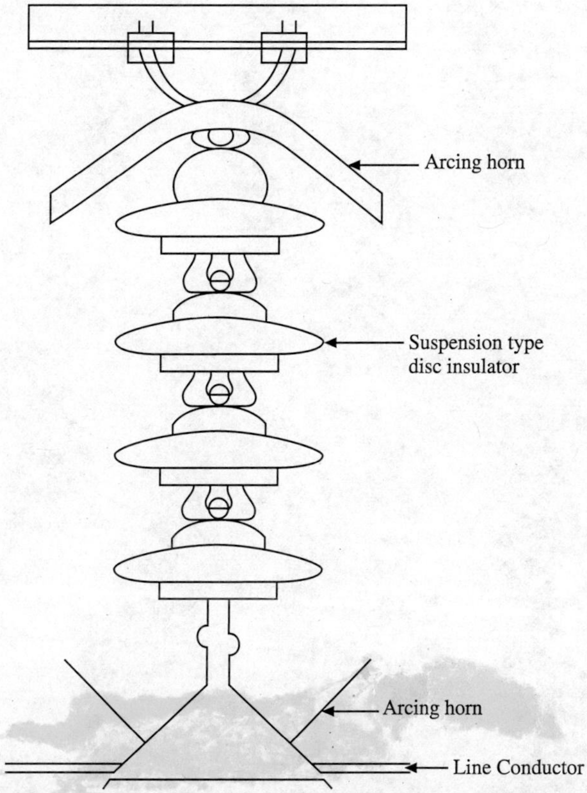

Figure 5.6 Arcing horn.

EXERCISES

1. What are the different types of insulators used in overhead transmission lines?
2. Write a short note on materials for insulators in overhead transmission lines.
3. What are the advantages of the suspension type insulator over the pin type insulator?
4. What are strain insulators and shackle insulators? Where are they used?

5. Why potential distribution over a string of suspension type insulator is not uniform? Derive the expression potential distribution over a string of suspension type insulators consisting of four discs.
6. What is string efficiency? Why is it necessary to have high string efficiency?
7. Discuss the methods of improving string efficiency.
8. What are the reasons of insulation failure? Why arcing horn is provided across the insulator?
9. What is a bushing?
10. Discuss the different tests carried out to prove the quality of an insulator.

CHAPTER 6

Power Cables

6.1 INTRODUCTION

Power cables are extensively used for the distribution of electrical energy in low or medium voltage grade in metropolitan cities, where population density is high. There are instances where power cables have also been used for the transmission of electrical energy in metropolitan cities between substation and load centres. In low tension (LT) systems, three phase three core or four core cables are mostly used whereas in high tension (HT) systems, the use of three single core cables is quite common. In comparison to overhead lines, cables are costlier but more reliable as there are fewer chances of faults being developed due to meteorological problems. In this chapter, the analytical aspects of power cables and their application are presented.

6.2 TYPES OF CABLES

The selection of a cable for a particular application depends upon system voltage and service requirement. On the basis of system voltage, cables may be classified as follows:

1. Low tension cables: for system voltage up to 1 kV
2. High tension cables: for system voltage up to 11 kV
3. Super tension cables: for system voltage up to 33 kV
4. Extra high tension cables: for system voltages beyond 33 kV.

6.2.1 Low Tension (LT) Cables

LT cables are extensively used for maximum operating voltage of 1.1 kV. Therefore, this type of cable does not require any special construction as the electrostatic stresses developed are small and the thermal conductivity is also not of very much concern. The insulation may be of paper, impregnated with oil. Usually, polyvinyl chloride (PVC) is used as insulation to increase the dielectric strength and to prevent damage. Sometimes, varnished cambric is also used for insulation. A lead sheath may be provided over the insulation in order to protect the cable from moisture and bad handling. It also makes the voltage stress more uniform in the insulation. Low tension cables may be divided into two categories: (i) single core cables and (ii) multi-core cables.

Single core cables

Single core cables consist of a copper/aluminum stranded circular core as conductor (as shown in Figure 6.1). This conductor is insulated with paper or PVC (Poly Vinyl Chloride) insulation and is covered with a lead sheath over it. There is a provision of a layer of compounded fibrous material or Hessian tape over the lead sheath in order to protect the metallic sheath against corrosion. Usually no armour is provided with these cables in order to avoid excessive power loss in the armour due to induced eddy currents during ac operation of these single core cables with single phase supply.

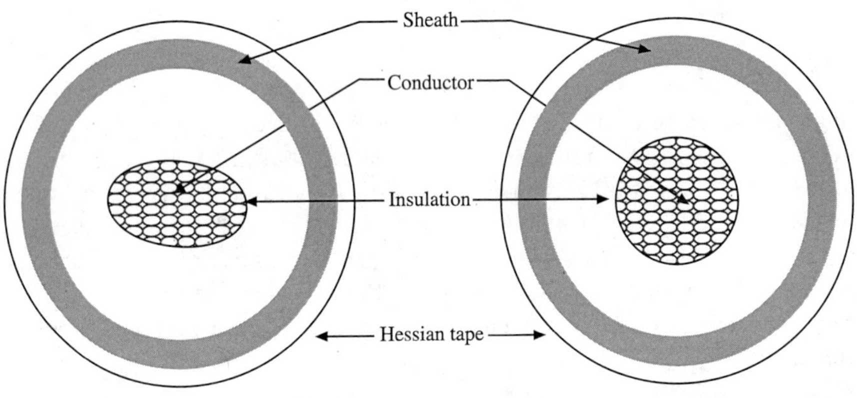

Figure 6.1 Single core cable with oval and round shaped conductor.

Multi-core cables

Multi-core cables are generally three core, four core or three and half core cables. Let us consider a three core cable. All three cores are insulated with each other and placed. The cores are either circular-shaped, oval or sector-shaped conductors wrapped around by proper insulation. The multi-core cables are of *belt type* or *sector shaped*. The shaped core improves the copper space factor. A belt of insulation is provided surrounding the three cores. A lead sheath may be provided within the cable to protect the cable from moisture. A coating of lime wash is provided outside the cable in order to prevent adhesion. A simple three core LT cable is shown in Figure 6.2.

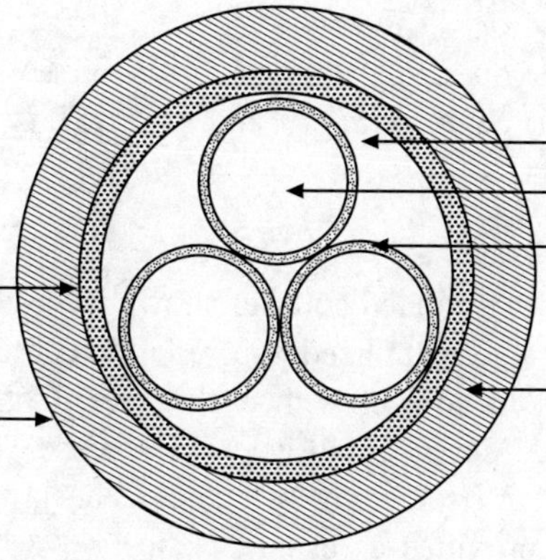

Figure 6.2 A three core LT cable.

6.2.2 High Tension (HT) Cables

High tension cables are used for three-phase medium and high voltage distribution. Here paper is still used as insulation through *cross linked polyethylene* (XLPE) insulation is becoming more popular due to better dielectric strength and longer life. XLPE is also resistant to corrosion. The three-core belted type cable may be an example of this type of cables. The construction is same as that of the low tension cables. The three-core belted type cable has three insulated cores and these three cores are again insulated with impregnated paper and covered with a lead sheath (as shown in Figure 6.3). Round conductors are also used frequently for three and four core cables.

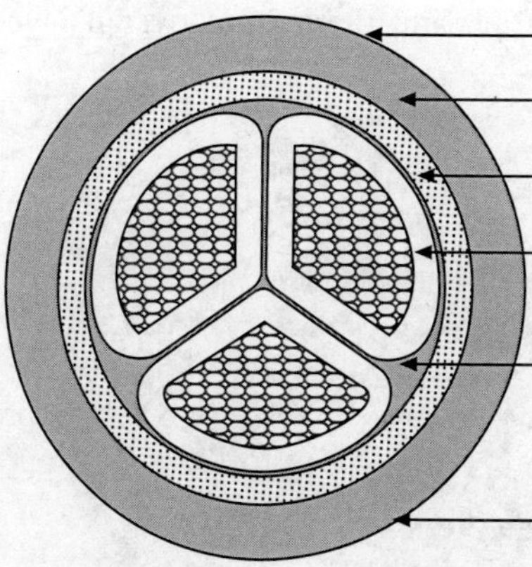
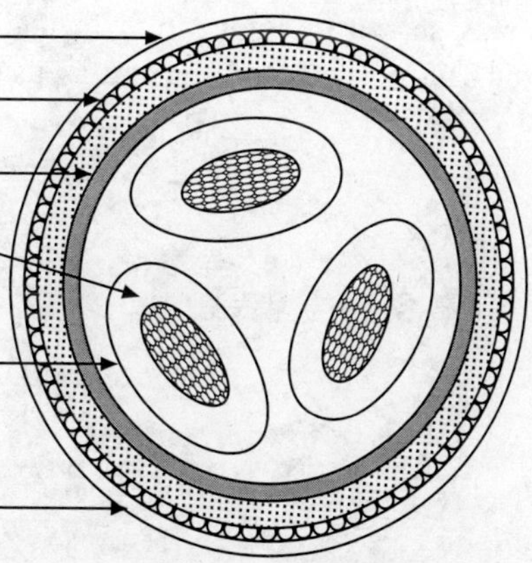

Figure 6.3 Three core high tension cable.

Figure 6.4 Three core H-type cable.

6.2.3 Super Tension (ST) Cables

Super tension cables are used for operating voltages up to 33 kV and these are classified as: (i) *H-type* or *screened cables*, (ii) *SL type cables* and (iii) *HSL type cables*.

H-type or screened cables

Screened cables contain three paper insulated cores as shown in Figure 6.4. Here, each conductor is insulated with a layer of perforated metallized paper. These cables are laid up in the ordinary way with the metallized screens in contact with each other. Cotton tape with fine wires of copper is wound around the three cores which act as a screen. The fine copper wires within the cotton insulation provide overall screening of cables. The layers of lead sheath, bedding, armouring, and serving are also provided as usual practice.

The advantages of this type of cables are as follows:

1. Electric stresses are entirely radial and reduce the dielectric loss since all four screens (three outside each core and one over the whole cable) and the lead sheath are at the earth potential.
2. The metal foil increases the heat dissipating power and hence there is little sheath loss.
3. There is an increase in the current carrying capacity of the cable due to reduction in loss.
4. There is least possibility of forming voids and vacuous spaces as these cables are screened.

SL type cables

In SL type of cables, each core has its own lead sheath over the insulation with an impregnated paper as shown in Figure 6.5. In addition to the overall sheath, there is an overall layer of bedding, armouring and serving as usual.

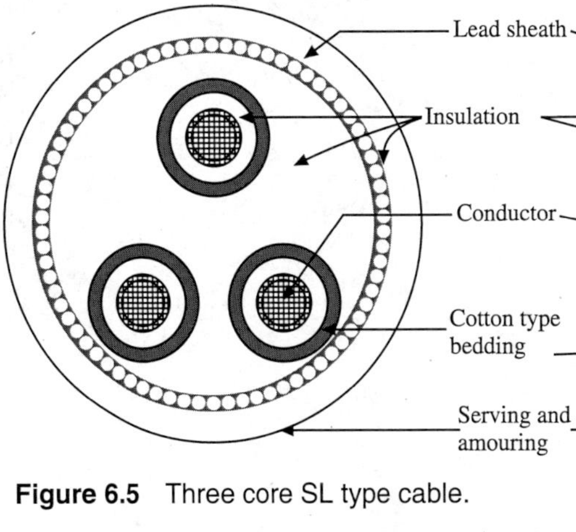

Figure 6.5 Three core SL type cable.

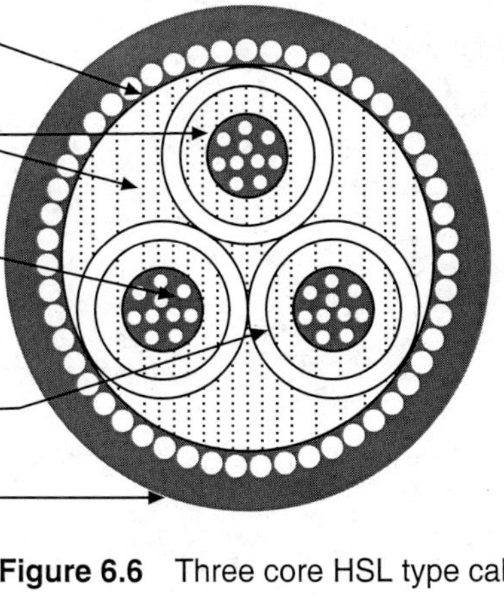

Figure 6.6 Three core HSL type cable.

SL type cables have the following advantages:

1. Bending of the cable is possible due to absence of overall lead sheath.
2. Electrical stresses are radial.

HSL type cables

HSL type cables are the combination of H type and SL type cables as shown in Figure 6.6. Here, each core is insulated with impregnated paper and a separate lead sheath is provided. All three cores are arranged in the required form and then insulated by providing the layers of bedding, armouring, and serving as usual.

The advantages of screened cables over belted type cables are as follows:

1. This has greater core-to-core thickness to a given overall diameter which reduces the possibility of core-to-core faults.
2. Electric stresses in the cables are uniformly radial in all sections.

6.2.4 Extra High Tension (EHT) Cables

To meet the increased operating voltage in metropolitan cities, super tension cables are used. The difficulties with these cables are:

1. Three-core cables are not economical and practicable due to their high weight and size and hence, single core cables are widely used.
2. There are greater sheath losses in single-core cables.
3. Due to the possibility of voids in extra high tension voltage cables, there is a danger of breakdown of the dielectric.
4. Due to the failure of the dielectric, there is a possibility of ionization and allied chemical reactions, which damage the insulation.

There are two possible solutions to minimize the formation of voids, that is, by using (i) oil-filled cables and (ii) gas pressure cables.

Oil-filled cables

In oil-filled cables, an oil channel is formed at the centre of the core by stranding the conductor wire built around a hollow cylindrical steel spiral. This channel formed at the centre of the core is filled with low viscosity mineral oil by means of oil reservoirs and feeding tanks placed at about every 500–600 m length along the route of the cable and maintained at a constant pressure (not below the atmospheric pressure) at any point along the cable.

Oil-filled cables are usually single phase type and these cables are so designed that when oil gets expanded due to rise in temperature of the cable, the extra oil gets collected in the external reservoir, which sends it back during contraction of oil when the temperature falls under light load conditions. An oil-filled single-core cable is shown in Figure 6.7.

There are two more varieties of oil filled cable, viz. single core sheath channel cable and there core filler space channel cable.

In single core sheath channel oil-filled cable, solid conductor is insulated with paper and the oil channel is provided by grooving the sheath (Figure 6.8).

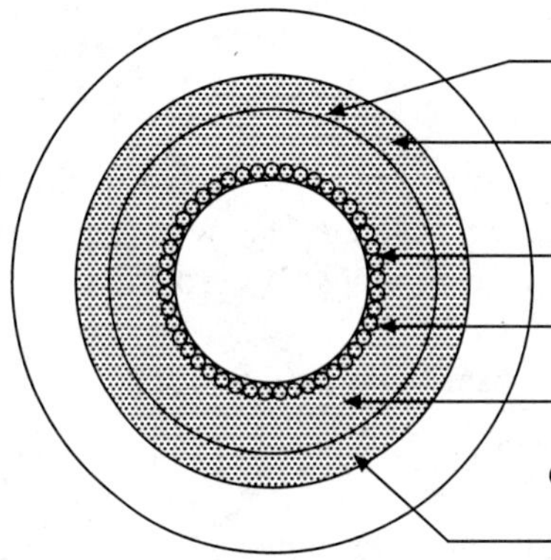

Figure 6.7 Single core oil filled cable.

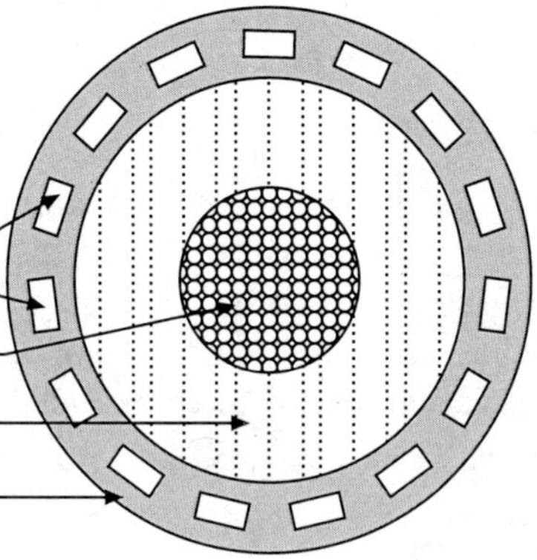

Figure 6.8 Single core sheath-channel cable.

In a three core filler space channel cable, the oil channels are located within the hollow filler space. The channels are of perforated metal ribbon tubing and are at earth potential. These channels are filled with oils in the factory and are dispatched by placing them lengthwise over drums provided with tanks having oil under pressure. A three core, oil-filled cable is shown in Figure 6.9.

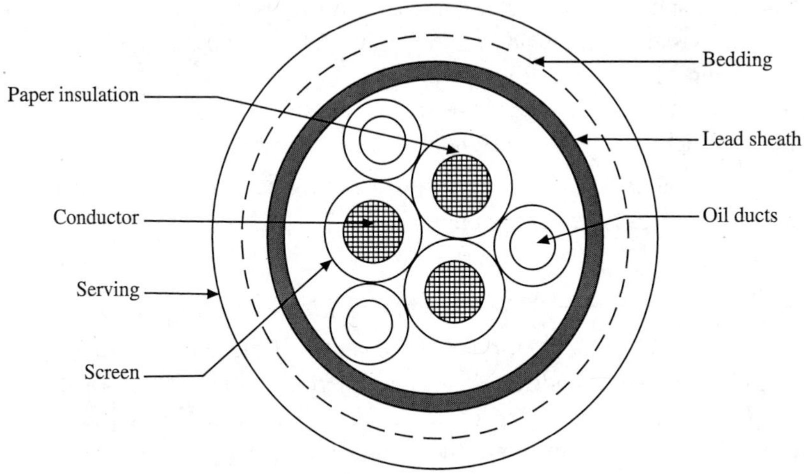

Figure 6.9 Three core filler space channel cable.

The advantages of oil-filled cables over solid cables are as follows:

1. The thickness of the required dielectric medium decreases, therefore the overall size and weight of the cable are lesser.
2. There is no possibility of formation of voids, oxidation and ionization within the cable.
3. Impregnation is more perfect.

4. Thermal resistance of the cable reduces and, as a result, current rating of the cable increases.
5. The maximum allowable stress increases.
6. There is possibility of increased working range of temperature for these cables.
7. As soon as the lead at any section deteriorates, the oil starts leaking. Therefore, the possibility of detection of fault is easy.

Oil-filled cables have the following disadvantages:

1. They involve high cost.
2. Installation and maintenance of the cable is complicated.

Gas-filled cables

Paper impregnated with petroleum jelly is normally used as dielectric in gas-filled cables. The space between layers of paper insulation is filled with dry nitrogen gas with a pressure of 1400 kN/m^2. This pressure is maintained constantly by pumps and by the lead sheath inside the cable. These cables are usually single core cables and a small clearance is left to allow the gas to flow axially. In the case of three core cables, clearance is not necessary because the filler spaces and strands allow the gas to flow. Gas-pressure cables are mainly of two types: (i) external-pressure cables and (ii) internal-pressure cables.

1. ***External-pressure gas-filled cables:*** In this type of cable, pressure is applied externally and raised to such an extent that no ionization can take place. Due to this increased pressure, radial compression takes place which tends to close voids (if any). The operating power factor of such a cable is better. The construction of the cable is similar to that of the ordinary solid type cable, except that it is triangular instead of a circular cross-section. The triangular section reduces its weight and has low thermal resistance. This type of cable has been developed recently for application in higher voltages. A simple external-pressure gas-filled cable is shown in Figure 6.10.

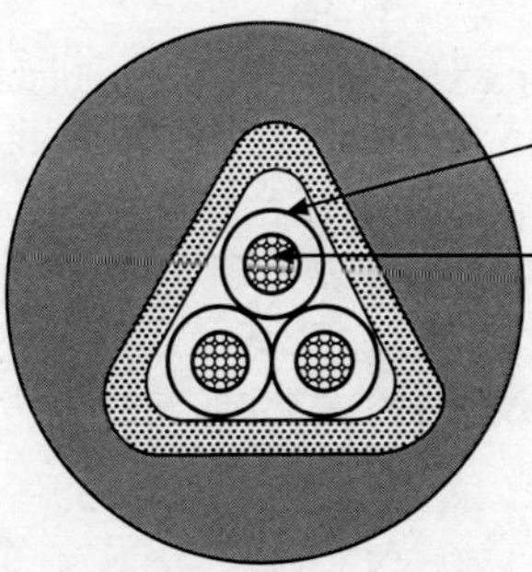

Figure 6.10 Three core external gas pressure cable.

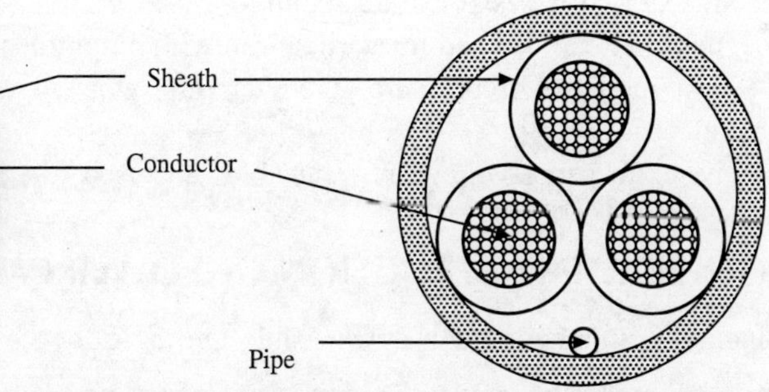

Figure 6.11 Three core impregnated pressure cable.

The advantages of this type of cables are:

(i) In comparison with the normal cable, these cables can carry from 1.4 to 1.6 times the load current. They can withstand double the operating voltage and transmit 2.4 to 3.2 times of the rated power.

(ii) The maximum potential gradient is 100 kV/cm and the power factor is much better.
(iii) The steel pipe provides mechanical protection to the cables and acts as an ideal method of cable laying.
(iv) The nitrogen in the steel pipes helps in quenching any flame.
(v) Reservoirs are not needed.
(vi) Maintenance cost is low.

The only disadvantage of these cables is their high initial cost.

2. ***Internal-pressure cables:*** These cables are of three types: (i) high-pressure gas-filled cables, (ii) gas-cushion cables and (iii) impregnated-pressure cables.

3. ***High-pressure gas-filled cables:*** For this type of cable, the space for the gas is provided in the dielectric itself, which is filled up by inert gases like nitrogen at a pressure of about 6 atmospheres for extra high-tension voltage cables and 12 atmospheres for super-voltage cables. This facilitates the axial flow of gas, which also passes along the un-impregnated strand. This clearance is not essential in the case of multi-core cables because the filler spaces and strands provide a low resistance path for the flow of gas.

4. ***Gas-cushion cables:*** Here a screened space is provided in between the lead sheath and the dielectric along the length of the cable. The screened space is subdivided by means of barriers and the inert gas is stored at various points throughout the length of the cable.

The advantage of this cable is that no separate arrangement is required for the transmission of pressure to the cables from outside. Hence, the cable is a complete unit with its own armouring, without requiring the external pipe protection

5. ***Impregnated pressure cable:*** It consists of mass-impregnated paper dielectric, which is maintained under a pressure of 14 atmospheres by means of nitrogen. Otherwise, the construction is same as that of a solid cable. A special reinforcement is provided here to cater to the large hoop and longitudinal-stress set-up. This type of cable is shown in Figure 6.11.

The advantages of internal-pressure cables are:

(a) No external accessories are required.
(b) The cable can be used for vertical run without any fear of leakage with suitable designs.
(c) With increase in pressure, there is improvement in the power factor of the cable dielectric.
(d) The extra super-voltage power cables are used for operating voltage beyond 132 kV.

6.3 GENERAL CONSTRUCTION OF POWER CABLES

The principal components of a power cable are as follows:

1. ***Core:*** All cables have one or more cores depending upon the service required (single phase or three phase). The material used in the core is generally electrical quality copper although, at times, aluminum is also used.

2. ***Insulation:*** The insulation over the core is generally provided with different layers to isolate it from the surroundings. The thickness of each layer depends upon the operating voltage. The insulation should have high resistance, high dielectric strength, good mechanical strength and high temperature withstand capability. The following material can be used for cable insulation:

(a) *Poly vinyl chloride (PVC):* It is a synthetic material and is obtained as white odourless, tasteless, chemically inert, non-inflammable and insoluble powder. It is chemically combined with a plastic compound and the gel is used over the conductor to form the insulation cover. PVC has dielectric strength of 17 kV per mm and dielectric constant of 5. Maximum continuous temperature rating is 75°C.

(b) *Paper insulation:* Paper is still one of the most important insulating materials used in high voltage power cable manufacturing. It is highly advantageous as insulating medium due to its higher thermal conductivity, dielectric strength, better thermal withstand capability, low cost and high durability. Paper is dried and impregnated with insulating oil and taped over the conductor to act as insulation on cable conductor. The dielectric strength is around 20 kV/mm while the dielectric constant is 3.5. It can be used for maximum continuous temperature rating of 80°C.

(c) *Cross linked polythene cable (XLPE):* Specially treated low density polyethylene results in cross linking of carbon atoms, and the compound is a new material having extremely high melting point with light weight, small dimension, low dielectric constant and high mechanical strength. XLPE cable has high maximum continuous temperature rating of 90°C with dielectric strength of 20 kV/mm. Due to high temperature withstand capability and very low moisture absorption, these cables can be laid directly on soil bed and easily work for voltages upto and including 33 kV.

(d) *Vulcanised India rubber (VIR):* This cable insulation was developed in the early stages of cable manufacture during 1880–1930. Though it has got reasonably good dielectric strength of 15 kV/mm, its use is limited because of its low melting point, low chemical resistance capability and short span of life.

3. ***Metallic sheath:*** A metallic sheath is provided over the insulation to protect the insulation material from moisture, gases and any other harmful liquids in the soil. It also protects the insulation from mechanical damage. The metallic sheath is usually made of lead or lead alloy. Nowadays, aluminium is also being used as metallic sheath because of its greater mechanical strength, low weight and cost as compared to lead sheath.

4. ***Bedding:*** A layer of 'bedding' consisting of paper tape compounded with a fibrous material like jute or Hessian tape is used over the metallic sheath to protect it from corrosion and mechanical injury.

5. ***Armouring and covering:*** Armouring is provided over the bedding to protect the cable from mechanical injury while laying and handling it. It consists of one or two layers of galvanized steel wires or two layers of steel tape. The steel tapes are coated with preservative compound and are helically wound on the cable insulation. Single core cables in AC systems are not provided with armouring because eddy currents induce in the steel armour producing high loss. In these cables, a plastic wrap is used for mechanical protection. Power cables are usually protected by bituminous compound wrapping (covering).

6. ***Serving:*** A layer of fibrous material is again used over the armouring which is same as bedding. This is called *serving*. Serving is useful to protect the armouring from atmospheric conditions.

Different parts of a cable are shown in Figure 6.12.

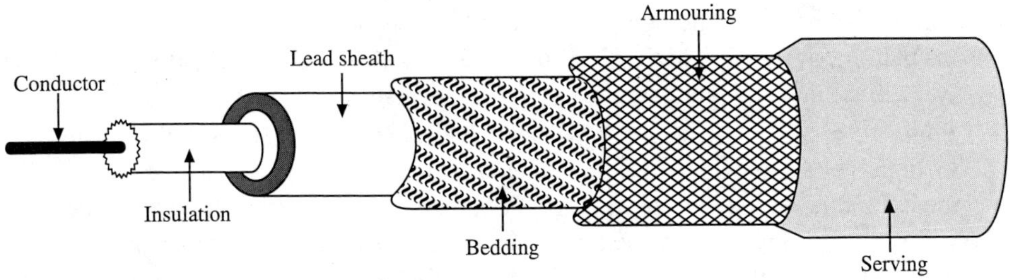

Figure 6.12 Different parts of a cable.

6.4 ADVANTAGES AND DISADVANTAGES OF UNDERGROUND CABLES OVER OVERHEAD TRANSMISSION LINES

The advantages of underground transmission through cables are as follows:

1. The operation of an underground cable remains unaffected by weather conditions like lightning, and thunderstorms. So, there is no interruption in service to consumers.
2. There is lesser chance of accident caused by the breaking of the conductors.
3. Since there is less chance of fault occurrence, the maintenance cost of the underground system is very low.
4. The spacing between the conductors in an underground system is much less as compared to overhead transmission line. Inductance of the underground cable is very low as compared to overhead lines and as a result, the voltage drop is less.
5. The sheath absorbs surge energy; therefore, surge effect is smoothened down in the underground system.
6. Interference with communication lines in the underground system in negligibly small.
7. The underground system of distribution or transmission is more aesthetic because no wiring is visible.
8. Cable system injects capacitive reactive power and hence improves voltage regulation.

The disadvantages of underground transmission through cables are as follows:

1. An underground system requires trenching, conduits, cables, manholes and other special equipment which involve high cost. Hence, the underground system with cables is more expensive than overhead lines.
2. The cost of the cable increases with increase in transmission voltage.
3. Addition of new loads is not possible in the underground system. It can only be met by installation of new lines.
4. Tapping of loads and service mains is not conveniently possible in the underground system because joining of underground cables is more difficult.

5. The cables have high capacitance due to less spacing between the conductors and insulation and also having high dielectric constant surrounding it. So, they draw high charging current. This may lead to overvoltage in the system when the cable is lightly loaded.
6. Though faults are rare in the underground system, it is very difficult to locate the fault point and its repair is difficult and expensive.

6.5 INSULATION RESISTANCE OF POWER CABLES

The cross section of a single core cable is shown in Figure 6.13. Let the core have a diameter d (= $2r$) and the cable have overall diameter D. Let us consider an elementary cylindrical section of the insulation having a radius x and thickness dx. The resistance of this cylindrical shell for length l of the cable is,

$$dR = \frac{\rho\, dx}{2\pi x l}$$

where ρ is the resistivity of the insulation.

The insulation resistance is therefore

$$R = \int_{d/2}^{D/2} \frac{\rho\, dx}{2\pi x l} = \frac{\rho}{2\pi l} \ln x \Big|_{d/2}^{D/2} = \frac{\rho}{2\pi l} \ln (D/d) \text{ ohm} \tag{6.1}$$

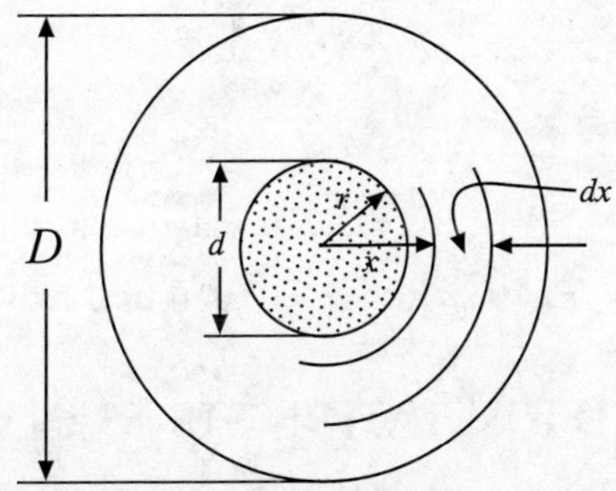

Figure 6.13 A single core cable cross section.

EXAMPLE 6.1 A 8 km long single-phase single core cable has an insulation resistance of 0.4 Mohm. The core diameter is 2.5 cm and the overall diameter of the cable is 6.5 cm. Calculate the resistivity of the insulation material.

Solution: Given:

Length of the cable, l = 8 km; Insulation resistance, R = 0.5 Mohm;

Diameter of the core, d = 2.5 cm; Diameter of the cable with insulation, D = 6.5 cm,

Radius of the core, $r = d/2$ = 1.25 cm

From equation (6.1), we have

$$R = \frac{\rho}{2\pi l} \ln (D/d)$$

or,
$$\rho = R \times \frac{2\pi l}{\ln (D/d)} = 0.4 \times 10^6 \times \frac{2\pi \times 8 \times 10^3}{\ln (6.5/2.5)}$$

$$= 21.042 \times 10^9 \text{ Ohm-m.} \quad \text{(Ans.)}$$

EXAMPLE 6.2 The insulation resistance of a single-core cable is 450 MΩ per km. If the core diameter is 4 cm and resistivity of insulation is 7.5×10^{12} Ohm-m, calculate the insulation thickness.

Solution: Length of the cable, l = 1 km; insulation resistance, R = 450 Mohm;
Resistivity of insulation, $\rho = 7.5 \times 10^{12}$ Ohm-m;
Diameter of core, d = 4 cm; Radius of core, $r_1 = d/2$ = 2 cm
Let the radius of the cable with insulation (overall radius) be r_2 in cm.
From equation (6.1), we have

$$R = \frac{\rho}{2\pi l} \ln (r_2/r_1) \qquad [\because d = 2r_1 \text{ and } D = 2r_2]$$

or,
$$\ln (r_2/r_1) = R \times \frac{2\pi l}{\rho} = 450 \times 10^6 \times \frac{2\pi \times 1 \times 10^3}{7.5 \times 10^{12}} = 0.377$$

or,
$$\frac{r_2}{r_1} = \varepsilon^{0.377} = 1.458$$

or,
$$r_2 = r_1 \times 1.458 = 2 \times 1.458$$

$$r_2 = 2.916 \text{ cm}$$

Insulation thickness = $r_2 - r_1$ = 2.916 – 2 = 0.916 cm. **Ans.**

6.6 CAPACITANCE AND INSULATION STRESS IN CABLES

Let us consider a single core cable with the conductor diameter d and overall diameter D as shown in Figure 6.13. Let the charge per metre axial length be Q. Then, the total electric flux passing through a cylinder of radius x metres concentric with the conductor, according to Gauss's theorem, is Q lines. The electric flux density due to this flux is $Q/2\pi x$.

The *electric stress* g at a distance x being $g = Q/2\pi\varepsilon x$, (where ε is the permittivity of the dielectric), the work done in moving a unit charge through a distance dx in the direction of the electric field is gdx. Therefore, the work done in moving a unit charge from the conductor to the outer covering (which is the potential difference between the conductor and outer cover) is given by

$$V = \int_{d/2}^{D/2} g dx = \int_{d/2}^{D/2} \frac{Q}{2\pi\varepsilon x} dx = \frac{Q}{2\pi\varepsilon} \ln x \Big|_{d/2}^{D/2} = \frac{Q}{2\pi\varepsilon} \ln(D/d) \text{ Volt} \qquad (6.2)$$

Capacitance, $$C = \frac{Q}{V} = \frac{2\pi \varepsilon_0 \varepsilon_r}{\ln (D/d)} \text{ F/m} \quad [\because \varepsilon = \varepsilon_0 \varepsilon_r] \tag{6.3}$$

Here $\varepsilon_0 = \dfrac{1}{36\pi \times 10^9}$ F/m.

$\therefore$ from equation (6.3), $$C = \frac{\varepsilon_r}{18 \ln (D/d)} \ \mu\text{F/m}.$$

We have, $$g = \frac{Q}{2\pi\varepsilon x} = \frac{V}{x \ln (D/d)} \tag{6.4}$$

The stress is maximum at the surface of the conductor, i.e. when $x = r$. Therefore, the maximum stress is

$$g_{\max} = \frac{2V}{d \ln (D/d)} \quad \left[\because r = \frac{d}{2}\right] \tag{6.5}$$

Let us now find the minimum value of $g_{\max}$, if d is varied. We can write $\dfrac{\partial g_{\max}}{\partial d} = 0$, for g to be minimum.

$$\frac{\partial g_{\max}}{\partial d} = -\frac{2V}{[d \ln (D/d)]^2} \times \left[\ln (D/d) + \frac{d}{D/d}\left(-\frac{D}{d^2}\right)\right]$$

$$= -\frac{2V[\ln (D/d) - 1]}{[d \ln (D/d)]^2}$$

$$\therefore \qquad \frac{2V[\ln (D/d) - 1]}{[d \ln (D/d)]^2} = 0$$

or, $$\ln (D/d) = 1, \quad \text{i.e. } D = 2.718 \text{ or, } d = \frac{D}{2.718}$$

In high voltage cables where thickness of insulation is considerable, the conductor diameter, if increased and made equal to $D/2.718$, it will give the minimum value of $g_{\max}$.

The minimum value of stress is $$g_{\max} = \frac{2V}{d} \tag{6.6}$$

The stress at the sheath $$= \frac{2V}{D \ln (D/d)} \tag{6.7}$$

Therefore, the stress is maximum at the conductor and minimum at the outer cover. In order to avoid breakdown of the insulation, it would be advisable to have a more uniform stress distribution. This may be done in two ways:

(i) By the use of intersheaths
(ii) By the use of insulating materials with different dielectric constants.

EXAMPLE 6.3 Consider a graded cable with permittivity as shown in Figure E6.1. Obtain the condition under which the maximum values of the electric fields in the two regions are equal.

Solution:

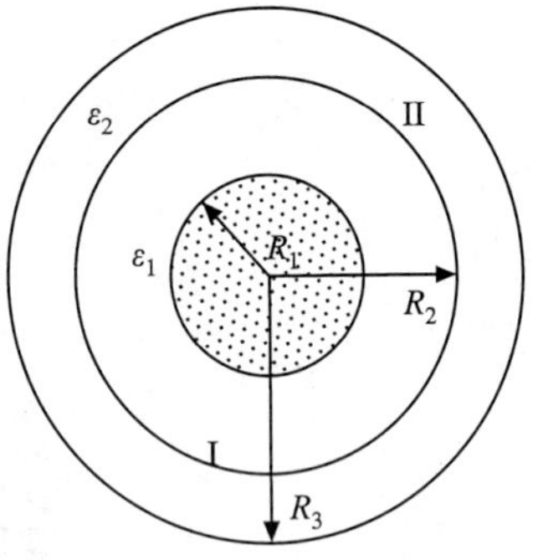

Figure E6.1

The electric fields in the two dielectrics are,

$$g_1 = \frac{Q}{2\pi\varepsilon_1 r} \text{ when } R_1 < r < R_2, \text{ and } g_2 = \frac{Q}{2\pi\varepsilon_2 r} \text{ when } R_2 < r < R_3$$

The maximum values of the electric fields are at, $r = R_1$ for g_1 and at $r = R_2$ for g_2.
For the maximum values of the electric fields in the two regions to be same, we must have

$$g_{max} = \frac{Q}{2\pi\varepsilon_1 R_1} = \frac{Q}{2\pi\varepsilon_2 R_2}$$

i.e.,

$$\varepsilon_1 R_1 = \varepsilon_2 R_2$$

since $R_1 < R_2$. The above equation implies that $\varepsilon_1 > \varepsilon_2$; that is, the dielectric closest to the conductor must have the higher permittivity than the outer one. (Ans.)

EXAMPLE 6.4 A single-core cable is working on 220 kV, three-phase system. The maximum permissible stress in the dielectric is not to exceed 250 kV/cm. Determine the economical-core diameter of the cable.

Solution: Operating voltage, $V = 220$ kV

Peak value of operating voltage, $\frac{220}{\sqrt{3}} \times \sqrt{2} = 179.6292$ kV

Maximum permissible stress, $g_{max} = 250$ kV
Now from equation (6.6), g_{max} to be minimum,

$$d = \frac{2V}{g_{max}} = \frac{2 \times 179.6292}{250} = 1.4370 \text{ cm}$$

Therefore, the most economical conductor diameter is 1.4370 cm. (Ans.)

6.7 SHEATHING AND GRADING IN CABLES

6.7.1 Use of Intersheaths

A metal covering is provided over the cable insulation such that the moisture ingress is low, voltage stress is uniformly distributed and the cable is saved from mechanical damage. Lead is used as the sheath material due to its high flexibility and high corrosion resistance. However, lead sheath has low mechanical strength and small resistance to vibrations. Lead alloys are commonly used in sheathing. Aluminium can also be used as a sheathing material. Though it has greater mechanical strength and low weight, its cost is higher.

Suppose two intersheaths having diameters d_1 and d_2 are used and kept at potentials V_1 and V_2 respectively. The electric stress g_1 between the conductor and the first intersheath at a distance x varies inversely as x.

Thus, $g_1 = \dfrac{Q}{2\pi\varepsilon x} = \dfrac{k}{x}$, where k is a constant.

The potential difference between the conductor and the first intersheath is

$$V - V_1 = \int_{d/2}^{d_1/2} g_1 dx = k \ln x \Big|_{d/2}^{d_1/2} = k \ln (d_1/d)$$

$$k = \frac{V - V_1}{\ln (d_1/d)}$$

and

$$g_1 = \frac{V - V_1}{x \ln (d_1/d)} \tag{6.8}$$

The maximum stress $g_{1\max}$ which occurs when $x = d/2$ is given by

$$g_{1\max} = \frac{V - V_1}{(d/2) \ln (d_1/2)} \tag{6.9a}$$

With a similar reasoning, the maximum stress $g_{2\max}$ between the first and second intersheaths is

$$g_{2\max} = \frac{V_1 - V_2}{(d_1/2) \ln (d_2/d_1)} \tag{6.9b}$$

The maximum stress $g_{3\max}$ between the second intersheath and the outermost cover is

$$g_{3\max} = \frac{V_2}{(d_2/2) \ln (D/d_2)} \tag{6.9c}$$

Note that the potential of the outermost cover is zero while D is the overall cable diameter. By choosing V_1 and V_2 properly, $g_{1\max}$ can be made equal to $g_{2\max}$, and so on. The improvement in electric stress distribution of a cable may be seen from Figs. 6.14 and 6.15 which show electric stress variation in insulation of a cable (a) without and (b) with intersheaths, respectively.

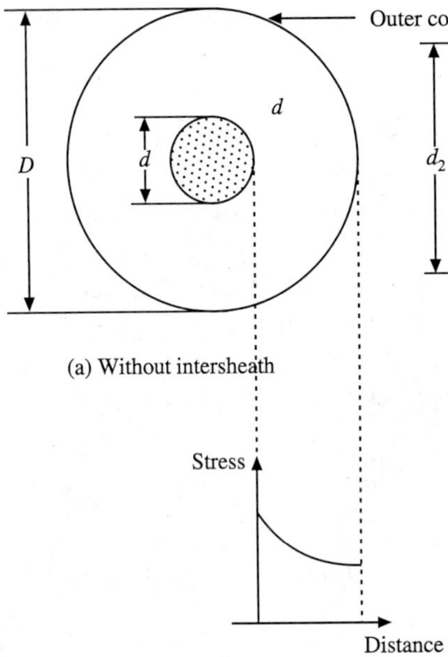

Figure 6.14 Voltage stress distribution without intersheath.

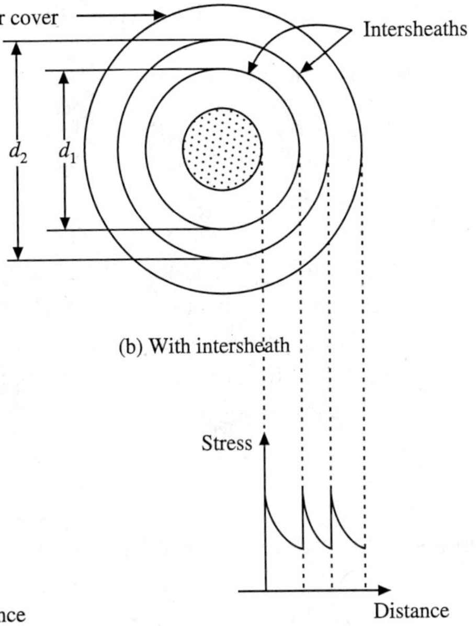

Figure 6.15 Voltage stress distribution with intersheath.

The stress can also be made to vary between the same maximum and minimum value by choosing d_1 and d_2 such that $d_1/d = d_2/d_1 = D/d_2 = \alpha$.

Since the maximum stresses are required to be made equal, we have

$$g_{3\max} = g_{2\max} = g_{1\max}$$

or,
$$\frac{V_2}{(d_2/2)\ln(D/d_2)} = \frac{V_1 - V_2}{(d_1/2)\ln(d_2/d_1)} = \frac{V - V_1}{(d/2)\ln(d_1/d)}$$

or,
$$\frac{V_2}{d_2 \ln \alpha} = \frac{V_1 - V_2}{d_1 \ln \alpha} = \frac{V - V_1}{d \ln \alpha}$$

or,
$$\frac{V_2}{d_2} = \frac{V_1 - V_2}{d_1} = \frac{V - V_1}{d}$$

or,
$$V_2 = \frac{d_2}{d_1}[V_1 - V_2] = \alpha V_1 - \alpha V_2$$

or,
$$V_2 = \frac{\alpha V_1}{1 + \alpha} = \frac{V_1}{1 + \dfrac{1}{\alpha}} \tag{6.10}$$

From the relationship,
$$\frac{V_1 - V_2}{d_1} = \frac{V - V_1}{d}$$

we have
$$V_1 - V_2 = \frac{d_1}{d}[V - V_1]$$

i.e., $$V_1 - V_2 = \alpha V - \alpha V_1$$

or, $$(1+\alpha)V_1 = \alpha V + V_2$$

or, $$(1+\alpha)V_1 = \alpha V + \frac{\alpha V_1}{1+\alpha} \quad \text{[Substituting } V_2 \text{ from equation (6.10)]}$$

or, $$\left(1+\alpha - \frac{\alpha}{1+\alpha}\right)V_1 = \alpha V$$

$$V_1 = \frac{\alpha(1+\alpha)V}{1+\alpha+\alpha^2} = \frac{V\left(1+\dfrac{1}{\alpha}\right)}{1+\dfrac{1}{\alpha}+\dfrac{1}{\alpha^2}} \tag{6.11a}$$

Substituting V_1 in equation (6.10), we get

$$V_2 = \frac{V_1}{1+\dfrac{1}{\alpha}} = \frac{V\left(1+\dfrac{1}{\alpha}\right)}{\left(1+\dfrac{1}{\alpha}\right)\left(1+\dfrac{1}{\alpha}+\dfrac{1}{\alpha^2}\right)}$$

$$V_2 = \frac{V}{1+\dfrac{1}{\alpha}+\dfrac{1}{\alpha^2}} \tag{6.11b}$$

The maximum stress is

$$g_{1\max} = \frac{V - V_1}{(d/2)\ln(d_1/d)} = \frac{V - V_1}{(d/2)\ln\alpha}$$

$$= \frac{V - V\left(1+\dfrac{1}{\alpha}\right)\Big/\left(1+\dfrac{1}{\alpha}+\dfrac{1}{\alpha^2}\right)}{(d/2)\ln\alpha} \quad \text{[Substituting } V_1 \text{ from equation (6.11a)]}$$

$$= \frac{V/\alpha^2}{\left(1+\dfrac{1}{\alpha}+\dfrac{1}{\alpha^2}\right)(d/2)\ln\alpha} = \frac{V}{(1+\alpha+\alpha^2)(d/2)\ln\alpha}$$

As $D/d = \alpha^3$ and $\ln(D/d) = \ln\alpha^3 = 3\ln\alpha$

$$g_{1\max} = \frac{V}{\dfrac{1}{3}(1+\alpha+\alpha^2)(d/2)\ln(D/d)} \tag{6.12a}$$

When no intersheath is used, the maximum stress is,

$$g_{\max} = \frac{V}{(d/2)\ln(D/d)} \tag{6.12b}$$

Therefore, the value of maximum stress with intersheaths is only $\frac{1}{3}(1+\alpha+\alpha^2)$ of the previous one ($\because\ \alpha > 1$).

6.7.2 Capacitance Grading

The insulation may be made of dielectrics of different permittivity. Such a cable is known as *graded cable*, and its arrangement results in a more uniform stress in the dielectric. Let d_1 be the diameter of the dielectric having permittivity and D the diameter of the dielectric having permittivity as shown in Figure 6.16.

The stress in the inner dielectric at a radius x is, $g_1 = \dfrac{Q}{2\pi\varepsilon_1 x}$

Dielectric stress in the outer dielectric is, $g_2 = \dfrac{Q}{2\pi\varepsilon_2 x}$

Therefore,
$$V = \int_{d/2}^{d_1/2} g_1 dx + \int_{d_1/2}^{D/2} g_2 dx = \frac{Q}{2\pi}\left[\frac{1}{\varepsilon_1}\ln(d_1/d) + \frac{1}{\varepsilon_2}\ln(D/d_1)\right] \tag{6.13}$$

and
$$C = \frac{Q}{V} = \frac{2\pi}{\left[\frac{1}{\varepsilon_1}\ln(d_1/d) + \frac{1}{\varepsilon_2}\ln(D/d_1)\right]} \tag{6.14}$$

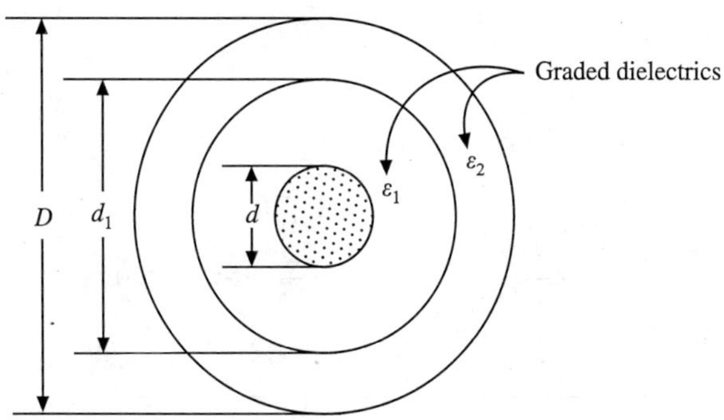

Figure 6.16 Capacitance graded cable.

Electric stress will be maximum at $x = \dfrac{d}{2}$ and at $x = \dfrac{d_1}{2}$.

i.e.
$$g_{max} = \frac{Q}{2\pi\varepsilon_1 (d/2)} = \frac{Q}{\pi\varepsilon_1 d}$$

Substituting the value of Q in terms of V from equation (6.14),

$$g_{1\,max} = \frac{2\pi V}{\pi\varepsilon_1 d\left[\frac{1}{\varepsilon_1}\ln(d_1/d) + \frac{1}{\varepsilon_2}\ln(D/d_1)\right]}$$

$$g_{1\max} = \frac{2V}{d\left[\ln(d_1/d) + \frac{\varepsilon_1}{\varepsilon_2}\ln(D/d_1)\right]} \tag{6.15a}$$

Similarly,

$$g_{2\max} = \frac{Q}{\pi\varepsilon_2 d_1} = \frac{2V}{d_1\left[\frac{\varepsilon_2}{\varepsilon_1}\ln(d_1/d) + \ln(D/d_1)\right]} \tag{6.15b}$$

Practical difficulties in grading

With capacitance grading, there is the practical difficulty of obtaining materials having the required values of permittivity. A second difficulty is the possibility of change of permittivity with time. Such a change could completely alter the distribution of dielectric stress and might even lead to breakdown at the normal working voltage. With the intersheath method, there is possibility of damage to the intersheath during laying of the cable since the intersheath has to be thin. The intersheath also has to carry a charging current which in a very long cable may be enough to cause overheating.

For these reasons, the present-day tendency is to avoid grading in favour of the oil-filled and gas filled cables.

EXAMPLE 6.5 A cable has intersheath grading that satisfies the equation $r/R_1 = R_1/R = \alpha$. The cable radii are r = 1.2 cm and R = 3 cm. Determine the location of the intersheath and calculate the ratio of maximum electric field strengths with and without intersheath.

Solution: The location of the intersheath radius R_2 is given by

$$\frac{R_1}{r} = \frac{R}{R_1} = \alpha \text{ or, } R_1^2 = rR = 1.2 \times 3$$

$$R_1 = 1.8974 \text{ cm}$$

$$\frac{R_1}{r} = \alpha = \frac{1.8974}{1.2} = 1.5812$$

For one intersheath, equation (6.12a) can be written as

$$g_{\max} = \frac{V}{\frac{1}{2}(1+\alpha)(d/2)\ln(D/d)}$$

$$\frac{g_{\max}\text{ (with intersheath)}}{g_{\max}\text{ (without intersheath)}} = \frac{2}{(1+\alpha)} = \frac{2}{1+1.5812} = 0.7748. \quad \text{(Ans.)}$$

EXAMPLE 6.6 A single core cable has a conductor diameter of 2.0 cm and a sheath of inside diameter 5 cm. Calculate the maximum stress. It is desired to reduce the maximum stress by using two intersheaths. Determine their best position, the maximum stress and the voltage on each intersheath. Assume system voltage as 66 kV, 3 phase.

Solution: Given: d = 2.0 cm and D = 5 cm, system voltage, V = 66 kV

$$\alpha^3 = D/d = 5/2 = 2.5$$

$$\alpha = 2.5^{1/3} = 1.3572$$

Peak voltage (= V) at core $= \dfrac{66}{\sqrt{3}} \times \sqrt{2} = 53.8887 \text{ kV}$

From equation (6.12b), the maximum stress is,

$$g_{max} = \frac{V}{(d/2)\ln(D/d)} = \frac{53.8887}{(2/2)\ln(5/2)} = 58.8118 \text{ kV/cm}$$

Now, $d_1/d = d_2/d1 = D/d_2 = \alpha.$

Position of first intersheath is, $d_1 = d\alpha = 2 \times 1.3572 = 2.7144$ cm.

Position of second intersheath is, $d_2 = d_1\alpha = 2.7144 \times 1.3572 = 3.6840$ cm.

From equation (6.11a), peak voltage on the first intersheath is

$$V_1 = \frac{V\left(1+\dfrac{1}{\alpha}\right)}{1+\dfrac{1}{\alpha}+\dfrac{1}{\alpha^2}} = \frac{53.8887\left(1+\dfrac{1}{1.3572}\right)}{1+\dfrac{1}{1.3572}+\dfrac{1}{1.3572^2}} = 41.0556 \text{ kV}$$

From equation (6.11b), peak voltage on the second intersheath is

$$V_2 = \frac{V}{1+\dfrac{1}{\alpha}+\dfrac{1}{\alpha^2}} = \frac{53.8887}{1+\dfrac{1}{1.3572}+\dfrac{1}{1.3572^2}} = 23.6385 \text{ kV}$$

From equation (6.12a), with two intersheaths, the maximum stress is

$$g'_{max} = \frac{V}{\dfrac{1}{3}(1+\alpha+\alpha^2)(d/2)\ln(D/d)}$$

$$= \frac{53.8887}{\dfrac{1}{3}(1+1.3572+1.3572^2)(2/2)\ln(5/2)} = 42.0165 \text{ kV/cm} \quad \text{(Ans.)}$$

So with the introduction of intersheaths, the maximum stress has been reduced $\left(g'_{max} < g_{max}\right)$.

EXAMPLE 6.7 A 3 phase, 66 kV single core lead sheathed cable is graded by using two dielectrics of relative permittivity 4.5 (inner) and 2.8 (outer), the thickness of each being 1 cm. The core diameter is 1 cm. Determine the maximum stress in the two dielectrics.

Solution: Given: $\varepsilon_1 = 2.8$ and $\varepsilon_2 = 4.5$, system voltage, V = 66 kV, d = 1.0 cm

$$d_1 = 1 + (2 \times 1) = 3 \text{ cm and } D = 3 + (2 \times 1) = 5 \text{ cm}$$

[Since thickness of dielectric is 1 cm.]

Peak voltage at core, $V = \frac{66}{\sqrt{3}} \times \sqrt{2} = 53.8887 \text{ kV}$

From equation (6.15a), the maximum stress at the first dielectric is

$$g_{1\max} = \frac{2V}{d\left[\ln(d_1/d) + \frac{\varepsilon_1}{\varepsilon_2}\ln(D/d_1)\right]} = \frac{2 \times 53.8887}{1 \times \left[\ln(3/1) + \frac{4.5}{2.8}\ln(5/3)\right]}$$

$$= 56.1463 \text{ kV/cm.} \quad \text{(Ans.)}$$

Similarly, from equation (4.15b), the maximum stress at the second dielectric is

$$g_{2\max} = \frac{2V}{d_1\left[\frac{\varepsilon_2}{\varepsilon_1}\ln(d_1/d) + \ln(D/d_1)\right]} = \frac{2 \times 53.8887}{3 \times \left[\frac{2.8}{4.5}\ln(3/1) + \ln(5/3)\right]}$$

$$= 30.0784 \text{ kV/cm.} \quad \text{(Ans.)}$$

6.8 POWER FACTOR OF CABLES

For a single-core cable, the insulation resistance is given by

$$R_i = \frac{\rho}{2\pi l}\ln\frac{D}{d} \tag{6.16}$$

Assuming ρ remaining constant, the impressed voltage V will send a current through the insulation and this current will be in phase with V.

On account of the capacity effect of the dielectric and the intersheaths, a charging current flows from the supply. The value of this charging current (I_c) is $\frac{V}{1/\omega C} = \omega CV$, where $\omega = 2\pi f$ and C has the values as found in earlier sections. The charging current leads the voltage V by 90° and the total current drawn from the supply is the phasor sum of currents V/R_i and ωCV, as shown in Figure 6.17. The total current I leads the voltage V by an angle ϕ, where cos ϕ gives *the power factor of the cable*. (In practice, ϕ is very nearly equal to $\pi/2$.)

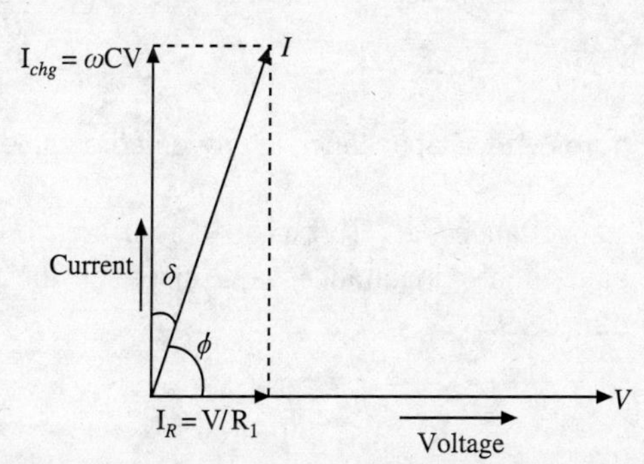

Figure 6.17 Phasor diagram of cable currents and voltage.

More correctly, $$\cos\phi = \cos\left(\frac{\pi}{2} - \delta\right) = \sin\delta \cong \delta \text{ (since } \delta \text{ is small)}$$

In other words, power factor of the cable = δ

Therefore, $$\delta = \frac{V}{R_i} \times \frac{1}{\omega CV} = \frac{1}{\omega CR_i} = \frac{1}{R_i} \times \frac{1}{\omega C} = \frac{G}{\omega C}$$

where $1/R_i = G$, conductance of the cable.

6.9 CAPACITANCE OF A THREE PHASE CABLE

The three-core cable has the following capacitances:

1. A capacitance between conductors due to the existence of an electrostatic field caused by the potential difference between conductors.
2. A capacitance between the conductor and the sheath due to the existence of an electrostatic field caused by the potential difference between the conductor and the sheath.

Let the capacitance formed between the core and the sheath = C_S
the capacitance formed between the two conductors = C_C

The three-core cable with capacitances formed is shown in Figure 6.18(a). Since the sheath is at same potential all over it, this can be represented in the other as shown in Figure 6.18(b).

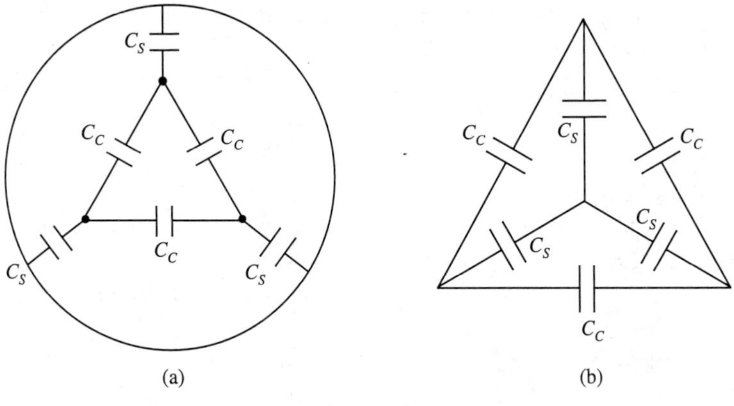

Figure 6.18 Capacitance in a three-core cable.

If the delta connected capacitances, C_C (between cores) are converted into an equivalent star arrangement, then the equivalent capacitance value between the star point and the core is, $C_1 = 3C_C$, as shown in Figure 6.19.

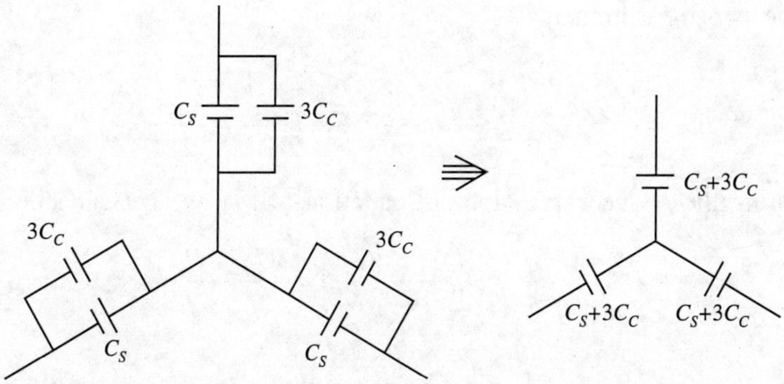

Figure 6.19 Equivalent capacitance in a three-core cable.

If the potential of the star point is assumed to be at sheath potential (i.e. at zero), then the capacitance of each conductor to neutral is given by

$$C_N = C_S + 3C_C \tag{6.17}$$

It is very difficult to calculate the capacitance of each conductor to neutral *CN* of the cable from geometry. Therefore, the empirical formula used to find the capacitance between the core and the neutral is given by

$$C_N = \frac{0.03\varepsilon_r}{\ln\left[1+\frac{T+t}{d}\left\{3.84-1.7\frac{t}{T}+0.52\frac{t^2}{T^2}\right\}\right]}\ \mu\text{F/Km} \tag{6.18}$$

where ε_r = relative permittivity of the dielectric material
d = conductor diameter
t = belt insulation thickness
T = conductor insulation thickness

All of the above parameters are in the same units. The main uncertainty in this formula is the value of ε_r.

Measurement of C_C and C_S

1. Bunch the three cores and measure the capacitance between the bunched cores and the sheath. Take this as C_A.

$$C_A = 3C_S;\ C_S = \frac{C_A}{3} \tag{6.19}$$

2. Connect any two cores to the sheath and measure the capacitance between the remaining core and the sheath. Take this as C_B.

$$C_B = C_S + 2C_C$$

From the two measurements,

$$C_S = \frac{C_A}{3};\; C_C = \frac{C_B - C_S}{2} = \frac{C_B - \frac{C_A}{3}}{2} = \frac{C_B}{2} - \frac{C_A}{6}$$

Substituting the above expressions in equation (4.17), we get capacitance/phase as:

$$C_N = \frac{C_A}{3} + 3\left(\frac{C_B}{2} - \frac{C_A}{6}\right) = \frac{3C_B}{2} - \frac{C_A}{6} \tag{6.20}$$

EXAMPLE 6.8 The capacitance per km of a three-phase belted cable is 0.5 μF between the two cores, with the third core connected to the lead sheath. Calculate the charging current taken by 1.0 km of this cable when connected to a three-phase, 50 Hz, 33 kV supply.

Solution: Capacitance between cores 1 and 2 with third core connected to the sheath is,

$$C_C + \frac{1}{2}C_C + \frac{1}{2}C_S = \frac{3}{2}C_C + \frac{1}{2}C_S = \frac{1}{2}(3C_C + C_S)$$

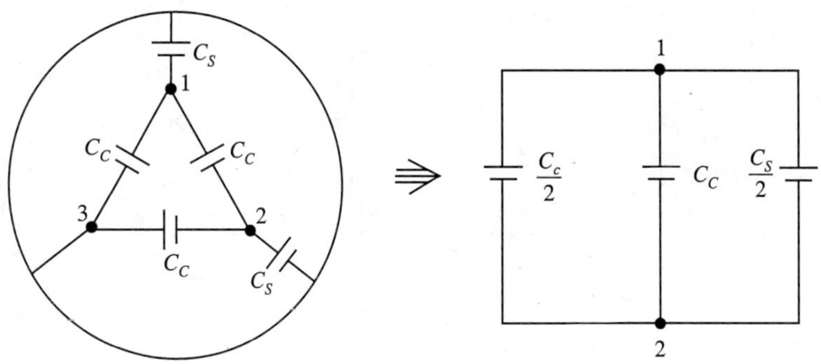

Figure E6.2

Given: $\frac{1}{2}(3C_C + C_S) = 0.5\ \mu\text{F}$

But $C_N = C_S + 3C_C$ [C_N = capacitance of each conductor to neutral]

∴ Capacitance per phase, $C_N = 2 \times 0.5 = 1.0\ \mu\text{F}$

Charging current per phase per kilometer is

$$= \omega C_N V_{ph} = 2\pi \times 50 \times 1.0 \times 10^{-6} \times \frac{33 \times 10^3}{\sqrt{3}} = 5.9855\ \text{A}.$$

EXAMPLE 6.9 A 33 kV, 3 phase, three core underground cable is 5 km long. Each of the conductors has a diameter of 2 cm and the radial thickness of insulation is 0.8 cm. The relative permittivity of the dielectric is 3.5. Determine

(i) Capacitance of the cable/phase
(ii) Charging current/phase
(iii) Total charging KVAR
(iv) Dielectric loss/phase if the power factor of the unloaded cable is 0.05
(v) Maximum stress in the cable.

Solution: Given: Length, l = 5 km; Insulation resistance, R = 0.5 Mohm;
Diameter of core, d = 2 cm; Diameter of cable with insulation, $D = d + 2R$ = 3 cm
Radius of core, $r = d/2$ = 1 cm
The relative permittivity of the dielectric, $\varepsilon_r = 3.5$

(i) From equation (6.3), capacitance of the cable/phase is

$$c = \frac{Q}{V} = \frac{2\pi\varepsilon_0\varepsilon_r}{\ln(D/d)} = \frac{2\pi \times 8.854 \times 10^{-12} \times 3.5}{\ln(3/2)} \quad \left[\because \varepsilon_r \simeq 8.854 \times 10^{-12}\,\text{F/m}\right]$$

$= 480.21 \times 10^{-12}$ farads/metre = 480.21 pF/m

The cable is 5 km long; therefore, total capacitance of the cable/phase is

$C = cl = 480.21 \times 10^{-12} \times 5 \times 10^3 = 2.4 \times 10^{-6}$ farads = 2.4 μF (Ans.)

(ii) Charging current/phase (I_c) = $\dfrac{V_{ph}}{X_C}$

Now $\quad V_{ph} = \dfrac{33 \times 10^3}{\sqrt{3}}$ and $X_C = \dfrac{1}{2\pi f C} = \dfrac{1}{2\pi \times 50 \times 2.4 \times 10^{-6}}$

∴ Charging current/phase (I_c) = $\dfrac{33 \times 10^3}{\sqrt{3}} \times 2\pi \times 50 \times 2.4 \times 10^{-6}$

= 14.3653 A/ph (= I_{ph}) (Ans.)

(iii) Total charging kVAR = $\sqrt{3}V_L I_c = 3V_{ph}I_{ph} = 3 \times \dfrac{33 \times 10^3}{\sqrt{3}} \times 14.3653$

= 821.0871 kVAR. (Ans.)

(iv) Given: Power factor of cable at no load, $\cos\phi = 0.05$

$$\phi = \cos^{-1} 0.05 = 87.1340°$$

∴ Loss angle, $\delta = 90° - 87.1340° = 2.866°$

Dielectric loss/phase,

$$P_{loss} = V_{ph}I_{ph}\cos\phi = V_{ph}I_{ph}\cos(90° - \delta) = V_{ph}I_{ph}\sin\delta$$

$$= V_{ph}^2\,\omega C \sin\delta = \left(\frac{33 \times 10^3}{\sqrt{3}}\right)^2 \times 2 \times \pi \times 50 \times 2.4 \times 10^{-6} \times \sin 2.866°$$

= 13.6848 kW/ph. (Ans.)

(v) From equation (6.5),

$$\text{Maximum stress in the cable, } g_{max} = \frac{2V_{ph}}{d \ln (D/d)} = \frac{2 \times \dfrac{33 \times 10^3}{\sqrt{3}}}{2 \ln (3/2)}, \text{ with rms value of } V_{ph}.$$

$$= 46.9894 \text{ kV/cm (rms).} \quad \text{(Ans.)}$$

6.10 HEATING OF CABLES

The temperature rise of a cable under working conditions depends upon the following factors:

- Heat generated inside the cable, i.e. upto its peripheral
- The rate of dissipation of heat upto peripheral of the cable
- Surrounding medium and temperature
- The current rating of the cable
- Nature of the load.

6.10.1 Generation of Heat within the Cable

The main sources of generation of heat inside the cables are:

1. Copper loss
2. Dielectric loss
3. Sheath and armouring losses

Copper loss

Copper loss occurs mainly due to the presence of resistance in the cables. The resistance of the conductor mainly depends upon the temperature by the following relation

$$R_t = R_0 (1 + \alpha t) \tag{6.21}$$

where

R_t = resistance at temperature t
R_0 = resistance at initial temperature
α = temperature coefficient
t = change in temperature

Hence, as the temperature increases, the value of the resistance of the conductor increases and hence more losses result in more heat.

Dielectric loss

Losses occurring in the dielectric of cables are called *dielectric losses*. These are mainly caused due to the leakage of charging currents.

Dielectric loss per phase is given by

$$P = VI\cos\phi = VI\sin\delta = V\frac{I_{chg}}{\cos\delta}\sin\delta = V\frac{\omega CV}{\cos\delta}\sin\delta = \omega CV^2\tan\delta \simeq \omega CV^2\delta \quad (6.22)$$

From phasor diagram charging current, we have, $I_c = I\cos\delta = \omega CV$

$$P = V\frac{\omega CV}{\cos\delta}\sin\delta = \omega CV^2\tan\delta; P = \omega CV^2\tan\delta \quad (6.22)$$

Sheath and armouring losses

The presence of lead sheath influences the electrical characteristics of the circuit when power is transmitted by the cable. This effect is considerably more for single-core cables than that of three-core cables. The flow of currents in the conductor produces magnetic fields around the conductor which link with the lead sheath and induce an e.m.f and current in it. This induced current depends upon the frequency, current in the conductor, sheath resistance, arrangement of cables and on the conditions that the sheath is bounded or unbounded.

The approximate *sheath losses* due to sheath eddy currents are given by

$$P_{\text{loss}}\text{ sheath} = I^2\left[\frac{78\omega^2}{R_S}\left(\frac{r_m}{d}\right)^2 \times 10^{-9}\right]\text{Watt/phase}$$

where

I = current per conductor
$\omega = 2\pi f$
r_m = mean radius of the sheath
R_S = sheath resistance in ohm
d = inter-axial spacing of conductors.

These losses are negligible as they form only about 2 per cent of the core losses.

6.11 BREAKDOWN OF CABLES

Breakdown of cables may be due to puncture, thermal instability and void formation. A cable punctures when a discharge occurs between the conductor and the sheath at one or more than one points within the dielectric forming a hole in it. This leads to cable breakdown.

Thermal instability occurs in the cable when the power dissipation in the cable is high due to dielectric loss beyond the recommended value. Additional ohmic loss may be cumulatively added such that the rate of heat within the cable is very high. This may lead to thermal breakdown of the cables. However, if the temperature rise can be halted, permanent damage can be stopped.

The formation of air space (*void*) in the dielectric medium, due to imperfect impregnation, faulty design or material and successive changing of load cycles may be detrimental for the cable. Ionization can take place within the void and the voltage gradient may be very high in dielectric. This may produce minute carbon within the insulation and near the conductive surface. In HV cables, the formation of carbonized track is gradually extended. This leads to breakdown of the cable.

6.12 SELECTION OF CABLES

The selection of LT cables is easier. Only continuous current rating, voltage grade and voltage drop are to be calculated to select the conductor size and type of cable. Obviously, thermal conditions and derating factors (i.e. the derated capacity of the cable due to temperature rise from the adjacent cable) are to be checked. Voltage drop at full load current should also be taken care of.

However, for HV cables the additional factor, being the most important, is short circuit current rating. It is required because in high voltage systems, the fault current is very high. Thus, even if the continuous current rating is satisfactory for the selected cable size, it may not be sufficient from the viewpoint of short circuit current rating. Higher size of cable may then be required such that it is capable of dissipating the short circuit fault energy.

EXERCISES

1. Classify the cables according to their voltage level. Discuss their application.
2. Write short notes on: (i) SL type cables and (ii) HSL type cables.
3. With a schematic diagram, discuss the construct of single core oil-filled cables. Discuss the advantages and disadvantages of oil-filled cables over the solid cables.
4. Discuss different types of oil-filled cables in brief.
5. Discuss different types of available gas-filled cables. Discuss advantages and disadvantages of this type of cable.
6. What is gas-cushion cable?
7. With a simplified diagram, describe different parts of a cable.
8. Discuss the commonly used cable insulations.
9. Why metallic sheath is provided over in cable? What is bedding?
10. What is the function of armour in a cable?
11. Discuss the advantages and disadvantages of underground cables over overhead transmission line.
12. Derive expression of the following in single core power cables: (i) insulation resistance, (ii) maximum stress and (iii) capacitance.
13. Why sheathing is required in cable? Derive the expression for maximum stress in a single core cable (i) without intersheath and (ii) with two intersheaths. Discuss the effect of introduction of intersheaths.
14. What is capacitance grading? Why is it required? What are the practical difficulties involved here? How can they be solved?

15. Derive the expressions for maximum stress in a single core cable graded by two dielectrics of permittivities ε_1 (inner) and ε_2 (outer) with diameters d_1 and D, respectively. The cable diameter is d cm.
16. What is loss angle of a cable? How is it related to the power factor angle of the cable? Derive the expression for power loss in a cable in terms of power factor angle as well as loss angle. V and I are the voltage and current of the cable.
17. Derive the expression for per phase capacitance of a three phase cable.
18. Why heat is generated in the cable? Discuss different types of losses that occur in cable.
19. Derive the expression for (i) copper loss and (ii) dielectric loss.
20. Why sheath and armouring losses occur in cable.
21. Discuss different possibilities for breakdown of a cable.
22. Discuss different factors required to be considered for selection of a proper cable for a particular purpose.

CHAPTER 7

Mechanical Design of Overhead Lines

7.1 INTRODUCTION

The transmission line design should be robust from the point of view of mechanical considerations, in addition to electrical requirements. The line conductors, supports, tower, etc. should have ample mechanical strength to cope with the worst possible weather conditions and should satisfy the requirements of safety standards. The tension of the conductor should be well below the breaking load. An appreciable amount of sag needs to be allowed for this purpose. The following sections discuss the analytical aspects in order to obtain the best possible mechanical design of the transmission line.

7.2 LINE SUPPORTS

The supports for an overhead transmission line should be capable of carrying the load of conductors (including loading of ice deposition and wind thrust) and insulators together with the wind load on the support itself. The various types of supports include:

(i) Wooden poles
(ii) Reinforced concrete poles
(iii) Tubular steel poles
(iv) Latticed steel towers.

7.2.1 Wooden Poles

Wooden poles are the simplest (and perhaps the cheapest). They are useful for voltages upto 33 kV and for span upto about 30 to 40 metres. If properly treated against decay, these poles may give a long useful life. Under special conditions, single and H-type poles may be employed for lines working with higher voltages and span length of 50 metres. Double pole structures are often used to obtain higher traverse strength. Wooden poles are elastic and lines employing this type of support are often designed for the transverse load at terminals and for anchor supports. Longitudinal strength is provided by means of guys. Wooden poles are protected from atmospheric affects by metallic cap at the top.

The disadvantages with this type of supports are that they have the tendency to rot at the ground level and their life cannot be predicted with certainty. Use of wooden poles is not encouraged for environmental reasons.

7.2.2 Reinforced Concrete Poles

The difficulty with the use of *reinforced concrete* poles is the high cost of transport owing to their great weight. To overcome this, poles may often be cast on site. The advantages of these poles include very long life and very low maintenance charges. They are extensively used in medium voltage transmission (upto 11 kV) and LT distribution.

7.2.3 Tubular Steel Poles

Steel poles are often used as a substitute for wood poles. They enable rather longer spans and long life. Along the road side or in congested localities, their use is generally more favoured. This type of support needs to be galvanised which prolongs the life of the steel pole. (A useful life of 30 years is common.) Steel poles may be *swaged* type, i.e. having different diameters throughout their length. They are costlier than wooden poles but possess specific advantages like light weight, long life and higher strength. Steel poles are widely used for supply systems upto 33 kV.

7.2.4 Latticed Steel Towers

Steel towers are extremely useful for system voltage of and above 33 kV with long distance transmission line running across open country, where long spans are required. Even though they are costlier than other poles, they are economical for tall supports and long spans. The steel may be galvanised or painted. Steel towers can be classified into (i) narrow base lattice type and (ii) broad base towers. The former is used for spans not so long while the latter is particularly useful when the span is very large. There are separate types of towers designed for single circuit and double circuit lines. Steel towers are fabricated from galvanised angle iron sections and transported separately to site for assembly and erection. They are finally painted to withstand severe weather conditions. The following types of towers are in use:

(i) Self-supporting steel towers
(ii) Guyed towers
(iii) Chainnet type towers

Lattice type structure being one of the most common forms of self-supporting tower, H-type frames are also used for voltages 345 kV and above. V-type and H-type frames are used in guyed and chainnet type of towers also. Guys are used to secure more mechanical strength of the tower. Figure 7.1(a–j) shows lattice type and guyed type of towers.

A typical tower consists of (a) main stem, (b) cross-arm (on which conductors with insulators are supported), (c) insulators, and (d) optional guy wire. There should be adequate clearance between conductor to conductor, conductor to ground and conductor to nearest tower member. The configuration of the tower is determined by the requirement of various clearances and mechanical loading on structure members.

Though self-supporting structures are most widely used in India, *guyed towers* are used in USA and Canada. The self-supporting feature of the tower is replaced by guys in the direction of the line and with an increased strength. Guyed towers are mostly supported on hinged foundation.

In *chainnet structure*, there are two lattice steel/aluminium legs on hinged foundation and a system of wire rope, guys, etc. is utilised to keep the legs in the inclined pattern, as well as to provide strength. In the V-shaped guyed structure, the two legs are pivoted on the same foundation while in the chainnet structure, they are spaced apart. The overall width and height of components of chainnet structure are higher.

In steel towers, if the span between successive two towers is increased, the requirement of total number of towers is reduced. This enables reduction in the cost of insulators, though the towers become taller and costlier. Obviously, the use of longer span reduces the number of towers and hence it increases the reliability of the line. However, from mechanical considerations, there is a maximum value of span for each conductor size. Lightning hazards also affect the tower height criterion. Modern HV lines have spans between 200 and 400 m. In river crossing, hilly terrain, a long span length of 500 m or more is also utilised.

Regarding conductor configuration, there is no special advantage for symmetrical conductor configuration. Flat horizontal or vertical configurations are frequently used from mechanical considerations. The tower height becomes shorter when horizontal type of conductors are utilised. A vertical configuration is used for tall towers.

IE rules specify the minimum clearances between conductor to conductor and conductor to ground.

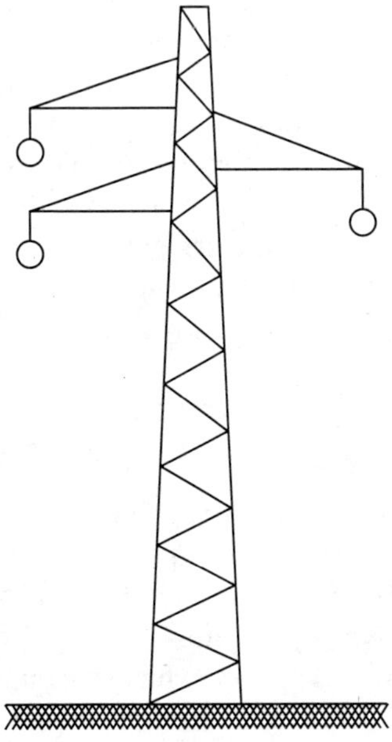

Figure 7.1(a) Narrow base single circuit (33 kV).

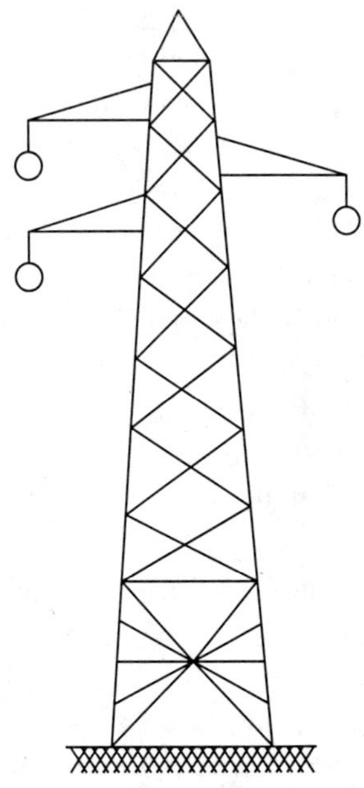

Figure 7.1(b) Broad base single circuit (33 kV/66 kV).

Figure 7.1(c) Broad base double circuit (66 kV/132 kV/220kV).

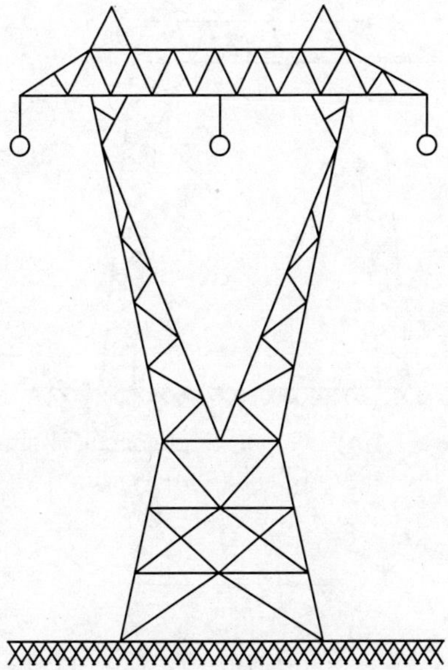

Figure 7.1(d) Broad base single circuit (220 kV/400 kV).

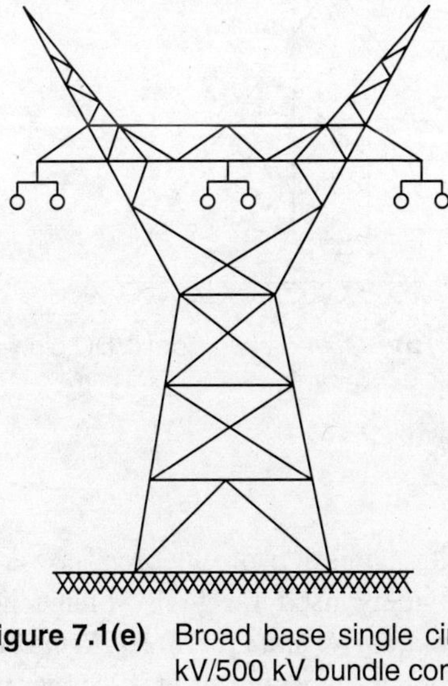

Figure 7.1(e) Broad base single circuit (400 kV/500 kV bundle conductor).

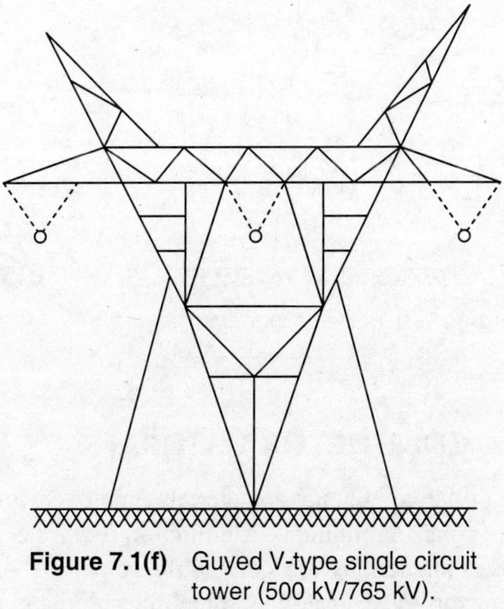

Figure 7.1(f) Guyed V-type single circuit tower (500 kV/765 kV).

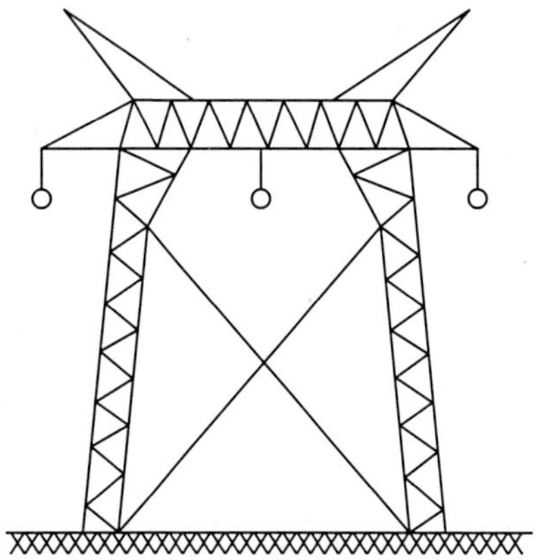

Figure 7.1(g) Guyed single circuit H-structure (765 kV/800 kV).

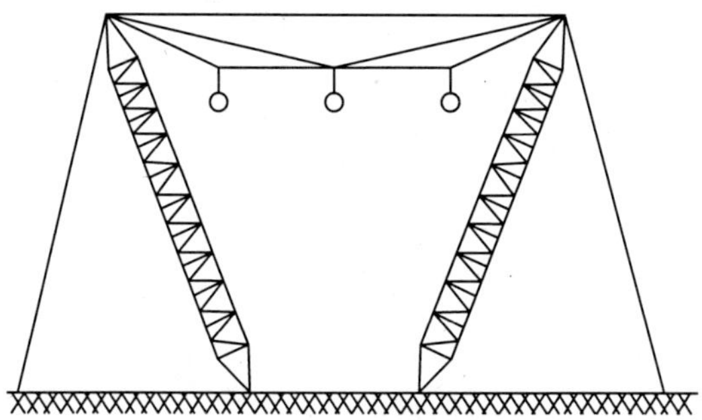

Figure 7.1(4) Chainned single circuit structure (765 kV/1100 kV/1500 kV).

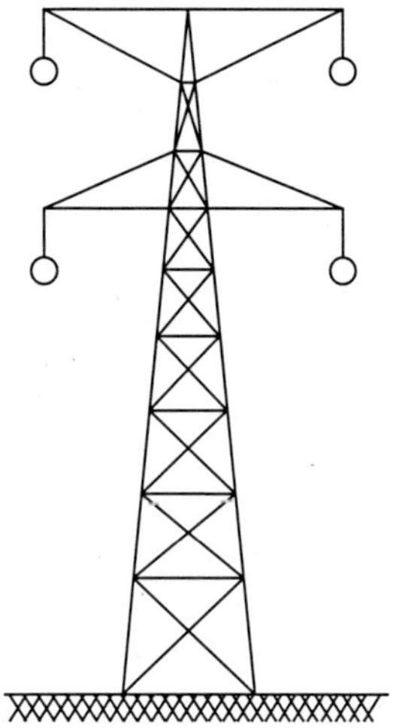

Figure 7.1(i) Self supporting tower for HVDC bipolar link.

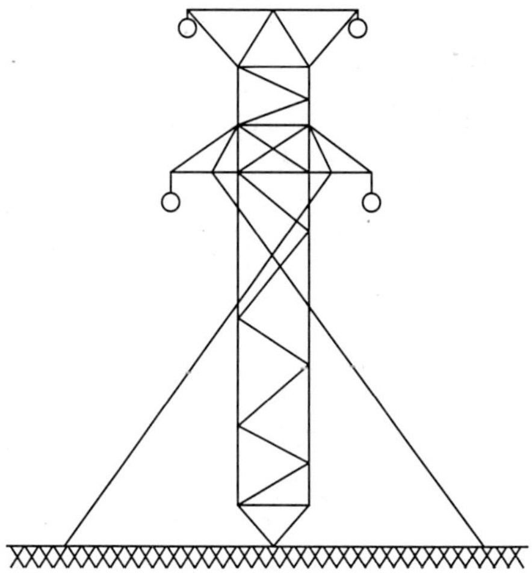

Figure 7.1(j) Guyed tower for HVDC bipolar link.

7.3 CONDUCTOR MATERIALS

The principal conductor materials employed for overhead transmission of electrical power are copper and aluminium. Aluminium (with steel core) is widely used for high voltage lines. These conductions are commonly named as ACSR (*Aluminium Conductor Steel Reinforced*) conductor. Other materials sometimes used are cadmium, copper, phosphor bronze, copper weld and galvanized steel. The choice in a particular conductor depends upon the cost and required electrical and mechanical properties.

7.4 CALCULATION OF SAG

7.4.1 Supports at Same Level

Let the conductor be suspended between points P_1 and P_2 with O as the lowest point as shown in Figure 7.2(a). The curve will be horizontal at O. Let w be the weight of conductor per unit length and let $P(x, y)$ be any point on the curve. Draw a tangent at P (it will indicate the direction tension/compression in conductor) as shown, this tangent making an angle θ with the horizontal.

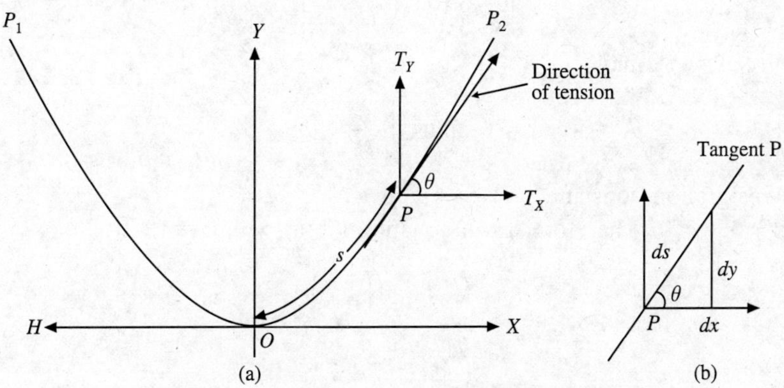

Figure 7.2 Sag of an overhead transmission line.

The forces acting on portion OP of the curve (length s) are:

(i) The horizontal tension at O, H

(ii) The weight of wire between OP acting vertically downwards through the centre of gravity, ws

(iii) Tensions T_X and T_Y which are the components of tensions in the wire $(T_X / T_Y = \tan\theta)$

For equilibrium, horizontal and vertical components of the forces must balance so that

$$H = T_X \tag{7.1a}$$

and

$$T_Y = ws \tag{7.1b}$$

Also,

$$\tan\theta = \frac{dy}{dx} \tag{7.2}$$

Combining equations (7.1) and (7.2), we can write

$$\frac{dy}{dx} = \frac{T_Y}{T_X} = \frac{ws}{H}$$

$\therefore$

$$\frac{H}{w} = s\frac{dx}{dy} \tag{7.3}$$

From Figure 7.1(b), we have

$$(ds)^2 = (dx)^2 + (dy)^2$$

or,

$$\frac{ds}{dx} = \sqrt{1 + \left(\frac{dy}{dx}\right)^2} = \sqrt{1 + \frac{w^2 s^2}{H^2}}$$

or,

$$dx = \frac{ds}{\sqrt{1 + \left(\frac{ws}{H}\right)^2}}$$

Integrating both sides, we have

$$x = \frac{H}{w}\sinh^{-1}\left(\frac{ws}{H}\right) + A \tag{7.4a}$$

where A is an integration constant.

Now, at $x = 0$, $s = 0$. Therefore, from equation (7.4a), we have

$$x = \frac{H}{w}\sinh^{-1}\left(\frac{ws}{H}\right) \tag{7.4b}$$

or,

$$\sinh\left(\frac{wx}{H}\right) = \frac{ws}{H}$$

or,

$$s = \frac{H}{w}\sinh\left(\frac{wx}{H}\right) \tag{7.5}$$

Replacing s in equation (7.5) with the help of equation (7.3), it yields

$$\frac{dy}{dx} = \sinh\left(\frac{wx}{H}\right)$$

Integrating both sides, we have

$$y = \frac{H}{w}\cosh\left(\frac{wx}{H}\right) + B \tag{7.6a}$$

where B is an integration constant.

Also, at $y = 0$, $s = 0$. Therefore, from equation (7.6a), we get

$$0 = \frac{H}{w} + B \Rightarrow B = -\frac{H}{w}$$

Substituting for B in equation (7.6a), we have

$$y = \frac{H}{w}\cosh\left(\frac{wx}{H}\right) - \frac{H}{w}$$

$\therefore$

$$y = \frac{H}{w}\left[\cosh\left(\frac{wx}{H}\right) - 1\right] \tag{7.7}$$

The tension at point P is

$$T^2 = T_X^2 + T_Y^2 = H^2 + w^2 s^2 = H^2 + H^2 \sinh^2\left(\frac{wx}{H}\right) \qquad \left[\because\ s = \frac{H}{w}\sin h\left(\frac{wx}{H}\right)\right]$$

$$= H^2 + H^2 \sinh^2\left(\frac{wx}{H}\right) \quad \text{[Substituting for } s \text{ from equation (7.5)]}$$

$$= H^2 \cosh^2\left(\frac{wx}{H}\right)$$

$$\therefore \qquad T = H\cosh\left(\frac{wx}{H}\right) \tag{7.8}$$

If we consider that P_1 and P_2 are at the same level and the *half span length* is ℓ, the tension at P_1 and P_2 is

$$T = H\cosh\left(\frac{w\ell}{H}\right) \tag{7.9}$$

And the length(s) of line in half span is

$$s = \frac{H}{w}\sinh\left(\frac{w\ell}{H}\right) \tag{7.10}$$

Also, from equation (7.7) ordinate y (sag) at P_1 and P_2 is

$$D = \frac{H}{w}\left[\cosh\left(\frac{w\ell}{H}\right) - 1\right] \tag{7.11}$$

From practical cases, when the sag is small compared to the span, equation (7.7) may be expanded as

$$y = \frac{H}{w}\left[1 + \frac{w^2x^2}{2H} + \frac{w^4x^4}{24H} + \cdots - 1\right] \cong \frac{wx^2}{2H} = \frac{wx^2}{2T} \qquad [\because\ T \cong H] \tag{7.12}$$

$$\therefore \qquad \text{Sag}(D) = \frac{w\ell^2}{2T} \tag{7.13a}$$

If the length of the entire span in the transmission line be L, then $\ell = \frac{L}{2}$.

$$\therefore \qquad \text{Sag}(D) = \frac{wL^2}{8T} \tag{7.13b}$$

From equation (7.5), arc length is given by

$$s = \frac{H}{w}\sinh\left(\frac{wx}{H}\right) = \frac{H}{w}\left[\frac{wx}{H} + \frac{w^3x^3}{6H^3} + \cdots\right] \cong x + \frac{w^2x^3}{6H^2} \tag{7.14}$$

Also, we can write from equation (7.14),

$$s \cong x\left(1 + \frac{w^2x^2}{6T^2}\right) \tag{7.15}$$

∴ The arc length in half span is, $$s = \ell\left(1+\frac{w^2\lambda^2}{6T^2}\right) \quad (7.16)$$

It will be seen that equation (7.12) is the curve of a *parabola*. Thus, if we assume that the conductor takes the form of a parabola, sag will be determined by equation (7.12) or (7.13), whereas if the conductor is assumed to take the form of a *catenary*, sag will be determined by equation (7.7) or (7.11).

7.4.2 Supports at Different Levels

Let P_1 and P_2 be at different levels from the ground as shown in Figure 7.3. Let O be the lowest point on the curve.

Here, $$D_1 = \frac{wx_1^2}{2T} \quad (7.17a)$$

and $$D_2 = \frac{wx_2^2}{2T} \quad (7.17b)$$

∴ $$D_2 - D_1 = \frac{wx_2^2}{2T} - \frac{wx_1^2}{2T} = \frac{w}{2T}(x_2^2 - x_1^2) = \frac{w}{2T}(x_2 + x_1)(x_2 - x_1) \quad (7.17c)$$

But $$x_2 + x_1 = L,\ L = \text{span length} \quad (7.18a)$$

and $$D_2 - D_1 = h \quad (7.19)$$

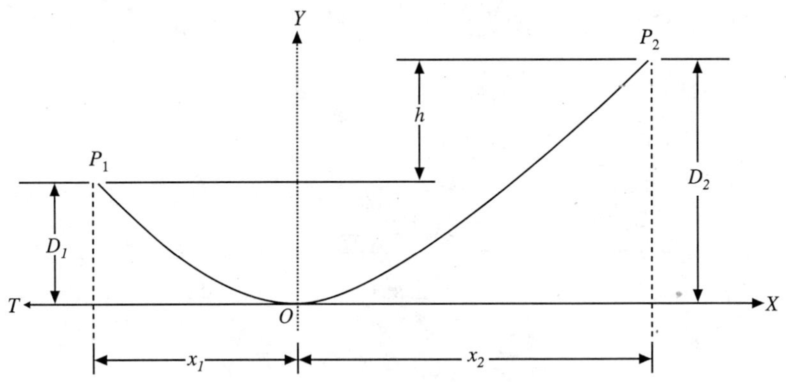

Figure 7.3 Two supports at different levels from ground.

Substituting equation (7.18a) in equation (7.17c), we get

∴ $$h = \frac{wL}{2T}(x_2 - x_1) \quad (7.20)$$

∴ $$x_2 - x_1 = \frac{2Th}{wL} \quad (7.18b)$$

Solving equations (7.18a) and (7.18b), we obtain

$$x_2 = \frac{L}{2} + \frac{Th}{wL} \tag{7.21a}$$

and

$$x_1 = \frac{L}{2} - \frac{Th}{wL} \tag{7.21b}$$

Thus, point O (the lowest point) can be found out.

Effect of ice loading

Transmission lines running through areas which experience severe winter and snowfalls would be covered with ice at the time of snowfall, the thickness of ice depending upon weather and also on the size of the conductor. The ice covering increases the weight of the conductor and thus increases the vertical sag. To guard against this, the tension on the conductors at the time of erection is suitably increased.

If d is the diameter of the conductor and t is the thickness of ice, then the overall diameter of the covered conductor as shown in Figure 7.4 becomes

$$D = d + 2t \tag{7.22}$$

The overall area of the conductor is $= \frac{\pi D^2}{4}$.

Therefore, the area of ice covering $= \frac{\pi D^2}{4} - \frac{\pi d^2}{4} = \frac{\pi}{4}(D^2 - d^2)$

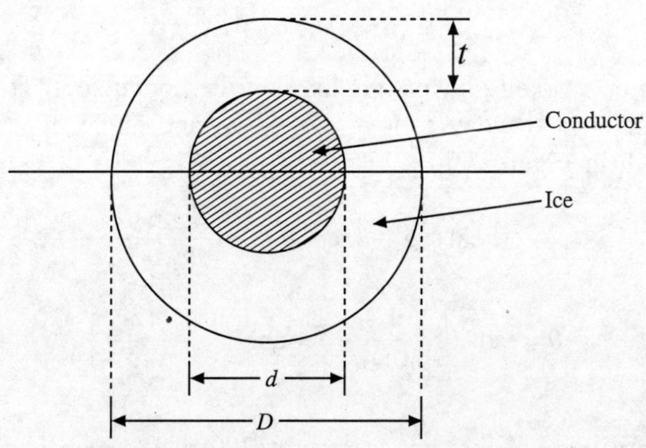

Figure 7.4 Conductor covered with ice.

If D and d are in metres, then this expression is also equal to the volume of ice in cubic metres per metre length of the conductor. Taking the weight of ice as 915 kg per cubic metre, the weight of ice becomes

$$w_i = \frac{\pi}{4}(D^2 - d^2) \times 915 \text{ kg/metre length} \tag{7.23}$$

It is obvious that this weight of ice acts vertically downward.

Effect of wind pressure

The wind pressure on the conductors of the transmission line acts perpendicular to the conductors and taken as pressure on the projected area of the conductors. In the case of towers and lattice type poles, the wind pressure is assumed to act on a surface area 1.5 times the projected area of one face. If d is the diameter of the conductor in metres and p the wind pressure in kg/m^2, then the wind load per metre of the conductor length (assuming the wind pressure to act on 2/3rd of the of the projected area) is $2/3p(d \times 1)$ kg. When the effect of both ice and wind is to be taken, d will be substituted by D, i.e. $(d + 2t)$, where t is the thickness of ice.

Since the wind pressure is assumed to act perpendicular to the conductor, its effect is only to increase the transverse loading on the conductors. The maximum wind load varies from 100 kg/m^2 to 150 kg/m^2 in India. The wind load can also be calculated by applying the formula $w_w = 0.006V^2$, where w_w is wind loading kg/m^2 and V is wind velocity in km/hr.

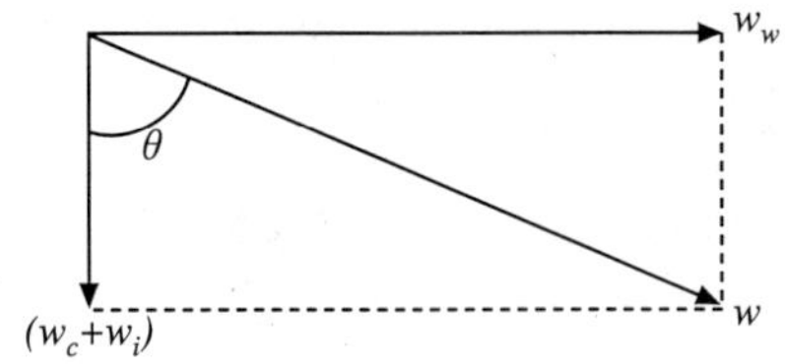

Figure 7.5 Effective conductor weight due to the presence of wind and ice.

Thus, if w_c is the weight of conductor per metre itself, w_i is the weight of ice per metre and w_w is the weight of wind per metre (shown in Figure 7.5), the resultant weight per metre is given by

$$w_r = \sqrt{(w_c + w_i)^2 + w_w^2} \tag{7.24}$$

The total sag can be worked out for this load on the conductors. To calculate the vertical sag, the component of w in the direction of $(w_c + w_i)$ and angle of w should be taken into consideration as shown in Figure 7.5.

$$\text{Vertical sag} = D\cos\theta = \frac{w_r L^2 \cos\theta}{8T} \tag{7.25}$$

where
$$\theta = \tan^{-1}\left(\frac{w_w}{w_c + w_i}\right) = \cos^{-1}\left(\frac{w_c + w_i}{w}\right) \tag{7.26}$$

EXAMPLE 7.1 A transmission line has a span of 200 m between level supports. The cross sectional area of conductor is 1.5 cm^2 and weighs 100 kg per 100 metres if breaking stress is 4500 kg/cm^2. Calculate the factor of safety if the sag of the line is 3 m. Assume a maximum wind pressure of 100 kg/m^2.

Solution: Given:

Cross sectional area of the conductor a = 1.5 cm^2. Span length = L = 200 m

Sag (= D) = 3 m. Diameter of the conductor (= d) = $\sqrt{\frac{4a}{\pi}} = \sqrt{\frac{4 \times 1.5}{\pi}} = 1.3820$ cm

Breaking stress = 4500 kg/cm^2

Weight of the conductor per metre $(= w_c) = \frac{100}{100} = 1\text{ kg}$

Wind force per metre $(= w_w) = 100 \times (1 \times 1.3820 \times 10^{-2}) = 1.3820\text{ kg}$

[$\because$ Cross sectional area of 1 m length of the conductor (1 × 1.3820 × 10^{-2}) m^2]

Resultant force per metre $(= w_r) = \sqrt{w_c^2 + w_w^2} = \sqrt{1^2 + 1.3820^2} = 1.7058\text{ kg}$

Using equation (7.13b), it can be written that

$$D = \frac{w_r L^2}{8T}$$

$$\therefore \qquad T = \frac{w_r L^2}{8D} = \frac{1.7058 \times 200^2}{8 \times 3} = 2843\text{ kg} \quad \text{(Ans.)}$$

Now, Ultimate stress = Breaking stress × Cross sectional area = $4500 \times 1.38 = 6210$ kg

[$\because$ Cross sectional area of 1 cm length of the conductor $(1 \times 1.3820)\text{ cm}^2$]

$$\therefore \text{ Factor of safety } = \frac{\text{Ultimate stress}}{\text{Working stress}} = \frac{6210}{2843} = 2.1843 \cong 2 \quad \text{(Ans.)}$$

EXAMPLE 7.2 An overhead transmission line has a span of 150 m between level supports. The conductor diameter is 1 cm and weighs 0.65 kg/m. The allowable tension is 650 kg. Determine sag when the wind pressure is 40 kg per square metre of the projected area.

Solution: Given:

Cross sectional area of the conductor (= a) = 1.5 cm^2, Span length (= L) = 150 m
Sag (= D) = 3 m, Diameter of the conductor (= d) = 1 cm, Allowable tension (= T) = 650 kg
Weight of the conductor per metre (w_c) = 0.65 kg

Wind force per metre $(= w_w) = 40 \times (1 \times 1 \times 10^{-2}) = 0.4\text{ kg}$

Resultant force per metre $(= w_r) = \sqrt{w_c^2 + w_w^2} = \sqrt{0.65^2 + 0.4^2} = 0.7632\text{ kg}$

Using equation (7.13b), sag $(D) = \frac{w_r L^2}{8T} = \frac{0.7632 \times 150^2}{8 \times 650} = 3.3023\text{ m}$ (Ans.)

Using equations (7.25) and (7.26), we obtain

$$\text{Vertical sag} = D\cos\theta = D \times \frac{w_c}{w_r} = 3.3023 \times \frac{0.65}{0.7632} = 2.8125\text{ m} \quad \text{(Ans.)}$$

EXAMPLE 7.3 An overhead transmission line has the span length of 180 m and maximum sag of 5 m. Calculate conductor loading per unit tension in the conductor. Also, find loading if the ultimate stress and conductor diameter are 4750 kg/cm^2 and 0.42 cm respectively. Consider safety factor as 5.

Solution: Given:
Span length (= L) = 180 m. Maximum sag (= D) = 5 m.
Diameter of the conductor (= d) = 0.42 cm. Ultimate stress = 4750 kg/cm^2
Factor of safety = 5
Loading per unit tension in the conductor [from equation (7.13b)] is given by.

$$\frac{w}{T} = \frac{8D}{L^2} = \frac{8\times 5}{180^2} = 1.2345\times 10^{-3}$$

$$f = \text{Working stress} = \frac{\text{Ultimate stress}}{\text{Factor of safety}} = \frac{4750}{5} = 950 \text{ kg/cm}^2$$

$$\text{Working tension} = f \times \text{Area of the conductor} = 950\times\frac{\pi(0.42)^2}{4} = 131.6170 \text{ kg}$$

$$\text{Conductor loading} = 1.2345\times 10^{-3}\times 131.6170 = 0.1625 \text{ kg/ m} \quad \text{(Ans.)}$$

EXAMPLE 7.4 An overhead line conductor at a river crossing is supported from two towers at a height of 40 m and 50 m respectively above water level. The span length is 400 m and the weight of the conductor 0.925 kg/m. Determine the minimum clearance between the conductor and water at a point midway between the towers if the tension in the conductor is 2400 kg.

Solution: Given: Total span length (= L) = 400 m,
Weight of the conductor per metre (= w_c) = 0.925 kg
Tension (= T) = 2400 kg, $h = h_2 - h_1 = 50 - 40 = 10$ m

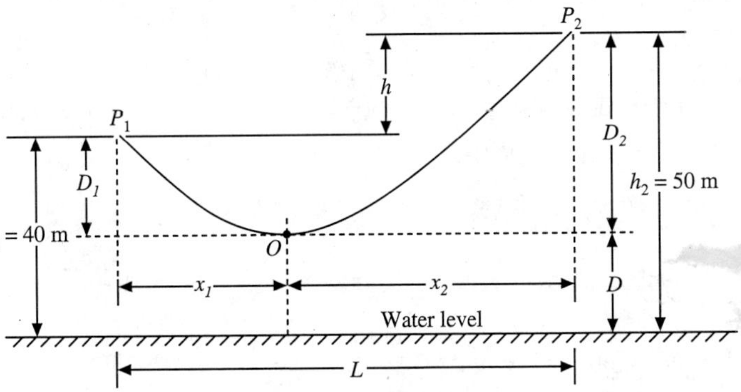

Figure E7.1

From equations (7.21a) and (7.21b),

$$x_2 = \frac{L}{2} + \frac{Th}{w_c L} = \frac{400}{2} + \frac{2400\times 10}{0.925\times 400} = 264.8649 \text{ m}$$

and

$$x_1 = \frac{L}{2} - \frac{Th}{w_c L} = \frac{400}{2} - \frac{2400\times 10}{0.925\times 400} = 135.1351 \text{ m}$$

From equation (7.17a), the height of P_1 above O is (O is the lowermost point of conductor),

$$D_1 = \frac{w_c x_1^2}{2T} = \frac{0.925 \times 135.1351^2}{2 \times 2400} = 3.5191 \text{ m}$$

From equation (7.17b), the height of P_2 above O is,

$$D_2 = \frac{w_c x_2^2}{2T} = \frac{0.925 \times 264.8649^2}{2 \times 2400} = 13.5191 \text{ m}$$

$\therefore$ Minimum clearance between the conductor and water is given by (Figure E7.1)

$$D = h_2 - D_2 = 50 - 13.5191 = 36.4809 \text{ m} \quad \text{(Ans.)}$$

[Alternatively, $D = h_1 - D_1 = 36.4809 \text{ m}$.]

EXAMPLE 7.5 A 132 kV transmission line uses ACSR conductors having a self weight of 844 kg/km. The wind force is 1.465 kg/m. If the ultimate strength of the conductor is 7950 g, the factor of safety being 2, find the sag. If the minimum ground clearance required beyond the lowest point is 7 m, determine the height at which the conductor needs to be supported. Assume the span length to be 300 m.

Solution: Given:

Weight of the conductor per metre (= w_c) = 0.844 kg

Wind force per metre (= w_w) = 1.465 kg, Ultimate strength = 7950 kg

$$\text{Resultant force per metre } (= w_r) = \sqrt{w_c^2 + w_w^2} = \sqrt{0.844^2 + 1.465^2} = 1.6907 \text{ kg}$$

$$\text{Working tension } (= T) = \frac{\text{Ultimate strength}}{\text{Factor of safety}} = \frac{7950}{2} = 3975 \text{ kg}$$

$$\text{Using equation (6.13b), sag } (= D) = \frac{w_r L^2}{8T} = \frac{1.6907 \times 300^2}{8 \times 3975} = 4.785 \text{ m}$$

Since minimum ground clearance of 7 m is needed over and above the sag, we should provide support at (7 + 4.785) m, i.e. 11.785 m above the ground. (Ans.)

7.5 STRINGING CHART

In still air, the conductor is acted on by its own weight only. If at the same time the temperature is high, the resulting sag will give a low tension. We thus find that this is the easiest operating condition for a transmission line. Compare this with the conditions arising from a low temperature (which reduces the sag) and effect of ice which increase the weight per unit length and also the projected area for wind. A high cross wind will make conditions all the more severe.

Further, the regulations insist on the factors of safety based upon a particular tension and sag occurring at specified temperature and ice and wind conditions. Obviously, these conditions are different from those at the time of erection of the line. A *stringing chart* is helpful, as it enables tension and sag to be determined at any temperature and specified loading conditions once similar figures are known at a given temperature and loading conditions.

It has already been deduced in section 7.4 that [from equations (7.12), (7.13a) and (7.16)]

$$T \cong H;\ y = \frac{wx^2}{2T};\ D = \frac{w\ell^2}{2T};\ s = \ell + \frac{w^2\ell^3}{6T^2} \tag{7.27}$$

Let w, f, s, D and t represent the load per unit length, the stress, the half span, sag and temperature, respectively. Suffix 1 will denote maximum load conditions (with ice and wind and low temperature usually –5.5°C); suffix 2 will similarly denote corresponding values under erection conditions (no ice and wind and actual temperature at the time of erection of line). Let, a = Area of cross-section of conductor; α = Coefficient of linear expansion and E =Young's modulus

Then

$$f = \frac{T}{a} \quad \text{or,} \quad s_1 = \ell + \frac{w_1^2\ell^3}{6f_1^2a^2} \tag{7.28}$$

Next consider the effect of increase of temperature from t_1 to t_2 over half span length s_1. This will result in increase in the length given by $(t_2 - t_1)\alpha s_1 = (t_2 - t_1)\alpha\ell$.

At the same time, another effect of increase of temperature from t_1 to t_2 is the reduction in stress from f_1 to f_2. The decrease in the length on this account is

$$\frac{(f_1 - f_2)}{E}s_1 = \frac{(f_1 - f_2)}{E}\ell \tag{7.29}$$

The new length s_2 is thus

$$s_2 = s_1 + (t_2 - t_1)\alpha\ell - \frac{(f_1 - f_2)}{E}\ell \tag{7.30}$$

We may now write an expression for s_2 similar to that given by equation (7.27).

$$s_2 = \ell + \frac{w_2^2\ell^3}{6f_2^2a^2} \tag{7.31}$$

Substituting the values of s_1 in equation (7.30) and equating with expression in equation (7.31), we have

$$\ell + \frac{w_1^2\ell^3}{6f_1^2a^2} + (t_2 - t_1)\alpha\ell - \frac{(f_1 - f_2)}{E}\ell = \ell + \frac{w_2^2\ell^3}{6f_2^2a^2}$$

or,

$$\frac{w_1^2\ell^2}{6f_1^2a^2} + (t_2 - t_1)\alpha - \frac{(f_1 - f_2)}{E} = \frac{w_2^2\ell^2}{6f_2^2a^2}$$

or,

$$\frac{w_1^2\ell^2E}{6f_1^2a^2} + (t_2 - t_1)\alpha E + (f_2 - f_1) = \frac{w_2^2\ell^2E}{6f_2^2a^2}$$

$\therefore$

$$f_2^2\left[(f_2 - f_1) + (t_2 - t_1)\alpha E + \frac{w_1^2\ell^2E}{6f_1^2a^2}\right] = \frac{w_2^2\ell^2E}{6a^2} \tag{7.32}$$

Equation (7.32) is a cubic equation of f_2 and D_2 can be determined from equation (7.5) as

$$D_2 = \frac{w_2 \ell^2}{2T_2} = \frac{w_2 \ell^2}{2 f_2 a} \tag{7.33}$$

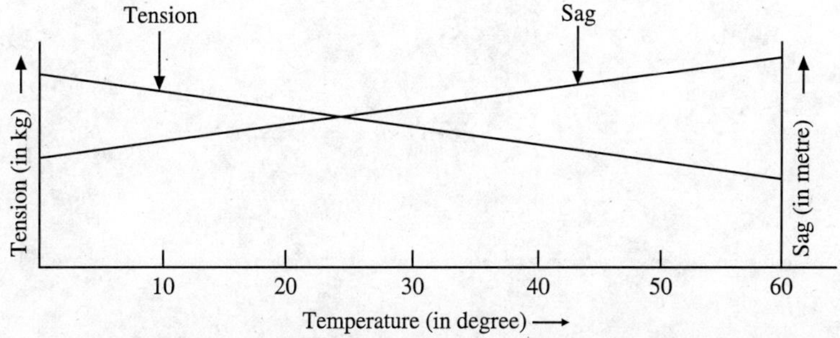

Figure 7.6 Sag versus temperature chart (Stringing chart).

Thus, it is possible to plot a graph belween temperature (independent variable) and sag (dependent variable). Also, f_2 can be plotted (as dependent variable) against temperature (independent variable). Figure 7.6 shows the plot of tension versus temperature and sag versus temperature. This is called the *stringing chart*.

EXERCISES

1. Discuss different types of supports used in overhead power transmission.
2. What are the different types of conductor materials employed for overhead transmission of electrical power?
3. Prove that a transmission line conductor between two supports at equal heights from ground takes the form of a catenary.
4. Deduce the expressions for (i) total length of conductor, (ii) tension at ends in terms of span length, horizontal tension, maximum sag and weight of the conductor per unit length.
5. Derive the expression for sag of an overhead line if supports are at different levels of heights from the ground.
6. How the effects of wind and ice loading are taken into account while determining sag and stress of an overhead line conductor?
7. How are the sag and tension of an overhead line conductor affected by an extreme fall of temperature?
8. What is stringing chart? Why it is necessary for an overhead transmission line? How such a chart can be constructed?

CHAPTER 8

Corona

8.1 MECHANISM OF CORONA

A small discharge current is created in the electric field between two parallel conductors of a transmission line with high potential gradient. Beyond a certain critical voltage gradient (typically 30 kV/cm), electric breakdown of the dielectric medium occurs between the two conductors, establishing an arc between these two conductors accompanied by a violet luminous glow associated with a hissing sound around the conductor. The brightest part of the glow is visible near the rough surface of the conductor. This phenomenon is known as *corona*. Corona produces power loss and is dependent upon weather condition. During humid and moist climate, corona loss is higher. It should be noted that if the spacing between conductors is not large with respect to their diameter, spark-over may take place before corona is observed. Corona discharge is further prominent as the electric field intensity goes up. In EHV AC transmission lines, the visible glow of corona is uniform about both the conductors whereas in HVDC transmission system, the glow is uniform about the positive wire but spotty about the negative wire.

8.2 CONCEPT OF ELECTRIC STRESS FOR CORONA DISCHARGE

Consider a single phase system as shown in Figure 8.1. The electric force or stress (g) at any point P is given by

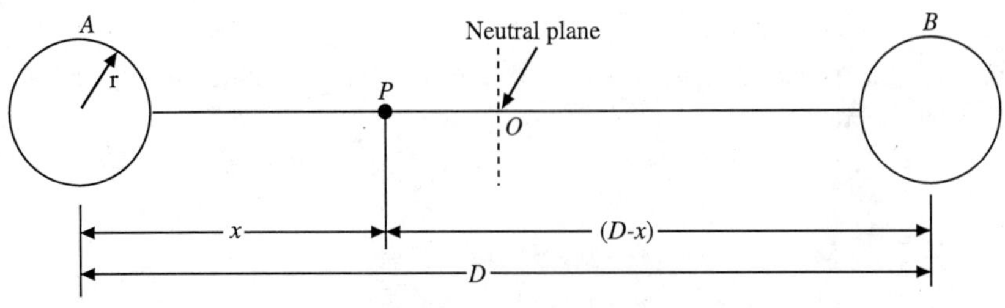

Figure 8.1 A single phase system, consisting of two conductors (A and B).

$$g = \frac{1}{2\pi\varepsilon}\left[\frac{q}{x} + \frac{q}{D-x}\right] \tag{8.1}$$

Therefore, the potential difference between the two conductors is

$$V = \frac{q}{2\pi\varepsilon}\left[\int_r^D \frac{1}{x} + \int_r^{D-r} \frac{1}{D-x}\right]$$

$$= \frac{q}{2\pi\varepsilon}\left[\int_r^D \frac{1}{x} + \int_r^{D-r} \frac{1}{D-x}\right]$$

$$= \frac{q}{2\pi\varepsilon}\left[\ln\frac{D}{r} - \ln\frac{r}{D}\right] = \frac{q}{2\pi\varepsilon} \times 2\ln\frac{D}{r}$$

or,

$$V = \frac{q}{\pi\varepsilon}\ln\frac{D}{r} \tag{8.2a}$$

$$q = \frac{\pi\varepsilon V}{\ln(D/r)} \tag{8.3}$$

Since the gradient is maximum at the surface of the conductor, the value of g will be minimum for $x = D/2$ and maximum for $x = r$. The maximum value of g is then given by

$$g_{max} = \frac{1}{2\pi\varepsilon}\left[\frac{q}{r} + \frac{q}{D-r}\right] \tag{8.4a}$$

Since D is very large in comparison to r, we may write approximately

$$\frac{q}{D-r} \cong \frac{q}{D} \text{ and } \frac{q}{D} \ll \frac{q}{r}$$

Therefore, equation (8.4a) can be written as

$$g_{max} \cong \frac{q}{2\pi\varepsilon r} \quad \left[\frac{q}{d} \text{ is neglected.}\right] \tag{8.4b}$$

The substitution of q from equation (8.3) to equation (8.4b) yields

$$g_{max} \cong \frac{V}{2r\ln(D/r)} \tag{8.4c}$$

Assuming E_0 be the voltage of each conductor with respect to neutral, we can write

$$V = 2E_0 \tag{8.2b}$$

$$\text{and } g_{max} \cong \frac{2E_0}{2r\ln(D/r)} \tag{8.4d}$$

The same relationship holds true for a 3-phase symmetrically spaced system also. It is to be noted that actually the gradient at the surface of a conductor is maximum at the inner point varying round the circumference. However, if (D/r) is very large, the two gradients may differ slightly.

8.3 CORONA DISCHARGE AND CRITICAL VOLTAGES

With a high operating voltage, the surface stress of a conductor causes the breakdown of surrounding air and produces corona. The occurrence of corona virtually increases the conductor diameter and as a consequence of this, there is a decrease in maximum stress.

The dielectric strength of air at barometric pressure of 76 cm of mercury and 25°C is 30 kV per cm or 21.1 kV r.m.s. per cm. If this is denoted by g_0, the dielectric strength of air at any other temperature t°C and barometric pressure b cm of mercury is δg_0 where

$$\delta = \frac{b}{76}\left(\frac{273+25}{273+t}\right) = \frac{3.92b}{273+t} \tag{8.5}$$

The *disruptive voltage gradient* of air is attained at the conductor surface when

$$g_0 = \frac{E_0}{r \ln (D/r)}$$

i.e.

$$E_0 = g_0 r \ln (D/r) \tag{8.6}$$

Allowing for surface irregularity factor of the wire, m, and air density factor δ, complete expression for disruptive critical voltage, E_0 is obtained as

$$m\delta g_0\, r \ln (D/r)\, E_0 = m\delta g_0\, r \ln (D/r) \text{ (to neutral)} \tag{8.7}$$

With

$$g_0 = 21.1 \text{ kV (r.m.s.)},$$

$$E_0 = 21.1\, m\delta r \ln (D/r) \text{ kV to neutral (r.m.s.)} \tag{8.8}$$

where δ = air density factor

D = distance between conductors in cm

r = radius of conductors in cm

m = surface roughness (irregulatory) factor

For stranded conductors, the value of m may be taken in the range of 0.8 to 0.85 in the above expression. For solid conductor, it is 1.0 while for polished conductors m is 0.95–0.98. Voltage E_0 is the *disruptive critical voltage* and holds for fair weather condition. In the case of rains and bad atmospheric conditions, the value of E_0 is considerably decreased.

It may be of interest to note that in the case of parallel wires, the visual corona does not begin at voltage E_0 but at a higher voltage E_v, called the *visual critical voltage*. The visual corona will occur when the breakdown value is attained at a distance $\left[r\left(1+0.3/\sqrt{\delta r}\right)\right]$ from the axis instead of r. This requires that voltage to neutral should be $\left[1+0.3/\sqrt{\delta r}\right]$ times the disruptive critical voltage.

Complete expression for visual critical voltage E_v (to neutral) is given by

$$E_v = 21.1\, m_v \delta r \left(1 + \frac{0.3}{\sqrt{\delta r}}\right) \ln (D/r)\, \text{kV to neutral (r.m.s.)} \tag{8.9}$$

[Authors like Peek stated that m_v (which is the roughness factor) may be taken as 0.72 for stranded wires for *local visual corona* while he recommends a figure of 0.82 for *general visual corona.*]

EXAMPLE 8.1 A three phase line has conductor 4 cm in diameter and is spaced equilaterally 1m apart. If air density factor is 0.962 and irregularity factor of conductor (m) is 0.8, find the disruptive critical voltage for the line.

Solution: Given: $\delta = 0.962$, $m = 0.8$, $r = 2$ cm, $D = 100$ cm

From equation (8.8), disruptive critical voltage is given by

$E_0 = 21.1\, m\delta r \ln (D/r) = 21.1 \times 0.8 \times 0.962 \times 2 \ln (100/2) = 127.0512$ kV/ph. (Ans.)

8.4 POWER LOSS DUE TO CORONA

When the surface voltage gradient at the line conductor exceeds the critical breakdown stress, corona appears and energy is dissipated in the form of light and heat. This is known as *corona power loss* (or simply corona loss) and this is derived from the power lines, reducing the transmission energy. Corona loss in an EHV transmission line can vary between a few kW/km in fair weather to hundred of kW/km during thunderstorm or bad weather conditions.

Under fair weather conditions, *power loss due to corona* in kW per phase per km (*Peek's formula*) is given by

$$P_c = \frac{244\,(f + 25)}{\delta} \sqrt{\frac{r}{D}}\,(E - E_0)^2 \times 10^{-5} \tag{8.10}$$

where f = frequency in Hz and E = voltage in kV to neutral (r.m.s.).

Factors like rain, sleet and fog may reduce disruptive critical voltage (E_0) to $0.8E_0$, with the result that corona loss (in kW per phase per km) under storm conditions is given by

$$= \frac{244\,(f + 25)}{\delta} \sqrt{\frac{r}{D}}\,(E - 0.8E_0)^2 \times 10^{-5} \tag{8.11}$$

Peek's formula is useful only when the corona losses are predominant and it is not applicable when the losses are low. Peterson's formula (another widely used empirical formula to determine corona loss) holds well when the ratio (E/E_0) is less than 1.8. *Peterson's formula* states that the power loss (P_c) is given by

$$P_c = \frac{21 \times 10^{-6} fE^2}{\left(\log_{10} \dfrac{D}{r}\right)^2} \times F \quad \text{(in kW per phase per km)} \tag{8.12}$$

where F is the factor which varies with the ratio E/E_0 as given below:

E/E_0	0.6	0.8	1.0	1.2	1.4	1.6	1.8	2.0	2.2
F	0.012	0.018	0.05	0.08	0.3	1.0	3.0	6.0	8.0

Corona loss in 3-phase EHV (≥ 400 kV) lines in rainy weather is given by

$$P'_{c(3\phi)} = P_{c(3\phi)} + \frac{E}{\sqrt{3}} Jr^2 \ln(1+KR) \sum_{i=1}^{m} E_i^{\rho} \text{ kW/km} \tag{8.13}$$

where, $P_{c(3\phi)}$ = 3-phase corona loss in fair weather, J (= Constant) = 4.37×10^{-10} at 400 kV, R = Rain rate in mm/hr, K = Wetting coefficient (= 10), i = subconductor, m = total number of subconductors, E_i = Voltage gradient on outer conductor (kV/cm), ρ = an exponent (usually 5).

EXAMPLE 8.2 A three phase overhead transmission line consisting of three stranded copper conductors is spread 2.65 metres apart at the corners of an equilateral triangle. Determine the disruptive critical voltage and the visual critical voltages for local and general corona when air temperature and pressure are 25°C and 76 cm of mercury, respectively. Conductor diameter is 1 cm. Take irregularity factor (m) = 0.85 and surface factors for local and general corona = 0.7 and 0.8, respectively. Breakdown strength of air is 21.1 kV (rms) per cm.

Solution: Given m = 0.85, m_v (for local corona) = 0.7, m_v (for general corona) = 0.8, r = 0.5 cm, D = 2.65 m = 265 cm, b = 76, t = 25°C, length of the line, g = 21.1 kV.
Using equation (8.5), we have

$$\delta = \frac{3.92b}{273+t} = \frac{3.92 \times 76}{273+25} = 0.9997$$

Disruptive critical voltage is obtained in equation (8.7) and is given by,

$$E_0 = gm\delta r \ln(D/r) = 21.1 \times 0.85 \times 0.9997 \times 0.5 \ln(265/0.5)$$

$$= 56.2351 \text{ kV/ph (r.m.s.)}.$$

Using equation (8.9), disruptive critical voltage for local visual corona is obtained as

$$E_{v\,(\text{local})} = gm_v\delta r\left(1+\frac{0.3}{\sqrt{\delta r}}\right)\ln(D/r)$$

$$= 21.1 \times 0.7 \times 0.9997 \times 0.5\left(1+\frac{0.3}{\sqrt{0.9997 \times 0.5}}\right)\ln(265/0.5)$$

[m_v = 0.7 for local visual corona]

$$= 65.9625 \text{ kV/ph (r.m.s.)}.$$

Also, using equation (8.9), disruptive critical voltage for general corona is obtained as

$$E_{v\,(\text{general})} = gm_v\delta r\left(1+\frac{0.3}{\sqrt{\delta r}}\right)\ln(D/r)$$

$$= 21.1 \times 0.8 \times 0.9997 \times 0.5\left(1+\frac{0.3}{\sqrt{0.9997 \times 0.5}}\right)\ln(265/0.5)$$

[m_v = 0.8 for general corona]

$$= 75.3857 \text{ kV/ph (r.m.s.)}.$$

EXAMPLE 8.3 Calculate the disruptive critical voltage and critical voltages for visual corona of a 3 phase, 170 km long line consisting of three stranded copper (7/4.75) mm, area 120 mm^2, spaced in a 4 m delta arrangement. Temperature of ambience is 30°C and barometric pressure is 73 cm. This line operates on 220 kV at 50 Hz. Calculate fair weather and storm corona power loss for the line. Assume additional data if required.

Solution: Here we consider that a seven strand conductor will have strands in two layers–one and six respectively as shown in Figure E8.1.

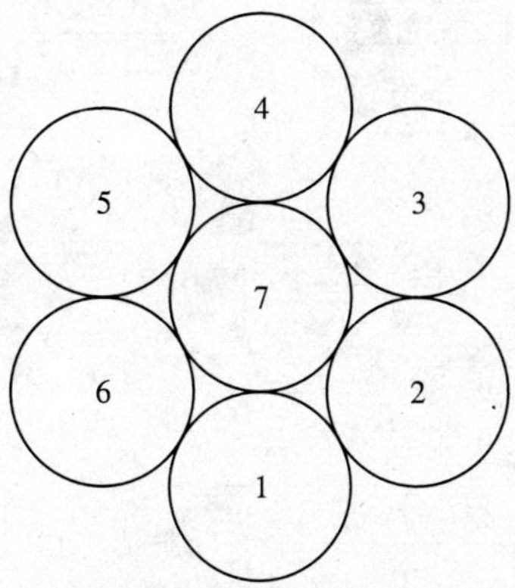

Figure E8.1 Seven strand conductor in each of the three stranded line conductors.

Diameter of the conductor = 3 × 4.75 = 14.25 mm

$$r = \frac{3 \times 4.75}{2} = \frac{14.25}{2} = 0.7125 \text{ cm, D} = 4 \text{ m} = 400 \text{ cm, } b = 73,\ t = 30°\text{C},$$

Length of the line, λ = 170 km

We assume that, g = 21.1 kV, m = 0.85, m_v (for local corona) = 0.72,

m_v (for general corona) = 0.82.

Using equation (8.5), we have

$$\delta = \frac{3.92b}{273 + t} = \frac{3.92 \times 73}{273 + 30} = 0.9444$$

Disruptive critical voltage is obtained in equation (8.7) and is given by,

$$E_0 = gm\delta r \ln (D/r) = 21.1 \times 0.85 \times 0.9444 \times 0.7125 \ln (400 / 0.7125)$$

$$= 76.3970 \text{ kV/ph (r.m.s.)}.$$

Using equation (8.9), disruptive critical voltage for local visual corona is obtained as

$$E_{v(\text{local})} = gm_v\delta r \left(1 + \frac{0.3}{\sqrt{\delta r}}\right) \ln (D/r)$$

$$= 21.1 \times 0.72 \times 0.9444 \times 0.7125 \left(1 + \frac{0.3}{\sqrt{0.9444 \times 0.7125}}\right) \ln (400 / 0.7125)$$

$$= 88.3796 \text{ kV/ph (r.m.s.).}$$

Also, using equation (8.9), disruptive critical voltage for general corona is obtained as

$$E_{v\,(\text{general})} = gm_v \delta r \left(1 + \frac{0.3}{\sqrt{\delta r}}\right) \ln (D/r)$$

$$= 21.1 \times 0.82 \times 0.9444 \times 0.7125 \left(1 + \frac{0.3}{\sqrt{0.9444 \times 0.7125}}\right) \ln (400 / 0.7125)$$

$$= 100.6545 \text{ kV/ph (r.m.s.).}$$

Now, phase to neutral rms voltage applied to the conductor is given by,

$$E = \frac{220}{\sqrt{3}} = 127.017 \text{ kV}$$

Let us calculate coronal loss using Peek's formula as follows:

(i) *For fair weather* [equation (8.10)]

$$P_c = \frac{244\,(f + 25)}{\delta} \sqrt{\frac{r}{D}} (E - E_0)^2 \times 10^{-5}$$

$$= \frac{244\,(50 + 25)}{0.9444} \sqrt{\frac{0.7125}{400}} (127.0171 - 76.3970)^2 \times 10^{-5}$$

$$= 20.9558 \text{ kW per km per phase}$$

$\therefore$ Total power loss due to corona, $P_{c\,(\text{total})} = 3 \times P_c \times \lambda$ [λ = length of the line]

$= 3 \times 20.9558 \times 170 = 10.6874$ MW (Ans.)

(ii) *For stormy weather* [equation (8.11)]

$$P_{c\,(st)} = \frac{244\,(f + 25)}{\delta} \sqrt{\frac{r}{D}} (E - 0.8E_0)^2 \times 10^{-5}$$

$$= \frac{244\,(50 + 25)}{0.9444} \sqrt{\frac{0.7125}{400}} (127.0171 - 0.8 \times 76.3970)^2 \times 10^{-5}$$

$$= 35.5158 \text{ kW per km per phase}$$

$\therefore$ Total power loss due to corona, $P_{c\,(\text{total})} = 3 \times P_{c\,(st)} \times \lambda$

$= 3 \times 35.5158 \times 170 = 18.1131$ MW (Ans.)

EXAMPLE 8.4 Find the corona characteristics of a 110 kV, 50 Hz, three phase, 200 km long transmission line consisting of three stranded copper conductors spaced with 2.5 m delta arrangement. Ambient temperature is 28°C and barometric pressure is 75 cm of mercury; $m = 0.85$, m_v for local corona = 0.72 and m_v for general corona = 0.82. The diameter of each stranded conductor is 1 cm while the spacing between any two of them is 2.5 metre.

Solution: Given m = 0.85, m_v (for local corona) = 0.72, m_v (for general corona) = 0.82, r = 0.5 cm, D (= 2.5 m) = 250 cm, b = 75, t = 28°C, length of the line, λ = 200 km.
Using equation (8.5), we have

$$\delta = \frac{3.92 \times 75}{273 + 28} = 0.9767$$

Determination of disruptive critical voltage (E_0)
Disruptive critical voltage is obtained in equation (8.8) and is given by,

$$E_0 = 21.1 m\delta r \ln (D/r)$$

$$= 21.1 \times 0.85 \times 0.9767 \times 0.5 \ln (250/0.5) \quad [\because D = 2.5 \text{ m} = 250 \text{ cm}]$$

$$= 54.4310 \text{ kV to neutral (r.m.s.)} \qquad \text{(i)}$$

Using equation (8.9), disruptive critical voltage for local visual corona is obtained as

$$E_{v1} = 21.1\, m_v \delta r \left(1 + \frac{0.3}{\sqrt{\delta r}}\right) \ln (D/r)$$

$$= 21.1 \times 0.72 \times 0.9767 \times 0.5 \left(1 + \frac{0.3}{\sqrt{0.9767 \times 0.5}}\right) \ln (250/0.5)$$

$$= 65.8994 \text{ kV to neutral (r.m.s.)} \qquad \text{(ii)}$$

Also, using equation (8.9), disruptive critical voltage for general visual corona is obtained as

$$E_{v2} = 21.1\, m_v \delta r \left(1 + \frac{0.3}{\sqrt{\delta r}}\right) \ln (D/r)$$

$$= 21.1 \times 0.82 \times 0.9767 \times 0.5 \left(1 + \frac{0.3}{\sqrt{0.9767 \times 0.5}}\right) \ln (250/0.5)$$

$$= 75.0521 \text{ kV to neutral (r.m.s.)} \qquad \text{(iii)}$$

Now, phase to neutral rms voltage applied to the conductor is,

$$E = \frac{110}{\sqrt{3}} = 63.5085 \text{ kV} \qquad \text{(iv)}$$

Comparing equations (i), (ii), (iii) and (iv), it can be written that

$$E_{v2} > E_{v1} > E > E_0$$

Therefore, corona will take place but it will not be visual as E_{v1} and $E_{v1} > E$.
Let us calculate coronal loss using Peek's formula as follows:

(i) Using equation (8.10), for fair weather condition

$$P_c = \frac{244\,(f + 25)}{\delta}\sqrt{\frac{r}{D}}(E - E_0)^2 \times 10^{-5}$$

$$= \frac{244\,(50 + 25)}{0.9767}\sqrt{\frac{0.5}{250}}(63.5085 - 54.4310)^2 \times 10^{-5}$$

$$= 0.6905 \text{ kW per km per phase}$$

Total power loss due to corona, $P_{c(\text{total})} = 3 \times P_c \times \lambda$ [λ = length of the line]

$$= 3 \times 0.6905 \times 200 = 414.3 \text{ kW (fair weather condition)}$$

(ii) Using equation (8.11), for stormy weather condition

$$P_c = \frac{244\,(f+25)}{\delta}\sqrt{\frac{r}{D}}(E - 0.8E_0)^2 \times 10^{-5}$$

$$= \frac{244\,(50+25)}{0.9767}\sqrt{\frac{0.5}{250}}(63.5085 - 0.8 \times 54.4310)^2 \times 10^{-5}$$

$$= 3.3395 \text{ kW per km per phase}$$

Total power loss due to corona, $P_{c(\text{total})} = 3 \times P_c' \times \lambda$

$$= 3 \times 3.3395 \times 200 = 2003.7 \text{ kW (stormy weather condition)}$$

Let us calculate coronal loss using Peterson's formula as follows:

(i) *For fair weather*

Here $$\frac{E}{E_0} = \frac{63.5085}{54.4310} = 1.1668$$

Using the method of interpolation, we can write from the table given in the text

$$\frac{1.1668 - 1.0}{F - 0.05} = \frac{1.2 - 1.1668}{0.08 - F}$$

$$F = \frac{(1.1668 - 1.0) \times 0.08 + (1.2 - 1.1668) \times 0.05}{1.2 - 1.0} = 0.075$$

Now using equation (8.12),

$$P_c = \frac{21 \times 10^{-6} fE^2}{\left(\log_{10}\dfrac{D}{r}\right)^2} \times F = \frac{21 \times 10^{-6} \times 50 \times 63.5085^2}{\left(\log_{10}\dfrac{250}{0.5}\right)^2} \times 0.075$$

$$= 0.0436 \text{ (in kW per phase per km)}$$

$\therefore$ Total power loss due to corona, $P_{c(\text{total})} = 3 \times P_c \times \lambda$

$$= 3 \times 0.0436 \times 200 = 26.16 \text{ kW (fair weather condition)}$$

(ii) *For stormy weather*

Here $$\frac{E}{E_0} = \frac{63.5085}{0.8 \times 54.4310} = 1.4585$$

Using the method of interpolation, we can write from the table given in the text

$$\frac{1.4585 - 1.4}{F - 0.3} = \frac{1.6 - 1.4585}{1.0 - F}$$

$$F = \frac{(1.4585 - 1.4) \times 1.0 + (1.6 - 1.4585) \times 0.3}{1.6 - 1.4} = 0.5048$$

Now using equation (8.12),

$$P'_c = \frac{21\times10^{-6} fE^2}{\left(\log_{10}\frac{D}{r}\right)^2}\times F = \frac{21\times10^{-6}\times50\times63.5085^2}{\left(\log_{10}\frac{250}{0.5}\right)^2}\times0.5048$$

$$= 0.2935 \text{ (in kW per phase per km)}$$

Total power loss due to corona, $P_{c(\text{total})} = 3\times P'_c\times\lambda$

$$= 3\times0.2935\times200 = 176.1 \text{ kW (in stormy weather condition)}$$

Peek's formula and Peterson's formula differ in results as the losses in insulators have not been accounted for in Peterson's formula.

EXAMPLE 8.5 A certain three phase equilateral transmission line has a total corona loss of 48 kW at 105 kV and a loss of 125 kW at 115 kV. What will be disruptive critical voltage and the corona loss at 110 kV?

Solution: Given:

For the first case, corona loss = P_{c1} = 55 kW

System voltage (per phase), $E_1 = \frac{105}{\sqrt{3}} = 60.6218$ kV

For the second case, corona loss = P_{c2} = 100 kW

System voltage (per phase), $E_2 = \frac{115}{\sqrt{3}} = 66.3953$ kV

Using Peek's formula for fair weather as given in equation (8.10), we know corona loss is given by

$$P_c = \frac{244\,(f+25)}{\delta}\sqrt{\frac{r}{D}}(E-E_0)^2\times10^{-5} \text{ kW per km per phase}$$

Therefore, if all the things remain unchanged, except system voltage, we can write

$$P_c \propto (E-E_0)^2$$

$$\frac{P_{c1}}{P_{c2}} = \frac{(E_1-E_0)^2}{(E_2-E_0)^2}$$

or,
$$\sqrt{\frac{P_{c1}}{P_{c2}}} = \frac{E_1-E_0}{E_2-E_0}$$

or,
$$E_2\sqrt{\frac{P_{c1}}{P_{c2}}} - E_0\sqrt{\frac{P_{c1}}{P_{c2}}} = E_1 - E_0$$

or,
$$E_0\left(1-\sqrt{\frac{P_{c1}}{P_{c2}}}\right) = E_1 - E_2\sqrt{\frac{P_{c1}}{P_{c2}}}$$

or,
$$E_0 = \frac{E_1 - E_2\sqrt{\frac{P_{c1}}{P_{c2}}}}{\left(1-\sqrt{\frac{P_{c1}}{P_{c2}}}\right)} = \frac{60.6218 - 66.3953\sqrt{\frac{48}{125}}}{1-\sqrt{\frac{48}{125}}}$$

$$E_0 = 51.2148 \text{ kV (line to neutral)}.$$

Let the coronal loss is P_c at 110 kV.

In this case, system voltage (per phase), $E = \frac{110}{\sqrt{3}} = 63.5085 \text{ kV}$

$$\frac{P_c}{P_{c2}} = \frac{(E - E_0)^2}{(E_2 - E_0)^2}$$

or,
$$P_c = \frac{(E-E_0)^2}{(E_2 - E_0)^2} \times P_{c2} = \frac{(63.5085 - 51.2148)^2}{(66.3953 - 51.2148)^2} \times 125$$

$$P_c = 81.9791 \text{ kW}.$$

8.5 FACTORS AFFECTING CORONA LOSS

1. *Effect of frequency:* Corona loss is directly proportional to system frequency.
2. *Effect of system voltage:* With increase in system voltage, the electric field increases and hence corona loss also increases in a steep way. Figure 8.2 represents the profile of a corona loss curve with change in system voltage.
3. *Effect of conductivity of air:* If the number of ions in air increases, the conductivity of air also increases. During rain and thunderstorms, ion content in the air enhances and thus corona loss becomes high during bad weather conditions.
4. *Effect of conductor diameter:* The electric field intensity reduces with increase in conductor size (i.e. diameter). Hence, with greater conductor diameters, the electric field intensity reduces, resulting in lower corona loss.
5. *Effect of load current:* The flow of the load current increases the temperature of the conductor. Thus, it prevents deposition of dew or snow on the conductor's surface. This reduces corona loss.
6. *Effect of conductor's surface:* Roughness of the conductor's surface results in field distortion and gives rise to high potential gradient, causing higher corona loss.
7. *Effect of atmospheric condition:* Rain drop and dust deposition increase corona loss, particularly local corona discharge. If the density of air decreases, corona loss increases. Figure 8.3 represents the effect of rain on corona loss.

8.6 RADIO INTERFERENCE OF CORONA

Wireless signals are adversely affected by corona discharges. Amplitude modulated broadcasting, carrier signal, aviation, navigation, etc. are affected due to interference of corona discharge on communication channel. The effects of corona discharge may create interference to communication line within even few kilometres. The corona interference is extremely high above a value of particular system voltage. A transmission line with smooth surface in good weather condition may cause little radio interference problem. The radio interference field created from transmission line conductors varies as the inverse of radio frequency. Hence, high frequency transmissions like television, radar communication, microwave channel, FM broadcasting, etc. are less likely to be affected. The radio interference due to corona is of importance for lines above 200 kV. Line design is thus largely affected due to corona.

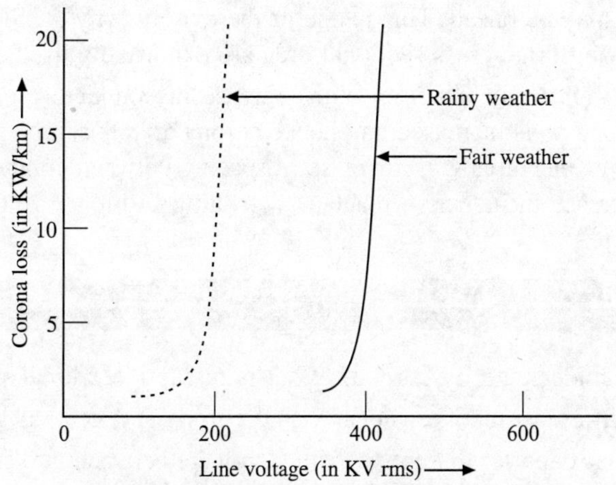

Figure 8.2 Effect of line voltage on corona loss.

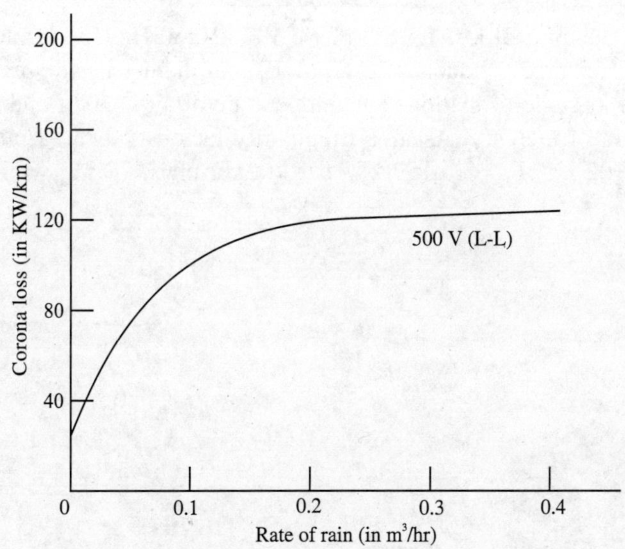

Figure 8.3 Effect of rate of rain on corona loss.

8.7 METHODS OF REDUCING CORONA

Intense corona effects are observed even at a working voltage of 33 kV or above. Therefore, design should be made carefully to avoid corona on the sub-stations or bus bars rated for 33 kV and higher voltage. Otherwise, highly ionised air may cause flash over in the insulator or between the phases, causing considerable damage to equipment. The corona effect may be reduced by the following methods:

1. *By increasing conductor size:* With increase in the conductor size (diameter), the potential gradient decreases. By increasing the conductor size, the voltage at which corona occurs may be raised. This is one of the reasons why ACSR conductor (bundled conductor) having large cross sectional area are extensively used in transmission lines.
2. *By increasing conductor spacing:* With increase in conductor spacing, the voltage at which corona occurs is increased and hence corona effects can be eliminated. However, the cost of tower increases with increase in spacing between conductors and as a result, there is an increase in inductive reactance resulting in higher voltage drop in the line.

EXERCISES

1. What do you understand by corona? What is disruptive critical voltage?
2. What is the difference between local visual corona and general visual corona?
3. Derive the expression for disruptive critical voltage between two smooth circular wires.
4. Discuss the factor which affects disruptive critical voltage of a transmission line.
5. How the power loss due to corona can be calculated? Discuss how it can be minimized.
6. Discuss the factor which affects corona loss.
7. Calculate the coronal loss for a 3 phase 132 kV, 50 Hz, 150 km long transmission line consisting of three conductors, each of 12 mm diameter and spread 2.5 m apart in an equilateral triangle formation. The temperature of air is 30°C and atmospheric pressure of 750 mm of mercury. Assume irregularity factor as 0.85. [*Hints:* E_0 = 62.9853 kV, E = 76.2102 kV, P_c = 1.6159 kW per km per phase, $P_{c(\text{total})}$ = 727.155 kW.]

CHAPTER 9

Transient Analysis and Wave Propagation in Transmission Lines

9.1 INTRODUCTION

When a single circuit transmission line is connected to a voltage source at $t = 0$, the whole length of the line is not energized all at once; there must be a time lag before the disturbance initiated at one point can be observed at the distant point in the line. This is due to the presence of distributed constants (inductance and capacitance in a lossfree line).The process is equivalent to launching a voltage wave, which travels along the length of the line at a certain velocity (almost equal to the velocity of light) and reaches the distant end after a period of time. The voltage wave is usually accompanied by a current wave. These two waves, which are related to each other by the characteristic impedance or resistance of the line, must reach the distant end in a finite time and, in general, must meet a discontinuity as a result of which total or partial reflection will take place. Then, for a finite line, the voltage and current at any point (including the ends) is the result of superimposed incident and reflected waves. In this chapter, we shall study the propagation of transient voltage and current waves along the transmission lines which may be caused by switching or lightning and which may subject the electrical systems to dangerous over voltages.

9.2 APPROXIMATE REPRESENTATION OF A TRANSMISSION LINE IN TRANSIENT STUDY

A transmission line can be regarded as cascade made up of a number of elements of short lengths as shown in Figure 9.1. In this figure, $r\Delta x$ represents the series resistance of an element of length Δx of a transmission line, $\ell\Delta x$ the series inductance, $g\Delta x$ the shunt conductance (to take into account the leakage of current from one conductor to the reference) and $c\Delta x$ the shunt capacitance. Although the inductance and resistance are shown in lumped form in one conductor in the figure, they actually represent the net effect of both conductors in the short element.

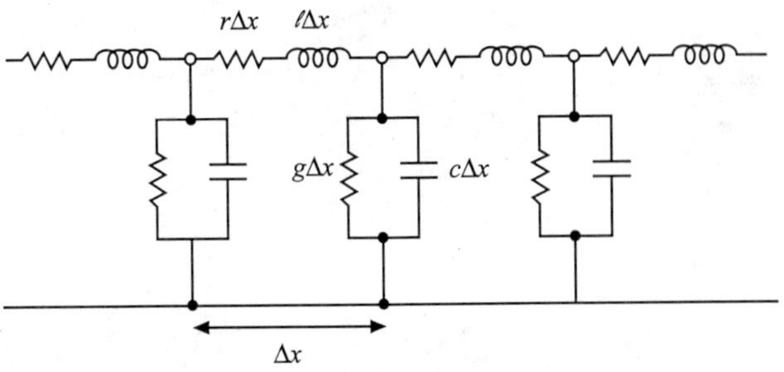

Figure 9.1 Approximate representation of a transmission line.

9.3 CONCEPT OF INFINITE LINE

Before going into the details of theory of transmission lines, the concept of the *infinite line* should be made clear.

Let the propagation of electricity along any uniform and symmetrical transmission line of infinite length having electrical constants per unit length be identical to those of the line under construction.

The infinite line will have the following important characteristics:

(a) Since the line is of infinite length, no electric wave will ever reach the distant end. Then, there will not be any possibility of reflection of electrical energy at the distant end and consequently reflected wave also.

(b) The voltage being applied at the sending end, the current will be depending only on the characteristics impedance Z_0, and will not be affected by the terminating impedance Z_R at the remote end. Considering the infinite line, the impedance will be Z_0 looking from input terminals. Even if a short section of the line at the end of the line is removed, the line will now start from the cut section and will again be Z_0. The removal of the short section does not affect in any way the infinite nature of the line. The currents and voltages at all the points along the length of the line would be exactly the same as if it were terminated in an infinite length of the line.

Therefore, from the logical point of view, it follows that a short line terminated in Z_0 behaves electrically, at all points along its length, as if it were an infinite line.

When an alternating voltage is applied at the sending end of an infinite line, a finite current flows (due to the capacitance effect and the leakage conductance) between the two wires constituting the transmission. The ratio of the voltage applied to the currents flowing gives the input impedance of the line (commonly termed the characteristic impedance Z_0, a vector quantity having a modulus $|Z_0|$). The magnitude of Z_0 for a open wire line usually lies between 200 ohms and 1600 ohms with an angle between (–45°) and (0°). The modulus and angle of the characteristic impedance will be dependent on frequency, and the frequency at which the impedance has been measured should be stated.

9.4 EXPRESSION OF CHARACTERISTICS IMPEDANCE OF A SHORT LINE IN TERMS OF OPEN CIRCUIT AND SHORT CIRCUIT PARAMETERS

Figure 9.2(a) shows an infinite line with input terminals *ab*. By definition, the impedance looking into the terminals will be Z_0. If a short section of line be removed from the infinite line, as shown in Figure 9.2(b), the impedance looking through terminals *c* and *d* will still be Z_0. Hence, the removed short section (say *AB*) may be assumed to be terminated in impedance Z_0 at *B* having the current and voltage at all points along its length to be exactly the same as if it were terminated in an infinite line. This type of line and termination is popularly known as *short line terminated* in Z_0 (Figure 9.2(c)).

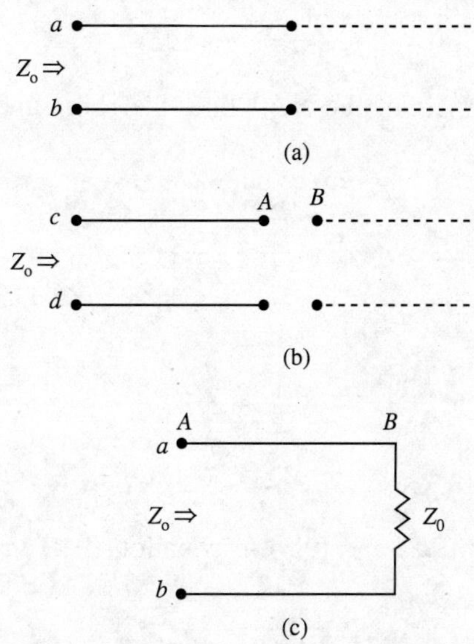

Figure 9.2 Short line terminated in Z_0.

The short line, being a complex electrical network, may be represented by a *T* section having input impedance Z_0, as shown in Figure 9.3.

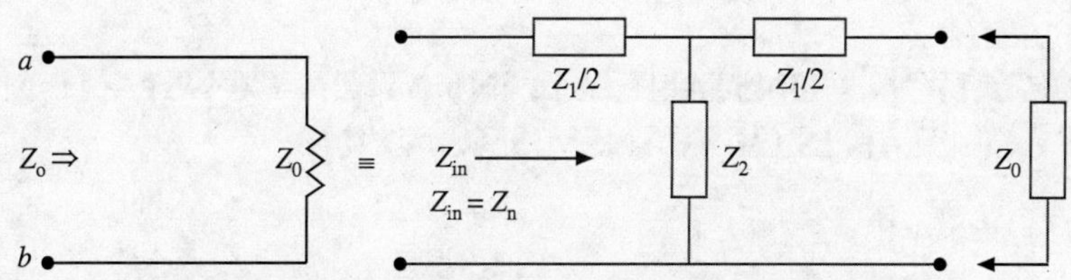

Figure 9.3 *T* section representation of a short line.

Here,

$$Z_{in} = \frac{Z_1}{2} + \frac{Z_2\left(\frac{Z_1}{2} + Z_0\right)}{\frac{Z_1}{2} + Z_2 + Z_0} \tag{9.1}$$

But $$Z_{in} \equiv Z_0$$

The simplification of equation (9.1) gives,

$$Z_0^2 = \frac{Z_1^2}{4} + Z_1 Z_2$$

or, $$Z_0 = \sqrt{\frac{Z_1^2}{4} + Z_1 Z_2} \tag{9.2}$$

However, the open circuit impedance of the short line looking from terminals *a-b* and removing Z_0 is given by

$$Z_{oc} = \frac{Z_1}{2} + Z_2 \tag{9.3}$$

Shorting the output terminals of the short line, the short line looking from terminals *a-b* is given by

$$Z_{sc} = \frac{Z_1}{2} + \frac{\frac{Z_1 Z_2}{2}}{\frac{Z_1}{2} + Z_2}$$

$$= \frac{\frac{Z_1^2}{4} + Z_1 Z_2}{\frac{Z_1}{2} + Z_2} \tag{9.4}$$

The substitution of equations (9.2) and (9.3) in equation (9.4) yields

$$Z_{sc} = \frac{Z_0^2}{Z_{oc}}$$

or, $$Z_0 = \sqrt{Z_{sc} Z_{oc}} \tag{9.5}$$

Expression (9.5) represents the characteristics impedance of a short line in terms of *open circuit* and *short circuit impedances.*

9.5 PROPAGATION CONSTANT, ATTENUATION CONSTANT AND PHASE SHIFT CONSTANTS OF TRANSMISSION LINE

Propagation constant

The *propagation constant* is defined as the natural logarithm of the ratio of the input to the output current and is usually denoted by γ.

Hence, $$\gamma = \log_e \frac{I_s}{I_R} \tag{9.6}$$

i.e. $$\frac{I_s}{I_R} = e^{\gamma} \tag{9.7}$$

γ, the propagation constant, is thus a phasor quantity and can be represented by $(\alpha + j\beta)$ where α is the *attenuation constant* and β the *phase shift constant*. When β is multiplied by the total length of the line, it gives the *electrical length* of the line. However, if α is multiplied by the length of the line, the product gives *total attenuation*. α is expressed in Neper or decibel and β is expressed in radian.

Attenuation constant and phase shift constants

The propagation constant γ being given by $\gamma = (\alpha + j\beta)$, α represents the real part determining the reduction or attenuation in voltage and current along the line while β represents the imaginary part determining the variation of the phase position of voltage and current along the line.

Also, $$\gamma = \sqrt{zy} = \sqrt{(r + j\omega\ell)(g + j\omega c)}$$

i.e. $$\alpha + j\beta = \sqrt{(r + j\omega\ell)(g + j\omega c)} \tag{9.8}$$

By squaring both sides, we get

$$\gamma^2 = (\alpha + j\beta)^2 = \left[\sqrt{(r + j\omega\ell)(g + j\omega c)}\right]^2$$
$$= (r + j\omega\ell)(g + j\omega c) \tag{9.9}$$

By equating the real part of equation (23.9), we get

$$\alpha^2 - \beta^2 = rg - \omega^2 \ell c \tag{9.10}$$

On the other hand, by equating the imaginary part of equation (9.9), we get

$$2\alpha\beta = \omega rc + \omega \ell g$$

Again, the magnitude of γ is given by

$$|\gamma| = \sqrt{\alpha^2 + \beta^2} \tag{9.11}$$

i.e. $$\alpha^2 + \beta^2 = \sqrt{r^2 + (\omega\ell)^2}\sqrt{g^2 + (\omega c)^2} \tag{9.12}$$

By adding equations (9.10) and (9.12), we have

$$2\alpha^2 = (rg - \omega^2 \ell c) + \sqrt{(r^2 + \omega^2 \ell^2)(g^2 + \omega^2 c^2)}$$

$$\alpha = \sqrt{\frac{1}{2}\left\{(rg - \omega^2 \ell c) + \sqrt{(r^2 + \omega^2 \ell^2)(g^2 + \omega^2 c^2)}\right\}} \tag{9.13}$$

Similarly, subtracting equation (9.12) from equation (9.10), simplification yields,

$$\beta = \sqrt{\frac{1}{2}\left\{(\omega^2 \ell c - rg) + \sqrt{(r^2 + \omega^2 \ell^2)(g^2 + \omega^2 c^2)}\right\}} \tag{9.14}$$

Expressions (9.13) and (9.14) represent the relation between the primary constants R, L, C and G in terms of α and β, the secondary constants. The profiles of α and β with respect to ω are shown in Figure 9.4 and 9.5.

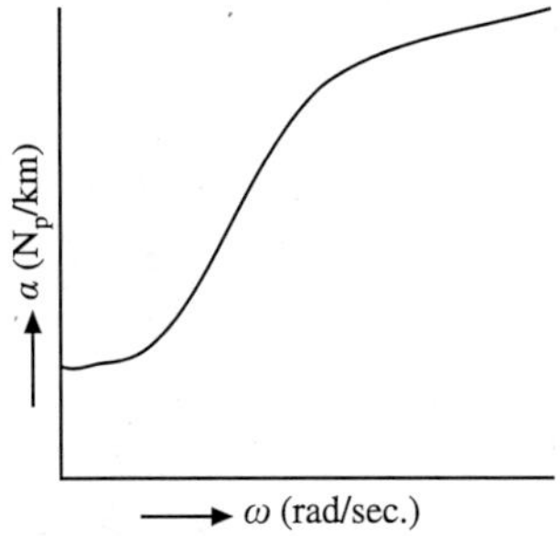

Figure 9.4 Profile of α vs ω.

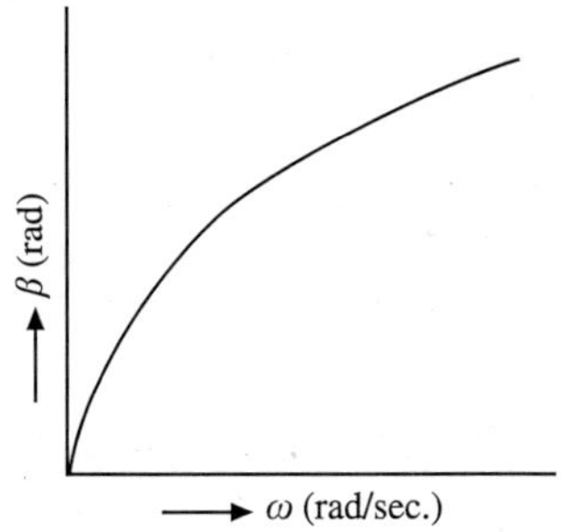

Figure 9.5 Profile of β vs ω.

9.6 WAVELENGTH AND VELOCITY OF PROPAGATION

The *wavelength* is defined as the distance that an electric wave traverses along a line in order that the total phase shift of input energy is 2π radians. The wavelength is generally denoted by "λ", the unit being metre (Figure 9.6).

$\therefore$ $$\beta\lambda = 2\pi, \text{ where } \beta \text{ is the phase shift constant}$$

i.e. $$\lambda = \frac{2\pi}{\beta}\text{ m}.$$

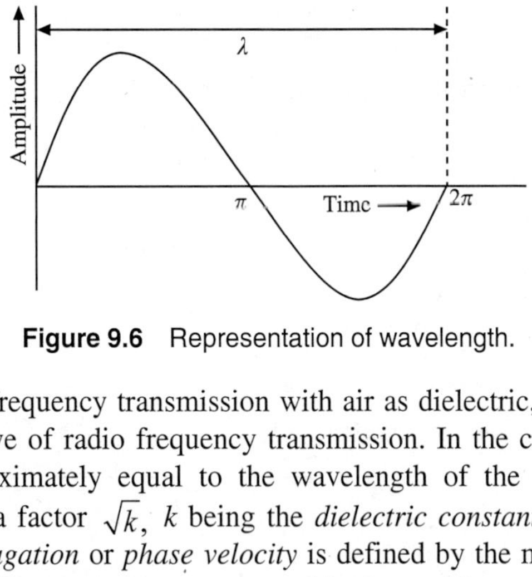

Figure 9.6 Representation of wavelength.

In the case of radio frequency transmission with air as dielectric, λ is approximately equal to the wavelength of wave of radio frequency transmission. In the case of cable transmission, the wavelength is approximately equal to the wavelength of the wave in radio frequency transmission divided by a factor $\sqrt{k}$, k being the *dielectric constant* of the medium.

The *velocity of propagation* or *phase velocity* is defined by the magnitude of velocity with which a signal of single frequency propagates and is denoted by v_p, its unit being km/sec.

It is evident that change of 2π represents one cycle operation and hence

$$\lambda = v_p \times t \tag{9.15a}$$

But, t, (the time period) $= \dfrac{1}{f}$ (9.15b)

Substituting the value of t from equation (9.15b) to equation (9.15a), we get

$$\lambda = v_p \times \frac{1}{f}$$

or,

$$v_p = \lambda \times f \tag{9.16}$$

Substituting the value of $\lambda \left(= \dfrac{2\pi}{\beta} \right)$ in equation (9.16) gives

$$v_p = \frac{2\pi}{\beta} \times f = \frac{\omega}{\beta} (\omega = 2\pi f) \tag{9.17}$$

In practical transmission line, with air as dielectric, v_p is approximately equal to the velocity of light. In lines with chemical or solid dielectric medium, v_p is the velocity of light divided by k; the velocity of the propagation cannot exceed the velocity of the light.

9.7 DIFFERENTIAL EQUATION FOR THE PROPAGATION WAVES IN A UNIFORM LINE

Let us consider an infinitesimal element Δx of a transmission line as shown in Figure 9.7. Let the instantaneous voltage and current be e and i respectively and they both are functions of distance and time. Applying Kirchhoff's voltage law around the loop $a\ b\ c\ d\ a$,

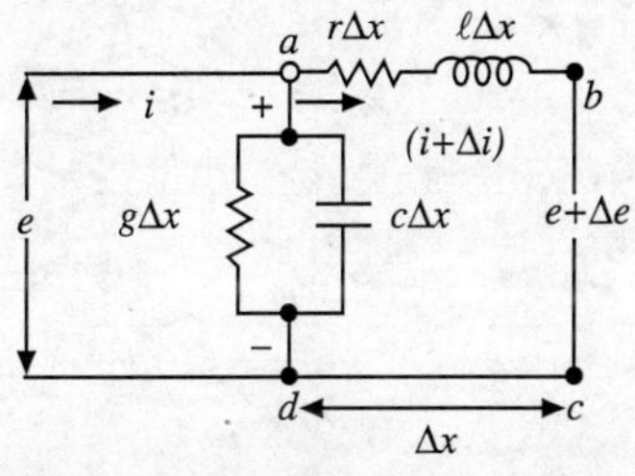

Figure 9.7 Infinitesimal element of a line.

$$(i + \Delta i)(r\Delta x) + (\ell\Delta x)\frac{\partial}{\partial t}(i + \Delta i) + (e + \Delta e) - e = 0$$

or

$$i(r\Delta x) + (\ell\Delta x)\frac{\partial i}{\partial t} + \Delta e = 0$$

[Neglecting differential quantities of the second and higher order]

or,

$$\left(ri + \ell\frac{\partial i}{\partial t} \right)\Delta x + \Delta e = 0$$

or,

$$\Delta e = -\left(ri + \ell\frac{\partial i}{\partial t} \right)\Delta x$$

or,
$$\frac{\Delta e}{\Delta x} = -ri - \ell \frac{\partial i}{\partial t}$$

Let $\Delta x \to 0$. This gives

$$\frac{\partial e}{\partial x} = -ri - \ell \frac{\partial i}{\partial t} \tag{9.18}$$

Similarly, applying Kirchhoff's Current law in the given circuit, we have

$$i - \left(g\Delta x e + c\Delta x \frac{\partial e}{\partial t} \right) - (i + \Delta i) = 0$$

$\therefore$
$$\Delta i = -\left(g\Delta x e + c\Delta x \frac{\partial e}{\partial t} \right)$$

i.e.
$$\frac{\Delta i}{\Delta x} = -\left(ge + c\frac{\partial e}{\partial t} \right)$$

with $\Delta x \to 0$,

$$\frac{\partial i}{\partial x} = -ge - c\frac{\partial e}{\partial t} \tag{9.19}$$

Equations (9.18) and (9.19) are two simultaneous partial differential equations which can be solved for *e* and *i*.

Differentiating equation (9.18) with respect to *x*, we get

$$\frac{\partial^2 e}{\partial x^2} = -r\frac{\partial i}{\partial x} - \ell \frac{\partial^2 i}{\partial t \partial x} \tag{9.20}$$

and differentiating equation (9.19) with respect to *t*, we get

$$\frac{\partial^2 i}{\partial t \partial x} = -g\frac{\partial e}{\partial t} - c\frac{\partial^2 e}{\partial t^2} \tag{9.21}$$

Substituting for $\frac{\partial i}{\partial x}$ and $\frac{\partial^2 i}{\partial t \partial x}$ in equation (9.20) from equations (9.19) and (9.21) respectively, it yields

$$\frac{\partial^2 e}{\partial x^2} = -r\left(-ge - c\frac{\partial e}{\partial t} \right) - \ell\left(-g\frac{\partial e}{\partial t} - c\frac{\partial^2 e}{\partial t^2} \right)$$

$$= rge + (rc + \ell g)\left(\frac{\partial e}{\partial t} \right) + \ell c\frac{\partial^2 e}{\partial t^2} \tag{9.22}$$

Similarly, by eliminating *e* the equation for *i* can be obtained as

$$\frac{\partial^2 i}{\partial x^2} = rgi + (rc + \ell g)\left(\frac{\partial i}{\partial t} \right) + \ell c\frac{\partial^2 i}{\partial t^2} \tag{9.23}$$

Equations (9.22) and (9.23) are called the *Telegrapher's equation.*

Let us now consider the case of a line without loss*, for which $R = 0$, $G = 0$. This approximation is reasonably good when the line losses are much smaller than the energy which travels along the line and it affords a simple and useful result. Therefore, substituting $R = 0$, $G = 0$ in equations (9.22) and (9.23), we get

$$\frac{\partial^2 e}{\partial x^2} = \ell c \frac{\partial^2 e}{\partial t^2} \tag{9.24}$$

and

$$\frac{\partial^2 i}{\partial x^2} = \ell c \frac{\partial^2 i}{\partial t^2} \tag{9.25a}$$

The general solution for equation (9.24) is given by**

---

* We shall limit our discussion to a lossless line. On such a line, waves travel without any attenuation or distortion. An actual line has, however, losses the effect of which is to attenuate the traveling waves and also to distort them. These losses comprise copper loss, dielectric loss and corona loss. A lightning surge, for example, is attenuated by corona to one half its initial voltages after it has traveled only a few kilometres.

** Let us write

$$\sqrt{\ell c}\, x - t = s \tag{i}$$

Therefore, the supposed solution $e = f(\sqrt{\ell c}\, x - t)$ of equation (23.24) can be written as

$$e = f(s) \tag{ii}$$

Differentiating (ii) with respect to x, we have $\dfrac{\partial e}{\partial x} = \dfrac{\partial f}{\partial s}\dfrac{\partial s}{\partial x}$ (iii)

But from (i) $\dfrac{\partial s}{\partial x} = \sqrt{\ell c}$, and so $\dfrac{\partial e}{\partial x} = \sqrt{\ell c}\dfrac{\partial f}{\partial s}$ (iv)

Taking the second derivative with respect to x, we obtain

$$\frac{\partial^2 e}{\partial x^2} = \ell c \frac{\partial^2 f}{\partial s^2} \tag{v}$$

In a similar way, it can be shown that $\dfrac{\partial^2 e}{\partial t^2} = \dfrac{\partial^2 f}{\partial s^2}$ (vi)

From (v) and (vi),

$$\frac{\partial^2 e}{\delta s^2} = \ell c \frac{\partial^2 e}{\partial t^2} \tag{vii}$$

This equation (23.24) therefore is to be satisfied by any function, say $e = f(\ell c\, x - t)$

It may at first seem strange that the quantity $\dfrac{1}{\sqrt{\ell c}}$ is a velocity. This is, however, true since

$$\frac{1}{\sqrt{\ell c}} = \frac{1}{\sqrt{\text{(henry/length)(farad/length)}}}$$

$$= \frac{\text{length}}{\sqrt{\text{henry} \times \text{farads}}} = \frac{\text{length}}{\text{seconds}}$$

which is the dimension of velocity.

$$e = f(\sqrt{\ell c}\, x - t) \tag{9.25b}$$

Similarly, it can be shown that $e = \phi(\sqrt{\ell c}\, x + t)$ is also a solution to equation (9.24).

Since the differential equation (23.24) is linear, the sum of any two solutions is also a solution. Hence, the complete solution for the voltage in the ideal line (i.e. a line without losses) has the form

$$e = f(\sqrt{\ell c}\, x - t) + \phi(\sqrt{\ell c}\, x + t) \tag{9.26}$$

This may be written in the following form:

$$e = e' + e'' \tag{9.27}$$

where $e' = f(\sqrt{\ell c}\, x - t)$ and it represents the *incident wave*, i.e. the wave travelling in the direction of x increasing and $e'' = \phi(\sqrt{\ell c}\, x + t)$ and it represents the *reflected wave*, i.e. the wave travelling in the direction of x decreasing. Thus, the transient voltage wave is seen to have two voltage wave components, e' travelling in the forward direction (i.e. when x is increasing) and e'' travelling in the backward direction (i.e. when x is decreasing).

The component in the forward direction is called *forward travelling wave* and the wave travelling in the backward direction is called *backward travelling wave.*

A similar solution may be written for the current.

Again from equation (9.19), we have

$$\frac{\partial i}{\partial x} = -ge - c\frac{\partial e}{\partial t} = -c\frac{\partial e}{\partial t} \quad \text{since } g = 0$$

Utilizing equation (9.26), we get

$$\therefore \quad \frac{\partial i}{\partial x} = -c\left[\frac{\partial f(\sqrt{\ell c}x - t)}{\partial(\sqrt{\ell c}x - t)}\cdot\frac{\partial(\sqrt{\ell c}x - t)}{\partial t} + \frac{\partial\phi(\sqrt{\ell c}x + t)}{\partial(\sqrt{\ell c}x + t)}\cdot\frac{\partial(\sqrt{\ell c}x + t)}{\partial t}\right]$$

$$= -c\left[-\frac{\partial f(\sqrt{\ell c}x - t)}{\partial(\sqrt{\ell c}x - t)} + \frac{\partial\phi(\sqrt{\ell c}x + t)}{\partial(\sqrt{\ell c}x + t)}\right]$$

$$= c\left[\frac{\partial f(\sqrt{\ell c}x - t)}{\partial(\sqrt{\ell c}x - t)} - \frac{\partial\phi(\sqrt{\ell c}x + t)}{\partial(\sqrt{\ell c}x + t)}\right]$$

$$= c\left[\frac{\partial f(\sqrt{\ell c}x - t)}{\partial(\sqrt{\ell c}x)} - \frac{\partial\phi(\sqrt{\ell c}x + t)}{\partial(\sqrt{\ell c}x)}\right]$$

$$= \frac{c}{\sqrt{\ell c}}\left[\frac{\partial f(\sqrt{\ell c}x - t)}{\partial x} - \frac{\partial\phi(\sqrt{\ell c}x + t)}{\partial x}\right]$$

and

$$i = \frac{1}{\sqrt{\ell/c}}\left[f(\sqrt{\ell c}\, x - t) - \phi(\sqrt{\ell c}\, x + t)\right] \tag{9.28}$$

The quantity $\sqrt{\ell / c}$ is defined as the *characteristic impedance* of the line and is denoted by symbol Z_0. The characteristic impedance of a lossless line is a real quantity, similar to a resistance, and is therefore the characteristic impedance is more commonly called the *surge impedance*. The surge impedance is denoted by Z_0.

$$\therefore \qquad Z_0 = \sqrt{\ell / c} \text{ for a lossless line.} \tag{9.29}$$

Substituting Z_0 in equation (9.28), we get

$$i = \frac{1}{Z_0}\left[f(\sqrt{\ell c}\, x - t) - \phi(\sqrt{\ell c}\, x + t)\right]$$

or,

$$i = \frac{1}{Z_0}[e' - e'']$$

$$\therefore \qquad i = \left[\frac{e'}{Z_0} - \frac{e''}{Z_0}\right] \tag{9.30}$$

This may also be written in the following form:

$$i = i' + i''$$

where $i' = \left(e'/Z_0\right)$ and it represents the *incident current wave* and $i'' = \left(-e''/Z_0\right)$ and it represents the *reflected current wave.*

9.8 EVALUATION OF SURGE IMPEDANCE

(i) *For 2 wire line with air as dielectric*

$$\ell = 4 \times 10–7 \ln (D/r) \text{ henry/meter}$$

where D is the distance between the centres of the conductors and R is the radius of the conductor and $D \gg r$.

$$c = \frac{\pi\varepsilon}{\ln(D/r)} = \frac{\pi \times (10^{-9}/36\pi)}{\ln(D/r)} = \frac{10^{-9}}{36\ln(D/r)} \text{ Farads/metre}$$

$$Z_0 = \sqrt{\left(\frac{\ell}{c}\right)} = \sqrt{\frac{4\times 10^{-7}\ln(D/r)}{10^{-9}/36\ln(D/r)}} = 120\ln(D/r)$$

$$= 276 \log_{10}(D/r) \text{ ohms} \tag{9.31}$$

(ii) *Coaxial cable*

$$\ell = 2 \times 10^{-7} \ln(R/r) \text{ henry/metre}$$

where R is the radius of the outer conductor and r is the radius of the inner conductor

$$c = \frac{2\pi\varepsilon}{\ln(R/r)}$$

$$= \frac{10^{-9} \varepsilon r}{18 \ln (R / r)}$$

Assuming a dielectric having a relative dielectric constant of ε_r

$$\therefore \qquad Z_0 = \sqrt{\left(\frac{\ell}{c}\right)} = \frac{138}{\sqrt{\varepsilon_r}} \tag{9.32}$$

A value of 200–600 ohms is usually assumed for the *surge resistance of an overhead line* while a value of 30–60 ohms is assumed for the *surge resistance of a cable.*

9.9 VELOCITY OF PROPAGATION OF TRAVELLING WAVES

We know, $$v = \frac{1}{\sqrt{\ell c}}$$

(i) for overhead line, $v = \sqrt{\dfrac{1}{(4 \times 10^{-7} \ln (D/r) \times 10^{-9})/36 \ln(D/r)}}$

$= 3 \times 10^8$ metres per second (186,000 miles per sec), which equals the velocity of light in vacuum.

(ii) for cables $v = \sqrt{\dfrac{1}{(2 \times 10^{-7} \ln (R / r) \times 10^{-9} \varepsilon_r) / 18 \ln(R / r)}}$

$= \dfrac{3 \times 10^8}{\sqrt{\varepsilon_r}}$ metre per second

9.10 ANALYTICAL EXPRESSION FOR VOLTAGE AND CURRENT OF A LINE AT ANY POINT OF LENGTH "X" FROM THE SENDING END

Figure 9.8 shows the transmission line of length D terminated by impedance Z_R. Let the voltage across the load be V_R.

Hence,

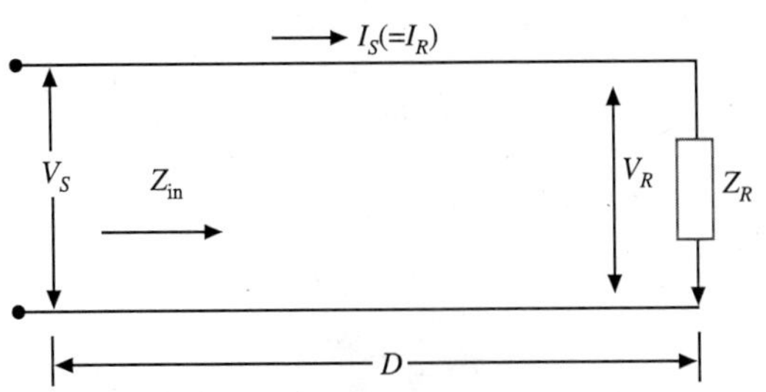

Figure 9.8 Transmission line terminated by Z_R.

or,
$$I_R = \frac{V_R}{Z_R} \tag{9.33}$$

$$V_R = V_s \cosh(\gamma D) - I_s Z_0 \sinh(\gamma D) \tag{9.34}$$

$$I_R = I_s \cosh(\gamma D) - \left(\frac{V_s}{Z_0}\right) \sinh(\gamma D) \tag{9.35}$$

Substituting the values of V_R and I_R from equations (9.34) and (9.35) to equation (9.33), we have

$$\left[I_s \cosh(\gamma D) - \left(\frac{V_s}{Z_0}\right)\sinh(\gamma D)\right] = \left[\frac{V_s}{Z_R}\cosh(\gamma D) - \frac{I_s Z_0}{Z_R}\sinh(\gamma D)\right]$$

or,
$$I_s \cosh(\gamma D) + \frac{I_s Z_0}{Z_R}\sinh(\gamma D) = \frac{V_s}{Z_R}\cosh(\gamma D) + \frac{V_s}{Z_0}\sinh(\gamma D)$$

or,
$$I_S = \frac{\frac{V_s}{Z_R}\cosh(\gamma D) + \frac{V_S}{Z_0}\sinh(\gamma D)}{\cosh(\gamma D) + \frac{Z_0}{Z_R}\sinh(\gamma D)} = \frac{V_s\left[Z_0\cosh(\gamma D) + Z_R\sinh(\gamma D)\right]}{Z_0\left[Z_R\cosh(\gamma D) + Z_0\sinh(\gamma D)\right]} \tag{9.36}$$

However, the sending end impedance or input impedance as viewed from the sending end is given by

$$Z_{in} = \frac{V_s}{I_s} = \frac{V_s}{\frac{V_S}{Z_0}\left[\frac{Z_0\cosh(\gamma D) + Z_R\sinh(\gamma D)}{Z_R\cosh(\gamma D) + Z_0\sinh(\gamma D)}\right]}$$

or,
$$Z_{in} = Z_0\left[\frac{Z_R\cosh(\gamma D) + Z_0\sinh(\gamma D)}{Z_0\cosh(\gamma D) + Z_R\sinh(\gamma D)}\right] \tag{9.37a}$$

$$= Z_0\left[\frac{Z_R + \tanh(\gamma D)}{Z_0 + \tanh(\gamma D)}\right] \tag{9.37b}$$

Let V_X and I_X represent the voltage and current at any point "P" at a distance x from the sending end (Figure 9.9).

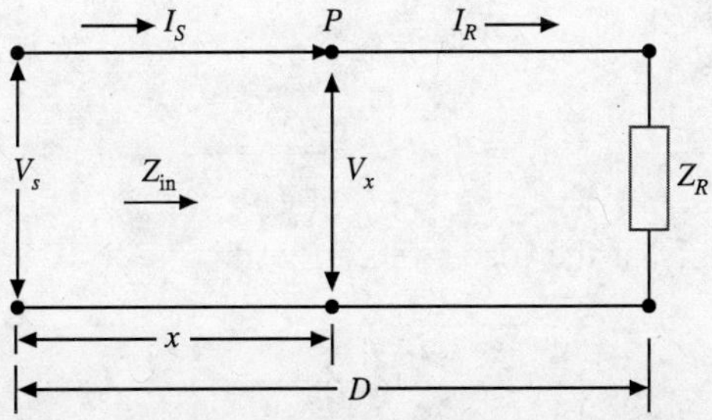

Figure 9.9 Voltage and current at any point P of the line.

Substituting the value of I_S from equation (9.36) to equation (9.34), we have

$$V_x = V_s \cosh \gamma x - Z_0 \left[\frac{V_S}{Z_0} \left\{ \frac{Z_0 \cosh \gamma D + Z_R \sinh \gamma D}{Z_R \cosh \gamma D + Z_0 \sinh \gamma D} \right\} \right] \times \sinh \gamma x$$

$$= V_s \left[\frac{(Z_R \cosh \gamma x \cosh \gamma D + Z_0 \sinh \gamma D \cosh \gamma x) - (Z_0 \sinh \gamma x \cosh \gamma D + Z_R \sinh \gamma x \sinh \gamma D)}{Z_R \cosh \gamma D + Z_0 \sinh \gamma D} \right]$$

or,

$$V_x = V_s \left[\frac{Z_R \cosh \gamma (D-x) + Z_0 \sinh \gamma (D-x)}{Z_R \cosh \gamma D + Z_0 \sinh \gamma D} \right] \tag{9.38}$$

Similarly, the current equation at a distance x from the sending end becomes,

$$I_x = \frac{V_s}{Z_0} \left[\frac{Z_R \cosh \gamma (D-x) + Z_R \sinh \gamma (D-x)}{Z_R \cosh \gamma D + Z_0 \sinh \gamma D} \right] \tag{9.39}$$

9.11 REFLECTION AT LOAD

When a travelling voltage or current wave meets any discontinuity on the line, the wave is wholly or partially reflected. Let us consider transmission line terminated by impedance Z_L as shown in Figure 9.10.

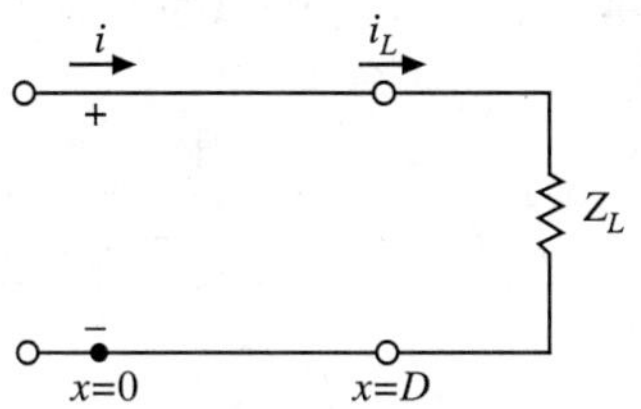

Figure 9.10 Transmission line terminated by Z_L.

At any value of x,

$$e = e' + e'' \tag{9.40}$$

$$i = i' + i''$$

$$= \frac{e'}{Z_0} - \frac{e''}{Z_0} \tag{9.41}$$

Therefore, at $x = L$

$$e_L = e_L' + e_L'' \tag{9.42}$$

$$i_L = i_L' + i_L'' = \frac{e_L'}{Z_0} - \frac{e_L''}{Z_0} \tag{9.43}$$

Since Ohm's law must apply at the load, we can write

∴

$$i_L = \frac{e_L}{Z_L} \tag{9.44}$$

$$= \frac{e'_L}{Z_L} + \frac{e''_L}{Z_L} \text{ [utilizing equation (9.42)]} \tag{9.45}$$

Equating (9.43) and (9.45), we get

$$\frac{e'_L + e''}{Z_L} = \frac{e'_L - e''_L}{Z_0}$$

whence

$$e''_L = \frac{Z_L - Z_0}{Z_L + Z_0} e'_L \tag{9.46}$$

$$= K_L e'_L \tag{9.47}$$

Here $K_L = \dfrac{Z_L - Z_0}{Z_L + Z_0}$ and is called the *reflection coefficient* for the voltage wave for the load.

Now since

$$i''_L = -\frac{e''_L}{Z_0}$$

$$i''_L = -\frac{K_L e'_L}{Z_0} = -K_L i'_L \tag{9.48}$$

The *reflection coefficient* for the current wave for the load is negative.

Thus, for the reflected voltage wave, $e''_L = K_L\, e'_L$ (9.49)

and for the reflected current wave, $i''_L = -K_L\, i'_L$ (9.50)

The following cases for different values of Z_L are interesting to note:

(i) ***Reflection at short circuit (i.e. the line is terminated by a short circuit)***

Here, $Z_L = 0$ $\therefore$ $K_L = -1$

$\therefore$ $e''_L = -e'_L$ and $i''_L = i'_L$

Thus, $e_L = e'_L + e''_L = 0$

and $i_L = i'_L + i''_L = 2i'_L$

Thus, for the short circuited condition, the reflected voltage wave is the negative of the incident voltage wave, the resultant being zero and the reflected current wave is equal to the incident current wave, the resultant being twice the incident wave.

(ii) ***Reflection at an open circuit (i.e. the line is terminated at open circuit)***

Here $Z_L = \infty$

$$\therefore \quad K_L = \frac{Z_L - Z_0}{Z_L + Z_0} = \frac{1 - (Z_0/Z_L)}{1 + (Z_0/Z_L)} = 1$$

$\therefore$ $e''_L = e'_L$ and $i''_L = -i'_L$

Hence, $e_L = e'_L + e''_L = 2e'_L$

and $i_L = i'_L + i''_L = 0$

Thus, the reflected voltage wave is equal to the incident voltage wave, the total voltage being twice the incident voltage, and the reflected current wave is the negative of the incident current wave, the total current being zero.

(iii) ***Reflection at $Z_L = Z_0$ (i.e. the line is terminated by Z_0)***

$$K_L = \frac{Z_L - Z_0}{Z_L + Z_0}$$

$$\therefore \qquad e''_L = 0 \qquad \text{and} \qquad i''_L = 0$$

Thus, there is no *reflected wave.* Since the line has the same characteristic as that of the load resistance, it is evident that electrically there is no discontinuity. The wave merely disappears as the end of the line is reached. The transmission line, terminated by Z_0, is thus called an infinite line.

(iv) ***Other cases.*** If Z_L is greater than Z_0, partial reflection will occur with *reversal of current* only. If Z_L is less than Z_0, partial reflection will occur with reversal of voltage only.

9.12 REFLECTION FROM TERMINAL INDUCTANCE

Referring to Figure 9.10, with Z_L replaced by an inductance of value L,

$$e_L = L\frac{di_L}{dt} = Lpi_L \left[p = \frac{d}{dt}\right]$$

i.e.

$$e'_L + e''_L = Lp\left(\frac{e'_L}{R_0} - \frac{e''_L}{R_0}\right)$$

or

$$Z_0 e''_L + Lpe''_L = -Z_0 e'_L + Lpe'_L$$

If we consider that the incident wave is flat-topped and of constant voltage E, then $pe'_L = 0$. Then we can write

$$Z_0 e''_L + Lpe''_L = -Z_0 E$$

Replacing by d/dt for p, we finally obtain

$$\frac{L}{-Z_0 E - Z_0 e''_L} de''_L = dt$$

Integration gives,

$$\frac{L}{-Z_0} \ln\left[-Z_0 E - Z_0 e''_L\right] = t + C$$

where C is a constant of integration. By changing from the logarithmic form to the exponential form, and at the same time writing as K the exponential factor that contains the constant of integration, we have

$$-Z_0 E - Z_0 e''_L = K\varepsilon^{(-Z_0/L)t} \tag{9.51}$$

The co-efficient K must be evaluated by means of known boundary conditions that relate and t. At the instant of arrival of the oncoming wave, the inductance cannot carry any current; consequently the wave is fully reflected and the voltage of the reflected wave at that instant is equal to the voltage of the incident wave. Thus, the line behaves as if it were open-circuited. Thus, for

$$t = 0, i = 0 \text{ and } e_L'' = E \tag{9.52}$$

Introducing this in equation (9.51), we get $-2Z_0E = K$.
Therefore, from equation (9.51), we have

$$e_L'' = E[-1 + 2\varepsilon^{(-Z_0/L)t}] \tag{9.53}$$

The reflected wave thus would vary from E, initially, to $-E$, finally, which can be physically interpreted as follows:

There is voltage across the terminal inductance from the instant of arrival of the wave, so the inductance will begin to carry current. It will carry more current as the time passes, until at least it is practically equivalent to a short circuit and carries steady current with no terminal voltage (i.e. $e_L'' = E$). When the voltage across the inductance has become zero, the current through it has become twice the current of the incident wave, as in the case of the short circuit line.

Now
$$e_L = e_L' + e_L''$$

$\therefore$
$$e_L = E + E(-1 + 2\varepsilon^{\frac{-Z_0 t}{L}}) = 2E\varepsilon^{\frac{-Z_0 t}{L}} \tag{9.54}$$

Also, since
$$i_L = \frac{e_L'}{Z_0} - \frac{e_L''}{Z_0}$$

$\therefore$
$$i_L = \frac{E}{Z_0} - \frac{E}{Z_0}(-1 + 2\varepsilon^{\frac{-Z_0 t}{L}})$$

$$i_L = \frac{2E}{Z_0}(1 - \varepsilon^{\frac{-Z_0 t}{L}}) \tag{9.55}$$

9.13 REFLECTION FROM TERMINAL CAPACITANCE

If there is a condenser connected to the terminals of the line, it acts as a short circuit at the moment of arrival of the incident wave. At this instant, the reflected voltage wave is negative, for the terminal voltage is momentarily zero, and the current of incident wave is momentarily doubled. As the condenser becomes charged, the terminal voltage rises. Finally, the condenser becomes fully charged; it then behaves as an open circuit. The terminal current is zero and the voltage of the incident wave is doubled. Mathematically,

$$e_L'' = E(1 - 2\varepsilon^{\frac{-t}{Z_0 C}}) \tag{9.56}$$

$$e_L = 2E(1 - 2\varepsilon^{\frac{-t}{Z_0 C}}) \tag{9.57}$$

$$i_L = \frac{2E}{R_0} \varepsilon^{\frac{-t}{Z_0 C}} \tag{9.58}$$

There is more practical importance of the effect of terminal capacitance in the case of a line with terminal transformer, owing to the paradox that for lightning waves (or other transient overvoltages), a transformer acts as a capacitance rather than as an inductance. Lightning voltage arrives at a transformer as a travelling wave, and the front of the wave is so steep and rises so suddenly to a maximum, that there is practically no time for current to start to flow through the large inductance of the transformer winding. But there is some small capacitance in the transformer from turn to turn, and from winding to core, and it is largely this capacitance rather than the inductance, that determines the transformer's reaction to lightning. In the case of termination of line by a HT motor, the motor winding being inductive, high voltage is developed across the winding as a travelling wave reaches the motor terminals. Also, due to winding capacitances, the leakage current flow increases and the motor becomes vulnerable to the travelling voltage wave.

9.14 CONCEPT OF STANDING WAVE AND VOLTAGE STANDING WAVE RATIO

Standing wave

Whenever reflection takes place in line transmission, at some points, the incident and reflected signals are in phase and both the components add together; on the other hand, at some other points, two components may oppose each other. The net or resultant graphical profile of both of these incident and reflected waves is called *standing wave*; the points where the resultant signal (voltage or current) is maximum are known as *voltage or current maxima.* On the other hand, the points where the resultant signal (voltage or current) is minimum are called *voltage or current minima.* There will be standing waves corresponding to the resultant voltage or current signal.

Thus, it is evident that

$$|V_{\max}| = |V_i| + |V_r|$$

$$|I_{\max}| = |I_i| + |I_r|$$

$$|V_{\min}| = |V_i| - |V_r|$$

$$|I_{\min}| = |I_i| - |I_r|$$

Voltage standing wave ratio (VSWR)

Voltage standing wave ratio (*VSWR*) is defined as the ratio of magnitudes of maximum and minimum voltage of the standing wave,

i.e.

$$VSWR(= S) = \frac{|V_{\max}|}{|V_{\min}|} \tag{9.59}$$

Here, the standing wave has been specified in terms of voltage. In high frequency transmission*, the standing wave problem is very common.

[In a similar way, *current standing wave ratio* may be defined by the ratio of maximum to the minimum current on the standing wave. Voltage standing wave ratio (VSWR) is more popular since it is measure of SWR at different points of the line by just a voltmeter while for the measurement of current standing wave ratio, we require an ammeter to be connected in series at different maximum and minima points which is cumbersome.]

Figure 9.11 represents the profiles of voltage and current standing waves. It may be noted that as the power of the wave must remain same at any point on the line, current and voltages are opposite in phase, i.e. at voltage maximum point the current is minimum and vice-versa. In Figure 9.11, λ represents wave length and it may be noted that the maximum impedance and minimum impedance points can be determined by the ratios of (V_{max} and I_{min}) and (V_{min} and I_{max}), respectively. It may also be noted that the current minimum and the voltage maximum or the current maximum and the voltage minimum points are $\left(\lambda/4\right)$ distance apart from each other.

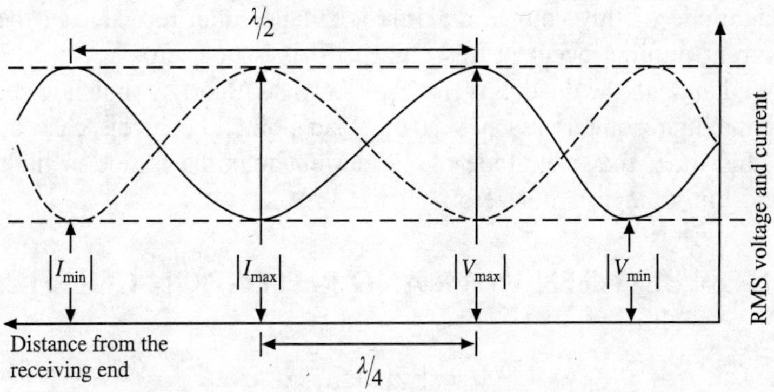

Figure 9.11 Standing wave patterns on an HF line.

---

* In high frequency lines, propagation constant γ is given by

$$\gamma = \sqrt{zy} = \sqrt{(r + j\omega\ell)(g + j\omega c)} \qquad \text{(i)}$$

However, at high frequency,

$$\omega\ell >> r \quad \text{and} \quad \omega c >> g$$

Hence equation (i) becomes

$$\gamma = \sqrt{j\omega\ell \times j\omega c} \qquad \text{(ii)}$$

$$\gamma = j\omega\sqrt{\ell c}$$

But,

$$\gamma = \alpha + j\beta \qquad \text{(iii)}$$

Comparing equations (i) and (ii), we get

$$\alpha = 0, \gamma = j\beta, \text{ where } \beta = \omega\sqrt{\ell c}$$

The velocity of signal is then given by

$$v_P = \frac{\omega}{\beta} = \frac{1}{\sqrt{\ell c}} \text{ km/sec.}$$

Also, when $\omega L >> R$ and $\omega C >> G$,

$$Z_0 = \sqrt{\frac{r + j\omega\ell}{g + j\omega c}} = \sqrt{\frac{\ell}{c}}$$

In the case of *open-circuited line*, the load current is zero. This clearly indicates that the incident and reflected currents at load are equal in magnitude but opposite in sign so that I_r and I_r cancel each other. Thus, at load the current minima point is found. Obviously, the voltage maxima point will be located here also and at a distance $\left(\lambda/4\right)$ from the load end along the line, the conditions are opposite and at $\left(\lambda/4\right)$ the current will be maximum while the voltage will be minimum.

On the other hand, when the load end of the line is *short circuited*, the load current is maximum at this point and the load voltage is zero. Thus, incident and reflected voltages cancel each other at load point and the voltage minima; the current maxima point is located at the load. At a distance $\left(\lambda/4\right)$ along the line from the load point, the condition is reversed and the voltage maxima current minima point is reached.

If the load at the termination of the line is very high, i.e. more than Z_0 (but not infinity like the open-circuited case), the voltage maxima-current minima point is also reached at the load but the magnitude of this voltage maxima is smaller than its value in the open-circuited line. Though current minima occur at this point, still it is not zero.

On the other hand, if the load is very low, i.e. less than Z_0 (but not zero), the voltage minima current maxima point is reached at the load point like in the case of short-circuited line. However, this time, the magnitude of the current maxima is not so high like the short-circuited one and the voltage is not zero at load.

9.15 RELATION BETWEEN VSWR AND REFLECTION COEFFICIENT *K*

It is obvious that

$$|V_{\max}| = |V_i| + |V_r|$$

$$|V_{\min}| = |V_i| - |V_r|$$

But

$$VSWR = \frac{|V_{\max}|}{|V_{\min}|} = \frac{|V_i| + |V_r|}{|V_i| - |V_r|} = \frac{1 + \frac{|V_r|}{|V_i|}}{1 - \frac{|V_r|}{|V_i|}}$$

$$= \frac{1 + |K|}{1 - |K|}$$

or,

$$S = \frac{1 + |K|}{1 - |K|} \tag{9.60}$$

or,

$$S(1 - |K|) = 1 + |K|$$

or,

$$S - S|K| = 1 + |K|$$

or,
$$S-1=|K|+S|K|$$

∴
$$|K|=\frac{S-1}{S+1} \tag{9.61}$$

EXAMPLE 9.1 A 30 V dc pulse is applied to a transmission line at t = 0. The parameters of the line are as follows:

Source resistance = 0 ohm
Characteristic resistance of line = 50 Ω
Load resistance = 100 ohm
Velocity of wave = 200 m/micro-sec
Length of the line = 400 m

Draw the distribution of voltage along the line at t = 1 μs, t = 2.5 μs, and t = 4.5 μs.

Solution:

Load reflection coefficient, $K_L=\frac{Z_L-Z_0}{Z_L+Z_0}=\frac{100-50}{100+50}=\frac{1}{3}$ [∵ Z_L = 100; Z_0 = 50 ohms]

Source reflection coefficient, $K_S=\frac{Z_s-Z_0}{Z_s+Z_0}=-1$ [∵ Z_s = 0; Z_0 = 50 ohms]

Time required to traverse distance L (= 400 m) is

$$\frac{L}{\text{Velocity}}=\frac{400\text{ m}}{200\text{ m/}\mu\text{s}}=2\ \mu\text{s}$$

At t = 0, the 30 V pulse is sent down the line. However, it is zero for the points on the line as it is just connected. After the arrival of the pulse, 30 V is imposed on the line.

At t = 1.0 ms, the pulse could travel upto 200 m. The distribution has been shown in Figure E9.1.

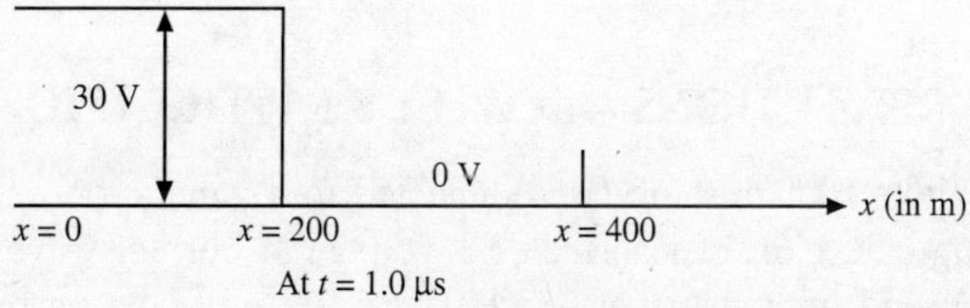

Figure E9.1

At t = 2.0 μs, the pulse arrives at the load. At t = 2.5 μs, the pulse is reflected partially and traverses backwards with a magnitude of 10 volts [∵ magnitude of reflected pulse = K_L × 30 = 10 V].

Figure E9.2 shows the distribution. The backward pulse will traverse a distance of 100 m from the load towards the source.

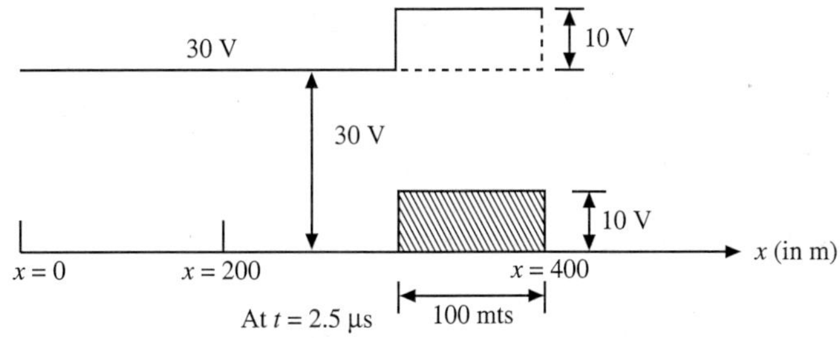

Figure E9.2

The distribution shows that for 100 m from the load side towards the source, the net voltage is (30 V + 10 V) = 40 V while for the rest of the line (300 m), the voltage is only 30 V.

At t = 4 μs, the reflected voltage of 10 V reaches the source [∵ in 2 μs, the pulse covers a distance of 200 m, hence within 4 μs, it will complete one to and fro trip] and starts coming back with the magnitude of (K_S × 10 V), i.e. –10 V towards the load. At t = 4.5 μs, this reflected pulse from source will traverse a distance of 100 m from the source towards the load. Figure E9.3 shows the voltage distribution where, for the 100 m from the source, the net voltage becomes (40 – 10) V = 30 V; the rest of the line is having 40 V.

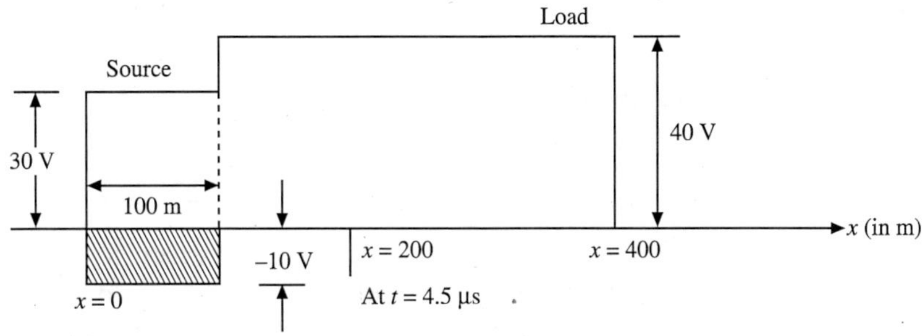

Figure E9.3

9.16 REPEATED REFLECTIONS—BEWLEY'S LATTICE (ZIGZAG) DIAGRAM

Figure 9.12 shows a finite transmission line connected to a generation on the sending end and a load on the receiving end. Consider an incident voltage or current wave on a line. When this wave reaches the load end, it will be reflected. The reflected wave on reaching the generator end will again be reflected due to the generator impedance Z_g, and this process will continue. To keep track of the reflections, it is convenient to draw a diagram showing successive reflections. This diagram is known as the Bewley's lattice diagram.

We shall illustrate the method of constructing the zigzag diagram with the help of the following example.

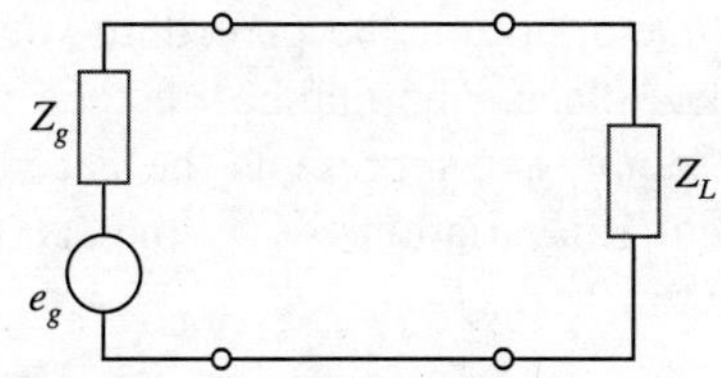

Figure 9.12 Transmission line connecting the generator to the load.

EXAMPLE 9.2 A battery with an e.m.f E and series resistance R is connected at $t = 0$ to the sending end of the transmission line which is short-circuited at the far end (Figure E9.4). Plot the sending end voltage and current as a function of time for $R = Z_0/3$. The time required for a wave to travel to full length for the line is T. Z_0 is the characteristic impedance of the line.

Solution: (i) The first step is to find the reflection coefficient for the two ends. Reflection coefficient for the load end is

$$K_L = \frac{Z_L - Z_0}{Z_L + Z_0} = -1$$

Reflection coefficient for the source end is $K_S = \dfrac{(Z_0/3) - Z_0}{(Z_0/3) + Z_0} = -\dfrac{1}{2}$

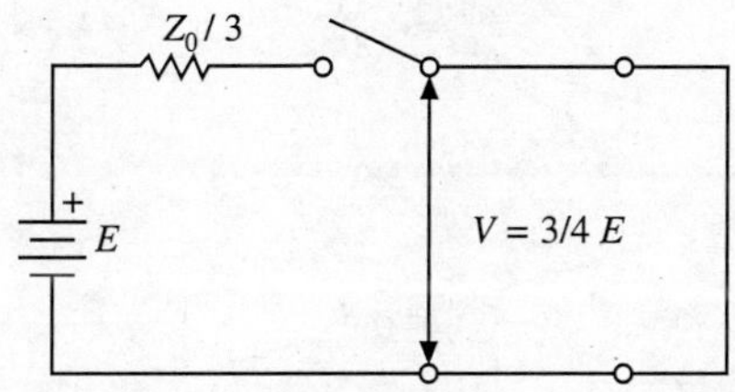

Figure E9.4

(ii) The second step is to find the sending end voltage and current

$$\text{Sending end voltage} = E - \frac{E}{Z_0 + (Z_0/3)} \times \frac{Z_0}{3} = \frac{3}{4}E \quad \text{[ref. Figure 9.4]}$$

$$\text{Sending end current} = \frac{3}{4}\frac{E}{Z_0}$$

(iii) The third step is to draw the zigzag diagrams, one for the voltage and the other for the current. In a zigzag diagram, distance is plotted horizontally and time is plotted downwards as shown in Figure E9.5(a) and (b).

The zigzag lines are traces of the wave fronts of the various reflections. The numbers attached to the line indicate the magnitude of individual waves. The magnitude of each reflection is obtained by multiplying the magnitude of the preceding wave by the reflection coefficient at the point where reflection takes place. The number in each intervening space is the sum of the individual waves above the point, and represents the net voltage or current in that region of the diagram. The voltage or current variations with time can then be easily plotted as shown in Figure E9.6(a) and Figure E9.6(b).

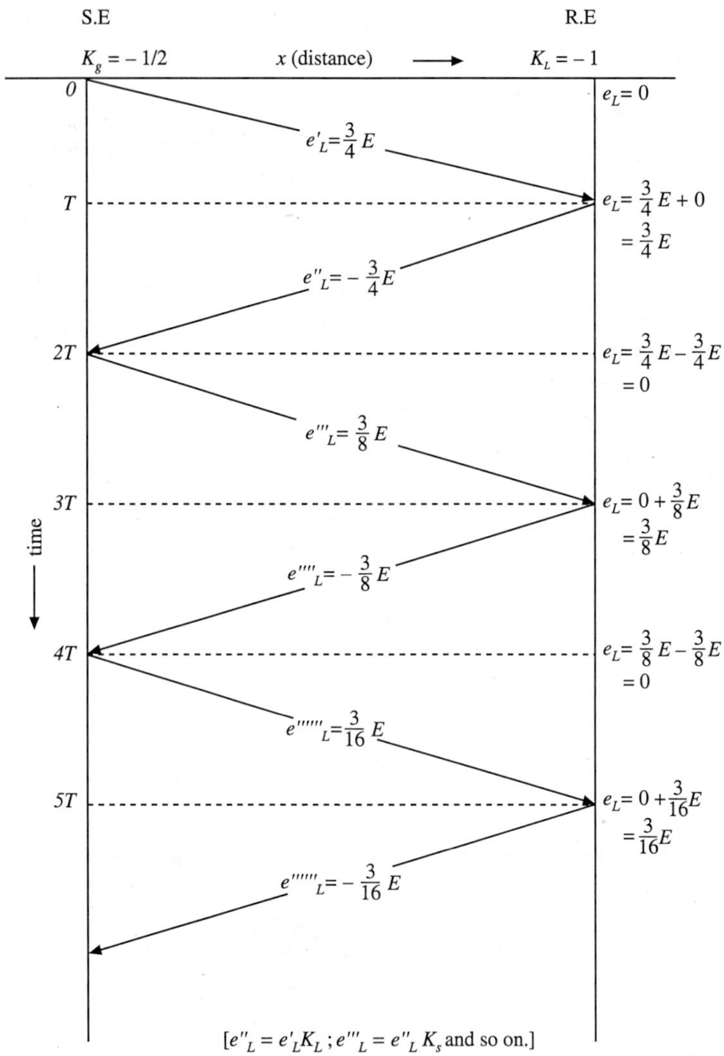

Figure E9.5(a)

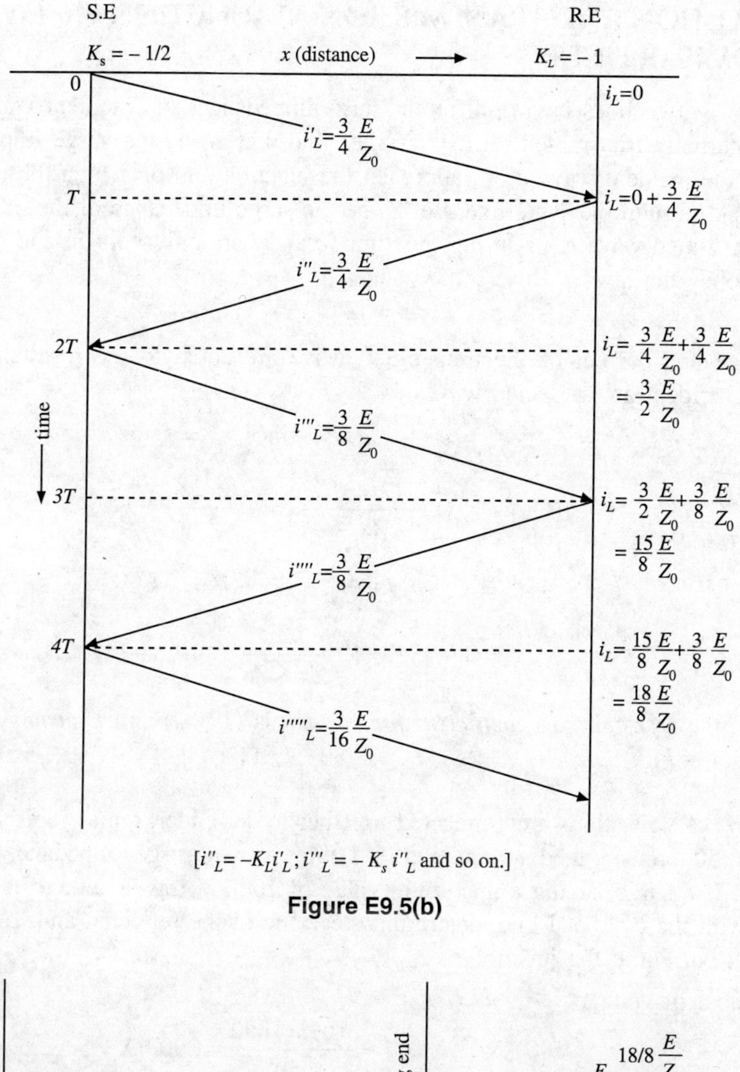

[$i''_L = -K_L i'_L$; $i'''_L = -K_s\, i''_L$ and so on.]

Figure E9.5(b)

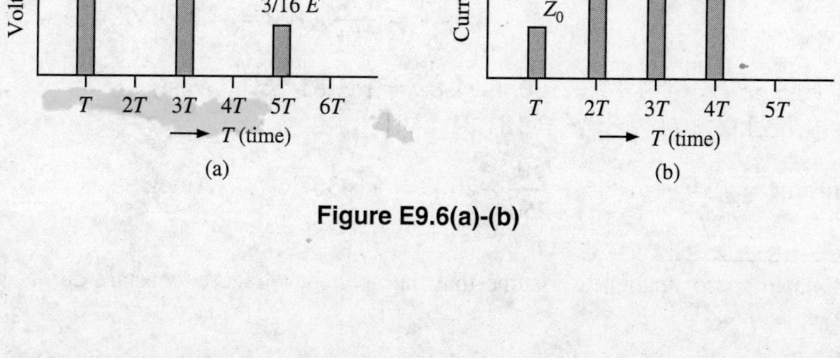

Figure E9.6(a)-(b)

9.17 REFLECTION AND TRANSMISSION AT JUNCTION OF TWO DISSIMILAR LINES

At the junction of two lines having different surge impedances, an incident wave is partially reflected and partially transmitted on to the other line. Let Z_1 be the surge impedance of the line on which the incident wave e', i', approach the junction and on which the reflected wave e'', i'', goes back through the same line. Let Z_2 be the surge impedance of the other line which carries the transmitted wave e, i. During this time, e and i are only voltage and current on this line and they obey the law

$$e = iZ_2 \tag{9.62}$$

The voltage and current of the transmitted wave are equal to the resultant voltage and current of the incident and reflected waves.

$$e = e' + e'' = e' + k\,e' \qquad \text{since } e'' = k\,e' \tag{9.63}$$

$$= e'(1+k) = e'\left(1+\frac{Z_2 - Z_1}{Z_2 + Z_1}\right) = e' \cdot \frac{2Z_2}{Z_2 + Z_1}$$

and

$$i = i' + i'' = i' + (-k)\,i' \qquad \text{since } i'' = -k\,i'$$

$$= (1-k)\,i' = i' \cdot \frac{2Z_1}{Z_2 + Z_1}$$

Therefore, the *transmission coefficient for voltage* is $(1 + k)$ and *transmission coefficient for current* is $(1 - k)$.

EXAMPLE 9.3 Two stations are connected together by an underground cable having a surge impedance of 50 ohms joined to an overhead line with a surge impedance of 500 ohms (Figure E9.7). If a surge having a maximum value of 100 kV travels along the cable towards the junction with the overhead line, determine the value of the reflected and transmitted wave of voltage and current at the junction.

Solution: Incident voltage

$$\therefore \qquad e' = 100 \text{ kV*} ; \qquad i' = \frac{100 \times 1000}{50} = 2000 \text{ A}$$

$$\text{Reflected voltage } e'' = ke' = \frac{Z_2 - Z_1}{Z_2 + Z_1} e'$$

$$= \frac{450}{550} \times 100 = 81.81 \text{ kV.} \quad \text{(Ans.)}$$

Transmitted voltage $e = e' + e'' = 100 + 81.82 = 181.81$ kV. (Ans.)

[Check $e = (1 + k)\,e' = 181.81$ kV.]

$$\text{Reflected current} = i'' = -k\,i' = \frac{450}{-550} \times 2000 = -1636.36 \text{ A. (Ans.)}$$

---

* It is a reasonable approximation to assume that the applied voltage is constant during the life of the travelling wave.

$$\text{Transmitted current } i = \frac{e}{R_2} = \frac{181.81 \times 1000}{500} = 363.2 \text{ A. (Ans.)}$$

[Check $i' = (1 - k)\ i' = (1 - 0.8181) \times 2000 = 363.8$ A.]

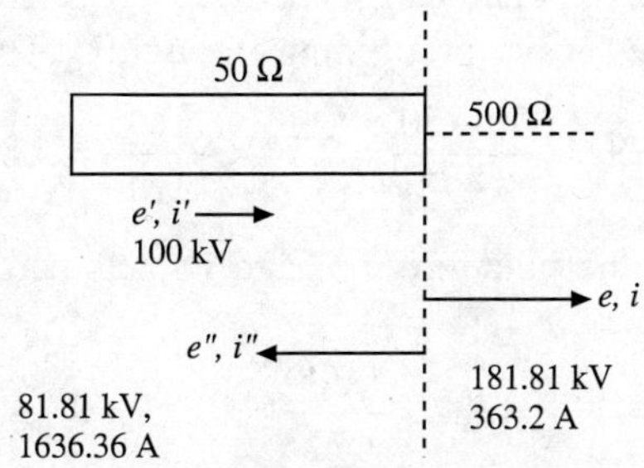

Figure E9.7

EXAMPLE 9.4 A cable C is connected in between two transmission lines A and B. The surge impedances A, B and C are 500 ohm, 600 ohm, and 70 ohm, respectively. A rectangular voltage wave of 40 kV magnitude and of infinite length is initiated in A and travels to B through C. Determine the first and final voltages impressed on B.

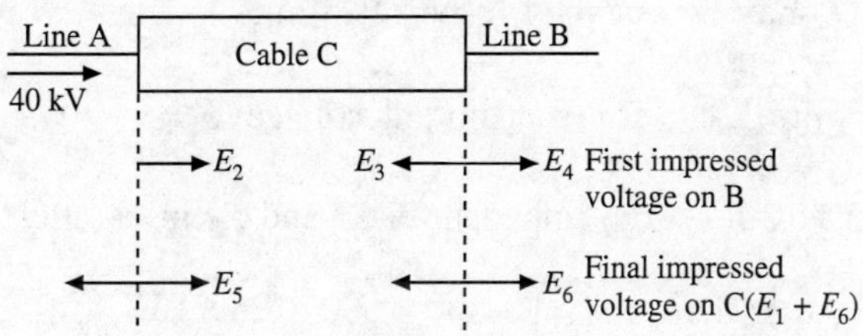

Figure E9.8

Solution: Refer to Figure E9.8.

The voltage wave (magnitude 40 kV) is initiated in A is partly reflected and partly transmitted on to cable C on reaching the junction of A and C.

$$\text{The transmitted wave } E_2 = 40 \times \left(1 + \frac{70 - 500}{70 + 500}\right)$$

$$= 40 \times \frac{140}{570} = 9.83 \text{ kV.}$$

This transmitted wave on reaching the junction of C and B again meets an irregularly so that a part of it is reflected and a part is transmitted on to B. This transmitted voltage wave is obtained as

$$E_4 = 9.83 \times (1 + \frac{600 - 70}{600 + 70}) = 9.83 \times \frac{1200}{670} = 17.61 \text{ kV.} \quad \text{(Ans.)}$$

This is also the first voltage impressed on B. The reflected wave E_3 is, however, obtained as

$$E_3 = 9.83 \times \frac{600 - 70}{600 + 70} = 9.83 \times \frac{530}{670} = 7.78 \text{ kV}.$$

E_3, on reaching the junction of A and C, gets partially reflected in C and partially transmitted on to A. Let E_5 be the reflected wave proceeding through C. Then

$$E_5 = 7.78 \times \frac{500 - 70}{500 + 70} = 7.78 \times \frac{430}{570} = 5.87 \text{ kV}.$$

However, E_5, on reaching the junction of C and B, gets partially transmitted on to B. Let this be E_6.

Then,
$$E_6 = 5.87 \times \frac{1,200}{670} = 10.51 \text{ kV}$$

∴ Then the second impressed voltage = $E_4 + E_6$ = 17.61 + 10.51 = 28.12 kV. (Ans.)

EXAMPLE 9.5 A cable with a surge impedance of 100 ohm is terminated in two parallel connected open wires having surge impedances of 400 ohm and 600 ohm, respectively. If a step fronted voltage wave of 5 kV travels along the cable, find the voltage and current in the cable and the open-wires lines immediately after the travelling wave has reached the transmission point. The voltage wave may be assumed to be of infinite length.

Solution: Refer to Figure E9.9, The transmitted voltage $E = 5 \times \left(1 + \frac{Z_t - Z}{Z_t + Z}\right)$

where Z_t is the resultant of the surge impedances Z_1 and Z_2 in parallel, i.e.

$$Z_t = \left(\frac{Z_1 \times Z_2}{Z_1 + Z_2}\right) = \frac{400 \times 600}{400 + 600} = \frac{24 \times 10^4}{10^3} = 240\Omega$$

∴
$$E = 5\left[1 + \frac{240 - 100}{240 + 100}\right] = 7.06 \text{ kV} \quad \text{(Ans.)}$$

$$I_1 = \frac{E}{Z_1} = \frac{7060}{400} = 17.65\text{A} \quad \text{(Ans.)}$$

$$I_2 = \frac{E}{Z_2} = \frac{7060}{600} = 11.77\text{A} \quad \text{(Ans.)}$$

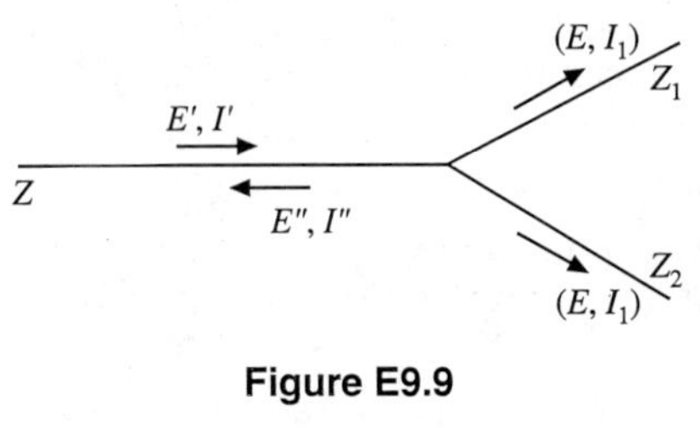

Figure E9.9

Reflected voltage $E'' = 5 \times \dfrac{240-100}{240+100} = 2.059$ kV (Ans.)

Reflected current $I'' = -\dfrac{2059}{100} = -20.59$ A (Ans.)

[Check $I' + I'' = \dfrac{5,000}{100} - 20.59\text{ A} = 29.41; I_1 + I_2 = 17.65 + 11.77 = 29.42$ A and $I' + I'' = I_1 + I_2$.]

EXAMPLE 9.6 A voltage wave $v = V\,(e^{-0.4t} - e^{-t})$ travelling along an overhead line has a surge impedance of 400 Ω. At a distant point, the overhead line terminates and the circuit is continued by two parallel overhead line transformer feeders, each of which has a surge impedance of 400 Ω. Each of the transformers is protected by surge–diverters having effective surge impedance of 50 Ω. Assuming the relative transformer surge impedance to be infinite, what is the maximum voltage which would initially appear across the feeder end of winding of each transformer. Assume $V = 600$ V.

Solution: Figure E9.10 shows the circut.

Voltage transmitted at juinction A is given by $e = v\left[1 + \dfrac{200-400}{200+400}\right] = 0.66$ v.

[∵ the net surge impedance of two parallel overload lines is 200 ohm.]

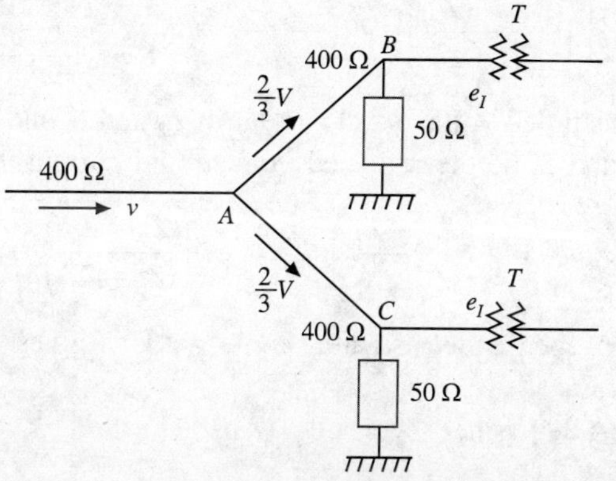

Figure E9.10

Voltage transmitted at junction B or C are obtained as

$$e_1 = 0.66v\left[1 + \frac{50-400}{50+400}\right]$$

$$= 0.66v \times \left[\frac{100}{450}\right] = 0.146v$$

Maximum voltage would appear when

$$\left[\frac{dv}{dt}\right] = 0, \quad \text{i.e.} \quad \frac{d}{dt}[600(e^{-0.4t} - e^{-t})] = 0$$

This gives t = 1.53 sec.

∴ Maximum voltage appearing across the feeder and winding of each transformer is obtained as

$$e_{1(m)} = 0.146 \times 600(e^{-0.6} - e^{-1.53}) = 38.3 \text{ kV.} \quad \text{(Ans.)}$$

EXAMPLE 9.7 A transmission line has surge impedance Z_1. A travelling wave is incident on this line and proceeds to meet another line of surge impedance Z_2 at point P. Derive expressions for transmitted and reflected voltages and currents beyond the point P. Assume the incident wave to be step fronted.

Using the expression of transmitted and reflected voltages, interprete the following situations:

(i) a surge meets a sub-station with an opened out isolator
(ii) a surge meets a sub-station from where 2 more lines are taken out
(iii) a surge travelling over the overhead line meets a cable.

Solution: We know,

$$e = e'\left(1 + \frac{Z_2 - Z_1}{Z_1 + Z_2}\right) = e'\frac{2Z_2}{Z_1 + Z_2} \quad \text{and} \quad i = i'\frac{2Z_1}{Z_1 + Z_2}$$

where e and i are the transmitted voltage and current in Z_2 and e' and i' are the incident voltage and current in Z_1. If e'' and i'' be the reflected voltage and current, we have

$$e'' = e'\left(\frac{Z_2 - Z_1}{Z_2 + Z_1}\right); \; i'' = -i'\left(\frac{Z_2 - Z_1}{Z_1 + Z_2}\right)$$

If the surge meets an open ended isolator, $Z_2 \to \infty$. This gives

(i) $$e = e' \times \frac{2}{\dfrac{Z_1}{Z_2} + 1} = 2e' \text{ (since } Z_2 = \infty\text{).} \quad \text{(Ans.)}$$

(ii) In this case, let Z_R be the resultant of two surge impedances of lines (each of Z) in parallel; this means $Z_2 = Z_R$ and hence

$$\therefore \; Z_2 = \left(\frac{Z_1 \times Z_1}{Z_1 + Z_1}\right) = \frac{Z_1}{2} \text{ and } e = e' \times \frac{2}{\left(\dfrac{Z_1}{Z_1/2}\right) + 1} = e' \times \frac{2}{2+1} = \frac{2}{3}e' \quad \text{(Ans.)}$$

(iii) $$e = e' \times \frac{2Z_2}{Z_1 + Z_2} = e'.\frac{2}{Z_1 / Z_2 + 1}$$

Assuming $Z_1 = 500\ \Omega$ and $Z_2 = 50\ \Omega$, we have for this case, $e = e' \times \frac{2}{11} = \frac{2}{11}e'$. (Ans.)

We may therefore say that in case (i), the transmitted voltage is double the incoming voltage. In case (ii) the transmitted voltage is reduced to two-thirds of the incoming voltage, and in case (iii), the transmitted voltage is very much reduced. Hence, cables play a beneficial role in reducing the effects of surge voltage.

EXAMPLE 9.8 A surge in the line with surge impedance Z_1 meets a substation bus having n number of lines. In such a case, one of the lines is considered to receive the incident wave and $(n - 1)$ lines are assumed to be in parallel carrying the transmitted waves. Under such a condition, obtain the expression for voltage and current.

Solution: $e = e'(1+k) = e'\left[1+\frac{Z_t - Z_1}{Z_t + Z_1}\right]$, where $Z_t = \frac{Z_1}{n-1}$

$\therefore$
$$e = e'\left[1+\frac{Z_1\left(\frac{1}{n-1}-1\right)}{Z_1\left(\frac{1}{n-1}+1\right)}\right] = \frac{2}{n}e'$$

and
$$i = i'(1-k) = \frac{e'}{Z_1}\left[\frac{2(n-1)}{n}\right]$$

[Check: Putting $(n - 1) = 2$ as given in part (ii) of the previous question, $e = \frac{2}{3}e'$.]

EXERCISES

1. Develop an expression of characteristic impedance of a short line in term of open and short circuit parameters.
2. Show that

$$\gamma = \sqrt{(r + j\omega\ell)(g + j\omega c)}$$

[**Hint:** In Figure P9.1,

$$I_R = \frac{Z_2}{Z_2 + Z_0 + \frac{Z_1}{2}} \times I_s$$

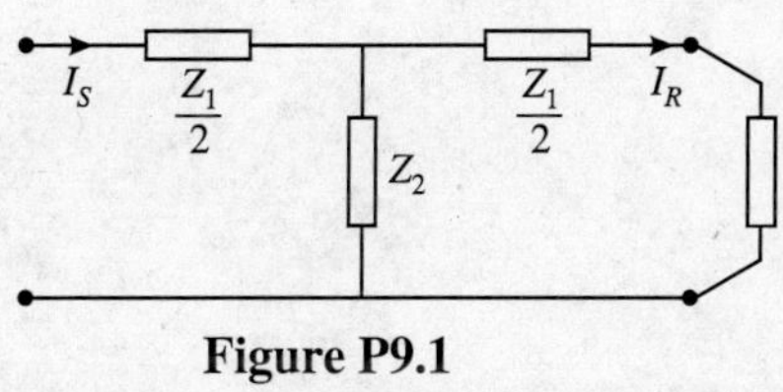

Figure P9.1

However, by definition,

$$e^{\gamma} = \left(\frac{I_S}{I_R}\right) = 1 + \frac{Z_1}{2Z_2} + \frac{Z_0}{Z_2}$$

$$\therefore \qquad e^{\gamma L} = 1 + \frac{Z_1}{2Z_2} + \frac{Z_0}{Z_2}$$

But $\quad Z_1 = (r + j\omega\ell)L \;\; ; Z_2 = \left(\frac{1}{(g + j\omega c)}\right)L$

$$\therefore \qquad e^{\gamma L} = 1 + \frac{(r + j\omega\ell)(g + j\omega c)L^2}{2} + \sqrt{\frac{r + j\omega\ell}{g + j\omega c}}\,(g + j\omega c)L$$

$$= 1 + \sqrt{(r + j\omega\ell)(g + j\omega c)} \cdot L + \frac{(r + j\omega\ell)(g + j\omega c)L^2}{2} \qquad \text{(i)}$$

As $L \to 0$, terms containing L^3 and higher may be neglected.

i.e.

$$e^{\gamma L} = 1 + \gamma L + \frac{\gamma^2 L^2}{L^2} \qquad \text{(ii)}$$

Comparison of equations (i) and (ii) yield,

$$\gamma = \sqrt{(r + j\omega\ell)(g + j\omega c)}\;]$$

3. Derive the telegrapher's equation and show that an incident voltage wave or current wave in a line can be assumed to be propagated as forward and backward waves.

4. Derive an expression for voltage and current in a line at any point x from the sending end.

5. Develop the expressions for reflection coefficients for voltages and currents when the line is terminated by R_L, the load resistance.

6. What are the effects of travelling waves in a line when the line is terminated by pure inductance or pure capacitance?

7. Illustrate Bewley's lattice diagram for travelling waves.

CHAPTER 10

Load Flow Study

10.1 INTRODUCTION

Load flow (or *Complex power flow*) techniques provide basic calculation procedure in order to determine the characteristics of power system under *steady state* operating mode. Load flow studies are usually conducted for planning purposes, or to obtain system behaviour in order to predict the loading of lines and equipment (as well as voltages and currents) of the entire network. Usually, these techniques are off-line and involve computers for solving *steady state load flow equations* (SLFE).

In load flow operations, active power generations are normally specified and the generator bus voltage magnitudes are normally maintained at a specified level (by AVR and machine excitation). The loads are usually specified as *constant power type* with real and reactive components being given. The loads are assumed to be unaffected by small variations of voltage and frequency expected during normal steady state operation.

The types of buses and corresponding parameters being specified are as under:

1. *Load bus* (or *PQ bus*): The total injected power ($P_i + jQ_i$) is specified and normally the loads are constant power type and remains unaffected by small variations in bus voltages.
2. *Slack bus* (or *swing bus*): This type of bus arises because the system losses are not known in advance before the load flow calculation. Any generating bus may be chosen as slack bus, which is assumed to supply line losses. Usually a generator bus with maximum capacity is considered as slack bus. The slack bus voltage is assigned as system reference; its complex bus voltage ($V + j.0$) is specified.
3. *Voltage controlled bus* (or *PV bus*): The total injected power (active) P_i is specified while the voltage magnitude is maintained at $|V|$ (using reactive power injection). This type of bus usually corresponds to a generator bus or may even a load bus where the bus voltage is fixed by supplying reactive power from reactive power compensators.

The power flow problem is formulated assuming the power system network to be linear, bilateral and balanced and having lumped parameters. However, the power and voltage constraints impose non-linearity in the power flow formulation and this invites the help of *iterative techniques* for solution. This load flow solution is expected to give the final result about voltage magnitudes with angles, active and reactive power flows in the individual lines, losses and the reactive power absorbed or generated at PV bus.

Load flow is required to be run for the following cases:

(i) In power system planning
(ii) In addition or outage of transmission lines or other equipment
(iii) When a new load centre is established or when the network remains unchanged but load grows.

A load flow (or complex power flow) solution should have the following solution properties:

- High computational speed
- Simplicity of the program
- Flexibility of the program
- Low computer storage
- Reliability of solution.

In computer-based power system analysis, the Bus *Admittance Matrix,* $[Y_{Bus}]$ is often used in solving *load flow* (or *complex power flow*) problems. Its widespread application in power system computations is due to its simplicity in data preparation and the ease with which it can be formed and modified for any network change (e.g. addition or tripping of line etc.). $[Y_{Bus}]$ matrix is highly *sparse* and facilitates minimum computer storage as well as reduces *computer operation time*. There are different methods of formulation of $[Y_{Bus}]$ matrix and the nodal method is reviewed here which is easily amenable to computer programming and easy to grasp.

10.2 NODAL METHOD FOR DEVELOPMENT OF $[Y_{BUS}]$

In this method of $[Y_{Bus}]$ formation, the variables include the *complex load voltages* being treated as *node voltages* (the reference is the 'ground' for designating the magnitudes of bus voltages and for *voltage angles*, the reference is the angle of one of the bus (or node) voltages which is usually fixed at a datum value (say, zero)). The *node current* being the other variable, it is the net current injected into the network at a given node (from a source and/or load external to the network). When the current enters the network from a node, the sign of the current is assumed to be positive, while for the current leaving the network, the sign is negative; the net nodal current being the algebraic sum of these node currents.

In the *nodal method* it is usual to use branch admittances rather than branch impedances. For an isolated branch y_{ij} (Figure 10.1), the node voltage being V_i and V_j at the buses i and j, respectively, current flowing from node i to node j is given by

$$I_{ij} = y_{ij}\left(V_i - V_j\right) \tag{10.1}$$

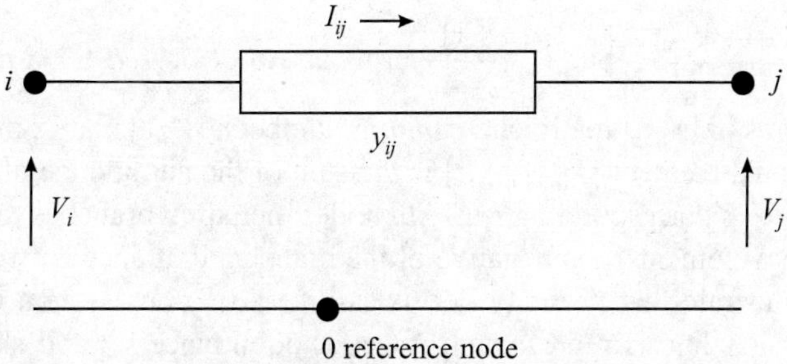

Figure 10.1 Nodal relationship between node voltages and branch currents.

In a complex network the nodes being numbered 0, 1, 2, …, n, where node 0 indicates the reference node, by Kirchoff's current law, the injected current I_i being equal to the sum of all currents leaving node i; thus, we can write

$$I_i = \sum_{j=0}^{n} I_{ij} = \sum_{j=0}^{n} y_{ij}\left(V_i - V_j\right) \tag{10.2}$$

With no ground potential (i.e. with zero reference voltage), for a linear system,

$$I_i = \sum_{\substack{j=0 \\ j\neq i}}^{n} y_{ij}V_i - \sum_{\substack{j=1 \\ j\neq i}}^{n} y_{ij}V_j \tag{10.3}$$

This equation, for a n-bus network, in matrix form can be represented as:

$$\begin{bmatrix} I_1 \\ I_2 \\ \vdots \\ I_n \end{bmatrix} = \begin{bmatrix} Y_{11} & Y_{12} & \cdots & Y_{1n} \\ Y_{21} & Y_{22} & \cdots & Y_{2n} \\ \vdots & \vdots & \vdots & \vdots \\ Y_{n1} & Y_{n2} & \cdots & Y_{nn} \end{bmatrix} \begin{bmatrix} V_1 \\ V_2 \\ \vdots \\ V_n \end{bmatrix} \tag{10.4}$$

or,

$$[I] = [Y_{Bus}]\,[V] \tag{10.5}$$

$[Y_{Bus}]$ is called *Bus Admittance matrix* and it has a well-defined structure. The elements of $[Y_{Bus}]$ are important and hence defined below:

Y_{ii}, the *diagonal element*, is called *self admittance* of node i, while Y_{ij}, the *off-diagonal element*, is called *mutual admittance* (or *transfer admittance*) between nodes i and j.

Obviously,

$$\left.\begin{aligned} Y_{ii} &= \sum_{\substack{j=0 \\ j\neq i}}^{n} y_{ij} \\ \text{and} \quad Y_{ij} &= y_{ij} \end{aligned}\right\} \tag{10.6}$$

The properties of the $[Y_{Bus}]$ matrix are as follows:

(i) $[Y_{Bus}]$ is a *square matrix* of order $n \times n$.

(ii) $[Y_{Bus}]$ is symmetrical, since $y_{ij} = y_{ji}$.

(iii) Only $\frac{(n \times n) - n}{2} + n$, i.e. $\frac{n(n+1)}{2}$ terms are to be *stored* for a n-bus power system.

(iv) The elements of $[Y_{Bus}]$ matrix are *complex* numbers; $[Y_{Bus}]$ matrix itself is thus complex.

(v) Each diagonal element $[Y_{ii}, Y_{jj}, \ldots]$ is the sum of the admittance of the branches which are linked with corresponding i-th, j-th nodes including branches to ground, while each off-diagonal element Y_{ij} is negative of the branch admittance between nodes i and j. In order to illustrate this property, let us assume a two-bus system (Figure 10.2) where a transmission line is represented by series admittance y_{se} and shunt admittance y_{sh}.

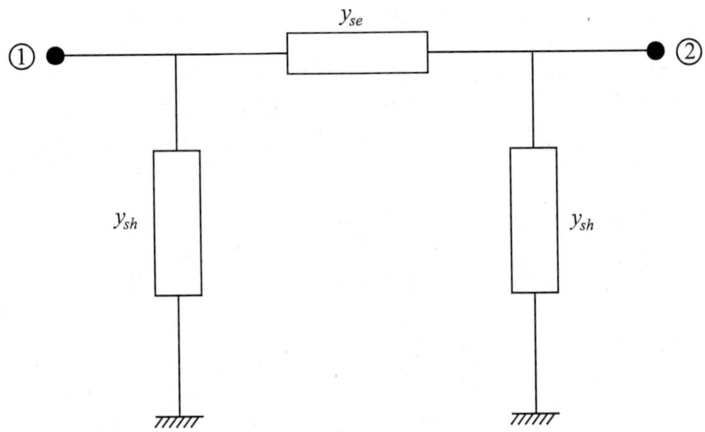

Figure 10.2 Schematic representation of a two-bus system.

In this case the diagonal elements of $[Y_{Bus}]$ are given as

$$Y_{12} = Y_{21} = -y_{se}$$

$$\therefore \quad [Y_{Bus}] = \begin{bmatrix} Y_{11} & Y_{12} \\ Y_{21} & Y_{22} \end{bmatrix} = \begin{bmatrix} y_{se} + y_{sh} & -y_{se} \\ -y_{se} & y_{se} + y_{sh} \end{bmatrix} \tag{10.7}$$

(vi) Y_{ij} $(i \neq j) = 0$ if i-th bus and j-th buses are not connected.

In actual systems lots of interconnections do not exist between a number of buses and hence the $[Y_{Bus}]$ matrix becomes highly *sparse* (containing number of zero elements in the matrix). This saves tremendous computer storage and memory requirements.

10.3 MODELLING OF A TRANSMISSION LINE DUE TO PRESENCE OF REGULATING TRANSFORMER BETWEEN TWO BUSES

In this section we will discuss about the formation of *admittance matrix* between two buses with inclusion of a two-winding transformer between them.

We take into account the p.u. transformer admittances (the series and shunt, using π model) which are actually reciprocal of the p.u. impedance of the transformer that has the complex transformation ratio $(a : 1)$; where a is the off nominal tap ratio; a is complex in general and

$|a| > 1$. We assume first that the regulating transformer (a:1) is situated with the π model included at the off-nominal tap side (non-unity side) of the transformer (Figure 10.3) and in the second case the regulating transformer is situated with the π model included in the same way at the unity tap side of the transformer (Figure 10.4).

Case A: When the regulating transformer (a:1) is present with π model of line at non-unity tap side.

Let us assume the regulating transformer is having a *complex transformation ratio* of a : 1 ($|a| \angle \alpha$). Figure 10.3 represents the voltages and currents at the line and bus side for the transformer being included in the line.

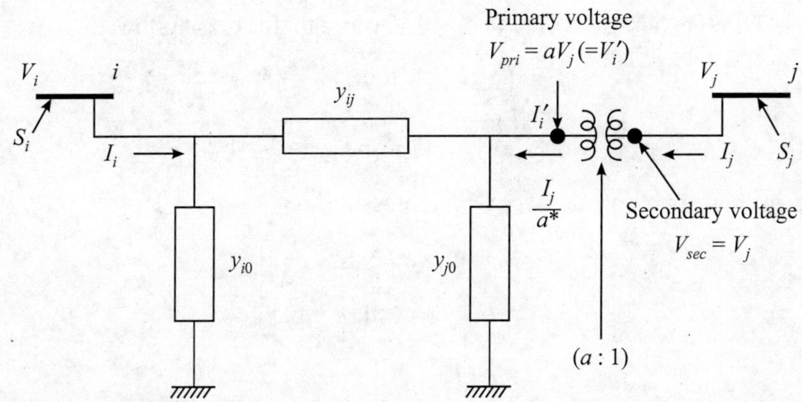

Figure 10.3 Equivalent circuit for a line containing regulating transformer between two buses. [S_i and S_j are the injected complex powers at the i-th and j-th buses respectively, while V_i and V_j are respective bus voltages]. *π model of line is placed at the non-unity side of the off-nominal tap ratio.*

Since the transformer is assumed having complex *off-nominal* tap ratio a : 1, hence assuming the transformer to be loss-less we can write,

$$\frac{V_i'}{V_j} = a = |a|\angle\alpha$$

or, $$V_i'\left(\text{i.e., } V_{pri}\right) = aV_j \tag{10.9}$$

Also, input power being equal to output power for the transformer,

$$V_i' I_i'^* = V_j I_j^* \text{ or, } \frac{V_i'}{V_j} = a = \frac{I_j^*}{I_i'^*} \quad [I_i' \text{ is the secondary current of transformer}]$$

$$\therefore \quad I_i'^* = \frac{I_j^*}{a} \text{ or } I_i' = \frac{I_j}{a^*} \tag{10.10}$$

Here transformer is assumed to be ideal. In case of non-ideal transformer, the transformer reactance is to be added in series with the corresponding line reactions.

Next we consider the current balance at two buses by the following two equations,

$$I_i = \frac{S_i^*}{V_i^*} = V_i y_{i0} + (V_i - aV_j) y_{ij} = V_i y_0 + (V_i - aV_j) y_{ij} \tag{10.11}$$

$$= (y_0 + y_{ij})V_i + (-ay_{ij})V_j \tag{10.11a}$$

and $\dfrac{I_j}{a^*} = \dfrac{S_i^*}{(aV_j)^*} = aV_j y_{j0} + (aV_j - V_i) y_{ij} = aV_j y_0 + (aV_j - V_i) y_{ij}$ [assuming $y_{i0} = y_{j0} = y_0$]

$$\therefore \quad I_j = (-a^* y_{ij})\ V_i + aa^*(y_0 + y_{ij})V_j \tag{10.12}$$

Let us now rewrite equations (10.11) and (10.12) in pair form as follows:

$$I_i = (y_0 + y_{ij})\ V_i + (-ay_{ij})V_j$$

and

$$I_j = (-a^* y_{ij})\ V_i + aa^*(y_0 + y_{ij})V_j$$

In matrix form these two equations can be represented as

$$\begin{bmatrix} I_i \\ I_j \end{bmatrix} = \begin{bmatrix} y_0 + y_{ij} & -ay_{ij} \\ -a^* y_{ij} & aa^*(y_0 + y_{ij}) \end{bmatrix} \begin{bmatrix} V_i \\ V_j \end{bmatrix}$$

or,

$$[I] = [Y][V]$$

where

$$[Y] = \begin{bmatrix} y_0 + y_{ij} & -ay_{ij} \\ -a^* y_{ij} & aa^*(y_0 + y_{ij}) \end{bmatrix} \tag{10.13}$$

It may be noted that a is *complex* and $[Y]$ is *not symmetric*. If a be a *real quantity*, i.e. $a = (\text{kV})_{\text{base}}/(\text{kV})_{\text{tap}}$, then

$$[Y] = \begin{bmatrix} y_0 + y_{ij} & -ay_{ij} \\ -ay_{ij} & a^2(y_0 + y_{ij}) \end{bmatrix} = \begin{bmatrix} Y_1 + Y_2 & -Y_1 \\ -Y_1 & Y_1 + Y_3 \end{bmatrix} \tag{10.14}$$

where,

$$Y_1 = ay_{ij};\ Y_2 = (1-a)y_{ij} + y_0;\ Y_3 = a(a-1)y_{ij} + a^2 y_0$$

The matrix $[Y]$ then becomes *symmetric*.

Case B: When the regulating transformer (a:1) is present with π model of line at unity tap side

With reference to Figure 10.4, $\dfrac{V_j'}{V_i} = \dfrac{1}{a} = \dfrac{1}{a\angle\alpha}$ i.e., $V_j' = \dfrac{V_i}{a}$ (10.15)

Also

$$V_i I_i^* = \left(\frac{V_i}{a}\right) I_j'^*$$

[Power being equal at the transformer input and output, while I_j' is the secondary current of transformer]

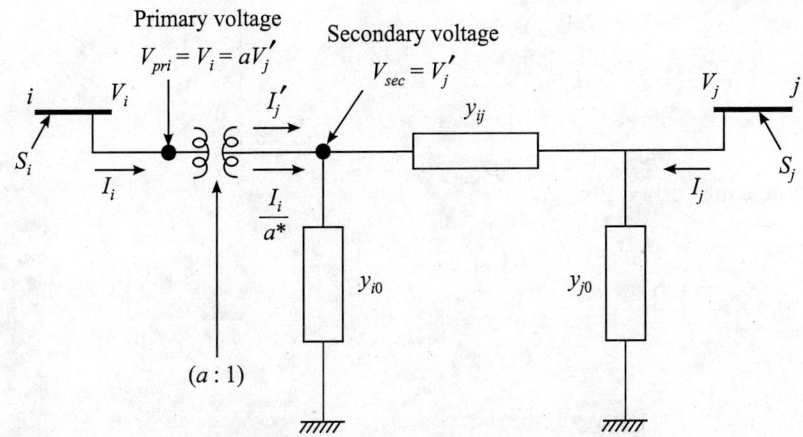

Figure 10.4 Equivalent circuit for a line containing regulating transformer between two buses. π model of line is placed at the unity side of the off-nominal tap ratio.

Also,
$$\frac{I_j'^*}{I_i^*} = a$$

or,
$$I_i = \frac{1}{a^*} I_j' \tag{10.16}$$

At bus i we can write,
$$I_i = \frac{1}{a^*} I_j' = \frac{1}{a^*}\left[\frac{V_i}{a} y_{i0} + \left(\frac{V_i}{a} - V_j\right) y_{ij}\right]$$

or,
$$I_i = \frac{1}{aa^*} V_i y_{i0} + \frac{1}{aa^*} V_i y_{ij} - \frac{1}{a^*} V_j y_{ij}$$

$$\therefore \quad I_i = \frac{1}{aa^*}\left(y_0 + y_{ij}\right) V_i - \frac{1}{a^*} y_{ij} V_j \quad \text{[assuming } y_{i0} = y_{j0} = y_0\text{]} \tag{10.17}$$

Also at bus j we can write, $I_j = y_0 V_j + \left(V_j - \frac{V_i}{a}\right) y_{ij}$

$$\therefore \quad I_j = -\frac{1}{a} V_i y_{ij} + V_j \left(y_0 + y_{ij}\right) \tag{10.18}$$

In matrix form equations (10.17) and (10.18) can be rearranged as

$$\begin{bmatrix} I_i \\ I_j \end{bmatrix} = \begin{bmatrix} \frac{1}{aa^*}\left(y_0 + y_{ij}\right) & -\frac{1}{a^*} y_{ij} \\ -\frac{1}{a} y_{ij} & y_0 + y_{ij} \end{bmatrix} \begin{bmatrix} V_i \\ V_j \end{bmatrix} \tag{10.19}$$

Here
$$[Y] = \begin{bmatrix} \frac{1}{aa^*}(y_0 + y_{ij}) & -\frac{1}{a^*}y_{ij} \\ -\frac{1}{a}y_{ij} & y_0 + y_{ij} \end{bmatrix} \tag{10.20}$$

It may be noted that a being *real*,

$$[Y] = \begin{bmatrix} \frac{1}{a^2}(y_0 + y_{ij}) & -\frac{1}{a}y_{ij} \\ -\frac{1}{a}y_{ij} & y_0 + y_{ij} \end{bmatrix} = \begin{bmatrix} Y_1 + Y_2 & -Y_1 \\ -Y_1 & Y_1 + Y_3 \end{bmatrix} \tag{10.21}$$

where, $Y_1 = \frac{1}{a}y_{ij}$; $Y_2 = \frac{(1-a)}{a^2}y_{ij} + \frac{y_0}{a^2}$ and $Y_3 = \frac{(a-1)}{a}y_{ij} + y_0$

the $[Y]$ matrix becomes *symmetrical.*

[In practical cases, the regulating transformer is designed for either *voltage magnitude* or *phase angle control.* In the former case $\alpha = 0$ and $|a|$ can be changed in discrete steps of $\Delta\, |a|$. In the latter case, $|a|$ is constant and α is changed in discrete steps of $\Delta\alpha$.]

10.4 FORMATION OF $[Y_{BUS}]$ WITH TRANSFORMER PRESENT IN THE LINE

Once the modelling of the branch with the transformer installed between two buses is done, we proceed to modify $[Y_{Bus}]$. We are now in a position to draw the equivalent π circuit of the line-transformer system connected between bus i and j. Figure 10.3(a) represents the π equivalent circuit when the transformer is placed with π line model placed at the non-unity side of off-nominal tap ratio of the transformer, while Figure 10.4(a) represents the π equivalent circuit when π line model is placed at the unity side of off-nominal tap ratio.

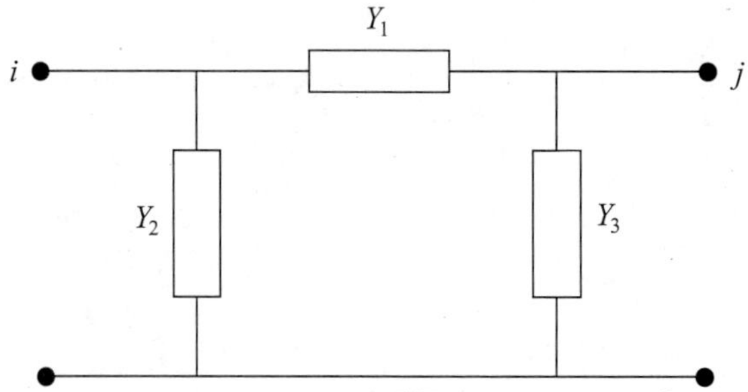

$Y_1 = ay_{ij}$; $Y_2 = (1-a)y_{ij} + y_0$; $Y_3 = a(a-1)y_{ij} + a^2 y_0$

Figure 10.3(a) Equivalent circuit of the transformer with off nominal tap ratio as shown in Figure 10.3.

The $[Y_{Bus}]$ matrix can be modified for inclusion of line with transformer with revised form of its self (diagonal) and transfer (off-diagonal) elements as given below:

$$Y_{ii_{(new)}} = Y_{ii_{(old)}} + Y_1 + Y_2 \tag{10.22}$$

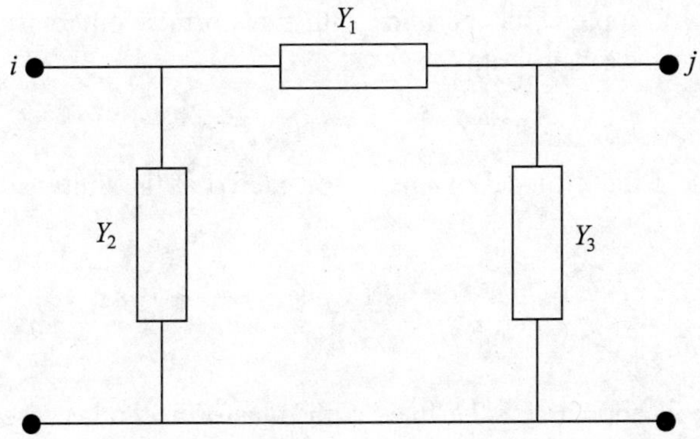

$$Y_1 = \frac{y_{ij}}{a}; Y_2 = \frac{(1-a)}{a^2} y_{ij} + \frac{y_0}{a^2} \text{ and } Y_3 = \frac{(a-1)}{a} y_{ij} + y_0$$

Figure 10.4(a) Equivalent circuit of the transformer with off nominal tap ratio as shown in Figure 10.4.

$$Y_{ij_{(new)}} = Y_{ji_{(new)}} = -Y_1 \tag{10.23}$$

$$Y_{jj_{(new)}} = Y_{jj_{(old)}} + Y_1 + Y_3 \tag{10.24}$$

where $Y_{ii_{(old)}}, Y_{jj_{(old)}}$ and $Y_{ij_{(old)}}$ are the elements of $[Y_{Bus}]$ formed without considering line transformers. The algorithm for obtaining $[Y_{Bus}]$ by nodal method is given in Article 10.4.1.

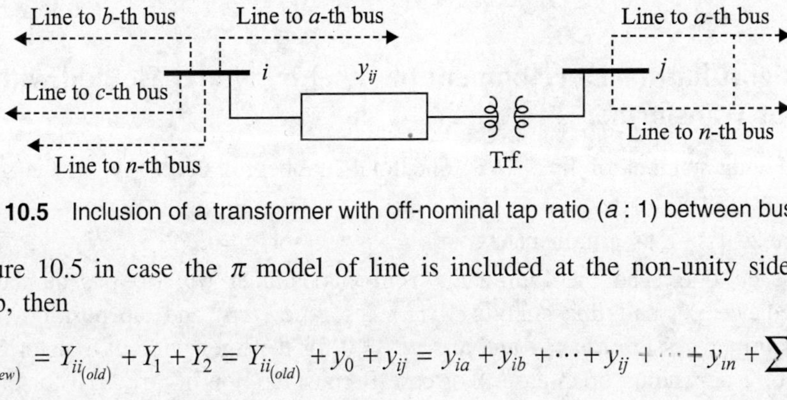

Figure 10.5 Inclusion of a transformer with off-nominal tap ratio (*a* : 1) between bus *i* and *j*.

In Figure 10.5 in case the π model of line is included at the non-unity side of the off-nominal tap, then

$$Y_{ii_{(new)}} = Y_{ii_{(old)}} + Y_1 + Y_2 = Y_{ii_{(old)}} + y_0 + y_{ij} = y_{ia} + y_{ib} + \cdots + y_{ij} + \cdots + y_{in} + \sum y_0 \tag{10.22a}$$

and if y_{i0} is neglected for all lines, we can write

$$Y_{ii_{(new)}} = y_{ia} + y_{ib} + \cdots + y_{ij} + \cdots + y_{in} \tag{10.22b}$$

Also,

$$Y_{ij_{(new)}} = Y_{ji_{(new)}} = -Y_1 = -ay_{ij} \tag{10.23a}$$

$$Y_{jj_{(new)}} = Y_{jj_{(old)}} + Y_1 + Y_3 = Y_{jj_{(old)}} + a^2(y_0 + y_{ij})$$

$$= y_{ja} + y_{jb} + \cdots + a^2(y_0 + y_{ij}) + \cdots + y_{jn} + \sum_{\substack{\text{excluding} \\ \text{line i-j}}} y_0 \tag{10.24a}$$

[Note that $a^2 y_0$ term will appear for the lines with transformer, otherwise it will be y_0]. If y_{i0} is neglected for all lines, we can write

$$Y_{jj(new)} = y_{ja} + y_{jb} + \cdots + a^2 y_{ij} + \cdots + y_{jn} \tag{10.24b}$$

On the other hand, if the π model of line is considered at the unity side of the off-nominal tap, then

$$Y_{ii\,(new)} = Y_{ii\,(old)} + Y_1 + Y_2 = Y_{ii\,(old)} + \frac{y_0}{a^2} + \frac{y_{ij}}{a^2} = y_{ia} + y_{ib} + \cdots + \frac{y_{ij}}{a^2} + \cdots + y_{in} + \frac{y_0}{a^2} + \sum_{\substack{\text{excluding} \\ \text{line i-j}}} y_0 \tag{10.22c}$$

[Note that $\frac{y_0}{a^2}$ term will apprear for the lines with transformer, otherwise it will be y_0]. If y_{i0} is neglected for all lines, we can write

$$Y_{ii(new)} = y_{ia} + y_{ib} + \cdots + \frac{y_{ij}}{a^2} + \cdots + y_{in} \tag{10.22d}$$

Also,

$$Y_{ij(new)} = Y_{ji(new)} = -Y_1 = -\frac{y_{ij}}{a} \tag{10.23b}$$

and

$$Y_{jj(new)} = Y_{jj(old)} + Y_1 + Y_3 = Y_{jj(old)} + (y_0 + y_{ij}) = y_{ja} + y_{jb} + \cdots + y_{ij} + \cdots + y_{jn} + \sum y_{i0} \tag{10.24c}$$

If y_{i0} is neglected for all lines, we can write

$$Y_{jj(new)} = y_{ja} + y_{jb} + \cdots + y_{ij} + \cdots + y_{jn} \tag{10.24d}$$

10.4.1 Algorithm for Development of $[Y_{Bus}]$ by Nodal Method with or without Transformer

1. Read total number of lines (*NL*) and total number of buses (*N*) in the given power system network.
2. Initialize $[Y_{Bus}]$ as a null matrix, i.e., $Y_{Bus_{ij}} = 0$, for, i = 1, 2, ..., *N*.
3. For each line, read the following: from bus number (*fb*), to bus number (*tb*), line impedance (*z*), half line charging (*ych*), tap ratio (*tap*) and tap position (*tap_pos*) of transformer. Assign *tap* = 1 and *tap_pos* = 0 for the lines without transformer. *tap_pos* is 1 or 2 depending upon case-A or case-B consideration of Art. 10.3 for the lines with transformer respectively.
4. Compute line admittance, i.e., $y = 1/z$.
5. Set line count $l = 1$.
6. Go to step 8 if the line has transformer.
7. Modify $[Y_{Bus}]$ for the line-l as follows:
   (a) Modification of diagonal elements,

$$Y_{Bus_{fb\,fb}} = \left(Y_{Bus_{fb\,fb}} + y + ych\right) \text{ and } Y_{Bus_{tb\,tb}} = \left(Y_{Bus_{tb\,tb}} + y + ych\right)$$

(b) Modification of off diagonal elements,

$$Y_{Bus_{fb\,tb}} = \left(Y_{Bus_{fb\,tb}} - y\right) \text{ and } Y_{Bus_{tb\,fb}} = Y_{Bus_{fb\,tb}}$$

(c) Go to step 9.

8. Modify $[Y_{Bus}]$ for the line-l as follows:
   (a) Compute Y_1, Y_2 and Y_3 as given in Figures 10.3(a) and 10.4(a) depending on the position of π model of line with the transformer (refer cases of Art. 10.3).
   (b) Modification of diagonal elements,

$$Y_{Bus_{fb\,fb}} = \left(Y_{Bus_{fb\,fb}} + Y_1 + Y_2\right) \text{ and } Y_{Bus_{tb\,tb}} = \left(Y_{Bus_{tb\,tb}} + Y_1 + Y_3\right)$$

   (c) Modification of off diagonal elements, $Y_{Bus_{fb\,tb}} = \left(Y_{Bus_{fb\,tb}} - Y_1\right)$ and $Y_{Bus_{tb\,fb}} = Y_{Bus_{fb\,tb}}$.

9. Increase line count by 1, i.e., $l \leftarrow l+1$. Go to step 6 if $l \le NL$.
10. Display and/or store final $[Y_{Bus}]$.
11. Stop.

EXAMPLE 10.1 A three-bus system is shown in Figure E10.1. Each line has a series impedance of (0.05 + *j*0.15) p.u. while the shunt admittance is neglected. Find $[Y_{Bus}]$.

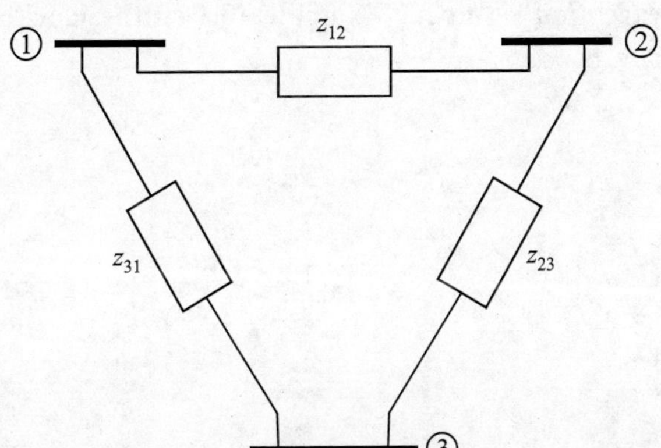

Figure E10.1 A three-bus three-line power system.

Solution: Given: p.u. $z_{12} = z_{23} = z_{31} = (0.05 + j0.15)$

$$\therefore \quad y_{12} = y_{23} = y_{31} \qquad \text{(series admittance of each line)}$$

$$= \frac{1}{(0.05 + j0.15)} = (2 - j6) \text{ p.u.}$$

Since the given problem is a three-bus system, hence $[Y_{Bus}]$ matrix would be a 3 × 3 matrix.

$$\left[Y_{Bus}\right] = \begin{bmatrix} Y_{11} & Y_{12} & Y_{13} \\ Y_{21} & Y_{22} & Y_{23} \\ Y_{31} & Y_{32} & Y_{33} \end{bmatrix}$$

where, $Y_{11} = y_{12} + y_{31};\ Y_{12} = Y_{21} = -y_{12};$

$Y_{22} = y_{12} + y_{23};\ Y_{13} = Y_{31} = -y_{31};$

$Y_{33} = y_{23} + y_{31};\ Y_{23} = Y_{32} = -y_{23}$

Since, $y_{12} = y_{23} = y_{31} = (2 - j6)$ p.u.

$\therefore$ $Y_{11} = 2 - j6 + 2 - j6 = (4 - j12)$ p.u.;

$Y_{12} = Y_{21} = (-2 + j6)$ p.u.;

$Y_{22} = 2 - j6 + 2 - j6 = (4 - j12)$ p.u.;

$Y_{23} = Y_{32} = (-2 + j6)$ p.u.;

$Y_{33} = (4 - j12)$ p.u.;

$Y_{13} = Y_{31} = (-2 + j6)$ p.u.

$$\therefore \quad \left[Y_{Bus}\right] = \begin{bmatrix} (4-j12) & (-2+j6) & (-2+j6) \\ (-2+j6) & (4-j12) & (-2+j6) \\ (-2+j6) & (-2+j6) & (4-j12) \end{bmatrix} \text{ p.u.} \qquad \text{(Ans.)}$$

Identical result is obtained for Example 10.1 by executing computer program following the algorithm given in Art. 10.4.1 and is presented below.

Computer program execution result for Example 10.1 (with algorithm of Art. 10.4.1)

```
All quantities are in p.u.
No. of buses = 3
Elements of [Ybus]:
Row  Column    Ybus            Row   Column    Ybus
 1      1     (4-12i)           2       3     (-2+6i)
 1      2     (-2+6i)           3       1     (-2+6i)
 1      3     (-2+6i)           3       2     (-2+6i)
 2      1     (-2+6i)           3       3     (4-12i)
 2      2     (4-12i)
```

EXAMPLE 10.2 In Example 10.1, for the same three-bus system (Figure E10.1) let a new bus (bus no. 4) be added with bus no. 3 through a transmission line of z (= 0.1 + j0.3) p.u. Obtain $[Y_{Bus}]$.

Solution: Let the bus no. 4 be added to bus no. 3 through a transmission line of $z = (0.1 + j0.3)$ p.u., i.e. $y_{32} = 1/(0.1 + j0.3) = (1 - j3)$ p.u. [Figure E10.2]. Since the new element y_{34} is added with bus 3, entries of Y_{33} will change and new entries of Y_{34} and Y_{44} will appear in the new bus admittance matrix.

Obviously, due to presence of 4-bus system, this bus admittance matrix will be a 4 × 4 matrix.

$Y_{33} = Y_{33(old)} + (1 - j3) = (5 - j15)$ p.u.; $Y_{34} = Y_{43} = -y_{34} = (-1 + j3)$ p.u.; $Y_{44} = (1 - j3)$ p.u.

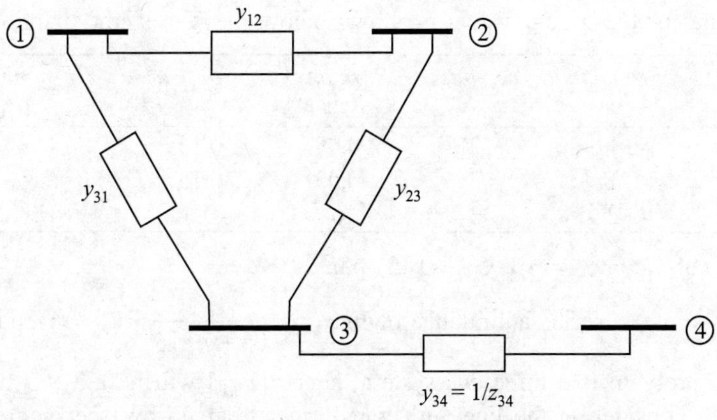

Figure E10.2 A new bus added to three-bus system.

Since there is no connection of bus 4 with any other bus, except bus no. 3, hence,

$$Y_{14} = Y_{41} = 0; \quad Y_{24} = Y_{42} = 0.$$

$[Y_{Bus}]$ is thus obtained as

$$[Y_{Bus}] = \begin{bmatrix} 4-j12 & -2+j6 & -2+j6 & 0 \\ -2+j6 & 4-j12 & -2+j6 & 0 \\ -2+j6 & -2+j6 & 5-j15 & -1+j3 \\ 0 & 0 & -1+j3 & 1-j3 \end{bmatrix} \text{p.u.} \quad \text{(Ans.)}$$

Computer execution for Example 10.2 is left for the reader as an exercise.

EXAMPLE 10.3 A three-bus system is shown in Figure E10.3(a). Assume an ideal transformer to be connected between buses 2 and 3 in series with a line reactance $j0.5$ p.u. If off-nominal tap ratio be 1:1.02, find $[Y_{Bus}]$. Assume line model at the non-unity side of the transformer.

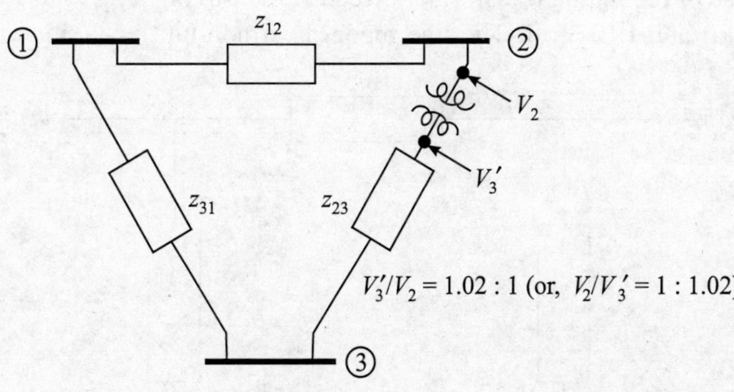

Figure E10.3(a) A three-bus three-line power system.

The line data for the given system is shown below (neglect shunt charging effect)

Line no.	*From bus*	*To bus*	*R (in p.u.)*	*X (in p.u.)*	*Off-nominal tap ratio of transformer*
1	1	2	0.05	0.15	
2	1	3	0.05	0.15	
3	2	3	0	0.5	1:1.02

Solution: Given: $z_{12} = z_{13} = (0.05 + j0.15)$ p.u.

$$\therefore \quad y_{12} = y_{13} = \text{series admittance of line} = \frac{1}{0.05 + j0.15} = (2 - j6) \text{ p.u.}$$

Since the given problem is a three-bus system, hence $[Y_{Bus}]$ will be a 3 × 3 matrix.
∴ $[Y_{Bus}]$ without considering the line with transformer is given by [see Figure E10.3(b)].

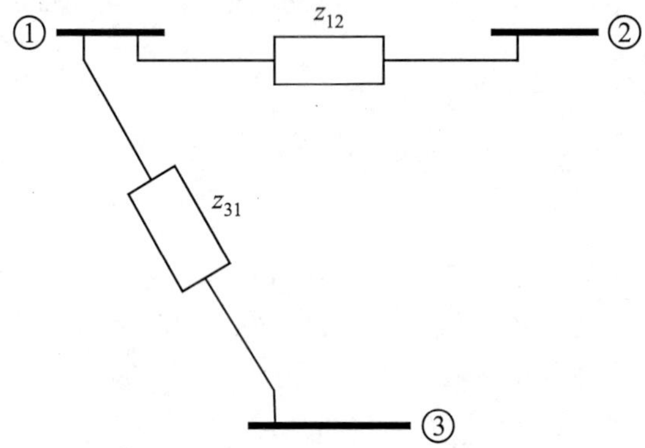

Figure E10.3(b) The given system before considering the line with transformer.

$$[Y_{Bus}] = \begin{bmatrix} (4 - j12) & (-2 + j6) & (-2 + j6) \\ (-2 + j6) & (2 - j6) & (0 + j0) \\ (-2 + j6) & (0 + j0) & (2 - j6) \end{bmatrix} \text{p.u.}$$

This is the case of regulating transformer placed at the i-th bus $(V_3'/V_2 = 1.02)$, i.e., $a = 1.02$ [see Figures 10.3(a) and E10.3(c)]. The line model is at non-unity tap side of the transformer.

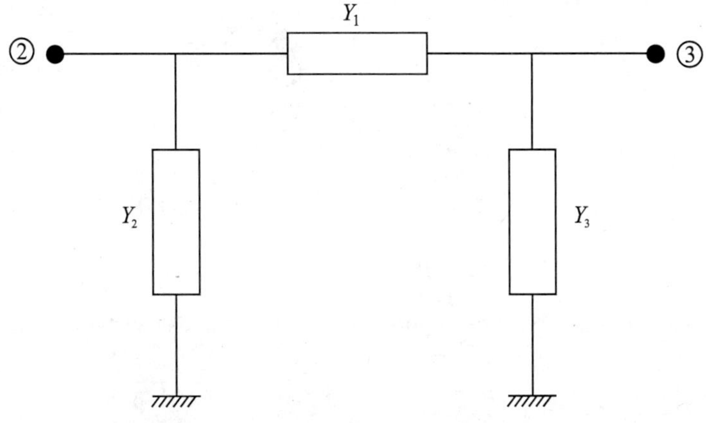

Figure E10.3(c) π model of the line with transformer.

$\because \quad y_{23} = \dfrac{1}{j0.5} = -j2$ p.u.

$\therefore \quad Y_1 = ay_{23} = 1.02 \times (-j2) = -j2.04$ p.u.

$Y_2 = (1-a)y_{23} + y_0 = (1-1.02)\times(-j2)\, [\because y_0 = 0] = j0.04$ p.u.

$Y_3 = a(a-1)y_{23} + a^2 y_0 = 1.02 \times (1.02-1) \times (-j2)\, [\because y_0 = 0] = -j0.0408$ p.u.

$\therefore$ For the transformer in line 2–3,

$$Y_{Bus_{22}} = Y_{Bus_{22_{old}}} + Y_1 + Y_2 = (2-j6) + (-j2.04) + (j0.04) = (2-j8) \text{ p.u.}$$

$$Y_{Bus_{33}} = Y_{Bus_{33_{old}}} + Y_1 + Y_3 = (2-j6) + (-j2.04) + (-j0.0408) = (2-j8.0808) \text{ p.u.}$$

$$Y_{Bus_{23}} = Y_{Bus_{32}} = -Y_1 = -(-j2.04) = j2.04 \text{ p.u.}$$

$$\therefore \quad [Y_{Bus}] = \begin{bmatrix} (4-j12) & (-2+j6) & (-2+j6) \\ (-2+j6) & (2-j8) & j2.04 \\ (-2+j6) & j2.04 & (2-j8.0808) \end{bmatrix} \text{ p.u.} \quad \text{(Ans.)}$$

Computer program execution result for Example 10.3 (following algorithm of Art. 10.4.1)

```
All quantities are in p.u.
No. of buses = 3
Elements of [Ybus] without considering line transformers:
Row   Column      Ybus          Row     Column       Ybus
 1      1       (4-12i)          2        3            0
 1      2       (-2+6i)          3        1         (-2+6i)
 1      3       (-2+6i)          3        2            0
 2      1       (-2+6i)          3        3          (2-6i)
 2      2        (2-6i)
The line-3 is in non-unity tap side of transformer (Case A of Art 10.3).
For line-3, Y1 =(0-2.04i); Y2 =(0+0.04i); Y3 =(0-0.0408i)
Elements of final [Ybus]:
Row   Column      Ybus          Row     Column       Ybus
 1      1       (4-12i)          2        3        (0+2.04i)
 1      2       (-2+6i)          3        1         (-2+6i)
 1      3       (-2+6i)          3        2        (0+2.04i)
 2      1       (-2+6i)          3        3       (2-8.0808i)
 2      2        (2-8i)
```

10.5 ANALYTICAL FORMULATION OF LOAD FLOW SOLUTION

In any power flow problem, it is required to have *four variables* at each bus i of the system, e.g. P_i (the active bus power), Q_i (the reactive bus power), V_i (the bus voltage magnitude), δ_i (the voltage phase angle). Only two of these four variables are known *a priori* and the load flow solution provides the solution of the remaining two variables at any bus.

The complex power injected at bus i is given by

$$S_i = V_i I_i^* \tag{10.25}$$

Here,
$$V_i = |V_i|\angle\delta_i$$

$$I_i = \sum_{k=1}^{N} Y_{ik} V_k$$

$$\left[I_i = \text{Net current injected into the network at bus } i = Y_{i1}V_1 + Y_{i2}V_2 + \cdots + Y_{iN}V_N = \sum_{k=1}^{N} Y_{ik}V_k \right]$$

$$S_i = P_i + jQ_i$$

Rewriting the load flow equation (10.25) as $S_i^* = V_i^* I_i$, we have

$$P_i - jQ_i = V_i^* \sum_{k=1}^{N} Y_{ik} V_k = \sum_{k=1}^{N} |Y_{ik} V_i V_k| \angle(\phi_{ik} + \delta_k - \delta_i) \tag{10.26}$$

[since $Y_{ik} = |Y_{ik}|\angle\phi_{ik} = |Y_{ik}|\cos\phi_{ik} + j|Y_{ik}|\sin\phi_{ik} = G_{ik} + jB_{ik}$]

Separating equation (10.26) into real and imaginary parts, we obtain

$$\left.\begin{aligned} P_i &= \sum_{k=1}^{N} |Y_{ik} V_i V_k| \cos(\phi_{ik} + \delta_k - \delta_i) \\ Q_i &= -\sum_{k=1}^{N} |Y_{ik} V_i V_k| \sin(\phi_{ik} + \delta_k - \delta_i) \end{aligned}\right\} \tag{10.27}$$

Equation (10.27) represents the polar form of the power flow also and is known as *steady state load flow equation (SLFE)*. It provides the calculated values of the network real and reactive power injected at any bus i.

If P_{gi} be the scheduled active power being generated at the i-th bus and P_{di} the scheduled power demand of the load at that bus, then

$$P_{i_{\text{scheduled}}} = P_{g_i} - P_{d_i} \tag{10.27a}$$

Similarly for reactive power
$$Q_{i_{\text{scheduled}}} = Q_{g_i} - Q_{d_i} \tag{10.27b}$$

Therefore, active power mismatch is given by

$$\Delta P_i = P_{i_{\text{scheduled}}} - P_{i_{\text{calculated}}} = (P_{g_i} - P_{d_i}) - P_i (= \theta) \tag{10.28a}$$

Similarly, for reactive power mismatch at the i-th bus

$$\Delta Q_i = Q_{i_{\text{scheduled}}} - Q_{i_{\text{calculated}}} = (Q_{g_i} - Q_{d_i}) - Q_i (= \psi) \tag{10.28b}$$

In solving power flow problems, mismatches do appear where P_i and Q_i (calculated values) do not coincide with scheduled values. The mismatches approach zero when the calculated values of P and Q coincide with the scheduled values.

Each bus i has two equations involving P and Q (equation 10.27). The power flow problem requires solution of equation (10.27) for values of the unknown bus voltages which makes equation (10.27) to be numerically satisfied at each bus. In case there is no scheduled value ($P_{i_{calculated}}$) at any bus i, then the mismatch (ΔP_i) cannot be defined and there is no need to satisfy equation (10.27a). Similarly once (ΔQ_i) is not specified, equation (10.27b) need not be solved (*for PV buses* ΔQ_i *is not specified*).

In solving the power flow problems, the bus type is first identified and at each bus, two of the four quantities $|V_i|$, δ_i, P_i and Q_i are specified and the remaining two are to be calculated. *For a PQ bus,* P_i *and* Q_i *are known and* $|V_i|$ and δ_i *are to be determined. In a PV bus,* P_i *and* $|V_i|$ *being known and it is required to obtain* Q_i *and* δ_i.

Table 10.1 provides configurations of each power system bus for complex power flow analysis:

Table 10.1 Summary of power flow variables.

Bus	*Bus type*	*Specified quantities*	*No. of SLFE*	*No. of variables* $(\delta, \lvert V\rvert)$
Slack (or swing bus)	0	$\lvert V\rvert, \delta$	0	0
Voltage controlled (or *PV*) bus	1	$P, \lvert V\rvert$	1	δ
Load bus (or *PQ* bus)	2	P, Q	2	$\delta, \lvert V\rvert$

Power flow solution techniques solve a set of simultaneous non-linear algebraic equations involving trigonometric functions. There are two unknown variables at each node in a system. A second set of equations, which are linear, are derived from the original set of equations and a method of iteration is applied to this second set.

In any of the power flow solution techniques, the system data (e.g., bus power, network connection, impedance/admittance) are read. Initial voltage being specified to all the buses, for the first iteration, *PQ* bus voltages are set to $(1 + j.0)$ while *PV* bus voltages are set to $(V + j.0)$. Next $[Y_{Bus}]$ matrix being formed, iterative methods are applied to update voltages and angles in order to satisfy the specified conditions of load and generation. The iteration cycle is terminated when the bus voltages and angles are such that the specified condition of load and generation are satisfied [this condition stipulates that the power mismatches ΔP and ΔQ for all buses are to be less than a tolerance value].

Power flow calculations usually employ conventional iterative techniques such as *Gauss-Seidal (G-S), Newton-Raphson (N-R), Fast Decoupled Load Flow (FDLF)* and *DC Load Flow (DCLF)* methods. Using the solution techniques, when a solution is obtained, complete terminal condition for all buses are computed. Line flows, system losses and slack bus power can then be calculated.

To incorporate the *generation limit* in any generator bus, we simply set $P_{g_{i_{min}}} \le P_{g_i} \le P_{g_{i_{max}}}$ to fix the real power generation. For inclusion of reactive power limits, first Q_{gi} is calculated and then it is to be checked whether $Q_{g_{i_{min}}} \le Q_{g_i} \le Q_{g_{i_{max}}}$. If Q_{gi} is not within limits, it is a usual practice to set at Q_{gi} appropriate limit [if $Q_{g_i} > Q_{g_{i_{max}}}$, the practice is to set $Q_{g_i} = Q_{g_{i_{max}}}$ releasing the constraint that $|V_i|$ is fixed. On the other hand, if $Q_{g_i} < Q_{g_{i_{min}}}$, set $Q_{g_i} = Q_{g_{i_{min}}}$ releasing again the constraint that $|V_i|$ is fixed]. If $|V_j|$ rises above $V_{i_{specified}}$ when $Q_{g_i} = Q_{g_{i_{max}}}$, it is possible to set the bus to be a generator bus again. Once Q_{gi} violates any of the limits, δ_i and $|V_j|$ of that i-th bus become variables.

10.6 GAUSS-SEIDAL (G-S) METHOD OF POWER FLOW

Let us rewrite the power flow equation (10.26)

$$S_i^* = P_i - jQ_i = V_i^* \sum_{k=1}^{N} Y_{ik} V_k$$

This equation may also be expressed in the following form

$$P_i - jQ_i = V_i^* Y_{ii} V_i + V_i^* \sum_{\substack{k=1 \\ k \neq i}}^{N} Y_{ik} V_k \text{, for } i = 2, 3, \ldots, N$$

[N being the total number of buses and bus-1 being the slack bus]

or,

$$V_i = \frac{1}{Y_{ii}} \left[\frac{P_i - jQ_i}{V_i^*} - \sum_{\substack{k=1 \\ k \neq i}}^{N} Y_{ik} V_k \right] \text{, for } i = 2, 3, \ldots, N \tag{10.29}$$

In *G-S algorithm*, equation (10.29) is utilised to find the final bus voltages using successive steps of iterations, where

$$V_i^{p+1} = \frac{1}{Y_{ii}} \left[\frac{P_i - jQ_i}{\left(V_i^p\right)^*} - \sum_{\substack{k=1 \\ k \neq i}}^{N} Y_{ik} V_k^p \right] \text{, for } i = 2, 3, \ldots, N$$

[p is the iteration count] (10.30)

i.e.,

$$V_i^{p+1} = \frac{1}{Y_{ii}} \left[\frac{P_i - jQ_i}{(V_i^p)^*} - \sum_{k=1}^{i-1} Y_{ik} V_k^{p+1} - \sum_{k=i+1}^{N} Y_{ik} V_k^p \right] \tag{10.31a}$$

As V_1 is known *a priori*, computations can to be performed for buses 2, 3, …, N.

G-S algorithm convergence is slower and it is conventional to use *acceleration factor* (α) for speeding up the convergence process. However, if α increases too much, the system may diverge. The acceleration factor is introduced in the algorithm for the p-th iteration as follows:

$x^{p+1} = x^p + \alpha\Delta x$, α being the acceleration factor.

[Usually, the choice of acceleration factor of 1.4 to 1.6 gives best convergence.]

Hence,

$$V_{i_{acc}}^{p+1} = V_i^p + \alpha\left(V_i^{p+1} - V_i^p\right) \tag{10.31b}$$

Advantages of G-S method

- It is one of the simplest iterative methods in power flow studies since early days of digital power flow analysis.
- Because of its simplicity, G-S method does have a definite tutorial value, particularly for the beginners.
- G-S method can be conveniently used for load flow studies in small power systems.
- G-S method may be used for even large systems to obtain first approximate solution, which can then be used as "initial solution" for Newton-Raphson method.

The algorithm of G-S method of power flow is given below. However, G-S method convergence becomes increasingly slower as the system size grows and hence it is not very much common to use for practical load flow studies or for general research studies involving power flows in large complex networks.

10.6.1 Algorithm to Calculate Bus Voltages by G-S Method of Power Flow

1. Read line data for the power system having total number of lines = *NL* and total number of buses = *N*.
2. Read bus type ($type_i$), initial bus voltage magnitude in p.u. $\left(|V_i|^0\right)$, initial bus voltage angle in radian (δ_i^0), real power generation (P_{g_i}), reactive power generation (Q_{g_i}), real power demand (P_{d_i}), reactive power demand (Q_{d_i}), lower and upper Q-limit at PV buses ($Q_{g_{i_{min}}}$ and $Q_{g_{i_{max}}}$) all in p.u. for each bus $i = 1, 2, ..., N$.
3. Specify acceleration factor, α (say 1.4) and tolerance ε (say 0.0001).
4. Form [Y_{Bus}] by nodal method [as described in Article 10.4.1].
5. Compute scheduled injected bus power for all buses using equations (10.27a) and (10.27b) $S_i = \left(P_{g_i} - P_{d_i}\right) + j\left(Q_{g_i} - Q_{d_i}\right)$, $\left(j = \sqrt{-1}\right)$ for $i = 1, 2, ..., N$.
6. Convert all the bus voltages into rectangular form, $V_i^0 = |V_i|^0 \angle\delta_i^0$
7. Store specified voltage for PV buses, i.e., $\left|V_{i_{specified}}\right| = \left|V_i^0\right|$, for $i = 1, 2, ..., N$ (only if $type_i = 1$).
8. Set iteration count $p = 0$
9. Fix slack bus voltage for next iteration, $V_1^{p+1} = V_1^0$
10. Set bus count $i = 2$
11. If $type_i = 2$ (i.e., load bus) go to step 16 else go to step 12.
12. Calculate Q_i for the *i*-th PV bus ($type_i = 1$), using equation (10.27) and assign reactive power generation as $Q_{g_i} = (Q_i + Q_{d_i})$.
13. If $Q_{g_{i_{min}}}$ and $Q_{g_{i_{max}}}$ are specified then go to step 14, otherwise go to step 15.
14. If $Q_{g_i} > Q_{g_{i_{max}}}$ or, $Q_{g_i} < Q_{g_{i_{min}}}$ for the PV bus-*i* (i.e., Q-limit is failed) then compute the following else go to step 15.

    (i) set $Q_{g_i} = Q_{g_{i_{max}}}$ for $Q_{g_i} > Q_{g_{i_{max}}}$,

    or, set $Q_{g_i} = Q_{g_{i_{min}}}$ for $Q_{g_i} < Q_{g_{i_{min}}}$

    and also update the reactive injection at bus-*i* as $Q_i = Q_{g_i} - Q_{d_i}$

    (ii) update S_i with Q_i as obtained in step 14(i) keeping real part unaltered, i.e., $S_i = (P_{g_i} - P_{d_i}) + jQ_i$

(iii) release the constraint of $|V_i|$ to be fixed and initialize it for a flat start, i.e., $V_i^p = 1\angle 0°$ and also change bus type from PV bus to PQ bus, i.e., set $type_i = 2$ (as Q-limit is failed).

(iv) calculate V_i^{p+1} using equation (10.31a) and accelerate it using equation (10.31b) as $V_i^{p+1} = V_i^{p+1} + \alpha(V_i^{p+1} - V_i^p)$.

(v) calculate $\Delta V_i = V_i^{p+1} - V_i^p$ and go to step 17.

15. Compute the following steps as $Q_{g_{i_{\min}}} \le Q_i \le Q_{g_{i_{\max}}}$ (i.e., Q-limit is maintained) or, $Q_{g_{i_{\min}}}$ and $Q_{g_{i_{\max}}}$ are not specified for the PV bus-i.

(i) update S_i with Q_i obtained in step 12 as $S_i = (P_{g_i} - P_{d_i}) + jQ_i$.

(ii) calculate accelerated V_i^{p+1} using equations (10.31a) and (10.31b) as in step 14 (iv).

(iii) express V_i^{p+1} in polar form as $V_i^{p+1} = |V_i^{p+1}|\angle\delta_i^{p+1}$.

(iv) update V_i^{p+1} as $V_i^{p+1} = |V_{i_{\text{specified}}}|\angle\delta_i^{p+1}$ and calculate $\Delta V_i = V_i^{p+1} - V_i^p$ and go to step 17.

16. Calculate accelerated V_i^{p+1} for PQ bus ($type_i = 2$) using equations (10.31a) and (10.31b) [as shown in step 14(iv)] and also calculate $\Delta V_i = V_i^{p+1} - V_i^p$.
17. Increase bus count by 1, i.e., $i \leftarrow i+1$. Go to step 11 if $i \le N$.
18. If max(ΔV) > ε then update iteration count $p \leftarrow p+1$, else go to step 20.
19. If p > 20 then break the iteration process and go to step 23. Otherwise go to step 9.
20. Calculate reactive power generations for those PV buses who have not crossed their Q-limit. $Q_{g_i} = Q_i + Q_{d_i}$ for i = 1, 2, ..., NB provided $type_i = 1$.
21. Display and/or store final bus voltages, V_i^{p+1} for i = 1, 2, ..., N.
22. Display and/or store reactive power generation, Q_{g_i} for PV buses ($type_i = 1$), for i = 1, 2, ..., N.
23. Stop.

Note that in order to generalize the algorithm in execution part $Q_{g_{i_{\min}}}$ and $Q_{g_{i_{\max}}}$ are considered to be zero for load (PQ) buses. Also, for the same reason reactive power generation (Q_{g_i}) for PV buses are considered to be zero initially. For convenience, it is assumed that if the program does not converge within 20 iterations then the convergence criterion is not possible to achieve.

EXAMPLE 10.4 A four-bus, five-line system (Figure E10.4) has the following line data

Line no.	*From bus*	*To bus*	*Line impedance (p.u.)*	*B/2 (p.u.)*
1	1	2	(0.15 + *j*0.4)	*j*0.04
2	1	3	(0.10 + *j*0.3)	*j*0.05
3	1	4	(0.15 + *j*0.6)	*j*0.04
4	2	3	(0.07 + *j*0.25)	*j*0.03
5	3	4	(0.09 + *j*0.3)	*j*0.04

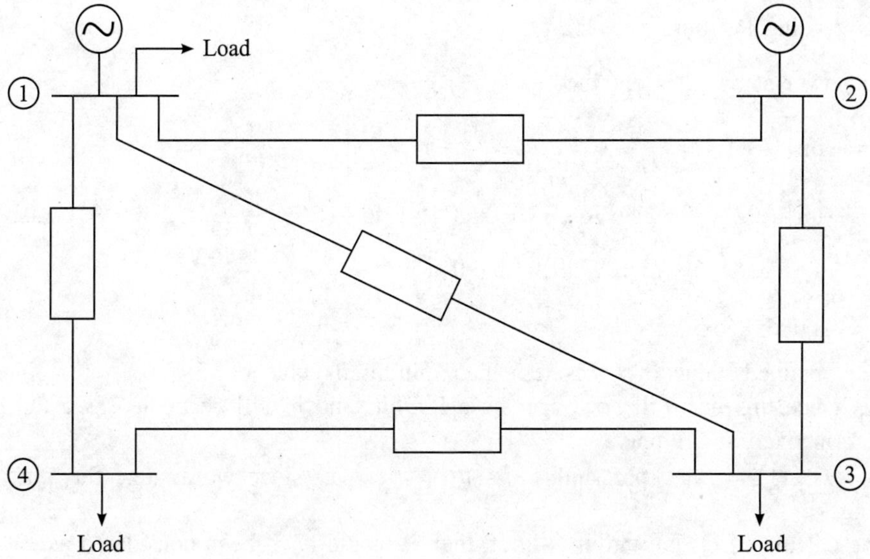

Figure E10.4 Four-bus, five-line system.

The power and bus voltages are as under (P_d, Q_d, P_g, Q_g and V are in p.u.),

Bus no.	P_d	Q_d	P_g	Q_g	V	*Bus type*
1	2	0.6	?	?	$1.06\angle 0°$	Slack bus
2	0	0	1.2	?	$\lvert 1.02 \rvert$	PV bus
3	0.7	0.5	0	0	?	PQ bus
4	0.6	0.3	0	0	?	PQ bus

[The suffix '*d*' represents demand while the suffix '*g*' represents generation].

Find bus voltages by *G-S* method, when bus-2 has *Q*-limit of $0.1 \le Q_2 \le 1.0$ p.u. Assume acceleration factor and specified tolerance as 1.4 and 0.0001 respectively. Write a MATLAB code for this problem.

Solution: $[Y_{bus}]$ is obtained using the techniques described in Art. 10.4.1 and the non-zero elements are given below. Note that only upper triangular off diagonal elements are provided.

```
Row  Column         Ybus                    Row  Column          Ybus
 1      1    (2.214075-6.630408i)            2      3     (-1.038576+3.709199i)
 1      2   (-0.821918+2.191781i)            3      3      (2.956007-9.647303i)
 1      3   (-1.000000+3.000000i)            3      4     (-0.917431+3.058104i)
 1      4   (-0.392157+1.568627i)            4      4      (1.309588-4.546731i)
 2      2    (1.860493-5.830980i)
```

In this solution all quantities are expressed in p.u. and angles are expressed in radian. Initial bus voltages are,

$$V_1^0 = (1.06 + j0)(\text{fixed}),\ V_2^0 = (1.02 + j0),\ V_3^0 = (1 + j0),\ V_4^0 = (1 + j0)$$

[Superscript zero stands for the initial guess.]

Here bus-2 being PV bus, $P_2 = 1.2, Q_{g_{2_{min}}} = 0.1$ and $Q_{g_{2_{max}}} = 1$.

Reactive power at bus-2 is obtained as,

$$Q_2 = -\operatorname{Im}\left[\left(V_2^0\right)^*\left\{Y_{21}V_1^0 + Y_{22}V_2^0 + Y_{23}V_3^0 + Y_{24}V_4^0\right\}\right]$$

$$= -\operatorname{Im}\left[1.02\{(-0.8219178 + j2.191281)\times 1.06 + (1.860493 - j5.830979)\times 1.02 + (-1.038576 + j3.709199)\times 1.0 + 0\}\right]$$

$$= -0.086586 \text{ p.u.}$$

Since obtained value of Q_2 is less than minimum value of Q_{g2} (0.1 p.u.) given in the problem, hence this bus will no longer be a PV bus and it will be treated as a PQ bus with Q-generation fixed at 0.1 p.u.

$\therefore$ $P_2 - jQ_2 = (1.2 - j0.1)$ p.u. and $V_2^0 = (1.0 + j0)$ p.u. [since we assume flat start for load bus voltages].

If the calculated Q_2 be within Q-limit, then it would have been considered as PV bus and then $P_2 - jQ_2 = (1.2 - jQ_2)$ p.u.

In G-S method, the bus voltage at i-th bus at $(p+1)$-th iteration equation (10.31a) is given by

$$V_i^{p+1} = \frac{1}{Y_{ii}}\left[\frac{P_i - jQ_i}{(V_i^p)^*} - \sum_{k=1}^{i-1} Y_{ik}V_k^{p+1} - \sum_{k=i+1}^{N} Y_{ik}V_k^p\right]$$

Net complex power injection being S_i, we can write

$$S_i^* = (-S_{load})^*, \text{ if } S_g = 0; \text{ otherwise,}$$

$$S_i^* = (S_g - S_{load})^* = (P_g + jQ_g - P_{load} - jQ_{load})^* = (P_g - P_{load}) - j(Q_g - jQ_{load})$$

Thus $P_3 - jQ_3 = (-0.7 + j0.5)$ p.u. and $P_4 - jQ_4 = (-0.6 + j0.3)$ p.u.

For bus-2,

$$V_2^1 = \frac{1}{Y_{22}}\left[\frac{P_2 - jQ_2}{\left(V_2^0\right)^*} - \sum_{j=1}^{2-1} Y_{2j}V_j^1 - \sum_{j=3}^{4} Y_{2j}V_j^0\right] = \frac{1}{Y_{22}}\left[\frac{P_2 - jQ_2}{\left(V_2^0\right)^*} - Y_{21}V_1^1 - \left\{Y_{23}V_3^0 + Y_{24}V_4^0\right\}\right]$$

$$= \frac{P_2 - jQ_2}{Y_{22}\left(V_2^0\right)^*} - \frac{Y_{21}}{Y_{22}}V_1^1 - \frac{Y_{23}}{Y_{22}}V_3^0 - \frac{Y_{24}}{Y_{22}}V_4^0$$

$$= (0.075162 + j0.18182) - (-0.40489 - j0.020226) - (-0.62892 + j0.022557) - 0$$

$$= (1.108976 + j0.179484) \text{ p.u.}$$

$\therefore$ The change in voltage is,

$$\Delta V_2^1 = V_2^1 - V_2^0 = (1.108976 + j0.179484) - (1 + j0) = (0.108976 + j0.179484) \text{ p.u.}$$

$$\therefore \quad V_{2_{accelerated}}^1 = V_2^0 + \alpha\Delta V_2^1 = 1.0 + 1.4\,(0.108976 + j0.179484)$$

$$= (1.152566 + j0.251277) = 1.179639\ \angle 0.214656 \text{ p.u.}$$

This value of voltage at bus-2 will be used in subsequent calculations of voltage for the remaining buses.

Again, $V_3^1 = \frac{1}{Y_{33}}\left[\frac{P_3 - jQ_3}{\left(V_3^0\right)^*} - \sum_{k=1}^{3-1} Y_{3k}V_k^1 - \sum_{j=4}^{4} Y_{3k}V_k^0\right] = \frac{P_3 - jQ_3}{Y_{33}\left(V_3^0\right)^*} - \frac{Y_{31}}{Y_{33}}V_1^1 - \frac{Y_{32}}{Y_{33}}V_2^1 - \frac{Y_{34}}{Y_{33}}V_4^0$

$$= (-0.067704 - j0.051814) - (-0.33211 - j0.0081137) - (-0.44219 - j0.085199)$$
$$-(-0.31642 + j\,0.0018568)$$
$$= (1.023023 + j0.039641) \text{ p.u.}$$

$$\Delta V_3^1 = V_3^1 - V_3^0 = (1.023023 + j0.039641) - (1.0 + j0)$$
$$= (0.023023 + j0.039641) \text{ p.u.}$$

$$\therefore \quad V_{3_{\text{accelerated}}}^1 = V_3^0 + \alpha\,\Delta V_3^1 = 1 + 1.4 \times (0.023023 + j0.039641)$$
$$= (1.032232 + j0.055498) = 1.033723\angle 0.053713 \text{ p.u.}$$

and $V_4^1 = \frac{1}{Y_{44}}\left[\frac{P_4 - jQ_4}{(V_4^0)^*} - \sum_{k=1}^{4-1} Y_{4k}V_k^1\right] = \frac{P_4 - jQ_3}{Y_{44}(V_4^0)^*} - \frac{Y_{41}}{Y_{44}}V_1^1 - \frac{Y_{42}}{Y_{44}}V_2^1 - \frac{Y_{43}}{Y_{44}}V_3^1$

$$= (-0.096024 - j0.10431) - (-0.362 + j0.012842) - 0 - (-0.69607 - j0.045121)$$
$$= (0.96205 - j0.072026) \text{ p.u.}$$

$$\Delta V_4^1 = V_4^1 - V_4^0 = (0.96205 - j0.072026) - (1 + j0) = (-0.03975 - j0.072026) \text{ p.u}$$

$$V_{4_{\text{accelerated}}}^1 = V_4^0 - \alpha\,\Delta V_4^1 = 1 + 1.4(-0.03975 - j0.072026)$$
$$= (0.946868 - j0.100836) = 0.952223\angle -0.106094 \text{ p.u.}$$

After first iteration, $\Delta V_2^1 = V_2^1 - V_2^0 = (1.152566 + j0.251277) - (1 + j0)$

$$= (0.152566 + j0.251277) = 0.29397\angle 1.0251 \text{ p.u.}$$
$$\Delta V_3^1 = (0.032232 + j0.055498) = 0.064179\angle 1.0446 \text{ p.u.}$$
$$\Delta V_4^1 = (-0.053132 - j0.100836) = 0.11398\angle -2.0557 \text{ p.u.}$$

It can be observed from the calculations shown above that the magnitude of maximum change in bus voltages exceeds the given tolerance value. Thus the iterative process has not converged. In the next step, the process of iteration will starts again with calculation of voltage at bus-2 in the prescribed manner as described in the text. Note that bus voltage magnitude and angle for slack bus will remain constant throughout the iterative process.

If bus-2 be a PV bus (i.e., its Q-limit being maintained) then Q_2 is to be calculated again in next iteration. Q-limit is also to be checked and if this criterion is maintained then only load angle of the voltage needs to be modified while the magnitude of bus voltage remains at 1.02 p.u.

The MATLAB code for Example 10.4 is as follows:

```
% Program to calculate bus voltages by G-S method of power flow for Example 10.4
clc
clear all
alp=1.4; epsil=0.0001; bus_no=4;line_no=5;type=[0 1 2 2];
v=[1.06 1.02 1 1]'; del=[0 0 0 0]'; pg=[0 1.2 0 0]; qg=[0 0 0 0];
pd=[2 0 0.7 0.6]; qd=[0.6 0 0.5 0.3]; qmin=0.1; qmax=1;
disp('All quantities are in p.u. and all angles are in radian.')
disp(sprintf('No. of buses = %d. No. of lines = %d.',bus_no,line_no));
Ybus_prog %Prog. to calculate [Ybus]
vsp=v; psch=pg-pd; qsch=qg-qd; ssch=psch+i*qsch; vr=v.*exp(i*del);
kp=1;
for kp =1:100
  vr(1,kp+1)=vr(1,1);
  for m=2:bus_no
   if type(m)==1  s(m)=0;
    for k=1:bus_no
       if (k < m) s(m)= s(m)+ conj(vr(m,kp))*Y(m,k)*vr(k,kp+1);
       else s(m)= s(m)+ conj(vr(m,kp))*Y(m,k)*vr(k,kp);end
    end
    q(m)=-imag(s(m));
    if (q(m)<qmin||q(m)>qmax)
    if q(m)<qmin q(m)=qmin; qg(m)=q(m);  else q(m)=qmax; qg(m)=q(m); end
    type(m)=2; vr(m,kp)=1.0; ssch(m)=psch(m)+i*q(m);
    vr(m,kp+1)=(conj(ssch(m)))/conj(vr(m,kp));yv1=0;
    for n=1:(m-1)  yv1= yv1-Y(m,n)*vr(n,kp+1); end
    yv2=0;
    for n=(m+1):bus_no yv2=yv2-Y(m,n)*vr(n,kp); end
    vr(m,kp+1)= (vr(m,kp+1)+yv1+yv2)/Y(m,m);
    vr(m,kp+1)=vr(m,kp)+alp*(vr(m,kp+1)-vr(m,kp));
    delv(m)=vr(m,kp+1)-vr(m,kp);
    else
    ssch(m)=psch(m)+i*q(m);qg(m)=q(m);
    vr(m,kp+1)=conj(ssch(m))/conj(vr(m,kp));
    for k=1:(m-1) vr(m,kp+1)=vr(m,kp+1)-Y(m,k)*vr(k,kp+1); end
     for k=m+1:bus_no vr(m,kp+1)=vr(m,kp+1)-Y(m,k)*vr(k,kp);end
     vr(m,kp+1)= vr(m,kp+1)/Y(m,m);
     vr(m,kp+1)=vr(m,kp)+alp*(vr(m,kp+1)-vr(m,kp));
     vr(m,kp+1)= vsp(m)*exp(j*angle(vr(m,kp+1)));
     delv(m)=vr(m,kp+1)-vr(m,kp);
    end
   else
     vr(m,kp+1)=(conj(ssch(m)))/(conj(vr(m,kp))*Y(m,m)); yv1=0;
     for n=1:(m-1)  yv1= yv1-Y(m,n)*vr(n,kp+1)/Y(m,m);  end
     yv2=0;
     for n=(m+1):bus_no  yv2=yv2-Y(m,n)*vr(n,kp)/Y(m,m); end
```

```
      vr(m,kp+1)= (vr(m,kp+1)+yv1+yv2);
      vr(m,kp+1)=vr(m,kp)+alp*(vr(m,kp+1)-vr(m,kp));
      delv(m)=vr(m,kp+1)-vr(m,kp);
    end
  end
    diff=abs(max(delv));
    if diff < epsil
      vf=vr(:,kp+1);
disp('G-S loadflow result considering Q-limits at PV buses.')
disp(sprintf('Specified  tolerance  =%7.5f  and  acceleration  factor
=%5.2f.',epsil,alp))
     disp(sprintf('Load flow converges with %d iterations and maximum delV
=%f.',kp,diff))
     disp('Final bus voltages are:')
     disp('Bus no.   Complex bus voltage     Magnitude   Angle')
     for n=1:bus_no
disp(sprintf('  %d \t\t(%s)\t%f\t%f',n,num2str(vf(n),'% 10.6f'),abs(vf(n
)),angle(vf(n))));
     end
     break
    end
end
```

Computer program execution result for Example 10.4 (Ans.)

```
All quantities are in p.u. and all angles are in radian.
No. of buses = 4. No. of lines = 5.
G-S loadflow result considering Q-limits at PV buses.
Specified tolerance = 0.00010 and acceleration factor = 1.40.
Lower bound of Q-limit is crossed for bus-2 in iteration=1.
Load flow converges with 12 iterations and maximum delV =0.000032.
Final bus voltages are:
Bus no.   Complex bus voltage      Magnitude      Angle
 1              (1.060000)          1.060000         0
 2         (1.027325+0.159872i)     1.039690     0.154381
 3         (0.920204-0.037384i)     0.920963    -0.040603
 4         (0.854963-0.133254i)     0.865286    -0.154616
```

From the computed final bus voltages (polar form) of Example 10.4, it is clear that although bus no. 2 is converted to a load bus (due to violation of Q-limit) its voltage magnitude is boosted up from 1.02 p.u. to 1.039690 p.u. This can be explained in the way that Q_g in bus-2 is more than actual Q_d of the bus. This excess Q_g helps raise the bus voltage.

10.7 CALCULATION OF LINE POWER FLOW

Once voltage solutions are attained, line power flows may be calculated using final bus voltages. The power flow from the i-th bus to j-th bus through the transmission line connected between these buses is given by

$$S_{ik} = V_i I_{ik}^* = V_i \left(\frac{V_i - V_k}{z_{ik}} + V_i y_{ik0} \right)^*$$

or,

$$S_{ik}^* = V_i^* I_{ik} = V_i^* \left(\frac{V_i - V_k}{z_{ik}} + V_i y_{ik_0} \right)$$

$$\therefore \quad \left. \begin{aligned} P_{ik} &= \text{Real}\left[V_i^* (V_i - V_k) y_{ik} + V_i^* V_i y_{ik_0} \right] \\ Q_{ik} &= -\text{Im}\left[V_i^* (V_i - V_k) y_{ik} + V_i^* V_i y_{ik_0} \right] \end{aligned} \right\} \left[\because y_{ik} = \frac{1}{z_{ik}} \right] \qquad (10.32a)$$

Similarly,

$$\left. \begin{aligned} P_{ki} &= \text{Re}\left[V_k^* (V_k - V_i) y_{ki} + V_k^* V_k y_{ki_0} \right] \\ Q_{ki} &= -\text{Im}\left[V_k^* (V_k - V_i) y_{ki} + V_k^* V_k y_{ki_0} \right] \end{aligned} \right\} \qquad (10.32b)$$

[where z_{ik} is the line impedance and y_{ik_0} is the half line charging of transmission line between buses i and k. Note that here positive reactive power flow indicates inductive VAR flow].

For the transmission lines with transformer, the line should be transformed to π-equivalent representation of the line with transformer (explained in Article 10.3; see Figures 10.3(a) and 10.4(a)). Let us assume that there is a transmission line with transformer between bus-i and bus-k.

Therefore

$$\therefore \quad \left. \begin{aligned} P_{ik} &= \text{Re}\left[V_i^* (V_i - V_k) Y_{1_{ik}} + V_i^* V_i Y_{2_{ik}} \right] \\ Q_{ik} &= -\text{Im}\left[V_i^* (V_i - V_k) Y_{1_{ik}} + V_i^* V_i Y_{2_{ik}} \right] \end{aligned} \right\} \qquad (10.32c)$$

and

$$\left. \begin{aligned} P_{ki} &= \text{Re}\left[V_k^* \left(V_k - V_i \right) Y_{1_{ki}} + V_k^* V_k Y_{3_{ki}} \right] \\ Q_{ki} &= -\text{Im}\left[V_k^* \left(V_k - V_i \right) Y_{1_{ki}} + V_k^* V_k Y_{3_{ki}} \right] \end{aligned} \right\} \qquad (10.32d)$$

Slack bus power is the total complex power supplied by it. Total load demand at slack bus is required to be known for calculation of slack bus power.

Slack bus power (SBP) = Summation of outgoing power flows from slack bus + load demand on slack bus (if any)

$$= \sum_{i=1}^{N} (P_{1i} + jQ_{1i}) + S_{d_1} \quad \text{[where } N = \text{total number of buses]} \qquad (10.32e)$$

For power flow computation of line i–k, we have

$$S_{ik} = V_i^* \{(V_i - V_k) y_{ik} + V_i y_{ik_0}\}; \; S_{ki} = V_k^* \{(V_k - V_i) y_{ki} + V_k y_{ki_0}\}$$

It may be noted that $y_{ik} = y_{ki}$ and $y_{ik_0} = y_{ki_0}$ (but for the lines with transformer, y_{ik_0} (i.e., Y_2) $\neq y_{ki_0}$ (i.e., Y_3) though $V_i \neq V_k$. Hence, $S_{ik} \neq S_{ki}$, i.e., the line flow from bus-i to bus-k is not equal to the line flow from bus-k to bus-i.

The real and reactive power system losses can be obtained as follows:

$$\text{Total real power loss } (P_L) = \underbrace{\sum_{i=1}^{N} P_{g_i}}_{\substack{\text{Total active}\\ \text{generation}}} - \underbrace{\sum_{i=1}^{N} P_{d_i}}_{\substack{\text{Total active}\\ \text{demand}}} = \sum_{i=1}^{N}\sum_{k=1}^{N}(P_{ik} + P_{ki})$$

$[\sum_{i=1}^{N}\sum_{k=1}^{N}(P_{ik} + P_{ki}) =$ Summation of real power flows]

$$\text{Total reactive power loss } (Q_L) = \underbrace{\sum_{i=1}^{N} Q_{g_i}}_{\substack{\text{Total reactive}\\ \text{generation}}} - \underbrace{\sum_{i=1}^{N} Q_{d_i}}_{\substack{\text{Total reactive}\\ \text{demand}}} = \sum_{i=1}^{N}\sum_{k=1}^{N}(Q_{ik} + Q_{ki})$$

$[\sum_{i=1}^{N}\sum_{k=1}^{N}(Q_{ik} + Q_{ki}) =$ Summation of reactive power flows]

Hence total complex line loss being the summation of all power flows, we can write,

$$\text{Total line loss } = S_{loss} = \sum_{i=1}^{N}\sum_{k=1}^{N}\left\{\left(P_{ikj} + jQ_{ik}\right) + \left(P_{ki} + jQ_{ki}\right)\right\} \tag{10.32f}$$

[where N = total number of buses]

The algorithm of line power flow calculation has been furnished in the following sub article. Line data and final bus voltages are required for the calculation of line flows and total line loss.

10.7.1 Algorithm to Calculate Line Power Flow, Line Loss and Slack Bus Power

1. Read line data for the power system having total number of lines = NL and total number of buses = N.
2. Read final bus voltages of each bus obtained from power flow calculation, i.e., V_i for $i = 1, 2, \ldots, N$.
3. Initialize slack bus power and total line loss as $SBP = 0$ and $S_{loss} = 0$ respectively.
4. Read the following for each line: from bus number (*fb*), to bus number (*tb*), line impedance (*z*), half line charging (*ych*), tap ratio (*tap*) and case type (*tap_pos*).
5. Compute line admittance of for each line using equation, $y = 1/z$.
6. Set line count $inl = 1$.
7. Calculate line flows S_{ij} and S_{ji} for line *inl*.
   (i) Use equations (10.32a) and (10.32b) respectively for $tap = 1$.
   (ii) Use equations (10.32c) and (10.32d) respectively for $tap \neq 1$.
8. Display and/or store line flows S_{ij} and S_{ji} for line *inl*.
9. Calculate slack bus power (*SBP*) and line loss (S_{loss}) using equations (10.32e) and (10.32f) respectively.
10. Increase line count by 1, i.e., $inl \leftarrow inl + 1$.
11. If $inl \leq NL$ then go to step 7 else go to next step.

12. Update slack bus power with its demand (if any), $SBP = SBP + (P_{d_1} + jQ_{d_1})$.
13. Display and/or store final SBP and S_{loss}.
14. Stop.

EXAMPLE 10.5 A six-bus, eight-line system has the following line data [schematic of the system is shown in Figure E10.5]. In case of transformers, the lines are in the non unity tap side.

Line no.	*From bus*	*To bus*	*Line impedance (p.u.)*	*B/2 (p.u.)*	*Off-nominal tap ratio in case of transformer*
1	1	3	$(0.04 + j0.3)$	$j0.01$	–
2	2	3	$(0.03 + j0.2)$	$j0.01$	–
3	3	4	$(0.04 + j0.2)$	$j0.01$	–
4	3	5	$(0.04 + j0.2)$	$j0.01$	–
5	4	5	$(0.03 + j0.15)$	$j0.01$	–
6	5	6	$(0.06 + j0.3)$	$j0.01$	–
7	2	1	$j0.4$	–	1.05
8	6	2	$j0.2$	–	1.02

The scheduled bus powers and initial bus voltages (in p.u.) are as under:

Bus no.	P_d	Q_d	P_g	Q_g	V	*Bus type*
1	0	0	?	?	$1.04\angle 0°$	Slack bus
2	1.0	0.1	0	0	$1.0\angle 0°$	PQ bus
3	0	0	1.5	0.75	$1.0\angle 0°$	PQ bus
4	0.45	0.25	0	0	$1.0\angle 0°$	PQ bus
5	0.40	0.25	0	0	$1.0\angle 0°$	PQ bus
6	0.35	0.1	0	0	$1.0\angle 0°$	PQ bus

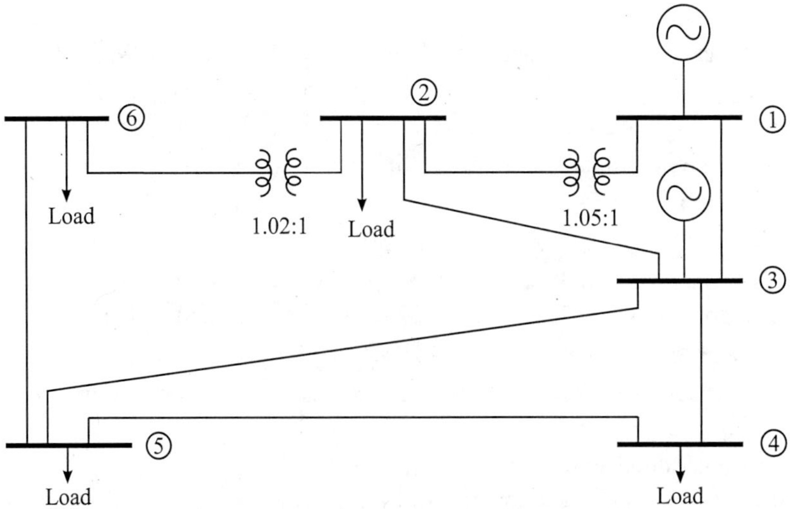

Figure E10.5 A six bus, eight line system.

Find final bus voltages by Gauss-Seidal method. Also calculate line flows, total line loss and slack bus power [all data are referred on 100 MVA base and in per unit]. Assume acceleration factor and specified tolerance as 1.4 and 0.0001 respectively. Write a MATLAB code for power flow calculation.

Solution: **The MATLAB code to calculate power flow for Example 10.5.**

```
%Program for Example 10.5
clc
clear all
bus_no=6; line_no=8;
vf= [1.04 (0.991619-0.187446i) (1.052868-0.069270i) (0.977224-0.150829i)
(0.977838-0.158874i) (0.973658-0.217581i)];
sloss=0; sbp=0; sdl=0;
%line data
fb=[1 2 3 3 4 5  2  6]; tb=[3 3 4 5 5 6  1  2];
z=[(0.04+ 0.3i) (0.03 +0.2i) (0.04+0.2i) (0.04+0.2i) (0.03+.15i) (0.06+0.3i)
0.4i 0.2i];
hlch=[0.01i  0.01i  0.01i  0.01i  0.01i  0.01i  0 0];
tap=[1 1 1 1 1 1 1.05 1.02]; tap_pos=[0 0 0 0 0 0 1 1];
y=1./z;
disp(sprintf('All quantities are in p.u. No. of buses = %d. No. of lines =
%d.',bus_no,line_no));
disp('LINE POWER FLOWS:')
disp('Line no. bus-i   bus-j        Sij           Sji') %for normal load flow
  for n=1:line_no
   y2=0; y3=0;%initialise for line not havinf trf
  if (tap(n)~=1)
   if tap_pos(n)==1
     y2=(1-tap(n))/z(n); y3=tap(n)*(tap(n)-1)/z(n); y(n)=tap(n)/z(n);
   else
      y(n)=1/(z(n)*tap(n));  y2=(1-tap(n))/(z(n)*tap(n)^2);
     y3=(tap(n)-1)/(z(n)*tap(n));
   end
  end
%Calculation of line flows
S1_ser(n)=conj(vf(fb(n)))*(vf(fb(n))-vf(tb(n)))*y(n);
S1_sht(n)=conj(vf(fb(n)))*(vf(fb(n))*(y2+hlch(n)));
S1(n)=S1_ser(n)+S1_sht(n);
S2_ser(n)=conj(vf(tb(n)))*(vf(tb(n))-vf(fb(n)))*y(n);
S2_sht(n)=conj(vf(tb(n)))*(vf(tb(n))*(y3+hlch(n)));
S2(n)=S2_ser(n)+S2_sht(n) ;
S1(n)=conj(S1(n)); S2(n)=conj(S2(n));
disp(sprintf(' %d\t\t%d \t%d\t(%s)\t(%s)',n,fb(n),tb(n),num2str(S1(n),'%
10.6f'),num2str(S2(n),'% 10.6f')));%for normal load flow
sloss=sloss+S1(n)+S2(n);
  if (fb(n)==1) sbp=sbp+S1(n); end
  if (tb(n)==1) sbp=sbp+S2(n);end
end
```

```
sbp=sbp+ sd1;
disp(sprintf('Slack bus power = (%s)', num2str(sbp,'% 10.6f')));
disp(sprintf('Total transmission line loss = (%s)', num2str(sloss,'%
10.6f')));
```

Computer program execution result for Example 10.5 (following algorithm of Articles 10.4.1, 10.6.1 and 10.7.1) (Ans.)

```
All quantities are in p.u. and all angles are in radian.
No. of buses = 6. No. of lines = 8.
Non zero elements of [Ybus]:
Note that only upper triangular off diagonal elements are provided.
Row  Column       Ybus                 Row  Column       Ybus
 1     1    (0.436681-6.021359i)        3     4    (-0.961538+4.807692i)
 1     2    (0.000000+2.625000i)        3     5    (-0.961538+4.807692i)
 1     3    (-0.436681+3.275109i)       4     4    (2.243590-11.197949i)
 2     2    (0.733496-12.581976i)       4     5    (-1.282051+6.410256i)
 2     3    (-0.733496+4.889976i)       5     5    (2.884615-14.393077i)
 2     6    (0.000000+5.100000i)        5     6    (-0.641026+3.205128i)
 3     3    (3.093254-17.740469i)       6     6    (0.641026-8.195128i)
G-S loadflow result without considering Q-limits at PV buses.
Specified tolerance =0.00010 and acceleration factor = 1.40.
Load flow converges with 20 iterations and maximum delV =0.000091.
Final bus voltages are:
Bus no. Complex bus voltage      Magnitude     Angle
  1          (1.040000)          1.040000        0
  2     (0.991619-0.187446i)     1.009180    -0.186826
  3     (1.052868-0.069270i)     1.055144    -0.065697
  4     (0.977224-0.150829i)     0.988796    -0.153136
  5     (0.977838-0.158874i)     0.990661    -0.161068
  6     (0.973658-0.217581i)     0.997673    -0.219855
LINE POWER FLOWS:
Line no. bus-i   bus-j          Sij                     Sji
  1        1       3    (0.230098-0.086104i)   (-0.227931+0.080412i)
  2        2       3    (-0.657474-0.104482i)  (0.670469+0.169799i)
  3        3       4    (0.509179+0.256995i)   (-0.497281-0.218416i)
  4        3       5    (0.548541+0.243104i)   (-0.535408-0.198387i)
  5        4       5    (0.047479-0.031363i)   (-0.047395+0.012189i)
  6        5       6    (0.182760-0.063832i)   (-0.180540+0.055166i)
  7        1       2    (0.511727+0.274040i)   (-0.511727-0.161009i)
  8        2       6    (0.169570+0.165905i)   (-0.169570-0.155282i)
Slack bus power = (0.741825+0.187936i)
Total transmission line loss = (0.042498+0.238736i)
```

10.8 NEWTON-RAPHSON (N-R) METHOD

10.8.1 Review of Newton-Raphson Method

Newton-Raphson (N-R) method is basically an iterative process for solving a set of simultaneous non-linear equation with an equal number of unknowns.

Let the set of simultaneous non-linear equations be written as

$$f_m(x_n) = 0 \text{ for } m = 1, 2, 3, \ldots, N \text{ and } n = 1, 2, 3, \ldots, N$$

The basic philosophy of N-R method of solution is that at each step of the iteration process, the non-linear problem is approximated by linear matrix equation. Let us show the linearising approximation in case of a single variable problem.

Let x^p be an approximation to the solution of the single variable equation while it possesses an error Δx^p at the p-th process of iteration (Figure 10.6). We can then write

$$f(x^p + \Delta x^p) = 0 \tag{10.33}$$

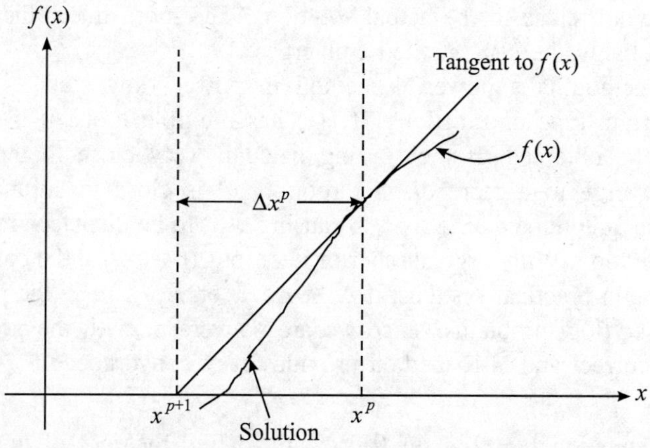

Figure 10.6 Linear approximation of a single variable equation.

Expanding in Taylor series, we have

$$f(x^p + \Delta x^p) = 0 = f(x^p) + \Delta x^p f'(x^p) + \frac{(\Delta x^p)^2}{2!} f''(x^p) + \cdots$$

The first derivative $f'(x^p)$ is called *Jacobian* while the second derivative $f''(x^p)$ is known as *Hessian*. Provided the initial estimate of the variable x^p approaches the solution value, Δx^p will be very small and the higher power of Δx^p can easily be neglected.

This gives

$$f(x^p) + \Delta x^p f'(x^p) = 0 \tag{10.34}$$

i.e.

$$\Delta x^p = \frac{-f(x^p)}{f'(x^p)} \tag{10.34a}$$

The new value of the variable can then be obtained from

$$x^{p+1} = x^p + \Delta x^p \tag{10.35}$$

Equation (10.34) can be further written as

$$f(x^p) + J\Delta x^p = 0 \quad \text{or} \quad f(x^p) = -J\Delta x^p \tag{10.36}$$

The method of solving the single variable problem can easily be extended to the set of N equations having N number of unknowns. J becomes the *square Jacobian matrix* of first order partial differentials of $f_m(x_n)$ function. Any element of J being given by $J_{mn} = \frac{\partial f_m}{\partial x_n}$, it represents the *slopes of the tangent hyperplanes that approximate the function $f_m(x_n)$ at each step of iteration.*

N-R method requires selection of a starting value $x^{(0)}$, from which Δx is calculated using equation (10.34a). If $\Delta x > \varepsilon$, where ε represents some error tolerance value. $x^{(1)}$ is calculated from equation (10.35). Again Δx is calculated and the process continues till $\Delta x \le \varepsilon$.

N-R method converges quadratically if the functions have continuous first derivative, whose values approach the solution. The Jacobian matrix becomes non-singular and the initial approximations of x are close to the actual solutions. The more linear the equations are, the more rapidly and reliably the N-R method converges.

Though N-R method is a powerful method in power flow study, it does have some difficulties with certain type of functions. If $f(x)$ has multiple roots, it is difficult to predict which root would be obtained. If the starting value is very close to the desired root, the procedure may converge to a more distinct root. If x^p is close to a function maximum or minimum, Δx can be quite large ($\Delta x \ge \Delta x_{max}$), causing x^{p+1} to be displaced far from the solution and adding more iteration. With large number of iterations ($p \ge p_{max}$), the computations become uneconomical. Thus, in practical systems, if $\Delta x \ge \Delta x_{max}$ or if $p \ge p_{max}$, the program is abruptly terminated. In power flow problems, since we are conversant with the problem, the starting values are usually correct and N-R method provides easy convergence.

An illustration

Let us assume an equation $\theta = 2 - \sin\theta$. We are to find the root of this equation by N-R method when ε is the tolerance ($\varepsilon = 0.01$, say for simplicity in calculation) assuming an initial guess $\theta^0 = 0$. We write $f(\theta) = \theta - 2 + \sin\theta$

$$\therefore \qquad \frac{df(\theta)}{d\theta} = 1 + \cos\theta$$

Starting now with the iterations, for the first iteration, ($p = 0$),

$$\theta^0 = 0; \quad f(\theta^0) = -2; \quad \frac{df(\theta^0)}{d\theta} = 2.$$

Hence,

$$\Delta\theta^0 = \frac{-f(\theta^0)}{f'(\theta^0)} = -\left(\frac{-2}{2}\right) = 1$$

For the second iteration, ($p = 1$),

$$\theta^1 = \theta^0 + \Delta\theta = 0 + 1 = 1; \ f(\theta^1) = 1 - 2 + \sin\,(1\text{ rad}) = -0.1585$$

$$\frac{df(\theta^1)}{d\theta} = 1 + \cos\,(1\text{ rad}) = 1.5403$$

$$\therefore \qquad \Delta\theta^1 = -\frac{f(\theta^1)}{f'(\theta^1)} = -\frac{-0.1585}{1.5403} = 0.1029$$

For the third iteration, ($p = 2$),

$$\theta^2 = \theta^1 + \Delta\theta = 1 + 0.1029 = 1.1029$$

$$f(\theta^2) = 1.1029 - 2 + \sin(1.1029 \text{ rad}) = -0.0046$$

$$\frac{df(\theta^2)}{d\theta} = 1 + \cos(1.1029) = 1.451$$

$$\Delta\theta^2 = -\frac{f(\theta^2)}{f'(\theta^2)} = -\frac{-0.0046}{1.451} = 0.0032$$

Since $\Delta\theta < 0.01$, we can stop here and the root of the equation then becomes 1.1029.

10.8.2 Application Procedure of N-R Method of Solution for Two Non-linear Equations with Two Unknowns

Let the two non-linear equations in two unknowns be

$$f_1(x,y) = 0 \quad \text{and} \quad f_2(x,y) = 0.$$

The initial guess of solutions being x^p and y^p with the errors being Δx^p and Δy^p, we can write

$$f_1(x^p + \Delta x^p, y^p + \Delta y^p) = 0$$

$$f_2(x^p + \Delta x^p, y^p + \Delta y^p) = 0, \qquad \text{for } p\text{th step of iteration.}$$

Expanding by Taylor series, the first equation gives

$$f_1(x^p, y^p) + \Delta x^p f_1'(x^p, y^p) + \Delta y^p f_1'(x^p, y^p) + \cdots = 0$$

or

$$f_1(x^p, y^p) + \frac{\partial f_1}{\partial x}(x^p, y^p)(x - x^p) + \frac{\partial f_1}{\partial y}(x^p, y^p)(y - y^p) + \cdots = 0$$

or

$$f_1^p + \frac{\partial f_1^p}{\partial x}\Delta x^p + \frac{\partial f_1^p}{\partial y}\Delta y^p + \cdots = 0 \tag{10.37}$$

[Simplifying the notation $f_1(x^p, y^p)$ as f_1^p, as $\frac{\partial f_1}{\partial x}(x^p, y^p)$ as $\frac{\partial f_1^p}{\partial x}$ as well as $(x^p - x)$ as Δx^p, as $\frac{\partial f_1}{\partial y}(x^p, y^p)$ as $\frac{\partial f_1^p}{\partial y}$ well as $(y^p - y)$ as Δy^p.]

Neglecting higher order terms in equation (10.37), transposition yields

$$\frac{\partial f_1^p}{\partial x}\Delta x^p + \frac{\partial f_1^p}{\partial y}\Delta y^p = -f_1^p \tag{10.38}$$

In a similar way, the second equation yields

$$\frac{\partial f_2^p}{\partial x}\Delta x^p + \frac{\partial f_2^p}{\partial y}\Delta y^p = -f_2^p \tag{10.39}$$

The final form of these two equations can be placed in matrix form as

$$\begin{bmatrix} \dfrac{\partial f_1^p}{\partial x} & \dfrac{\partial f_1^p}{\partial y} \\ \dfrac{\partial f_2^p}{\partial x} & \dfrac{\partial f_2^p}{\partial y} \end{bmatrix} \begin{bmatrix} \Delta x^p \\ \Delta y^p \end{bmatrix} = - \begin{bmatrix} f_1^p \\ f_2^p \end{bmatrix} \tag{10.40}$$

The coefficient matrix being called *Jacobian matrix*, and symbolised as $[J^p]$, we can write

$$[J^p] = \begin{bmatrix} \dfrac{\partial f_1^p}{\partial x} & \dfrac{\partial f_1^p}{\partial y} \\ \dfrac{\partial f_2^p}{\partial x} & \dfrac{\partial f_2^p}{\partial y} \end{bmatrix} \tag{10.41}$$

$$\therefore \qquad \begin{bmatrix} \Delta x^p \\ \Delta y^p \end{bmatrix} = -\left[J^p\right]^{-1} \begin{bmatrix} f_1^p \\ f_2^p \end{bmatrix} \tag{10.42}$$

The variables x and y can be updated as follows:

$$\left.\begin{aligned} x^{p+1} &= x^p + \Delta x^p \\ y^{p+1} &= y^p + \Delta y^p \end{aligned}\right\} \tag{10.43}$$

Iteration continues till $|\Delta x^p| \le \varepsilon_x$ and $|\Delta y^p| \le \varepsilon_y$

10.8.3 Application Procedure of N-R Method for Solving 2*n* Equations

N-R method can also be extended to 2n equations having 2n unknowns. Let the equations be

$$\begin{aligned}
f_{11}\left(x_1, x_2, \cdots, x_n, y_1, y_2, \cdots, y_n\right) &= 0 \\
f_{12}\left(x_1, x_2, \cdots, x_n, y_1, y_2, \cdots, y_n\right) &= 0 \\
\vdots \quad \vdots \quad \vdots \\
f_{1i}\left(x_1, x_2, \cdots, x_n, y_1, y_2, \cdots, y_n\right) &= 0 \\
\vdots \quad \vdots \quad \vdots \\
f_{1n}\left(x_1, x_2, \cdots, x_n, y_1, y_2, \cdots, y_n\right) &= 0 \\
f_{21}\left(x_1, x_2, \cdots, x_n, y_1, y_2, \cdots, y_n\right) &= 0 \\
f_{22}\left(x_1, x_2, \cdots, x_n, y_1, y_2, \cdots, y_n\right) &= 0 \\
\vdots \quad \vdots \quad \vdots \\
f_{2i}\left(x_1, x_2, \cdots, x_n, y_1, y_2, \cdots, y_n\right) &= 0 \\
\vdots \quad \vdots \quad \vdots \\
f_{2n}\left(x_1, x_2, \cdots, x_n, y_1, y_2, \cdots, y_n\right) &= 0
\end{aligned}$$

At pth stage of iteration,

$$\left.\begin{aligned} f_{1i}(\tilde{x}^p, \tilde{y}^p) = 0 \\ f_{2i}(\tilde{x}^p, \tilde{y}^p) = 0 \end{aligned}\right\} \quad (10.44)$$

where,

$$\tilde{x}^p = \begin{bmatrix} x_1^p \\ x_2^p \\ \vdots \\ x_i^p \\ \vdots \\ x_n^p \end{bmatrix}; \tilde{y}^p = \begin{bmatrix} y_1^p \\ y_2^p \\ \vdots \\ y_i^p \\ \vdots \\ y_n^p \end{bmatrix}; \tilde{f}_1 = \begin{bmatrix} f_{11}^p \\ f_{12}^p \\ \vdots \\ f_{1i}^p \\ \vdots \\ f_{1n}^p \end{bmatrix}; \tilde{f}_2 = \begin{bmatrix} f_{21}^p \\ f_{22}^p \\ \vdots \\ f_{2i}^p \\ \vdots \\ f_{2n}^p \end{bmatrix}$$

Also,

$$\left[J^p\right] = \begin{bmatrix} \left[\dfrac{\partial f_1^p}{\partial x}\right] & \left[\dfrac{\partial f_1^p}{\partial y}\right] \\ \left[\dfrac{\partial f_2^p}{\partial x}\right] & \left[\dfrac{\partial f_2^p}{\partial y}\right] \end{bmatrix} = \begin{bmatrix} \text{Quad I} & \text{Quad II} \\ \text{Quad III} & \text{Quad IV} \end{bmatrix}$$

$[J^p]$ is $2n \times 2n$ matrix when each quadrant gives a $n \times n$ matrix as shown below:

$$\text{Quad I} = \left[\frac{\partial f_1^p}{\partial x}\right]; \quad \text{Quad II} = \left[\frac{\partial f_1^p}{\partial y}\right]$$

$$\text{Quad III} = \left[\frac{\partial f_2^p}{\partial x}\right]; \quad \text{Quad IV} = \left[\frac{\partial f_2^p}{\partial y}\right]$$

The Jacobian matrix thus becomes an array of all first order partial derivatives.

Also,

$$\Delta\tilde{x} = \begin{bmatrix} \Delta x_1 \\ \Delta x_2 \\ \vdots \\ \Delta x_i \\ \vdots \\ \Delta x_n \end{bmatrix} \text{ and } \Delta\tilde{y} = \begin{bmatrix} \Delta y_1 \\ \Delta y_2 \\ \vdots \\ \Delta y_i \\ \vdots \\ \Delta y_n \end{bmatrix}$$

Thus,

$$\begin{bmatrix} \Delta\tilde{x} \\ \Delta\tilde{y} \end{bmatrix} = -\left[J^p\right]^{-1} \begin{bmatrix} \tilde{f}_1^p \\ \tilde{f}_2^p \end{bmatrix} \quad (10.45)$$

and finally,

$$\left.\begin{aligned} \tilde{x}^{p+1} &= \tilde{x}^p + \Delta\tilde{x}^p \\ \tilde{y}^{p+1} &= \tilde{y}^p + \Delta\tilde{y}^p \end{aligned}\right\} \quad (10.46)$$

N-R method converges when $\Delta x_i \leq \varepsilon_x$ and $\Delta y_i \leq \varepsilon_y$ for $i = 1, 2, 3, ..., n$.

10.9 APPLICATION OF N-R METHOD IN POWER FLOW STUDIES

In power system, let r_i represent the vector function $f(x^p)$ while the derivative $f'(x^p)$ is replaced by the Jacobian [J] with elements i, k and defined as

$$J_{ik} = \frac{\partial r_i}{\partial x_k} = J_{ik}(x)$$

Element J_{ik} is obtained from differentiating function r_i partially with respect to the unknown x_k and inserting the "present" value of x.

From equations (10.35) and (10.36) we can then write

$$x^{p+1} = x^p - \frac{f(x^p)}{J} = x^p - \frac{r}{J} \tag{10.47}$$

Since no division of matrix is possible, we can use multiplication by the inverse of J, written as J^{-1}. Thus equation (10.47) converts to

$$x^{p+1} = x^p - [J^p]^{-1} r^p, \text{ i.e. } [J^p]^{-1} r^p = x^p - x^{p+1}$$

Multiplying both sides by J^p we get

$$J^p[J^p]^{-1}r^p = J^p (x^p - x^{p+1})$$

Since J^p $[J^p]^{-1}$ yields a unity matrix I and Ir is r, hence we obtain

$$r^p = J^p (x^p - x^{p+1}) \tag{10.48}$$

Equation (10.48) represents a set of linear equations for the unknown vector x^{p+1}.

In order to have the solution for these equations, let us at first approximately guess the solution as x^p. Then it is inserted into the equations defining r^p and J^p. The linear set of these equations can then be solved using numerical methods to obtain x^{p+1}, the better solution value. Next, the value of x^{p+1} is used as an approximation and the equations are further computed after defining r^{p+1} and J^{p+1}. A better value of x^{p+2} is the obtained value, which again is used to yield another better value x^{p+3}. The process continues till the problem *converges*, i.e. till $x^p - x^{p+1} \leq \varepsilon$, ε being the tolerance.

10.10 APPLICATION OF N-R METHOD TO SOLVE POWER FLOW EQUATIONS IN RECTANGULAR FORM

After outlining the solution procedure, let us turn to power system equations. The governing equation in any network bus, using nodal admittance matrix form, is given by

$$I_i = \sum_{k=1}^{n} Y_{ik} V_k$$

[I_i being the injected current vector into i-th bus; N is the total number of buses]

The power at the i-th bus is then given by

$$S_i^* = P_i - jQ_i = V_i^* I_i = V_i^* \sum_{k=1}^{N} Y_{ik} V_k \tag{10.49}$$

However, in rectangular form of presentation,

$$V_i = (e_i + jf_i) \text{ and } Y_{ik} = (G_{ik} + jB_{ik})$$

Thus from equation (10.49)

$$P_i - jQ_i = (e_i - jf_i) \sum_{k=1}^{N} (G_{ik} + jB_{ik})(e_k + jf_k) \tag{10.50}$$

Separating real and imaginary parts, we have

$$\left.\begin{aligned} P_i &= \sum_{k=1}^{N} \{e_i (e_k G_{ik} - f_k B_{ik}) + f_i (f_k G_{ik} + e_k B_{ik})\} \\ Q_i &= \sum_{k=1}^{N} \{f_i (e_k G_{ik} - f_k B_{ik}) - e_i (f_k G_{ik} + e_k B_{ik})\} \end{aligned}\right\} \tag{10.51}$$

Thus, we have obtained a set of non-linear simultaneous equations in rectangular form of expression, two for each bus in the power system network. It may be noted that for any node (bus) the active and reactive power injections P_i and Q_i are known but the real and imaginary components of voltage (e_i and f_i) are unknown, except for the slack bus. In slack or swing bus, the voltage is specified and fixed. Thus, the number of equations to be solved is $2(N - 1)$.

Linear relationships are obtained for small variations in e and f by forming the total differentials as follows:

$$\Delta P_i = \sum_{k=1}^{N} \frac{\partial P_i}{\partial e_k} \Delta e_k + \sum_{k=1}^{N} \frac{\partial P_i}{\partial f_k} \Delta f_k = \sum_{k=1}^{N} J_1 \Delta e_k + \sum_{k=1}^{N} J_2 \Delta f_k \tag{10.52}$$

[*Equation* (10.52) *is valid for all buses except the slack bus.*]

Also,

$$\Delta Q_i = \sum_{k=1}^{N} \frac{\partial Q_i}{\partial e_k} \Delta e_k + \sum_{k=1}^{N} \frac{\partial Q_i}{\partial f_k} \Delta f_k = \sum_{k=1}^{N} J_3 \Delta e_k + \sum_{k=1}^{N} J_4 \Delta f_k \tag{10.53}$$

[*Equation* (10.53) *is valid for all non-voltage controlled PQ buses.*]

Thus, we can write,

$$\begin{bmatrix} \Delta P \\ \Delta Q \end{bmatrix} = \begin{bmatrix} J_1 & J_2 \\ J_3 & J_4 \end{bmatrix} \begin{bmatrix} \Delta e \\ \Delta f \end{bmatrix} \tag{10.54}$$

The Jacobian matrix is $2(N - 1) \times 2(N - 1)$ for power injections specified at $(N - 1)$ buses. The next task is to obtain J_1, J_2, J_3 and J_4.

The real power equation obtained in equation (10.51) can further be written as

$$P_i = e_i (e_i G_{ii} - f_i B_{ii}) + f_i (f_i G_{ii} + e_i B_{ii}) + \sum_{\substack{k=1 \\ k \neq i}}^{N} \{e_i (e_k G_{ik} - f_k B_{ik}) + f_i (f_k G_{ik} + e_k B_{ik})\} \tag{10.55}$$

$$i = 1, 2, 3, ..., (N - 1)$$

By differentiation of equation (10.55) the *off-diagonal elements* of $[J_1]$ are obtained as

$$\frac{\partial P_i}{\partial e_k} = e_i G_{ik} + f_i G_{ik}, \quad k \neq i$$

while the *diagonal elements* of $[J_1]$ are

$$\frac{\partial P_i}{\partial e_i} = 2e_iG_{ii} - f_iB_{ii} + f_iB_{ii} + \sum_{\substack{k=1\\k\neq i}}^{N} \left(e_kG_{ik} - f_kB_{ik}\right) \quad (10.56)$$

Next, representing the i-th bus *current vector* as $I_i = c_i + jd_i$, the equation of current at bus-i is given by

$$I_i = c_i + jd_i = \left(G_{ii} + jB_{ii}\right)\left(e_i + jf_i\right) + \sum_{\substack{k=1\\k\neq i}}^{N}\left(G_{ik} + jB_{ik}\right)\left(e_k + jf_k\right)$$

Separating into real and imaginary parts, we obtain

$$c_i = e_iG_{ii} - f_iB_{ii} + \sum_{\substack{k=1\\k\neq i}}^{N}\left(e_kG_{ik} - f_kB_{ik}\right) \quad (10.57)$$

and

$$d_i = f_iG_{ii} + e_iB_{ii} + \sum_{\substack{k=1\\k\neq i}}^{N}\left(f_kG_{ik} + e_kjB_{ik}\right) \quad (10.58)$$

$$[i = 1, 2, 3, ..., (N-1)]$$

Substitution of equation (10.57) in equation (10.56) yields

$$\frac{\partial P_i}{\partial e_i} = e_iG_{ii} + f_iB_{ii} + c_i \quad (10.59)$$

Again, from equation (10.55), the *off-diagonal elements* of $[J_2]$ are obtained as

$$\frac{\partial P_i}{\partial f_k} = -\,e_iB_{ik} + f_iG_{ik}, i \neq k; \quad (10.60)$$

while *diagonal elements* of $[J_2]$ are given as

$$\frac{\partial P_i}{\partial f_i} = -\,e_iB_{ii} + 2f_iG_{ii} + e_iB_{ii} + \sum_{\substack{k=1\\k\neq i}}^{N}\left(f_kG_{ik} + e_kB_{ik}\right) \quad (10.61)$$

Combining equations (10.58) and (10.61), we have

$$\frac{\partial P_i}{\partial f_i} = -\,e_iB_{ii} + f_iG_{ii} + d_i \quad (10.62)$$

Again, the reactive power equation for bus-i is given as

$$Q_i = f_i\left(e_iG_{ii} - f_iB_{ii}\right) - e_i\left(f_iG_{ii} + e_iB_{ii}\right) + \sum_{\substack{k=1\\k\neq i}}^{N}\left\{f_i\left(e_kG_{ik} - f_kB_{ik}\right) - e_i\left(f_kG_{ik} + e_kB_{ik}\right)\right\} \quad (10.63)$$

$$\left[i = 1, 2, 3, ..., (N-1)\right]$$

By differentiation of equation (10.63), *off-diagonal elements* of $[J_3]$ are obtained as

$$\frac{\partial Q_i}{\partial e_k} = -e_i B_{ik} + f_i G_{ik}, \quad k \neq i; \tag{10.64}$$

while *diagonal elements* of $[J_3]$ are

$$\frac{\partial Q_i}{\partial e_i} = f_i G_{ii} - f_i G_{ii} - 2e_i B_{ii} + \sum_{\substack{k=1 \\ k \neq i}}^{N} \left(f_k G_{ik} + e_k B_{ik} \right) \tag{10.65}$$

Combining equations (10.58) and (10.65), we have

$$\frac{\partial Q_i}{\partial e_i} = -e_i B_{ii} + f_i G_{ii} - d_i \tag{10.66}$$

The *off-diagonal elements* of $[J_4]$ can be obtained from equation (10.63) as

$$\frac{\partial Q_i}{\partial f_k} = -e_i G_{ik} - f_i B_{ik}, \quad i \neq k; \tag{10.67}$$

while the *diagonal elements* of $[J_4]$ can also be obtained as

$$\frac{\partial Q_i}{\partial f_i} = e_i G_{ii} - 2 f_i B_{ii} - e_i G_{ii} + \sum_{\substack{k=1 \\ k \neq i}}^{N} \left(e_k G_{ik} - f_k B_{ik} \right) \tag{10.68}$$

Combining equations (10.57) and (10.68), we have

$$\frac{\partial Q_i}{\partial f_i} = -e_i G_{ii} - f_i B_{ii} + c_i \tag{10.69}$$

Thus, for the system Jacobian matrix we can now arrange the terms in the four quadrants as follows:

Quadrant-1 $[J_1]$

Diagonal elements: $$\frac{\partial P_i}{\partial e_i} = e_i G_{ii} + f_i B_{ii} + c_i$$

Off-diagonal elements: $$\frac{\partial P_i}{\partial e_k} = e_i G_{ik} + f_i B_{ik}$$

Quadrant-2 $[J_2]$

Diagonal elements: $$\frac{\partial P_i}{\partial f_i} = -e_i B_{ii} + f_i G_{ii} + d_i$$

Off-diagonal elements: $$\frac{\partial P_i}{\partial f_k} = -e_i B_{ik} + f_i G_{ik}$$

Quadrant-3 [J_3]

Diagonal elements: $$\frac{\partial Q_i}{\partial e_i} = -e_i B_{ii} + f_i G_{ii} - d_i$$

Off-diagonal elements: $$\frac{\partial Q_i}{\partial f_k} = -e_i B_{ik} + f_i G_{ik}$$

Quadrant-4 [J_4]

Diagonal elements: $$\frac{\partial Q_i}{\partial f_i} = -e_i G_{ii} - f_i B_{ii} + c_i$$

Off-diagonal elements: $$\frac{\partial Q_i}{\partial f_k} = -e_i G_{ik} - f_i B_{ik}$$

Consideration of voltage controlled bus

In case there is presence of voltage-controlled bus in the system, the bus voltage magnitude being fixed (but not the phase angle), both e_i and f_i would vary at each iteration. It is then necessary to have one more equation

$$V_i^2 = e_i^2 + f_i^2 \tag{10.70}$$

to be solved with the real power equation for the buses concerned.

Here, $$\Delta V_i^2 = \frac{\partial V_i^2}{\partial e_i}\Delta e_i + \frac{\partial V_i^2}{\partial f_i}\Delta f_i = J_5 \Delta e_i + J_6 \Delta f_i$$

the *diagonal elements* of [J_5] are $$\frac{\partial |V_i|^2}{\partial e_i} = 2e_i \tag{10.71}$$

while *off-diagonal elements* of [J_5] are $$\frac{\partial |V_i|^2}{\partial e_k} = 0 \tag{10.72}$$

the *diagonal elements* of [J_6] are $$\frac{\partial |V_i|^2}{\partial f_i} = 2f_i \tag{10.73}$$

while *off-diagonal elements* of [J_6] $$\frac{\partial |V_i|^2}{\partial f_k} = 0 \tag{10.74}$$

The modified nodal relationship is then

$$\begin{bmatrix} \Delta P \\ \Delta Q \\ \Delta V^2 \end{bmatrix} = \begin{bmatrix} J_1 & J_2 \\ J_3 & J_4 \\ J_5 & J_6 \end{bmatrix} \begin{bmatrix} \Delta e \\ \Delta f \end{bmatrix} \tag{10.75}$$

[Since $|V|$ is specified only (but not the real and imaginary components of V (i.e. e and f) for voltage-controlled buses, hence the voltage of such type of buses can be considered to be $V = (1 + j0)$ p.u. in computations.]

10.10.1 Algorithm to Calculate Bus Voltages in Rectangular Form by N-R Method of Power Flow

Once the Jacobian elements are formulated, we can now develop the process of solving the power flow problem using N-R method using following steps of computation, the unknowns being expressed in rectangular form.

1. Read line data for the power system having total number of lines = *NL* and total number of buses = *N*.
2. Read bus type ($type_i$), initial bus voltage magnitude in p.u. $\left(|V_i|^0\right)$, initial bus voltage angle in rad $\left(\delta_i^0\right)$, real power generation $\left(P_{g_i}\right)$, reactive power generation $\left(Q_{g_i}\right)$, real power demand $\left(P_{d_i}\right)$, reactive power demand $\left(Q_{d_i}\right)$, lower and upper *Q*-limit at PV buses ($Q_{g_{i_{\min}}}$ and $Q_{g_{i_{\max}}}$) in p.u. for each bus i = 1, 2, ..., *N*.
3. Specify tolerance ε (say 0.0001).
4. Form $\left[Y_{Bus}\right]$.
5. Compute scheduled active and reactive bus powers for all buses using equations (10.27a) and (10.27b), i.e., $P_{i_{\text{scheduled}}}$ and $Q_{i_{\text{scheduled}}}$ for i = 1, 2, ..., *N*.
6. Set iteration count p = 0.
7. Fix slack bus voltage magnitude and angle for next iteration, $|V_i|^{p+1} = |V_i|^0$, $\delta_i^{p+1} = 0$ if $type_i = 0$.
8. Fix specified voltage magnitude for PV buses. $|V_i|^p = |V_i|^0$, for i = 1, 2, ..., *N* (only if $type_i = 1$).
9. Separate real and imaginary parts of bus voltages $e_i^p = |V_i|^p \cos\delta_i^p$ and $f_i^p = |V_i|^p \sin\delta_i^p$.
10. Calculate active and reactive power (P_i^p and Q_i^p) at all buses (i = 1, 2, …, *N*) using equation (10.51).
11. Calculate reactive power generation for PV buses as $Q_{g_i} = (Q_i^p + Q_{di})$ for i = 1, 2, ..., *N* (only if $type_i = 1$).
12. If $Q_{g_{\min}}$ and $Q_{g_{\max}}$ are specified for all PV buses then go to step 13, otherwise go to step 14.
13. Check the following for all PV buses (for i = 1, 2, ..., *N* and consider bus-i only if $type_i = 1$) $Q_{g_i} > Q_{g_{i_{\max}}}$ or, $Q_{g_i} < Q_{g_{i_{\min}}}$ for the PV bus-i (i.e., *Q*-limit is failed),

    (i) set $Q_{g_i} = Q_{g_{i_{\max}}}$ for $Q_{g_i} > Q_{g_{i_{\max}}}$,

    or, set $Q_{g_i} = Q_{g_{i_{\min}}}$ for $Q_{g_i} < Q_{g_{i_{\min}}}$.

    (ii) release the constraint of $|V_i|^p$ to be fixed and initialize it for a flat start, i.e., $|V_i|^p = 1$ and also change bus type from PV bus ($type_i = 1$) to PQ bus ($type_i = 1$) as *Q*-limit is failed.

14. If $Q_{g_{i\min}} \le Q_i \le Q_{g_{i\max}}$ (i.e., Q-limit is maintained) or, $Q_{g_{i\min}}$ and $Q_{g_{i\max}}$ are not specified for the PV bus-i.

    (i) express V_i^{p+1} in polar form as $V_i^{p+1} = |V_i^{p+1}| \angle \delta_i^{p+1}$

    (ii) update V_i^{p+1} as $V_i^{p+1} = |V_{i_{\text{specified}}}| \angle \delta_i^{p+1}$

15. Form ΔP^p matrix considering all buses excluding slack bus using equation (10.28a) where $\Delta P_i^p = P_{i_{\text{scheduled}}} - P_i^p$, for i = 1, 2, ..., N (i.e., exclude bus-i, if $type_i$ = 0).
16. Form ΔQ^p matrix considering all buses excluding slack bus using equation (10.28b) $\Delta Q_i^p = Q_{i_{\text{scheduled}}} - Q_i^p$ for i = 1, 2, ..., N (i.e., exclude bus-i, if $type_i$ = 0).
17. If $\max\left(\Delta P^p \text{ and } \Delta Q^p\right) < \varepsilon$ then go to step 23 else go to step 18.
18. Form the Jacobian matrix $[J^p]$ using equation (10.54). Compute the elements of the matrix using equations (10.56)–(10.68).
19. Solve equation $\begin{bmatrix} \Delta e_i^p \\ \Delta f_i^p \end{bmatrix} = \left[J^p\right]^{-1} \begin{bmatrix} \Delta P_i^p \\ \Delta Q_i^p \end{bmatrix}$.
20. Update bus voltages angles for all buses except slack bus

    $e_i^{p+1} = e_i^p + \Delta e_i^p$, $f_i^{p+1} = f_i^p + \Delta f_i^p$ for i = 1, 2, ..., N except slack bus (i.e., exclude bus-i, if $type_i$ = 0).
21. Update iteration count $p \leftarrow p+1$.
22. If p > 20 then break the iteration process and go to step 27. Otherwise go to step 7.
23. Convert final bus voltages into polar form, $V_i^p = |V_i|^p \angle \delta_i^p$ for i = 1, 2, ..., N.
24. Calculate reactive power generations for those PV buses who have not crossed their Q-limit. $Q_{g_i} = Q_i^p + Q_{d_i}$ for i = 1, 2, ..., N provided $type_i = 1$.
25. Display and/or store final bus voltages, V_i^p, for i = 1, 2, ..., N.
26. Display and/or store reactive power generation, Q_{g_i} for PV buses ($type_i$ = 1), for i = 1, 2, ..., N,
27. Stop.

An additional note for the formation of Jacobian matrix

The elements of Jacobian matrix can also be obtained using exponential forms of the system variables (V_i, V_k and Y_{ik}) and using equation (10.70) of the text for the purpose of computational convenience. Proceeding with exponential type of representation, we can now write

$$P_i - jQ_i = \sum_{k=1}^{N} \left| V_i V_k Y_{ik} \right| \varepsilon^{j(\phi_{ik} + \delta_k - \delta_i)} \tag{10.76}$$

Differentiating partially with respect to δ_k, $k \neq i$;

$$\frac{\partial P_i}{\partial \delta_k} - j\frac{\partial Q_i}{\partial \delta_k} = j\,|V_i|\,\varepsilon^{-j\delta_i}\,|Y_{ik}|\,\varepsilon^{j\phi_{ik}}\,|V_k|\,\varepsilon^{j\delta_k} = j\,(e_i - jf_i)(a_k + jb_k)$$

$$= (a_k f_i - b_k e_i) + j\,(a_k e_i + b_k f_i) \tag{10.77}$$

where $(a_k + jb_k) = Y_{ik}V_k = (G_{ik} + jB_{ik})(e_k + jf_k)$.

Therefore,

$$\frac{\partial P_i}{\partial \delta_k} = (a_k f_i - b_k e_i) = J_{1ik};\ i \neq k \tag{10.78}$$

$$\frac{\partial Q_i}{\partial \delta_k} = -(a_k e_i + b_k f_i) = J_{3ik};\ i \neq k \tag{10.79}$$

Again, differentiating equation (10.76) partially with respect to $|V_k|$, $k \neq i$;

$$\frac{\partial P_i}{\partial |V_k|} - j\frac{\partial Q_i}{\partial |V_k|} = |V_i|\,\varepsilon^{-j\delta_i}\,\varepsilon^{j\delta_k}\,|Y_{ik}|\,\varepsilon^{j\phi_{ik}}$$

multiplying both sides by $|V_k|$ we have

$$|V_k|\frac{\partial P_i}{\partial |V_k|} - j\,|V_k|\frac{\partial Q_i}{\partial |V_k|} = V_i|\,\varepsilon^{-j\delta_i}\,|Y_{ik}|\,\varepsilon^{j\phi_{ik}}\,|V_k|\,\varepsilon^{j\delta_k} = (e_i - jf_i)(a_k + jb_k)$$

$$= (a_k e_i + b_k f_i) + j\,(b_k e_i - a_k f_i) \tag{10.80}$$

$$|V_k|\frac{\partial P_i}{\partial |V_k|} = (a_k e_i + b_k f_i) = J_{2ik};\ i \neq k \tag{10.81}$$

$$|V_k|\frac{\partial Q_i}{\partial |V_k|} = (a_k f_i + b_k e_i) = J_{4ik};\ i \neq k \tag{10.82}$$

We can now summarise the elements of $[J_1]$, $[J_2]$, $[J_3]$ and $[J_4]$ as follows:

Quadrant-1 $[J_1]$

Diagonal elements:
$$J_{1ii} = \frac{\partial P_i}{\partial \delta_i} = -Q_i - |V_i|^2\,B_{ii} \tag{10.83}$$

Off-diagonal elements:
$$J_{1ik} = \frac{\partial P_i}{\partial \delta_k} = (a_k f_i - b_k e_i) = J_{4ik} \tag{10.84}$$

Quadrant-2 $[J_2]$

Diagonal elements:
$$J_{2ii} = |V_i|\frac{\partial P_i}{\partial |V_i|} = P_i + |V_i|^2\,G_{ii} \tag{10.85}$$

Off-diagonal elements:
$$J_{2ik} = |V_k|\frac{\partial P_i}{\partial |V_k|} = (a_k e_i + b_k f_i) = -\,J_{3ik} \tag{10.86}$$

Quadrant-3 [J_3]

Diagonal elements: $$J_{3ii} = \frac{\partial Q_i}{\partial \delta_i} = -|V_i|^2 G_{ii} + P_i \quad (10.87)$$

Off-diagonal elements: $$J_{3ik} = \frac{\partial Q_i}{\partial \delta_k} = -\left(a_k e_i + b_k f_i\right) = -J_{2ik} \quad (10.88)$$

Quadrant-4 [J_4]

Diagonal elements: $$J_{4ii} = |V_i| \frac{\partial Q_i}{\partial |V_i|} = -|V_i|^2 B_{ii} + Q_i \quad (10.89)$$

Off-diagonal elements: $$J_{4ik} = |V_k| \frac{\partial Q_i}{\partial |V_k|} = \left(a_k f_i - b_k e_i\right) = J_{1ik} \quad (10.90)$$

10.11 APPLICATION OF N-R METHOD TO SOLVE POWER FLOW EQUATION IN POLAR FORM

Next, the N-R method is applied to solve the complex power flow equations when the equations are expressed in *polar coordinates.* In polar coordinates we first express the i-th bus voltage and transfer admittance in the following form:

$$V_i = |V_i| \angle \delta_i = |V_i| \left(\cos \delta_i + j \sin \delta_i\right) = |V_i| \, \varepsilon^{j\delta_i}$$

and $$Y_{ik} = |Y_{ik}| \left(\cos \phi_{ik} + j \sin \phi_{ik}\right) = |Y_{ik}| \, \varepsilon^{j\phi_{ik}}$$

[It may be recalled that $|Y_{ik}| \cos \phi_{ik}$ is G_{ik}, the transfer conductance and $|Y_{ik}| \sin \phi_{ik}$ is B_{ik}, the transfer susceptance and hence $Y_{ik} = (G_{ik} + jB_{ik})$.]

At the very beginning of formulating N-R method in the load flow equations, we have shown the bus power at bus-i is given by

$$P_i - jQ_i = V_i^* I_i = V_i^* \sum_{k=1}^{n} Y_{ik} V_k$$

Substitution of the polar forms of V and Y, we find from the complex power equation

$$P_i - jQ_i = \sum_{k=1}^{N} |V_i V_k Y_{ik}| \, \varepsilon^{j(\phi_{ik} + \delta_k - \delta_i)} \quad (10.91a)$$

Since from trigonometry,

$$\varepsilon^{j(\phi_{ik} + \delta_k - \delta_i)} = \cos\left(\phi_{ik} + \delta_k - \delta_i\right) + j \sin\left(\phi_{ik} + \delta_k - \delta_i\right) \quad (10.91b)$$

hence substitution of equation (10.91b) in equation (10.91a), after separating the real and imaginary components yields

$$\left.\begin{aligned} P_i &= \sum_{k=1}^{N} |V_i V_k Y_{ik}| \cos(\phi_{ik} + \delta_k - \delta_i) \\ Q_i &= -\sum_{k=1}^{N} |V_i V_k Y_{ik}| \sin(\phi_{ik} + \delta_k - \delta_i) \end{aligned}\right\}; \; i = 1, 2, 3, \ldots, (N-1) \tag{10.92}$$

It may be noted that equations (10.92) and (10.27) are identical. The elements of the Jacobian matrix are then obtained as follows:

Quadrant-1 [J_1]

Diagonal elements: $$\frac{\partial P_i}{\partial \delta_i} = \sum_{\substack{k=1 \\ k \neq i}}^{n} |V_i V_k Y_{ik}| \sin(\phi_{ik} + \delta_k - \delta_i)$$

Off-diagonal elements: $$\frac{\partial P_i}{\partial \delta_k} = -|V_i V_k Y_{ik}| \sin(\phi_{ik} + \delta_k - \delta_i); i \neq k$$

Quadrant-2 [J_2]

Diagonal elements: $$\frac{\partial P_i}{\partial |V_i|} = 2|V_i Y_{ii}| \cos\phi_{ii} + \sum_{\substack{k=1 \\ k \neq i}}^{N} |V_k Y_{ik}| \cos(\phi_{ik} + \delta_k - \delta_i)$$

Off-diagonal elements: $$\frac{\partial P_i}{\partial |V_k|} = |V_i Y_{ik}| \cos(\phi_{ik} + \delta_k - \delta_i); \; i \neq k$$

Quadrant-3 [J_3]

Diagonal elements: $$\frac{\partial Q_i}{\partial \delta_i} = \sum_{\substack{k=1 \\ k \neq i}}^{N} |V_i V_k Y_{ik}| \cos(\phi_{ik} + \delta_k - \delta_i)$$

Off-diagonal elements: $$\frac{\partial Q_i}{\partial \delta_k} = -|V_i V_k Y_{ik}| \cos(\phi_{ik} + \delta_k - \delta_i); \; i \neq k$$

Quadrant-4 [J_4]

Diagonal elements: $$\frac{\partial Q_i}{\partial |V_i|} = -2|V_i Y_{ii}| \sin\phi_{ii} + \sum_{\substack{k-1 \\ k \neq i}}^{N} |V_k Y_{ik}| \sin(\phi_{ik} + \delta_k - \delta_i)$$

Off-diagonal elements: $$\frac{\partial Q_i}{\partial |V_k|} = -|V_i Y_{ik}| \sin(\phi_{ik} + \delta_k - \delta_i); \; i \neq k$$

Again, rewriting the power equations

$$P_i = \sum_{k=1}^{N} |V_i V_k Y_{ik}| \cos(\phi_{ik} + \delta_k - \delta_i)$$

and $$Q_i = -\sum_{k=1}^{N} |V_i V_k Y_{ik}| \sin(\phi_{ik} + \delta_k - \delta_i); \; i = 1, 2, 3, \ldots, (N-1)$$

as
$$P_i = |V_i|^2 G_{ii} + \sum_{\substack{k=1\\k\neq i}}^{N} |V_i V_k Y_{ik}| \cos(\phi_{ik} + \delta_k - \delta_i) \tag{10.93}$$

and
$$Q_i = -|V_i|^2 B_{ii} - \sum_{\substack{k=1\\k\neq i}}^{N} |V_i V_k Y_{ik}| \sin(\phi_{ik} + \delta_k - \delta_i) \tag{10.94}$$

the elements of the Jacobian can also be shown to have the following form of representations:

For quadrant-1 [J_1]

Diagonal elements: $$\frac{\partial P_i}{\partial \delta_i} = \sum_{\substack{k=1\\k\neq i}}^{N} |V_i V_k Y_{ik}| \sin(\phi_{ik} + \delta_k - \delta_i) = -\sum_{\substack{k=1\\k\neq i}}^{N} \frac{\partial P_i}{\partial \delta_k} \tag{10.95}$$

While the *off-diagonal elements* remain the same

$$\frac{\partial P_i}{\partial \delta_k} = -|V_i V_k Y_{ik}| \sin(\phi_{ik} + \delta_k - \delta_i);\ i \neq k$$

Comparing equation (10.95) with equation (10.94), we find

$$\frac{\partial P_i}{\partial \delta_i} = -Q_i - |V_i|^2 B_{ii} \tag{10.96}$$

For quadrant-2 [J_2]

$$\frac{\partial P_i}{\partial |V_k|} = |V_i Y_{ik}| \cos(\phi_{ik} + \delta_k - \delta_i)$$

or,
$$|V_k| \frac{\partial P_i}{\partial |V_k|} = |V_k||V_i Y_{ik}| \cos(\phi_{ik} + \delta_k - \delta_i);\ i \neq k$$

$$= -\frac{\partial Q_i}{\partial \delta_k}, \text{ the } \textit{off-diagonal element} \text{ of } [J_3] \tag{10.97}$$

Again for the same quadrant (quadrant-2), [J_2], the *diagonal element* is given by

$$\frac{\partial P_i}{\partial |V_i|} = 2|V_i Y_{ii}| \cos\phi_{ii} + \sum_{\substack{k=1\\k\neq i}}^{N} |V_k Y_{ik}| \cos(\phi_{ik} + \delta_k - \delta_i)$$

or,
$$|V_i| \frac{\partial P_i}{\partial |V_i|} = |V_i| \left[2|V_i Y_{ii}| \cos\phi_{ii} + \sum_{\substack{k=1\\k\neq i}}^{N} |V_k Y_{ik}| \cos(\phi_{ik} + \delta_k - \delta_i) \right]$$

$$= |V_i| \left[2|V_i| G_{ii} + \sum_{\substack{k=1\\k\neq i}}^{N} |V_k Y_{ik}| \cos(\phi_{ik} + \delta_k - \delta_i) \right] = 2|V_i|^2 G_{ii} + \frac{\partial Q_i}{\partial \delta_i}$$

or,
$$|V_i| \frac{\partial P_i}{\partial |V_i|} = 2|V_i|^2 G_{ii} + \frac{\partial Q_i}{\partial \delta_i} = P_i + |V_i|^2 G_{ii} \tag{10.98}$$

For quadrant-3 [J_3]

Diagonal element: $$\frac{\partial Q_i}{\partial \delta_i} = \sum_{\substack{k=1\\k\neq i}}^{N} |V_i V_k Y_{ik}| \cos(\phi_{ik} + \delta_k - \delta_i) = -\sum_{\substack{k=1\\k\neq i}}^{N} \frac{\partial Q_i}{\partial \delta_k} \tag{10.99}$$

The *off-diagonal element* is given by

$$\frac{\partial Q_i}{\partial \delta_k} = -|V_i V_k Y_{ik}| \cos(\phi_{ik} + \delta_k - \delta_i);\ i \neq k \tag{10.100}$$

The diagonal element $\frac{\partial Q_i}{\partial \delta_i}$ (equation (10.99)) can further be written as

$$\frac{\partial Q_i}{\partial \delta_i} = -|V_i|^2 G_{ii} + P_i \text{ [comparing equation (10.99) with (10.93)]} \tag{10.101}$$

For quadrant-4 [J_4]

Finally, we express the *off-diagonal* and *diagonal elements* of [J_4]

$$|V_k| \frac{\partial Q_i}{\partial |V_k|} = -|V_i V_k Y_{ik}| \sin(\phi_{ik} + \delta_k - \delta_i) = \frac{\partial P_i}{\partial \delta_k};\ i \neq k \tag{10.102}$$

and $$|V_i| \frac{\partial Q_i}{\partial |V_i|} = -2|V_i|^2 |Y_{ii}| \sin\phi_{ii} - \sum_{\substack{k=1\\k\neq i}}^{N} |V_i V_k Y_{ik}| \sin(\phi_{ik} + \delta_k - \delta_i)$$

$$= -|V_i|^2 B_{ii} + Q_i \tag{10.103}$$

We can now summarise the elements of [J_1], [J_2], [J_3] and [J_4] as follows:

Quadrant-1 [J_1]

Diagonal elements: $$\frac{\partial P_i}{\partial \delta_i} = -Q_i - |V_i|^2 B_{ii}$$

Off-diagonal elements: $$\frac{\partial P_i}{\partial \delta_k} = -|V_i V_k Y_{ik}| \sin(\phi_{ik} + \delta_k - \delta_i)$$

Quadrant-2 [J_2]

Diagonal elements: $$|V_i| \frac{\partial P_i}{\partial |V_i|} = P_i + |V_i|^2 G_{ii}$$

Off-diagonal elements: $|V_i| \frac{\partial P_i}{\partial |V_k|} = \frac{\partial Q_i}{\partial \delta_k}$ (i.e., the negative of *off-diagonal element* of [J_3])

Quadrant-3 [J_3]

Diagonal elements: $$\frac{\partial Q_i}{\partial \delta_i} = -|V_i|^2 G_{ii} + P_i$$

Off-diagonal elements: $$\frac{\partial Q_i}{\partial \delta_k} = -|V_i V_k Y_{ik}| \cos(\phi_{ik} + \delta_k - \delta_i)$$

[*It may be noted here that the off-diagonal elements of* [J_2] *are simply the negatives of the off-diagonal element of* [J_3]. *It reduces the computation involved for the formulation of Jacobian and its evaluation in large systems.*]

Quadrant-4 [J_4]

Diagonal elements: $$|V_i| \frac{\partial Q_i}{\partial |V_i|} = -|V_i|^2 B_{ii} + Q_i$$

Off-diagonal elements: $$|V_k| \frac{\partial Q_i}{\partial |V_k|} = \frac{\partial P_i}{\partial \delta_k}$$

The equations relating the power mismatch to the voltage and power angle changes ($\Delta|V|$ and $\Delta\delta$) can then be represented in N-R method as

$$\begin{bmatrix} \Delta P \\ \Delta Q \end{bmatrix} = \begin{bmatrix} [J_1] & [J_2] \\ [J_3] & [J_4] \end{bmatrix} \begin{bmatrix} \Delta\delta \\ \dfrac{\Delta|V|}{|V|} \end{bmatrix} \tag{10.104}$$

The Jacobian [J] here is $2(N - 1) \times 2(N - 1)$ for power injection specified at $(N - 1)$ buses. Each quadrant of Jacobian is $(N - 1) \times (N - 1)$. The Jacobian is usually *non-singular due to diagonal dominance* in each quadrant.

For any PQ bus,

$$\Delta P_i = \sum_{k=1}^{N} \frac{\partial P_i}{\partial \delta_k} \Delta\delta_k + \sum_{k=1}^{N} \frac{\partial P_i}{\partial |V_k|} \Delta|V_k|$$

$$= [J_1]\Delta\delta_k + [J_2]\frac{\Delta|V_k|}{|V_k|} \left[\because [J_2] = |V_k| \frac{\partial P_i}{\partial |V_k|}\right] \tag{10.105}$$

$$\Delta Q_i = \sum_{k=1}^{N} \frac{\partial Q_i}{\partial \delta_k} \Delta\delta_k + \sum_{k=1}^{N} |V_k| \frac{\partial Q_i}{\partial |V_k|} \frac{\Delta|V_k|}{|V_k|}$$

$$= [J_3]\Delta\delta_k + [J_4]\frac{\Delta|V_k|}{|V_k|} \tag{10.106}$$

It may be noted here that for a PV bus, only equation (10.105) *is to be used since Q_i is not specified and P_i is only specified. On the other hand, for the slack bus, since both P_i and*

Q_i are unspecified, no equation will appear for the slack bus. It may also be noted that the voltage magnitudes for PV bus, voltage-controlled bus and for slack bus are not variables but are fixed at their specified values. Power angle (δ) at slack bus is also fixed.

An illustration

Let us now apply the N-R method for a typical four-bus system (Figure 10.7) using polar version.

We do not have any voltage-controlled bus here and all the buses except bus no. 1 (the slack bus) are regarded as load buses (i.e., demand P_{d_i} and Q_{d_i} are known). The slack bus would have specified values of δ_i and $|V_i|$ corresponding to estimates of first iteration.

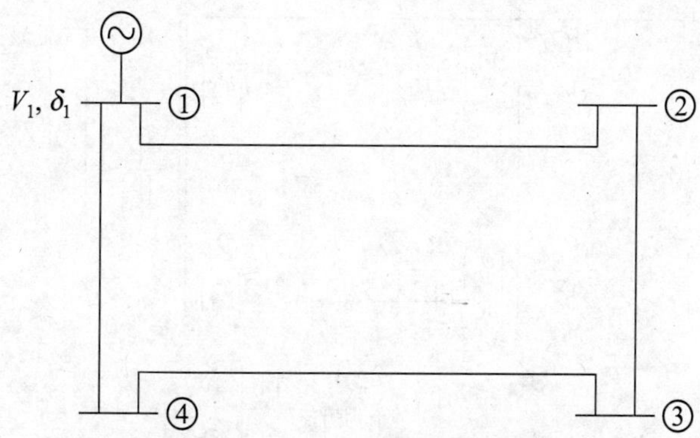

Figure 10.7 A four bus system.

Let us now first observe the mismatch power equations:

$$\Delta P_i = \frac{\partial P_i}{\partial \delta_2}\Delta\delta_2 + \frac{\partial P_i}{\partial \delta_3}\Delta\delta_3 + \frac{\partial P_i}{\partial \delta_4}\Delta\delta_4 + \frac{\partial P_i}{\partial |V_2|}\Delta|V_2| + \frac{\partial P_i}{\partial |V_3|}\Delta|V_3| + \frac{\partial P_i}{\partial |V_4|}\Delta|V_4| \quad (10.107)$$

The last three terms of the above equation can be multiplied and divided by their respective voltage magnitudes and this is not likely to alter their values. We thus obtain

$$\Delta P_i = \frac{\partial P_i}{\partial \delta_2}\Delta\delta_2 + \frac{\partial P_i}{\partial \delta_3}\Delta\delta_3 + \frac{\partial P_i}{\partial \delta_4}\Delta\delta_4$$
$$+|V_2|\frac{\partial P_i}{\partial |V_2|}\cdot\frac{\Delta|V_2|}{|V_2|} + |V_3|\frac{\partial P_i}{\partial |V_3|}\cdot\frac{\Delta|V_3|}{|V_3|} + |V_4|\frac{\partial P_i}{\partial |V_4|}\cdot\frac{\Delta|V_4|}{|V_4|} \quad (10.108a)$$

[Since $|V_i|\frac{\partial P_i}{\partial |V_i|}\times\frac{\Delta|V_i|}{|V_i|} \equiv \frac{\partial P_i}{\partial |V_i|}\Delta|V_i|$ hence it is evident that equations (10.107) and (10.108a) do have same dimension and values.] Similarly,

$$\Delta Q_i = \frac{\partial Q_i}{\partial \delta_2}\Delta\delta_2 + \frac{\partial Q_i}{\partial \delta_3}\Delta\delta_3 + \frac{\partial Q_i}{\partial \delta_4}\Delta\delta_4$$
$$+|V_2|\frac{\partial Q_i}{\partial |V_2|}\cdot\frac{\Delta|V_2|}{|V_2|} + |V_3|\frac{\partial Q_i}{\partial |V_3|}\cdot\frac{\Delta|V_3|}{|V_3|} + |V_4|\frac{\partial Q_i}{\partial |V_4|}\cdot\frac{\Delta|V_4|}{|V_4|} \quad (10.108b)$$

Each load bus thus has two mismatch equations. Combining these equations for the four buses, in matrix form we have

$$\left[\begin{array}{ccc|ccc} \frac{\partial P_2}{\partial \delta_2} & \cdots & \frac{\partial P_2}{\partial \delta_4} & |V_2|\frac{\partial P_2}{\partial |V_2|} & \cdots & |V_4|\frac{\partial P_2}{\partial |V_4|} \\ \vdots & \vdots & \vdots & \vdots & \vdots & \vdots \\ \frac{\partial P_4}{\partial \delta_2} & \cdots & \frac{\partial P_4}{\partial \delta_4} & |V_2|\frac{\partial P_4}{\partial |V_2|} & \cdots & |V_4|\frac{\partial P_4}{\partial |V_4|} \\ \hline \frac{\partial Q_2}{\partial \delta_2} & \cdots & \frac{\partial Q_2}{\partial \delta_4} & |V_2|\frac{\partial Q_2}{\partial |V_2|} & \cdots & |V_4|\frac{\partial Q_2}{\partial |V_4|} \\ \vdots & \vdots & \vdots & \vdots & \vdots & \vdots \\ \frac{\partial Q_4}{\partial \delta_2} & \cdots & \frac{\partial Q_4}{\partial \delta_4} & |V_2|\frac{\partial Q_4}{\partial |V_2|} & \cdots & |V_4|\frac{\partial Q_4}{\partial |V_4|} \end{array}\right] \begin{bmatrix} \Delta\delta_2 \\ \vdots \\ \Delta\delta_4 \\ \frac{\Delta|V_2|}{|V_2|} \\ \vdots \\ \frac{\Delta|V_4|}{|V_4|} \end{bmatrix} = \begin{bmatrix} \Delta P_2 \\ \vdots \\ \Delta P_4 \\ \Delta Q_2 \\ \vdots \\ \Delta Q_4 \end{bmatrix}$$

or,

$$\begin{bmatrix} J_1 & J_2 \\ J_3 & J_4 \end{bmatrix} \begin{bmatrix} \Delta\delta_i \\ \Delta|V_i|/|V_i| \end{bmatrix} = \begin{bmatrix} \Delta P_i \\ \Delta Q_i \end{bmatrix} \tag{10.109}$$

Next equation (10.109) is to be solved iteratively to find $\Delta\delta_i^p$ and $\Delta|V_i|^p$. The variables can then be updated as

$$\left.\begin{aligned} \delta_i^{p+1} &= \delta_i^p + \Delta\delta_i^p \\ |V_i|^{p+1} &= |V_i|^p + \Delta|V_i|^p = |V_i|^p\left(1 + \frac{\Delta|V_i|^p}{|V_i|^p}\right) \end{aligned}\right\} \tag{10.110}$$

10.11.1 Algorithm to Calculate Bus Voltages in Polar Form by N-R Method of Power Flow

1. Read line data for the power system having total number of lines = NL and total number of buses = N.
2. Read bus type ($type_i$), initial bus voltage magnitude in p.u. ($|V_i|^0$), initial bus voltage angle in rad (δ_i^0), real power generation (P_{g_i}), reactive power generation (Q_{g_i}), real power demand (P_{d_i}), reactive power demand (Q_{d_i}), lower and upper Q-limit at PV buses ($Q_{g_{i\min}}$ and $Q_{g_{i\max}}$) in p.u. for each bus, for i = 1, 2, ..., N.
3. Specify tolerance ε (say 0.0001).
4. Form $[Y_{Bus}]$.

5. Compute scheduled active and reactive bus powers for all buses using equation (10.27a) and (10.27b), i.e., $P_{i_{\text{scheduled}}}$ and $Q_{i_{\text{scheduled}}}$ for i = 1, 2, ..., N.
6. Set iteration count p = 0.
7. Fix slack bus voltage magnitude and angle for next iteration, $|V_i|^{p+1} = |V_i|^0$, $\delta_i^{p+1} = 0$ if $type_i = 0$.
8. Fix specified voltage magnitude for PV buses. $|V_i|^p = |V_i|^0$, for i = 1, 2, ..., N (only if $type_i = 1$).
9. Calculate active and reactive power at all buses (i = 1, 2, ..., N) using equation (10.92) or (10.27).
10. Calculate reactive power generation for PV buses as $Q_{g_i} = \left(Q_i^p + Q_{di}\right)$ for i = 1, 2, ..., N (only if $type_i = 1$).
11. If $Q_{g_{\min}}$ and $Q_{g_{\max}}$ are specified for all PV buses then go to next step, otherwise go to step 13.
12. Compute the following for all PV buses (for i = 1, 2, ..., N and consider bus-i only if $type_i = 1$).

    if $Q_{g_i} > Q_{g_{i\max}}$ or, $Q_{g_i} < Q_{g_{i\min}}$ for the PV bus-i (i.e., Q-limit is failed),

    (i) set $Q_{g_i} = Q_{g_{i\max}}$ for $Q_{g_i} > Q_{g_{i\max}}$,

    or, set $Q_{g_i} = Q_{g_{i\min}}$ for $Q_{g_i} < Q_{g_{i\min}}$.

    (ii) release the constraint of $|V_i|^p$ to be fixed and initialize it for a flat start, i.e., $|V_i|^p$ = 1 and also change bus type from PV bus ($type_i$ = 1) to PQ bus ($type_i$ = 2) as Q-limit is failed.
13. Form ΔP^p matrix considering all buses excluding slack bus using equation (10.28a) where $\Delta P_i^p = P_{i_{\text{scheduled}}} - P_i^p$ for i = 1, 2, ..., N (i.e., exclude bus-i if $type_i$ = 0)
14. Form ΔQ^p matrix considering only for the PQ buses using equation (10.28b). $\Delta Q_i^p = Q_{i_{\text{scheduled}}} - Q_i^p$ for i = 1, 2, ..., N (i.e., include bus-i, only if $type_i$ = 2)
15. If max (ΔP^p and ΔQ^p) < ε then go to step 21 else go to step 16.
16. Form the Jacobian matrix [J^p]. Compute the elements of the matrix using equations (10.95)–(10.103).
17. Solve equation $$\begin{bmatrix} \Delta\delta_i^p \\ \dfrac{\Delta|V_k|^p}{|V_k|^p} \end{bmatrix} = \left[J^p\right]^{-1} \begin{bmatrix} \Delta P_i^p \\ \Delta Q_k^p \end{bmatrix}$$

    where, i = 1, 2, ..., N except slack bus (i.e., exclude bus-i, if $type_i$ = 0) and k = 1, 2, ..., N only for PQ buses (i.e., include bus-k, only if $type_k$ = 0).

18. (i) Update bus voltage angles for all buses except slack bus $\delta_i^{p+1} = \delta_i^p + \Delta\delta_i^p$, for i = 1, 2, ..., N. Exclude bus-i if $type_i = 0$.

    (ii) Update bus voltage magnitudes for PQ buses only $|V_k|^{p+1} = |V_k|^p \left(1 + \frac{\Delta|V_k|^p}{|V_k|^p}\right)$, for k = 1, 2, ..., N Include bus-k, only if $type_k = 2$.
19. Update iteration count $p \leftarrow p+1$.
20. If $p > 20$ then break the iteration process and go to step 25. Otherwise go to step 7.
21. Convert final bus voltages into rectangular form, $V_i^p = |V_i|^p \angle \delta_i^p$ for i = 1, 2, ..., N.
22. Calculate reactive power generations for those PV buses who have not crossed their Q-limit. $Q_{g_i} = Q_i^p + Q_{d_i}$ for i = 1, 2, ..., N provided $type_i = 1$.
23. Display and/or store final bus voltages, V_i^p, for i = 1, 2, ..., N.
24. Display and/or store reactive power generation, Q_{g_i} for PV buses ($type_i = 1$), for i = 1, 2, ..., N.
25. Stop.

10.12 DISCUSSION ABOUT N-R METHOD

The complete set of governing equations consists of two numbers for each PQ bus and one for each PV or voltage-controlled bus. For any PQ bus, the variables are V and δ while it is δ for any PV bus. This makes the number of variables to be equal to the number of equations.

The algorithm for N-R method, applicable for polar expression of equations of load flow, can be further simplified by dividing each $\Delta|V_i|^p$ by $|V_i|^p$ and then we can write

$$\begin{bmatrix} \Delta P^p \\ \Delta Q^p \end{bmatrix} = \begin{bmatrix} J_1^p & J_2^p \\ J_3^p & J_4^p \end{bmatrix} \begin{bmatrix} \Delta\delta^p \\ \dfrac{\Delta|V|^p}{|V|^p} \end{bmatrix}$$

The division of each $\Delta|V_i|^p$ by $|V_i|^p$ does not numerically affect the algorithm, but only simplifies some of the Jacobian terms. $\Delta\delta^p$ is the correction to PQ and PV buses while $\Delta|V_i|^p / |V_i|^p$ is the correction to PQ buses.

To some users, the polar coordinate representation of equations and subsequent computations involving N-R method are preferable to rectangular version in computational aspects. Real power mismatch equations are present for all buses except the slack bus, while reactive power mismatch equations are needed for load buses only.

The form of Jacobian shown in the application of N-R method for solving power flow equations when they are expressed in polar forms, is used for evaluating the Jacobian and for comparison of magnitudes. In the power systems if M buses are voltage controlled, successive iterated bus voltage change at each of these buses must be zero, i.e., for m-th bus

$$\Delta V_m = V_m^{p+1} - V_m^p = 0; \frac{\partial Q_m}{\partial V_m} = 0 \quad [\text{for } m = 1, 2, 3, ..., M]$$

In the absence of voltage-controlled buses, power injections being specified at $(N - 1)$ buses, the Jacobian is $2(N - 1) \times 2(N - 1)$, while for the presence of M number of voltage-controlled buses, the Jacobian would then obviously be $(2N - 2 - M) \times (2N - 2 - M)$; the real power constraint equations are $(N - 1)$ numbers while the reactive power constraint equations are $(N - 1 - M)$ numbers.

Also, Q-limit being specified for PV buses, if at any step of iteration $Q_{\text{calculated}}$ goes beyond the limit of Q (limit) range, $Q_{\text{calculated}}$ is then assigned to the value of Q_{limit} or $Q_{\text{specified}}$ at the nearest limit and PV bus types are then changed to PQ types of buses. The number of equations to be solved is then again $(2N - 2)$.

EXAMPLE 10.6 A simple two-bus power system is shown in Figure E10.6:

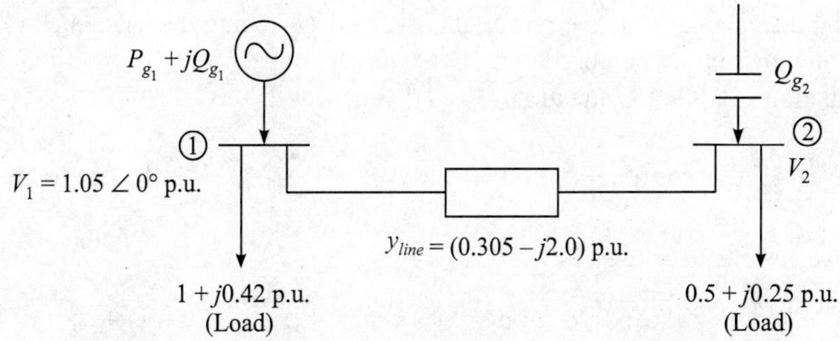

Figure E10.6 A two bus power system [$b_{line}/2 = j0.064$ p.u.].

(a) Use N-R method to find V_2 and δ_2. Assuming bus no. 2 as PQ bus

(b) Repeat (a) when the specified bus voltages are $|V_1| = 1.05$ p.u. and $|V_2| = 1.0$ p.u. Obtain δ_2 and Q_{g_2} (assume specified tolerance, $\varepsilon = 0.0001$.)

Solution: (a) Let us assume bus-1 be the slack bus ($type_1 = 0$) in the given simple two bus system. Note that slack bus load is not required for calculation of bus voltage magnitude and angle of bus-2.

In the first part of the given problem there is only one PQ bus and there is no PV bus. The matrix equation for power mismatches by N-R method is given by

$$\begin{bmatrix} \Delta P_2 \\ \Delta Q_2 \end{bmatrix} = \begin{bmatrix} J_1 & J_2 \\ J_3 & J_4 \end{bmatrix} \begin{bmatrix} \Delta\delta_2 \\ \Delta|V_2|/|V_2| \end{bmatrix}$$

In the algorithm for computer simulation, powers are calculated using polar form of voltage and $[Y_{Bus}]$. $[Y_{Bus}]$ is here obtained in rectangular form (by inspection in this simple two bus problem) and then it is converted to polar form as shown below (all quantities are in p.u. and angles are in radian).

$$[Y_{Bus}] = \begin{bmatrix} 0.305 - j1.936 & -0.305 + j2 \\ -0.305 + j2 & 0.305 - j1.936 \end{bmatrix} = \begin{bmatrix} 1.9599\angle -1.4145 & 2.0231\angle 1.7221 \\ 2.0231\angle 1.7221 & 1.9599\angle -1.4145 \end{bmatrix}$$

In order to start the iterative process, initial guess of bus voltage magnitudes and angles are as follows:

$$|V_1|^0 = 1.05 \text{ p.u. (fixed)}, \ \delta_1^0 = 0 \text{ rad (fixed)}, \ |V_2|^0 = 1.0 \text{ p.u.}, \ \delta_2^0 = 0 \text{ rad},$$

$P_{2_{\text{scheduled}}} = -0.5$ p.u and $Q_{2_{\text{scheduled}}} = -0.25$ p.u. [The inductive reactive power generation is assumed to be positive while the inductive reactive demand is negative. This assumption will be maintained in all the subsequent load flow problems throughout this chapter.]

For p-th iteration,
$$\begin{bmatrix} \Delta P_2^p \\ \Delta Q_2^p \end{bmatrix} = \begin{bmatrix} J_1^p & J_2^p \\ J_3^p & J_4^p \end{bmatrix} \begin{bmatrix} \Delta \delta_2^p \\ \Delta |V_2|^p / |V_2|^p \end{bmatrix}$$

and
$$\Delta P_2^p = P_{2_{\text{scheduled}}} - P_2^p \ ; \ \Delta Q_2^p = Q_{2_{\text{scheduled}}} - Q_2^p$$

[P_2^p being the calculated active power at bus-2 and Q_2^p being the calculated reactive power at same bus at p-th step of iteration].

From equation (10.92) [or equation (10.27)] for the i-th bus,

$$P_i^p = \sum_{k=1}^{N} |V_i|^p |V_k|^p |Y_{ik}| \cos(\phi_{ik} + \delta_k^p - \delta_i^p) \text{ and } Q_i^p = -\sum_{k=1}^{N} |V_i|^p |V_k|^p |Y_{ik}| \sin(\phi_{ik} + \delta_k^p - \delta_i^p)$$

With $p = 0$ we have

$$\therefore \quad P_2^0 = |V_2|^0 |V_1|^0 |Y_{21}| \cos(\phi_{21} + \delta_1^0 - \delta_2^0) + |V_2|^0 |V_2|^0 |Y_{22}| \cos(\phi_{22} + \delta_2^0 - \delta_2^0)$$
$$= 1.0 \times 1.05 \times 2.0231 \cos(1.7221 + 0 - 0) + 1.0 \times 1.0 \times 1.9599 \cos(-1.4145 + 0 - 0)$$
$$= -0.32025 + 0.30501 = -0.01525 \text{ p.u.}$$

and
$$Q_2^0 = -|V_2|^0 |V_1|^0 |Y_{21}| \sin(\phi_{21} + \delta_1^0 - \delta_2^0) - |V_2|^0 |V_2|^0 |Y_{22}| \sin(\phi_{22} + \delta_2^0 - \delta_2^0)$$
$$= -1.0 \times 1.05 \times 2.0231 \sin(1.7221 + 0 - 0) - 1.0 \times 1.0 \times 1.9599 \sin(-1.4145 + 0 - 0)$$
$$= -2.1 + 1.936 = -0.164 \text{ p.u.}$$

$\therefore \ \Delta P_2^0 = -0.5 - (-0.01525) = -0.48475$ p.u. and $\Delta Q_2^0 = -0.25 - (-0.164) = -0.086$ p.u.

Since changes in bus powers (both active and reactive) are more than tolerance ($\varepsilon = 0.0001$), hence it is required to find the elements of Jacobian matrix and then to find the bus voltage magnitudes and angles in next step of iterative process. If the changes in bus power be remain within the given tolerance then bus voltage at this iteration (here first iteration) would be the final value of bus voltages, otherwise the iterative process will continue. Note that bus voltage magnitude and angle for slack bus will remain constant throughout the iterative process.

Formation of Jacobian matrix

The Jacobian is given by, $\left[J^p\right] = \begin{bmatrix} J_1^p & J_2^p \\ J_3^p & J_4^p \end{bmatrix}$; in this problem since there are only two buses in the system out of which one is slack bus, hence there is no off-diagonal element of J_1, J_2, J_3 and J_4. The problem thus gets simpler with diagonal elements being given by (using equations (10.96), (10.98), (10.101) and (10.103) respectively)

$$J^0_{1_{22}} = -Q_2^0 - B_{22}\left(\left|V_2\right|^0\right)^2 = -(-0.164) - (-1.936)\,(1.0)^2 = 2.1$$

$$J^0_{2_{22}} = P_2^0 + G_{22}\left(\left|V_2\right|^0\right)^2 = (-0.01525) + 0.305(1.0)^2 = 0.28975$$

$$J^0_{3_{22}} = P_2^0 - G_{22}\left(\left|V_2\right|^0\right)^2 = (-0.01525) - 0.305\,(1.0)^2 = -0.32025$$

$$J^0_{4_{22}} = Q_2^0 - B_{22}\left(\left|V_2\right|^0\right)^2 = -0.164 - (-1.936)\,(1.0)^2 = 1.772$$

Thus Jacobian matrix becomes a (2 × 2) square matrix and is as follows:

$$\left[J^0\right] = \begin{bmatrix} 2.1 & 0.28975 \\ -0.32025 & 1.772 \end{bmatrix}$$

$[J^0]$ being the Jacobian matrix for first iteration (i.e., 0th iteration), the mismatches can be written as

$$\therefore \qquad \begin{bmatrix} \Delta\delta_2^0 \\ \dfrac{\Delta\left|V_2\right|^0}{\left|V_2\right|^0} \end{bmatrix} = \begin{bmatrix} 2.1 & 0.28975 \\ -0.32025 & 1.772 \end{bmatrix}^{-1} \begin{bmatrix} -0.48475 \\ -0.086 \end{bmatrix}$$

$$= \begin{bmatrix} 0.464605 & -0.075970 \\ 0.083967 & 0.550604 \end{bmatrix} \begin{bmatrix} -0.48475 \\ -0.086 \end{bmatrix} = \begin{bmatrix} -0.218684 \\ -0.088055 \end{bmatrix}$$

$$\therefore \qquad \Delta\delta_2^0 = -0.218684 \text{ rad and } \frac{\Delta\left|V_2\right|^0}{\left|V_2\right|^0} = -0.088055 \text{ p.u.}$$

Next, new bus voltage magnitude and load angle are obtained using the following equations:

$$\delta_i^{p+1} = \delta_i^p + \Delta\delta_i^p\,;\; \left|V_i\right|^{p+1} = \left|V_i\right|^p\left(1 + \frac{\Delta\left|V_i\right|^p}{\left|V_i\right|^p}\right)$$

Here, $\delta_2^1 = \delta_2^0 + \Delta\delta_2^0 = 0.0 + (-0.218684) = -0.218684$ rad

$$\left|V_2\right|^1 = \left|V_2\right|^0\left(1 + \frac{\Delta\left|V_2\right|^0}{\left|V_2\right|^0}\right) = 1.0\,\{1 + (-0.88055)\} = 0.911945 \text{ p.u.}$$

ΔP_2^1 and ΔQ_2^1 are again to be calculated and iterative process will continue until the changes in bus powers are within given tolerance (ε = 0.0001). Continuing with the iterative process, the final bus voltages are obtained as

$$V_1 = 1.05\angle 0\,;\; V_2 = 0.855801\angle -0.257869$$

[Bus voltage magnitudes are in p.u. and angles are in radian]

(b) In the second part of the given problem there is only one PV bus (bus no. 2), obviously excluding slack bus in the system, with Q-limit. Hence Jacobian matrix is a (1 × 1) matrix.

The equation required to be solved is then given by

$$\left[\Delta P_2^p\right]=\left[J^p\right]\left[\Delta\delta_2^p\right]$$

However, $P_2^0 = -0.01525$ and $\Delta P_2^0 = -0.48475$ (as obtained in Problem no. 10.6(a)). It is observed that ΔP_2^0 is beyond the given tolerance. Hence iterative process is required to be continued.

Now, $$[J^0]=[J^0_{2_{22}}]=2.1 \qquad \therefore \qquad [J^0]^{-1}=0.47619$$

and $$\Delta\delta_2^0 = 0.47619\,(-0.48475) = -0.230833 \text{ rad.}$$

Hence, $$\delta_2^1 = 0+(-0.230833) = -0.230833 \text{ rad,}$$

ΔP_2^1 is again to be calculated, and iterative process will continue until change in bus powers remains within given tolerance (ε = 0.0001).

In this problem reactive power demand of bus-2 (i.e., Q_{d_2}) is not required for determination of δ_2, but this is required for calculation of reactive power generation (i.e., Q_{g_2}) of bus-2 as follows:

$$Q_{g_2} = Q_2 + Q_{d_2}$$

where, Q_2 = calculated reactive power at bus-2 after iterative solution i.e. after the convergence of iterative process and Q_{d_2} = reactive power demand at bus-2.

After iterative solution, the final bus voltages are,

$$V_1 = 1.05\angle 0 \quad \text{and} \quad V_2 = 1.0\angle -0.237321$$

[bus voltages are in p.u. while load angles are in radians]

$$\therefore \qquad Q_2 = -|V_2||V_1||Y_{21}|\sin(\phi_{21}+\delta_1-\delta_2) - |V_2||V_2||Y_{22}|\sin\phi_{22}$$

$$= -1.0\times 1.05\times 2.0231\sin\{1.7221+0-(-0.237321)\} - 1.0\times 1.0\times 1.9599\sin(-1.4145)$$

$$= -0.029849 \text{ p.u.}$$

and $$Q_{g_2} = -0.029849+0.25 = 0.220151 \text{ p.u.}$$

Computer program execution result for Example 10.6 (following algorithm of Article 10.11.1)

Part-(a)

```
All quantities are in p.u. and all angles are in radian.
No. of buses = 2. No. of lines = 1.
Non zero elements of [Ybus]:
Note that only upper triangular off diagonal elements are provided.
Row  Column          Ybus              Row  Column          Ybus
 1      1    (0.305000-1.936000i)      2      2    (0.305000-1.936000i)
 1      2   (-0.305000+2.000000i)
N-R load flow result without considering Q-limits at PV buses.
Specified tolerance =0.00010.
```

```
Load flow converges with 3 iterations and maximum tolerance =0.000054.
Final bus voltages are:
Bus no.  Complex bus voltage   Magnitude     Angle
 1          (1.050000)          1.050000      0
 2       (0.827504-0.218247i)   0.855801   -0.257869        (Ans.)
```

Part-(b) Here $[Y_{Bus}]$ is same as in Problem 10.6 (a). In this problem bus no. 2 is considered as PV bus whereas bus no. 2 was considered as PQ bus (load bus) in problem (a).

```
All quantities are in p.u. and all angles are in radian.
No. of buses = 2. No. of lines = 1.
N-R load flow result without considering Q-limits at PV buses.
Specified tolerance = 0.00010.
Load flow converges with 2 iterations and maximum tolerance =0.000017.
Final bus voltages are:
Bus no.   Complex bus voltage    Magnitude     Angle
 1            (1.050000)          1.000000       0
 2        (0.971971-0.235100i)    1.000000   -0.237321        (Ans.)
Reactive power generation at PV buses:
PV bus no.    Reactive Power
 2               0.220151                                     (Ans.)
```

EXAMPLE 10.7 Use *N-R* method to obtain V_2 and V_3 for the following data set referring to the three-bus, three-line system as shown in Figure E10.7 [all data are in p.u.]. Assume each line to have series impedance (z_s) of $(0.8 + j0.2)$ p.u. and a total shunt admittance of 0.04 p.u. Also calculate line flows, slack bus power and total line loss. Assume specified tolerance, $\varepsilon = 0.0001$.

Bus no.	P_d	Q_d	P_g	Q_g	V
1	1	0.75	?	?	1.05
2	0.5	1.0	0	0	–
3	1.2	0.5	0	?	$\lvert 1.02 \rvert \angle ?$

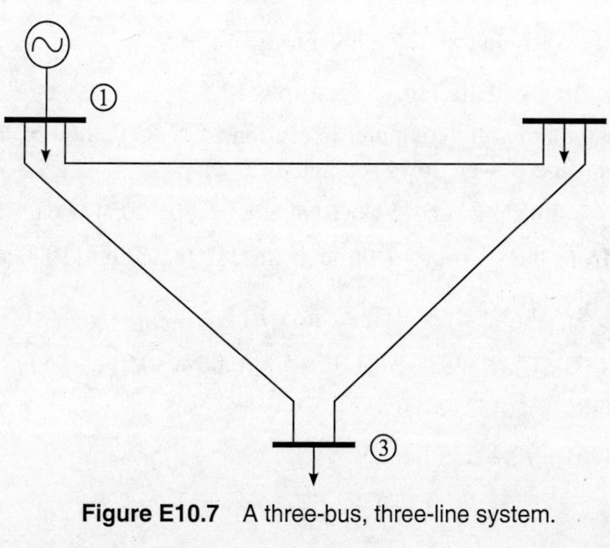

Figure E10.7 A three-bus, three-line system.

Solution: It is evident that bus no. 1 is the slack bus, bus no. 2 is the PQ bus and bus no. 3 is the voltage controlled bus.

Computer program execution result for Example 10.7 (following algorithm of Article 10.11.1) (Ans.)

```
All quantities are in p.u. and all angles are in radian.
No. of buses = 3. No. of lines = 3.
Non zero elements of [Ybus]:
Note that only upper triangular off diagonal elements are provided.
Row  Column        Ybus                  Row  Column        Ybus
 1     1    (3.448276-8.540690i)          2     2    (3.448276-8.540690i)
 1     2    (-1.724138+4.310345i)         2     3    (-1.724138+4.310345i)
 1     3    (-1.724138+4.310345i)         3     3    (3.448276-8.540690i)
N-R load flow result without considering Q-limits at PV buses.
Specified tolerance =0.00010.
Load flow converges with 3 iterations and maximum tolerance =0.000001.
Final bus voltages are:
Bus no.     Complex bus voltage      Magnitude      Angle
 1              (1.050000)           1.050000         0
 2         (0.896083-0.110547i)      0.902877     -0.122746
 3         (0.995066-0.224151i)      1.020000     -0.221564
LINE POWER FLOWS:
Line no.  bus-i   bus-j            Sij                      Sji
 1          1       2      (0.778961+0.452378i)   (-0.717046-0.374297i)
 2          1       3      (1.113927-0.201267i)   (-1.022097+0.345127i)
 3          2       3      (0.217047-0.625702i)   (-0.177903+0.649339i)
Slack bus power = (2.892888+1.001110i)
Total transmission line loss = (0.192890+0.245577i)
```

Line flow from bus-1 to bus-2 is not equal to the line flow from bus-2 to bus-1. For power flow in lines 1-2,

$$S_{12} = V_1^* \{(V_1 - V_2)\, y_{12} + V_1 y_{12_0}\} \text{ and } S_{21} = V_2^* \{(V_2 - V_1)\, y_{21} + V_2 y_{21_0}\}$$

whereas $z_{12} = z_{21}$, $y_{12_0} = y_{21_0}$ and $V_1 \neq V_2$, therefore, $S_{12} \neq S_{21}$.

A sample line power flow calculation of Example 10.7

The final bus voltages obtained by computer execution of N-R method of load flow are given above. Let us calculate line power flows for line 1-2.

$$V_1 = 1.05 \text{ p.u. and } V_2 = (0.896083 - j0.110547) \text{ p.u.}$$

Therefore, line flow from bus-1 to bus-2 through line1-2 (equation (10.32a)),

$$S_{12}^* = P_{12} - jQ_{12} = V_1^* \{(V_1 - V_2)/z_{12} + V_1 y_{12_0}\}$$

$$= 1.05\,[\{1.05 - (0.896083 - j0.110547)\}/(0.08 + j0.2) + 1.05\,(j0.04)]$$

$$= (0.778961 - j0.452378) \text{ p.u.}$$

$$\therefore \quad S_{12} = (0.778961 + j0.452378) \text{ p.u.}$$

∴ P_{12} = Active power flow from bus-1 to bus-2 = Re (S_{12}^*) = 0.778961 p.u.

Q_{12} = Reactive power flow from bus-1 to bus-2 = – Im (S_{12}^*) = 0.452378 p.u. (inductive)

Similarly, P_{21} = Active power flow from bus-1 to bus-2 = –0.717046 p.u.

Q_{21} = Reactive power flow from bus-1 to bus-2 = – 0.374297 p.u. (negative sign means VAR is capacitive)

Power loss in line 1–2,

$$S_{loss_{12}} = S_{12} + S_{21} = (0.778961 + j0.452378) + (-0.717046 - j0.374297)$$
$$= (0.061915 + j0.078081) \text{ p.u.}$$

Slack bus power is given by (equation (10.32e)),

$$SBP = S_{12} + S_{13} + S_{d_1} = (0.778961 + j0.452378) + (1.113927 - j0.201267) + (1 + j0.75)$$
$$= (2.8929 + j1.0011) \text{ p.u. (Here } S_{d_1} = (1 + j0.75) \text{ p.u.)}$$

EXAMPLE 10.8 Find final bus voltages, line flows, total line loss, slack bus power by N-R method for the six-bus, eight-line system shown in Example 10.5.

Solution:

Computer program execution result for Example 10.8 (following algorithm of Article 10.11.1) (Ans.)

```
All quantities are in p.u. and all angles are in radian.
No. of buses = 6. No. of lines = 8.
N-R load flow result without considering Q-limits at PV buses.
Specified tolerance =0.00010.
Load flow converges with 3 iterations and maximum tolerance =0.000032.
Final bus voltages are:
Bus no.    Complex bus voltage      Magnitude      Angle
  1           (1.040000)             1.040000        0
  2         (0.991418-0.187553i)     1.009002    -0.186967
  3         (1.052691-0.069361i)     1.054973    -0.065794
  4         (0.976992-0.150937i)     0.988583    -0.153280
  5         (0.977628-0.158974i)     0.990469    -0.161201
  6         (0.973459-0.217681i)     0.997501    -0.219996
LINE POWER FLOWS:
Line no. bus-i  bus-j           Sij                     Sji
  1        1      3     (0.230487-0.085541i)   (-0.228316+0.079879i)
  2        2      3    (-0.657493-0.104454i)    (0.670493+0.169812i)
  3        3      4     (0.509285+0.257155i)   (-0.497376-0.218515i)
  4        3      5     (0.548570+0.243168i)   (-0.535430-0.198411i)
  5        4      5     (0.047369-0.031471i)   (-0.047285+0.012304i)
  6        5      6     (0.182705-0.063876i)   (-0.180484+0.055218i)
  7        1      2     (0.512021+0.274589i)   (-0.512021-0.161357i)
  8        2      6     (0.169507+0.165831i)   (-0.169507-0.155214i)
Slack bus power = (0.742508+0.189048i)
Total transmission line loss = (0.042523+0.239120i)
```

EXAMPLE 10.9 Find final bus voltages by *N-R* method for the four-bus, five-line system given in Example 10.4, where bus-2 has *Q*-limit of $0.1 \le Q_2 \le 1.0$ p.u. Write a MATLAB code for this problem.

Solution:

The MATLAB code for Example 10.9

```
% Program to calculate bus voltages by N-R method of power flow for Example 10.9
clc
clear all
epsil=0.0001;bus_no=4;line_no=5;
type=[0 1 2 2]; v=[1.06 1.02 1 1]; del=[0 0 0 0]; pg=[0 1.2 0 0];
qg=[0 0 0 0]; pd=[2 0 0.7 0.6]; qd=[0.6 0 0.5 0.3];qmin=0.1;qmax=1;
disp('All quantities are in p.u. and all angles are in radian.')
disp(sprintf('No. of buses = %d. No. of lines = %d.',bus_no,line_no));
Ybus_prog %Prog. to calculate [Ybus]
G=real(Y); B=imag(Y);psch=pg-pd;qsch=qg-qd;vmag=v; vang=del;
for it=1:20,
for n=1:bus_no,
pcal(n)=0;
qcal(n)=0;
for k=1:bus_no,
pcal(n)=pcal(n)+vmag(n)*vmag(k)*abs(Y(n,k))*cos(vang(k)+angle(Y(n,k))
vang(n));
qcal(n)=qcal(n)-vmag(n)*vmag(k)*abs(Y(n,k))*sin(vang(k)+angle(Y(n,k))-
vang(n));
end
end
for k=1:bus_no
  if(type(k)==1);
     qg(k)=qcal(k);
   if (qg(k)<qmin||qg(k)>qmax)
      if qg(k)<qmin qg(k)=qmin; else qg(k)=qmax; end
      type(k)=2; vmag(k)=1.0; qsch(k)=qg(k);
   end
  end
end
pchng=psch-pcal; qchng=qsch-qcal; m=1;
for k=1:bus_no,
   if type(k)~=0  pqchng(m)=pchng(k); m=m+1; end
end
for k=1:bus_no,
   if type(k)==2 pqchng(m)=qchng(k); m=m+1;  end
end
   max(abs(pqchng));
   diff=max(abs(pqchng));
   if diff<epsil
      break
   end
```

```
%Form Jacobian matrix for NRLF_Final
J1=[];J2=[];J3=[];J4=[];
for k=1:bus_no
 if k~=1
  for n=1:bus_no
   if type(n)~=0
    if n~=k
G(k,n)=i*vmag(k)*exp(-i*vang(k))*abs(Y(k,n))*exp(i*angle(Y(k,n)))*vmag
(n)*exp(i*vang(n));
     J1(k-1,n-1)=real(G(k,n));
     else
    D(k,n)=-qcal(n)-(vmag(n))^2*imag(Y(n,n))-i*(pcal(n)
(vmag(n))^2*real(Y(n,n)));
       J1(k-1,n-1)=real(D(k,n));
     end
    end
   end
 end
end
for k=1:bus_no
  if k~=1
   for n=1:bus_no
    if type(n)==2
     if n~=k J2(k-1,n-1)=imag(G(k,n));
     else J2(k-1,n-1)=(pcal(n)+(vmag(n))^2*real(Y(n,n)));end
    end
   end
  end
end
for k=1:bus_no,
  if k>1 %type(k)==2  %PQ bus
   for n=1:bus_no
    if n~=1
   if n~=k  J3(k-1,n-1)=-imag(G(k,n)); else  J3(k-1,n-1)=-imag(D(k,n));end
    end
   end
  end
end
for k=1:bus_no,
  if k>1 %type(k)==2
   for n=1:bus_no,
    if n>1 %type(n)==2
     if n~=k  J4(k-1,n-1)=real(G(k,n));
     else J4(k-1,n-1)=qcal(n)-(vmag(n))^2*imag(Y(n,n));  end
    end
   end
  end
end
```

```
jac=[J1 J2;J3 J4]; ddeldv=(inv(jac))*pqchng'; n=1;
for k=1:bus_no,
if k~=1 ddel(k)=ddeldv(n); n=n+1; else ddel(k)=0; end
end
for k=1:bus_no
   if k>1 dv(k)=ddeldv(n); n=n+1; else  dv(k)=0; end
end
vang=vang+ddel; del=vang';
for n=1:bus_no  vmag(n)=vmag(n)+dv(n)*vmag(n); end
if (it>20)
   break
end
end
if diff<epsil
vf=vmag.*exp(i*vang);
disp('N-R load flow result considering Q-limits at PV buses.')
disp(sprintf('Specified tolerance =%7.5f.',epsil))
disp(sprintf('Load flow converges with %d iterations and maximum tolerance
=%f.',it-1,diff))
disp('Final bus voltages are:')
disp('Bus no.   Complex bus voltage      Magnitude   Angle')
for n=1:bus_no
disp(sprintf('  %d \t\t(%s)\t%f\t%f',n,num2str(vf(n),'% 10.6f'),abs(vf(n
)),angle(vf(n))));
end
end
```

Computer program execution result for Example 10.9 (following algorithm of Article 10.11.1) (Ans.)

```
All quantities are in p.u. and all angles are in radian.
No. of buses = 4. No. of lines = 5.
N-R load flow result considering Q-limits at PV buses.
Specified tolerance =0.00010.
Lower bound of Q-limit is crossed for bus-2 in iteration=1.
Load flow converges with 3 iterations and maximum tolerance =0.000083.
Final bus voltages are:
Bus no.     Complex bus voltage      Magnitude      Angle
  1               (1.060000)          1.060000        0
  2         (1.027349+0.159855i)      1.039711     0.154362
  3         (0.920233-0.037391i)      0.920993    -0.040610
  4         (0.855003-0.133230i)      0.865321    -0.154580
```

10.13 APPLICATION ASPECT OF N-R METHOD IN MULTI-BUS SYSTEM

The system equations to be solved in complex power flow problem can be written in the form:

$$P_i = F_1(\delta, V)$$

$$Q_i = F_2(\delta, V)$$

where δ and V are vectors and components are δ_i and V_i respectively. Then it is possible to select one of the nodes as the reference of phase angles (i.e. δ_i can be set as zero). This node then acts as the *reference node*.

The Jacobian matrix of the system can then be written as

$$[J] = \begin{bmatrix} \dfrac{\partial F_1}{\partial \delta} & \dfrac{\partial F_1}{\partial |V|} \\ \dfrac{\partial F_2}{\partial \delta} & \dfrac{\partial F_2}{\partial |V|} \end{bmatrix}$$

where $[J]$ is a $(2N \times 2N)$ square matrix. Without including slack bus, $[J]$ will have dimension of $(2N - 2) \times (2N - 2)$.

In power systems $[J]$ is a *sparse matrix* such that the magnitude of

$$\frac{\partial F_{1_i}}{\partial \delta_j} = \left(\frac{\partial F_{1_i}}{\partial |V_j|} = \frac{\partial F_{2_i}}{\partial \delta_j} = \frac{\partial F_{2_i}}{\partial |V_j|} \right) = 0, \text{ if } j \neq i \,;$$

also, the non-zero elements are located symmetrically with respect to the diagonal elements. Ordinarily, corresponding to the stable state of system operation we can write

$$\sin(\delta_j - \delta_i) \approx (\delta_j - \delta_i)\,; |V_i| \approx |V_j| \approx V_{\text{nominal}}.$$

Since the resistance r_{ij} of an EHV transmission line is negligibly small in comparison to the inductive reactance x_{ij} (obviously, not for cables), further simplification can be obtained by setting shunt admittance (or line charging) y_{ij_0} to be zero. The non-zero elements of the submatrices $(\partial F_1/\partial |V|)$ and $(\partial F_2/\partial \delta)$ become very small in comparison to those of submatrices $(\partial F_1/\partial \delta)$ and $(\partial F_2/\partial |V|)$. We can then write for the power mismatches ΔP and ΔQ,

$$\Delta P = \left(\frac{\partial F_1}{\partial \delta} \right) \Delta \delta; \; \Delta Q = \left(\frac{\partial F_2}{\partial |V|} \right) \Delta |V|$$

where $\Delta\delta$ and $\Delta|V|$ are the mismatches of δ and $|V|$ with respect to the solution.

In the Jacobian, first N column are not linearly dependent and for $i = 1, 2, 3, ..., N$,

$$\sum_{j=1}^{N} \frac{\partial F_{1_i}}{\partial \delta_j} = 0; \; \sum_{j=1}^{N} \frac{\partial F_{2_i}}{\partial \delta_j} = 0$$

Then the matrix $[J]$ approaches singularity irrespective of the values of δ and V under these special sets of assumptions. For the usual values of δ and V, during stable operation of the system, $[J]$ has rank $(2N - 1)$. Rank of $[J]$ becomes less when the system approaches instability.

In the presence of tap changer transformer, $[J]$ is no longer a topologically symmetric matrix but it contains more zero elements. If V_1 and V_2 be the primary and secondary voltages while tap changer is at primary, off-nominal tap ratio being α, in the Jacobian, $\dfrac{\partial F_{1_i}}{\partial \alpha} = \dfrac{\partial F_{2_i}}{\partial \alpha} = 0$, although in the absence of the tap changing transformer, $\dfrac{\partial F_{1_i}}{\partial V_2} \neq 0$ and $\dfrac{\partial F_{2_i}}{\partial V_2} \neq 0$.

The Jacobian matrix is *sparse* like the bus admittance matrix $[Y_{Bus}]$ and does have positional (but not numerical) symmetry. The Jacobian matrix ordinarily is also *non-singular*. Also, use of $[\Delta\delta, \Delta |V|/|V|]^T$ is normally used for the variable voltage vector reducing computational effort. The values of $[J]$ are all functions of the variables $|V|$ and δ and are to be calculated after each step of iteration.

A further improvement in computational effort can be achieved by replacing the reactive power residual (ΔQ) in the Jacobian matrix by $\Delta Q/|V|$. [In the equation $\Delta Q = Q_{\text{specified}} - Q_{\text{calculated}}$ be divided throughout by $|V|$ the term $(Q_{\text{specified}}/|V|)$ come on the right hand side of load flow equation instead of ΔQ. Usually, for normal load flow problems $Q_{\text{specified}}/|V|$ is numerically small. Since the performance of NR method is very much affected by the degree of non-linearity, hence with very small values of non-linearity of $Q_{\text{specified}}/|V|$, the obvious advantage is achieved.] Division of ΔP by $|V|$ is also, to some extent, advantageous but due to weak coupling between active power and voltage, this advantage is less effective.

The N-R method may *diverge* very rapidly or may *converge* to give a *wrong* solution if the *starting voltages* are not properly chosen. A method to solve these problems is the imposition of limits on the size of $\Delta\delta$ and $\Delta|V|$ correction at each step of iteration. Another useful method is obviously to provide good starting values of $|V|$ and δ. It is thus a usual practice to set bus voltage to $(1 + j0)$ p.u. as starting voltage for load buses and setting $(V + j0)$ p.u. as starting voltage for voltage-controlled buses. However, there are still cases where these starts may even give a poor starting point. A different method involves use of one iteration of a dc load flow in order to provide a reasonable estimate of the voltage angles. This step may be followed by another iteration to get the voltage magnitudes.

The convergence of N-R method is *quadratic* while that of G-S method is *linear*. Researches reveal that quadratic convergence is much better if and only if starting values of $|V|$ and δ are properly chosen. A usual technique nowadays being adapted, in which it is required to set the voltage magnitudes to given values, obviously if the values are given, otherwise the voltage magnitudes are set to that of slack bus. All angles are set equal to the slack bus angle. This is known as *flat start*. Even if N-R method does have less number of iterations, it may still not be faster than G-S method as each iteration in the N-R method may take more time than each iteration in G-S method. Another disadvantage lies in the storage requirement of the N-R method. This method takes comparatively greater computer space than the G-S method.

In N-R method, the number of iterations is almost independent of the system size due to convergence characteristic. Mostly system convergence is obtained within five iterations. Due to quadratic convergence, high accuracy near the exact solution is available within a few iterations. Though the *coefficients* of the Jacobian matrix are functions of the voltage and power angle variables and they vary with each iteration, still in the N-R method, after a few iterations, the coefficients tend towards constant values ($|V|$ and δ also approach to their final values). Some researchers show that it is possible to *calculate the Jacobian for the first two or three iterations only and then use the final value for all subsequent iterations*. It is also possible to update the Jacobian for every two or three iterations throughout the iterative process. These simplified procedures do not greatly affect the convergence of the algorithm, though much time is saved.

N-R method is very much reliable provided the starting approximations are good. By N-R method, even heavily loaded systems can be solved. N-R method may not be even disturbed by ill-conditioned system and the location or selection of slack bus is not critical.

10.14 FAST DECOUPLED LOAD FLOW (FDLF)

In any practical power system, operating in steady state condition, there is *strong interdependence* between active power (P) and bus voltage angle (δ) and between reactive powers (Q) and bus voltage magnitude $|V|$, whereas couplings between P–V and Q–δ are relatively weak. This weak-coupling effect may be neglected without introducing much inaccuracy in the load flow solution. This is popularly known as *decoupled load flow*. The most successful decoupled load flow is derived from Jacobian matrix equations of the NR method, N-R method being represented as

$$\begin{bmatrix} \Delta P \\ \Delta Q \end{bmatrix} = \begin{bmatrix} J_1 & J_2 \\ J_3 & J_4 \end{bmatrix} \begin{bmatrix} \Delta\delta \\ \dfrac{\Delta|V|}{|V|} \end{bmatrix} \tag{10.111}$$

Corresponding to the concept of decoupling effect, it may further be observed that in the Jacobian, $\dfrac{\partial P_i}{\partial \delta_k}$ terms are much larger than $\dfrac{\partial Q_i}{\partial \delta_k}$ terms and $\dfrac{\partial P_i}{\partial |V_k|}$ terms are much larger than $\dfrac{\partial P_i}{\partial |V_k|}$ terms. Thus, neglecting the $\dfrac{\partial Q_i}{\partial \delta_k}$ and $\dfrac{\partial P_i}{\partial |V_k|}$ terms, we finally obtain from (10.111)

$$\begin{bmatrix} \Delta P \\ \Delta Q \end{bmatrix} = \begin{bmatrix} J_1 & 0 \\ 0 & J_4 \end{bmatrix} \begin{bmatrix} \Delta\delta \\ \dfrac{\Delta|V|}{|V|} \end{bmatrix} \tag{10.112}$$

We are thus left with two separate systems of equation (equations (10.93)and (10.94)) for an N-bus power system, (bus no. 1 being the slack bus.) as described below

$$\begin{bmatrix} \Delta P_2 \\ \vdots \\ \Delta P_N \end{bmatrix} = \underbrace{\begin{bmatrix} \dfrac{\partial P_2}{\partial \delta_2} & \cdots & \dfrac{\partial P_2}{\partial \delta_N} \\ \vdots & \vdots & \vdots \\ \dfrac{\partial P_N}{\partial \delta_2} & \cdots & \dfrac{\partial P_N}{\partial \delta_N} \end{bmatrix}}_{[J_1]} \begin{bmatrix} \Delta\delta_2 \\ \vdots \\ \Delta\delta_N \end{bmatrix} \tag{10.113}$$

$$\begin{bmatrix} \Delta Q_2 \\ \vdots \\ \Delta Q_N \end{bmatrix} = \underbrace{\begin{bmatrix} |V_2|\dfrac{\partial Q_2}{\partial |V_2|} & \cdots & |V_N|\dfrac{\partial Q_2}{\partial |V_N|} \\ \vdots & \cdots & \vdots \\ |V_2|\dfrac{\partial Q_N}{\partial |V_2|} & \cdots & |V_N|\dfrac{\partial Q_N}{\partial |V_N|} \end{bmatrix}}_{[J_4]} \begin{bmatrix} \dfrac{\Delta|V_2|}{|V_2|} \\ \vdots \\ \dfrac{\Delta|V_N|}{|V_N|} \end{bmatrix} \tag{10.114}$$

These two sets of system equations are decoupled in the sense that the voltage-angle corrections ($\Delta\delta$) are obtained using the real power mismatches (ΔP) only while the voltage-magnitude correction ($\Delta|V|$) are obtained using only the reactive power mismatches (ΔQ). The Jacobian matrix elements $[J_1]$ and $[J_4]$ are still inter-dependent because the elements of $[J_1]$ depend on voltage magnitudes being solved in equation (10.114), whereas the elements of $[J_4]$ depend on voltage angles of equation (10.113). However, the two sets of equations represented in equations (10.113) and (10.114) can be solved alternately using in one set the latest solution from the other set, but this will require evaluation and factorisation of $[J_1]$ and $[J_4]$ at each step of iterations.

Let us now consider some practical approximations in power transmission.

(i) The angular differences between voltage angles are typically small, i.e. $(\delta_k - \delta_i)$ being very small,

$$\cos(\delta_k - \delta_i) \simeq 1\,;\ \sin(\delta_k - \delta_i) \simeq (\delta_k - \delta_i)$$

(ii) Since transmission lines have high X/R ratio, line susceptances B_{ik} become much larger than the line conductances Gik and hence

$$G_{ik}\sin(\delta_k - \delta_i) << B_{ik}\cos(\delta_k - \delta_i)$$

(iii) The reactive power injected into any bus (Q_i) during steady state operation is much less than the reactive power that would flow if all the lines connected with that bus were short-circuited to reference. This gives

$$Q_i << |V_i|^2 B_{ii}$$

Again, the off-diagonal element of $[J_1]$ and $[J_4]$ (derived in N-R method) being given by

$$\frac{\partial P_i}{\partial \delta_k} = -|V_i V_k Y_{ik}|\sin\left(\phi_{ik} + \delta_k - \delta_i\right) = |V_k|\frac{\partial Q_i}{\partial |V_k|}$$

We can write now

$$-|V_i V_k Y_{ik}|\sin\left(\phi_{ik} + \delta_k - \delta_i\right) = -|V_i V_k|\left\{B_{ik}\cos\left(\delta_k - \delta_i\right) + G_{ik}\sin\left(\delta_k - \delta_i\right)\right\}$$

where $\quad B_{ik} = |Y_{ik}|\sin\phi_{ik}$ and $G_{ik} = |Y_{ik}|\cos\phi_{ik}$.

Using the practical approximations listed above, the elements of $[J_1]$ and $[J_4]$ become

$$\frac{\partial P_i}{\partial \delta_k} = |V_k|\frac{\partial Q_i}{\partial |V_i|} = -|V_i V_k| B_{ik} \tag{10.115}$$

Next, the diagonal elements of $|J_1|$ and $|J_4|$ being represented as

$$\frac{\partial P_i}{\partial \delta_i} = -Q_i - |V_i|^2 B_{ii} \tag{10.116}$$

and $\quad |V_i|\dfrac{\partial Q_i}{\partial |V_i|} = Q_i - |V_i|^2 B_{ii}$

above listed approximation ($Q_i << |V_i|^2 B_{ii}$) reduce the diagonal elements of $[J_1]$ and $[J_4]$ shown in equation (10.116) as

$$\frac{\partial P_i}{\partial \delta_i} = |V_i| \frac{\partial Q_i}{\partial |V_i|} = -|V_i|^2 B_{ii} \tag{10.116a}$$

Thus, the decoupled matrix representation of the load flow equations [(10.113) and (10.114)] can further be simplified with off-diagonal elements having terms given by $\{-|V_i V_k| B_{ik}\}$ and diagonal elements given by $\{-|V_i|^2 B_{ii}\}$.

∴ For an N-bus system with bus no. 1 as swing (or slack) bus, we obtain the decoupled equation as

$$\begin{bmatrix} \Delta P_2 \\ \vdots \\ \Delta P_N \end{bmatrix} = \begin{bmatrix} -|V_2V_2|B_{22} & -|V_2V_3|B_{23} & \cdots & -|V_2V_N|B_{2N} \\ \vdots & \vdots & & \vdots \\ -|V_2V_N|B_{N2} & -|V_3V_N|B_{N3} & \cdots & -|V_NV_N|B_{NN} \end{bmatrix} \begin{bmatrix} \Delta\delta_2 \\ \vdots \\ \Delta\delta_N \end{bmatrix} \tag{10.117}$$

$$\begin{bmatrix} \Delta Q_2 \\ \vdots \\ \Delta Q_N \end{bmatrix} = \begin{bmatrix} -|V_2V_2|B_{22} & -|V_2V_3|B_{23} & \cdots & -|V_2V_N|B_{2N} \\ \vdots & \vdots & & \vdots \\ -|V_2V_N|B_{N2} & -|V_3V_N|B_{N3} & \cdots & -|V_NV_N|B_{NN} \end{bmatrix} \begin{bmatrix} \dfrac{\Delta|V_2|}{|V_2|} \\ \vdots \\ \dfrac{\Delta|V_N|}{|V_N|} \end{bmatrix} \tag{10.118}$$

It has been found that the second set of equations [equation (10.118)] become *relatively unstable* at some distance from the exact solution (due to non-linear defining functions). An improvement is possible by *removing* the voltage from the entries of the Jacobian matrix. This is possible by multiplying each row of the second set of equations in the decoupled method by *correction vector* and then dividing the resultant equation by $|V_i|$ to obtain

$$\begin{aligned} \frac{\Delta Q_2}{|V_2|} &= -B_{22}\Delta|V_2| - B_{23}\Delta|V_3| - \cdots -B_{2N}\Delta|V_N| \\ &\vdots \\ \frac{\Delta Q_N}{|V_N|} &= -B_{N2}\Delta|V_2| - B_{N3}\Delta|V_3| - \cdots -B_{NN}\Delta|V_N| \end{aligned} \tag{10.118a}$$

It is thus evident that the coefficients in equation (10.118a) are equal to the negative of the susceptances in the row of $[Y_{Bus}]$ corresponding to the respective bus. All the entries in the coefficient matrix of equation (10.118a) become constants given by the known susceptances of $[Y_{Bus}]$.

Next, we modify the first set of the decoupled equations by multiplying each row by the *vector of angle correction* and division by $|V_i|$.

We then obtain

$$\begin{aligned}\frac{\Delta P_2}{|V_2|} &= -B_{22}\Delta\delta_2 - B_{23}\Delta\delta_3 - \cdots - B_{2N}\Delta\delta_N \\ \vdots \quad & \quad \vdots \qquad\qquad\qquad \vdots \\ \frac{\Delta P_N}{|V_N|} &= -B_{N2}\Delta\delta_2 - B_{N3}\Delta\delta_3 - \cdots - B_{NN}\Delta\delta_N\end{aligned} \tag{10.118b}$$

The coefficients in the above equation can be made the same as those of equation (10.118a) by setting the magnitudes of V_2, V_3, ..., V_N equal to 1.0 p.u. in the left-hand side of equation (10.118b). The decoupled equations can now be arranged as follows [equations (10.119) and (10.120)]:

$$\begin{bmatrix} \dfrac{\Delta Q_2}{|V_2|} \\ \vdots \\ \dfrac{\Delta Q_N}{|V_N|} \end{bmatrix} = \begin{bmatrix} -B_{22} & -B_{23} & \cdots & -B_{2N} \\ \vdots & \vdots & & \vdots \\ -B_{N2} & -B_{N3} & \cdots & -B_{NN} \end{bmatrix} \begin{bmatrix} \Delta|V_2| \\ \vdots \\ \Delta|V_N| \end{bmatrix} \tag{10.119}$$

$$\begin{bmatrix} \dfrac{\Delta P_2}{|V_2|} \\ \vdots \\ \dfrac{\Delta P_N}{|V_N|} \end{bmatrix} = \begin{bmatrix} -B_{22} & -B_{23} & \cdots & -B_{2N} \\ \vdots & \vdots & & \vdots \\ -B_{N2} & -B_{N3} & \cdots & -B_{NN} \end{bmatrix} \begin{bmatrix} \Delta\delta_2 \\ \vdots \\ \Delta\delta_N \end{bmatrix} \tag{10.120}$$

It may be noted here that $\Delta Q_i/|V_i|$ represents the reactive power mismatch at bus-i while $\Delta P_i/|V_i|$ represents the real power mismatch at the same bus. In general, both sets of the decoupled equations can then be written as

$$[\Delta P/|V|] = [B^*][\Delta\delta] \tag{10.121}$$

$$[\Delta Q/|V|] = [B^*][\Delta|V|] \tag{10.122}$$

Matrix B^* is generally *symmetrical* and *sparse* with non-zero elements. These elements are constant, real number and exactly equal to the negative of susceptances of $[Y_{Bus}]$. Matrix B^* can be easily formed and once computed at the beginning of the solution, it does not require any further computation. This leads to fast convergence. At voltage-controlled buses Q is not specified, while $\Delta|V| = 0$. The rows and columns corresponding to such buses are then deleted from the decoupled equation.

The decoupled load flow equation can be further modified using some more practical modifications as follows:

(a) From the calculation of Jacobian of active power mismatch equation $[\Delta P/|V|] = [B^*][\Delta\delta]$, series resistances of lines are neglected and the elements that predominantly affect reactive power (i.e. shunt reactance, off-nominal in-phase transformer taps etc.) are omitted.

(b) From the calculation of Jacobian component of reactive power flow the angle-shifting effects are neglected.

The resulting *decoupled load flow equation* then takes the shape of

$$[\Delta P/|V|] = [B'][\Delta\delta] \tag{10.123}$$

and

$$[\Delta Q/|V|] = [B''][\Delta|V|] \tag{10.124}$$

Here,
$$B'_{ik} = -(1/X_{ik}) \text{ for } i \neq k;\; B''_{ii} = \sum_{i=1}^{k}(1/X_{ik})$$

$$B''_{ik} = -B_{ik} \text{ for } i \neq k;\; B''_{ii} = \sum_{i=1}^{k} B_{ik}$$

The matrix B' and B'' are real and of order $(N - 1)$ and $(M - 1)$, respectively, N being the number of buses while M is the number of load (PQ) buses including the slack bus. Normally (i.e. when there is no phase-shifting transformer), B' and B'' are *symmetric*. The convergence is *geometric* and may ordinarily be obtained for normal power system within five iterations. The convergence of decoupled method has been found to be more reliable than Newton-Raphson method in a number of applications, though this achievement may not be universally true for any multi-bus system.

With efficient computation of $\Delta P/|V|$ and $\Delta Q/|V|$, the speed of iterations of the decoupled method is much faster than the N-R method. Also, storage requirements are less. This method is conventionally called *fast decoupled load flow* (*FDLF*) method. If the PV buses have any Q-limit then, it should be taken into care in the similar way of N-R method of load flow. The algorithm of FDLF method is given below.

10.14.1 Algorithm to Calculate Bus Voltages by Fast Decoupled Load Flow Method

1. Read line data for the power system having total number of lines = NL and total number of buses = N.
2. Read bus type ($type_i$), initial bus voltage magnitude in p.u. ($|V_i|^0$), initial bus voltage angle in rad (δ_i^0), real power generation (P_{g_i}), reactive power generation (Q_{g_i}), real power demand (P_{d_i}), reactive power demand (Q_{d_i}), lower and upper Q-limit at PV buses ($Q_{g_{i_{min}}}$ and $Q_{g_{i_{max}}}$) in p.u. for each bus i = 1, 2, ..., N.
3. Specify tolerance ε (say 0.0001).
4. Form [Y_{Bus}].
5. Form the [B'] matrix from [Y_{Bus}] excluding slack bus.

   $B'_{ij} = -B_{ij} = -\text{Im}\,(Y_{ij})$

   where i = 1, 2, ..., N and j = 1, 2, ..., N provided $type_i \neq 0$ as well as $type_j \neq 0$.

6. Compute scheduled active and reactive bus powers for all buses using equation (10.27a) and (10.27b); i.e., $P_{i_{\text{scheduled}}}$ and $Q_{i_{\text{scheduled}}}$ for i = 1, 2, ..., N.
7. Set iteration count $p = 0$.
8. Fix slack bus voltage magnitude and angle for next iteration, $|V_i|^{p+1} = |V_i|^0$, $\delta_i^{p+1} = 0$ if $type_i = 0$.
9. Fix specified voltage magnitude for PV buses. $|V_i|^p = |V_i|^0$, for i = 1, 2, ..., N (only if $type_i = 1$).
10. Calculate active and reactive power at all buses (i = 1, 2, ..., N) using equation (10.92 or 10.27).
11. Calculate reactive power generation for PV buses as $Q_{g_i} = (Q_i^p + Q_{d_i})$ for i = 1, 2, ..., N (only if $type_i = 1$).
12. If $Q_{g_{\min}}$ and $Q_{g_{\max}}$ are specified for all PV buses then go to next step, otherwise go to step 14.
13. Check the following for all PV buses (for i = 1, 2, ..., N and consider bus-i only if $type_i = 1$)

    if $Q_{g_i} > Q_{g_{i\max}}$ or, $Q_{g_i} < Q_{g_{i\min}}$ for the PV bus-i (i.e., Q-limit is failed),

    (i) set $Q_{g_i} = Q_{g_{i\max}}$ for $Q_{g_i} > Q_{g_{i\max}}$,

    or, set $Q_{g_i} = Q_{g_{i\min}}$ for $Q_{g_i} < Q_{g_{i\min}}$.

    (ii) release the constraint of $|V_i|^p$ to be fixed and initialize it for a flat start, i.e., $|V_i|^p = 1$ and also change bus type from PV bus ($type_i = 1$) to PQ bus ($type_i = 2$) as Q-limit is failed.
14. Form ΔP^p matrix considering all buses excluding slack bus using equation (10.28a) with updated values of $|V|$, where $\Delta P_i^p = P_{i_{\text{scheduled}}} - P_i^p$ for i = 1, 2, ..., N (but $type_i \neq 0$).
15. If $\max(\Delta P^p) > \varepsilon$ then compute the following for all buses except slack bus (i.e., for i = 1, 2, ..., N but $type_i \neq 0$) else go to step 16:

    (i) Form $\left[\dfrac{\Delta P^p}{|V|^p}\right]$ matrix

    (ii) Solve equation $\left[\Delta\delta^p\right] = [B']^{-1}\left[\dfrac{\Delta P^p}{|V|^p}\right]$

    (iii) Update bus voltage angles: $\delta_i^{p+1} = \delta_i^p + \Delta\delta_i^p$
16. Form ΔQ^p matrix considering only the PQ buses using equation (10.28b) with updated values of δ.

    $\Delta Q_i^p = Q_{i_{\text{scheduled}}} - Q_i^p$ for i = 1, 2, ..., N(only for $type_i = 2$).

17. If $\max(\Delta P^p \text{ and } \Delta Q^p) < \varepsilon$ then go to step 21 else go to step 18.

18. If $\max(\Delta Q^p) > \varepsilon$ then compute the following for PQ buses (i.e., for i = 1, 2, ..., N provided $type_i = 2$).

    (i) Form the $[B'']$ matrix from $[Y_{Bus}]$

    $B''_{ij} = -B_{ij} = -\text{Im}(-Y_{ij})$ where i = 1, 2,..., N and j = 1, 2, ..., N provided $type_i$ = $type_j$ = 2.

    (ii) Form $\left[\dfrac{\Delta Q^p}{|V|^p}\right]$ matrix

    (iii) Solve equation $\left[\Delta|V|^p\right] = [B'']^{-1}\left[\dfrac{\Delta Q^p}{|V|^p}\right]$

    (iv) Update bus voltage magnitudes: $|V_i|^{p+1} = |V_i|^p + \Delta|V_i|^p$

19. Update iteration count $p \leftarrow p+1$.

20. If $p > 20$ then break the iteration process and go to step 25. Otherwise go to step 8.

21. Convert final bus voltages into rectangular form, $V_i^p = |V_i|^p \angle \delta_i^p$ for i = 1, 2, ..., N.

22. Calculate reactive power generations for those PV buses who have not crossed their Q-limit. $Q_{g_i} = Q_i^p + Q_{d_i}$ for i = 1, 2, ..., N provided $type_i = 1$.

23. Display and/or store final bus voltages, V_i^p, for i = 1, 2, ..., N.

24. Display and/or store reactive power generation, Q_{g_i} for PV buses ($type_i = 1$), for i = 1, 2, ..., N.

25. Stop.

EXAMPLE 10.10 A three-bus, three-line system has been shown in Figure E10.10. Each line has series impedance of (0.02+ j0.1) p.u. and shunt admittance of j0.04 p.u. in 100 MVA base. Obtain, bus voltages, line flows and total line loss using FDLF method. Also obtain real and reactive power generation at bus-1 and reactive power generation at bus-2. Assume specified tolerance, ε = 0.0001.

Bus no.	P_d	Q_d	P_g	Q_g	V
1	1	0.5	?	?	$1.03 + j0$
2	1.5	0.75	0	0	?
3	1	0.2	0.5	?	$\|1.0\|\angle ?$

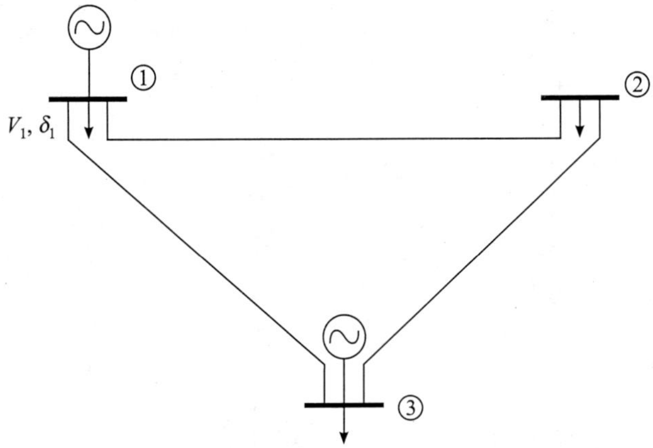

Figure E10.10 A three-bus, three-line system.

Solution: The $[Y_{Bus}]$ of the given power system, both in rectangular form and in polar form as shown below (all quantities are in p.u. and angles are in radian).

$$[Y_{Bus}] = \begin{bmatrix} (3.8462 - j19.151) & (-1.9231 + j9.6154) & (-1.9231 + j9.6154) \\ (-1.9231 + j9.6154) & (3.8462 - j19.151) & (-1.9231 + j9.6154) \\ (-1.9231 + j9.6154) & 9.805806\angle 1.768192 & (3.8462 - j19.151) \end{bmatrix} \text{p.u.}$$

$$= \begin{bmatrix} 19.533170\angle -1.372598 & 9.805806\angle 1.768192 & 9.805806\angle 1.768192 \\ 9.805806\angle 1.768192 & 19.533170\angle -1.372598 & 9.805806\angle 1.768192 \\ 9.805806\angle 1.768192 & 9.805806\angle 1.768192 & 19.533170\angle -1.372598 \end{bmatrix} \text{p.u.}$$

In FDLF method bus voltage magnitudes and load angles are obtained by solving following matrix equations

$$\left[\frac{\Delta P}{|V|}\right] = [B'][\Delta\delta] \text{ and } \left[\frac{\Delta Q}{|V|}\right] = [B''][\Delta|V|]$$

Where, ΔP = Change in active bus power, ΔQ = Change in reactive bus power, $|V|$ = Magnitude of bus voltage in the preceding iteration.

Let us assume bus-1 is the slack bus ($type_1 = 0$). Bus-2 is PQ bus ($type_2 = 2$) and bus-3 is PV bus ($type_3 = 1$). Then bus voltage angles are to be calculated for bus 2 and 3 though bus voltage magnitude is to be calculated only for bus-2. FDLF matrix equations can then be represented as

$$\begin{bmatrix} \dfrac{\Delta P_2}{|V_2|} \\ \dfrac{\Delta P_3}{|V_3|} \end{bmatrix} = \begin{bmatrix} B'_{22} & B'_{23} \\ B'_{32} & B'_{33} \end{bmatrix} \begin{bmatrix} \Delta\delta_2 \\ \Delta\delta_3 \end{bmatrix} \text{ and } \left[\frac{\Delta Q_2}{|V_2|}\right] = \left[B''_{22}\right]\left[\Delta|V_2|\right]$$

Obviously, B' and B'' matrices are formed by susceptance part of $[Y_{bus}]$ matrix, following step-5 and 18(i) of the FDLF algorithm respectively. There is no Q-limit in PV bus in this problem, hence both B' and B'' matrices are fixed. (Note that dimension B'' matrix will change if Q-limit be violated in PV buses of a power system).
Therefore, in this problem

$$[B'] = \begin{bmatrix} -B_{22} & -B_{23} \\ -B_{32} & -B_{33} \end{bmatrix} = \begin{bmatrix} 19.150770 & -9.615384 \\ -9.615384 & 19.150770 \end{bmatrix} \text{ while } [B''] = [-B_{22}] = [19.150770]$$

[B' and B'' matrices are formed by following step-5 and 18(i) of the algorithm respectively.]

In fast decoupled load flow, bus voltage angles (δ) are calculated first, then bus voltage magnitude ($|V|$) are calculated using updated bus voltage angles.

In order to start the iterative process, initial guess of bus voltage magnitudes and angles are,

$$|V_1| = 1.03 \text{ p.u. (fixed)},\ \delta_1 = 0 \text{ (fixed)}, |V_2|^0 = 1.0 \text{ p.u.},\ \delta_2^0 = 0$$

$$|V_3|^0 = 1.0 \text{ p.u. (fixed)},\ \delta_3^0 = 0$$

For p-th iteration,

$$\begin{bmatrix} \dfrac{\Delta P_2^p}{|V_2|^p} \\ \dfrac{\Delta P_3^p}{|V_3|^p} \end{bmatrix} = \begin{bmatrix} B'_{22} & B'_{23} \\ B'_{32} & B'_{33} \end{bmatrix} \begin{bmatrix} \Delta\delta_2^p \\ \Delta\delta_3^p \end{bmatrix} \tag{A1}$$

and $\Delta P_2^p = (P_{2_{\text{scheduled}}} - P_2^p)$; $\Delta P_3^p = (P_{3_{\text{scheduled}}} - P_3^p)$. ($P_2^p$ and P_3^p are the calculated active powers at bus-2 and bus-3 respectively at p-th step of iteration).

From equation (10.92) [or equation (10.27)] for the i-th bus,

$$P_i^p = \sum_{k=1}^{N} |V_i|^p |V_k|^p |Y_{ik}| \cos(\phi_{ik} + \delta_k^p - \delta_i^p)$$

With $p = 0$,

$$P_2^0 = |V_2|^0 |V_1|^0 |Y_{21}| \cos\left(\phi_{21} + \delta_1^0 - \delta_2^0\right) + |V_2|^0 |V_2|^0 |Y_{22}| \cos\left(\phi_{22} + \delta_2^0 - \delta_2^0\right)$$
$$+ |V_2|^0 |V_3|^0 |Y_{23}| \cos\left(\phi_{23} + \delta_3^0 - \delta_2^0\right)$$
$$= 1 \times 1.03 \times 9.805806 \cos(1.768192 - 0 + 0) + 1 \times 1 \times 19.533170 \cos(-1.372598 - 0 + 0)$$
$$+ 1 \times 1 \times 9.805806 \cos(1.768192 - 0 + 0)$$
$$= -0.057692 \text{ p.u.}$$

$$P_3^0 = |V_3|^0 |V_1|^0 |Y_{31}| \cos\left(\phi_{31} + \delta_1^0 - \delta_3^0\right) + |V_3|^0 |V_2|^0 |Y_{32}| \cos\left(\phi_{32} + \delta_2^0 - \delta_3^0\right)$$
$$+ \left(|V_3|^0\right)^2 |Y_{33}| \cos\phi_{33}$$

$$= 1 \times 1.03 \times 9.805806 \cos(1.768192 - 0 + 0) + 1 \times 1 \times 9.805806 \cos(1.768192 - 0 + 0)$$
$$+ 1^2 \times 19.533170 \cos(-1.372598)$$
$$= -0.057694 \text{ p.u.}$$

$$\Delta P_2^0 = P_{2_{\text{scheduled}}} - P_2^0 = -1.5 - (-0.057694) = -1.442306 \text{ p.u.}$$

Similarly, $$\Delta P_3^0 = P_{3_{\text{scheduled}}} - P_3^0 = -0.5 - (-0.057694) = -0.442306 \text{ p.u.}$$

Since both ΔP_2^0 and ΔP_3^0 exceeds given tolerance value (ε = 0.0001), load angles are required to be calculated. Once the active power mismatch is satisfied, it is then required to check the reactive power mismatch. If both the active and reactive power mismatches be satisfied then the iterative process is said to be converged.

$$\frac{\Delta P_2^0}{|V_2|^0} = -1.442306; \frac{\Delta P_3^0}{|V_3|^0} = -0.442306$$

∴ We can then write from equation (A1) for $p = 0$,

$$\begin{bmatrix} \Delta\delta_2^0 \\ \Delta\delta_3^0 \end{bmatrix} = \begin{bmatrix} 19.150770 & -9.615384 \\ -9.615384 & 19.150770 \end{bmatrix}^{-1} \begin{bmatrix} -1.442306 \\ -0.442306 \end{bmatrix}$$

$$= \begin{bmatrix} 0.0698178 & 0.0350547 \\ 0.0350547 & 0.0698178 \end{bmatrix} \begin{bmatrix} -1.442306 \\ -0.442306 \end{bmatrix} = \begin{bmatrix} -0.116204 \\ -0.081441 \end{bmatrix}$$

New bus voltage angles are obtained in the next step as follows:

$$\Delta\delta_2^0 = -0.116204 \text{ rad} \quad \text{and} \quad \Delta\delta_3^0 = 0.081441 \text{ rad.}$$

$$\therefore \; \delta_2^1 = \delta_2^0 + \Delta\delta_2^0 = 0 - 0.116204 = -0.116204 \text{ rad.}$$

and, $$\delta_3^1 = \delta_3^0 + \Delta\delta_3^0 = 0 - 0.081441 = -0.081441 \text{ rad.}$$

Note that for slack bus $\delta_1^1 = \delta_1^0 = 0$ rad.

Next Q_2^0 will be calculated by using updated value of δ_2^1 and δ_3^1

$$Q_2^0 = -|V_2|^0 |V_1|^0 |Y_{21}| \sin(\theta_{21} + \delta_1^1 - \delta_2^1) - |V_2|^0 |V_2|^0 |Y_{22}| \sin\phi_{22}$$
$$-|V_2|^0 |V_3|^0 |Y_{23}| \sin(\theta_{21} + \delta_3^1 - \delta_2^1)$$
$$= -1.0 \times 1.03 \times 9.805806 \sin(1.768192 + 0 + 0.116204)$$
$$-1^2 \times 19.533170 \sin(-1.372598 - 0 + 0)$$
$$-1.0 \times 1.0 \times 9.805806 \sin(1.768192 - 0.081441 + 0.116204)$$
$$= 0.000637 \text{ p.u.}$$

∴ $$\Delta Q_2^0 = Q_{2_{\text{scheduled}}} - Q_2^0 = -0.75 - 0.000637 = -0.750637 \text{ p.u.}$$

Since ΔQ_2^0 exceeds the given tolerance value (ε = 0.0001), hence bus voltage magnitude of the PQ bus (here bus-2 only) is required to be calculated.

Since $$\left[\frac{\Delta Q_2^0}{|V_2|^0}\right] = \left[B''_{22}\right]\left[\Delta|V_2|^0\right] \quad \text{and} \quad \left[\frac{\Delta Q_2^0}{|V_2|^0}\right] = -0.750637$$

Hence, $\Delta|V_2|^0 = [19.150770]^{-1}[-0.750637] = -0.039196$ p.u. $\left(\because \left[B''_{22}\right] = [19.150770]\right)$

Thus, $$|V_2|^1 = |V_2|^0 + \Delta|V_2|^0 = 1.0 + (-0.039196) = 0.960804 \text{ p.u.}$$

The results obtained after first iteration, are as follows

Bus no.	$\|V\|$ *(p.u.)*	δ (rad)
1	1.03	0
2	0.960804	– 0.116204
3	1.0	– 0.081441

Both of the active and reactive mismatches exceed the specified tolerance (ε = 0.0001) in the first iteration. Hence the iterative process will continue. ΔP_2^1 and ΔP_3^1 are first calculated using updated values of $|V|$ and δ. Then $\delta's$ are updated again. Next ΔQ_2^1 is calculated with updated $\delta's$ and new value of $|V_2|$ are obtained. The process is repeated until changes in bus power (both active and reactive) remain within the given tolerance (ε = 0.0001). Calculations of power flows have already been explained earlier in the text.

Computer program execution result for Example 10.10 (following algorithm of Art. 10.14.1) (Ans.)

```
All quantities are in p.u. and all angles are in radian.
No. of buses = 3. No. of lines = 3.
Non zero elements of [Ybus]:
Note that only upper triangular off diagonal elements are provided.
Row   Column          Ybus                Row   Column          Ybus
 1       1    (3.846154-19.150769i)        2       2    (3.846154-19.150769i)
 1       2    (-1.923077+9.615385i)        2       3    (-1.923077+9.615385i)
 1       3    (-1.923077+9.615385i)        3       3    (3.846154-19.150769i)
Fast Decoupled load flow result without considering Q-limits at PV buses.
Specified tolerance =0.00010.
Iteration completed with 3 iterations with delp_max=0.000011 and delQ_max
=0.000087.
Final bus voltages are:
Bus no.   Complex bus voltage     Magnitude      Angle
 1            (1.030000)          1.030000         0
 2        (0.955129-0.105182i)    0.960903    -0.109681
 3        (0.996771-0.080302i)    1.000000    -0.080389
Reactive power generation at PV buses:
PV bus no.     Reactive Power
  3                0.348359
```

```
LINE POWER FLOWS:
Line no.  bus-i  bus-j           Sij                    Sji
  1         1      2    (1.190005+0.490736i)  (-1.157950-0.409828i)
  2         1      3    (0.861120+0.127603i)  (-0.846595-0.137418i)
  3         2      3   (-0.342061-0.340085i)   (0.346586+0.285777i)
Slack bus power = (3.051125+1.118339i)
Total transmission line loss = (0.051105+0.016785i)
```

EXAMPLE 10.11 Find final bus voltages, line flows, total line loss, slack bus power by FDLF method for the six-bus, eight-line system shown in Example 10.5.

Solution:

Computer program execution result for Example 10.11 (following algorithm of Art. 10.14.1) (Ans.)

```
All quantities are in p.u. and all angles are in radian.
No. of buses = 6. No. of lines = 8.
Fast Decoupled load flow result without considering Q-limits at PV buses.
Specified tolerance =0.00010.
Iteration completed with 4 iterations with delp_max=0.000068 and delQ_max
=0.000006.
Final bus voltages are:
Bus no.   Complex bus voltage    Magnitude    Angle
 1            (1.040000)          1.040000      0
 2       (0.991409-0.187543i)     1.008992   -0.186959
 3       (1.052682-0.069346i)     1.054964   -0.065781
 4       (0.976982-0.150922i)     0.988570   -0.153266
 5       (0.977617-0.158959i)     0.990456   -0.161188
 6       (0.973450-0.217667i)     0.997488   -0.219985
LINE POWER FLOWS:
Line no.  bus-i  bus-j           Sij                    Sji
  1         1      3    (0.230441-0.085506i)  (-0.228271+0.079836i)
  2         2      3   (-0.657505-0.104451i)   (0.670506+0.169815i)
  3         3      4    (0.509274+0.257171i)  (-0.497366-0.218529i)
  4         3      5    (0.548559+0.243184i)  (-0.535419-0.198425i)
  5         4      5    (0.047368-0.031470i)  (-0.047285+0.012304i)
  6         5      6    (0.182709-0.063878i)  (-0.180488+0.055221i)
  7         1      2    (0.511992+0.274612i)  (-0.511992-0.161385i)
  8         2      6    (0.169491+0.165837i)  (-0.169491-0.155220i)
Slack bus power = (0.742433+0.189107i)
Total transmission line loss = (0.042523+0.239116i)
```

EXAMPLE 10.12 Find final bus voltages by FDLF method for the four-bus, five-line system given in Example 10.4 when bus-2 has a Q-limit of $0.1 \le Q_2 \le 1.0$ p.u. Write a MATLAB code for this example.

Solution:

The MATLAB code for Example 10.12:

```
% Program to calculate bus voltages by Fast Decoupled load flow method for
Example 10.12
```

```
clc
clear all
bus_no=4;line_no=5;epsil=0.0001;
disp('All quantities are in p.u. and all angles are in radian.')
disp(sprintf('No. of buses = %d. No. of lines = %d.',bus_no,line_no));
v=[1.06 1.02 1 1]; del=[0 0 0 0]; type=[0 1 2 2];pg=[0 1.2 0 0];
qg=[0 0 0 0]; pd=[2 0 0.7 0.6]; qd=[0.6 0 0.5 0.3];
qmin=0.1; qmax=1;
psch=pg-pd; qsch=qg-qd; vmag=v; vang=del;
Ybus_prog % Prog. to calculate [Ybus]
G=real(Y); B=imag(Y);
for k=1:bus_no,
 if k~=1
   for n=1:bus_no,
      if n~=1  B1(k-1,n-1)=-B(k,n);  end
   end
  end
end
for it=1:20
for n=1:bus_no
  pcal(n)=0;
  for k=1:bus_no
pcal(n)=pcal(n)+vmag(n)*vmag(k)*abs(Y(n,k))*cos(vang(k)+angle(Y(n,k))-
vang(n));
  end
end
pchng=psch-pcal;
for k=1:bus_no,
 if k~=1  delp(k-1)=pchng(k);  pvchng(k-1)=pchng(k)/vmag(k); end
end
delp_max=max(abs(delp));
if delp_max>epsil
ddel_it=(inv(B1))*pvchng';
for k=1:bus_no,
 if k~=1 ddel(k)=ddel_it(k-1);  else  ddel(k)=0;  end
end
vang=vang+ddel;del=vang';
end
for n=1:bus_no,
qcal(n)=0;
 for k=1:bus_no,
  qcal(n)=qcal(n)-vmag(n)*vmag(k)*abs(Y(n,k))*sin(vang(k)+angle(Y(n,k))-
vang(n));
 end
end
for k=1:bus_no
 if(type(k)==1)
 qg(k)=qcal(k);
```

```
 if (qg(k)<qmin||qg(k)>qmax)
 if qg(k)<qmin  qg(k)=qmin; else qg(k)=qmax; end
  type(k)=2; vmag(k)=1.0; qsch(k)=qg(k);
 end
 end
end
qchng=qsch-qcal;
for k=1:bus_no,
  if k>1  delq(k-1)=qchng(k);  qvchng(k-1)=qchng(k)/vmag(k);  end
end
for k=1:bus_no,
 if k>1
   for n=1:bus_no   if n>1  B2(k-1,n-1)=-B(k,n);  end
   end
 end
end
delq_max=max(abs(delq));
if ((delp_max<epsil)&& (delq_max<epsil))
break
end
if delq_max>epsil
dv_it=(inv(B2))*qvchng';
for k=1:bus_no,
if k>1 dv(k)=dv_it(k-1); else dv(k)=0; end
end
for n=1:bus_no  vmag(n)=vmag(n)+dv(n); end
end
if (it>10)
disp(sprintf('Iteration  incomplete  for  p  =%d  and  delp_max=%f,  delq_
max=%f.',it-1,delp_max,delq_max))
break
end
end
if ((delp_max<epsil)&& (delq_max<epsil))
vf=vmag.*exp(i*vang);
disp('Fast Decoupled load flow result considering Q-limits at PV buses.')
disp(sprintf('Specified tolerance =%7.5f.',epsil))
disp(sprintf('Iteration completed with %d iterations with delP_max=%f and
delQ_max  =%f.',it-1,delp_max,delq_max))
disp('Final bus voltages are:')
disp('Bus no.  Complex bus voltage     Magnitude   Angle')
for n=1:bus_no,
disp(sprintf('  %d \t\t(%s)\t%f\t%f',n,num2str(vf(n),'% 10.6f'),abs(vf(n
)),angle(vf(n))));
end
end
```

Computer program execution result for Example 10.12 (following algorithm of Art. 10.14.1)

```
All quantities are in p.u. and all angles are in radian.
No. of buses = 4. No. of lines = 5.
Fast Decoupled load flow result considering Q-limits at PV buses.
Specified tolerance =0.00010.
Iteration completed with 5 iterations with delp_max=0.000100 and delQ_max
=0.000020.
Final bus voltages are:
Bus no.   Complex bus voltage      Magnitude       Angle
 1           (1.060000)            1.060000          0
 2        (1.027338+0.159838i)     1.039698      0.154347
 3        (0.920215-0.037402i)     0.920975     -0.040622
 4        (0.854978-0.133248i)     0.865299     -0.154606
```

10.15 DC LOAD FLOW

We rewrite the decoupled power flow equations (equations (10.123) and (10.124))

$$\left[\frac{\Delta P}{|V|}\right] = [B'][\Delta\delta] \quad \text{and} \quad \left[\frac{\Delta Q}{|V|}\right] = [B''][\Delta|V|]$$

Let us further simplify the load flow solution technique by dropping the Q-V iteration. It results in a completely linear, non-iterative power flow method. Assuming all bus voltage magnitudes to be unity, the real power decoupled equation becomes

$$[\Delta P] = [B'][\Delta\delta]$$

The "DC" power flow can then be calculated as

$$P_{ij} = \frac{1}{X_{ij}}(\delta_i - \delta_j)\,;\; P_i = \sum^{N} P_{\text{buses connected to } i}$$

However, this DC load flow method is good only for the calculation of MW flows in lines but does not give any indication of MVAR or MVA flows. the algorithm to calculate bus voltages by DC load flow method is given in Article 10.14.1.

10.15.1 Algorithm to Calculate Bus Voltages by DC Load Flow Method

1. Read line data for the power system having total number of lines = NL and total number of buses = N.
2. Read bus type ($type_i$), initial bus voltage magnitude in p.u. ($|V_i|^0$), initial bus voltage angle in rad (δ^0_i), real power generation (P_{g_i}), real power demand (P_{d_i}) in p.u. for each bus i = 1, 2, ..., N.
3. Specify tolerance ε (say 0.0001).
4. Form [Y_{Bus}] matrix.
5. Form the [B'] matrix from [Y_{Bus}] excluding slack bus.

$B'_{ij} = -B_{ij} = -\text{Im}(Y_{ij})$

where $i = 1, 2, ..., N$ and $j = 1, 2, ..., N$ provided $type_i \neq 0$ as well as $type_j \neq 0$.

6. Compute scheduled active bus powers for all buses using equation (10.27a), i.e., $P_{i_{\text{scheduled}}}$ for $i = 1, 2, ..., N$.
7. Set iteration count $p = 0$.
8. Calculate active power at all buses (for $i = 1, 2, ..., N$) using equation (10.92) or (10.27).
9. Form ΔP^p matrix considering all buses excluding slack bus using equation (10.28a).

   where $\Delta P_i^p = P_{i_{\text{scheduled}}} - P_i^p$ for $i = 1, 2, ..., N$ (but $type_i \neq 0$).
10. If $\max(\Delta P^p) < \varepsilon$ then go to step 15 else go to step 11.
11. Solve equation $[\Delta\delta^p] = [B']^{-1}[\Delta P^p]$.
12. Update bus voltage angles: $\delta_i^{p+1} = \delta_i^p + \Delta\delta_i^p$
13. Update iteration count, $p \leftarrow p+1$.
14. If $p > 20$ then break the iteration process and go to step 18. Otherwise go to step 8.
15. Convert final bus voltages into rectangular form, $V_i^p = |V_i|^0 \angle\delta_i^p$ for $i = 1, 2, ..., N$.
16. Display and/or store final bus voltages, V_i^p, for $i = 1, 2, ..., N$.
17. Display and/or store reactive power generation, Q_{g_i} for PV buses ($type_i = 1$), for $i = 1, 2, ..., N$.
18. Stop.

EXAMPLE 10.13 Develop DC Load Flow for the problem furnished in Example 10.6 in order to obtain bus voltages.

Solution: In this example all quantities are expressed in p.u. and angles are expressed in radian.

$$[Y_{Bus}] = \begin{bmatrix} 0.305 - j1.936 & -0.305 + j2 \\ -0.305 + j2 & 0.305 - j1.936 \end{bmatrix} = \begin{bmatrix} 1.9599\angle -1.4145 & 2.0231\angle 1.7221 \\ 2.0231\angle 1.7221 & 1.9599\angle -1.4145 \end{bmatrix}$$

In DC load flow method bus voltage magnitudes are constant at its rated value and load angles are to be obtained by solving following matrix equation

$$[\Delta P] = [B'][\Delta\delta], \text{ where, } \Delta P = \text{Change in active bus power}$$

Here, $$[\Delta P_2] = [B'_{22}][\Delta\delta_2] \quad \text{and} \quad [B'] = [-B_{22}] = [1.936]$$

[B' matrix is obtained by following step-5 of the algorithm.]

In order to start the iterative process, initial guess of bus voltage magnitudes and angles are,

$$|V_1|^0 = 1.05 \text{ p.u. (fixed)}, \delta_1 = 0 \text{ (fixed)}, |V_2|^0 = 1.0 \text{ p.u., (fixed) and } \delta_2^0 = 0$$

[Superscript zero stands for the initial guess.]

For p-th iteration, $$[\Delta P_2^p]=[B'_{22}][\Delta\delta_2^p]$$

and $$\Delta P_2^p = P_{2_{\text{scheduled}}} - P_2^p.$$

P_2^p is the calculated active power at bus-2 at p-th step of iteration.

From equation (10.92) or equation (10.27) for the i-th bus,

$$P_i^p = \sum_{k=1}^{N} |V_i|^p |V_k|^p |Y_{ik}| \cos\left(\phi_{ik} + \delta_k^p - \delta_i^p\right)$$

With $p = 0$,

$$\therefore \quad P_2^0 = |V_2|^0 |V_1|^0 |Y_{21}| \cos\left(\phi_{21} + \delta_1^0 - \delta_2^0\right) + |V_2|^0 |V_2|^0 |Y_{22}| \cos\left(\phi_{22} + \delta_2^0 - \delta_2^0\right)$$

$$= 1.0 \times 1.05 \times 2.0231 \cos(1.7221 + 0 - 0) + 1.0 \times 1.0 \times 1.9599 \cos(-1.4145 + 0 - 0)$$

$$= -0.32025 + 0.30501 = -0.01525 \text{ p.u.}$$

$$\therefore \quad \Delta P_2^0 = -0.5 - (-0.01525) = -0.48475 \text{ p.u.}$$

Since this change in active bus power exceeds the given tolerance (ε = 0.0001), load angle (δ_2) is required to be calculated again. If the active power mismatch be satisfied then the iterative process is said to be converged.

$$[\Delta\delta_2^0] = [B'_{22}]^{-1}[\Delta P_2^p] = [1.936]^{-1}[-0.48475] = [0.51653][-0.48475] = -0.25039$$

Next, new bus voltage angle is obtained as follows:

$$\delta_2^1 = \delta_2^0 + \Delta\delta_2^0 = 0 - 0.25039 = -0.25039 \text{ rad.}$$

Note that for slack bus $\delta_1^1 = \delta_1^0 = 0$ rad.

After first iteration, the obtained results are shown below in tabular form

Bus no.	$\lvert V\rvert$ (p.u.)	δ (rad)
1	1.05	0
2	1.0	– 0.25039

The active power mismatch exceeds specified tolerance (ε = 0.0001) in the first iteration. So the iterative process will continue. ΔP_2^1 is first calculated using updated values of δ. The process is repeated until changes in active bus power remain within the given tolerance (ε = 0.0001).

The MATLAB code for Example 10.13

```
% Program to calculate bus voltages by DC load flow method for Example 10.13
clc
clear all
bus_no=2;line_no=1; epsil=0.0001;
disp('All quantities are in p.u. and all angles are in radian.')
disp(sprintf('No. of buses = %d. No. of lines = %d.',bus_no,line_no));
v=[1.05 1]; del=[0 0]; type=[0 2];pg=[0 0];
```

```
qg=[0 0]; pd=[1 0.5]; qd=[0.42 0.25];
psch=pg-pd; vmag=v; vang=del;
Ybus_prog %Prog. to calculate [Ybus]
G=real(Y);B=imag(Y);
for k=1:bus_no,
 if k~=1
  for n=1:bus_no,
  if n~=1   B1(k-1,n-1)=-B(k,n);end
  end
 end
end
for it=1:20
for n=1:bus_no
 pcal(n)=0;
 for k=1:bus_no
  pcal(n)=pcal(n)+vmag(n)*vmag(k)*abs(Y(n,k))*cos(vang(k)+angle(Y(n,k))-
vang(n));
 end
end
pchng=psch-pcal;
for k=1:bus_no
if k~=1  delp(k-1)=pchng(k); pchng_dc(k-1)=pchng(k); end
end
delp_max=max(abs(delp));
if delp_max>epsil ddel_it=(inv(B1))*pchng_dc';
for k=1:bus_no,
 if type(k)~=0
 ddel(k)=ddel_it(k-1);  else  ddel(k)=0;
 end
end
vang=vang+ddel;del=vang';
end
if delp_max<epsil
break
end
if (it>10)
disp(sprintf('Iteration  incomplete  for  p  =%d  and  delp_max=%f,  delp_
max=%f',it-1,delp_max))
break
end
end
if delp_max<epsil
vf=vmag.*exp(i*vang);
disp(sprintf('DC loadflow result with specified tolerance =%7.5f.',epsil))
disp(sprintf('Load flow converges with %d iterations and delP_max =%f.',it-
1,delp_max))
disp('Final bus voltages are:')
disp('Bus no.  Complex bus voltage     Magnitude   Angle')
```

```
for n=1:bus_no
disp(sprintf('  %d \t\t(%s)\t%f\t%f',n,num2str(vf(n),'% 10.6f'),abs(vf(n
)),angle(vf(n))));
end
end
```

Computer program execution result for Example 10.13 (following algorithm of Art. 10.14.1) (Ans.)

```
No. of buses = 2. No. of lines = 1.
DC loadflow result with specified tolerance =0.00010.
Load flow converges with 3 iterations and maximum delP =0.000005.
Final bus voltages are:
Bus no.   Complex bus voltage     Magnitude      Angle
 1              (1.050000)          1.050000         0
 2         (0.971969-0.235111i)     1.000000    -0.237332
```

EXAMPLE 10.14 A 14 bus 20 line power system has the following line data:

Line no.	*From bus*	*To bus*	*Line impedance (p.u.)*	*B/2 (p.u.)*
1	1	2	(0.02 + *j*0.26)	*j*0.06
2	1	3	(0.08 + *j*0.24)	*j*0.05
3	2	3	0.06 + *j*0.15)	*j*0.04
4	2	4	(0.05 + *j*0.14)	*j*0.04
5	2	5	(0.04 + *j*0.15)	*j*0.03
6	2	6	(0.06 + *j*0.15)	*j*0.05
7	3	4	(0.02 + *j*0.20)	*j*0.02
8	4	5	(0.08 + *j*0.25)	*j*0.05
9	4	7	(0.06 + *j*0.18)	*j*0.04
10	5	10	(0.02 + *j*0.25)	*j*0.025
11	6	8	(0.05 + *j*0.12)	*j*0.01
12	6	12	(0.04 + *j*0.35)	*j*0.005
13	7	10	(0.05 + *j*0.435)	*j*0.015
14	7	12	(0.056 + *j*0.315)	*j*0.025
15	9	11	(0.03 + *j*0.15)	*j*0.015
16	9	14	(0.04 + *j*0.2)	*j*0.02
17	10	12	(0.02 + *j*0.55)	*j*0.013
18	10	14	(0.05 + *j*0.95)	*j*0.02
19	11	13	(0.05 + *j*0.1)	*j*0.02
20	12	14	(0.015 + *j*0.35)	*j*0.03

The scheduled bus powers and initial bus voltages (in p.u.) are as under:

Bus no.	P_d	Q_d	P_g	Q_g	V	*Bus type*
1	0	0	?	?	$1.04\angle 0°$	Slack bus
2	0.4	0.2	0	0	$1.0\angle 0°$	PQ bus
3	0.45	0.15	0	0	$1.0\angle 0°$	PQ bus
4	0.4	0.05	0	0	$1.0\angle 0°$	PQ bus
5	0.6	0.1	0	0	$1.0\angle 0°$	PQ bus
6	0.5	0.2	0	0	$1.0\angle 0°$	PQ bus
7	0.15	0.2	0	0	$1.0\angle 0°$	PQ bus
8	0.1	0.1	0	0	$1.0\angle 0°$	PQ bus
9	0.15	0.3	0	0	$1.0\angle 0°$	PQ bus
10	0.15	0.25	0	0	$1.0\angle 0°$	PQ bus
11	0.25	0.25	0	0	$1.0\angle 0°$	PQ bus
12	0	0.1	0.25	?	$1.01\angle 0°$	PV bus
13	0	0.15	0.25	?	$1.02\angle 0°$	PV bus
14	0	0.25	0.25	?	$1.025\angle 0°$	PV bus

Find final bus voltages, line flows, total line loss, slack bus power by N-R method. Verify the results by Fast decoupled and DC load flow methods (all data are referred on 100 MVA base and in per unit). For DC load flow, calculations are to be done for real parts only. Assume specified tolerance, $\varepsilon = 0.0001$.

Solution: Execution of N-R method algorithm for Example 10.14.

```
All quantities are in p.u. and all angles are in radian.
No. of buses = 14. No. of lines = 20.
Non zero elements of [Ybus]:
Note that only upper triangular off diagonal elements are provided.
Row      Column          Ybus
 1         1       (1.544118-7.463529i)
 1         2       (-0.294118+3.823529i)
 Row     Column          Ybus
 1         3       (-1.250000+3.750000i)
 2         2       (8.814013-27.656690i)
 2         3       (-2.298851+5.747126i)
 2         4       (-2.262443+6.334842i)
 2         5       (-1.659751+6.224066i)
 2         6       (-2.298851+5.747126i)
 3         3       (4.043900-14.337621i)
 3         4       (-0.495050+4.950495i)
 4         4       (5.585263-19.763784i)
 4         5       (-1.161103+3.628447i)
 4         7       (-1.666667+5.000000i)
 5         5       (3.138819-13.722076i)
```

```
  5      10     (-0.317965+3.974563i)
  6       6     (5.579751-15.603024i)
  6       8     (-2.958580+7.100592i)
  6      12     (-0.322321+2.820306i)
  7       7     (2.474540-10.266219i)
  7      10     (-0.260790+2.268875i)
  7      12     (-0.547083+3.077344i)
  8       8      (2.958580-7.090592i)
  9       9     (2.243590-11.182949i)
  9      11     (-1.282051+6.410256i)
  9      14     (-0.961538+4.807692i)
 10      10      (0.700032-9.035942i)
 10      12     (-0.066028+1.815781i)
 10      14     (-0.055249+1.049724i)
 11      11     (5.282051-14.375256i)
 11      13     (-4.000000+8.000000i)
 12      12     (1.057657-10.492336i)
 12      14     (-0.122224+2.851905i)
 13      13      (4.000000-7.980000i)
 14      14      (1.139012-8.639321i
```

By N-R load flow method:

```
N-R load flow result without considering Q-limits at PV buses.
Specified tolerance =0.00010.
Load flow converges with 4 iterations and maximum tolerance =0.000000.
Final bus voltages are:
Bus no.    Complex bus voltage      Magnitude      Angle
 1            (1.040000)            1.040000        0
 2        (0.861182-0.326342i)      0.920941    -0.362226
 3        (0.875530-0.273887i)      0.917369    -0.303180
 4        (0.845289-0.361509i)      0.919349    -0.404134
 5        (0.816512-0.391300i)      0.905432    -0.446897
 6        (0.802770-0.389393i)      0.892226    -0.451625
 7        (0.843905-0.387665i)      0.928687    -0.430619
 8        (0.782527-0.388769i)      0.873779    -0.461094
 9        (0.886854-0.401083i)      0.973333    -0.424727
10        (0.830984-0.399758i)      0.922140    -0.448386
11        (0.892367-0.406351i)      0.980530    -0.427305
12        (0.924510-0.406670i)      1.010000    -0.414404
13        (0.932918-0.412389i)      1.020000    -0.416217
14        (0.943262-0.401102i)      1.025000    -0.402064
LINE POWER FLOWS:
Line no.  bus-i  bus-j           Sij                        Sji
  1        1      2    (1.352385+0.546347i)   (-1.311657-0.132669i)
  2        1      3    (1.281969+0.231300i)   (-1.154389+0.055283i)
  3        2      3   (-0.275580+0.108052i)    (0.282378-0.158643i)
  4        2      4    (0.229709-0.100177i)   (-0.226339+0.041878i)
  5        2      5    (0.467578-0.034997i)   (-0.457263+0.023642i)
```

```
 6      2     6   (0.489950-0.040209i)  (-0.472967+0.000455i)
 7      3     4   (0.422011-0.046639i)  (-0.417757+0.055439i)
 8      4     5   (0.144859-0.034395i)  (-0.142867-0.042630i)
 9      4     7   (0.099237-0.112922i)  (-0.098094+0.048046i)
10      5    10   (0.000130-0.081012i)  (-0.000041+0.040375i)
11      6     8   (0.101214+0.087317i)  (-0.100000-0.100000i)
12      6    12  (-0.128246-0.287772i)   (0.133120+0.321332i)
13      7    10   (0.036141-0.002802i)  (-0.036060-0.022180i)
14      7    12  (-0.088047-0.245244i)   (0.091799+0.219286i)
15      9    11   (0.006798-0.062249i)  (-0.006723+0.033989i)
16      9    14  (-0.156798-0.237751i)   (0.159857+0.213088i)
17     10    12  (-0.062772-0.155102i)   (0.063353+0.146756i)
18     10    14  (-0.051128-0.113092i)   (0.051824+0.088308i)
19     11    13  (-0.243277-0.283989i)   (0.250000+0.257399i)
20     12    14  (-0.038272-0.072023i)   (0.038318+0.010993i)
Slack bus power = (2.634354+0.777647i)
Total transmission line loss = (0.234354+0.162786i)
```

By FDLF method:

```
Fast Decoupled load flow result without considering Q-limits at PV buses.
Specified tolerance =0.00010.
Iteration completed with 5 iterations with delp_max=0.000094 and delQ_max
=0.000005.
Final bus voltages are:
Bus no.    Complex bus voltage      Magnitude      Angle
 1              (1.040000)           1.040000         0
 2         (0.861179-0.326345i)      0.920940     -0.362231
 3         (0.875527-0.273890i)      0.917368     -0.303184
 4         (0.845285-0.361513i)      0.919347     -0.404140
 5         (0.816508-0.391305i)      0.905430     -0.446904
 6         (0.802767-0.389397i)      0.892224     -0.451632
 7         (0.843901-0.387670i)      0.928686     -0.430626
 8         (0.782523-0.388773i)      0.873777     -0.461100
 9         (0.886842-0.401105i)      0.973332     -0.424753
10         (0.830980-0.399764i)      0.922139     -0.448394
11         (0.892352-0.406376i)      0.980527     -0.427334
12         (0.924506-0.406679i)      1.010000     -0.414413
13         (0.932909-0.412409i)      1.020000     -0.416238
14         (0.943254-0.401119i)      1.025000     -0.402083
LINE POWER FLOWS:
Line no. bus-i  bus-j           Sij                      Sji
  1        1      2    (1.352400+0.546359i)  (-1.311671-0.132667i)
  2        1      3    (1.281984+0.231307i)  (-1.154401+0.055285i)
  3        2      3   (-0.275583+0.108054i)   (0.282381-0.158645i)
  4        2      4    (0.229713-0.100178i)  (-0.226344+0.041880i)
  5        2      5    (0.467586-0.034998i)  (-0.457270+0.023644i)
  6        2      6    (0.489954-0.040212i)  (-0.472971+0.000459i)
  7        3      4    (0.422017-0.046640i)  (-0.417763+0.055441i)
```

```
  8      4      5    (0.144861-0.034395i)   (-0.142869-0.042629i)
  9      4      7    (0.099243-0.112926i)   (-0.098099+0.048050i)
 10      5     10    (0.000135-0.081014i)   (-0.000046+0.040378i)
 11      6      8    (0.101212+0.087317i)   (-0.099999-0.100000i)
 12      6     12   (-0.128239-0.287777i)    (0.133112+0.321337i)
 13      7     10    (0.036143-0.002802i)   (-0.036061-0.022179i)
 14      7     12   (-0.088042-0.245248i)    (0.091794+0.219290i)
 15      9     11    (0.006820-0.062245i)   (-0.006746+0.033985i)
 16      9     14   (-0.156835-0.237751i)    (0.159895+0.213091i)
 17     10     12   (-0.062770-0.155104i)    (0.063351+0.146758i)
 18     10     14   (-0.051118-0.113094i)    (0.051814+0.088309i)
 19     11     13   (-0.243348-0.283980i)    (0.250073+0.257393i)
 20     12     14   (-0.038245-0.072025i)    (0.038292+0.010994i)
Slack bus power = (2.634385+0.777666i)
Total transmission line loss = (0.234362+0.162822i)
```

By DC load flow method:

```
DC loadflow result with specified tolerance =0.00010.
Load flow converges with 5 iterations and maximum delP =0.000070.
Final bus voltages are:
Bus no.    Complex bus voltage      Magnitude      Angle
 1             (1.040000)            1.040000        0
 2        (0.942795-0.333373i)      1.000000     -0.339879
 3        (0.957026-0.290003i)      1.000000     -0.294230
 4        (0.929805-0.368052i)      1.000000     -0.376913
 5        (0.915460-0.402408i)      1.000000     -0.414146
 6        (0.912400-0.409299i)      1.000000     -0.421686
 7        (0.921956-0.387294i)      1.000000     -0.397695
 8        (0.906528-0.422146i)      1.000000     -0.435811
 9        (0.925794-0.378029i)      1.000000     -0.387666
10        (0.916320-0.400446i)      1.000000     -0.412004
11        (0.925595-0.378516i)      1.000000     -0.388192
12        (0.940185-0.368989i)      1.010000     -0.373993
13        (0.951834-0.366623i)      1.020000     -0.367662
14        (0.958810-0.362365i)      1.025000     -0.361339
LINE POWER FLOWS:
Line no. bus-i   bus-j        Pij          Pji
  1        1       2        1.355382    -1.319915
  2        1       3        1.238877    -1.125144
  3        2       3       -0.259867     0.264657
  4        2       4        0.236100    -0.232997
  5        2       5        0.466393    -0.457243
  6        2       6        0.477315    -0.461939
  7        3       4        0.410547    -0.407164
  8        4       5        0.135873    -0.134263
  9        4       7        0.104261    -0.103542
 10        5      10       -0.008515     0.008516
 11        6       8        0.100594    -0.100003
```

```
 12      6     12    -0.138655     0.139427
 13      7     10     0.032491    -0.032438
 14      7     12    -0.078976     0.079341
 15      9     11     0.003376    -0.003375
 16      9     14    -0.153419     0.154703
 17     10     12    -0.070305     0.070408
 18     10     14    -0.055799     0.055979
 19     11     13    -0.246656     0.249976
 20     12     14    -0.039201     0.039249
Slack bus active power = 2.594259
Total active power loss in transmission line = 0.194047
```

EXERCISES

1. In Figure P10.1 using bus no. 1 as slack bus, find the bus voltages for all bus bars using G-S and N-R methods. Assume initial voltage of all buses to be 1.01 p.u. The line data and load schedule are shown in Tables P10.1 and P10.2 (line data has been furnished on 100 MVA basis). Consider acceleration factor and specified tolerance as 1.4 and 0.0001 respectively.

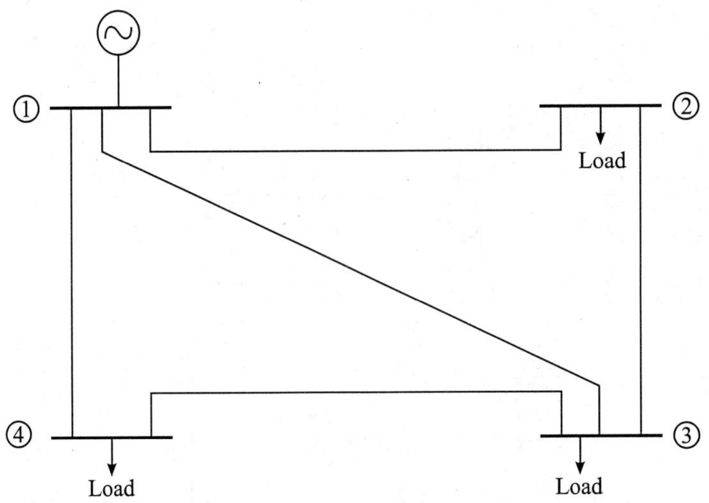

Figure P10.1 A four-bus, five-line system.

Table P10.1

Line no.	*From bus*	*To bus*	*Line impedance (p.u.)*
1	1	2	$j0.2$
2	1	3	$j0.4$
3	1	4	$j0.1$
4	2	3	$j0.1$
5	3	4	$j0.2$

Table P10.2

Bus no.	Generation		Load	
	MW	*MVAR*	*MW*	*MVAR*
1	?	?	0	0
2	0	0	50	20
3	0	0	40	10
4	0	0	20	0

2. (a) In Example 10.4 in the text, the line and bus data remaining the same, compute bus voltage using G-S method for Q_{limit} of bus-2 as $-0.1 \le Q_2 \le 0.2$.

(b) Repeat 2(*a*) but for flat start and assume $|V_2| = 1.0$ p.u. Apply G-S method and check the results using N-R method. Consider acceleration factor and specified tolerance as 1.4 and 0.0001 respectively in each case.

3. Figure P10.2 represents a five-bus power system feeding constant power loads. The line data and load schedule are shown in Tables P10.3 and P10.4 (line data has been furnished on 100 MVA basis).

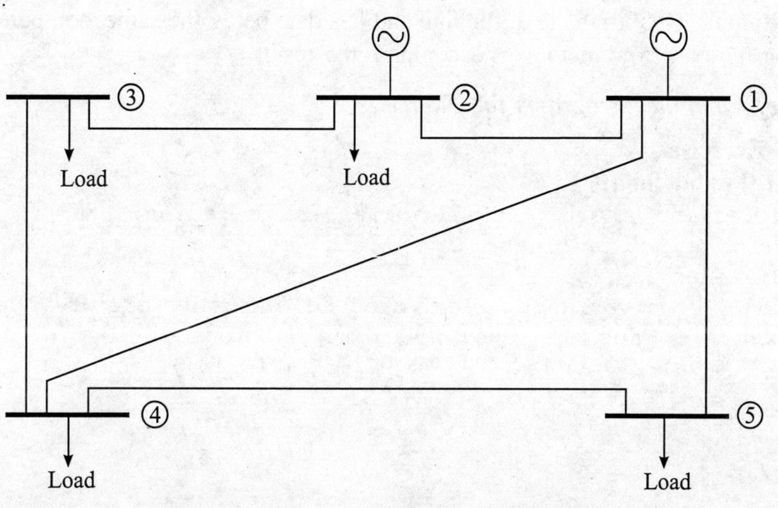

Figure P10.2 A five bus, six line system.

Table 10.3

Line no.	*From bus*	*To bus*	*Line impedance (p.u.)*	*Shunt Susceptance(B/2) (p.u.)*
1	1	2	$(0.03 + j0.25)$	$j0.02$
2	1	5	$(0.1 + j0.8)$	$j0.025$
3	1	4	$(0.08 + j0.62)$	$j0.025$
4	2	3	$(0.05 + j0.45)$	$j0.02$
5	3	4	$(0.08 + j0.75)$	$j0.02$
6	4	5	$(0.02 + j0.15)$	$j0.01$

Table 10.4

Bus no	*Generation*		*Load*	
	MW	*MVAR*	*MW*	*MVAR*
1	?	?	0	0
2	50	20	15	9
3	0	0	20	6
4	0	0	25	12
5	0	0	30	20

Assume slack bus voltage = $1\angle 0°$ p.u. Perform G-S, N-R and FDLF methods of load flow study and compare the results.

4. In Figure P10.2, assume a transformer between buses 2 and 4 having p.u. reactance of $j0.3$

(a) Find bus voltages and line flows using N-R method, if $V_2 : V_4 = 1.02:1$, $[V_i' : V_i = 1.02:1]$.

(b) Repeat the same exercise for $V_2 : V_4 = 1:1.02$ [i.e., $V_i : V_i' = 1.02:1$].

5. Repeat problem no 4, but with FDLF method.

6. In Example 10.8 in the text, the line and bus data being the same, compute bus voltage using dc load flow method and compare the result.

Esxecution of Computer Programs for Exercises

1. Result of Exercise-1:

By G-S load flow method:

```
All quantities are in p.u. and all angles are in radian.
No. of buses = 4. No. of lines = 5.
Non zero elements of [Ybus]:
Note that only upper triangular off diagonal elements are provided.
Row   Column     Ybus          Row   Column     Ybus
 1      1      (0-17.5i)        2      3       (0+10i)
 1      2       (0+5i)          3      3      (0-17.5i)
 1      3      (0+2.5i)         3      4       (0+5i)
 1      4      (0+10i)          4      4       (0-15i)
 2      2      (0-15i)
G-S loadflow result without considering Q-limits at PV buses.
Specified tolerance =0.00010 and acceleration factor = 1.40.
Load flow converges with 9 iterations and maximum delV =0.000042.
Final bus voltages are:
Bus no.    Complex bus voltage      Magnitude      Angle
 1            (1.010000)            1.010000         0
 2        (0.970765-0.090333i)      0.974959     -0.092787
 3        (0.976358-0.086367i)      0.980171     -0.088228
 4        (0.998230-0.042133i)      0.999119     -0.042182
```

By N-R load flow method:

```
N-R load flow result without considering Q-limits at PV buses.
Specified tolerance =0.00010.
```

```
Load flow converges with 3 iterations and maximum tolerance =0.000000.
Final bus voltages are:
Bus no.    Complex bus voltage       Magnitude      Angle
 1              (1.010000)           1.010000         0
 2         (0.970796-0.090367i)      0.974993    -0.092818
 3         (0.976372-0.086390i)      0.980186    -0.088251
 4         (0.998228-0.042130i)      0.999117    -0.042180
```

2. Result of Exercise-2:

Result of Exercise 2(a):

```
All quantities are in p.u. and all angles are in radian.
No. of buses = 4. No. of lines = 5.
G-S loadflow result considering Q-limits at PV buses.
Lower bound of Q-limit is crossed for bus-2 in iteration=2.
Specified tolerance =0.00010 and acceleration factor = 1.40.
Load flow converges with 17 iterations and maximum delV =0.000061.
Final bus voltages are:
Bus no.    Complex bus voltage       Magnitude      Angle
 1              (1.060000)           1.060000         0
 2         (0.971652+0.179397i)      0.988074     0.182575
 3         (0.888642-0.031305i)      0.889193    -0.035213
 4         (0.829394-0.132048i)      0.839840    -0.157885
```

Result of Exercise 2(b):

By G-S load flow method:

```
All quantities are in p.u. and all angles are in radian.
No. of buses = 4. No. of lines = 5.
G-S loadflow result considering Q-limits at PV buses.
Lower bound of Q-limit is crossed for bus-2 in iteration=1.
Specified tolerance =0.00010 and acceleration factor = 1.40.
Load flow converges with 12 iterations and maximum delV =0.000075.
Final bus voltages are:
Bus no.    Complex bus voltage       Magnitude      Angle
 1              (1.060000)           1.060000         0
 2         (0.971643+0.179429i)      0.988071     0.182609
 3         (0.888644-0.031299i)      0.889195    -0.035207
 4         (0.829399-0.132059i)      0.839847    -0.157897
```

By N-R load flow method:

```
N-R load flow result considering Q-limits at PV buses.
Lower bound of Q-limit is crossed for bus-2 in iteration=1.
Specified tolerance =0.00010.
Load flow converges with 4 iterations and maximum tolerance =0.000000.
Final bus voltages are:
Bus no.    Complex bus voltage       Magnitude      Angle
 1              (1.060000)           1.060000         0
 2         (0.971659+0.179418i)      0.988085     0.182594
 3         (0.888646-0.031305i)      0.889198    -0.035213
 4         (0.829397-0.132034i)      0.839840    -0.157868
```

3. Result of Exercise 3

By G-S load flow method:

```
All quantities are in p.u. and all angles are in radian.
No. of buses = 5. No. of lines = 6.
Non zero elements of [Ybus]:
Note that only upper triangular off diagonal elements are provided.
Row  Column        Ybus                Row  Column         Ybus
 1     1    (0.831741-6.690476i)        3     3    (0.384525-3.473456i)
 1     2   (-0.473186+3.943218i)        3     4   (-0.140622+1.318334i)
 1     4   (-0.204708+1.586489i)        4     4    (1.218693-9.400041i)
 1     5   (-0.153846+1.230769i)        4     5   (-0.873362+6.550218i)
 2     2    (0.717089-6.098340i)        5     5    (1.027209-7.745988i)
 2     3   (-0.243902+2.195122i)
G-S loadflow result without considering Q-limits at PV buses.s
Specified tolerance =0.00010 and acceleration factor = 1.40.
Load flow converges with 11 iterations and maximum delV =0.000069.
Final bus voltages are:
Bus no.    Complex bus voltage      Magnitude      Angle
 1            (1.000000)            1.000000         0
 2        (1.006996+0.016549i)      1.007132     0.016433
 3        (0.937107-0.103814i)      0.942840    -0.110332
 4        (0.875050-0.152790i)      0.888289    -0.172865
 5        (0.857195-0.162453i)      0.872454    -0.187296
```

By N-R load flow method:

```
N-R load flow result without considering Q-limits at PV buses.
Specified tolerance =0.00010.
Load flow converges withs 3 iterations and maximum tolerance =0.000008.
Final bus voltages are:
Bus no.    Complex bus voltage      Magnitude      Angle
 1            (1.000000)            1.000000         0
 2        (1.006992+0.016568i)      1.007129     0.016452
 3        (0.937111-0.103812i)      0.942844    -0.110329
 4        (0.875060-0.152803i)      0.888301    -0.172877
 5        (0.857207-0.162468i)      0.872468    -0.187310
```

By FDLF method:

```
Fast Decoupled load flow result without considering Q-limits at PV buses.
Specified tolerance =0.00010.
Iteration completed with 5 iterations with delp_max=0.000020 and delQ_max
=0.000013.
Final bus voltages are:
Bus no.    Complex bus voltage      Magnitude      Angle
 1            (1.000000)            1.000000         0
 2        (1.006993+0.016566i)      1.007130     0.016449
 3        (0.937115-0.103812i)      0.942847    -0.110329
 4        (0.875066-0.152801i)      0.888307    -0.172874
 5        (0.857215-0.162465i)      0.872475    -0.187305
```

4. Result of Exercise 4

Result of Exercise 4(a)

All quantities are in p.u. and all angles are in radian.
No. of buses = 5. No. of lines = 7.
The transformer position in line- 7 is like Case A of Art 12.3.
N-R load flow result without considering Q-limits at PV buses.
Specified tolerance =0.00010.
Load flow converges with 3 iterations and maximum tolerance =0.000001.
Final bus voltages are:

Bus no.	Complex bus voltage	Magnitude	Angle
1	(1.000000)	1.000000	0
2	(0.988646-0.02ss1828i)	0.988886	-0.022075
3	(0.947860-0.105410i)	0.953704	-0.110753
4	(0.932828-0.094502i)	0.937603	-0.100962
5	(0.910079-0.113964i)	0.917186	-0.124576

LINE POWER FLOWS:

Line no.	bus-i	bus-j	Sij	Sji
1	1	2	(0.091447+0.014444i)	(-0.091160-0.051615i)
2	1	5	(0.154097+0.068140i)	(-0.150855-0.088234i)
3	1	4	(0.163677+0.062222i)	(-0.160925-0.087873i)
4	2	3	(0.192733+0.044578i)	(-0.190623-0.063340i)
5	3	4	(-0.009377+0.003340i)	(0.009425-0.038658i
6	4	5	(0.149927+0.100434i)	(-0.149144-0.111766i)
7	2	4	(0.248427+0.117038i)	(-0.248427-0.093902i)

Slack bus power = (0.409221+0.144806i)
Total transmission line loss = (0.009221-0.125193i)

Result of Exercise 4(b)

All quantities are in p.u. and all angles are in radian.
No. of buses = 5. No. of lines = 7.
The transformer position in line- 7 is like Case B of Art 12.3.
N-R load flow result without considering Q-limits at PV buses.
Specified tolerance =0.00010.
Load flow converges with 3 iterations and maximum tolerance =0.000000.
Final bus voltages are:

Bus no.	Complex bus voltage	Magnitude	Angle
1	(1.000000)	1.000000	0
2	(0.979674-0.021015i)	0.979899	-0.021448
3	(0.947396-0.105463i)	0.953248	-0.110862
4	(0.946612-0.095874i)	0.951454	-0.100937
5	(0.922256-0.114709i)	0.929363	-0.123743

LINE POWER FLOWS:

Line no.	bus-i	bus-j	Sij	Sji
1	1	2	(0.092486+0.050207i)	(-0.092081-0.086040i)
2	1	5	(0.153141+0.053037i)	(-0.150187-0.075996i)
3	1	4	(0.163032+0.040074i)	(-0.160567-0.068601i)
4	2	3	(0.190373+0.025969i)	(-0.188379-0.045406i)
5	3	4	(-0.011620-0.014594i)	(0.011633-0.021563i)

```
  6        4      5    (0.150641+0.112523i)   (-0.149813-0.124003i)
  7        2      4    (0.251708+0.170071i)   (-0.251708-0.142359i)
Slack bus power = (0.408658+0.143318i)
Total transmission line loss = (0.008658-0.126682i)
```

5. Result of Exercise 5

Result of Exercise 5(a)

```
All quantities are in p.u. and all angles are in radian.
No. of buses = 5. No. of lines = 7.
Fast Decoupled load flow result without considering Q-limits at PV buses.
Specified tolerance =0.00010.
Iteration completed with 4 iterations with delp_max=0.000050 and delQ_max
=0.000013.
Final bus voltages are:
Bus no.    Complex bus voltage     Magnitude      Angle
 1             (1.000000)           1.000000        0
 2         (0.988648-0.021834i)     0.988889     -0.022081
 3         (0.947866-0.105410i)     0.953709     -0.110752
 4         (0.932834-0.094502i)     0.937608     -0.100962
 5         (0.910086-0.113962i)     0.917193     -0.124573
LINE POWER FLOWS:
Line no. bus-i  bus-j           Sij                      Sji
  1        1      2    (0.091468+0.014432i)   (-0.091181-0.051602i)
  2        1      5    (0.154094+0.068131i)   (-0.150852-0.088227i)
  3        1      4    (0.163676+0.062213i)   (-0.160924-0.087866i)
  4        2      3    (0.192720+0.044572i)   (-0.190611-0.063338i)
  5        3      4   (-0.009376+0.003340i)    (0.009425-0.038659i)
  6        4      5    (0.149915+0.100426i)   (-0.149132-0.111760i)
  7        2      4    (0.248410+0.117027i)   (-0.248410-0.093895i)
Slack bus power = (0.409238+0.144776i)
Total transmission line loss = (0.009221-0.125204i)
```

Result of Exercise 5(b)

```
All quantities are in p.u. and all angles are in radian.
No. of buses = 5. No. of lines = 7.
Fast Decoupled load flow result without considering Q-limits at PV buses.
Specified tolerance =0.00010.
Iteration completed with 4 iterations with delp_max=0.000041 and delQ_max
=0.000009.
Final bus voltages are:
Bus no.    Complex bus voltage     Magnitude      Angle
 1             (1.000000)           1.000000        0
 2         (0.979676-0.021020i)     0.979901     -0.021452
 3         (0.947401-0.105462i)     0.953252     -0.110861
 4         (0.946616-0.095874i)     0.951458     -0.100937
 5         (0.922262-0.114708i)     0.929368     -0.123741
```

```
LINE POWER FLOWS:
Line no.  bus-i  bus-j          Sij                        Sji
  1         1      2   (0.092502+0.050197i)  (-0.092098-0.086030i)
  2         1      5   (0.153139+0.053031i)  (-0.150185-0.075992i)
  3         1      4   (0.163031+0.040068i)  (-0.160566-0.068595i)
  4         2      3   (0.190363+0.025965i)  (-0.188369-0.045404i)
  5         3      4  (-0.011620-0.014595i)   (0.011633-0.021563i)
  6         4      5   (0.150633+0.112518i)  (-0.149805-0.123999i)
  7         2      4   (0.251694+0.170064i)  (-0.251694-0.142354i)
Slack bus power = (0.408672+0.143296i)
Total transmission line loss = (0.008658-0.126690i)
```

6. Result of Exercise 6

```
All quantities are in p.u. and all angles are in radian.
No. of buses = 6. No. of lines = 8.
DC loadflow result with specified tolerance =0.00010.
Load flow converges with 3 iterations and maximum delP =0.000035.
Final bus voltages are:
Bus no.    Complex bus voltage      Magnitude      Angle
 1             (1.040000)            1.040000         0
 2         (0.981481-0.191560i)      1.000000    -0.192751
 3         (0.998302-0.058249i)      1.000000    -0.058282
 4         (0.986729-0.162376i)      1.000000    -0.163098
 5         (0.985494-0.169711i)      1.000000    -0.170537
 6         (0.974467-0.224530i)      1.000000    -0.226461
```

CHAPTER 11

Economic Operation

11.1 INTRODUCTION

Electric power system is required to operate economically to supply power efficiently as well as to make electrical energy cost-effective to the consumers. The main objective is to reduce the generation cost while maintaining the voltage magnitude and nominal frequency throughout the electrical network. Cost saving of generation denotes a significant reduction in the operating cost which includes cost of fuel and hence this area requires a great deal of attention from operating and planning engineers throughout the years. A power plant has to cater to variable load conditions throughout the day, round the year. It is therefore illogical to assume that the same level of power must be generated at all time. The power generation must vary according to the load pattern, which may in turn vary with season. Therefore the *economic operation* must take into account the aspect of loading throughout the interval of operation and hence the power dispatch is always associated with the economics of generation and transmission. Various computational methods are normally used to solve the economic dispatch problem of power generating systems.

The economic operation can be subdivided into two parts: the *economic dispatch* problem and the *optimal power flow* problem. Economic dispatch problem deals with determining the power output of each plant to meet the specified load, such that the overall fuel cost is minimized. Optimal power flow problem deals with delivery of active power and or reactive power with minimum real or reactive power loss and accordingly the power flow is optimized through the network of transmission lines.

The factors influencing the cost of generation are the generator efficiency, fuel cost and transmission losses. The most efficient generator may not give minimum cost, since it may be located in a place where fuel cost is high. Further, if the plant is located far from the load centers, transmission losses may be high and running the plant may not be economical as desired. The economic scheduling of generators basically determines the optimal generation of different generating units to minimize total operating cost. This cost has got two components viz.,

(i) *fixed cost* incorporating the capital invested, interest charged on money borrowed, tax paid, labor wages, salary given to staff and other expenses that continue irrespective of the load on the power system, and

(ii) *variable cost* which is a function of loading on generating units, losses, daily load requirement and purchase or sale of power. The economic operation of a power system is concerned about minimizing the variable cost factor mostly, as we have little control over the fixed costs.

Mixture of different modes of generation exists in power system out of which thermal, hydro and nuclear are the conventional power generating processes. Normally, economic operation is considered by proper scheduling of thermal or hydrogenation only or of both. Nuclear stations for their safety are required to be operated at a fixed load only and the operation may be approximated similar to thermal generation and hence, no separate analytical tool is presented for scheduling of nuclear generating plants.

11.2 INPUT-OUTPUT (I-O) OPERATIONAL CHARACTERISTICS OF CONVENTIONAL THERMAL GENERATING PLANTS

The input output (I-O) characteristic is the fundamental cost profile for a thermal plant and is a plot of the input in British thermal units (Btu) or kcal unit per hour versus the power output of the plant in MW as shown in Figure 11.1. The overall efficiency of a thermal unit is then determined by measuring the heat input i.e. the electrical energy output. This depicts *input-output* curve of the thermal plant and can be developed for each thermal generating unit involved. The efficient generating unit develops a given amount of power with lesser fuel input and hence it has become the usual practice to load the more efficient unit prior to loading the comparatively lesser efficient unit.

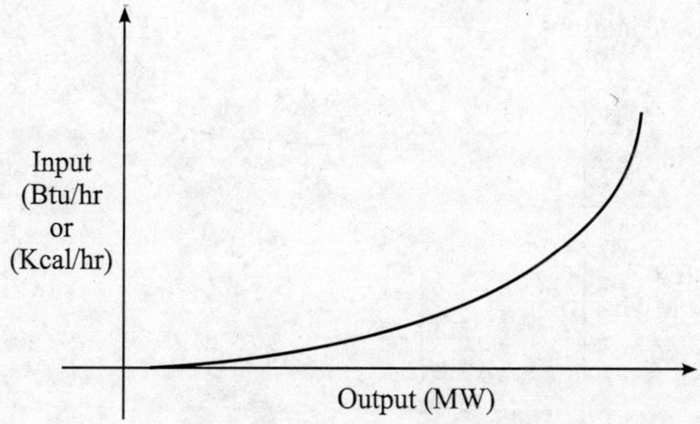

Figure 11.1 Input-output curve of a thermal generating unit.

11.2.1 Input-Output (I-O) Operational Characteristics of Hydel Power Plant

Figure 11.2 shows the input-output characteristic for a typical hydro-electric unit. It seems that the input of a hydel unit does not cost anything but with a deeper insight will reveal that capacity of the storage, agricultural requirement and cost of running the plant are governed by the water flow that includes the associated costs. The input of the plant is then basically this cost involved. In addition, artificial storage requirement imposes a cost on water input to turbine as well as the cost of control for the water output from the turbine due to agricultural needs virtually give rise to the cost factor involved in the input of the hydel plant.

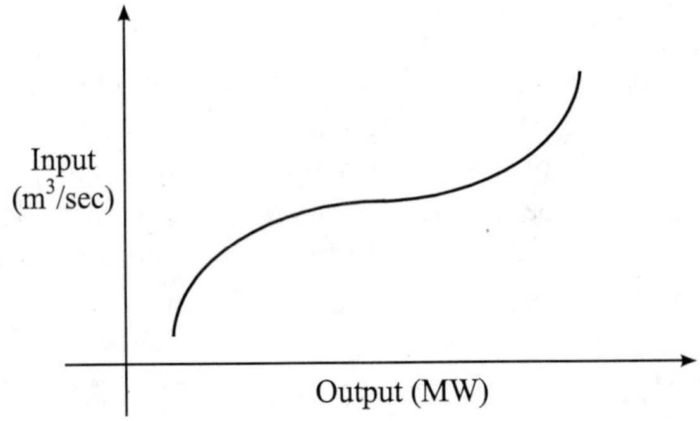

Figure 11.2 Input-output curve for a typical hydel power station.

11.2.2 Incremental Fuel Rate Curves

The input output curves, determined from the operating data of power station, can be used to get the *incremental fuel rate* (*IFR* or simply *IR*) curve. Incremental fuel rate (IFR) curve is given by the ratio of incremental change in input to the incremental change in output for a conventional power generating unit.

Thus, by calculating the slope of the input-output curves at various points of operation, the profile of IFR can be obtained. Figure 11.3(a) and (b) give the IFR profiles for typical thermal and hydro power stations.

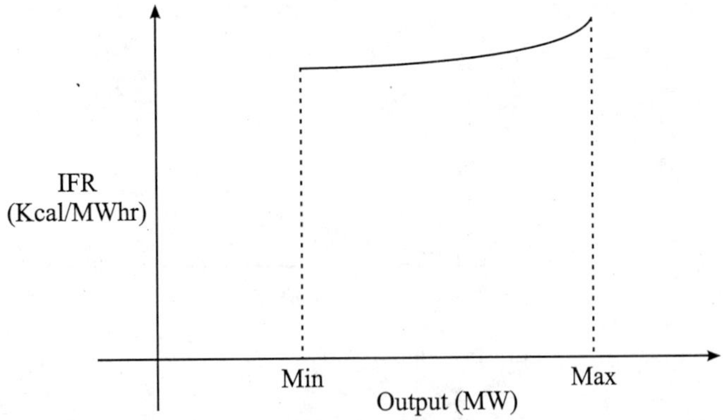

Figure 11.3(a) Typical IFR curves for thermal power station.

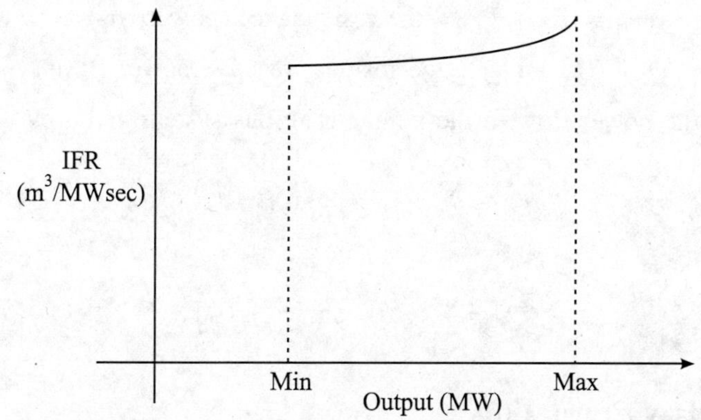

Figure 11.3(b) Typical IFR curve for hydro power station.

11.2.3 Incremental Fuel Cost (IFR) Curve

The *incremental fuel cost* is a measure of cost to produce an increment of power input with corresponding increment in power demand on the generating unit. The incremental production cost is made up of incremental fuel cost plus the incremental cost of labour, water, maintenance etc. which can be taken to be some percentage of the incremental fuel cost. This curve can be obtained from IFR curve by multiplying the 'IFR' by the cost of fuel per Btu (or kcal). In a power station fuel cost governs the actual total cost and hence IFR is very significant in economic loading of the generating units. The IFR curves will mostly be similar to the pattern of IFR characteristics, configuration wise. Normally, the cost curve is approximated by a linear curve. The slopes of the input-output curve and incremental fuel rate curve do not change for different fuels or for changes in the cost of same fuel. A multiplying factor may be used so that the actual cost is a realistic one. The unit of *IFC* (or simply the *IC*) is unit of cost/MWhr, (₹/MWhr or $/MWhr or £/MWhr or so).

11.3 CONSTRAINTS IN ECONOMIC OPERATION OF POWER SYSTEM

11.3.1 Primary Constraints

Equality constraints: Primary constraints, which are basically the *equality constraints,* associated with real and reactive power status for the power system and represent the balance of the load demand and total generation. If P_i and Q_i are the scheduled electrical generations in the system, P_R and Q_R are the respective load demands (Figure 11.4); the following equality needs to be satisfied in the system.

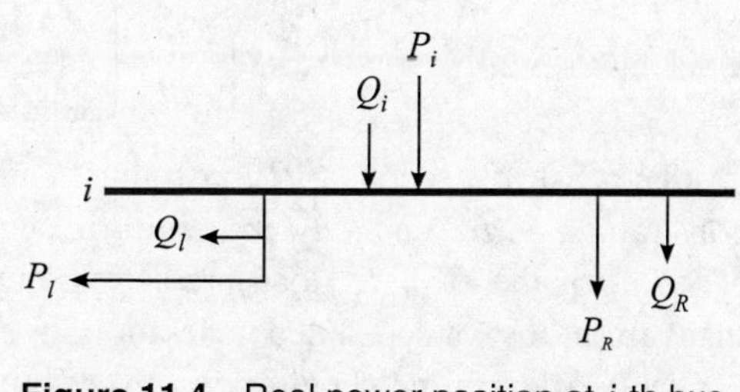

Figure 11.4 Real power position at *i*-th bus.

$$P_i - (P_R + P_L) = \text{the residual real power } (\mu_P) \tag{11.1}$$

$$Q_i - (Q_R + Q_L) = \text{the residual reactive power } (\mu_Q) \tag{11.2}$$

where, P_L and Q_L the power flow to the neighboring bus system given by

$$P_L = \sum_{j=1}^{N} V_i V_j Y_{ij} \cos\left(\delta_{ij} - \theta_{ij}\right) \tag{11.3}$$

$$Q_L = \sum_{j=1}^{N} V_i V_j Y_{ij} \sin\left(\delta_{ij} - \theta_{ij}\right) \tag{11.4}$$

11.3.2 Secondary Constraints

Secondary constraints are *inequality constraints* that arise due to physical and operational limitations of respective units and components. These constraints are listed below:

Power inequality constraint is a secondary constraint which is responsible for smooth operation of power units. For each generator there should be a minimum and maximum permissible output and the power production should be constrained to ensure that

$$P_i|^{\min} \le P_i \le P_i|^{\max}, \; i = 1, 2, 3, \ldots, N_P$$

and

$$Q_i|^{\min} \le Q_i \le Q_i|^{\max}, \; i = 1, 2, 3, \ldots., N_Q$$

N_P and N_Q being the total number of real and reactive sources in the system.

Moreover, *power injection constraint* $P_i^2 + Q_i^2 \le \left(S_{i_{rated}}\right)^2$ must be satisfied where $S_{i_{rated}}$ denotes the complex power capacity of the generating unit without any overloading.

Dynamic constraints are involved whenever fast charges in generation are required for picking up the increasing load demand. Here, for real power generation the dynamic constraint can be represented as

$$\left|\frac{dP_i(t)}{dt}\right|_{\min} \le \left|\frac{dP_i(t)}{dt}\right| \le \left|\frac{dP_i(t)}{dt}\right|_{\max}$$

while, for reactive power generation the dynamic constraint becomes

$$\left|\frac{dQ_i(t)}{dt}\right|_{\min} \le \left|\frac{dQ_i(t)}{dt}\right| \le \left|\frac{dQ_i(t)}{dt}\right|_{\max}$$

The list of secondary constraints can be extended to include some more inequality constraints. *Spare capacity constraint* is such type of inequality constraint which is utilized to account for the errors in load prediction, any sudden and fast change in load demands and the inadvertent loss of scheduled generation. In case of spare capacity constraint, the total generation available at any time should be in excess of the total anticipated load demand and any system loss by an amount not less then a specified *minimum* spare capacity P_{SP}.

$$P_i \ge \sum_{i=1}^{N} P_L + P_{SP} + P_R$$

For a bunch of generators, when all plants are not equally operationally suitable for taking up additional load, this constraint is then given by

$$P_i \geq \sum P_L + P_{SPG} + P_R$$

where, P_{SPG} is the spare capacity generation for the specified generators.

Thermal constraint of transmission line is also a typical inequality constraint. In transmission lines, when power injection ($+S_{i_{\text{Max}}}$) or power withdrawal ($-S_{i_{\text{Max}}}$) is allowed, then such constraint is represented as

$$\left|S_{i_{\max}}\right| \leq S_{tr} \leq \left|S_{i_{\min}}\right|, i = 1, 2, ...(t_r)_n$$

where, $(t_r)_n$ represents the number of branches and S_{tr} is the branch power transfer in MVA.

The next lot of inequality constraints is *bus voltage and angle constraint.* In order to maintain voltage profile at load bus and for limiting the overload capacity, these constraints appear. Here,

$$V_{i_{\min}} \leq V_i \leq V_{\max}, i = 1, 2, ..., N$$

and

$$\delta_{j_{\min}} \leq \delta_j \leq \delta_{j_{\max}}, j \neq 1, j = 2, 3, ..., M$$

where, N represents the number of units and M the number of loads in the system.

In addition to these inequality constraints *operational constraint* appears in case the transformer tap position needs to be included for optimization, the tap position "A_i" should lie within the range available in the transformer, i.e., $A_{i_{\min}}$ $A_{i_{\max}}$.

Emission constraint: The economic dispatch and emission dispatch are considerably different. The economic dispatch reduces the total fuel cost (operating cost) of the system at an increased rate of NO_x. On the other hand emission dispatch reduces the total emission from the system by an increase in the system operating cost. Therefore it is necessary to find out an operating point, that strikes a balance between cost and emission. This is achieved by *combined economic and emission dispatch* (CEED).

11.4 ECONOMIC OPERATION OF IDEAL THERMAL PLANTS (WITHOUT CONSIDERING ELECTRICAL SYSTEM LOSS)

Let us assume that a power system consists of N number of thermal generating units. These units are connected to a single bus having connected load of P_R (Figure 11.5). It has further been assumed that the electrical losses in the entire system are neglected. For the assumed system, the total *cost rate* (C) being the *summation* of cost rates of each of the individual units (C_1, C_2, C_3, ..., C_N), it is given by

$$C = C_1 + C_2 + C_3 + \cdots + C_i + \cdots + C_N = \sum_{i=1}^{N} C_i \tag{11.5}$$

where C_i is a function of P_i and it represents the input of the i-th unit in terms of cost rate while the electrical output of the i-th unit is P_i. The constraint equation in this analysis denotes the balance between that the summation of output generated $\left(\sum P_i\right)$ and the total load demand (P_R).

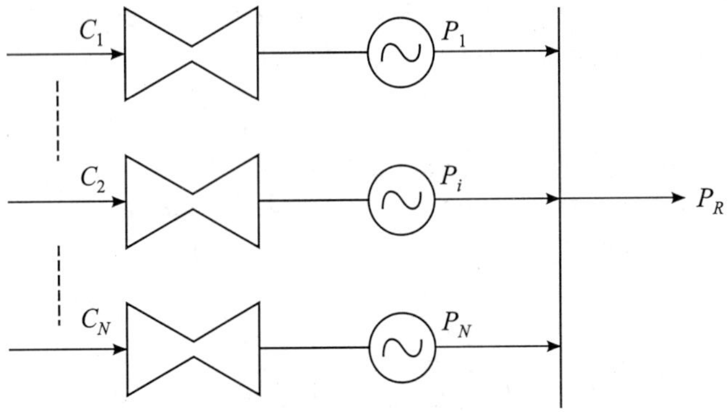

Figure 11.5 Block diagram of the power system comprising of thermal generators.

i.e.,

$$-\sum_{i=1}^{N} P_i + P_R = \Phi \tag{11.6}$$

The objective function being the total cost rate (C), economic operation is only possible when the objective function is minimized subject to the specified constraints.

The use of the *Lagrange function* (£) is conventional in order to achieve the minimum value of the objective function (C). Here, the objective function (the *cost function* C) is added to the constraint function (Φ) after the constrained function is multiplied by an undetermined *multiplier* (λ). The Lagrangian function is then obtained as

$$£ = C + \lambda\Phi = C + \lambda\left(-\sum_{i=1}^{N} P_i + P_R\right) \tag{11.7}$$

The substitution of the first partial derivative of £, with respect to each of the independent variable to zero gives the necessary condition for the *minimum value of the objective function.* This condition is commonly known as Kuhn-Tucker's condition of optimality. Here the partial derivative of the Lagrangian function with respect to power output values (one at a time) is computed in order to optimize the cost rate function. This results in formation of N number of equations (known as co-ordination equations) of the N-number of plants operating in the system. Generalizing, for the i-th plant, we have

$$\frac{\partial £}{\partial P_i} = \frac{dC_i\,(P_i)}{dP_i} - \lambda = 0$$

or,

$$\lambda = \frac{dC_i\,(P_i)}{dP_i} \quad \text{here} \quad i = 1, 2, 3, \ldots, N \tag{11.8}$$

Thus equation (11.8) depicts the necessary condition for the existence of a *minimum* cost of thermal power generation, i.e., cost minimization in thermal units in a no loss power system is achieved when the *incremental cost rate* of all the units is equal to, the Lagrangian multiplier. Hence for N number of thermal units, economic operation is obtained when

$$\frac{dC_1(P_1)}{dP_1} = \frac{dC_2(P_2)}{dP_2} = \frac{dC_3(P_3)}{dP_3} = \ldots = \frac{dC_i(P_i)}{dP_i} \ldots = \frac{dC_N(P_N)}{dP_N} = \lambda \qquad (11.9)$$

Equations (11.7) and (11.9) suffice for finding $(N + 1)$ unknowns (P_1, P_2, ..., P_i, ..., P_N and λ) in this optimization problem.

EXAMPLE 11.1 In a 100 MW thermal generating unit, the cost function is obtained as:

$$C(P_i) = (300 + 75P_i + 0.01P_i^2) \times 10^3 \text{ ₹/hr}$$

where P_i is the generator loading.

Find (a) IFC when 100% loading is applied (b) IFC when 50% loading is applied.

Solution: It is given that,

$$C(P_i) = (300 + 75P_i + 0.01P_i^2) \times 10^3 \text{ ₹/hr}$$

∴ IFC expression is given by

$$\frac{dC_i(P_i)}{dP_i} = (75 + 0.02P_i) \times 10^3 \text{ ₹/MWhr}$$

For 100% loading, IFC becomes

$$\text{IFC} = (75 + 0.02 \times 100) \times 10^3 = 77 \times 10^3 \text{ ₹/MWhr} \quad \text{(Ans. of (a))}$$

For 50% loading, IFC becomes

$$\text{IFC} = (75 + 0.02 \times 50) \times 10^3 = 76 \times 10^3 \text{ ₹/MWhr} \quad \text{(Ans. of (b))}$$

It may be observed here that incremental fuel cost alters a little even if loading is reduced to 50%.

EXAMPLE 11.2 In a generating plant there are two turbo-generators which feed a load bus with following fuel cost characteristics:

$$F_1(P_1) = 120 + 3P_1 + 0.0075P_1^2 \text{ ₹/hr}$$

$$F_2(P_2) = 100 + 2P_2 + 0.009P_2^2 \text{ ₹/hr}$$

(a) Find the economic schedule if total load is 160 MW. Assume no generator limits.

(b) Repeat (a) with following generation limits considered:

Unit-1: $P_{max} = 100$ MW, $P_{min} = 20$ MW

Unit-2: $P_{max} = 100$ MW, $P_{min} = 10$ MW

Solution: From given data we first calculate the IFC values as follows:

$$\frac{dF_1(P_1)}{dP_1} = 3 + 0.015P_1 \text{ ₹/MWhr}$$

N.B. The cost figures given in the numerical examples given in this chapter are indicative only and not the actual cost figures of a typical unit.

$$\frac{dF_2(P_2)}{dP_2} = 2 + 0.018P_2 \text{ ₹/MWhr}$$

For economic operation, we have

$$\frac{dF_1(P_1)}{dP_1} = \frac{dF_2(P_2)}{dP_2}$$

or, $$3 + 0.015P_1 = 2 + 0.018P_2$$

∴ $$0.015P_1 - 0.018P_2 = -1 \quad \text{(i)}$$

and $$P_1 + P_2 = 160 \text{ [since total load = 160 MW]} \quad \text{(ii)}$$

Solving for (i) and (ii),

$$P_1 = 56.97 \text{ MW}; \quad P_2 = 103.03 \text{ MW} \quad \text{(Ans. of (a))}$$

Here $\lambda = 3.854$.

(b) If generator limit is considered, $P_{2_{max}} = 100\,\text{MW}$; but for economic operation, P_2 should have been 103.03 MW

∴ $$P_2 = 100 \text{ MW } [P_2 \text{ is fixed at its upper limit } P_{2_{max}}]$$

Obviously, $$P_1 = (160 - 100) \text{ MW} = 60 \text{ MW}$$

Finally, in this case, $$P_1 = 60 \text{ MW}; P_2 = 100 \text{ MW (Ans. of (b))}$$

Here, $$\frac{dF_1(P_1)}{dP_1} \neq \frac{dF_2(P_2)}{dP_2}$$

Therefore the load scheduling is not economic [also note that $\lambda_1 = 3 + 0.015 \times 60 = 3.9$ and $\lambda_2 = 2 + 0.018 \times 100 = 3.8$]

EXAMPLE 11.3 The fuel cost characteristics in \$/hr of two thermal plants are given as under:

$$C_1 = 8500 + 52.8P_1 + 5.5 \times 10^{-3}P_1^2$$

$$C_2 = 3700 + 15P_2 + 0.05P_2^2$$

The limits of generation for the two units are $200 \le P \le 800$ MW. The load curve is shown in the Figure E11.1 below. Find the daily operating schedule to minimize the operating costs. The cost of taking a unit off and then putting it on is 1000.00 \$.

Solution: For 1200 MW load period, we must run both the units. So the respective incremental costs must be obtained and equated.

$$(IC)_1 = 52.8 + 11\times 10^{-3} P_1 = \lambda$$

$$(IC)_2 = 15 + 0.1P_2 = \lambda$$

For economic loading incremental costs being equal, we can write,

$$52.8 + 11 \times 10^{-3} P_1 = 15 + 0.1\,(1200 - P_1) \text{ [assuming loads to be } P_1 \text{ and } (1200 - P_1)]$$

Solving, $$P_1 = 740.54 \text{ MW and } P_2 = 429.73 \text{ MW}$$

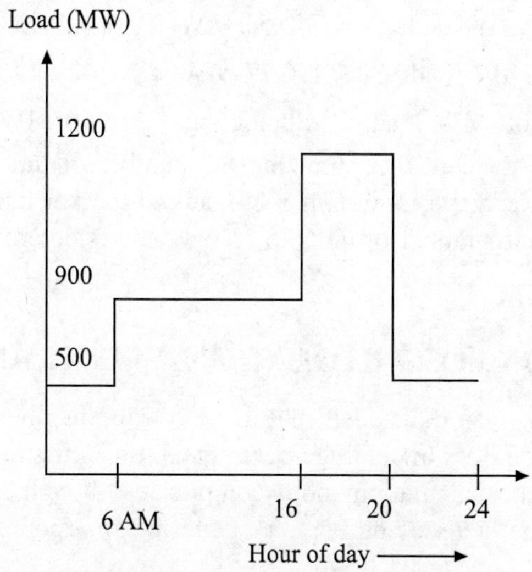

Figure E11.1 Load duration curve.

Again, for the 900 MW load also it is necessary to run both the units. So assuming the loads to be P_1 and $(900 - P_1)$ and equating the incremental costs, we get,

$$52.8 + 11 \times 10^{-3} P_1 = 15 + 0.1 (900 - P_1)$$

Solving, $$P_1 = 470.27 \text{ MW and } P_2 = 429.73 \text{ MW}$$

For the 500 MW load, we may run a single unit or both the units depending upon the costs involved. For this purpose we must determine the corresponding costs when both units are on and when a single 500 MW unit is operating.

First let us consider both units are on. Assuming the loads are P_1 and $(500 - P_1)$ and equating the incremental costs for economic operation,

We get, $$52.8 + 11\times 10^{-3} P_1 = 15 + 0.1 (500 - P_1)$$

Solving, $$P_1 = 109.9 \text{ MW} \quad \text{and} \quad P_2 = 390.1$$

The operating costs of each unit for the 500 MW load period are

$$C_1 = 8500 + 52.8 \times 109.9 + 5.5 \times 10^{-3} \times 109.9^2 = 14369.15 \text{ \$}$$

$$C_2 = 3700 + 15 \times 390.1 + 0.05 \times 390.1^2 = 17160.4 \text{ \$}$$

Total operating cost for the 10 hour period of 500 MW is $10 (C_1 + C_2) = 31529.55$ \$
The other option is to operate only one unit of 500 MW during 500 MW load period.
Then the cost for 10 our period is $10[3700 + 15 \times 500 + 0.05 \times 500^2]$ \$ = 237000 \$

Add the cost of shutting down and starting unit 1 which is 1000 \$. Therefore, total cost for 500 MW load duration is (237000 + 1000) \$ = 238000 \$ for one unit while \$(31529.55 + 2000), i.e., \$33529.55 for two units.

It may be observed that it is cheaper to run unit 2, as compared to this, the cost of operation if both units are operating is 319600\$. Hence it is economical to run only unit 2 during 500 MW load. Hence the complete operating schedule is

$$\left.\begin{aligned} 1200 \text{ MW load: } & P_1 = 740.54 \text{ MW; } P_2 = 459.73 \text{ MW} \\ 900 \text{ MW load: } & P_1 = 470.27 \text{ MW; } P_2 = 429.73 \text{ MW} \\ 500 \text{ MW load: } & P_1 \text{ Shutdown; } P_2 = 500 \text{ MW} \end{aligned}\right\} \text{Ans.}$$

The extension of this concept of optimizing the number of units to be operated during a particular time interval to achieve minimum cost leads to the concept of "Unit Commitment". Here for a N-unit sytem, the number of units to be operated is determined using the economics of operation of the units.

11.5 TRANSMISSION LOSS ALLOCATION AND ITS ANALYTICAL MODEL

Transmission loss-formula was developed in early of 1950s to obtain a practical method for calculation of transmission loss including incremental transmission loss in economic plant scheduling. This makes the calculation more complicated though it is well accepted in the power utility industry. The steps of derivation of the transmission loss formula are provided as follows:

Let us consider two thermal generators G_1 and G_2, in Figure 11.6, be connected to bus no-1 and 2 injecting currents I_1 and I_2 into the network of lines and load bus (3 and 4). Let n be the neutral bus where the no load current is I_n^0. Let I_3 and I_4 be the load current injections at bus 3 and 4 such that the net load current

$$I_D = I_3 + I_4 \tag{11.10}$$

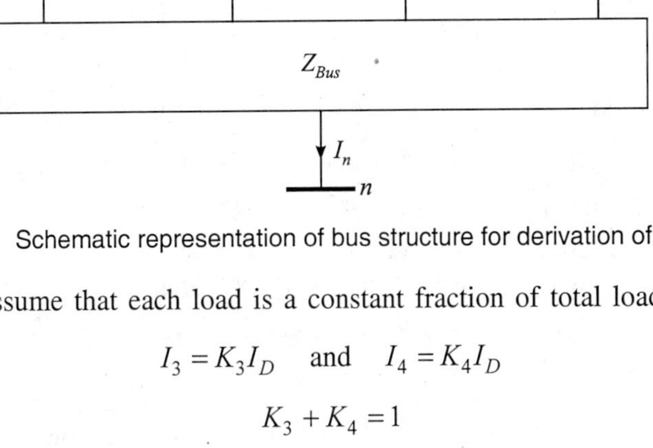

Figure 11.6 Schematic representation of bus structure for derivation of loss formulae.

Next let us now assume that each load is a constant fraction of total load such that

$$I_3 = K_3 I_D \quad \text{and} \quad I_4 = K_4 I_D \tag{11.11}$$

also,

$$K_3 + K_4 = 1 \tag{11.12}$$

For convenience, let us choose now the neutral bus as the *reference* bus. Since in a generalized network $[V]=[Z][I]$, hence for the given network, we have

$$\begin{bmatrix} V_{1n} \\ V_{2n} \\ V_{3n} \\ V_{4n} \end{bmatrix} = \begin{bmatrix} Z_{11} & Z_{12} & Z_{13} & Z_{14} \\ Z_{21} & Z_{22} & Z_{23} & Z_{24} \\ Z_{31} & Z_{32} & Z_{33} & Z_{34} \\ Z_{41} & Z_{42} & Z_{43} & Z_{44} \end{bmatrix} \begin{bmatrix} I_1 \\ I_2 \\ I_3 \\ I_4 \end{bmatrix}$$

where $\begin{bmatrix} Z_{11} & Z_{12} & Z_{13} & Z_{14} \\ Z_{21} & Z_{22} & Z_{23} & Z_{24} \\ Z_{31} & Z_{32} & Z_{33} & Z_{34} \\ Z_{41} & Z_{42} & Z_{43} & Z_{44} \end{bmatrix}$ is impedance matrix.

By expanding only the first now, we get

$$V_{1n} = Z_{11}I_1 + Z_{12}I_2 + Z_{13}I_3 + Z_{14}I_4 \tag{11.13}$$

Substituting for I_3 and I_4 from equation (11.11) we can write,

$$I_D = \frac{-Z_{11}}{K_3z_{13} + K_4z_{14}} I_1 + \frac{-Z_{11}}{K_3z_{13} + K_4z_{14}} I_2 + \frac{-Z_{11}}{K_3z_{13} + K_4z_{14}} I_n^0 \tag{11.14}$$

where, $\quad I_n^0 = -\dfrac{V_{1n}}{Z_{11}}$

We next write equation (11.14) as

$$I_D = -m_1I_1 - m_2I_2 - m_1I_n^0 \tag{11.14a}$$

where
$$\left.\begin{aligned} m_1 &= \frac{-Z_{11}}{K_3z_{13} + K_4z_{14}} \\ m_2 &= \frac{-Z_{12}}{K_3z_{13} + K_4z_{14}} \end{aligned}\right\} \tag{11.15}$$

Substitution of (11.14a) in (11.11) yields

$$I_3 = -K_3m_1I_1 - K_3m_2I_2 - K_3m_1I_n^0 \tag{11.16a}$$

and

$$I_4 = -K_4m_1I_1 - K_4m_2I_2 - K_4m_1I_n^0 \tag{11.16b}$$

In the next step, let us now relate currents I_1, I_2 and I_n^0 with the bus current through the *connection matrix* utilizing equations (11.16a) and (11.16b) assuming I_1 and I_2 remain *invariant*. This gives

$$\begin{bmatrix} I_1 \\ I_2 \\ I_3 \\ I_4 \end{bmatrix} = \begin{bmatrix} 1 & & \\ & 1 & \\ -K_3m_1 & -K_3m_2 & -K_3m_1 \\ -K_4m_1 & -K_4m_2 & -K_4m_1 \end{bmatrix} \begin{bmatrix} I_1 \\ I_2 \\ I_n^0 \end{bmatrix} = C \begin{bmatrix} I_1 \\ I_2 \\ I_n^0 \end{bmatrix} \tag{11.17)}$$

However, we know from the concepts of power invariant transformation that load bus power S_L, in terms of transformed (new) current, can be expressed as

$$S_L = I_{new}^T \, Z_{Bus(new)} \, I_{new}^* \tag{11.18}$$

But,

$$Z_{Bus(new)} = C^T Z_{Bus} C^* \tag{11.19}$$

where, $Z_{Bus} = (R_{Bus} + jX_{Bus})$. Thus equation (11.18) can be written for real load power as

$$P_L = I_{new}^T C^T R_{Bus} C^* I_{new}^* \tag{11.20}$$

Following equations (11.17), (11.20) becomes

$$P_L = [I_1 \quad I_2 \quad I_n^0][C^T R_{Bus} C^*]\begin{bmatrix} I_1 \\ I_2 \\ I_n^0 \end{bmatrix}^* \tag{11.21}$$

Let as now assume that at each generator bus the reactive power is a fraction 'f' of real power;

i.e.,

$$P_1 + jQ_1 = (1 + jf_1)P_1$$

and

$$P_2 + jQ_2 = (1 + jf_2)P_2$$

where, $f_1 = \dfrac{Q_1}{P_1}$ and $f_2 = \dfrac{Q_2}{P_2}$ [f_1 and f_2 are real numbers]

$\therefore$

$$I_1 = \frac{(1 - jf_1)P_1}{V_1^*} = \alpha_1 P_1 \tag{11.22a}$$

and

$$I_2 = \frac{(1 - jf_2)P_2}{V_2^*} = \alpha_2 P_2 \tag{11.22b}$$

$\therefore$ from equations (11.22a) and (11.22b) we have

$$\begin{bmatrix} I_1 \\ I_2 \\ I_n^0 \end{bmatrix} = \begin{bmatrix} \alpha_1 & & \\ & \alpha_2 & \\ & & I_n^0 \end{bmatrix}\begin{bmatrix} P_1 \\ P_2 \\ 1 \end{bmatrix} \tag{11.22(c)}$$

Equation (11.22(c)) can further be written as

$$P_\ell = \begin{bmatrix} P_1 \\ P_2 \\ 1 \end{bmatrix}^T \begin{bmatrix} \alpha_1 & & \\ & \alpha_2 & \\ & & I_n^0 \end{bmatrix} C^T R_{Bus} C^* \begin{bmatrix} \alpha_1 & & \\ & \alpha_2 & \\ & & I_n^0 \end{bmatrix}^* \begin{bmatrix} P_1 \\ P_2 \\ 1 \end{bmatrix}^* \tag{11.22(d)}$$

This finally gives,

$$P_\ell = [P_1 \quad P_2 \quad 1]\begin{bmatrix} B_{11} & B_{12} & B_{10}/2 \\ B_{21} & B_{22} & B_{20}/2 \\ B_{10}/2 & B_{20}/2 & B_{00} \end{bmatrix}\begin{bmatrix} P_1 \\ P_2 \\ 1 \end{bmatrix} \tag{11.23}$$

where $\begin{bmatrix} B_{11} & B_{12} & B_{10}/2 \\ B_{21} & B_{22} & B_{20}/2 \\ B_{10}/2 & B_{20}/2 & B_{00} \end{bmatrix} \begin{bmatrix} P_1 \\ P_2 \\ 1 \end{bmatrix} = \begin{bmatrix} \alpha_1 & & \\ & \alpha_2 & \\ & & I_n^0 \end{bmatrix} C^T R_{Bus} C^* \begin{bmatrix} \alpha_1 & & \\ & \alpha_2 & \\ & & I_n^0 \end{bmatrix}^* \begin{bmatrix} P_1 \\ P_2 \\ 1 \end{bmatrix}^*$

where B terms are called *loss coefficients* or B-coefficients ($N \times N$ matrix) and is always symmetrical.

In the symmetrical network, $B_{12} = B_{21}$, this gives,

$$P_\ell = B_{11}P_1^2 + 2B_{12}P_1P_2 + B_{22}P_2^2 + B_{10}P_1 + B_{20}P_2 + B_{00} \tag{11.24a}$$

$$= \sum_{i=1}^{2}\sum_{j=1}^{2} P_iB_{ij}P_j + \sum_{i=1}^{2} B_{i0}P_i + B_{00} \tag{11.24b}$$

In general form, expression (11.24b) may be written as

$$P_\ell = P^T BP + P^T B_0 + B_{00} \tag{11.25}$$

Expression (11.25) is the generalized *transmission loss formula.*

In a system of N sources, then we have

$$P_\ell = \sum_{i=1}^{N}\sum_{j=1}^{N} P_iB_{ij}P_j + \sum_{i=1}^{N} B_{i0}P_i + B_{00} \tag{11.26}$$

11.6 ECONOMIC DISPATCH WITH TRANSMISSION LOSS CONSIDERED

In this article the analytical model for economic allocation of generation between different generating units in a plant with transmission losses is developed. In an integrated power system, the transmission losses need to be considered and the dispatch of power thus involves these losses. The objective of this analysis is to model the economic operation of the generation in such a way so as to make the overall cost minimum.

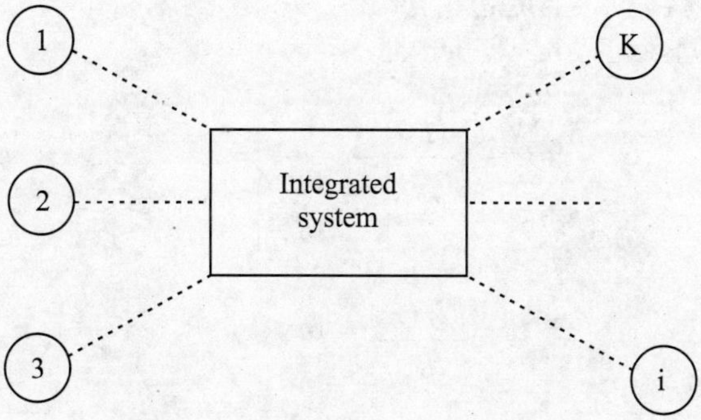

Figure 11.7 An integrated power pool.

The integrated power system is considered to have 'N' number of thermal plants interconnected by transmission lines and tie lines (Figure 11.7). Let P_1, P_2,..., P_N refer to the generation of each of the N plants respectively in MW while the assumed total load is P_R and ohmic losses in the transmission lines of the assumed system are P_ℓ. The power balance constraint equation is given by

$$-\sum_{i=1}^{N} P_i + P_R + P_\ell = \text{The residual power } (\Phi) \tag{11.27}$$

The total fuel input cost per hour for the entire generating system in the assumed network is taken as C (which is the summation of fuel costs per hour of each of the units) governed by the generated real power P.

i.e., $$C = \sum_{i=1}^{N} C_i(P_i) \text{ Unit of currency/hr.}$$

Application of Lagrangian optimization technique in order to optimize real power generation gives $£ = C + \lambda\Phi$, λ being the multiplier. For cost function to be minimum, the necessary condition can be obtained by setting the first derivative of the Lagrangian with respect to each of the independent variable to zero, i.e., $\dfrac{\partial £}{\partial P_i} = 0.$

Here, $£ = C + \lambda\left(-\sum_{i=1}^{N} P_i + P_R + P_\ell\right)$ and with $\dfrac{\partial £}{\partial P_i} = 0$ we can write

$$\frac{dC_i(P_i)}{dP_i} - \lambda\left[1 - \frac{\partial P_\ell}{\partial P_i}\right] = 0$$

or, $$\frac{dC_i(P_i)}{dP_i} + \lambda\frac{\partial P_\ell}{\partial P_i} = \lambda \tag{11.28}$$

Expression (11.28) represents the *modified criterion of economic operation for the thermal plants with transmission losses considered.*

Equation (11.28) may also be written as

$$\frac{dC_i(P_i)}{dP_i}\left[\frac{1}{1 - \dfrac{\partial P_\ell}{\partial P_i}}\right] = \lambda \tag{11.29}$$

or, $$\frac{dC_i(P_i)}{dP_i} \times PF_i = \lambda \tag{11.29a}$$

where, $$PF_i = \left[1 - \frac{\partial P_\ell}{\partial P_i}\right]^{-1}$$

PF_i is commonly known as *Penalty Factor* in this model of economic operation of thermal units with network losses considered. Equation (11.29 or 11.29a) is termed as *exact co-ordination equations.* In this model $\frac{dC_i\,(P_i)}{dP_i}$ is the incremental cost while $\frac{\partial P_\ell}{\partial P_i}$ is known as *incremental transmission loss.* λ stands here for the incremental cost of the received power in unit of currency/MWhr and hence during economic operation of plants with losses being considered, we have

$$\lambda = \frac{\text{Incremental fuel cost}}{1 - \text{Incremental transmission loss}}$$

For N numbers of thermal plants the co-ordination equations are listed below:

$$\left.\begin{aligned} \frac{dC_1\,(P_1)}{dP_1} - \lambda\left[1 - \frac{\partial P_\ell}{\partial P_1}\right] &= 0 \\ \frac{dC_2\,(P_2)}{dP_2} - \lambda\left[1 - \frac{\partial P_\ell}{\partial P_2}\right] &= 0 \\ &\vdots \\ \frac{dC_N\,(P_N)}{dP_N} - \lambda\left[1 - \frac{\partial P_\ell}{\partial P_N}\right] &= 0 \end{aligned}\right\} \tag{11.30}$$

while the constraint equation is given by

$$-\sum_{i=1}^{N} P_i + P_R + P_\ell = 0 \tag{11.31}$$

N number of optimum dispatch equations along with the constraint suffice for determining $(N + 1)$ unknowns P_1, P_2, ... P_N and λ.

If the numerical value of the partial derivative of the line loss with respect to each generator output $\left(\frac{\partial P_\ell}{\partial P_i}\right)$ is known, the generator output power may be adjusted to satisfy the following equation

$$\frac{dC_i\,(P_i)}{dP_i}\left[\frac{1}{1 - \frac{\partial P_\ell}{\partial P_i}}\right] = \lambda \qquad \text{where, } i - 1, 2, \ldots, N.$$

The optimum economy is thus achieved when the product of the incremental fuel cost times the penalty factor is the same for all plants.

i.e.
$$\frac{dC_i\,(P_i)}{dP_i} PF_1 = \frac{dC_2\,(P_2)}{dP_2} PF_2 = \cdots = \frac{dC_N(P_N)}{dP_N} PF_N = \lambda \tag{11.32}$$

EXAMPLE 11.4 In a two area power system, the bus voltages are kept fixed and the line power flow is a function of line angle. Power loss is a function of generation of area B only. Determine the incremental transmission losses for the two area power system.

Solution: It is evident that, $P_\ell = f\left(P_{g_B}\right)$

This also suggests that the incremental transmission loss for grid A will be zero and the incremental transmission loss of the line will be governed by the grid B only. Thus,

$$(ITL)_A = 0$$

$$(ITL)_B = \frac{\partial P_\ell}{\partial P_{g_B}} = \frac{\partial P_\ell}{\partial \delta} \times \frac{\partial \delta}{\partial P_{g_B}}$$

Economic operation being dictated by the criterion

$$\lambda = \frac{(IFC)_A}{1-(ITL)_A} = \frac{(IFC)_B}{1-(ITL)_B}$$

∴ We have for economic operation,

$$\lambda = \frac{(IFC)_A}{1-0} = \frac{(IFC)_B}{1-(ITL)_B} \quad \text{(Ans.)}$$

EXAMPLE 11.5 There are two generating units in a thermal power plant having cost characteristics

$$C_1 = 200 + 2.5P_1 + 0.01P_1^2$$

$$C_2 = 400 + 2P_2 + 0.02P_2^2$$

where P_1 and P_2 are in MW.

The system is operating with economic dispatch with P_1 = 325 MW and P_2 = 175 MW and $\frac{\partial P_\ell}{\partial P_2} = 0.2$. Find the penalty factor of plant 1 and plant 2.

Solution: With economic dispatch,

$$PF_1 \frac{dC_1}{dP_1} = PF_2 \frac{dC_2}{dP_2}, \text{ where, } \frac{dC_1}{dP_1} = 2.5 + 0.02P_1 \text{ and } \frac{dC_2}{dP_2} = 2 + 0.04P_2$$

Now,

$$PF_2 = \left[\frac{1}{1-\frac{\partial P_\ell}{\partial P_2}}\right] = \frac{1}{1-0.2} = 1.25 \text{ (Ans.)}$$

It is given that $P_1 = 325$ MW and $P_2 = 175$ MW

Substituting the above values of P_1, P_2 and PF_2

$$PF_1 (2.5 + 0.02 \times 325) = 1.25 (2 + 0.04 \times 175)$$

∴

$$PF_1 = 1.25 \text{ (Ans.)}$$

EXAMPLE 11.6 A system consists of two plants connected by a transmission line as shown in Figure E11.2.

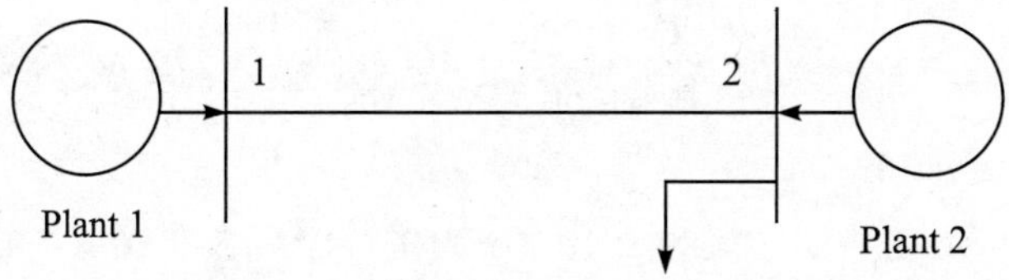

Figure E11.2

The load is at plant 2. If a load of 175 kW is transmitted from plant 1 to the load, there is a loss of 15 MW. Determine the generation schedule and the load demand if the cost of the received power is ₹ 90/MWh. Assume that the incremental costs of the two plants as:

$$\frac{dC_1}{dP_1} = 0.30\ P_1 + 50\ \text{₹/MWh}$$

And $$\frac{dC_2}{dP_2} = 0.40\ P_2 + 30\ \text{₹/MWh}$$

Solve the problem using (a) Co-ordination equations and (b) Penalty factor method.

Solution: Since the load is at the bus of plant 2, therefore the line losses will not be affected by variation of P_2.

Thus $$B_0 = B_{21} = 0 \text{ and } B_{22} = 0$$

For a 2-plant system, $$P_\ell = B_{11}P_1^2 + 2B_{12}P_1P_2 + B_{22}P_2^2$$

Here, $$P_\ell = B_{11}P_1^2$$

When $$P_1 = 175\ \text{MW},\ P_\ell = 15\ \text{MW}$$

We have, $$15 = (175)^2\ B_{11}$$

i.e., $$B_{11} = \frac{15}{(175)^2} = (4.89)\ 10^{-4}\ \text{MW}^{-1}$$

$$P_\ell = (4.89)\ 10^{-4}\ P_1^{\ 2}\ \text{MW}$$

(a) Using co-ordination equation:
It is also given, $\lambda = 90$ ₹/MWh.
The co-ordination equation for plant 1,

$$\frac{dC_1}{dP_1} + \lambda \frac{\partial P_{\ell oss}}{\partial P_1} = \lambda$$

or, $$0.30P_1 + 50 + 90 \times 9.78 \times 10^{-4}P_1 = 90$$

∴ $$P_1 = 103.087 \simeq 103.09\ \text{MW (Ans.)}$$

Similarly, the co-ordination equation for plant 2 is given by:

$$\frac{dC_2}{dP_2} + \lambda \frac{\partial P_{\ell oss}}{\partial P_2} = \lambda$$

or, $0.40P_2 + 30 + 90 \times 0 = 90$

$\therefore$ $P_2 = 150$ MW (Ans.)

The line losses are given by

$$P_\ell = 4.89 \times 10^{-4} P_1^2 = 4.89 \times 10^{-4} \times (103.09)^2 = 5.1966 \approx 5.197 \text{ MW}$$

$\therefore$ Total load $P_R = P_1 + P_2 - P_\ell = (103.09 + 150 - 5.197)$ MW $= 247.893$ MW (Ans.)

(b) Using of penalty factor:

The penalty factor for plant 1 is given by

$$PF_1 = \left[\frac{1}{1 - \dfrac{\partial P_{\ell oss}}{\partial P_1}}\right] = \frac{1}{1 - 9.78 \times 10^{-4} P_1}$$

For optimal dispatch, $PF_1 \dfrac{dC_1}{dP_1} = \lambda$

or,

$$\frac{1}{1 - 9.78 \times 10^{-4} P_1} \times (0.30 P_1 + 50) = 90$$

$\therefore$ $P_1 = 103.09$ MW (Ans.)

Similarly, the penalty factor for plant 2, $PF_2 = \left[\dfrac{1}{1 - \dfrac{\partial P_{\ell oss}}{\partial P_2}}\right] = \dfrac{1}{1 - 0} = 1$

For optimal dispatch, $PF_2 \times \dfrac{dC_2}{dP_2} = \lambda$

or, $1 \times (0.40 P_2 + 30) = 90$

$\therefore$ $P_2 = 150$ MW (Ans.)

It is seen that both the methods give identical results. This is due to the fact that the penalty factor has been derived from the co-ordination equations.

11.7 HYDROTHERMAL SCHEDULING

So far we were limited to the problem of optimal scheduling of thermal power plants. In case of thermal generating units, it was a static optimization problem where scheduling was obtained at any instant without reference to operation of any other time. In a power system combining both thermal and hydro units connected at various load centers through transmission network, the problem of scheduling is more complicated. For scheduling of hydrothermal systems, the aim is to generate and transmit power to load centers by an optimal mix of various types of plants at minimum fuel cost. On one hand, the hydel resources are limited and depending on climatic condition and on the other hand the thermal generation involves high operating cost. Meeting these two aspects is of primary importance in *hydrothermal scheduling*. Hence optimum level must be maintained so that compromise is made between limited resources in conjunction with saving in fuel and operating cost of thermal generation. Hydel resources vary in their amount depending on location and season. Specially in rainy seasons, resources are abundant while in winter and summer, water reservoirs dry up.

In conventional hydro thermal scheduling problem the approach incorporates two concepts– the *long range* and the *short range hydrothermal scheduling*. The long range optimal generation scheduling interval is normally for one year duration, while for short range scheduling the interval may be for a single day or one week. Cost optimized hydro thermal scheduling can be achieved by assuming the water heads of the hydel plants constant and converting the incremental water (i.e., fuel) rate characteristics into incremental fuel cost curves by multiplying it with cost of water per cubic meter and applying the conventional technique of minimizing the cost function.

The long range scheduling problem is normally solved with determination of water outflow from the reservoir in each subinterval and involves mainly the scheduling of water release. The corresponding thermal generation is required to meet the load demand over each subinterval utilizing the entire quantity of water available for power generation during the total Interval subjected to fulfillment of specified constraints. In the short-range hydrothermal scheduling problem usually the optimization interval is a day or a week. This optimization interval is normally divided into subintervals for scheduling purposes. In this optimization process the load, water inflows and unit availabilities are assumed to be known and a set of starting conditions (i.e., reservoir water storage) are given. The short range schedule can be obtained with minimization of the cost function and with presence of specified constraints. In both the approach, the load demand for the entire interval is assumed to be constant.

The concept of *cost of water* being important in a hydrothermal mix system, it is employed in both long range and short range hydrothermal scheduling solution. The incremental water value C_w is defined as

$$C_w = \frac{\text{Incremant in cost}}{\text{Change in generation}} = \frac{\Delta C_i}{\Delta P_i};$$

The following section illustrates the analytical aspects of long and short range hydrothermal scheduling problem in a mix system (i.e., hydro-thermal system). In long range system, the share of hydropower is high but due to some reasons, unable to cater the entire load. Thermal generation is operated for an optimum period to supplement the deficiency in hydro generation. N.B. In an only hydro system, the incremental water value is modified as

$$C_{w\,hydro} = C_w \times \frac{\eta}{\eta_{\max}}$$

where η is the efficiency of the plant as a function of output and $\eta_{\max}$ is the maximum value of η.

11.7.1 Long Range Hydro-thermal Scheduling

Figure 11.8 represents a simple hydrothermal power generation system where the capacity of hydro plant is larger.

The hydro-electric generation is normally capable enough to cater the entire load demand but in this analysis it is assumed that the energy output available from the hydro plant is insufficient to cater the total load during the interval t. Here, P_h and P_{th} represent the operating power capacities (in MW) of hydro-plant and thermal plant respectively and N_t represent number of hours in t-th interval.

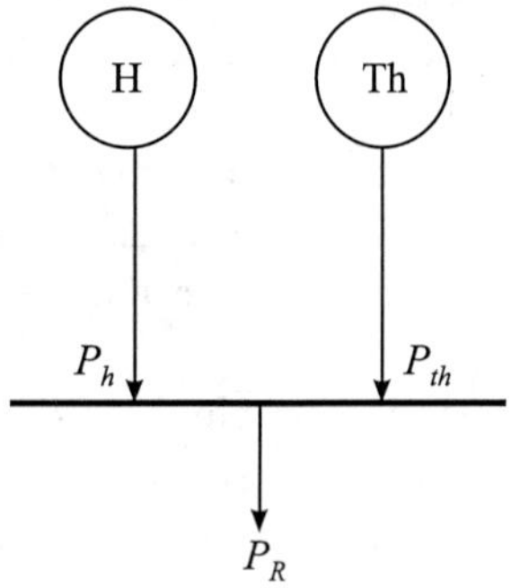

Figure 11.8 A simple hydrothermal system.

The deficit in hydro generation to cater the load is supplied by the thermal generation. All losses have been neglected.

Since $\sum_{t=1}^{t_{max}} P_h N_t$ represents the total hydro-energy and $\sum_{t=1}^{t_{max}} P_R N_t$ the total load energy, in this case obviously.

$$\sum_{t=1}^{t} P_h N_t < \sum_{t=1}^{t_{max}} P_R N_\tau$$

Also, $\sum_{t=1}^{t_{max}} N_t = T$ where T represents total operational time interval for which the generation scheduling is to be done. (Figure 11.9).

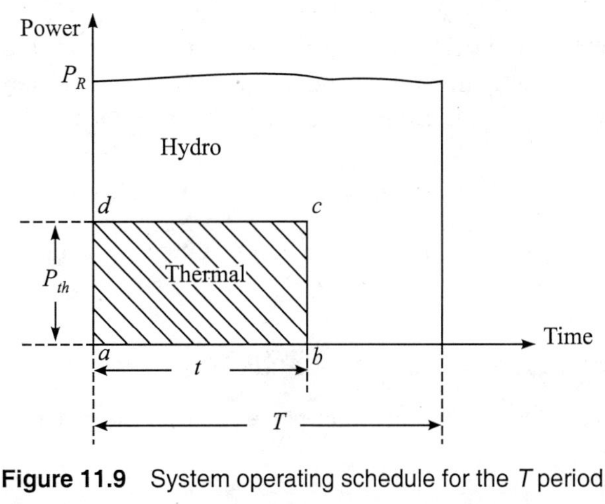

Figure 11.9 System operating schedule for the T period.

Since, (Load energy – Hydro energy) = Thermal energy (ε_{th})

Hence,
$$\sum_{t=1}^{t_{\max}} P_R N_t - \sum_{t=1}^{t_{\max}} P_h N_t = \varepsilon_{th} \tag{11.33}$$

However, the thermal energy must be equal to the thermal generators' electrical output. Hence, the constraint optimization can be done by utilizing the thermal power constraint equation which can be represented as

$$-\sum_{t=1}^{N} P_{th} N_t + \varepsilon_{th} = \Phi \tag{11.34}$$

Here N_{th} represents the number of periods the steam plant remains in service. Since, the thermal plant is not required to run for the entire time interval T, hence,

$\sum_{t=1}^{N} N_t = T_{th}$, ($T_{th} < T$) where T_{th} is the total operational time interval of thermal generation (hydro generation is also available during this time).

The cost function of thermal power plant is represented by

$$C_c = \sum_{t=1}^{N} C(P_{th})\, N_t \tag{11.35}$$

For economic operation, application of Lagrangian technique results

$$£ = \sum_{t=1}^{N} C(P_{th})\, N_t + \lambda\left(\varepsilon_{th} - \sum_{t=1}^{N} P_{th} N_t\right) \tag{11.36}$$

λ being the Lagrangian multiplier. In order to have economic operation, the first derivative of £ with respect to the independent variable is set to zero.

This gives,
$$\frac{dC(P_{th})}{dP_{th}} - \lambda = 0$$

or,
$$\lambda = \frac{dC(P_{th})}{dP_{th}} \tag{11.37}$$

Thus the in a hydro thermal system, economic operation is achieved when the thermal unit is loaded as per the economic loading criterion given by equation (11.37). It simply means that for the period 'T_{th}' when both the generations are available, the thermal plant must be operated with its economic loading criterion.

Let us now assume that P_{th_0} be the optimum value of the power generated by the thermal unit following the economic operating criterion. Here, the thermal energy is given by

$$\varepsilon_{th} = \sum_{t=1}^{N} P_{th_0} N_t \tag{11.38}$$

or,
$$\varepsilon_{th} = P_{th_0} \sum_{\tau=1}^{N} N_t$$

$$\therefore \qquad T_{th} = \frac{\varepsilon_{th}}{P_{th_0}} \tag{11.39}$$

The cost function of thermal plant during its optimum operation being represented by $C_c = \sum_{t=1}^{N} C(P_{th_0}) N_{th}$ and $C(P_{th}) = [\gamma + \beta P_{th_0} + \alpha P_{th}^2]$, the cost function can further be represented as $C_c = C(P_{th_0}) \sum_{t=1}^{N} N_{th} = C(P_{th}) T_{th} = \left(\gamma + \beta P_{th_0} + \alpha P_{th}^2\right) \frac{\varepsilon_{th}}{P_{th_0}}$

$$\frac{dC_c}{dP_{th_0}} = -\frac{\gamma \varepsilon_{th}}{P_{th_0}^2} + \alpha \varepsilon_{th} \tag{11.40}$$

For minimum cost function, $\qquad \frac{dC_c}{dP_{th}} = 0$

This gives from equation (11.40), $\qquad P_{th_0} = \sqrt{\frac{\gamma}{\alpha}} \qquad (11.41)$

Equation (11.39) represents the time for which the thermal generation is to be operated while equation (11.41) represents the amount of optimal thermal generation.

11.7.2 Short Range Hydrothermal Scheduling

In this analysis, the same simple hydrothermal model has been considered similar to the preceding article but it is assumed that the hydro plant is not capable enough to supply the entire load demand during the period of scheduling and that the thermal plant operates in conjunction with the hydro plant during the entire interval (T) of the short term operation. It has further has been assumed that the load is remaining constant during the assumed period of scheduling, the starting water volume is v (i.e., $v_s = v_t$ at $t = 0$) and the ending volume is v_e (i.e., $v_e = v_t$ at $t = t_{\max}$) t represents the subinterval of the scheduling period while N_t represents number of such sub intervals for entire period of scheduling (T).

Here, $\qquad \sum_{t=1}^{t_{\max}} N_t = T$

Also, flow limits are maintained such that $W_{\min} \le W_t \le W_{\max}$ and discharge remains fixed for any particular interval while water discharge is a function of hydro power generation. Since there is no spillage ($S_j = 0$), we would have a maximum total volume of water outflow throughout the period of scheduling interval (T).

The constraint equation (with *no spillage*) for the total water discharge W_{tot} is given by

$$\left[\sum_{t=1}^{t_{\max}} N_t W_t \left(P_{h_t}\right) - W_{tot}\right] = \Psi \tag{11.42}$$

where, W_t is the inflow of water during t interval.

The electrical constraint equation is given by $(P_R - P_{h_t} - P_{th_t}) = \Phi$.

Application of Lagrangian technique (Minimize £ = $C + \lambda\,\Phi + \gamma_h\,\Psi$) here yields

$$£ = \sum_{\tau=1}^{\tau_{max}} \left[N_t C\left(P_{th_t}\right) + \lambda\left(P_R - P_{h_t} - P_{th_t}\right)\right] + \gamma_h \left[\sum_{t=1}^{t_{max}} N_t W_t\left(P_{h_t}\right) - W_{tot}\right] \qquad (11.43)$$

where γ_h is a constant that converts an incremental water into an equivalent incremental cost (IC).

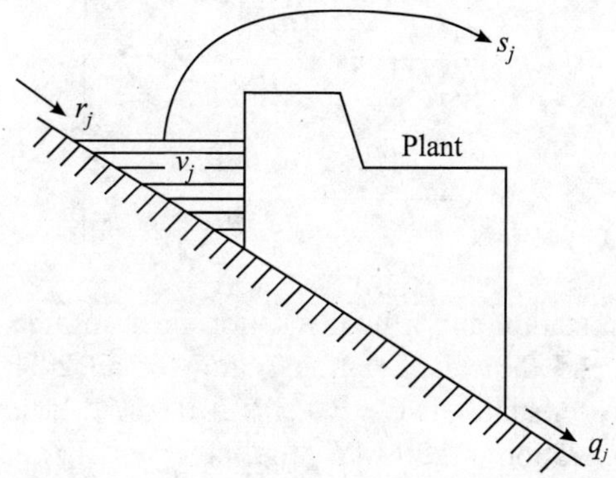

Figure 11.10 Scheduling parameters in the block diagram of hydel plant.

For any sub interval (t) scheduling of the thermal plant for economic operation is possible, if

$$\frac{\partial £}{\partial P_{th_t}} = 0$$

i.e., $\quad N_t \dfrac{dC(P_{th})}{dP_{th}} - \lambda = 0$ [obtained from equation (11.43), for $\dfrac{\partial £}{\partial P_{th_t}} = 0$]

or, $$\lambda = N_t \frac{dC(P_{th})}{dP_{th}} \qquad (11.44)$$

Similarly scheduling of the hydro plant is possible when $\dfrac{\partial £}{\partial P_{h_t}}$ = 0; i.e., from equation (11.43) we have

$$\gamma_h N_t \frac{dW_t(P_h)}{dP_{h_t}} - \lambda = 0$$

or, $$\lambda = \gamma_h N_t \frac{dW_t(P_h)}{dP_{h_t}} \qquad (11.45)$$

Hence in a hydrothermal system, a *short term* hydrothermal coordination can be obtained when

$$\lambda = N_t \frac{dC(P_{th})}{dP_{th}} = \gamma_h N_t \frac{dC(P_h)}{dP_h} \qquad (11.46)$$

With network losses (P_l) considered, the co-ordination equations for short range scheduling problem can be modified with application of Lagrangian technique in the following equation where the electrical constraint equation has been modified to include the network losses. Thus we have,

$$£ = \sum_{\tau=1}^{\tau_{max}} N_t C(P_{th_t}) + \lambda(P_R + P_l - P_{h_t} - P_{th_t} + \gamma_h \left[\sum_{t=1}^{t_{max}} N_t W_t(P_{h_t}) - W_{tot} \right] \quad (11.47)$$

Thus for any specific interval $t = t_i$ the co-ordination equation of a hydrothermal system, with network losses considered, are then given by

$$\left.\begin{aligned} N_{t_i} \frac{dC\left(P_{th_{t_i}}\right)}{dP_{th_{t_i}}} + \lambda_{t_i} \frac{\partial P_{\ell_{t_i}}}{\partial P_{th_{t_i}}} = \lambda_{t_i} \\ \gamma_h N_{t_i} \frac{dW_{t_i}\left(P_{h_{t_i}}\right)}{dP_{h_{t_i}}} + \lambda_{t_i} \frac{\partial P_{\ell_{t_i}}}{\partial P_{h_{t_i}}} = \lambda_{t_i} \end{aligned}\right\} \quad (11.48)$$

EXAMPLE 11.7 A steam station and a hydro station feed an area jointly. The hydro station is run for 14 hours daily and the steam station is run for all 24 hours. The production cost characteristics for steam power station is $C = 3 + 5\,P_s + 0.05\,P_s^2$. If the load on the steam station, when both plants are in operation is 220 MW, the incremental water rate of hydro-plant is

$$\frac{dW}{dP_h} = 40 + 0.03\,P_h \text{ m}^3\text{/MW sec.}$$

The total quantity of water used during the 14 hours is 400 million cubic meters. Determine the load on hydro-plant and cost of water use. Assume that load on hydro plant is constant for the 14 hour period.

Solution: $C = 3 + 5P_s + 0.05P_s^2$

$$\frac{dC}{dP_s} = 5 + 0.1P_s = \lambda$$

Since, $P_s = 220$ MW, $\lambda = 5 + 0.1 \times 220 = 27$.

Total quantity of water used during the 14 hours $= [P_h(40 + 0.03P_h) \times 3600 \times 14]$ m^3

$\therefore$ $$P_h(40 + 0.03\,P_h) \times 3600 \times 14 = 400 \times 10^6$$

or, $$P_h = 175.3514 \text{ MW (Ans.)}$$

If the cost of water is r, then,

$$r\,(40 + 0.03 \times 175.3514) = \lambda = 27$$

$\therefore$ $$r = 1.66 \times 10^{-4} \text{ ₹/m}^3 \text{ (Ans.)}$$

EXAMPLE 11.8 In a two plant system, one of the plants is thermal and the other is hydro electric. The fuel cost characteristics of thermal unit is $C = 264 + 22P + 0.0125P^2$ ₹/MWh and that of the water rate of hydro unit is $(11P + 0.0075P^2)$ m^3/sec. The units together supply a peak load of 700 MW for 12 hours in a day. During off load period of 12 hrs, the units together supplies 270 MW. Neglecting transmission losses, determine the economic scheduling of the plants both during peak and off-peak period, provided that water used is 459.47×10^6 m^3 during 24 hour period.

Solution:

For thermal unit, incremental fuel cost = $\frac{d}{dP}(264 + 22P + 0.0125P^2) = 22 + 0.025P$

and for hydro unit, incremental water rate = $\frac{d}{dP}(11P + 0.0075P^2) = 11 + 0.015P$

For off-peak period, let P_{so} and P_{ho} be the load supplied by the thermal and hydro unit respectively, while P_{sp} and P_{hp} are the corresponding loads shared during peak period.

Given, $\quad P_{so} + P_{ho} = 270$ and $P_{sp} + P_{hp} = 700$

The optimum conditions are:

For off-peak period, $22 + 0.025P_{so} = R(11 + 0.015P_{ho}) = \lambda_o$

where R = Factor which converts incremental water rate into incremental cost in ₹per hr./m^3 per sec.

Substituting the values of P_{so}, we get,

$$\lambda_o = 22 + 0.025(270 - P_{ho}) = R(11 + 0.015P_{ho}) \quad (1)$$

For peak period, $22 + 0.025\,P_{sp} = R(11 + 0.015P_{hp}) = \lambda_p$

or, $$\lambda_p = 22 + 0.025(700 - P_{hp}) = R(11 + 0.015P_{hp}) \quad (2)$$

The water used over 24 hours is given by

$$[(11 \times P_{ho} + 0.0075P_{ho}^2) + (11 \times P_{hp} + 0.0075P_{hp}^2)] \times 3000 \times 12 = 459.47 \times 10^6 \quad (3)$$

By solving these three equations (1), (2) and (3), we will get P_{ho}, P_{hp}, and R.

Here, $\quad P_{ho} = 109.9$ MW; $P_{hp} = 302.8$ MW; $R = 2.052$ ₹ per hr/m^3 per sec.
$P_{so} = 160.1$ MW; $P_{sp} = 397.2$ MW;
$\lambda_o = 26$; $\lambda_p = 32$ (Ans.)

11.7.3 Hydro-thermal Scheduling of Pumped Storage Plants

In its basic form, a *pumped storage plant* consists of an upper water reservoir along with a water reservoir in the lower level with arrangements for pumping water from lower reservoir to the upper reservoir during light loading condition and discharging water from the upper reservoir to the lower one through a turbine during peak loading condition. The utility thus saves the fuel cost by catering the peak load with hydro-energy during daytime while earning a greater tariff as it purchases power at low cost while pumping back the water into the upper reservoir when the load demand is low during highest time. Usually the pumped storage plant is operated on daily schedule. To minimize the total capital cost, the same electrical machine is used as a generator during peak load period and also as a motor for pumping of water during lean period. Reversible pump turbines are used to act as a generator during generating interval or as a pump during pumping interval. The pumped storage plant is operated until added pumping cost exceeds the savings in thermal cost.

Let us assume that the hydro-thermal system is composed of a pumped storage plant and a thermal plant. It is assumed here that t_1 and t_2 are the intervals of generation and pumping respectively while t is the scheduling interval. Also, we assume that P_R is the net load demand while P_{th} and P_h are the respective thermal and hydro generation. Suffixes 1 and 2 represent the respective values of assumed variables during the period of generation and the period of pumping. Here, v_t is water volume at the pumped storage reservoir at the end of t interval, W_t is the water discharge when in generating mode, ω_t water discharge when in pumping mode, and q_t is the water inflow.

The object is to minimize the hourly cost (N_t = 1) of thermal generation for the entire day ($t_{\max}$ = 24) subject to different constraints. C being the cost function, the total fuel cost for the entire day is given by

$$C = \sum_{\tau=1}^{t_{\max}} N_t C_t \left(P_{th_{(t)}}\right) = \sum_{t=1}^{24} C_t \left(P_{th_{(t)}}\right) \tag{11.49}$$

In order to have the economic operation for the pumped storage based hydrothermal system, the generation and pumping intervals are considered separately. For generation interval (period t_1), the electrical constraint equation is given by

The electric constraint equation is given by,

$$P_{R(t_1)} + P_{\ell oss(t_1)} - P_{th(t_1)} - P_{h(t_1)} = \Phi_1 \tag{11.50}$$

while the hydraulic constraint equation is given by

$$v_{t_1} - v_{t_1\text{-}1} - q_{t_1} + w_{t_1} = \Psi_1 \tag{11.51}$$

The Lagrangian function (for the generation interval) is then given by

$$£_{t1} = C_{(t_1)} + \lambda_{t_1}\left(P_{R(t_1)} + P_{\ell oss(t_1)} - P_{th(t_1)} - P_{h(t_1)}\right) + \gamma_{t_1}\left(v_{t_1} - v_{t_1\text{-}1} - q_{t_1} + W_{t_1}\right) \tag{11.52}$$

To minimize the cost function, the partial derivative of $£_{t1}$ must be zero for each independent variable. This gives

$$-\lambda_{t_1}\left(1 - \frac{\partial P_{\ell oss(t_1)}}{\partial P_{th(t_1)}}\right) + \frac{dC_{(t_1)}}{dP_{th(t_1)}} = 0 \tag{11.53a}$$

and

$$-\lambda_{t_1}\left(1 - \frac{\partial P_{\ell oss(t_1)}}{\partial P_{h(t_1)}}\right) + \gamma_{t_1}\frac{dW_{(t_1)}}{dP_{h(t_1)}} = 0 \tag{11.53b}$$

In the next step, the minimization of the objective function is done for the pumping interval (period t_2). The electrical constraint equation is now given by

$$P_{R(t_2)} + P_{\ell oss(t_2)} - P_{th(t_2)} + P_{h(t_2)} = \Phi_2 \tag{11.54}$$

[$P_{h(t_2)}$ is now demand on the system as the plant is in pumping mode].

The hydraulic constraint equation is obtained as

$$v_{t_2} - v_{t_2\text{-}1} - q_{t_2} - w_{t_2} = 0 \tag{11.55}$$

Application of Lagrangian technique during the pumping interval yields the following equation,

$$£_{t_2} = C_{(t_2)} + \lambda_{t_2}\left(P_{R(t_2)} + P_{\ell oss(t_2)} - P_{th(t_2)} + P_{h(t_2)}\right) + \gamma_{t_2}\left(v_{t_2} - v_{t_2\text{-}1} - q_{t_2} - w_{t_2}\right) \tag{11.56}$$

As in the case of generation interval, the condition for minimum cost function can now be obtained by setting partial derivative of $£_{t_2}$ to zero. This gives,

$$-\lambda_{t_2}\left(1-\frac{\partial P_{\ell oss(t_2)}}{\partial P_{th(t_2)}}\right)+\frac{dC_{(t_2)}}{dP_{th(t_2)}}=0 \quad (11.57a)$$

and

$$\lambda_{(t_2)}\left(1+\frac{\partial P_{\ell oss(t_2)}}{\partial P_{h(t_2)}}\right)-\gamma_{t_2}\frac{dw_{(t_2)}}{dP_{h(t_2)}}=0 \quad (11.57b)$$

In order to have economic operation of the pumped storage hydrothermal system, the total Lagrangian function £ is obtained for the for the entire period of scheduling and is given by,

$£ = £_{t_1} + £_{t_2}$ where $£_{t_1}$ is the optimized Lagrangian function during generation interval and $£_{t_2}$ is the optimized Lagrangian function during the pumping interval.

EXERCISES

1. What do you understand by economic operation of energy generating units?
2. Draw input-output characteristics of thermal power plant and hydel power plant and explain them.
3. What are incremental fuel rate and incremental fuel cost?
4. Discuss about the constraints in economic operation of a power system?
5. Develop the condition of economic operation of ideal thermal power plants.

   (**Hints:** Ideal thermal power plants do not contribute in transmission line loss.)
6. Determine the optimum distribution of a load of 250 MW to be shared by the three generators in a power plant. Assume operating cost (in ₹/hr) of the three generators as,

$$C_1 = 150 + 0.16P_1 + 0.002P_1^2$$

$$C_2 = 100 + 0.2P_2 + 0.001P_2^2$$

$$C_3 = 200 + 0.1P_1 + 0.001P_3^2$$

   [**Ans.** $P_1 = 48$ MW, $P_2 = 76$ MW, $P_3 = 126$ MW]
7. Assume that the rating of the three generators as 75, 100 and 120 MW respectively. Determine the optimum distribution of the load of 250 MW to be shared by the three generators in a power plant considering these ratings.

   [**Ans.** $P_1 = 50$, $P_2 = 80$ MW, $P_3 = 120$ MW]
8. Distinguish between the terms 'economic operation' and 'unit commitment'.
9. Derive 'Transmission Loss Formulae'.
10. Derive the condition of economic operation of thermal power plants considering transmission loss.

11. In a three-plant system the cost functions (in ₹/hr) are given by

$$C_1 = 200 + 6.8P_1 + 0.002P_1^2$$

$$C_2 = 150 + 7.2P_2 + 0.003P_2^2$$

while the transmission loss is expressed as, $P_\ell = 0.00002P_1^2 + 0.00004P_2^2$ MW. Find the economic dispatch schedule assuming total load to be 400 MW.

[**Ans.** P_1= 280.15 MW, P_2 = 122.01 MW, P_ℓ = 2.16 MW]

12. Write a short note on long range hydrothermal scheduling.

13. Discuss about short range hydrothermal scheduling.

14. Discuss about hydrothermal scheduling in presence of pumped storage plants.

CHAPTER 12

EHV AC and HVDC Power Transmission

12.1 INTRODUCTION

With tremendous growth in power transmission, there is an essential need for installation, commissioning and operation of extra high voltage AC (EHV AC) and high voltage DC (HVDC) transmission. It is advantageous to have high voltage transmission lines because of lower loss and facility to have bulk power transmitted over long distances. HVDC transmission system acts as a highly reliable asynchronous link between two grids and when operated in conjunction with EHV AC lines, the overall stability is improved. The following articles discuss the performance, operation and modelling of EHV AC and DC system.

12.2 NECESSITY FOR EHV AC POWER TRANSMISSION

The power transmission lines are usually operated in the EHV (extra high voltage) or UHV (ultra high voltage) range due to the following reasons:

(a) The transmission efficiency increases as the transmission voltage increases for a given amount of power to be transmitted over a given distance.

(b) The p.u. resistance drop and volume of conductor material decreases in the EHV/UHV voltage lines.

(c) The power transmitting capacity of the transmission line, at the EHV/UHV range, increases tremendously as the transmission capacity is proportional to the square of the operating voltages. Though the costs of the tower, insulation and terminal equipments increase but in general these costs are proportional to the voltages rather than square of voltage. As a consequence, overall capital cost of transmission decreases as the voltage increases. Transmission of large blocks of power is thus economically feasible at EHV/UHV.

(d) Economic considerations have led to the construction of large capacity power stations and hence the corresponding transmission lines need to evacuate bulk power. As bulk power transmission is economically feasible at EHV/UHV range hence with these higher voltage ranges of operation, it has become possible to achieve economies in generation also.
(e) The installation cost of the transmission line per km decreases with an increase in voltage level.
(f) It is a conventional practice to express the load carrying capability of lines in terms of *surge impedance loading* (SIL). At this loading the reactive power consumed by the inductive part of the line (i.e. ωLI^2) is theoretically equal to the reactive power charging (i.e., ωCV^2) in the line. For a transmission line, the surge impedance loading, P_{SIL}, is proportional to the square of voltage hence, with an increase in voltage level SIL itself increases which indicates that power transfer increases ($P_{SIL} = V^2/Z_0$; where, Z_0 is surge impedance of the line).
(g) It is practically not possible to have interconnections of the power systems on a large scale without EHV transmission.

12.3 PROBLEMS WITH EHV AC POWER TRANSMISSION

The major problems coupled with EHV transmission are:

(a) *Corona loss and radio interference with communication lines:* Corona loss is more acute in EHV transmission as system voltage level is a governing factor of corona loss. It results in interference with communication lines which create noise in communication. Under bad weather condition, corona loss further increases. To overcome this problem, bundled conductors are used which involve extra cost along with difficulty in large wind and snow loading of such heavy conductors.
(b) *Line supports:* EHV lines have large mechanical loadings on towers due to bundle conductors, large air and ground clearances, considerable dynamic forces due to broken conductors, etc. Transmission line towers with fabricated steel members have mostly been used which involves high cost.
(c) *Wind load:* The terrific wind pressure during storms and cyclones causes incalculable damage to towers, conductors and insulators. The transmission line must, therefore, be made more wind-resistant. If, however, the amount of metal in the tower is increased the cost goes up appreciably.
(d) *Erection difficulties:* The erection of EHV lines poses a wide range of problems requiring extremely high standard of workmanship and good transportation system as the supporting structures are to be transported across every corner of country. It also becomes necessary to evolve and employ new erection and conductor handling techniques.
(e) *Insulation requirements:* The transmission line requires insulation to protect against surges which may be due to internal causes (switching operations) or due to external causes (atmospheric disturbances like lightning). The protection against lightning may be obtained by the ground wire and rapid autoreclosing circuit breakers alongwith

careful insulation coordination. It is possible to control and minimise switching over-voltages with proper design of relay-breaker systems.

(f) *Power station and sub-station equipment:* The design and manufacture of power station equipment, i.e. turbogenerators, three phase two-winding transformers, circuit breaker, etc have also undergone some revolutionary changes in terms of rating and hazardless heavy duty. Oil-less type breakers have also been developed which are more efficient, are light in weight, are of higher voltage rating and which possess high speed.

12.4 ASPECTS OF EHV AC POWER TRANSMISSION

The limitations of three-phase EHV AC power transmission system are discussed below:

(i) ***Line Loading:*** It is a well-known fact that at surge impedance loading (SIL) of transmission line the lagging volt-amperes of line is equal to the leading volt-amperes of that line which keeps sending end and receiving end voltages equal. Surge impedance loading (also called *natural loading*) is given by

$$SIL = \frac{V^2}{Z_0} \tag{12.1}$$

where V = system voltage

Z_0 = characteristic impedance of transmission line

For lossless line $Z_0 = \sqrt{L/C}$

And for lossy line $Z_0 = \sqrt{Z/Y}$

L = line inductance

C = line capacitance

Z = series impedance of line

Y = shunt admittance of line

However, in practice a line is not always operated at its natural load. If the line is transferring power more than its natural loading then there will be reactive power deficit in the system and as a consequence there will be a voltage drop along the line but if the power transmitted is lower than its natural load then there will be reactive power surplus in the system and as a result there will be a rise of voltage along the line. In other words, if the line is transferring power at lagging power factor then there will be voltage drop along the line but if the power is transmitted with leading power factor there will be a rise of voltage along the line.

(ii) *Stability Considerations:* The power transmitted by a transmission line with negligible resistance and shunt leakage is given by

$$P = \left|\frac{V_s V_r}{X}\right| \sin\delta$$

where $|V_s|$ and $|V_r|$ are voltage magnitudes at sending end and receiving end respectively, $|X|$ is the line reactance and δ is the power angle in degree.

Though theoretically, at δ = 90° the maximum steady state transfer of power takes place but in actual practice δ is not allowed to vary more than 20° to 30° on full load so that the machines should not loose synchronism during transient disturbances. This consideration imposes restriction on the transmission distance in ac systems. The length determined accordingly, however, is too small for many transmission systems. In EHV systems, for lines exceeding 100 km in length additional and costly equipments like series/shunt capacitors and shunt reactors must be incorporated.

(iii) ***Current Carrying Capacity of Conductors:*** In the case of overhead line conductors the maximum power that can be transmitted is never dependent on thermal considerations but for underground cables both thermal current limit and charging current are important. Due to these considerations a cable is not operated at a loading as high as its natural load. Also for given length the cost of a cable is 10 to 20 times that of an overhead line and it puts serious limitation on the use of cables in high voltage schemes.

(iv) ***Ferranti Effect:*** If the line is loaded with leading reactive volt-amperes (i.e. capacitive load) or whenever the load on the generator is thrown off suddenly, there will be a rise of voltage at the receiving end with respect to sending end. This rise is of the order of 1.5% for 160 km, 13% for 500 km and 100 per cent for 960 km length of overhead transmission line. Shunt reactors in the load end are generally used to control this voltage rise. As sending end voltage remains constant, receiving end voltage then increases and this effect is popularly known as *Ferranti effect.*

(v) ***Power Transfer between Two Areas through an EHV Line:*** The maximum power transfer (i.e. *static transmission capacity*) of a EHV power link is given by, $P_{\max} = \dfrac{|V_s||V_r|}{|X|}$. This static transmission capacity can be enhanced by increasing the voltage level or by decreasing the line reactance. The incremental enhancement in transmitted power, say ΔP_{12} between two areas 1 and 2 can be caused by a small increment in the power angle δ (i.e. $\Delta\delta$). This incremental enhancement in transmitted power is known as "electrical stiffness" of the transmission link and the stiffness coefficient (T_{12}) is given by

$$T_{12} = \frac{\Delta P_{12}}{\Delta\delta} = \frac{dP_{12}}{d\delta} = \frac{d}{d\delta}(P_{\max}\sin\delta) = P_{\max}\cos\delta \tag{12.2}$$

As δ tends to 90°, the maximum power angle for steady state operation, stiffness goes to zero. Thus with higher power angle, T_{12}, stiffness coefficient reduces.

Let us now consider that area 1 is connected to area 2 by n number of transmission lines.

If ΔP_1 = total incremental power exported from area 1 to area 2

ΔP_{line} = incremental outflow of interconnected line power

$$\Delta P_1 = \sum_{line=1}^{n} \Delta P_{line}$$ [i.e. the summation of all incremental outflow of tie lines]

However,
$$\Delta P_{line} = T_{line}(\Delta\delta_{12}) = T_{line}(\Delta\delta_1 - \Delta\delta_2)$$

Therefore $$\Delta P_1 = \sum_{line=1}^{n} T_{line}(\Delta\delta_1 - \Delta\delta_2) = P_{\max_{12}} \cos(\delta_1^0 - \delta_2^0)(\Delta\delta_1 - \Delta\delta_2) \tag{12.3}$$

where, superscript "0" indicates initial condition of power angles. Here, $P_{\max 12}$ is the static transmission capacity of the tie line, δ_1^0 and δ_2^0 are the nominal angle of the voltage at both ends and $\Delta\delta_1$ and $\Delta\delta_2$ are the incremental changes in power angles of the respective sides.

(vi) ***Voltage Drop in an Interconnector:*** The voltage drop in an interconnector is given by

$$\Delta V = ZI^* = (R + jX)(I_p + jI_q)^* = (RI_p + XI_q) + j(XI_p - RI_q)$$

where I_p and I_q are real and reactive component of current respectively.

If $$\Delta V = V_i + jV_q$$

where V_i and V_q are *inphase* and *quadrature* component of voltage drops respectively.
In EHV transmission lines, $X >> R$.

$$\therefore \quad \Delta V = V_i + jV_q \cong XI_q + jXI_p \tag{12.4}$$

Hence it can be concluded that the quadrature voltage drop is governed by real power whereas inphase voltage drop is dependent on reactive power. Therefore EHV AC transmission system is very much reactive power sensitive.

(vii) ***Series Reactive Loss in EHV Transmission Line:*** Let S be the MVA power being fed at a load bus by a radial EHV line of per phase reactance X ohms (line resistance R is neglected as in practical EHV lines, R/X ratio is too small). Let Q_{se} be the series reactive loss in MVAR and is given by

$$Q_{se} = I^2X$$

where I be the magnitude line current and for constant MVA load, $I = \dfrac{S}{V}$

$$\therefore \quad Q_{se} = \left(\frac{S}{V}\right)^2 X$$

Thus the change in series reactive loss for change in voltage is given by

$$\therefore \quad \frac{dQ_{se}}{dV} = -\frac{2}{V^3}S^2X$$

Assume S and X are constant,

$$\therefore \quad \left|\frac{dQ_{se}}{dV}\right| \propto \left|\frac{1}{V^3}\right| \tag{12.5}$$

i.e. the *series reactive loss* increases sharply with the reciprocal of voltage cubed.

(viii) ***Line and Cable Charging in EHV Lines:*** Assuming Q_c to be the charging MVAR of the lines and cables in EHV system,

$$Q_c = \omega CV^2$$

where, C is the distributed line/cable capacitance and V is the system voltage.

Thus, the change in charging due to voltage change is given by

$$\left|\frac{dQ_c}{dV}\right| = |2\omega CV| \tag{12.5}$$

Hence it is seen that the cable and line charging increments are directly proportional to voltage.

[*Implication:* It has been revealed that the rate of series reactive loss is proportional to the reciprocal of voltage cubed while the charging rate is proportional to system voltage. Thus, in case of drop of voltage, reactive power loss increases while the charging capability decreases. This leads to further drop in voltage. This effect is cumulative and if no additional reactive power source is installed, this cumulative drop in voltage may lead to voltage instability].

(ix) ***Loadability of Transmission Lines:*** *Loadability* of a transmission line is defined as the optimum power transfer capability of a transmission line under specific operating criteria like prescribed voltage drop, etc.

The power transfer P is given by

$$P = \left|\frac{EV}{X}\right| \sin\delta$$

Now, $X = xL$; ℓ = the line inductance per unit length and L = line length

$$\therefore \qquad X = xL = \omega L\sqrt{\frac{\ell}{c}}\sqrt{\ell c}$$

where c is the distributed line capacitance per unit length.

Now, $\sqrt{\frac{\ell}{c}} = Z_0$ = (the surge impedance of transmission line)

and $\omega\sqrt{\ell c}$ = the phase shift constant (β) of transmission line

Therefore $X = Z_0 \beta L = Z_0 \theta = Z_0 \sin\theta$ where $\beta L = \theta \approx \sin\theta$ [$\because$ θ is very small]

Now,
$$P = \frac{EV}{X}\sin\delta = \frac{EV}{Z_0}\frac{\sin\delta}{\sin\theta}$$

Assuming
$$E = V = 1.0 \text{ p.u.,} \quad P = \frac{1}{Z_0}\frac{\sin\delta}{\sin\theta}$$

Thus it is seen that the power transfer is inversely proportional to the line length (θ) [θ is the line length expressed in radians]. The power transfer decreases with an increase in line length.

As the surge impedance loading is given by, $\text{SIL} = \frac{V^2}{Z_0} = \frac{1}{Z_0}$ with V = 1.0 p.u.

Thus
$$\frac{P}{\text{SIL}} = \frac{\sin\delta}{\sin\theta} \tag{12.6}$$

The power transfer P with respect to SIL has been shown in Figure 12.1. Under a specific set of operating criterion of voltage and power angle limit, this power (P) is denoted as loadability.

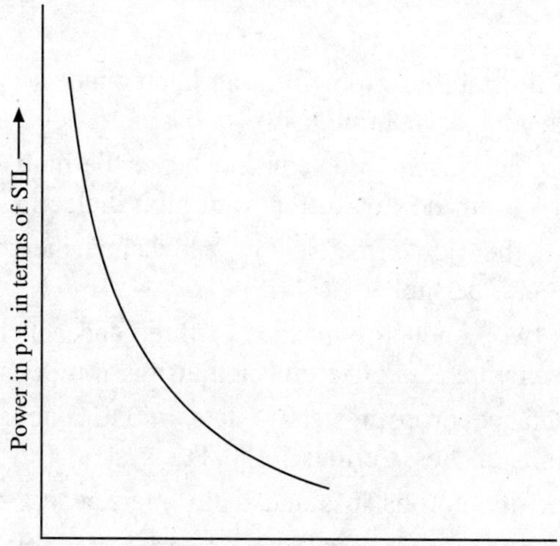

Figure 12.1 Loadability of a transmission line.

(x) ***Insulation Co-ordination:*** Insulation strength of transmission lines must be decided after a careful analysis of the different kinds of overvoltages. In general, there are three types of transient overvoltages, namely switching overvoltages, overvoltages due to lightning and temporary overvoltages.

(xi) ***Mechanical Vibrations and Oscillations:*** As the number of subconductors increases in bundle conductor transmission, hence there may be an enormous effect of vibrations of conductors. To encounter this problem, mechanical design of the system, i.e. tower dimension, phase spacings, subconductor spacing, conductor height, etc. are to be modified. These modifications affect the electrical design, particularly the voltage gradient.

Other problems include *Aeolian Vibration* (i.e. when a conductor is under tension and a steady wind blows across it, small vorties are formed which detach themselves creating minute vertical force on the conductor. The resulting vibrational force causes fatigue of conductor and supporting structures. This is known as Aeolian Vibration).

Another problem is *Galloping* (it is a very high amplitude, low-frequency conductor vibration occurring mainly in the areas of flat terrain under freezing rain and icing of conductors). Galloping can be controlled by applying weights at different locations of the span of the line.

(xii) ***Environmental and biological aspects:*** Recent researches in this field show that EHV and UHV lines generate electrostatic and electromagnetic fields. These fields can induce current and voltage in animals, human beings and birds. However, fortunately these effects are minimum and within tolerable limits.

Transmission lines also produce audible noise in line conductors in EHV range. However, these noises do not exceed the satisfactory noise level up to 500 kV.

12.5 HVDC TRANSMISSION—ADVANTAGES AND DISADVANTAGES

The high voltage DC (HVDC) transmission has been developed to overcome the associated problems of EHV AC transmission of bulk power. The advantages of HVDC transmission are:

(i) HVDC links has no stability problem in an interconnected power system as they can operate independently (asynchronously).

(ii) There is no charging current problem and hence Ferranti effect does not arise and cables in HVDC system do not suffer from high dielectric loss.

(iii) In HVDC system the skin effect is low and hence the current density in the DC transmission line can be higher.

(iv) DC line requires two conductors instead of three and hence costs on line conductor, insulator and towers are less. The power transmission per conductor is higher.

(v) There is no reactance drop at steady state in DC line and therefore the voltage regulation problems are less serious in HVDC system.

(vi) In HVDC systems the corona loss and radio interference are lesser.

(vii) The line construction is simpler and ground return is possible by using single conductor in HVDC transmission. Line loss is also lesser in DC transmission.

(viii) The interconnections of AC grids through AC tie lines increase the fault level whereas the interconnection of AC grids with DC links does not increases fault level to that extent.

(ix) HVAC lines and HVDC links can be used in parallel to improve the transient stability of the network.

(x) There is easy reversibility and controllability of power flow through an HVDC link.

(xi) During fault in the DC line, grid/gate control of the converter can drastically reduce the fault current.

(xii) Frequently long distance AC transmission lines require shunt compensation. Such compensation is not required in DC lines.

(xiii) These systems are economical for transmitting bulk power over long distances, say above 550 km.

In spite of the above advantages, HVDC transmission system has a few disadvantages also. These are:

(i) The overall cost increases, particularly for line lengths below 500 km, due to additional requirement of converters, filters, reactive power compensators.

(ii) Circuit breaking in multi-terminal DC systems is difficult and expensive.

(iii) Converter stations require considerable reactive power.

(iv) Harmonic generation and subsequent filtration pose problems in DC systems.

(v) Overload capacity of HVDC converters is low.

(vi) Voltage transformation is to be provided on the AC side only.

(vii) Reactive power required by the load is to be supplied locally as no reactive power can be transmitted.

(viii) Maintenance of insulators in HVDC transmission lines is more.

The present trend is to use HVDC lines for longer transmission lengths associated with EHV AC transmission lines. Terminal convertors utilize thyristors of ratings 50 KV and 100A for conversion of AC to DC at the sending end and DC to AC at the receiving end.

12.6 HVDC SYSTEM CONFIGURATION

The HVDC transmission lines are known as HVDC link. Different types of HVDC links used in HVDC transmission are discussed below:

1. **Monopolar link:** A monopolar link is shown in Figure 12.2. In this configuration, only one conductor is used (usually negative) and the return path is ground or water (sea). Negative polarity is generally utilised as the transmission conductor because it has a comparatively lesser radio interference. Instead of ground return a metallic return may be used in situations where earth resistivity is too high or possible interference with underground/underwater metallic structure is objectionable. The blocks in Figure 12.2 represent thyristorised rectifier and inverter stations.

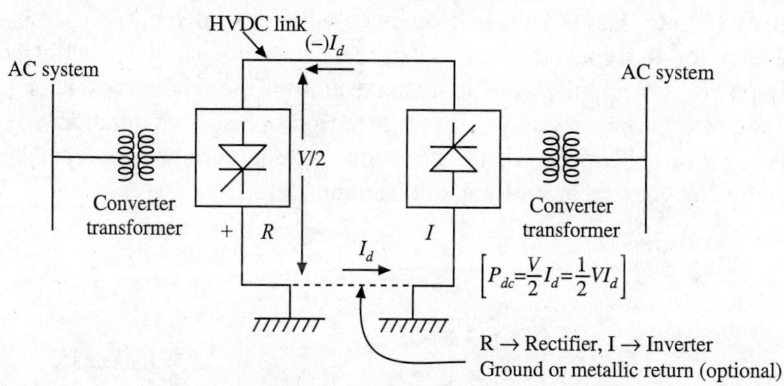

Figure 12.2 A monopolar link.

2. **Bipolar link:** This configuration has two conductors, one positive and another negative. In each terminal, two converters of equal rated voltages are connected in series, neutral points being grounded (Figure 12.3). Two poles can operate independently when both the neutrals are grounded. When the currents in the two conductors are equal, the ground current is zero. In case of a fault in one conductor, the remaining conductor alongwith ground return can supply 50% of the rated load. The rated voltage of a bipolar link is usually expressed as ±V volts.

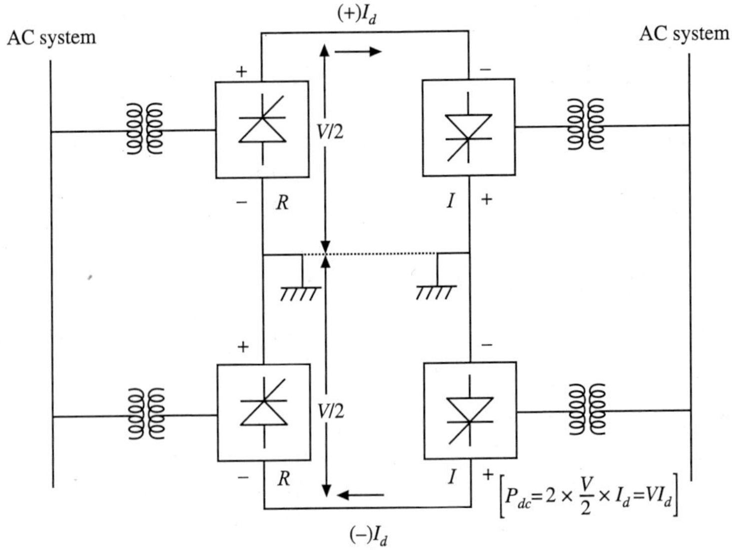

Figure 12.3 A bipolar link.

3. **Homopolar Link:** It has two conductors but having the same polarity (usually negative) (Figure 12.4). This link operates with ground return. In case of a fault in one conductor the converter can be connected such that the healthy conductor can supply power. For bipolar scheme, reconnection of the whole conductor in one pole is more complicated and usually not feasible. Homopolar configuration offers advantage in this regard in situations where continuous ground current is acceptable. The ground connection may be replaced by a suitable conductor.

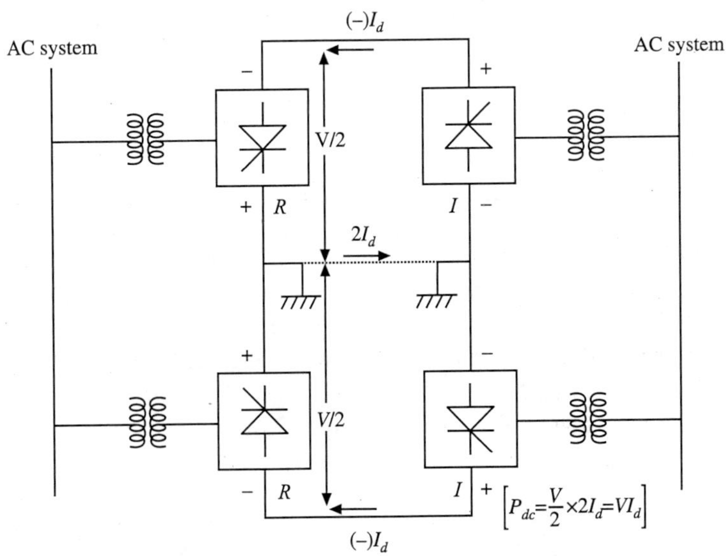

Figure 12.4 A homopolar link.

12.7 PRINCIPAL COMPONENTS OF HVDC SYSTEM

Any HVDC link is associated with a number of components as discussed below. A bipolar scheme with all of its main components is shown in Figure 12.5.

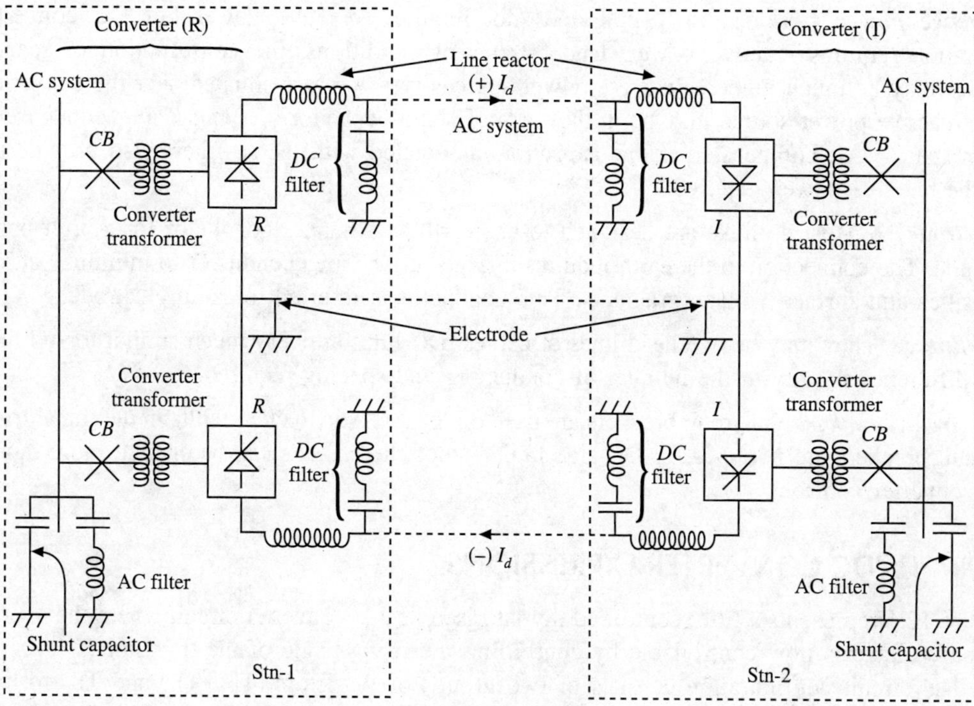

Figure 12.5 A bipolar link with all of its components.

Converters: Converters may be a rectifier (to convert AC to DC) or an inverter (to convert DC to AC). Converter consists of thyristor bridges and transformers with tap changers. The thyristor bridges consist of high-voltage semiconductor valves connected in a 6-pulse or 12-pulse arrangement. The converter transformers provide isolation between AC source and AC input to the converter and therefore the DC system will be able to establish its own reference to ground (usually by grounding the positive or negative end of the valve converter). The converter transformer is also capable of altering the voltage level of any side of the AC–DC system.

Smoothing reactors: These are large reactors having high value of inductance connected in series with each pole of each converter station. They serve the following purposes:

(i) Prevent commutation failure in inverters.
(ii) Reduce harmonic voltage and current in the DC line.
(iii) Restrict fault current in the rectifier during short circuit on DC line.
(iv) Decrease the incidence of commutation failures in the inverter during dips in the alternating voltage.
(v) Smoothen the ripple in the direct current output significantly to prevent the current from becoming discontinuous at light loads.

Harmonic filters: Converters generate harmonic voltages and currents due to switching action. Filters are used on both AC and DC side to eliminate these harmonics as these harmonics may cause overheating of capacitors and nearby generator and interference with telecommunication systems.

Reactive power supplies: DC link does not require reactive power but DC converters operation requires reactive power. Under transient conditions, the consumption of reactive power may be much higher. Reactive power sources are therefore situated near the converters. The reactive power source may be in the form of shunt capacitor or synchronous condenser or even static VAR compensators. The capacitors associated with the ac filters also provide part of the reactive power required.

Electrodes: Most dc links are designed to use earth as a neutral conductor for a brief period of time. The connection to the earth requires a large surface area conductor to minimize current densities and surface voltage gradients. This conductor is referred to as an electrode.

DC lines: They may be overhead lines or cables. DC lines are very much similar to AC lines, the difference is only in the number of conductors and spacing required.

AC circuit breakers: Circuit breakers are used on the AC side to clear faults in the transformer. Circuit breakers are not used for DC side faults, since these faults can be cleared more rapidly by converter control.

12.8 HVDC CONVERTER EXPRESSIONS

Figure 12.6 represents a fully-controlled thyristorised bridge converter circuit where the average output voltage is now controllable by controlling the firing angle of the thyristors.

The circuit configuration consists of two groups of thyristors (Th_1, Th_3 and Th_5 forming the positive group while thyristors Th_4, Th_6, and Th_2 form the negative group). The positive group of thyristors are turned on when the supply voltages are positive and negative group of thyristors are turned on when the supply voltages are negative.

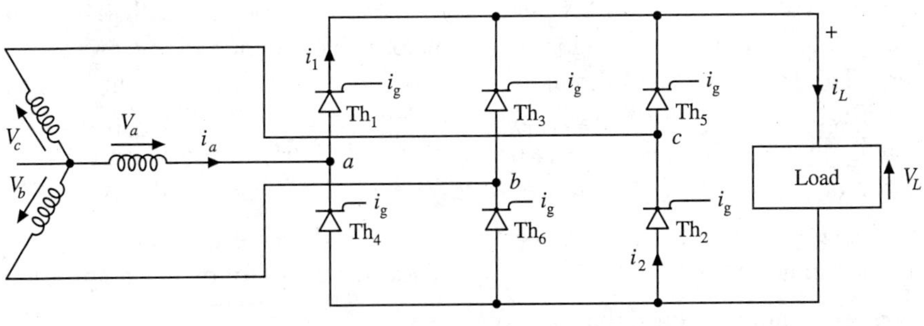

Figure 12.6 The connection diagram of three phase fully controlled bridge. ($i_1 \equiv i_2$ when Th_1 and Th_2 conduct)

In order to start the circuit functioning, two thyristors (one from each group) must be fired at the same time otherwise there will be no flow of current. Let us assume that initially voltage of phase "*a*" appears which is most positive. Thyristor Th_1 is triggered at an angle α (firing angle or triggering angle). Since phase "*b*" is most negative at this time, the return path of current should be through phase "b" and thus, Th_6 must be fired simultaneously with Th_1 and at the same firing angle. Similarly, when phase "b" will be having most positive, Th_3 is to be fired at an angle α and allow current to return through the most negative phase "c", Th_2 must be fired simultaneously. In the next sequence, phase "c" attains the most positive value while phase "a" attains the most negative value. This time Th_5 and Th_4 must be fired simultaneously at an angle α.

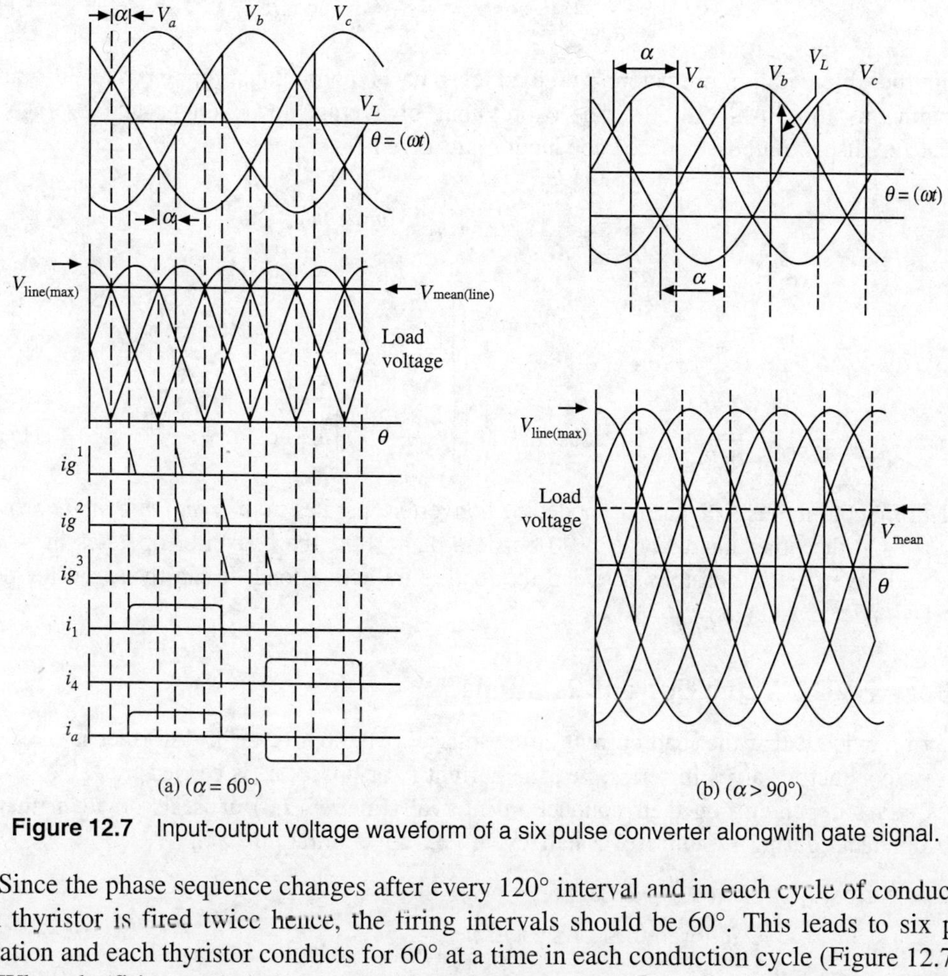

Figure 12.7 Input-output voltage waveform of a six pulse converter alongwith gate signal.

Since the phase sequence changes after every 120° interval and in each cycle of conduction, each thyristor is fired twice hence, the firing intervals should be 60°. This leads to six pulse operation and each thyristor conducts for 60° at a time in each conduction cycle (Figure 12.7(a)).

When the firing angle is large, with the load voltage having negative periods, it is difficult to visualise the load voltage waveform and here the six line voltages $(V_{ab}, V_{ac}, V_{bc}, V_{ba}, V_{ca}, V_{cb})$ give the direct waveform of the load voltage as shown in Figure 12.7(b).

12.8.1 Expression of Average Load Voltage and Current

$$V_{dc} = \frac{3}{\pi} \int_{\alpha+\pi/3}^{\alpha+2\pi/3} \sqrt{3} V_{m_{ph}} \sin \omega t \, d(\omega t)$$

where, $V_{m_{ph}}$ = peak value of phase voltage.

$$\therefore \qquad V_{dc} = \frac{3\sqrt{3}}{\pi} V_{m_{ph}} \cos\alpha = \frac{3\sqrt{2}}{\pi} V_{L\,(rms)} \cos\alpha \tag{12.7}$$

where, $V_{L(rms)}$ = rms value of line voltage.

and
$$I_{dc} = \frac{3\sqrt{3}}{\pi R} V_{m_{ph}} \cos\alpha = \frac{3\sqrt{2}}{\pi R} V_{L\,(rms)} \cos\alpha \tag{12.8}$$

In the three phase operation, the load current is continuous for $0° \le \alpha \le 30°$ and is discontinuous for $30° \le \alpha \le 150°$. The expressions of average load voltage and average load current for discontinuous mode of conduction are given by

$$V_{dc} = \frac{3}{\pi} \int_{\alpha+\pi/3}^{\pi} \sqrt{3} V_{m_{ph}} \sin\omega t \, d(\omega t)$$

$$\therefore \qquad V_{dc} = \frac{3\sqrt{3}}{\pi} V_{m_{ph}} \left[1 + \cos\left(\alpha + \frac{\pi}{3}\right)\right] = \frac{3\sqrt{2}}{\pi} V_L \left[1 + \cos\left(\alpha + \frac{\pi}{3}\right)\right] \tag{12.9}$$

while
$$I_{dc} = \frac{3\sqrt{3}}{\pi R} V_{m_{ph}} \left[1 + \cos\left(\alpha + \frac{\pi}{3}\right)\right] = \frac{3\sqrt{2}}{\pi R} V_L \left[1 + \cos\left(\alpha + \frac{\pi}{3}\right)\right] \tag{12.10}$$

During continuous load current mode, the converter operates in the *rectifying mode* provided $\alpha < 90°$. On the other hand, for $\alpha > 90$ but less than 180°, the converter operates in *inverter mode* and inversion is possible provided the load is an active load. It can then transfer power from DC side to AC side.

12.8.2 HVDC Valve Current Relations

It may be evident that the load current at any phase of the converter transformer is carried by two semiconductor valves in series, i.e., the current i_1 of phase '*a*' is carried by valve Th_1 and Th_6 is series depending on their conduction interval (Figure 12.6). it is also evident that eah valve conducts during twi-third of a half cycle i.e., up tp duration $2\pi/3$.

$\therefore$ valve current $i_1(t) = 0$ for $0 \le \omega t \le \pi/6$

$= I_d$ for $\pi/6 \le \omega t \le 5\pi/6$

$= 0$ for $5\pi/6 \le \omega t \le \pi$

$= 0$ for $\pi \le \omega t \le 2\pi$

The expression for average valve current can be obtained as follows:

$$I_{1(av)} = \frac{1}{2\pi} \int_0^{2\pi} i_1 \, d(\omega t) = \frac{1}{2\pi} \left[0 \times \frac{\pi}{6} + I_d \times \frac{2\pi}{3} + 0 \times \frac{\pi}{6}\right] = \frac{I_d}{3} \tag{12.11}$$

The expression for rms valve current can be obtained as follows:

$$I_{1\,\text{rms}}^2 = \frac{1}{2\pi}\int_0^{2\pi} i_1^2\, d(\omega t) = \frac{1}{2\pi}\left[0\times\frac{\pi}{6} + I_d^2\times\frac{2\pi}{3} + 0\times\frac{\pi}{6}\right] = \frac{I_d^2}{3}$$

$$\therefore \quad I_{1\,rms} = \frac{I_d}{\sqrt{3}} \qquad (12.12)$$

12.8.3 Transformer Currents

The current in phase 'a' flows towards load when valve Th_1 conducts (i.e., current is positive)

$$\begin{aligned} i_a(t) &= 0 \text{ for } 0 \le \omega t \le \pi/6 \\ &= I_d \text{ for } \pi/6 \le \omega t \le 5\pi/6 \\ &= 0 \text{ for } 5\pi/6 \le \omega t \le \pi \end{aligned}$$

The current in phase 'a' flows from load side to source side when valve Th_6 conducts (i.e., current is negative)

$$\begin{aligned} i_a(t) &= 0 \text{ for } \pi \le \omega t \le 7\pi/6 \\ &= -\,I_d \text{ for } 7\pi/6 \le \omega t \le 11\pi/6 \\ &= 0 \text{ for } 11\pi/6 \le \omega t \le 2\pi \end{aligned}$$

The rms value of phase current i_a is

$$I_{a\,\text{rms}}^2 = \frac{1}{2\pi}\int_0^{2\pi} i_a^2\, d(\omega t) = \frac{1}{2\pi}\left[0\times\frac{\pi}{6} + I_d^2\times\frac{2\pi}{3} + 0\times\frac{\pi}{6} + 0\times\frac{\pi}{6} + I_d^2\times\frac{2\pi}{3} + 0\times\frac{\pi}{6}\right]$$

$$= \frac{I_d^2}{2\pi}\times\frac{4\pi}{3} = \frac{2I_d^2}{3}$$

$$\therefore \quad I_{a\,\text{rms}}(=I_s) = \sqrt{\frac{2}{3}}\,I_d \qquad (12.13)$$

where I_s is the secondary side line current of the converter transformer.

12.9 SIX PULSE DOUBLE STAR CIRCUIT OF THREE PHASE CONVERTER

The converter transformer being connected in delta in the primary, it is having six phases in the secondary and is connected in star connection (Figure 12.8).

The secondary voltages are thus displaced from each other by 60° (Figure 12.9). Each of the thyristors conduct for 60° during each cycle. The average output DC voltage is given by

$$V_{dc} = \frac{1}{\frac{2\pi}{6}}\int_{\alpha+\pi/3}^{\alpha+2\pi/3} V_{m_{ph}} \sin\omega t\, d(\omega t) = \frac{6V_{m_{ph}}}{2\pi}\left[-\cos\omega t\right]_{\alpha+\pi/3}^{\alpha+2\pi/3}$$

$$\therefore \quad = \frac{3V_{m(ph)}}{\pi}\cos\alpha \qquad (12.14)$$

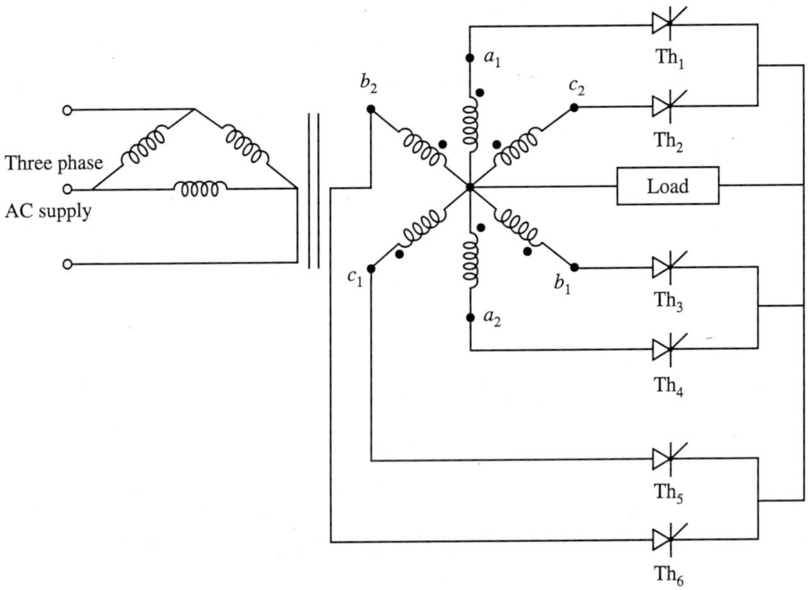

Figure 12.8 Schematic of six pulse double star converter.

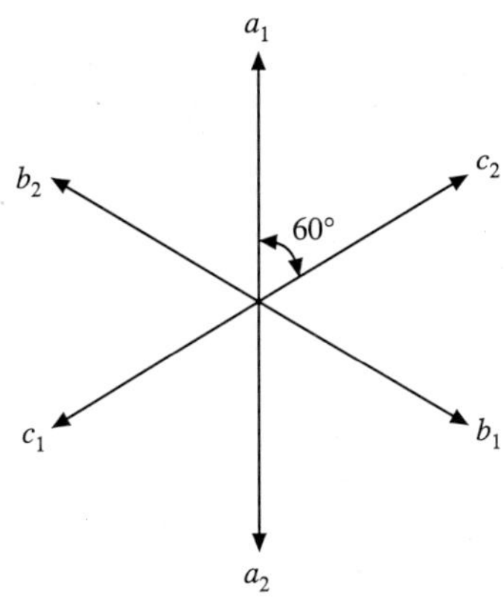

Figure 12.9 Vector relation of secondary voltages.

It may be observed that the output average value of the DC voltage in this case is more than that for the three pulse converter. It may also be observed that for $0° \leq \alpha \leq 90°$, this six pulse converter acts as a rectifier while for $90° \leq \alpha \leq 180°$, the converter acts as an inverter. This converter provides continuous load current for $0° \leq \alpha \leq \pi/3$ and discontinuous load current for $\pi/3 \leq \alpha \leq 2\pi/3$ for resistive loads.

Though the average DC output voltage for this converter is higher than that of three pulse converter, this converter is not very common in industry because of high ratio of RMS to average current in thyristors and in transformer secondary windings. This leads to poor utilisation factor of the transformer. Also, the flow of current in each limb of the primary winding for only 60° interval introduces a high level of harmonic current. In order to remove this difficulty, the six pulse double star circuit converter is used along with an interphase reactor. By interposing the interphase reactor between the DC terminals of the two three pulse circuits, a better utilisation of thyristors and transformer winding can be achieved. Such three pulse group now operates independently of the other and hence each thyristor conducts for 120°. The direct load current is being shared equally between the two groups and hence there is no magnetization of the core of the interphase reactor.

12.10 TWELVE PULSE CIRCUITS

Using two six pulse bridge converters in series, a twelve pulse circuit can be achieved (Figure 12.10). A transformer with two secondary windings with phase displacement of 30° can be used. These types of converters are used in HVDC transmission lines. Converters of higher pulse number can also be designed but are only of theoretical importance.

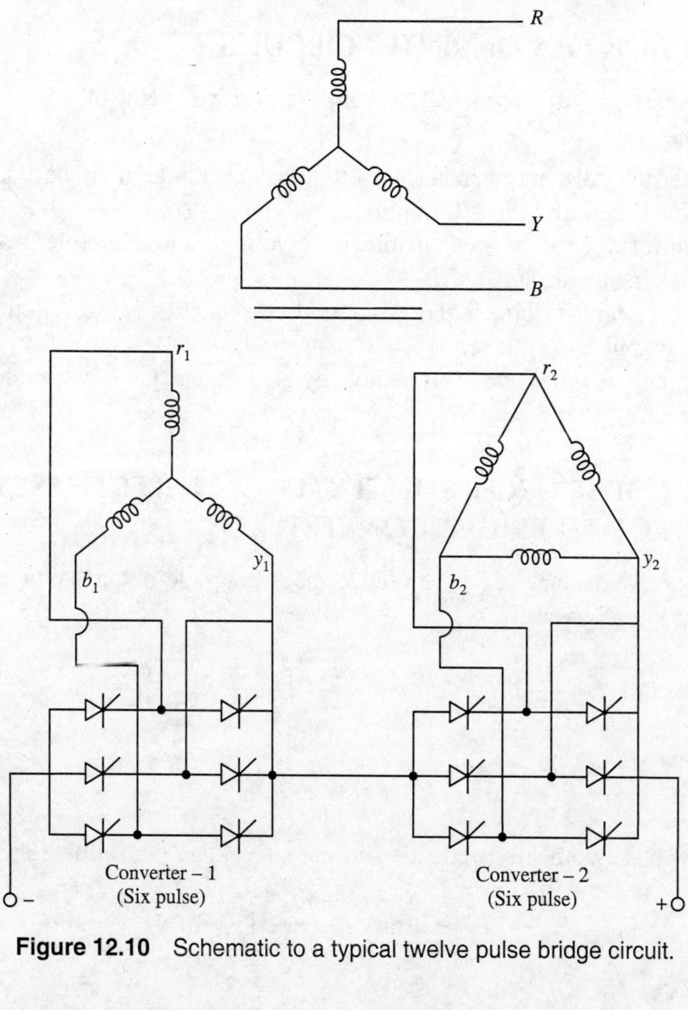

Figure 12.10 Schematic to a typical twelve pulse bridge circuit.

12.11 COMPARISON BETWEEN THREE PULSE AND SIX PULSE OPERATION

1. Ripple factor with full rectifier operation in a three pulse connection is 18% and may go up to a maximum value of 55%, which is very high.
2. Further with a three pulse connection, even second and fourth order harmonics in addition to 5th and 7th and other higher order harmonics are present.
3. Three pulse converters require special types of converter transformers to prevent demagnetisation.

The six pulse converter has less ripples than a three pulse connection. This results in the following advantages compared to three pulse converters:

1. Commutation is made easier.
2. Inductance required in series is considerably reduced.
3. Distortion on the AC side is reduced due to the reduction in lower order harmonics.

12.12 SELECTION OF CONVERTER CIRCUITS

The following points are to be considered from the designer point of view to have a correct choice of a circuit:

1. The higher the pulse number, the lower is the magnitude of ripple voltage and current in the load-circuit and the AC supply.
2. Fully-controlled circuits are capable of inversion and therefore re-generation, half controlled circuits are not.
3. Half-wave circuits require higher voltage thyristors than bridge circuits for producing the same output DC voltage.
4. Bridge circuit requires twice as many thyristors than half-wave circuits to carry the same current.

12.13 EFFECT OF SOURCE INDUCTANCE ON THREE PHASE FULLY CONTROLLED BRIDGE CONVERTER

The effect of source inductance L_s is to reduce the average dc output voltage due to overlap. This reduction (ΔV_R) is given by,

$$\Delta V_R = \frac{1}{\pi/3}\int_0^{\mu} L_s \frac{di}{dt}\, d(\omega t) = \frac{3L_s}{\pi}\int_0^{\mu/\omega} \omega \frac{di}{dt}\, dt = \frac{3\omega L_s}{\pi}\int_0^{I_d} di$$

$$\therefore \qquad \Delta V_R = \frac{3\omega L_s}{\pi} I_d \tag{12.15}$$

[In general, for p-pulse converter, reduction in output voltage due to overlap is obtained as

$$\Delta V_{R_p} = \frac{1}{2\pi/p}\int_0^{\mu} L_s \frac{di}{dt}\, d(\omega t) = \frac{p\omega L_s}{2\pi}\int_0^{\mu/\omega} \frac{di}{dt}\, dt = \frac{p\omega L_s}{2\pi}\int_0^{I_d} di$$

$$\therefore \qquad \Delta V_{R_0} = \frac{p\omega L_s}{2\pi} I_d \qquad (12.16)]$$

However, since output voltage with overlap is equal to internal voltage of three phase fully controlled rectifier minus drop in output voltage due to overlap, hence we have

$$V_{dc} = \frac{3\sqrt{3}}{\pi} V_{m_{ph}} \cos\alpha - \frac{3\omega L_s}{\pi} I_d \qquad (12.17)$$

The output voltage for a three phase fully-controlled converter with source inductance L_s and overlap angle μ, is then given by

$$V_{dc} = \frac{3\sqrt{3}}{\pi} V_{m_{ph}} \cos(\alpha + \mu) + \frac{3\omega L_s}{\pi} I_d = \frac{3\sqrt{2}}{\pi} V_L \cos(\alpha + \mu) + \frac{3\omega L_s}{\pi} I_d \qquad (12.18)$$

In a similar way, for *half-controlled three phase bridge rectifier*, with source inductance, we have

$$V_{dc} = \frac{3\sqrt{3}}{2\pi} V_{m_{ph}} \cos\alpha - \frac{3\omega L_s}{2\pi} I_d$$

With an overlap of μ, we then obtain the output voltage as

$$V_{dc} = \frac{3\sqrt{3}}{2\pi} V_{m_{ph}} \cos(\alpha + \mu) + \frac{3\omega L_s}{2\pi} I_d \qquad (12.19)$$

General expression for p-pulse converter for dc output voltage with source inductance and overlap

With reference to Figure 12.11,

$$V_{dc} = \frac{1}{\frac{2\pi}{p}} \left[\int_{-(\pi/p)+\alpha+\mu}^{(\pi/p)+\alpha} V_{m_{ph}} \cos\theta \, d\theta + \int_{\alpha}^{\alpha+\mu} V_{m_{ph}} \cos\left(\frac{\pi}{p}\right) \cos\theta \, d\theta \right]$$

$$V_{dc} = \frac{1}{\frac{2\pi}{p}} \left[\int_{-(\pi/p)+\alpha+\mu}^{(\pi/p)+\alpha} V_{m_{ph}} \cos\theta \, d\theta + \int_{\alpha}^{\alpha+\mu} V_{m_{ph}} \cos\left(\frac{\pi}{p}\right) \cos\theta \, d\theta \right]$$

$$= \frac{pV_{m_{ph}}}{2\pi} \left[\sin\left(\frac{\pi}{p} + \alpha\right) - \sin\left\{-\frac{\pi}{p} + (\alpha + \mu)\right\} \right.$$

$$\left. + \cos\left(\frac{\pi}{p}\right) \sin(\alpha + \mu) - \cos\left(\frac{\pi}{p}\right) \sin\alpha \right]$$

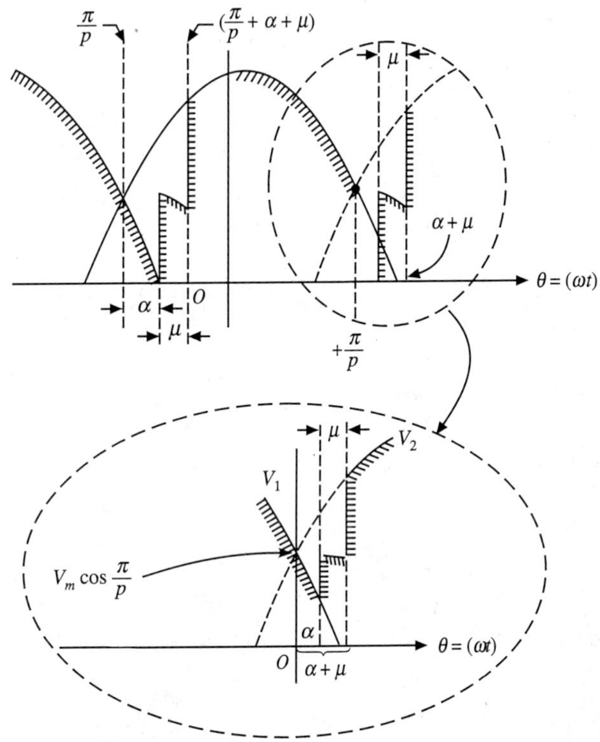

Figure 12.11 General waveform of commutation overlap due to source inductance.

Therefore for a p-pulse controlled converter,

$$V_{dc} = \frac{pV_{m_{ph}}}{2\pi}\sin\left(\frac{\pi}{p}\right)[\cos\alpha + \cos(\alpha+\mu)] \tag{12.20}$$

∴ For three pulse rectifier,

$$V_{dc} = \frac{3V_{m_{ph}}}{2\pi}\sin\left(\frac{\pi}{3}\right)[\cos\alpha + \cos(\alpha+\mu)]$$

i.e.

$$V_{dc} = \frac{3\sqrt{3}V_{m_{ph}}}{4\pi}[\cos\alpha + \cos(\alpha+\mu)] \tag{12.21}$$

Also, V_{dc} can be obtained in terms of commutating reactance X_c as follows:

$$V_{dc} = \frac{1}{2\pi/p\omega}\left[\int_{-(\pi/p\omega)+\alpha/\omega}^{(\pi/p\omega)+\alpha/\omega} V_{m_{ph}} \cos\omega t\, dt - L_c I_d\right]$$

i.e.

$$V_{dc} = \frac{pV_{m_{ph}}}{\pi}\sin\frac{\pi}{p}\cos\alpha - \frac{pX_c}{2\pi}I_d \tag{12.22}$$

Thus for three pulse rectifier,

$$V_{dc} = \frac{3V_{mph}}{\pi} \sin\frac{\pi}{3} \cos\alpha - \frac{3X_c}{2\pi} I_d$$

i.e.
$$V_{dc} = \frac{3\sqrt{3}V_{mph}}{2\pi} \cos\alpha - \frac{3X_c}{2\pi} I_d = \frac{3\sqrt{2}V_L}{2\pi} \cos\alpha - \frac{3X_c}{2\pi} I_d \quad (12.23)$$

12.14 DISCUSSION ABOUT RECTIFIER TRANSFORMER RATINGS

In the rectifier circuits, the supply transformers carry currents which are not sinusoidal. The connection of windings in the secondary requires different limbs of the transformer interconnected at different intervals. The rating of the winding of a transformer is the sum of products of rms voltage across the windings with corresponding rms currents times the number of windings. The primary rating of the winding may even differ from the secondary (for half-wave circuits). In the transformer where the number of secondary winding is more than one, the winding design must ensure mechanical balance in the windings. Leakage flux between the primary and each secondary winding must be the same. The secondary winding should also extend for the same length as the primary in order to ensure mmf balance.

12.14.1 VA Rating of Converter Transformer

$VA = \sqrt{3}V_L I_L \times 10^{-3}$ kVA, where V_L and I_L are the ac side line voltage and current of converter transformer. However, for rectifier

$$V_{dr(0)} = \frac{3\sqrt{2}}{\pi} V_L$$

i.e.,
$$V_L = \frac{\pi}{3\sqrt{2}} V_{dr(0)}$$

and
$$I_s = \sqrt{\frac{2}{3}} I_d$$

$$VA = \sqrt{3} \times \frac{\pi}{3\sqrt{2}} V_{dr(0)} \times \sqrt{\frac{2}{3}} I_d \times 10^{-3} \text{ kVA} = \frac{\pi}{3} V_{dr(0)} I_d \times 10^{-3}$$

∴
$$\text{kVA} = 1.047\, V_{dr(0)} I_d \times 10^{-3} \text{ kVA} \quad (12.24)$$

where $V_{dr(0)}$ is the no load dc output voltage of a six pulse converter with $\alpha = 0$.

I_d = dc output current of a six pulse converter.

Finally, we can write,

∴
$$VA = 1.047 P_d \times 10^{-3} \text{ kVA as } P_d = \text{dc power} = V_d I_d \quad (12.24a)$$

12.14.2 VA Rating of Valve

The VA rating of such semiconductor valve is the product of the average current through it and the PIV (Peak Inverse Voltage) rating of the valve.

$$VA_{(\text{Valve})} = \frac{I_d}{3} \times 1.047 V_{dr(0)} = 0.349 V_{dr(0)} I_d \qquad [\because \text{PIV}_{(\text{Valve})} = 1.047 V_{dr(0)}] \qquad (12.25)$$

∴ VA rating of each valve in the six pulse converter circuit is $0.349 V_{dr(0)} I_d$.

12.15 INVERSION

A fully-controlled converter with an active load can be used as an inverter. With continuous load current, considering the firing delay angle to be extended from a low value, it will give interesting results. Up to a delay of 90° (i.e. $\alpha < 90°$), the converter operates in the rectifying mode at $\alpha = 90°$, the DC mean output voltage is zero (remember the mean output voltage V_{dc} of the converter is $V_{dc} = \frac{E_m}{\pi} \cos\alpha$, where *Em* is the peak input AC voltage for the single phase operation) and delay beyond $\alpha = 90°$ would result in the waveform having a net negative value till it approaches 180°. Waveform at $\alpha = 180°$ will again have net zero value though it is reversed. In case we have an active DC source at the output side of the converter (say a battery), similarly we have an active AC source at the input, for $90° < \alpha < 180°$, the DC mean voltage is negative which means that the current flows in reversed direction for each phase through the converter and the power is being fed back to the AC source from the DC source. This is called inversion.

In order to illustrate the process of inversion, a three phase converter is assumed in Figure 12.12.

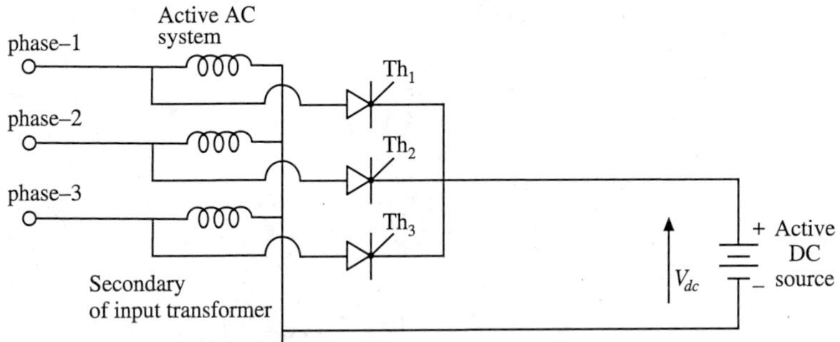

Figure 12.12 Converter operational schematic.

The commutation of thyristor Th_1 (Figure 12.12) is possible and the shifting of load current from Th_1 to Th_2 occurs when the instantaneous voltage of phase 2 is higher than that of phase 1. At $\alpha = 180°$, the instantaneous voltage of both the phases are equal and the net output voltage is zero. The relative voltage between the two phases, after this (i.e. $\alpha = 180°$), reverses. This makes the commutation possible. The operation is thus limited up to $\alpha = 180°$.

In rectification, firing delay angle being α (Figure 12.13), in inverting mode it is very much common to define the firing instant as firing advance angle is β (Figure 12.14). β and α are related by the following relation: $\beta = \pi - \alpha$.

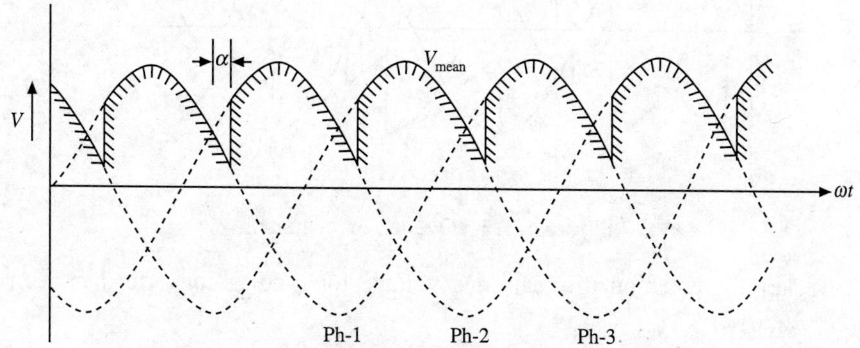

Figure 12.13 Rectification ($\alpha < 90°$) (V_{mean} is +ve).

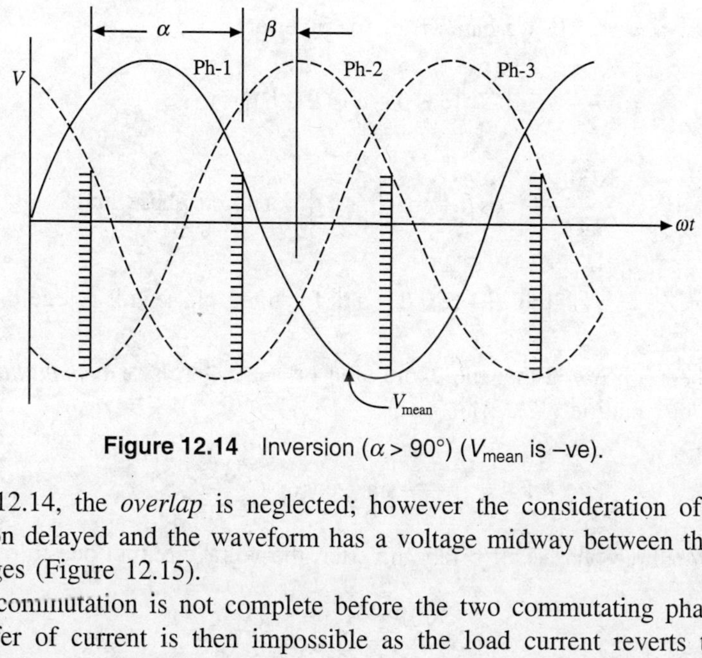

Figure 12.14 Inversion ($\alpha > 90°$) (V_{mean} is –ve).

In Figure 12.14, the *overlap* is neglected; however the consideration of overlap makes the commutation delayed and the waveform has a voltage midway between the incoming and outgoing voltages (Figure 12.15).

When the commutation is not complete before the two commutating phases reach equal voltages, transfer of current is then impossible as the load current reverts to the outgoing thyristor. The *Overlap angle* (μ) being less than the angle of firing advance β, therefore, we can write

$$\delta = \beta - \mu$$

where δ is the angle equivalent to the time available to the outgoing thyristor to regain the blocking state after commutation. δ is the *extinction angle*. In case $\mu = 0$, i.e., no overlap $\delta = \beta = \pi - \alpha$.

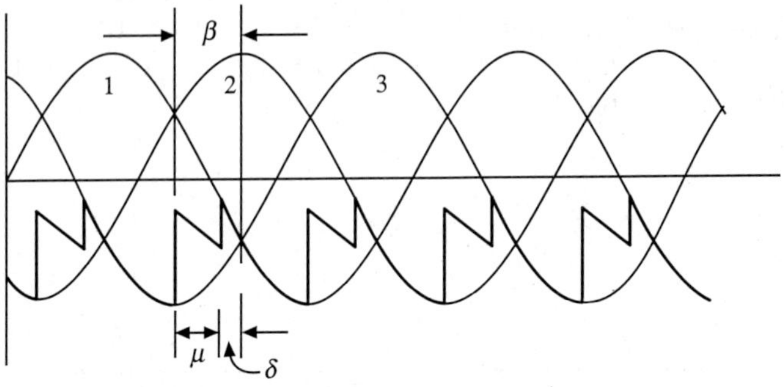

Figure 12.15 Overlap in inversion.

With overlap, the equation of mean load voltage for p-pulse converter has been obtained in equation (12.20) as

$$V_{dc} = \frac{pV_{m_{ph}}}{2\pi}\sin\left(\frac{\pi}{p}\right)[\cos\alpha + \cos(\alpha + \mu)]$$

For inverter since $\alpha = \pi - \beta$, we can write (for inverter),

$$V_{di} = \frac{pV_{m_{ph}}}{2\pi}\sin\left(\frac{\pi}{p}\right)[\cos\beta + \cos(\beta - \mu)]$$

i.e. $$V_{di} = \frac{3\sqrt{3}V_{m_{ph}}}{4\pi}[\cos\beta + \cos(\beta - \mu)] \text{ for three phase half bridge} \tag{12.26}$$

or $$V_{di} = \frac{3V_{m_{ph}}}{4\pi}[\cos\beta + \cos(\beta - \mu)] \text{ for three phase full bridge} \tag{12.27}$$

It has also been observed that the expression of V_{dc} in terms of *commutating reactance*X_c is given by (From equation (12.22))

$$V_{dc} = \frac{pV_{m_{ph}}}{\pi}\sin\frac{\pi}{p}\cos\alpha - \frac{pX_c}{2\pi}I_d$$

Therefore, for three phase half bridge inverter, the voltage output due to overlap would be

$$V_{di} = \frac{3\sqrt{3}V_{m_{ph}}}{2\pi}\cos\alpha + \frac{3X_c}{2\pi}I_d \tag{12.28}$$

[Here, the commutating voltage drop is added to the main voltage.]

On the other hand, for three phase full bridge inverter, the voltage drop due to overlap would be

$$V_{di} = \frac{3V_{m_{ph}}}{2\pi}\cos\alpha + \frac{3X_c}{2\pi}I_d \tag{12.29}$$

12.16 CONTROL OF HVDC SYSTEM

An HVDC transmission system is made controllable by using various levels of control in different segments of the HVDC system (in a hierarchical manner) in order to provide efficient and stable operation and to gain flexibility of power control. The following article discusses in depth about the principles of operation of these controls, their implementation and the performance during normal and abnormal system conditions.

12.16.1 Basic Principles of Control

Consider the monopolar HVDC link as shown in Figure 12.16(a). The corresponding equivalent circuit and voltage profile are shown in Figure 12.16(b) and (c), respectively.

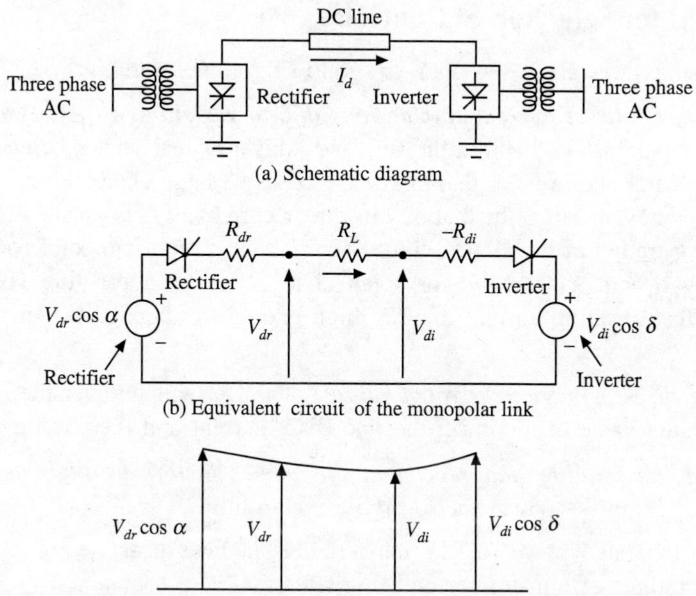

[R_{dr} = Commutating resistance of rectifier, R_{di} = Commutating resistance of inverter]
(c) Voltage profile

Figure 12.16 A monopolar HVDC transmission link.

The direct current flowing from the rectifier to the inverter is

$$I_d = \frac{V_{dr}\cos\alpha - V_{di}\cos\delta}{R_{dr} + R_L - R_{di}} \tag{12.30}$$

and the power at the rectifier terminals is given by

$$P_{dr} = V_{dr} I_d \tag{12.31}$$

Power at the inverter terminal becomes

$$P_{di} = V_{di} I_d = P_{dr} - I_d^2 R_L \tag{12.32}$$

12.16.2 Basic Means of Control

The direct voltage, current or power at any point on the line can be controlled by controlling the internal voltages of rectifier ($V_{dr}\cos\alpha$) and inverter ($V_{di}\cos\delta$). This can be achieved by gate control of the thyristor firing angle or by controlling input voltage through tap changing of the converter transformer. Gate control is a fast (1 to 10 ms) process whereas tap changing is comparatively a slower (5 to 6s per step) procedure. They are used in complementary manner (i.e. initially gate control is adopted for rapid action and then by using tap changing transformer the converter quantities (α for rectifier and δ for inverter) are restored to their normal range. Power reversal is obtained by reversal of polarity of direct voltages at both ends (i.e. transforming rectifier end as inverter and inverter end as rectifier).

12.16.3 Basis for Selection of Controls

The controlling measures are based upon the following considerations:

(i) ***Controlling large direct current fluctuations due to variations in AC system voltage:*** It is evident from equation (12.20) that the line and converter resistances being small, a minute change of the rectifier and inverter output voltages cause a large change in I_d. Rapid control of the converters to prevent large fluctuations in direct current is thus an important requirement for satisfactory operation of the HVDC link. Even if both α and δ are kept constant, the direct current can vary over a wide range for small changes in the alternating voltage magnitude at either end. Thus resulting current may be high enough to damage the thyristors and other equipment.

(ii) ***Maintaining direct voltage near rated value:*** The DC voltage over the HVDC line must be close to its rated value to minimize the line (DC) current and thereby the line losses.

(iii) ***Controlling the sending and receiving end power factors as high as possible:*** The reasons for maintaining the power factor high are as follows:

(a) to keep the rated power of the converter as high as possible for given current and voltage ratings of transformer and thyristors;
(b) to minimize losses and current rating of equipment in the AC system to which the converter is connected;
(c) to minimize voltage drops at the ac terminals as loading increases
(d) to minimize the burden of reactive power supply to the converters.

To achieve a high power factor, α for a rectifier and δ for an inverter should be kept as low as possible.

(iv) ***Prevention of commutation failure in inverters and rectifiers:*** The rectifier usually has a *minimum limit* (about 5°) to ensure adequate voltage across the thyristor before firing. The rectifier normally operates at a value of awithin the range of 15° to 20° so as to leave some margin for increasing rectifier voltage in order to control DC power flow.

It is necessary for an inverter to maintain a certain minimum extinction angle to avoid commutation failure and to ensure that commutation is completed with sufficient margin to allow for de-ionization before commutating voltage reverses at α = 180° (or δ = 0°). The extinction angle δ is equal to $(\beta - \mu)$ with the overlap μ. Because of the possibility of changes in direct current and alternating voltage even after commutation has begun, sufficient *commutation margin above the minimum δ limit* must be maintained. Typically, the value of δ with acceptable margin is around 15° to 18° for 50 Hz/60 Hz systems.

12.16.4 Control Characteristics

Ideal charateristics

The steady-state voltage-current (V-I) characteristics of a rectifier and an inverter are shown in Figure 12.16. Here the rectifier and inverter characteristics include the voltage drop across the line. During normal operation, the rectifier maintains *constant current* (CC) and the inverter operates with *constant extinction angle* (CEA). This facilitates adequate commutation margin and offers a better voltage regulation and current regulation at both the terminals.

The constant current V-I characteristic of rectifier is shown by a vertical line *PQ* in Figure 12.16. From Figure 12.16(b), we can write

$$V_{dr} = V_{di}\cos\delta + (R_L - R_{di})I_d \tag{12.33}$$

This indicates that δ must be maintained at a fixed value on the inverter side. If the commutating resistance R_{di} is slightly larger than the line resistance R_L, the characteristic of the inverter will be a straight line which has a small negative slope as shown by *RS* in Figure 12.17. The operating point is the point of intersection of the two characteristics (*T*).

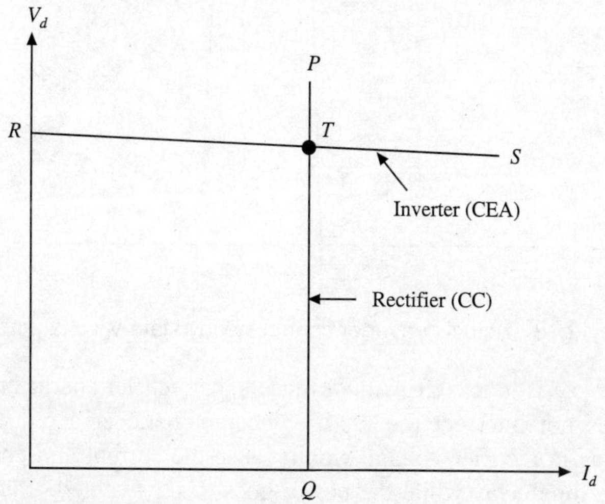

V_d measured at the rectifier side.

Inverter charateristics include line voltage drop ($R_L I_d$).

Figure 12.17 Ideal steady state V-I characteristics.

The rectifier characteristic can be shifted horizontally by adjusting the *reference current* of current regulator (if *measured current* is less than *reference current*, the regulator increase the converter output current by decreasing α).

The inverter characteristic can be adjusted (raised or lowered) by means of its transformer tap changer. When the tap changer operates, the CEA regulator quickly restores the desired d and as a result, the direct current changes, which is then quickly restored by the current regulator of the rectifier.

Actual characteristics

The rectifier operates in CC mode by controlling α which has a minimum limit (α_{min}). If the limit is reached (or violated) then the rectifier will operate at *constant ignition angle (CIA)* mode. This means that the rectifier characteristic has two segments (*OP* and *PQ*) as shown in Figure 12.18. The segment *OP* represents the CIA control mode whereas the segment *PQ* represents the normal CC mode. Also the CC characteristic (the *PQ* line) may not be exactly vertical but it may have a negative slope due to the finite gain of the current regulator with a proportional controller. But with a proportional integral regulator, the CC characteristic is again vertical. At a reduced voltage it shifts as shown by *O′P′Q*.

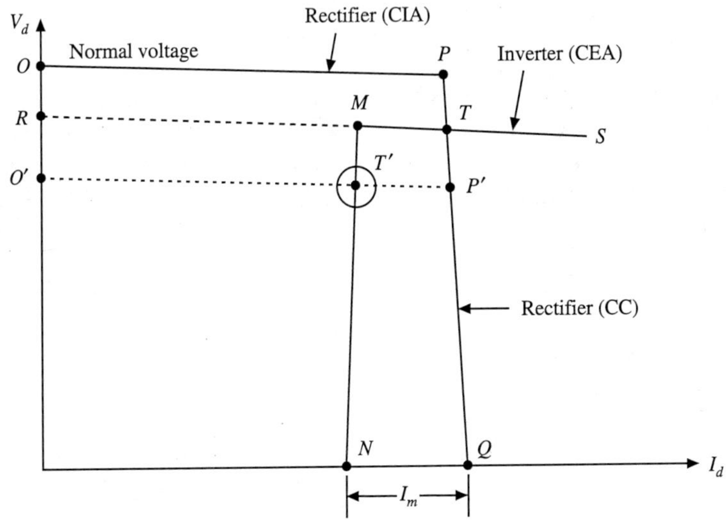

Figure 12.18 Actual converter control steady state V-I characteristics.

The inverter characteristic *RS* (CEA mode) meets the rectifier characteristics at *T* for normal voltage while it does not intersect the rectifier characteristic at a reduced voltage (*O′P′Q*). Hence, a big reduction in rectifier voltage would cause the current and power to be reduced to zero. There may be a time delay while the power goes to zero value because of the presence of DC reactor. In order to avoid this problem, the inverter is also provided with a current controller. This controller is set at a lower value than the current setting for the rectifier. The complete inverter characteristic is given by *SMN*, consisting of two segments: one of CEA mode (*SM*) and one of constant current mode (*MN*). The difference between the rectifier current order and the inverter current order is termed the current margin and is denoted by I_m in Figure 12.18. It is usually set at 8–16% of the rated current such that the two constant current characteristics do not cross each other due to errors or abnormal operation.

Under normal operating conditions the rectifier is in CC mode and controls the direct current whereas the inverter is in CIA mode and controls the direct voltage (the operating point is *T*). With a reduced rectifier voltage, the operating point shifts to *T*′. With the roles of the rectifier and the inverter being reversed, the rectifier controls dc voltage (through CIA mode) and the inverter controls the DC current (CC control mode).

Combined rectifier and inverter characteristics

In most HVDC systems, each converter is required to function as a rectifier as well as an inverter. Consequently, each converter is provided with a combined characteristic as shown in Figure 12.19.

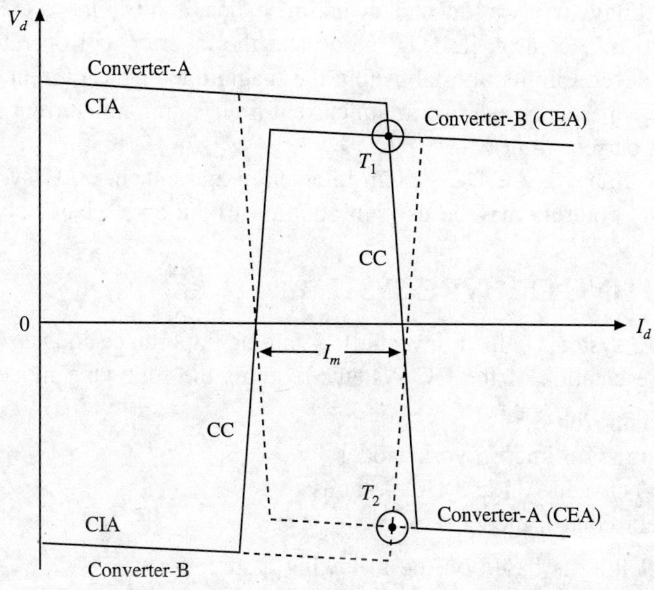

Figure 12.19 Operation with each converter operating both in inverter and rectifier mode.

The characteristic of each converter consists of three segments: constant ignition angle (CIA) corresponding to α_{min}, constant current (CC), and constant extinction angle (CEA). When the power transfer is from converter-A to converter-B, the characteristics are shown by solid lines and the corresponding operating point is T_1. When the power flow is reversed, the characteristics are shown by the dotted lines and the corresponding operating point is T_2. The line current I_d remains the same as before, but the voltage polarity has changed.

A HVDC system must have a high-speed constant current control characteristic for stable operation and to limit overcurrent and minimize damage due to faults. Also it is required to prevent the system from running down due to fluctuations of the AC voltages.

12.16.5 A Review of Basic Control Principles

The major aspects of the basic control system are:

(a) The rectifier is provided with a current control and α limit control (α_{min}) to ensure successful commutation. In the current control mode, a closed-loop regulator controls the firing angle α to control the direct current. Tap changer control of the rectifier transformer keeps α within the range of 10° to 20° A time delay is used to prevent unnecessary tap movements during transient excursions of α.

(b) The inverter is provided with a constant extinction angle (CEA) control and a current control. α is regulated to a value of about 15° to have less chance of commutation failure. Also, it reduces VAR demand. Tap changer control is used to bring the value of δ close to the desired range of 15° to 20°.

(c) The rectifier is in current control mode and the inverter is in CEA control mode under normal operating conditions. If there is a reduction in AC voltage at the rectifier end, the firing angle α is raised and at a time when it hits the α_{min} limit, the rectifier switches to α_{min} control, i.e. CIA mode and the inverter will operate in CC mode.

(d) There are several limits in establishing the magnitude of DC current in order to ensure satisfactory operation (e.g. maximum current limit, minimum current limit, and voltage-dependent current limit).

(e) In order to improve AC/DC system interaction and enhance AC system performance, higher level controls may be used in addition to the above basic controls.

12.17 MODELLING OF HVDC SYSTEM

Modelling of HVDC systems in load flow study is gaining importance and needs brief discussion. The complete representation of the DC systems requires the following modelling:

(i) Converter model
(ii) DC transmission line/network model
(iii) Interface between AC and DC systems
(iv) DC system control model

The converter modelling is based on the following assumptions:

(a) The direct current I_d is ripple-free.
(b) The AC systems at the inverter and the rectifier have perfectly sinusoidal balanced voltage sources of constant frequency behind balanced impedances.
(c) The converter transformers do not saturate under normal operating condition.

12.17.1 Expressions of Power at the Rectifier and Inverter End of HVDC Link

DC power at the rectifier end is given by

$$P_{dc_{(r)}} = V_{dc_{(r)}} I_d = V_{dr} \frac{V_{dr} - V_{di}}{R_{dc}} = \frac{V_{dr}^2}{R_{dc}} - \frac{V_{dr} V_{di}}{R_{dc}} \tag{12.34}$$

where, $R_{dc} = R_{dr} + R_L - R_{di}$

$\therefore$ Power at the rectifier end can be controlled by controlling the difference $(V_{dr} - V_{di})$ while keeping V_{dr} at a fixed value.

DC power at the inverter end is given by

$$P_{dc_{(i)}} = V_{dc_{(i)}} I_d = V_{di} \frac{V_{dr} - V_{di}}{R_{dc}} = -\frac{V_{di}^2}{R_{dc}} + \frac{V_{dr} V_{di}}{R_{dc}} \tag{12.35}$$

The power loss in the dc link is given by,

$$P_L = I_d^2 R_L = \left[\frac{V_{dr} - V_{di}}{R_{dc}} \right]^2 R_L \tag{12.36}$$

The power at the middle of the line is given by

$$P_{dc_{(m)}} = \frac{P_{dr} + P_{di}}{2} = \frac{V_{dr}^2 - V_{di}^2}{2R_{dc}} (= P_{d(\text{middle})}, \text{ say}) \tag{12.37}$$

The power at the rectifier end can also be expressed as

$$P_{dr} = P_{d(\text{middle})} + \frac{1}{2} P_L \tag{12.38}$$

While the power at the inverter end can be expressed as

$$P_{di} = P_{d(\text{middle})} - \frac{1}{2} P_L \tag{12.39}$$

It may also be noted that the power flow in the ac side is, $P_{ac} = \sqrt{3} V_L I_L \cos\phi$

While that in the dc side is, $P_{dc} = V_d I_d$

Neglecting conversion losses,

$$P_{ac} = P_{dc}$$

i.e.,

$$\sqrt{3} V_L I_L \cos\phi = V_d I_d$$

∴

$$V_d = \sqrt{3} V_L \frac{I_L}{I_d} \cos\phi = K_c V_L \cos\phi \tag{12.40}$$

Where K_c is the current ratio and is equal to $\frac{\sqrt{3} I_L}{I_d}$.

12.17.2 Representation of Power-flow

Let us assume,

V_{line} = rms value of line voltage on HT bus

N = number of bridges in series

K = transformer turns ratio

P = active power

Q = reactive power

X_c = commutating reactance per bridge/phase ($= \omega L_s$), L_s is source inductance.

V_d and I_d = direct voltage and current per pole.

From equation (12.17), we can write,

$$V_d = \frac{3\sqrt{3}}{\pi} NKV_{m_{ph}} \cos\alpha - \frac{3X_c}{\pi} NI_d$$

i.e. $$V_d = \frac{3\sqrt{2}}{\pi} NKV_{line} \cos\alpha - \frac{3X_c}{\pi} NI_d \left[\because \sqrt{3}V_{m_{ph}} / \sqrt{2} = V_{line\,(rms)}\right]$$

When $\alpha = 0°$ $$V_d = V_{d0} = \frac{3\sqrt{2}}{\pi} NKV_{line}$$

$\therefore$ $$V_{dr} = V_{d0r} \cos\alpha - \frac{3X_{cr}}{\pi} N_r I_d \text{ (Rectifier mode)} \quad (12.41a)$$

and $$V_{di} = V_{d0i} \cos\delta - \frac{3X_{ci}}{\pi} N_i I_d \text{ (Inverter mode)} \quad (12.41b)$$

$$\phi = \cos^{-1}\left(\frac{V_d}{V_{d0}}\right); P_{dr} = V_{dr} I_d = P_{ac} \text{ (if the line is loss less)}$$

$$Q_{ac} = P_{ac} \tan\phi$$

In a two-terminal DC link having a resistance R_L, the voltage-current relation is given by

$$V_{dr} = V_{di} + R_L I_d \quad (12.42)$$

Where the subscripts r and i denote rectifier and inverter quantities respectively.

EXAMPLE 12.1 A bipolar HVDC link operates at ±500 kV to deliver 1250 MW of power at the receiving end. Neglecting valve losses, the losses in the dc link is 40 MW. Obtain the following

(a) sending end power
(b) sending end voltage
(c) power at the middle of the dc link
(d) voltage at the middle of the dc link.

Solution: Sending end power $\left(P_{s(dc)}\right)$ = Receiving end power + Line loss = 1250 + 40 = 1290 MW. (Ans.(a))

Sending end power, $$(V_{dr}) = \frac{P_{s(dc)}}{I_{dc}} \quad (1)$$

Here, $$I_{dc} = \frac{\text{Receiving end dc power}}{\text{Voltage across receiving end valves}}$$

$$= \frac{1250 \times 10^6}{2 \times 500 \times 10^3}$$ [$\because$ it is bipolar link, voltage across valves is 2 × 500 = 1000 kV]

= 1250 A

$\therefore$ From equation (1), $V_{dr} = \dfrac{1290 \times 10^6}{1250}$ V $= 1232$ kV

i.e., between poles it is, $\pm \dfrac{1032}{2}$ kV or, 516 kV per pole. (Ans. (b))

Power at the middle of the dc link = Power flow through the mid-point of the line $+\dfrac{1}{2}$ line loss

$= 1250 + 20 = 1270$ MW (Ans. (c))

$$\text{Voltage at the mid point of the dc link} = \frac{\text{Power at the middle of the dc link}}{I_d}$$

$$= \frac{1270 \times 10^6}{1250}\text{V} = 1016 \text{ kV}$$

i.e., $\pm \dfrac{1016}{2}$ kV or, 508 kV between HVDC link conductor and ground. (Ans. (d))

EXAMPLE 12.2 At the sending end side of a HVDC link, the dc side voltage and currents are 100 kV and 500 A respectively. The sending end side rectifier has a commutating resistance of 8 Ω while the line resistance is 20 Ω. A dc voltage of 90 kV appears at the input terminal of the inverter. What is extinction angle (δ)? The inverter has no load voltage of 100 kV at extinction (δ) of 0°.

Solution:

Using equation (12.20) we can write, $I_d = \dfrac{V_{dr}\cos\alpha - V_{di}\cos\delta}{R_{dr} + R_L - R_{di}}$

Given: $V_{dr}\cos\alpha = 100 \text{ kV} = 100 \times 10^3 \text{ V}$; $V_{di}\cos\delta = 90 \text{ kV} = 90 \times 10^3 \text{ V}$;

$R_{dr} = 8\ \Omega$; $R_L = 20\ \Omega$ and $I_d = 500$ A.

$$\therefore \quad 500 = \frac{(100 - 90) \times 10^3}{8 + 20 - R_{di}}$$

$$\text{or,} \quad 500(128 - R_{di}) - 10^4$$

$$\therefore \quad R_{di} = 8\ \Omega$$

In inverter operation, from the concepts of power electronics

$$V_{di} = V_{di\,\text{(no load)}}\cos\delta + I_d R_{di}$$

$$\text{or,} \quad V_{di\,\text{(no load)}}\cos\delta = V_{di} - I_d R_{di} = (90 \times 10^3 - 500 \times 8)\text{ V} = 86 \text{ kV}$$

$$\therefore \quad \cos\delta = \frac{86}{V_{di\,\text{(no load)}}} = \frac{86}{100} = 0.86$$

$$\therefore \quad \delta = 30.68^\circ$$

(Ans.)

EXAMPLE 12.3 A HVDC link operates with firing angle (α) of rectifier end as 12° while the extinction angle (δ) at the inverter end is 20°. AC side line voltage at the rectifier end is 390 kV and that at the inverter end is 375 kV. Calculate

(a) the dc voltage at both ends of the link
(b) the dc line current if the net dc circuit resistance is 20 Ω

Solution:

$V_{dr(0)}$ = No load dc voltage at the rectifier end $= \dfrac{3\sqrt{2}}{\pi} V_{L_1} = \dfrac{3\sqrt{2}}{\pi} \times 390 \times 10^3 = 526.69$ kV

Similarly, the no load voltage at the inverter side $= \dfrac{3\sqrt{2}}{\pi} V_{L2} = \dfrac{3\sqrt{2}}{\pi} \times 375 \times 10^3 = 506.43$ kV

Since there is no overlap, $V_{dr} = V_{dr(0)} \cos\alpha = 526.69 \cos 12° = 515.18$ kV

and $V_{di} = V_{di\,(0)} \cos\delta = 506.43 \cos 20° = 474$ kV.

∴ DC side voltages are: 515.18 kV at the rectifier side and 474 kV at the inverter side (Ans.)

$$I_d = \frac{V_{dr} - V_{di}}{R_{dc}} = \frac{(506.43 - 474) \times 10^3}{20} = 1621.5 \text{ kA} \qquad \text{(Ans.)}$$

12.18 AC/DC INTERFACE AT THE GRID SIDE

Power-flow analysis requires simultaneous solution of the DC and AC system equations by some iterative approach as shown in Figure 12.20. The converter transformer HT bus (AC side) acts as an interface between the AC and DC equations.

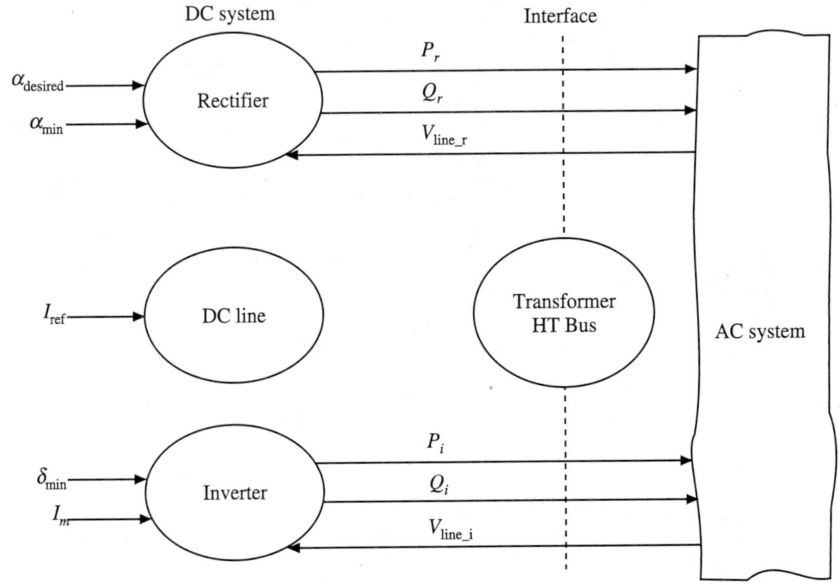

Figure 12.20 AC/DC inerface.

Here V_{line_r} and V_{line_i} are considered to be input quantities for the solution of DC equations as they are known from the previous step in AC solution, whereas P_{dr}, Q_{dr}, P_{di} and Q_{di} are the outputs of DC system solution and they are used in the next iteration for solving AC system equations.

The dependent and independent variables in the solution of DC equation depends on rectifier and inverter control modes. There are two possible modes of operation as follows:

Mode 1: Rectifier on CC control; inverter on CEA control

Mode 2: Inverter on CC control; rectifier on CIA control

When rectifier is on CIA control mode, another mode appears when the inverter is on CC mode.

Mode 1: Converter on CC control and inverter on CEA control

In mode 1, the operating conditions are:

$$\delta = \delta_{\min} \text{ and } \alpha \text{ is controlled to give } I_d = I_{ref}$$

Also, rectifier transformer tap is adjusted to give α within a desired range and inverter transformer tap adjusted to give the desired voltage.

With $I_d = I_{ref}$, the inverter equations may be written from equation (12.41b) as

$$\left.\begin{aligned} V_{d0i} &= \frac{3\sqrt{2}}{\pi} N_i K_i V_{line\_i} \\ V_{di} &= V_{d0i} \cos\delta_{\min} - \frac{3}{\pi} X_{ci} N_i I_{ref} \\ \phi_i &= \cos^{-1}(V_{di}/V_{d0i}) \\ P_i &= V_{di} I_{ref} \\ Q_i &= P_i \tan\phi_i \end{aligned}\right\} \quad (12.43)$$

Here $\alpha_{\min}$ and I_{ref} are known and V_{line_i} (ac line voltage at the inverter side) is also known from the previous AC solution. Therefore, V_{di}, P_i and Q_i can be computed. The transformer tap may be required to be controlled to adjust V_{di} within the desired range.

The rectifier equations are

$$\left.\begin{aligned} V_{d0r} &= \frac{3\sqrt{2}}{\pi} N_r K_r V_{line\_r} \\ V_{dr} &= V_{di} + R_L I_{ref} \\ \alpha &= \cos^{-1}\left(\frac{V_{dr}}{V_{d0r}} + \frac{X_{cr} I_{ref}}{\sqrt{2} V_{line\_r} K_r}\right) \\ \phi_r &= \cos^{-1}(V_{dr}/V_{d0r}) \\ P_r &= V_{dr} I_{ref} \\ Q_r &= P_r \tan\phi_r \end{aligned}\right\} \quad (12.44)$$

In this case I_{ref} is defined and V_{line_i} (ac line voltage at the rectifer side) is also known from the previous AC solution and therefore α, P_r and Q_r can be computed from equation (12.44). The transformer tap may be adjusted to maintain α within the desired range. Here P_i, Q_i, P_r and Q_r are the outputs to be used in the next iteration of the AC solution.

Mode 2: Converter on CIA control and inverter on CC control

In mode 2, the HVDC controls are:

$$\alpha = \alpha_{\min}$$

The inverter firing angle adjusted to give $I_d = I_{ref} - I_m$ (refer to Figure 12.19).

Rectifier transformer tap is adjusted to maximize DC voltage while inverter transformer tap adjusted so that $\delta > \delta_{\min}$ and VAR requirement is minimized.

The rectrifier equations are as under

$$\left.\begin{aligned} V_{d0r} &= \frac{3\sqrt{2}}{\pi} N_r K_r V_{line\_r} \\ V_{dr} &= V_{d0r} \cos\alpha_{\min} - \frac{3}{\pi} X_{cr}(I_{ref} - I_m) N_r \\ \phi_r &= \cos^{-1}(V_{dr}/V_{d0r}) \\ P_r &= V_{dr}(I_{ref} - I_m) \\ Q_r &= P_r \tan\phi_r \end{aligned}\right\} \qquad (12.45)$$

Here I_d is maintained at $(I_{ref} - I_m)$ by the inverter. V_{line_r} is also known from the previous AC solution. The rectifier output voltage V_{dr} may be maximized by adjusting transformer tap ratio K_r.

As V_{line_i} is known from the previous ac solution and the inverter equations are obtained as

$$\left.\begin{aligned} V_{d0i} &= \frac{3\sqrt{2}}{\pi} N_i K_i V_{line\_i} \\ V_{di} &= V_{dr} - R_L I_d = V_{dr} - R_L(I_{ref} - I_m) \\ \delta &= \cos^{-1}\left(\frac{V_{di}}{V_{d0i}} + \frac{X_{ci}(I_{ref} - I_m)}{\sqrt{2} K_i V_{line\_i}}\right) \\ \phi_i &= \cos^{-1}(V_{di}/V_{d0i}) \\ P_i &= V_{di} I_d = V_{di}(I_{ref} - I_m) \\ Q_i &= P_i \tan\phi_i \end{aligned}\right\} \qquad (12.46)$$

The transformer tap ratio K_i may be adjusted to ensure $\delta > \delta_{\min}$ and minimize VAR requirement. Here also P_i, Q_i, P_r and Q_r are the outputs to be used in the next iteration of the AC solution.

Appearance of special mode—rectifier on CIA control and inverter on modified characteristic

When a power flow solution is related with stability studies, it is necessary to consider the transient condition and therefore the transition between modes 1 and 2 need to be considered. During this transition period the converter is on CIA control and inverter is on modified characteristic. One method of realizing this modified characteristic is to operate the inverter in the constant β mode (β is inverter ignition advance angle).

In this mode the operating conditions are as under:

Rectifier ignition delay angle = $\alpha_{\min}$, inverter ignition advance angle = β_c (β_c is constant)

$I_d = I'_d$ such that $I_{ref} > I'_d > (I_{ref} - I_m)$

As line voltages (V_{line_r} and V_{line_i}) are known from the previous AC solution, following DC system equations are solved to compute the line current (I'_d).

$$\left.\begin{aligned} V_{d0r} &= \frac{3\sqrt{2}}{\pi} N_r K_r V_{line\_r} \\ V_{d0i} &= \frac{3\sqrt{2}}{\pi} N_i K_i V_{line\_i} \end{aligned}\right\} \tag{12.47}$$

Now,

$$\left.\begin{aligned} V_{dr} &= V_{d0r} \cos\alpha_{\min} - \frac{3}{\pi} X_{cr} I'_d N_r \\ V_{di} &= V_{d0i} \cos\beta_c + \frac{3}{\pi} X_{ci} I'_d N_i \end{aligned}\right\} \tag{12.48}$$

The V-I characteristic equation is

$$V_{dr} = V_{di} + R_L I'_d$$

or,

$$I'_d = \frac{V_{dr} - V_{di}}{R_L}$$

Substituting V_{dr} and V_{di} from equation (12.48), we obtain

$$I'_d = \frac{\left(V_{d0r} \cos\alpha_{\min} - \frac{3}{\pi} X_{cr} I'_d N_r\right) - \left(V_{d0i} \cos\beta_c + \frac{3}{\pi} X_{ci} I'_d N_i\right)}{R_L}$$

$$\therefore \quad I'_d = \frac{V_{d0r} \cos\alpha_{\min} - V_{d0i} \cos\beta_c}{R_L + \frac{3}{\pi} X_{cr} I'_d N_r + \frac{3}{\pi} X_{ci} I'_d N_i} \tag{12.49}$$

Next V_{dr} and V_{di} can be calculated by substituting the value of I'_d in equation (12.48). Following equations can be solved to compute the AC quantities:

$$\left.\begin{aligned} \phi_r &= \cos^{-1}(V_{dr}/V_{d0r}) \\ \phi_i &= \cos^{-1}(V_{di}/V_{d0i}) \\ P_r &= V_{dr}I'_d\,, Q_r = P_r \tan\phi_r \\ P_i &= V_{di}I'_d\,, Q_i = P_i \tan\phi_i \end{aligned}\right\} \tag{12.50}$$

The tap-changer action is too slow as compared to transient phenomena and hence not considered for transient stability simulations. A simple algorithm is furnished below to identify the operating modes and to solve the AC and DC system equations.

Step 1: Set iteration count p and tolerance ε.

Step 2: Solve for AC system equations and obtain V_{line_r} and V_{line_i}.

Step 3: (i) If $\alpha > \alpha_{min}$ then it is Mode-1 operating condition. Go to Step 4.

(ii) If $\alpha \le \alpha_{min}$ and $\delta > \delta_{min}$ then it is Mode-2 operating condition. Go to Step 4.

(iii) If $\alpha = \alpha_{min}$ and $\delta \le \delta_{min}$ then it is a special operating condition. Go to Step 4.

Step 4: Calculate P_r, Q_r, P_i and Q_i. If the mismatch in solution is more than tolerance ε, advance iteration count p and go to Step 1 to solve AC equations.

Step 5: If the mismatch is less than tolerance ε, the solution is complete.

Step 6: Stop.

EXERCISES

1. Why is it absolutely necessary to transmit power in EHV/UHV range?

2. What are the basic aspects of EHV AC transmission?

3. Why does the in-phase voltage drop in a line is more dependent on reactive power demand?

4. Why is HVDC transmission advantageous over EHV AC transmission?

5. What are the constraints in HVDC transmission?

6. What are the different types of links in HVDC transmission system? State relative advantages and disadvantages.

7. Why do you require a series inductive reactor in HVDC transmission system?

8. Briefly discuss the HVDC converter characteristics.

9. What are the basic control principles in HVDC system?

10. Write down the conditions and equations of the HVDC transmission model in different modes of operation.

CHAPTER 13

Power System Stability

13.1 INTRODUCTION

Power system stability is that property of a power system which ensures that the operation of the system is within the specified limits of voltage and power angle during normal and abnormal changes in operating conditions. It ensures that the operating point remains in equilibrium position despite slow, steady or abrupt changes in operating conditions.

In today's large power network, with many synchronous machines interconnected by complicated transmission networks, stability studies are performed by computer methods. In this chapter, we will analyse the aspects of power system stability following its classification.

13.2 TYPES OF STABILITY

Power system stability can be classified into four types: *transient stability*, *steady state stability*, *dynamic stability* and *voltage stability*. However, the first three types deal with power angle stability while the last one deals with stability of load bus voltage. Transient stability involves the response to large disturbance and is associated with appreciable change of rotor speeds, power angle and power transfers. The system response to such a disturbance is usually considered within one second. In most of the cases, transient stability is determined during the first swing of machine power angles following a disturbance. Steady state stability relates to the response of a synchronous machine to a gradually increasing load demand whereas dynamic stability involves the response of the synchronous machine to small perturbations. These perturbations may create oscillations of the system operating point around its equilibrium. If the oscillations are large, it may create dynamic instability. Dynamic stability is nowadays referred as *small signal stability*. Voltage stability is related with deviation of load bus voltage beyond a specified value determined by the reactive power status of the load bus, source reactance, type of load and sensitivity of a load bus voltage with change in reactive power demand of the load bus. In this chapter, the aspects of transient and steady state stability are discussed in detail.

13.3 TRANSIENT STABILITY

13.3.1 Representation of Transmission Lines, Loads and Generators in Transient Stability

Let us assume that a transmission network is represented by a n-port network (Figure 13.1), where the transmission system has n number of buses (excluding the reference or ground bus) with m number of generator buses ($m < n$), while the bus admittance matrix is denoted by $[Y_{Bus}]$. The transmission network model is basically the prefault ($t = 0^-$) model of the network which has been employed in the load flow study. The buses in this network have been numbered in the sequence as follows:

Bus number 2 to $m \rightarrow$ generator (PV) buses
Bus number m +1 to $n \rightarrow$ load (PQ) buses
Bus number 1 $\rightarrow$ slack bus.

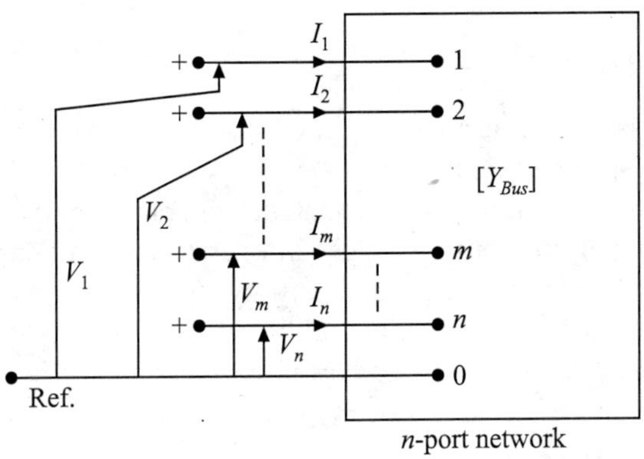

Figure 13.1 Port representation of a transmission network.
Bus number 2 to $m \rightarrow$ generator (PV) buses
Bus number m +1 to $n \rightarrow$ load (PQ) buses
Bus number 1 $\rightarrow$ slack bus.

The following information is known from load flow study

(i) P_{gi}, Q_{gi}, P_{di}, Q_{di}; for $i = 2, 3, \ldots, m$

(ii) P_{di}, Q_{di}; for $i = m + 1, m + 2, \ldots, n$

(iii) P_i and Q_i; for slack bus ($i = 1$)

(iv) $|V_i| \angle\delta_i$; for $i = 2, 3, \ldots, n$.

In the transient stability study, loads are taken into account and are converted into equivalent admittances y_{di} between i-th bus and reference bus.

$$y_{di} = (P_{di} - jQ_{di})/|V_i|^2; \; i = 2, 3, \ldots, n \tag{13.1}$$

In case any bus does not have any connected load, obviously $y_{di} = 0$.

Thus, $[Y_{Bus}]_{old}$ in the load flow study gets modified as

$$[Y_{Bus}] = [Y_{Bus}]_{old} + \text{diagonal } (y_{di}) \tag{13.2}$$

(since the loads are connected at the buses with reference to ground, only the diagonal elements of the $[Y_{Bus}]_{old}$ get modified.)

The generator in transient stability studies is represented by an equivalent voltage source in series with d-axis transient reactance (x'_{di}) of the machine.

Here,
$$E_i = V_i + jx'_{di} I_i$$

Where E_i is the internal voltage while V_i is the terminal voltage of the generator and I_i is the injected current at i-th bus from the generator.

and
$$I_i = \frac{P_{gi} - jQ_{gi}}{V_i^*}, \; i = 1, 2, 3, \ldots, m \tag{13.3a}$$

$\therefore$
$$E_i = V_i + jx'_{di}\left(\frac{P_{gi} - jQ_{gi}}{V_i^*}\right) = |E_i| \angle \delta_i \tag{13.3b}$$

13.3.2 Assumptions for Transient Stability Study

The conventional approach to study the transient stability aspect of a power system is used in classical model. The classical model is employed to study the transient stability of a power system for a period of time during which the dynamic behaviour of the system is dependent largely on stored energies of the machine in its rotor. This is a simple model with reasonable accuracy and having requirement of minimum data. The assumptions made in developing the classical model are as follows:

(i) P_m (mechanical power input to each generator) is constant.

(ii) D (damping power) is neglected.

(iii) Synchronous machines are represented electrically by constant voltage behind transient reactances, the excitation is also constant.

(iv) Loads are represented by constant impedances.

(v) Machine losses and saliency are neglected.

13.3.3 Derivation of Swing Equation

Let

G = MVA rating of the synchronous machine,

H = Its inertia constant (sec),

J = Polar moment of inertia of all rotating parts in kg-m^2,

ω_0 = Synchronous angular velocity in electrical radian/sec,

M = Corresponding angular momentum of the machine,

f = Electrical synchronous frequency (Hz),

T_e and P_e = Electromagnetic torque (N-m) and corresponding electromagnetic power in watts,

T_m and P_m = Mechanical torque provided at the generator shaft in N-m and corresponding electrical equivalent power (P_m) at shaft in watts,
[It may be noted though $P = \omega T$, in p.u. expression P is numerically same as T.]

T_a and P_a = Accelerating torque in N-m at shaft while P_a is the corresponding accelerating power in watts,

θ = Angular position of the rotor in electrical rad/sec,

δ = Power angle of the machine in rad,

E = Induced emf/ph of the machine in volts,

V = Terminal voltage/ph of the machine in volts, and

x'_{ds} = d-axis transient reactance in Ω/ph.

We can write for the synchronous machine, from the first principle,

$$GH = \frac{1}{2} J\omega_0^2$$

and
$$M = J\omega_0$$

∴
$$GH = \frac{1}{2} M\omega_0 = \frac{1}{2} M(360\,f) \tag{13.4a}$$

[∵ $\omega_0 = 360\,f$ electrical degrees/sec.]

i.e.,
$$M = \frac{GH}{180\,f} \text{ Megajoule-sec/degree.} \tag{13.4b}$$

At steady state, with no disturbance

$$T_e = T_m$$

i.e.,
$$T_e\omega_0 = T_m\omega_0$$

or,
$$T_m\omega_0 - T_e\omega_0 = 0$$

In case of disturbance (say, a fault) in the system,

$$T_m\omega_0 \neq T_e\omega_0$$

(in the transient period till the rotor will adjust with change of energy demand by the system); hence we can write

$$T_m\omega_0 - T_e\omega_0 = T_a\omega_0$$

or,
$$P_m - P_e = P_a \tag{13.5}$$

where, P_a is the *accelerating power*. In case of sudden load throw, the rotor will accelerate as it is now having surplus mechanical energy while for increase of sudden load demand, the rotor will decelerate and P_a would be *decelerating power* (–ve P_a).

Also,
$$\frac{d\theta}{dt} = \omega_0$$

as
$$\theta = \omega_0 t + \delta \tag{13.6}$$

[assuming increase of rotor angle uniformly with time.]

During rotor acceleration, from mechanics we can write

$$J\alpha = T_a \tag{13.7}$$

where α is the rotor angular acceleration in rad/sec^2 (or degree (elect.)/sec^2)

But $\alpha = \dfrac{d\omega}{dt}$ [ω being rotor angular velocity in rad/sec (or degree (elect.)/sec)]

$$= \frac{d^2\delta}{dt^2}$$

$\therefore$ From equation (13.7), we have

$$J\frac{d^2\delta}{dt^2} = T_a = T_m - T_e$$

or,

$$\omega_0 J\frac{d^2\delta}{dt^2} = P_a = P_m - P_e$$

or,

$$M\frac{d^2\delta}{dt^2} = P_a = P_m - P_e \tag{13.8a}$$

or,

$$\frac{GH}{180\,f}\frac{d^2\delta}{dt^2} = P_m - P_e = P_a \tag{13.8b}$$

Dividing both sides of equation (13.8b) by G (the machine rating), we get

$$\frac{H}{180\,f}\frac{d^2\delta}{dt^2} = P_m - P_e = P_a \tag{13.9a}$$

or,

$$M\frac{d^2\delta}{dt^2} = P_m - P_e = P_a \tag{13.9b}$$

[P_m, P_e and P_a in equations (13.9a) and (13.9b), as well as M in equation (13.9b) are expressed in p.u.]

Equation (13.9a) or (13.9b) represents the *swing equation* in transient stability study.

Here,

$$P_e = \frac{|E||E_i|}{X}\sin\delta\ (= P_{max}\sin\delta) \tag{13.10}$$

where the generated power is delivered by the machine having internal voltage $E\angle -\delta°$ to a bus having voltage $E_i\angle 0°$ through total impedance X (= $X_{line} + x'_d$). If $|E|$, $|E_i|$ and $|X|$ are expressed in p.u., P_e and P_{max} are also obtained in p.u. In the above analysis, we assumed no damping in the machine. Strictly speaking, this is not so and a damping term is to be included which is proportional to $(d\delta/dt)$.

The swing equation is then modified as

$$M\frac{d^2\delta}{dt^2} + D\frac{d\delta}{dt} = P_m - P_e = P_a \tag{13.11}$$

where D is the *damping constant* ($= 2k\omega_0/G$, k representing the constant of proportionality when the friction torque is proportional to angular velocity, and damping in the rotor speed is caused due to this friction).

EXAMPLE 13.1 The kinetic energy (KE) stored in the rotor of a 20 MVA, 4 pole, 50 Hz alternator is 60 MJ. The input to the machine is 15 MW at a developed power of 13.5 MW. Calculate the accelerating power assuming that acceleration remains constant for 10 cycles. Determine the value of δ at the end of 10 cycles.

Solution: Here $P_m = 15$ MW; $P_e = 13.5$ MW

The accelerating power, $P_a = P_m - P_e = 15 - 13.5 = 1.5$ MW

$$H = \frac{KE}{\text{rating in MVA}} = \frac{60}{20} = 3 \text{ MJ/MVA}$$

$$M = \frac{GH}{180\,f} = \frac{20 \times 3}{180 \times 50} = \frac{1}{150} \text{ Megajoule-sec/elect.degree.}$$

$$\left[\text{or, } M = \frac{GH}{\pi f} = \frac{20 \times 3}{\pi \times 50} = 0.382 \; MJ\text{-sec/rad}\right]$$

Using swing equation (equation (13.9b)), we can write,

$$M\frac{d^2\delta}{dt^2} = P_a$$

$$\therefore \quad \frac{d^2\delta}{dt^2} = \frac{P_a}{M} = \frac{1.5}{0.382} = 3.927 \text{ rad/sec}^2$$

Integrating both sides, we get

$$\frac{d\delta}{dt} = 3.927t + C \text{ [} C \text{ is an integration constant]}$$

At $t = 0$, $\dfrac{d\delta}{dt} = 0$, $\therefore C = 0$

Again integrating both sides, we get

$$\delta = 3.927\frac{t^2}{2} + \delta_0 \text{ [}\delta_0 \text{ is an integration constant (= initial power angle)]}$$

$$\therefore \quad \delta = 1.9635\, t^2 + \delta_0$$

Now, $t = 10 \text{ cycles} \left(= \dfrac{10}{50}\right) = 0.2 \text{ sec}$

$$\therefore \quad \delta = 1.9635 \times 0.2^2 + \delta_0 = 0.07854 + \delta_0 \text{ rad}$$

Hence, the new power angle after 10 cycles becomes

$$\delta = (0.07854 + \delta_0) \text{ rad.} \quad \text{(Ans.)}$$

EXAMPLE 13.2 A 30 MVA, 15 KV, 4 pole, 50 Hz turbo-alternator has an inertia constant of 10 KW sec/kVA.

(i) Compute the kinetic energy stored by the rotor at synchronous speed.
(ii) If the input power and the output power are 25 MW and 20 MW respectively, find the acceleration in degrees per sec^2 and in rpm per sec.

Solution: Here, G (machine rating in MVA) = 30

H (inertia constant) = 10 kW-sec/kVA

$\therefore$ Stored energy $(GH) = (30 \times 10)$ MJ

From equation (13.4b),

$$M = \frac{30 \times 10}{180 \times 50} = \frac{1}{30} \text{ Megajoule-sec/degree}$$

Now, $P_m = 25$ MW; $P_e = 20$ MW.

The accelerating power,

$$P_a = P_m - P_e = 25 - 20 = 5 \text{ MW}$$

Using swing equation (equation (13.9b)), we can write,

$$M\frac{d^2\delta}{dt^2} = P_a$$

or,

$$M\alpha = P_a$$

$\therefore$ The rotor angular acceleration,

$$\alpha = \frac{P_a}{M} = 5 \times 30 = 150 \text{ electrical degree/sec}^2$$

$= 75$ degree/sec^2 [since the machine is a 4 pole machine] Ans. (i)

$= 75 \times \dfrac{60}{360} = 12.5$ rpm/sec. Ans. (ii)

EXAMPLE 13.3 An alternator is delivering power of 0.8 p.u. at a terminal voltage of 1.05 p.u. This alternator is suddenly synchronized to an infinite bus through a transformer of reactance 0.1 p.u. and a double circuit transmission line of 0.4 p.u. reactances each as shown in Figure E13.1(a). Determine the swing equation assuming the infinite bus voltage magnitude to be 1.0 p.u. The alternator has the following data: $x'_d = 0.2$ p.u.; $H = 4$ MJ/MVA.

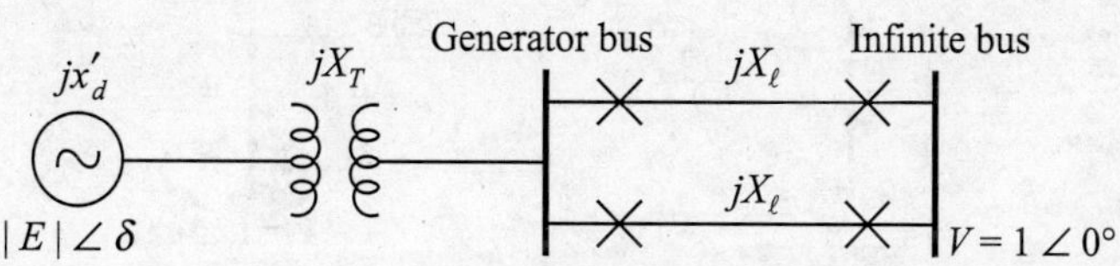

Figure E13.1(a)

Solution: The generator on the infinite bus has been drawn in Figure E13.1(a) and the equivalent circuit has been shown in Figure E13.1(b).

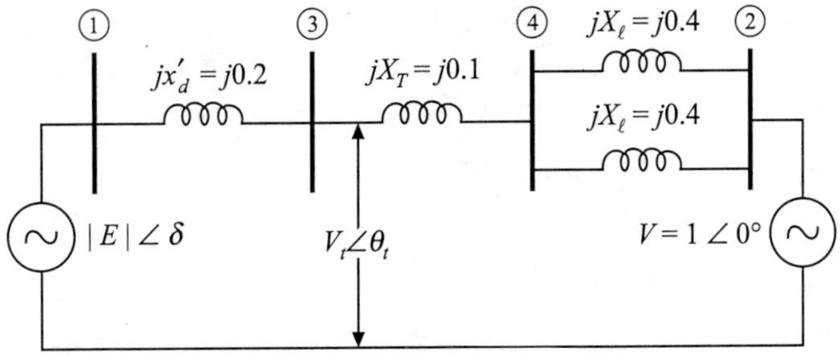

Figure E13.1(b)

From the equivalent circuit we can write,

$$x_{12} = x'_d + X_T + \frac{X_\ell \times X_\ell}{X_\ell + X_\ell}$$

where,

x'_d = transient d-axis reactance of alternator,
X_T = reactance of transformer,
X_ℓ = reactance of transmission line.

$$\therefore \quad x_{12} = j0.2 + j0.1 + \frac{j0.4 \times j0.4}{j0.4 + j0.4} = j0.5 \text{ p.u.}$$

$$\therefore \quad y_{12} = \frac{1}{j0.5} = -j2 \text{ p.u.}$$

Here, y_{12} = admittances connected between bus 1 and 2,
y_{10} = shunt admittance connected with bus-1,
y_{20} = shunt admittance connected with bus-2.

Obviously, $y_{10} = 0$; $y_{20} = 0$.

$$\therefore \quad [Y] = \begin{bmatrix} Y_{11} & Y_{12} \\ Y_{21} & Y_{22} \end{bmatrix} = \begin{bmatrix} (y_{12} + y_{10}) & -y_{12} \\ -y_{12} & (y_{12} + y_{20}) \end{bmatrix} = \begin{bmatrix} -j2 & j2 \\ j2 & -j2 \end{bmatrix}$$

Now, $P_e = EVY_{12} \sin \delta = E \times 1.0 \times 2 \sin \delta = 2E \sin \delta$ (1)

The equivalent circuit shown in Figure E13.1(b) is reduced to that shown in Figure E13.1(c).

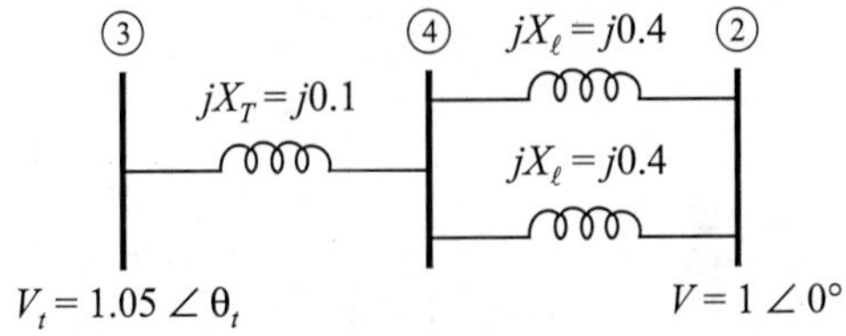

Figure E13.1(c)

From Figure E13.1(c),

$$V = 1 - \angle 0° \text{ p.u.};\ V_t = 1.05 - \angle \theta_t° \text{ p.u.};\ P_e = 0.8 \text{ p.u.};$$

and
$$x_{32} = j0.1 + \frac{j0.4 \times j0.4}{j0.4 + j0.4} = j0.3 \text{ p.u.}$$

Now,
$$P_e^0 = 0.8 = \frac{V_t V}{x_{32}} \sin \theta_t \quad \text{[here superscript 0 indicates initial condition]}$$

$\therefore$
$$\theta_t = \sin^{-1}\left(\frac{0.3 \times 0.8}{1.05 \times 1}\right) = 13.21°$$

$\therefore$
$$\text{Current } (I) = \frac{V_t - V}{Z} = \frac{1.05 \angle 13.21° - 1 \angle 0°}{j0.3} = 0.80 \angle{-5.28°} \text{ p.u.}$$

However, from Figure E13.1(b)

$$E \angle \delta = V_t \angle \theta_t + Ix'_d = 1.05 \angle 13.21° + 0.80 \angle{-5.28°} \times 0.2 \angle 90°$$

$$= 1.05 \angle 13.21° + 0.16 \angle 84.72°$$

$\therefore$
$$E \angle \delta = 1.11 \angle 21.06° \text{ p.u.}$$

From equation (1) we have,

$$P_e = 2E \sin \delta = 2 \times 1.11 \sin \delta = 2.22 \sin \delta$$

[This is the electrical output during transient process].

$\therefore$ Swing equation (13.9a) becomes

$$\frac{H}{\pi f}\frac{d^2\delta}{dt^2} = P_e^0 - P_e$$

i.e.,
$$\frac{4}{\pi \times 50}\frac{d^2\delta}{dt^2} = 0.8 - 2.22 \sin \delta$$

[Since P_m remains unchanged during transient process and is equal to initial electrical power output P_e^0.]

or,
$$\frac{d^2\delta}{dt^2} = \frac{\pi \times 50}{4} = (0.8 - 2.22 \sin \delta)$$

$\therefore$
$$\frac{d^2\delta}{dt^2} = 39.27\ (0.8 - 2.22 \sin \delta) \text{ rad/sec}^2$$

13.3.4 Swing Equation for Synchronous Machine Connected to Infinite Bus

Following Figure 13.2, it is evident that for the alternator injecting real power P_e to the infinite bus ($|V| \angle 0°$) through a line having reactance x_ℓ will have the swing equation given by equation (13.11a).

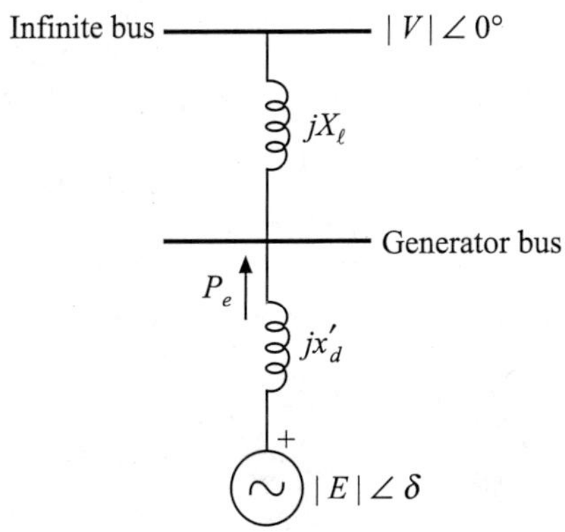

Figure 13.2 A synchronous generator connected to infinite bus.

$$\frac{H}{\pi f}\frac{d^2\delta}{dt^2} = P_m - P_{max} \sin \delta \tag{13.11a}$$

where,

$$P_e = \frac{|E'||V|}{X} \sin \delta = P_{max} \sin \delta$$

and

$$X = x'_d + x_\ell$$

All quantities except δ are expressed in p.u. δ is expressed in rad or degree.

13.3.5 Swing Equation for a Two Machine System

In illustrating the swing equation for a two machine system, we assume the following:

(i) One machine (No. 1) is generating while the other machine (No. 2) is motoring.
(ii) Interconnecting line has only reactance (x_ℓ) and hence the line is lossless.
(iii) Mechanical inputs (P_m) remain constant during the process and whatever be the electrical power output of the generator, it will be absorbed by the motor.

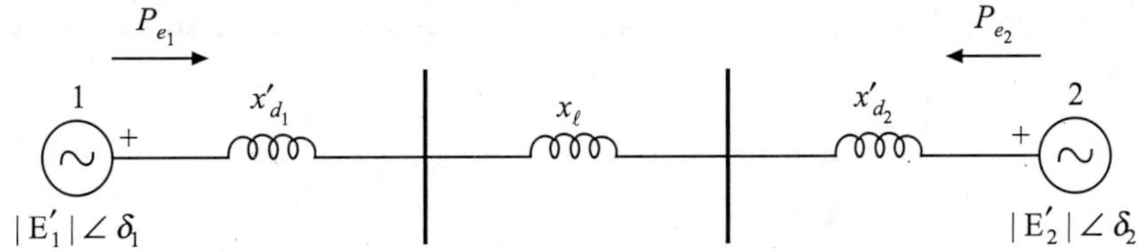

Figure 13.3 Two machine system.

Obviously,

$$P_{e_1} = P_{e_2} = P_e$$

The swing equation for a two machine system can now be written as

$$\frac{d^2\delta_1}{dt^2} = \pi f\left[\frac{P_{m_1} - P_{e_1}}{H_1}\right] = \pi f\left(\frac{P_m - P_e}{H_1}\right)$$

$$\frac{d^2\delta_2}{dt^2} = \pi f\left[\frac{P_{m_2} - P_{e_2}}{H_2}\right] = \pi f\left(\frac{P_e - P_m}{H_2}\right)$$

Subtracting, we get

$$\frac{d^2(\delta_1 - \delta_2)}{dt^2} = \pi f\left[\frac{H_1 + H_2}{H_1 H_2}\right](P_m - P_e)$$

or,
$$\frac{H_{eq}}{\pi f}\frac{d^2\delta}{dt^2} = (P_m - P_e) \tag{13.11b}$$

where,
$$H_{eq} = \left[\frac{H_1 H_2}{H_1 + H_2}\right];\ \delta = (\delta_1 - \delta_2)$$

and
$$P_e = \frac{|E_1'||E_2'|}{x'_{d_1} + x_\ell + x'_{d_2}}\sin\delta$$

Expression (13.11b) represents the swing equation for the two machine system. It may be observed that a two machine system is equivalent to a single machine connected to the infinite bus (SMIB). Thus, in most cases study of SMIB system can be studied as a generalisation.

13.3.6 Linearization of Swing Equation

Let $\delta = \delta^0 + \Delta\delta$ ($\Delta\delta$ being small).

Also, at initial state, $P_m^0 = P_e^0$. Hence, we can write the linearised expression of swing equation as

$$M\Delta\ddot{\delta} + D\ddot{\delta} + P_S\Delta\delta = 0 \tag{13.12}$$

where, P_s is the *synchronizing power coefficient* and is equal to $[\partial P_e(\delta)/\partial\delta]_{\delta = \delta_0}$. P_s represents the slope of the power output at the operating point (δ^0). As the load increases (δ^0 increasing), P_s decreases monotonically and for some value of δ^0 less than (or equal to) $\pi/2$, P_s approaches zero and then becomes negative. The characteristic equation is thus obtained as

$$Ms^2 + Ds + P_s = 0$$

The roots are given as

$$s_1, s_2 = \frac{-D \pm \sqrt{D^2 - 4MP_S}}{2M} \tag{13.13}$$

s_1, s_2 represent the *natural frequencies* of the mechanical system.

Under normal operating condition,

$$MP_S >> D^2 \text{ and we get } s_1, s_2 = \sigma \pm j\omega$$

Here $\sigma < 0$ and $\omega \cong \left(\frac{P_S}{M}\right)^{\frac{1}{2}} = \left(\frac{P_S \pi f}{H}\right)^{\frac{1}{2}}$ [$\because$ in p.u. $M = H/\pi f$]. With D being small, lightly damped oscillations (with amplitude of oscillation being $\Delta\delta$) are feasible and these oscillations will decay out exponentially. Thus with small D, $\delta = \delta^0 + \Delta\delta \cong \delta^0$. However with D negative, one of the natural frequency is positive and system instability may occur.

13.3.7 Swing Equation of Non-coherent and Coherent Machines

Let us consider two generators 1 and 2 which are connected to a common system base. The swing equations are given by

$$\frac{H_1}{180 f}\frac{d^2\delta}{dt^2} = P_{m_1} - P_{e_1} \tag{13.14}$$

and

$$\frac{H_2}{180 f}\frac{d^2\delta}{dt^2} = P_{m_2} - P_{e_2} \tag{13.15}$$

Due to *coherent swing* of the rotors of the two machines (say two generators),

$$\delta_1 = \delta_2 \Rightarrow \delta$$

Adding equations (13.14) and (13.15) we get,

$$\frac{H_{eq}}{180 f}\frac{d^2\delta_{eq}}{dt^2} = P_{m_{eq}} - P_{e_{eq}} \tag{13.16}$$

where,

$$\left.\begin{aligned} P_{m_{eq}} &= P_{m_1} + P_{m_2} \\ P_{e_{eq}} &= P_{e_1} + P_{e_2} \\ \delta_{eq} &= \delta \\ H_{eq} &= H_1 + H_2 = H_{1m}\frac{G_{1m}}{G_{sys}} + H_{2m}\frac{G_{2m}}{G_{sys}} \end{aligned}\right\} \tag{13.16a}$$

(suffixes 'm' and 'sys' stand for individual machine and for system (base), respectively).

Machines which swing together are called *coherent machines.* Two synchronous generators can swing together. Even though their speeds are different, the power angles may be same and it may be possible to represent them on a single base. The net effect of such a swing is known as *coherent swing* and is additive (as shown above). On the other hand, for a two machine system if one is a synchronous generator and the other is a synchronous motor, the swing is *non-coherent* and the net effect is obtained by subtracting the respective swing equations.

Let us assume that in the case of non-coherent swing, the swing equations are as follows:

$$\frac{2H_1}{\omega_S}\frac{d^2\delta_1}{dt^2} = P_{m_1} - P_{e_1}$$

and
$$\frac{2H_2}{\omega_S}\frac{d^2\delta_2}{dt^2} = P_{m_2} - P_{e_2}$$

Subtraction of these two equations gives us

$$\frac{d^2\delta_1}{dt^2} - \frac{d^2\delta_2}{dt^2} = \frac{\omega_S}{2}\left(\frac{P_{m_1} - P_{e_1}}{H_1} - \frac{P_{m_2} - P_{e_2}}{H_2}\right)$$

Multiplying both sides by $\frac{H_1 H_2}{H_1 + H_2}$, simplification yields

$$\frac{2}{\omega_S}\left(\frac{H_1 H_2}{H_1 + H_2}\right)\frac{d^2(\delta_1 - \delta_2)}{dt^2} = \left(\frac{P_{m_1}H_2 - P_{m_2}H_1}{H_1 + H_2} - \frac{P_{e_1}H_2 - P_{e_2}H_1}{H_1 + H_2}\right)$$

or,
$$\frac{2H_{eq}}{\omega_S}\frac{d^2\delta_{eq}}{dt^2} = (P_{m_{eq}} - P_{e_{eq}}) \tag{13.17}$$

where,
$$\left.\begin{aligned} P_{m_{eq}} &= \frac{P_{m_1}H_2 - P_{m_2}H_1}{H_1 + H_2} \\ P_{e_{eq}} &= \frac{P_{e_1}H_2 - P_{e_2}H_1}{H_1 + H_2} \\ H_{eq} &= \frac{H_1 H_2}{H_1 + H_2} \end{aligned}\right\} \tag{13.17a}$$

Equation (13.17) represents the swing equation of a non-coherent system of synchronous machine. For a multi-machine system involving several swing equations, no formal solution of the equation is feasible. In classical systems, a 'point by point' solution was made and in recent days digital computer methods are applied. Even for a simplest case of single machine system on infinite bus, with resistance neglected, the formal solution of the swing equation is possible only when $P_m = 0$.

The solution of the swing equation gives the values of δ *vs. t.* Such a graph is known as *swing curve*. If the swing curve diverges, the system goes to instability while in case of convergence of the swing curve, the system stability is maintained. Some popular techniques of solution of swing equation will be represented later in this chapter.

13.4 EQUAL AREA CRITERION

Let us neglect D, the damping coefficient. The linearised swing equation expressed in equation (13.12) is written as (for $D = 0$)

$$M\frac{d^2\Delta\delta}{dt^2} + \left[\frac{\Delta P_s}{\Delta\delta}\right]_0 \Delta\delta = 0 \tag{13.18}$$

The characteristic equation being given by

$$Ms^2 + P_S\Delta\delta = 0,$$

The roots are, $$s_1, s_2 = \pm\left[\frac{-P_S}{M}\right]^{\frac{1}{2}} = \left[\frac{\left(-\dfrac{\partial P_e}{\partial \delta}\right)_0}{M}\right]^{\frac{1}{2}} \tag{13.19}$$

If P_s is positive, we get a pair of conjugate imaginary roots and the system is stable. If P_s is negative, we get one root having negative real part while the another has positive real part. The system is then unstable. In an unstable system, δ increases indefinitely with time and the alternator loses synchronisation. *For a system to be stable, at the final instant* $\frac{d\delta}{dt} = 0$.

We have seen earlier (equation 13.9b) that $M\frac{d^2\delta}{dt^2} = Pa$.

i.e.,
$$2\frac{d\delta}{dt}\cdot\frac{d^2\delta}{dt^2} = 2\frac{d\delta}{dt}\cdot\frac{P_a}{M}$$

Upon integration, it gives $\left(\frac{d\delta}{dt}\right)^2 = \frac{2}{M}\int_{\delta_0}^{\delta} P_a d\delta$

∴
$$\frac{d\delta}{dt} = \sqrt{\frac{2}{M}\int_{\delta_0}^{\delta} P_a d\delta} \tag{13.20}$$

[where d_0 is the initial power angle before the rotor swing starts following a disturbance]

Application of stability criterion $\left[\frac{d\delta}{dt} = 0\right]$ leads

$$\sqrt{\frac{2}{M}\int_{\delta_0}^{\delta} P_a d\delta} = 0$$

i.e.,
$$\int_{\delta_0}^{\delta} P_a d\delta = 0 \tag{13.21}$$

In words, it means that *the area under the graph of accelerating power (P_a) versus δ must be zero for some value of δ.* It simply means that positive (or accelerating) area under the graph must be equal to the negative (or decelerating) area. This condition is, therefore, known as *equal area criterion* for stability.

13.5 INTERPRETATION OF EQUAL AREA CRITERION

Let us suppose, a generator is operating at steady state with initial operating angle δ_0 and delivering an electrical power $P_{e_0}(= P_{m_0})$, the $P_e - \delta$ profile being exhibited in Figure 13.4.

Let us further assume that there be a step change in mechanical power input from P_{m_0} to P_{m_1} at $t = 0$. Due to rotor inertia, the rotor position cannot change instantaneously (i.e., $\delta(0^-) = \delta(0) = \delta(0^+)$). This means that electrical power output remains unchanged and hence $P_e(0^-) = P_{e_0} = P_e(0^+)$. However, due to change (say increase) of P_{m_0} to P_{m_1} (we say that $P_{m_1} > P_{e_0}$) rotor will lead to consequent acceleration of the rotor ($P_e(0^+)$ is +ve) leading to $d^2\delta/dt^2(0^+)$ positive. With acceleration of the rotor, δ will increase to δ_1 till $P_{e_1} = P_{m_1}$. At this point (when $P_{e_1} = P_{m_1}$), $d^2\delta/dt^2$ is zero though $d\delta/dt$ is still positive and δ continuous to increase due to inertia of the rotor. It (δ) may even overshoot its final steady state operating point. When δ exceeds δ_1, P_{m_1} becomes less than P_e and obviously P_a (accelerating power) becomes negative. The rotor is then decelerating. Assuming absence of any damping, δ will continuously oscillate around δ_1. In practical machines, damping due to mechanical and electrical losses causes δ to stabilize at its final steady state operating point δ_1. It may be noted here that if δ exceeds δ_3, P_m would exceed P_e and rotor would accelerate again. This will lead to further increase in δ and loss of stability.

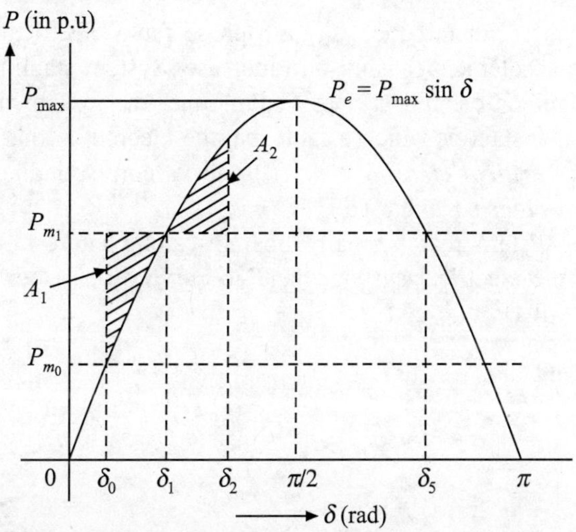

Figure 13.4 Profile of P_e vs δ (at $t = t_0$, $\delta = \delta_0$).

In Figure 13.4, $P_m > P_e$ during the interval $\delta_0 < \delta < \delta_1$ and the rotor accelerates during this time. The shaded area A_1 between P_m and P_e curves is called the *accelerating area* (A_1). During the interval $\delta_1 < \delta < \delta_2$, P_m is less than P_e and the rotor decelerates. The shaded area A_2 between P_m and P_e curves during this period is called *decelerating power*. At both the initial value ($\delta = \delta_0$) and the maximum value ($\delta = \delta_2$), $d\delta/dt = 0$. The equal area criterion states that A_1 must be equal to A_2 to have transient stabilty, i.e.

$$\int_{\delta_0}^{\delta_2} (P_m - P_e)d\delta = 0$$

or,

$$\int_{\delta_0}^{\delta_1} (P_m - P_e)d\delta + \int_{\delta_1}^{\delta_2} (P_m - P_e)d\delta = 0 \qquad [\because P_a = P_m - P_e]$$

or,
$$\int_{\delta_0}^{\delta_1} (P_m - P_e)d\delta = \int_{\delta_1}^{\delta_2} (P_e - P_m)d\delta \tag{13.22}$$

Stability problems similar to that described above may occur from sudden changes in electrical power, particularly during load throw, line switching action and faults in electrical network.

If we proceed further and perform integrations of both the sides of equation (13.22), we can write

$$P_m(\delta_1 - \delta_0) + P_{\max}(\cos\delta_1 - \cos\delta_0) = P_m(\delta_1 - \delta_2) + P_{\max}(\cos\delta_1 - \cos\delta_2) \tag{13.23}$$

But because of $P_m = P_e = P_{\max} \sin \delta_1$,
equation (13.23) becomes

$$(\delta_2 - \delta_0)\sin\delta_1 + \cos\delta_2 - \cos\delta_0 = 0 \tag{13.24}$$

If the quantities δ_0 and δ_1 are known, we can solve equation (13.24) to find δ_2.

13.6 CRITICAL CLEARING ANGLE AND ITS EXPRESSION

In case of a major electrical disturbance, say a 3-phase fault, electrical power abruptly drops down and the machine accelerates causing δ to increase. System stability is lost if δ increases too much, but if the fault is cleared at a particular time, the system may retain its transient stability. This particular instant at which a fault may be cleared so that the system stability is maintained is known as *critical clearing time* (t_c) and its corresponding angle in radian or in degree is called *critical clearing angle* (δ_c).

Let us consider a system that normally operates along curve-I. If a three phase short circuit fault occurs in the electrical system, the P–δ curve drops from its normal profile and is represented by curve-II (Figure 13.5).

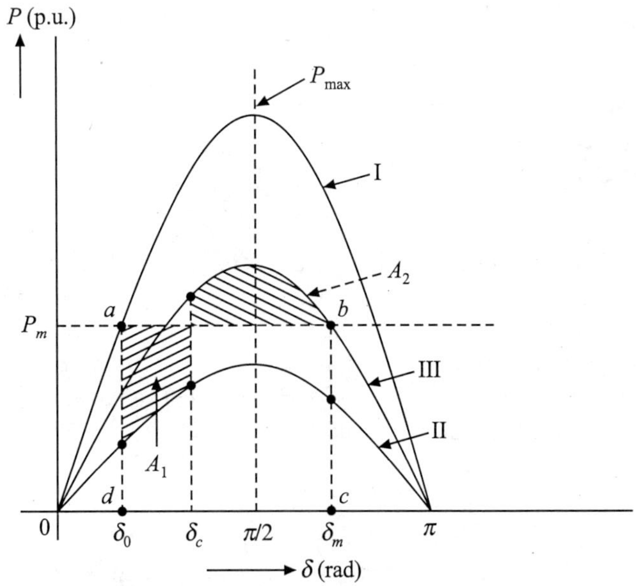

Figure 13.5 Power angle curve for different circuit conditions.

Once the fault is cleared, electrical power output is improved and the P–δ curve is represented by that given in curve-III.

Let $$P_{e(\text{curve-I})} = P_{\max_1} \sin\delta$$

$$P_{e(\text{curve-II})} = r_1 P_{e(\text{curve-I})} = P_{\max_2} \sin\delta$$

and $$P_{e(\text{curve-III})} = r_2 P_{e(\text{curve-I})} = P_{\max_3} \sin\delta$$

Also, $$r_1 < r_2 \text{ and } \delta_m = 180° - \sin^{-1}\left(\frac{P_m}{r_2 P_{\max_1}}\right) = \left[\pi - \sin^{-1}\left(\frac{P_m}{P_{\max_3}}\right)\right]$$

To have transient stability in Figure 13.5, we should have

$$\text{Area } A_1 = \text{Area } A_2$$

i.e., $$\int_{\delta_0}^{\delta_c} (P_m - P_{\max_2} \sin\delta)d\delta = \int_{\delta_c}^{\delta_m} (P_{\max_3} \sin\delta - P_m)d\delta$$

Integrating, we have

$$(P_m\delta + P_{\max_2} \cos\delta)\Big|_{\delta_0}^{\delta_c} + (P_{\max_3} \cos\delta + P_m\delta)\Big|_{\delta_c}^{\delta_m} = 0$$

or, $P_m(\delta_c - \delta_0) + P_{\max_2}(\cos\delta_c - \cos\delta_0) + P_m(\delta_m - \delta_c) + P_{\max_3}(\cos\delta_m - \cos\delta_c) = 0$

or, $$\cos\delta_c = \frac{P_m(\delta_m - \delta_0) - P_{\max_2}\cos\delta_0 + P_{\max_3}\cos\delta_m}{P_{\max_3} - P_{\max_2}} \tag{13.25a}$$

or, $$\delta_c = \cos^{-1}\left[\frac{P_m(\delta_m - \delta_0) - P_{\max_2}\cos\delta_0 + P_{\max_3}\cos\delta_m}{P_{\max_3} - P_{\max_2}}\right] \tag{13.25b}$$

Also, we can show that dividing equation (13.25b) by $P_{\max_1}$ at its numerator and denominator, we have

$$\delta_c = \cos^{-1}\left[\frac{1}{(r_2 - r_1)}\{(\delta_m - \delta_0)\sin\delta_0 - r_1\cos\delta_0 + r_2\cos\delta_m\}\right] \tag{13.25c}$$

where $r_1 = \dfrac{P_{\max_2}}{P_{\max_1}}$; $r_2 = \dfrac{P_{\max_3}}{P_{\max_1}}$, since $P_m = P_{e_1} = P_{\max_1}\sin\delta_0$, replacing $(\sin\delta_0)$ in equation (13.25c), we get

$$\delta_c = \cos^{-1}\left[\frac{1}{(r_2 - r_1)}\left\{\frac{P_m}{P_{\max_1}}(\delta_m - \delta_0) - r_1\cos\delta_0 + r_2\cos\delta_m\right\}\right] \tag{13.25d}$$

Expressions (13.25b), (13.25c) and (13.25d) give the value of δ_c in radian. Multiplying the value of δ_c by $(180°/\pi)$ we get the value of δ_c in degree.

In case there is *autoreclosure* and the fault is cleared, the post fault power approaches the prefault value (Figure 13.6), i.e., $P_{e(1V)} \equiv P_{e(1)} = P_{\max_1}\sin\delta$.

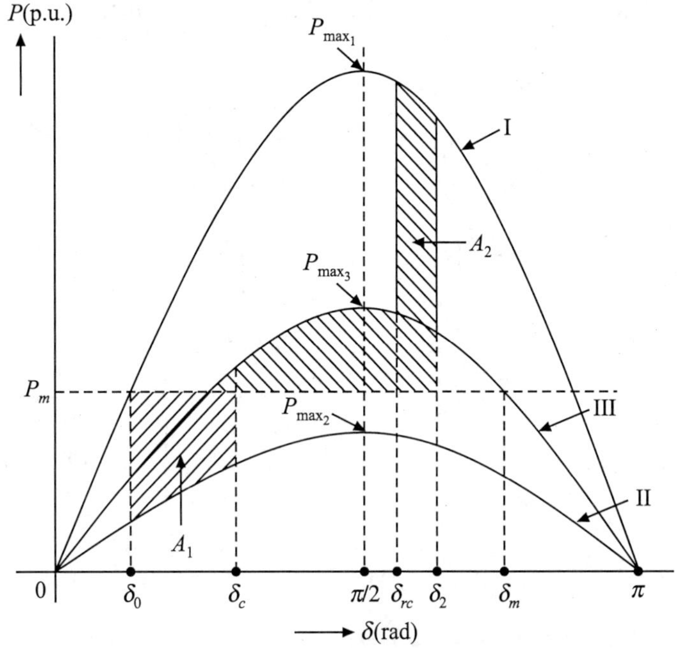

Figure 13.6 Fault clearing by autoreclosure.

Here $$\delta_m = \pi - \sin^{-1}\left(\frac{P_m}{P_{\max 4}}\right) = \pi - \sin^{-1}\left(\frac{P_m}{P_{\max 1}}\right)$$

To have transient stability, the following equation must be satisfied.

$$\text{Area } A_1 = \text{Area } A_2$$

i.e., $$\int_{\delta_0}^{\delta_c} (P_m - P_{\max_2} \sin\delta)d\delta = \int_{\delta_c}^{\delta_{rc}} (P_{\max_3} \sin\delta - P_m)d\delta + \int_{\delta_{rc}}^{\delta_m} (P_{\max_1} \sin\delta - P_m)d\delta \quad (13.26)$$

where δ_{rc} is the *angle of reclosure* and δ_m is the maximum value of δ_2.

Also, $$t_{rc} = t_c + t_0$$

where, t_{rc} = time of reclosure; t_c = time of critical clearing
and t_0 = time between clearing and reclosure.

[In order to get the expression of t_c, the critical clearing time, let us assume that following the short circuit, the electrical power drops down to zero.]

∴ From swing equation, we can write

$$\frac{2H}{\omega_0}\cdot\frac{d^2\delta}{dt^2} = P_m - 0 = P_m$$

or, $$\frac{d\delta}{dt} = \int_0^t \frac{\omega_0}{2H} P_m dt = \frac{\omega_0}{2H} P_m t$$

i.e., $$\delta = \int_0^t \frac{\omega_0}{2H} P_m t = \frac{\omega_0}{4H} P_m t^2 + \delta_0 \qquad (13.27)$$

If δ is replaced by δ_c, the critical clearing angle, we can write from equation (13.27)

$$\delta_c = \frac{\omega_0 P_m}{4H} t_c^2 + \delta_0$$

i.e., $$t_c = \sqrt{\frac{4H(\delta_c - \delta_0)}{\omega_0 P_m}} \qquad (13.28)]$$

EXAMPLE 13.4 A 50 Hz, three-phase synchronous generator delivers 0.8 p.u. power to an infinite busbar through a network of negligible resistance. A fault occurs which reduces the maximum transferable power to 0.50 p.u from 1.5 p.u. at post fault condition and after the clearance of the fault the maximum transferable power is 1.20 p.u. Determine the critical angle using the equal area criterion.

Solution: For this example,

$$P_m = 0.8 \text{ p.u.}, \; P_{\max_1} = 1.5 \text{ p.u.},$$

$$P_{\max_2} = 0.5 \text{ p.u.}, \; P_{\max_3} = 1.2 \text{ p.u.}$$

$$r_1 = \frac{P_{\max_2}}{P_{\max_1}} = \frac{0.5}{1.5} = 0.33; \; r_2 = \frac{P_{\max_3}}{P_{\max_1}} = \frac{1.2}{1.5} = 0.8$$

Now, $$\delta_0 = \sin^{-1}\left(\frac{P_m}{P_{\max_1}}\right) = \sin^{-1}\left(\frac{0.8}{1.5}\right) = \sin^{-1}(0.53) = 32.23° = 0.56 \text{ rad}$$

and $$\delta_m = 180° - \sin^{-1}\left(\frac{P_m}{P_{\max_3}}\right) = 180° - \sin^{-1}\left(\frac{0.8}{1.2}\right) = 138.19° = 2.41 \text{ rad}$$

∴ From equation (13.25c), we get

$$\delta_c = \cos^{-1}\left[\frac{1}{(r_2 - r_1)}\{(\delta_m - \delta_0)\sin\delta_0 - r_1\cos\delta_0 + r_2\cos\delta_m\}\right]$$

$$= \cos^{-1}\left[\frac{1}{(0.8 - 0.33)}\{(2.41 - 0.56)\sin 0.56 - 0.33\cos 0.56 + 0.8\cos 2.41\}\right]$$

[Here angles are in radian.]

$$= \cos^{-1}\left[\frac{1}{0.47}\{0.98 - 0.28 - 0.6\}\right]$$

$$= \cos^{-1}(0.21) = 1.36 \text{ rad}$$

∴ $\delta_c = 77.72°$ (Ans.)

EXAMPLE 13.5 An alternator, operating at 50% of its peak power capacity, is connected at an infinite bus. As soon as a fault occurs, the reactance between the generator and the infinite bus becomes four times its value before the fault. The peak power that can be delivered after clearance of fault is 80% of the original maximum value. Calculate the critical clearing angle.

Solution: Here, $P_m = 0.5\,P_{max_1},\ P_{max_3} = 0.8$

Now, $$\delta_0 = \sin^{-1}\left(\frac{P_m}{P_{max_1}}\right) = \sin^{-1}(0.5) = 0.52 \text{ rad}$$

Again, $$P_{max_1} = \frac{EV}{X}$$

[X represents the reactance between the generator and the infinite bus.]
During fault, X becomes $4X$.

$$\therefore \quad P_{max_2} = \frac{E'V}{4X}$$

$$r_1 = \frac{P_{max_2}}{P_{max_1}} = \frac{E'V/4X}{EV/X} = 0.25$$ [Due to constant flux linkage of the machine, $E' = E$.]

$$r_2 = \frac{P_{max_3}}{P_{max_1}} = 0.8$$

and $$\delta_m = \pi - \sin^{-1}\left(\frac{P_m}{P_{max_3}}\right) = \pi - \sin^{-1}\left(\frac{0.5\,P_{max_1}}{0.8\,P_{max_1}}\right) = 2.47 \text{ rad}$$

$$\delta_c = \cos^{-1}\left[\frac{1}{(r_2 - r_1)}\{(\delta_m - \delta_0)\sin\delta_0 - r_1\cos\delta_0 + r_2\cos\delta_m\}\right]$$

$$= \cos^{-1}\left[\frac{1}{(0.8 - 0.25)}\{(2.47 - 0.52)\sin 0.52 - 0.25\cos 0.52 + 0.8\cos 2.47\}\right]$$

[Here angles are in radian.]

$$= \cos^{-1}\left[\frac{1}{0.55}\{0.97 - 0.22 - 0.63\}\right]$$

$$= \cos^{-1}(0.22) = 1.35 \text{ rad}$$

$$\delta_c = 77.40^\circ \quad \text{(Ans.)}$$

13.7 APPLICATION OF EQUAL AREA CRITERION TO TRANSIENT STABILITY OF SYNCHRONOUS MOTOR

Let us consider a synchronous motor connected to an infinite busbar. It supplies a load of P_d and is operating at synchronous speed with torque angle δ_0, corresponding to this output power (P_e) on the power-angle diagram (sinusoidal) as shown in Figure 13.7. The power-angle equation is given by

$$P_e = \frac{|E_m||V|}{X}\sin\delta$$

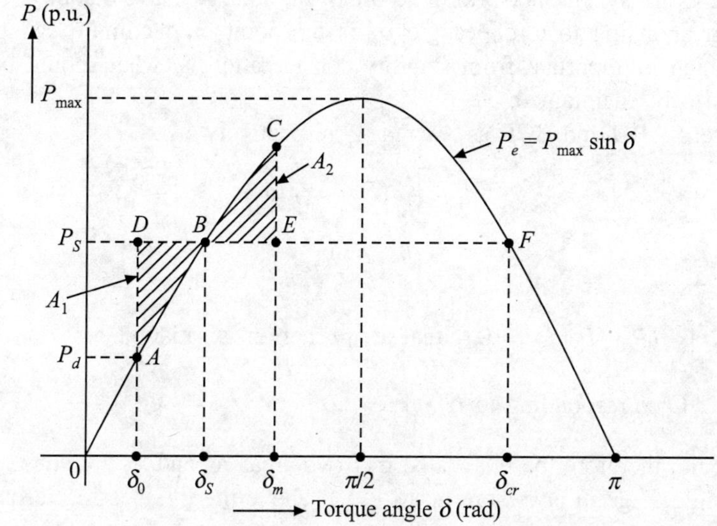

Figure 13.7 *P-δ* curve of a synchronous motor.

where,

$|V|$ = voltage of the infinite bus,

$|E_m|$ = voltage behind the transient reactance of the motor,

X = transient reactance of the motor including reactance of the transmission lines, transformers, etc., if any, between the infinite bus and the motor terminal.

It is clear that under these conditions the mechanical power output P_d and the electrical power input P_e are equal. If the load on the motor is suddenly increased to P_s, the electrical input power will initially be the same. The sudden excess of load must be available from the kinetic energy stored in the rotating system. This is possible only if the motor slows down. Immediately after the motor starts slowing down, the torque angle δ increases and with the increase in δ, there is an increase in electrical input power. This goes on till P_e and P_s are equal (corresponding to new increased load), at which point the deceleration ceases. The operating point on the power-angle diagram is then B. However, from A to B the motor has been decelerating at a finite rate and at B its speed is below synchronous speed. The result is that δ continues to increase until point C is reached, at which point the speed is again synchronous. Torque angle corresponding to C is δ_m. During the interval, B to C electrical power input P_e has been more than P_s and speed starts increasing till synchronous speed is reached. δ now starts decreasing till decreases point B is reached again with the speed above the synchronous speed. At B, the acceleration ceases and the rotor moves towards A with a finite deceleration; the cycle of oscillations then repeats itself till the oscillations are damped (due to losses). According to the equal area criterion of stability (ignoring losses), areas ABD and BCE must be equal.

The decelerating torque at any instant between A and B is $(T_s - T_e)$. Kinetic energy lost by the rotor is the area ABD. Similarly, the accelerating torque at any instant between B and C is $(T_e - T_s)$ and the kinetic energy gained by the rotor in this interval is the area BCE. Since speeds at A and C are synchronous, the two areas are equal to have stability.

Point C corresponding to torque angle δ_m is the point of maximum swing of the rotor. The question which is important from stability consideration is, what should be the value of δ_m if stability is to be maintained.

Denoting areas ABD and BCE as A_1 and A_2 respectively

$$A_1 = \int_{\delta_0}^{\delta_s} (P_s - P_e)d\delta \tag{13.29}$$

$$A_2 = \int_{\delta_s}^{\delta_m} (P_e - P_s)d\delta \tag{13.30}$$

With $A_1 - A_2 = \int_{\delta_0}^{\delta_m} (P_s - P_e)d\delta = 0$, equal area criterion is satisfied.

Also, $\dfrac{d\delta}{dt} = 0$ at C corresponding to δ_m.

Point C should, therefore, be so located that two areas A_1 and A_2 are equal. If point C lies on the power angle diagram anywhere up to F, i.e. above the line passing through point P_s, it will indicate transient stability.

EXAMPLE 13.6 A synchronous motor receives 25% of its peak power rating from an infinite bus. If the load on the motor is suddenly doubled, calculate the maximum value of power angle during the oscillation of the motor around its new equilibrium position. Neglect any loss in the motor.

Solution: Electrical input power to the motor is given by

$$P_e = P_{\max} \sin \delta$$

where $P_{\max}$ is the maximum power which the motor is capable to receive and is equal to 1.0 p.u. For this problem, according to the given data,

$$0.25\ P_{\max} = P_{\max} \sin \delta_0$$

$$\therefore \qquad \delta_0 = 14.47° = 0.25 \text{ rad.}$$

If the load on the motor is suddenly doubled, the new power angle be δ'; then we can write

$$2 \times 0.25\ P_{\max} = P_{\max} \sin \delta'$$

$$\therefore \qquad \delta' = 30° = 0.52 \text{ rad}$$

Let the maximum power angle be $\delta_{\max}$ [see Figure E13.2] such that stability is maintained.

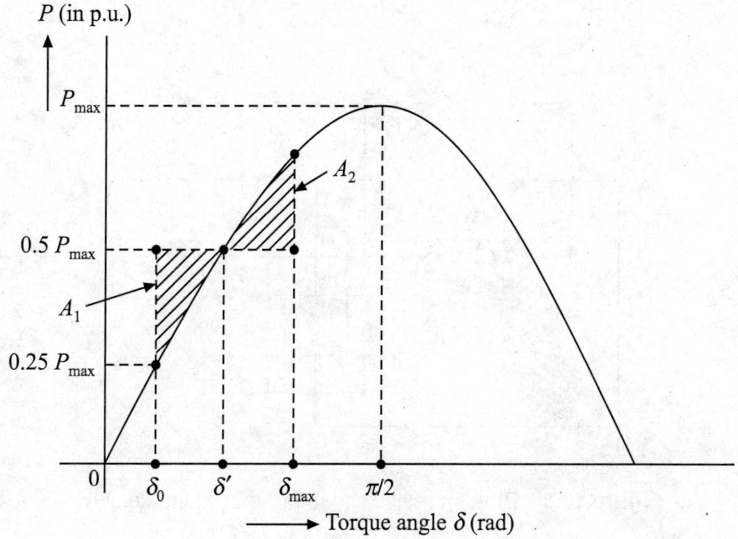

Figure E13.2

Applying equal area criterion, we have

$$-A_1 = A_2 \text{ [for motor]}$$

or, $$-\int_{\delta_0}^{\delta'} (0.5\, P_{max} - P_{max} \sin \delta) d\delta = \int_{\delta'}^{\delta_{max}} (P_{max} \sin \delta - 0.5\, P_{max}) d\delta$$

or, $$-P_{max} \left[0.5\right]_{0.25}^{0.52} - P_{max} \left[\cos \delta\right]_{0.25}^{0.52}$$

$$= -P_{max} \left[\cos \delta\right]_{0.52}^{\delta_{max}} - 0.5\, P_{max} \left[\delta\right]_{0.52}^{\delta_{max}}$$

or, $$-(0.26 - 0.13 + 0.87 - 0.97) + 0.5\delta_{max} + \cos \delta_{max} - 0.26 - 0.87 = 0$$

or, $$-0.03 = 0.5\delta_{max} + \cos \delta_{max} - 1.13$$

$\therefore$ $$\cos \delta_{max} + 0.5\delta_{max} - 1.1 = 0$$

This gives (by trial and error method)

$$\delta_{max} = 0.78 \text{ elect. rad} = 44.69°.$$

13.8 METHODS OF IMPROVING TRANSIENT STABILITY

13.8.1 Electrical Braking

The circuit breaker poles are normally closed and the resistances R in series with each generator winding are thus shorted under the normal working condition (see Figure 13.8). They are opened under fault condition, and absorb the accelerating energy during fault. Thus, the transient stability is improved. The use of braking resistors increases the generator output when the fault is on and even after it is cleared. It does not make the system inherently more reliable. It improves the stability at the cost of other characteristics of the system. Present-day quick-operating circuit breakers and relays leave little gain to be realized by braking resistors.

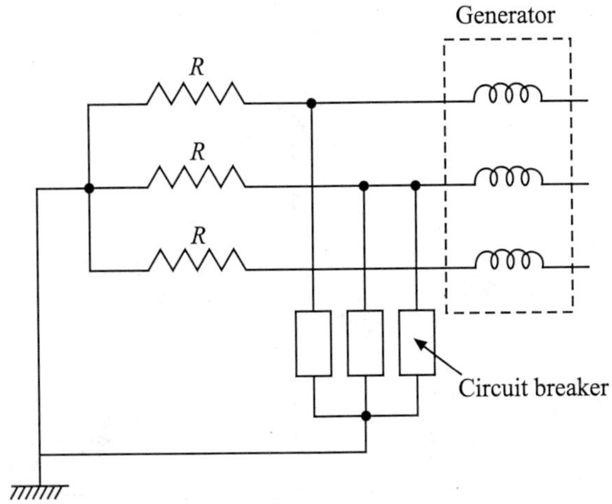

Figure 13.8 Braking resistor to improve system stability.

13.8.2 Effect of Voltage Regulators

The maximum power which generator can deliver is governed by field excitation of the generator. The excitation again depends upon the system operating conditions. Generators usually have their excitation adjusted to hold their terminal voltage at some predetermined value. With the help of an automatic voltage regulator, the terminal voltage can be held practically constant. If the load change is slow (such that the steady-state conditions may be assumed to exist), the field excitation would change slowly to adjust the terminal or system voltage to the nominal value. The so-called steady state stability limit corresponds to the maximum power development capability of the generator for slow changes of load. It implies infinitesimal change in load with constant field current. If the change in excitation takes place immediately after each small change in load, the stability limit under such a condition is known as the *dynamic stability limit*. This dynamic stability limit is always higher than the steady stability limit and is obtainable only with the help of automatic voltage regulators. The regulator has to be capable of making continuous and sufficiently rapid changes in the field excitation in order to maintain the dynamic stability. The use of voltage regulators helps to design large capacity generators when power is to be transmitted over fairly long distance. It is easier to maintain the dynamic stability with the help of regulators.

During or after a disturbance, the terminal voltage usually tends to be lower than normal. This means that even with a good excitation system, the current would be drawn at highly lagging power factor. The stability can be maintained and even the relays may be allowed to operate only and only if the voltage is not allowed falling appreciably below the prescribed value of the terminal voltage under the normal condition of operation. This condition can be assured only through the fast acting voltage regulators.

Though the use of automatic voltage regulators increases the transient stability of the system after the first swing, the stability limit during the first swing can be improved only by clearing the faults as early as possible by means of the circuit breakers and the protective devices used in the system.

13.8.3 Fast Governor Action

The fast changes occurring in the power angle during system disturbances can be arrested by varying the prime mover output with the help of the fast acting governor. If the prime mover output can be changed in time, the accelerating energy of the rotor can be decreased. Experiences have revealed that the earliest action of the governor can be brought in the system after about 15 cycles (i.e. 0.30 sec for 50 Hz system) of the occurrence of the fault. This is quite a long period from the severity of the faults and the system stability might have been lost within this period.

In the hydro power stations, the problem is still more difficult because of difficulty to control the flow of water than the steam flow. (The maximum instantaneous over speed has been about 35 to 50% upon sudden loss of load on the water-wheel generators while it is only 10% for the turbo-generators). Also, the governors and hydraulic conditions do not always permit rapid reduction of prime mover output to a water-wheel generator, though it is, to some extent, possible for a turbo-generator. This is the reason why governor action has not been so much advocated by the transient stability considerations.

The system stability can also be improved by inserting a permanent or switched series capacitor, or disconnecting a shunt-reactor and connecting a shunt capacitor and by closing throttle or gate valve in the sending area as soon as possible or by opening throttle or gate valve in the receiving area also as early as possible. The use of generators with higher H values also improves transient stability, though it is a costly method. The reduction of transfer reactance by reducing the conductor spacing or by increasing conductor diameter also improves transient stability. The use of bundled conductors is also an alternative method for improving transient stability. Another method to improve system stability is to increase the number of interconnecting EHV lines. Recent trends of improving transient stability include lowering of x_s (synchronous reactance) of alternator (i.e. enhancement of SCR, short circuit ratio), use of HVDC links in between regional grids and use of FACTS devices at load buses.

EXAMPLE 13.7 Assume a synchronous generator supplying power to an infinite bus through two parallel lines (Figure E13.3(a)). Study the transient stability using equal area criterion for sudden loss of line-2. Assume steady state operation before the assumed transient disturbance.

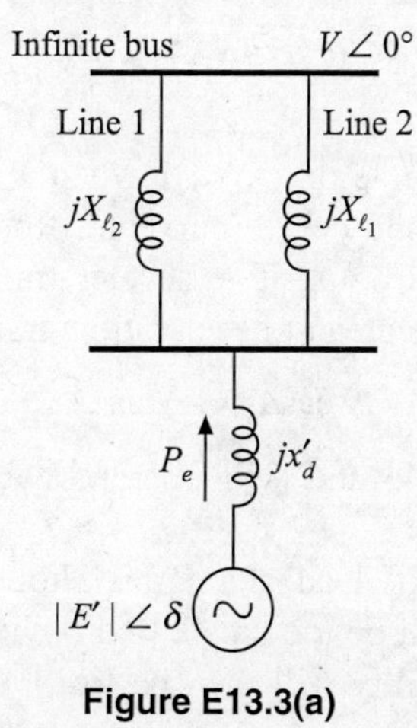

Figure E13.3(a)

Solution: During steady state operation (before switching of line 2), we have

$$P_{e_1} = \frac{|E'||V|\sin\delta}{x'_d + \{x_{\ell_1}x_{\ell_2}/(x_{\ell_1} + x_{\ell_1})\}} = P_{\max_1}\sin\delta$$

With removal of line 2, we have

$$P_{e_2} = \frac{|E'||V|\sin\delta}{x'_d + x_{\ell_1}} = P_{\max_2}\sin\delta$$

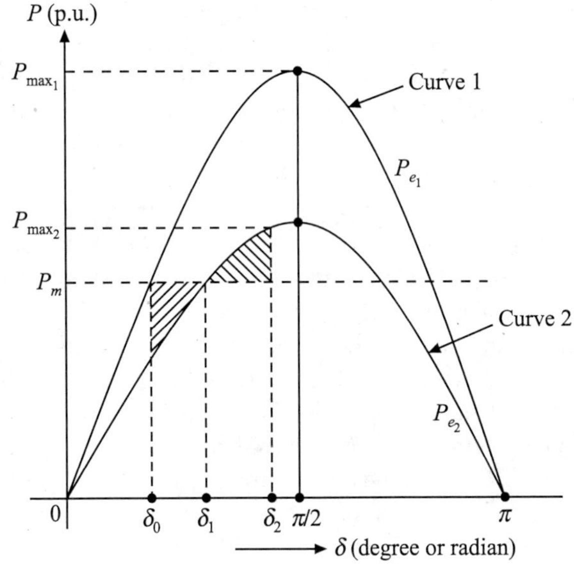

Figure E13.3(b)

Both these curves are plotted as shown in Figure E13.3(b). Obviously,

$$P_{\max_1} > P_{\max_2} \text{ as } (x'_d + x_{\ell_1}) > \left\{x'_d + \frac{x_{\ell_1}x_{\ell_2}}{x_{\ell_1} + x_{\ell_2}}\right\}.$$

If δ_0 be the initial operating angle on curve 1 when $P_e = P_m$ (at steady state), the critical operating point would shift to second curve (curve 2). Rotor will start accelerating as for P_m constant and P_e reducing there will be positive accelerating power P_a. Assuming A_1 to be the area of acceleration and A_2 to be the area of deceleration, transient stability will be maintained if

$$\text{Area } A_1 = \text{Area } A_2$$

The system will then be stable and will operate stably with $P_e = P_m$ and at angle δ_1 (obviously $\delta_1 > \delta_0$). (Ans.)

[In case there is an increase of load and P_m is shifted upwards, a limit will be finally achieved beyond which area A_2 will be too small, and obviously it will not be possible to have Area A_1 = Area A_2. Transient stability will then be lost.]

EXAMPLE 13.8 In the previous example (Example 13.7) without assuming sudden removal of line, let there be a three phase solid fault at the generator end of the line 2. All other conditions being same, examine the transient stability case with the help of equal area criterion.

Solution: Here as before,

$$P_{e_1} = \frac{|E'||V|\sin\delta}{x'_d + (x_{\ell_1} \| x_{\ell_2})} = P_{\max_1} \sin\delta$$

But $P_{e_2} = 0$, since there is a solid three phase fault at line 2 near the generator end. The circuit breakers at both the ends of line-2 will act on occurrence of the fault and they will isolate the line. The power flow will be immediately restored with healthy line (line 1).

Obviously,
$$P_{e_3} = \frac{|E'||V|\sin\delta}{x'_d + x_{\ell_1}} = P_{\max_3} \sin\delta$$

Also
$$P_{\max_1} > P_{\max_3}.$$

The rotor will start deceleration and the deceleration area is A_2 (A_1 was the accelerating area during fault). With reference to Figure E13.4(a), we conclude that transient stability will be maintained if Area A_1 = Area A_2. This is only possible if $\delta_2 < \delta_m$ and the fault is cleared within such a time so that there is ample scope to have Area A_1 = Area A_2. Thus (Figure E13.4(a) the fault clearing angle (or time) δ_{fc} is very important. Actually, δ_{fc} is called *critical clearing angle*.

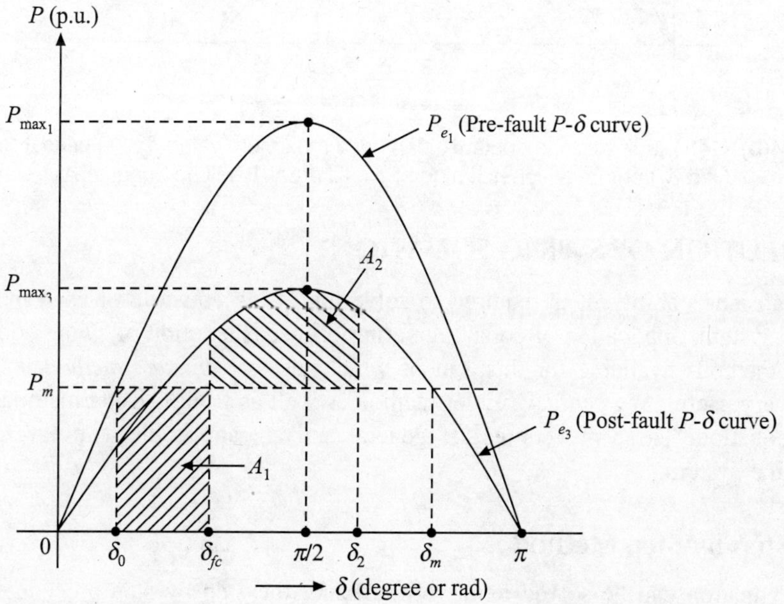

Figure E13.4(a) [During fault, the electrical power output is assumed to be zero, thus touches *x*-axis.]

[In case the fault on line 2 has not occurred near the generator end, and somewhere in between, then we need to use circuit reduction technique to find transfer reactance in the lines. This time

$$P_{e_2} = \frac{|E'||V|\sin\delta}{x'_d + x_t} = P_{\max_2} \sin\delta$$

where x_t is obtained using circuit reduction technique and is the transfer reactance in the lines. We will illustrate this problem through examples in the following part. Figure E13.4(b) represents the application of equal area criterion.

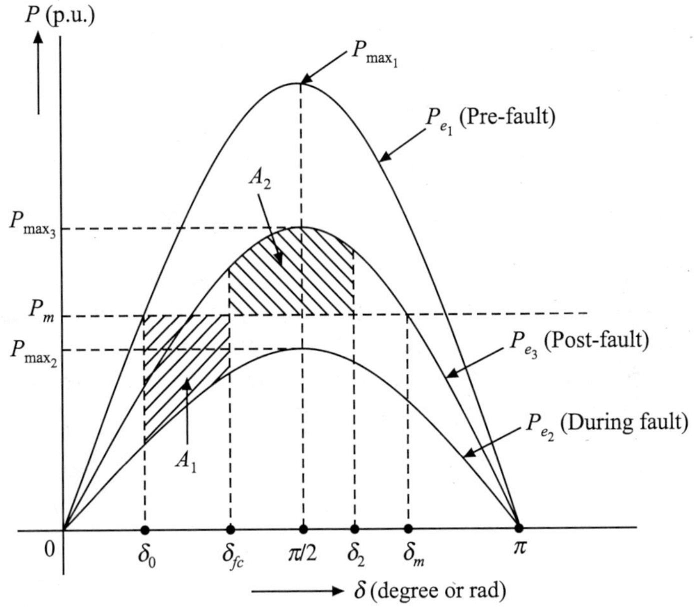

Figure E13.4(b) [The fault clearing angle δ_{fc} can have maximum value δ_c, the critical clearing angle. Also δ_2 can have highest value of δ_m, till area A_1 will be equal to A_2.]

13.9 SOLUTION OF SWING EQUATION

To study transient stability, it is required to solve the swing equation of each machine. Even for a SMIB system, application of equal area criterion needs solution of swing equation. There are several methods available for this, out of which the *step by step method* is conventional one and is very useful. We can have Euler method as well as Runge–Kutta methods of solution for swing equation. These procedures are general and are applicable to every machine of a multi-machine system.

13.9.1 Step by Step Method

The swing equation can be solved iteratively utilizing the step by step procedure. Here, it is assumed that the accelerating power P_a and the rotor angular velocity ω_r are constant within each of a succession of intervals as shown in Figure 13.9. These values are used to find the change in the power angle δ during each interval.

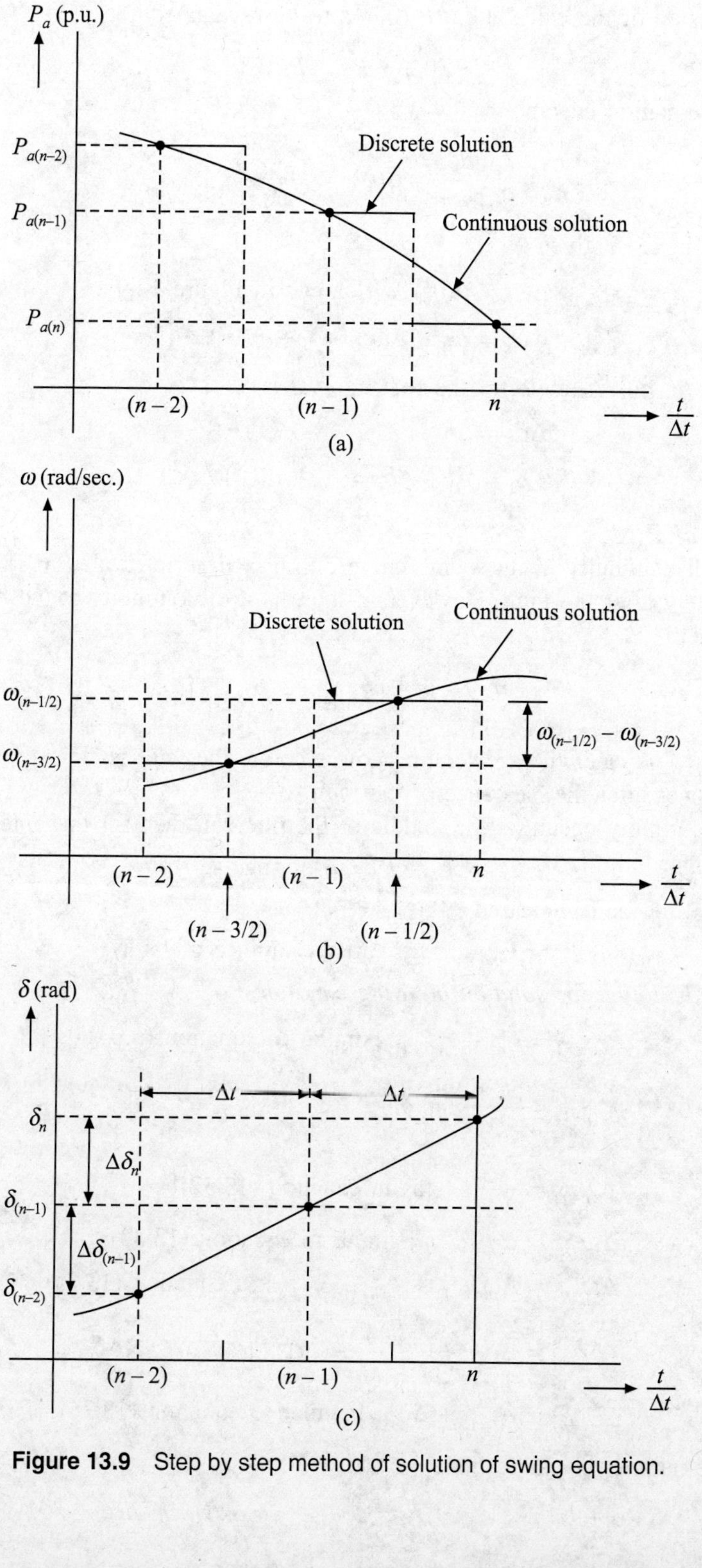

Figure 13.9 Step by step method of solution of swing equation.

In the first step of the calculation, $P_a(0^+)$ is to be revaluated.

$$P_a(0^+) = P_m - P_e(0^+) \tag{13.31}$$

The swing equation becomes

$$\frac{d^2\delta}{dt^2} = \alpha(0^+) = \frac{P_a(0^+)}{M} \tag{13.32}$$

and the change in ω_r is given by

$$\Delta\omega_r = \alpha(0^+)\Delta t \quad \text{[shown in Figure 13.9]} \tag{13.33}$$

Then,

$$\omega_r = \omega_0 + \Delta\omega_r = \omega_0 + \alpha(0^+)\ \Delta t \tag{13.34}$$

The change in power angle for the first interval is

$$\Delta\delta_1 = \Delta\omega_r\Delta t \tag{13.35}$$

And hence

$$\delta_1 = \delta_0 + \Delta\delta_1 = \delta_0 + \alpha(0^+)(\Delta t)^2 \tag{13.36}$$

Determination of P_a

If there is no discontinuity in the swing curve during a time interval, $P_a(0^+)$ is equal to half of P_a immediately after the fault. However, if there is a discontinuity at the beginning of the j-th interval, then

$$P_{a(j-1)} = \frac{1}{2}[P_{a(j-1)-} + P_{a(j-1)+}] \tag{13.37}$$

Here, $P_{a(j-1)-}$ indicates the accelerating power before the fault is cleared and $P_{a(j-1)+}$ is the accelerating power after the clearance of the fault.

If the discontinuity occurs at the middle of the interval, then for that interval,

$$P_a = P_m - P_{\text{output}} \text{ (during fault)} \tag{13.38}$$

As soon so the fault is cleared,

$$P_a = P_m - P_{\text{output}} \text{ (after clearance of fault)} \tag{13.39}$$

Steps of calculation for solution of swing equation

$$P_{a(n-1)} = P_m - P_{e(n-1)} \text{ [similar to equation (13.31)]} \tag{13.40}$$

$$P_{e(n-1)} = \frac{|E'V|}{X}\sin\delta \tag{13.41}$$

$$\alpha_{(n-1)} = \frac{P_{a(n-1)}}{M} \text{ [as in equation (13.32)]} \tag{13.42}$$

$$\Delta\omega_{r(n)} = \alpha_{(n-1)}\ \Delta t \text{ [similar to equation (13.33)]} \tag{13.43}$$

$$\Delta\omega_{r(n)} = \Delta\omega_{r(n-1)} + \alpha_{(n-1)}\ \Delta t \text{ [as in equation (13.34)]} \tag{13.44}$$

$$\Delta\delta_{(n)} = \Delta\delta_{(n-1)} + \frac{P_{a(n-1)}}{M}(\Delta t)^2 \quad [\because\ \Delta\delta_{(\gamma)} = \Delta\delta_{(n-1)} + \Delta\omega_{r(n)}\Delta t] \tag{13.45}$$

and

$$\delta_{(n)} = \delta_{(n-1)} + \Delta\delta_{(n)} \text{ [similar to equation (13.36)]} \tag{13.46}$$

where n is the number of calculation.

EXAMPLE 13.9 A synchronous generator having $H = 8$ MJ/MVA is connected to an infinite bus and supplying power of 1 p.u. with initial power angle (δ_0) as 25°. Assume a three phase fault occurring at $t = 0$ and cleared at $t = 0.2$ sec. The power equations expressed in p.u. are as under:

Power transfer in pre-fault condition = 2.5 sin δ
Power transfer during fault condition = 0.6 sin δ
Power transfer in post-fault condition = 1.5 sin δ

The system frequency is 50 Hz

Use step by step method to solve the swing equation with step size 0.05 till the fault is cleared. Write a MATLAB code for this problem.

Solution: Let us first segregate the swing equation in two first order differential equations as under:

$$\frac{d\delta}{dt} = \Delta\omega; \frac{d\Delta\omega}{dt} = \frac{P_a}{M} = \frac{\pi f}{H}(P_m - P_e) = \alpha \text{ (say)}$$

where $$M = \frac{H}{\pi f} \text{ MW-sec(p.u.)/rad}$$

Now let us use step by step method to solve these equations. At any state i, the following equations need to be solved.

$$\Delta\omega_i = \Delta\omega_{i-1} + \left.\frac{d\Delta\omega}{dt}\right|_{\delta_{i-1}} \times \Delta t = \Delta\omega_{i-1} + \alpha_{i-1} \times \Delta t$$

and $$\delta_i = \delta_{i-1} + \left.\frac{d\delta}{dt}\right|_{\Delta\omega_i} \times \Delta t = \delta_{i-1} + \Delta\omega_i + \Delta t$$

We are interested in getting δ and ω as function of t. Here, we assume that the machine has constant flux linkage and hence the induced emf is constant during the process of analysis. This leads to the assumption that E' (voltage behind transient reactance) is constant.

Let us now start calculation to find δ and $\Delta\omega$ for $t = 0.05$ sec

Here, $P_m = 1$ p.u.; $\delta_0(= 25°) = 0.4363$ rad; $H = 8$ MJ/MVA and $\Delta\omega_0 = 0$

$\therefore$ $P_a = 1 - 0.6 \sin \delta$ ($\because$ $P_a = P_m - P_e$ and $P_e = 0.6 \sin \delta$, during fault)

$$\alpha_0 = \left.\frac{d\Delta\omega}{dt}\right|_{\delta_0} = \frac{P_a}{M} = \frac{\pi \times 50}{8}(1 - 0.6 \sin 0.4363) = 14.6564 \text{ rad/sec}^2$$

and $$\Delta\omega_1 = \Delta\omega_0 + \alpha_0 \times \Delta t = 0 + 14.4564 \times 0.05 = 0.7328 \text{ rad/sec}$$

$\therefore$ $$\left.\frac{d\delta}{dt}\right|_{\Delta\omega_1} = \Delta\omega_1 = 0.7328 \text{ rad/sec}$$

Hence, $$\delta_1 = \delta_0 + \left.\frac{d\delta}{dt}\right|_{\Delta\omega_1} \times \Delta t = 0.4363 + 0.7328 \times 0.05 = 0.4729 \text{ rad}$$

$$= 27.0975°$$

Next, we will find δ and $\Delta\omega$ for t = 0.10 sec and for this we will perform the second calculation.

$$\alpha_1 = \left.\frac{d\Delta\omega}{dt}\right|_{\delta_1} = \frac{P_a}{M} = \frac{\pi \times 50}{8}(1 - 0.6 \sin 0.4729) = 14.2690 \text{ rad/sec}^2$$

And $\Delta\omega_2 = \Delta\omega_1 + \alpha_1 \times \Delta t = 0.7328 + 14.2690 \times 0.05 = 1.4463$ rad/sec

$$\therefore \quad \left.\frac{d\delta}{dt}\right|_{\Delta\omega_2} = \Delta\omega_2 = 1.4463 \text{ rad/sec}$$

Hence, $\delta_2 = \delta_1 + \left.\frac{d\delta}{dt}\right|_{\Delta\omega_2} \times \Delta t = 0.4729 + 1.4463 \times 0.05 = 0.5452$ rad

$= 31.2385°$

In the next step, we are to find δ and $\Delta\omega$ for t = 0.15 sec and for this, we are to perform the third calculation.

$$\alpha_2 = \left.\frac{d\Delta\omega}{dt}\right|_{\delta_1} = \frac{P_a}{M} = \frac{\pi \times 50}{8}(1 - 0.6 \sin 0.5452) = 13.5254 \text{ rad/sec}^2$$

And $\Delta\omega_3 = \Delta\omega_2 + \alpha_2 \times \Delta t = 1.4463 + 13.5254 \times 0.05 = 2.1226$ rad/sec

$$\therefore \quad \left.\frac{d\delta}{dt}\right|_{\Delta\omega_3} = \Delta\omega_3 = 2.1226 \text{ rad/sec}$$

Hence, $\delta_3 = \delta_2 + \left.\frac{d\delta}{dt}\right|_{\Delta\omega_3} \times \Delta t = 0.5452 + 2.1226 \times 0.05 = 0.6513$ rad

$= 37.3185°$

After 0.05 sec, i.e. at 0.2 sec, the fault is cleared. Hence, we are to find the value of δ and $\Delta\omega$ at t = 0.2 sec with post-fault swing equation.

Here, $P_a = 1 - 1.5 \sin \delta$; $\Delta t = 0.05$ sec

Also, $\left.\frac{d\delta}{dt}\right|_{\Delta\omega_3} = \Delta\omega_3 = 2.1226$ rad/sec

and $\alpha_3 = \left.\frac{d\Delta\omega}{dt}\right|_{\delta_3} = \frac{P_a}{M} = \frac{\pi \times 50}{8}(1 - 1.5 \sin 0.6513) = 1.7803 \text{ rad/sec}^2$

Also, $\Delta\omega_4 = \Delta\omega_3 + \alpha_3 \times \Delta t = 2.1226 + 1.7803 \times 0.05 = 2.2116$ rad/sec

$$\therefore \quad \left.\frac{d\delta}{dt}\right|_{\Delta\omega_4} = \Delta\omega_4 = 2.2116 \text{ rad/sec}$$

Hence, $\delta_4 = \delta_3 + \left.\frac{d\delta}{dt}\right|_{\Delta\omega_4} \times \Delta t = 0.6513 + 2.2116 \times 0.05 = 0.7619$ rad

$= 43.6526°$ (Ans.)

We can write the results obtained in a tabular form

Time (in sec)	*δ (in radian)*	*δ (in degree)*
0	0.4363	25
0.05	0.4729	27.0975
0.1	0.5452	31.2385
0.15	0.6513	37.3185
0.2	0.7619	43.6526

The MATLAB code for Example 13.9 is as follows:

```
%Prog to solve Example 13.9
clc; clear all
f=50;  h=8;  pm=1;pem_pre=2.5;  pem_f=0.6;  pem_post=1.5;  delta=0.4363;
delta=delta';
swcf=pi*f*pm/h; swvf=pi*f*pem_f/h; swcp=pi*f*pm/h; swvp=pi*f*pem_post/h;
tclr=0.2; t(1)=0;  dt=0.05; tstep=tclr/dt; dw(1)=0;
%solution of Swing Equation by step by step method
 for k=1:tstep
  t(k+1)=t(k)+dt;
   if(t(k+1)<tclr)
     sw(k)=swcf-swvf*sin(delta(k));
   else
     sw(k)=swcp-swvp*sin(delta(k));
   end
 alpha(k)=sw(k); ddw(k)=alpha(k)*dt; dw(k+1)=dw(k)+ddw(k);
 ddelta(k)=dw(k+1)*dt; delta(k+1)=delta(k)+ddelta(k);
 t(k+1)=t(k)+dt;
 end
disp('Solution of swing equation by Step by Step method.')
disp(sprintf('time interval=%4.3fs, total time step=%d, fault clearance
time=%4.3fs',dt,tstep,tclr))
 disp(sprintf('t(in sec)   delta(in radian)   delta(in degree)'))
 for k=1:tstep+1
 deltad(k)=delta(k)*180/pi;
 disp(sprintf('%4.3f     %10.6f       %10.6f  ',t(k),delta(k),deltad(k)))
 end
```

Execution of the computer program to solve swing equation by step by step method for Example 13.9:

```
Solution of swing equation by Step by Step method.
time interval=0.05s, total time step=4, fault clearance time=0.2s
t(in sec)   delta(in radian)   delta(in degree)
0.000          0.436300           24.998149
0.050          0.472941           27.097530
0.100          0.545254           31.240744
0.150          0.651379           37.321261
0.200          0.761950           43.656519
```

13.9.2 Modified Euler's Method

The modified Euler's method is extensively used here to solve the swing equation. Let us first segregate the swing equation in two first order differential equations as under:

$$\frac{d\delta}{dt} = \Delta\omega \,; \quad \frac{d\Delta\omega}{dt} = \frac{P_a}{M} = \frac{\pi f}{H}(P_m - P_e)$$

where $M = \dfrac{H}{\pi f}$ MW-sec (p.u.)/rad.

The modified Euler's method in transient stability problem solution is employed to solve these equations by predicting and subsequently correcting the variables in each calculation step. It is a two-step procedure for solving each of the solution state. At any state i, the following equations need to be solved.

$$\delta_i^p = \delta_{i-1} + \left.\frac{d\delta}{dt}\right|_{\Delta\omega_{i-1}} \times \Delta t$$

and
$$\Delta\omega_i^p = \Delta\omega_{i-1} + \left.\frac{d\Delta\omega}{dt}\right|_{\delta_{i-1}} \times \Delta t$$

With predicted values (represented as subscript p), we can write for the derivatives

$$\left.\frac{d\delta}{dt}\right|_{\Delta\omega_i^p} = \Delta\omega_i^p; \quad \left.\frac{d\Delta\omega}{dt}\right|_{\delta_i^p} = \left.\frac{P_a}{M}\right|_{\delta_i^p}$$

The corrected values (represented as subscript c) are then obtained as

$$\delta_i^c = \delta_{i-1} + \left[\frac{\left.\frac{d\delta}{dt}\right|_{\Delta\omega_{i-1}} + \left.\frac{d\delta}{dt}\right|_{\Delta\omega_i^p}}{2}\right] \times \Delta t$$

and
$$\Delta\omega_i^c = \Delta\omega_{i-1} + \left[\frac{\left.\frac{d\Delta\omega}{dt}\right|_{\delta_{i-1}} + \left.\frac{d\Delta\omega}{dt}\right|_{\delta_i^p}}{2}\right] \times \Delta t$$

For the next step of calculation, $\delta_i^c = \delta_i$ and $\Delta\omega_i^c = \Delta\omega_i$. The process continues and in each step, it is possible to get values of δ_i^c against the corresponding state.

EXAMPLE 13.10 Use modified Euler's method to solve the swing equation with step size 0.05 till the fault is cleared for the system described in Example 13.9. Write a MATLAB code to solve this problem.

Solution: The two first order differential equations are as under:

$$\frac{d\delta}{dt} = \Delta\omega \,; \quad \frac{d\Delta\omega}{dt} = \frac{P_a}{M} = \frac{\pi f}{H}(P_m - P_e)$$

Let us now start first step of calculation to find δ and $\Delta\omega$ for $t = 0.05$ sec

Here, $P_m = 1$ p.u.; δ_0 (= 25°) = 0.4363 rad and $H = 8$ MJ/MVA

Also, $P_a = 1 - 0.6 \sin\delta$

Since the fault is at $t = 0$,

$$\left.\frac{d\delta}{dt}\right|_{\Delta\omega_0} = \Delta\omega_0 = 0 \text{ rad/sec}$$

and $$\left.\frac{d\Delta\omega}{dt}\right|_{\delta_0} = \frac{P_a}{M} = \frac{\pi \times 50}{8}(1 - 0.6 \sin 0.4363) = 14.6564 \text{ rad/sec}^2$$

The predicted values of δ and $\Delta\omega$ at first step of calculation are:

$$\delta_i^p = \delta_0 + \left.\frac{d\delta}{dt}\right|_{\Delta\omega_0} \times \Delta t = 0.4363 + 0 = 0.4363 \text{ rad}$$

and $\Delta\omega_1^p = \Delta\omega_0 + \left.\frac{d\Delta\omega}{dt}\right|_{\delta_0} \times \Delta t = 0 + 14.4564 \times 0.05 = 0.7328$ rad/sec [step size $\Delta t = 0.05$s]

$$\therefore \quad \left.\frac{d\delta}{dt}\right|_{\Delta\omega_1^p} = \Delta\omega_1^p = 0.7328 \text{ rad/sec}$$

Also, $$\left.\frac{d\Delta\omega}{dt}\right|_{\delta_1^p} = \left.\frac{P_a}{M}\right|_{\delta_1^p} = \frac{\pi \times 50}{8}(1 - 0.6 \sin 0.4363) = 14.6564 \text{ rad/sec}^2$$

The corrected values ($\Delta t = 0.05$s) are now obtained as

$$\delta_1 (\equiv \delta_1^c) = \delta_0 + \left[\frac{\left.\frac{d\delta}{dt}\right|_{\Delta\omega_0} + \left.\frac{d\delta}{dt}\right|_{\Delta\omega_1^p}}{2}\right] \times \Delta t = 0.4363 + \left[\frac{0 + 0.7328}{2}\right] \times 0.05$$

$$= 0.4546 \text{ rad} = 26.0467°$$

and $$\Delta\omega_1 (\equiv \Delta\omega_1^c) = \Delta\omega_0 + \left[\frac{\left.\frac{d\Delta\omega}{dt}\right|_{\delta_0} + \left.\frac{d\Delta\omega}{dt}\right|_{\delta_1^p}}{2}\right] \times \Delta t$$

$$= 0 + \left[\frac{14.6564 + 14.6564}{2}\right] \times 0.05 = 0.7328 \text{ rad/sec.}$$

$\therefore$ At $t = 0.05$ sec, $\delta_1 = 0.4546$ rad = 26.0467°

Next, in the second step of calculation, we are to find δ and $\Delta\omega$ for $t = 0.1$ sec.

Here, $$\left.\frac{d\delta}{dt}\right|_{\Delta\omega_1} = \Delta\omega_1 = 0.7328 \text{ rad/sec}$$

and $$\left.\frac{d\Delta\omega}{dt}\right|_{\delta_1} = \frac{P_a}{M} = \frac{\pi \times 50}{8} (1 - 0.6 \sin 0.4546) = 14.4618 \text{ rad/sec}^2$$

The predicted values of δ and $\Delta\omega$ in second step of calculation are:

$$\delta_2^p = \delta_1 + \left.\frac{d\delta}{dt}\right|_{\Delta\omega_1} \times \Delta t = 0.4546 + 0.7328 \times 0.05 = 0.4912 \text{ rad}$$

and $$\Delta\omega_2^p = \Delta\omega_1 + \left.\frac{d\Delta\omega}{\partial t}\right|_{\delta_1} \times \Delta t = 0.7328 + 14.4618 \times 0.05 = 1.4559 \text{ rad/sec}$$

$\therefore$ $$\left.\frac{d\delta}{dt}\right|_{\Delta\omega_2^p} = \Delta\omega_2^p = 1.4559 \text{ rad/sec.}$$

$$\left.\frac{d\Delta\omega}{dt}\right|_{\delta_1^p} = \left.\frac{P_a}{M}\right|_{\delta_1^p} = \frac{\pi \times 50}{8} (1 - 0.6 \sin 0.4912) = 14.0781 \text{ rad/sec}^2$$

The corrected values ($t = 0.1$s and $\Delta t = 0.05$s) are thus obtained as

$$\delta_2 (\equiv \delta_2^c) = \delta_0 + \left[\frac{\left.\frac{d\delta}{dt}\right|_{\Delta\omega_1} + \left.\frac{d\delta}{dt}\right|_{\Delta\omega_2^p}}{2}\right] \times \Delta t$$

$$= 0.4546 + \left[\frac{0.7328 + 1.4559}{2}\right] \times 0.05$$

$$= 0.5093 \text{ rad} = 29.1817°$$

and $$\Delta\omega_2 (\equiv \Delta\omega_2^c) = \Delta\omega_1 + \left[\frac{\left.\frac{d\Delta\omega}{dt}\right|_{\delta_1} + \left.\frac{d\Delta\omega}{dt}\right|_{\delta_2^p}}{2}\right] \times \Delta t$$

$$= 0.7328 + \left[\frac{14.4618 + 14.0781}{2}\right] \times 0.05 = 1.4463 \text{ rad/sec}$$

$\therefore$ At $t = 0.1$ sec, $\delta_2 = 0.5093$ rad $= 29.1817°$

Next, we are to find the values of δ and $\Delta\omega$ for $t = 0.15$ sec and for this, we will perform the third step of calculation.

Here, $$\left.\frac{d\delta}{dt}\right|_{\Delta\omega_2} = \Delta\omega_2 = 1.4463 \text{ rad/sec}$$

and $$\left.\frac{d\Delta\omega}{dt}\right|_{\delta_2} = \frac{P_a}{M} = \frac{\pi \times 50}{8} (1 - 0.6 \sin 0.5093) = 13.8909 \text{ rad/sec}^2$$

The predicted values of δ and $\Delta\omega$ at the third step of calculation are:

$$\delta_3^p = \delta_2 + \left.\frac{d\delta}{dt}\right|_{\Delta\omega_2} \times \Delta t = 0.5093 + 1.4463 \times 0.05 = 0.5816 \text{ rad}$$

and $$\Delta\omega_3^p = \Delta\omega_2 + \left.\frac{d\Delta\omega}{dt}\right|_{\delta_2} \times \Delta t = 1.4463 + 13.8909 \times 0.05 = 2.1408 \text{ rad/sec.}$$

$\therefore$ $$\left.\frac{d\delta}{dt}\right|_{\Delta\omega_3^p} = \Delta\omega_3^p = 2.1408 \text{ rad/sec}$$

and $$\left.\frac{d\Delta\omega}{dt}\right|_{\delta_3^p} = \left.\frac{P_a}{M}\right|_{\delta_3^p} = \frac{\pi \times 50}{8}(1 - 0.6 \sin 0.5816) = 13.1629 \text{ rad/sec}^2$$

The corrected values (t = 0.15s and Δt = 0.05s) are thus

$$\delta_3 (\equiv \delta_3^c) = \delta_2 + \left[\frac{\left.\frac{d\delta}{dt}\right|_{\Delta\omega_2} + \left.\frac{d\delta}{dt}\right|_{\Delta\omega_3^p}}{2}\right] \times \Delta t$$

$$= 0.5093 + \left[\frac{1.4463 + 2.1408}{2}\right] \times 0.05$$

$$= 0.5990 \text{ rad} = 34.3189°$$

and $$\Delta\omega_3 (\equiv \Delta\omega_3^c) = \Delta\omega_2 + \left[\frac{\left.\frac{d\Delta\omega}{dt}\right|_{\delta_2} + \left.\frac{d\Delta\omega}{dt}\right|_{\delta_3^p}}{2}\right] \times \Delta t$$

$$= 1.4463 + \left[\frac{13.8909 + 13.1629}{2}\right] \times 0.05 = 2.1226 \text{ rad/sec}$$

$\therefore$ At t = 0.15 sec, δ_3 = 0.599 rad = 34.3189°

Again after 0.05 sec, i.e. at 0.2 sec, the fault is cleared. We will now find the value of δ and $\Delta\omega$ at t = 0.2 sec. For this, the fourth step of calculation is to be performed.

Here, $$P_a = 1 - 1.5 \sin \delta;\ \Delta t = 0.05 \text{ sec}$$

Also, $$\left.\frac{d\delta}{dt}\right|_{\Delta\omega_3} = \Delta\omega_3 = 2.1226 \text{ rad/sec}$$

and $$\left.\frac{d\Delta\omega}{dt}\right|_{\delta_3} = \frac{P_a}{M} = \frac{\pi \times 50}{8}(1 - 1.5 \sin 0.599) = 3.0292 \text{ rad/sec}^2$$

The predicted values of δ and $\Delta\omega$ at the fourth step of calculation are:

$$\delta_4^p = \delta_3 + \left.\frac{d\delta}{dt}\right|_{\Delta\omega_3} \times \Delta t = 0.599 + 2.1226 \times 0.05 = 0.7051 \text{ rad}$$

and $$\Delta\omega_4^p = \Delta\omega_3 + \left.\frac{d\Delta\omega}{dt}\right|_{\delta_3} \times \Delta t = 2.1226 + 3.0292 \times 0.05 = 2.2741 \text{ rad/sec}$$

$\therefore$ $$\left.\frac{d\delta}{dt}\right|_{\Delta\omega_4^p} = \Delta\omega_4^p = 2.2741 \text{ rad/sec}$$

and $$\left.\frac{d\Delta\omega}{dt}\right|_{\delta_4^p} = \left.\frac{P_a}{M}\right|_{\delta_4^p} = \frac{\pi \times 50}{8}(1 - 1.5 \sin 0.7051) = 0.5465 \text{ rad/sec}^2$$

The corrected values (t = 0.2s and Δt = 0.05s) are thus obtained as

$$\delta_4 (\equiv \delta_4^c) = \delta_3 + \left[\frac{\left.\frac{d\delta}{dt}\right|_{\Delta\omega_3} + \left.\frac{d\delta}{dt}\right|_{\Delta\omega_4^p}}{2}\right] \times \Delta t$$

$$= 0.599 + \left[\frac{2.1226 + 2.2741}{2}\right] \times 0.05 = 0.7089 \text{ rad} = 40.6170°$$

and $$\Delta\omega_4 (\equiv \Delta\omega_4^c) = \Delta\omega_3 + \left[\frac{\left.\frac{d\Delta\omega}{dt}\right|_{\delta_3} + \left.\frac{d\Delta\omega}{dt}\right|_{\delta_4^p}}{2}\right] \times \Delta t$$

$$= 2.1226 + \left[\frac{3.0292 + 0.5465}{2}\right] \times 0.05 = 2.2120 \text{ rad/sec}$$

$\therefore$ At t = 0.2 sec, δ_4 = 0.7089 rad = 40.6170°

We can write the results obtained in a tabular form:

Time (in sec)	*δ (in radian)*	*δ (in degree)*
0	0.4363	25
0.05	0.4546	26.0467
0.1	0.5093	29.1817
0.15	0.5990	34.3189
0.2	0.7089	40.6170

The MATLAB code for Example 13.10 is as follows:

```
%Prog to solve Example 13.10
clc; clear all
f=50;  h=8;  pm=1;pem_pre=2.5;  pem_f=0.6;  pem_post=1.5;  delta=0.4363;
```

```
delta=delta';
swcf=pi*f*pm/h; swvf=pi*f*pem_f/h; swcp=pi*f*pm/h; swvp=pi*f*pem_post/h;
tclr=0.2; t(1)=0;  dt=0.05; tstep=tclr/dt; dw(1)=0;
%Solution of swing equation by Modifier Euler method.
for k=1:tstep
t(k+1)=t(k)+dt;
 if(t(k+1)<tclr)
  sw(k)=swcf-swvf*sin(delta(k));
 else
  sw(k)=swcp-swvp*sin(delta(k));
 end
 alpha(k)=sw(k); ddw(k)=alpha(k)*dt; ddelta(k)=dw(k)*dt;
 deltap(k+1)=delta(k)+ddelta(k); dwp(k+1)=dw(k)+ddw(k);
 if(t(k+1)<tclr)
   sw(k+1)=swcf-swvf*sin(deltap(k+1));%sw during fault
 else
   sw(k+1)=swcp-swvp*sin(deltap(k+1)); %sw after fault
 end
 alpha(k+1)=sw(k+1);  ddw(k+1)=alpha(k+1)*dt;
 ddelta(k+1)=dwp(k+1)*dt;  dw(k+1)=dw(k)+(ddw(k)+ddw(k+1))/2;
 delta(k+1)=delta(k)+(ddelta(k)+ddelta(k+1))/2;
end
disp('Solution of swing equation by Modifier Euler method.')
disp(sprintf('time interval=%4.3fs, total time step=%d, fault clearance
time=%4.3fs',dt,tstep,tclr))
 disp(sprintf('t(in sec)  delta(in radian)   delta(in degree)'))
for k=1:tstep+1
 deltad(k)=delta(k)*180/pi;
 disp(sprintf('%4.3f    %10.6f       %10.6f  ',t(k),delta(k),deltad(k)))
end
```

Execution of the computer program to solve swing equation by modified Euler's method for Example 13.10:

```
Solution of swing equation by Modifier Euler method.
time interval=0.05s, total time step=4, fault clearance time=0.2s
t(in sec)   delta(in radian)   delta(in degree)
0.000           0.436300          24.998149
0.050           0.454621          26.047839
0.100           0.509339          29.182962
0.150           0.599017          34.321142
0.200           0.708934          40.618930
```

Note: Let us now assume that post-fault power equation is 2.0 sin δ instead of given value 1.5 sin δ. Coming back to calculation of step-4, we now find, $P_a = 1 - 2 \sin \delta$;

Also we have obtained earlier, $\left.\dfrac{d\delta}{dt}\right|_{\Delta\omega_3} = \Delta\omega_3 = 2.1226$ rad/sec.

With new post-fault power equation, we get

$$\left.\frac{d\Delta\omega}{dt}\right|_{\delta_3} = \frac{P_a}{M} = \frac{\pi \times 50}{8}(1 - 2 \sin 0.599) = -2.5061 \text{ rad/sec}^2$$

It may be noted here that in the post-fault period, the machine starts deceleration instead of acceleration in the original case. And this should improve transient stability. Let us continue to get the final values of δ and $\Delta\omega$.

The predicted values of δ and $\Delta\omega$ at this step of calculation are:

$$\delta_4^p(\text{new}) = \delta_3 + \left.\frac{d\delta}{dt}\right|_{\Delta\omega_3} \times \Delta t = 0.599 + 2.1226 \times 0.05 = 0.7051 \text{ rad}$$

and $$\Delta\omega_4^p(\text{new}) = \Delta\omega_3 + \left.\frac{d\Delta\omega}{dt}\right|_{\delta_3} \times \Delta t = 2.1226 - 2.5061 \times 0.05 = 1.9973 \text{ rad/sec}$$

Note that $\Delta\omega$ reduces than the value $\Delta\omega$ at $t = 0.15$ sec (with the new post-fault power equation).

$$\therefore \quad \left.\frac{d\delta}{dt}\right|_{\Delta\omega_4^p} = \Delta\omega_4^p = 1.9973 \text{ rad/sec}$$

Here $$\left.\frac{d\Delta\omega}{dt}\right|_{\delta_4^p} = \left.\frac{P_a}{M}\right|_{\delta_4^p} = \frac{\pi \times 50}{8}(1 - 2 \sin 0.7051) = -5.8163 \text{ rad/sec}^2$$

The corrected values of δ and $\Delta\delta$ ($t = 0.2$s and $\Delta t = 0.05$s in the new case) are

$$\delta_4 (\equiv \delta_4^c) = \delta_3 + \left[\frac{\left.\frac{d\delta}{dt}\right|_{\Delta\omega_3} + \left.\frac{d\delta}{dt}\right|_{\Delta\omega_4^p}}{2}\right] \times \Delta t$$

$$= 0.599 + \left[\frac{2.1226 + 1.9973}{2}\right] \times 0.05 = 0.7020 \text{ rad} = 40.2215°$$

and $$\Delta\omega_4 (\equiv \Delta\omega_4^c) = \Delta\omega_3 + \left[\frac{\left.\frac{d\Delta\omega}{dt}\right|_{\delta_3} + \left.\frac{d\Delta\omega}{\partial t}\right|_{\delta_4^p}}{2}\right] \times \Delta t$$

$$= 2.1226 + \left[\frac{-2.5061 - 5.8163}{2}\right] \times 0.05 = 1.9145 \text{ rad/sec.}$$

$\therefore$ At $t = 0.2$ sec, $\quad \delta_4 = 0.7089$ rad $= 40.6170°$

Comparing these results with the original case (when post-fault power equation was 1.5 sin δ), we find that in the later case the post-fault power equation being 2 sin δ, the value of δ starts reducing and $\Delta\omega$ also reduces. Obviously, we can say the transient stability is improved with post-fault power equation being 2 sin δ, i.e. when the post-fault power transfer is higher.

13.9.3 Runge–Kutta Method

The fourth order Runge–Kutta method can be applied here to solve the swing equation.

The swing equation is given by

$$\frac{H}{\pi f}\frac{d^2\delta}{dt^2} = P_m - P_e$$

where P_m is assumed to be constant input power. Here we assume that the machine has constant flux linkage during the study period of transient stability and hence the induced emf E' is constant.

The above equation can be split into two single order equations as

and $$\frac{d\Delta\omega}{dt} = \frac{\pi f}{H}(P_m - P_e) = f(\delta)$$

and $$\frac{d\delta}{dt} = \Delta\omega = f(\omega)$$

We are interested in getting δ and ω as function of t.

Let us assume, $x = t,\ y = \delta,\ z = \omega$

Hence $$\frac{d\delta}{dt} = f(\omega) = \frac{dy}{dx} = f(x, y, z) \tag{13.47a}$$

and $$\frac{d\Delta\omega}{dt} = f(\delta) = \frac{dz}{dx} = g(x, y, z) \tag{13.47b}$$

According to Runge–Kutta method, it is required to use equations (13.48a) and (13.48b) to solve two simultaneous differential equations given in equations (13.47a) and (13.47b)

[In this method, $$y_1 = y_0 + \frac{1}{6}(k_1 + 2k_2 + 2k_3 + k_4) \tag{13.48a}$$

and $$z_1 = z_0 + \frac{1}{6}(\ell_1 + 2\ell_2 + 2\ell_3 + \ell_4) \tag{13.48b}$$

where,

$$k_1 = f(x_0, y_0, z_0)h \tag{13.49a}$$

$$\ell_1 = g(x_0, y_0, z_0)h \tag{13.50a}$$

$$k_2 = f\{(x_0 + h/2), (y_0 + k_1/2), (z_0 + \ell_1/2)\}h \tag{13.49b}$$

$$\ell_2 = g\{(x_0 + h/2), (y_0 + k_1/2), (z_0 + \ell_1/2)\}h \tag{13.50b}$$

$$k_3 = f\{(x_0 + h/2), (y_0 + k_2/2), (z_0 + \ell_2/2)\}h \tag{13.49c}$$

$$\ell_3 = g\{(x_0 + h/2), (y_0 + k_2/2), (z_0 + l\ell_2/2)\}h \tag{13.50c}$$

$$k_4 = f\{(x_0 + h), (y_0 + k_3), (z_0 + \ell_3)\}h \tag{13.49d}$$

$$\ell_4 = g\{(x_0 + h), (y_0 + k_3), (z_0 + \ell_3)\}h \tag{13.50d}$$]

Let us apply this fourth order Runge–Kutta method to solve the swing equation. It is a four-step procedure for solving each of the solution state. At any state i, the following equations need to be solved.

$$\therefore \qquad \delta_i = \delta_{i-1} + \frac{1}{6}(k_{1i} + 2k_{2i} + 2k_{3i} + k_{4i}) \tag{13.51a}$$

and
$$\Delta\omega_i = \Delta\omega_{i-1} + \frac{1}{6}(\ell_{1i} + 2\ell_{2i} + 2\ell_{3i} + \ell_{4i}) \tag{13.52a}$$

where,
$$k_{1i} = f(\omega_{i-1})\Delta t = \left.\frac{d\delta}{dt}\right|_{\Delta\omega_{i-1}} \times \Delta t \text{ and } \ell_{1i} = g(\delta_{i-1})\Delta t = \left.\frac{d\Delta\omega}{dt}\right|_{\delta_{i-1}} \times \Delta t \tag{13.53a}$$

$$k_{2i} = f(\omega_{i-1} + \ell_{1i}/2)\Delta t = \left.\frac{d\delta}{dt}\right|_{\Delta\omega_{i-1} + \ell_1/2} \times \Delta t; \ \ell_{2i} = g(\delta_{i-1} + k_{1i}/2)\Delta t = \left.\frac{d\Delta\omega}{dt}\right|_{\delta_{i-1} + k_1/2} \times \Delta t \tag{13.53b}$$

$$k_{3i} = f(\omega_{i-1} + \ell_{2i}/2)\Delta t = \left.\frac{d\delta}{dt}\right|_{\Delta\omega_{i-1} + \ell_{2i}/2} \times \Delta t; \ \ell_{3i} = g(\delta_{i-1} + k_{2i}/2)\Delta t = \left.\frac{d\Delta\omega}{dt}\right|_{\delta_{i-1} + k_{2i}/2} \times \Delta t \tag{13.53c}$$

$$k_{4i} = f(\omega_{i-1} + \ell_{3i})\Delta t = \left.\frac{d\delta}{dt}\right|_{\Delta\omega_{i-1} + \ell_{3i}} \times \Delta t; \ \ell_4 = g(\delta_{i-1} + k_{3i})\Delta t = \left.\frac{d\Delta\omega}{dt}\right|_{\delta_0 + k_{3i}} \times \Delta t \tag{13.53d}$$

k's and ℓ's are the changes in δ and $\Delta\omega$ respectively obtained using derivatives evaluated at predetermined points. Units of k's and ℓ's radian and rad/sec respectively.

EXAMPLE 13.11 Use fourth order Runge–Kutta method to solve the swing equation with step size 0.05 till the fault is cleared for the system described in Example 13.9. Write a MATLAB code to solve this problem.

Solution: The two first order differential equations are as under:

$$\frac{d\delta}{dt} = \Delta\omega\,; \quad \frac{d\Delta\omega}{dt} = \frac{P_a}{M} = \frac{\pi f}{H}(P_m - P_e)$$

Let us now start first step of calculation to find δ and $\Delta\omega$ for $t = 0.05$ sec.

Here, $P_m = 1$ p.u.; $\delta_0(= 25°) = 0.4363$ rad; $H = 8$ MJ/MVA and $\Delta\omega_0 = 0$

Also, $P_a = 1 - 0.6 \sin \delta.$

Step 1: The initial estimates of changes in δ and $\Delta\omega$ are obtained from

$$k_{10} = \Delta\omega_0 \times \Delta t = 0$$

where ω_0 = synchronous angular velocity = $2\pi f$

$$\ell_{10} = \left.\frac{d\Delta\omega}{dt}\right|_{\delta_0} \times \Delta t = \frac{P_a}{M} = \frac{\pi f}{H}[P_m - P_e]\big|_{\delta_0} \times \Delta t$$

$$= \frac{\pi \times 50}{8}(1 - 0.6 \sin 0.4363) \times 0.05 = 0.7328$$

(Since the fault is at $t = 0$, here $\Delta t = 0.05$ sec.)

Step 2: The second set of estimate of changes in δ and $\Delta\omega$ are obtained as

$$k_{20} = \left.\frac{d\delta}{dt}\right|_{\Delta\omega_0 + \ell_{10}/2} \times \Delta t = \left(\Delta\omega_0 + \frac{\ell_1}{2}\right) \times \Delta t = \left(0 + \frac{0.7328}{2}\right) \times 0.05 = 0.0183$$

$$\ell_{20} = \left.\frac{d\Delta\omega}{dt}\right|_{\delta_0 + k_{10}/2} \times \Delta t = \frac{\pi \times 50}{8}\{1 - 0.6 \sin(0.4363 + 0)\} \times 0.05 = 0.7328$$

Step 3: The third set of estimates are obtained as

$$k_{30} = \left.\frac{d\delta}{dt}\right|_{\Delta\omega_0 + \ell_{20}/2} \times \Delta t = \left(\Delta\omega_0 + \frac{\ell_2}{2}\right) \times \Delta t = \left(0 + \frac{0.7328}{2}\right) \times 0.05 = 0.0183$$

$$\ell_{30} = \left.\frac{d\Delta\omega}{dt}\right|_{\delta_0 + k_{20}/2} \times \Delta t = \frac{\pi \times 50}{8}\left\{1 - 0.6 \sin\left(0.4363 + \frac{0.0183}{2}\right)\right\} \times 0.05$$

$$= 0.7279$$

Step 4: The fourth set of estimates are obtained as

$$k_{40} = \left.\frac{d\delta}{dt}\right|_{\Delta\omega_0 + \ell_{30}} \times \Delta t = (0 + 0.7279) \times 0.05 = 0.0364$$

$$\ell_{40} = \left.\frac{d\Delta\omega}{dt}\right|_{\delta_0 + k_{30}} \times \Delta t = \frac{\pi \times 50}{8} \{1 - 0.6 \sin(0.4363 + 0.0183)\} \times 0.05$$

$$= 0.7231$$

$$\therefore \quad \delta_1 = \frac{1}{6}\delta_0 + (k_{10} + 2k_{20} + 2k_{30} + k_{40})$$

$$= 0.4363 + \frac{1}{6}(0 + 2 \times 0.0183 + 2 \times 0.0183 + 0.0364)$$

$$= 0.4546 \text{ rad} = 26.0467°$$

And
$$\Delta\omega_1 = \Delta\omega_0 + \frac{1}{6}(\ell_{10} + 2\ell_{20} + 2\ell_{30} + \ell_{40})$$

$$= 0 + \frac{1}{6}(0.7328 + 2 \times 0.7328 + 2 \times 0.7279 + 0.7231) = 0.7296 \text{ rad/sec}$$

Next, we will find δ and $\Delta\omega$ for $t = 0.10$ sec and for this purpose, we will perform second step of calculation for next time step. From the first set of estimates in the second step of calculation we have,

$$k_{11} = \Delta\omega_1 \times \Delta t = 0.7296 \times 0.05 = 0.0365$$

$$\ell_{11} = \left.\frac{d\Delta\omega}{dt}\right|_{\delta_1} \times \Delta t = \frac{\pi \times 50}{8}(1 - 0.6 \sin 0.4546) \times 0.05 = 0.7231$$

The second set of estimates in the second step of calculation are obtained as

$$k_{21} = \left.\frac{d\delta}{dt}\right|_{\Delta\omega_1 + \ell_{11}/2} \times \Delta t = (\Delta\omega_1 + \ell_{11}/2) \times \Delta t$$

$$= \left(0.7296 + \frac{0.7231}{2}\right) \times 0.05 = 0.0546$$

$$\ell_{21} = \left.\frac{d\Delta\omega}{dt}\right|_{\delta_1 + k_{11}/2} \times \Delta t = \frac{\pi \times 50}{8}\left\{1 - 0.6 \sin\left(0.4546 + \frac{0.0365}{2}\right)\right\} \times 0.05$$

$$= 0.7135$$

The third set of estimates in the second step of calculation are as follows:

$$k_{31} = \left.\frac{d\delta}{dt}\right|_{\Delta\omega_1 + \ell_{21}/2} \times \Delta t = \left(\Delta\omega_1 + \frac{\ell_2}{2}\right) \times \Delta t = \left(0.7296 + \frac{0.7135}{2}\right) \times 0.05$$

$$= 0.0543$$

$$\ell_{31} = \left.\frac{d\Delta\omega}{dt}\right|_{\delta_1 + k_{21}/2} \times \Delta t = \frac{\pi \times 50}{8}\left\{1 - 0.6 \sin\left(0.4546 + \frac{0.0546}{2}\right)\right\} \times 0.05$$

$$= 0.7087$$

The fourth set of estimates in the second step of calculation are given as

$$k_{41} = \left.\frac{d\delta}{dt}\right|_{\Delta\omega_1 + \ell_{31}} \times \Delta t = (0.7296 + 0.7087) \times 0.05 = 0.0719$$

$$\ell_{41} = \left.\frac{d\Delta\omega}{dt}\right|_{\delta_1 + k_{31}} \times \Delta t = \frac{\pi \times 50}{8} \{1 - 0.6 \sin (0.4546 + 0.0543)\} \times 0.05$$

$$= 0.6948$$

$\therefore$

$$\delta_2 = \delta_1 + \frac{1}{6}(k_{11} + 2k_{21} + 2k_{31} + k_{41})$$

$$= 0.4546 + \frac{1}{6}(0.0365 + 2 \times 0.0546 + 2 \times 0.0543 + 0.0719)$$

$$= 0.5089 \text{ rad} = 29.1578°$$

and

$$\Delta\omega_2 = \Delta\omega_1 + \frac{1}{6}(\ell_{11} + 2\ell_{21} + 2\ell_{31} + \ell_{41})$$

$$= 0.7296 + \frac{1}{6}(0.7231 + 2 \times 0.7135 + 2 \times 0.7087 + 0.6948) = 1.44 \text{ rad/sec}$$

Next, we are to find σ and $\Delta\omega$ for $t = 0.15$ sec and for this, we need to perform third step of calculation. The first set of estimates in this third step of calculation are given as

$$k_{12} = \Delta\omega_2 \times \Delta t = 1.44 \times 0.05 = 0.072$$

$$\ell_{12} = \left.\frac{d\Delta\omega}{dt}\right|_{\delta_2} \times \Delta t = \frac{\pi \times 50}{8}(1 - 0.6 \sin 0.5089) \times 0.05 = 0.6947$$

The second set of estimates are obtained as

$$k_{22} = \left.\frac{d\delta}{dt}\right|_{\Delta\omega_2 + \ell_{12}/2} \times \Delta t = \left(1.44 + \frac{0.6947}{2}\right) \times 0.05 = 0.0894$$

$$\ell_{22} = \left.\frac{d\Delta\omega}{dt}\right|_{\delta_2 + k_{12}/2} \times \Delta t = \frac{\pi \times 50}{8}\left\{1 - 0.6 \sin\left(0.5089 + \frac{0.072}{2}\right)\right\} \times 0.05 = 0.6764$$

For the third set of estimates we have,

$$k_{32} = \left.\frac{d\delta}{dt}\right|_{\Delta\omega_2 + \ell_{22}/2} \times \Delta t = \left(1.44 + \frac{0.6764}{2}\right) \times 0.05 = 0.0889$$

$$\ell_{32} = \left.\frac{d\Delta\omega}{dt}\right|_{\delta_2 + k_{22}/2} \times \Delta t = \frac{\pi \times 50}{8}\left\{1 - 0.6 \sin\left(0.5089 + \frac{0.0894}{2}\right)\right\} \times 0.05 = 0.6721$$

For the third set of estimates we have,

$$k_{32} = \left.\frac{d\delta}{dt}\right|_{\Delta\omega_2 + \ell_{22}/2} \times \Delta t = \left(1.44 + \frac{0.6764}{2}\right) \times 0.05 = 0.0889$$

$$\ell_{32} = \left.\frac{d\Delta\omega}{dt}\right|_{\delta_2 + k_{22}/2} \times \Delta t = \frac{\pi \times 50}{8}\left\{1 - 0.6 \sin\left(0.5089 + \frac{0.0894}{2}\right)\right\} \times 0.05 = 0.6721$$

For the fourth set of estimates we get,

$$k_{42} = \left.\frac{d\delta}{dt}\right|_{\Delta\omega_2 + \ell_{32}} \times \Delta t = (1.44 + 0.6721) \times 0.05 = 0.1056$$

$$\ell_{42} = \left.\frac{d\Delta\omega}{dt}\right|_{\delta_2 + k_{32}} \times \Delta t = \frac{\pi \times 50}{8}\{1 - 0.6 \sin (0.5089 + 0.0889)\} \times 0.05 = 0.6502$$

$$\therefore \quad \delta_3 = \delta_2 + \frac{1}{6}(k_{12} + 2k_{22} + 2k_{32} + k_{42})$$

$$= 0.5089 + \frac{1}{6}(0.072 + 2 \times 0.0894 + 2 \times 0.0889 + 0.1056)$$

$$= 0.5979 \text{ rad} = 34.2591°$$

and $\Delta\omega_3 = \Delta\omega_2 + \frac{1}{6}(\ell_{12} + 2\ell_{22} + 2\ell_{32} + \ell_{42})$

$$= 1.44 + \frac{1}{6}(0.6947 + 2 \times 0.6764 + 2 \times 0.6721 + 0.6502) = 2.1136 \text{ rad/sec}$$

Again after 0.05 sec (i.e. at 0.2 sec), the fault is cleared. We are to find the value of δ and $\Delta\omega$ at t = 0.2 sec with post-fault swing equation.

Here, $P_a = 1 - 1.5 \sin \delta$; $\Delta t = 0.05$ sec

Also, $$\left.\frac{d\delta}{dt}\right|_{\Delta\omega_3} = \Delta\omega_3 = 2.1136 \text{ rad/sec}$$

and $$\left.\frac{d\Delta\omega}{dt}\right|_{\delta_3} = \frac{P_a}{M} = \frac{\pi \times 50}{8}(1 - 1.5 \sin 0.5979) = 3.0559 \text{ rad/sec}^2$$

In this fourth step of calculation, the first set of estimates are obtained as

$$k_{13} = \Delta\omega_3 \times \Delta t = 2.1136 \times 0.05 = 0.1057.$$

$$\ell_{13} = \left.\frac{d\Delta\omega}{dt}\right|_{\delta_3} \times \Delta t = \frac{\pi \times 50}{8}(1 - 1.5 \sin 0.5979) \times 0.05 = 0.1528$$

While the second set of estimates is given by

$$k_{23} = \left.\frac{d\delta}{dt}\right|_{\Delta\omega_3 + \ell_{13}/2} \times \Delta t = \left(2.1136 + \frac{0.1528}{2}\right) \times 0.05 = 0.1095$$

$$\ell_{23} = \left.\frac{d\Delta\omega}{dt}\right|_{\delta_3 + k_{13}/2} \times \Delta t = \frac{\pi \times 50}{8}\left\{1 - 1.5 \sin\left(0.5979 + \frac{0.1057}{2}\right)\right\} \times 0.05 = 0.0897$$

For the third set of estimates we have,

$$k_{33} = \left.\frac{d\delta}{dt}\right|_{\Delta\omega_3 + \ell_{23}/2} \times \Delta t = \left(2.1136 + \frac{0.0897}{2}\right) \times 0.05 = 0.1079$$

$$\ell_{33} = \left.\frac{d\Delta\omega}{dt}\right|_{\delta_3 + k_{23}/2} \times \Delta t = \frac{\pi \times 50}{8}\left\{1 - 1.5 \sin\left(0.5979 + \frac{0.1095}{2}\right)\right\} \times 0.05 = 0.0874$$

While for the fourth set of estimates we have,

$$k_{43} = \left.\frac{d\delta}{dt}\right|_{\Delta\omega_3 + \ell_{33}} \times \Delta t = (2.1136 + 0.0874) \times 0.05 = 0.11$$

$$\ell_{43} = \left.\frac{d\Delta\omega}{dt}\right|_{\delta_3 + k_{33}} \times \Delta t = \frac{\pi \times 50}{8}\{1 - 1.5 \sin (0.5979 + 0.1079)\} \times 0.05$$

$$= 0.0265$$

$\therefore$ $$\delta_4 = \delta_3 + \frac{1}{6}(k_{13} + 2k_{23} + 2k_{33} + k_{43})$$

$$= 0.5979 + \frac{1}{6}(0.1057 + 2 \times 0.1095 + 2 \times 0.1079 + 0.11)$$

$$= 0.7063 \text{ rad} = 40.4690°$$

and $$\Delta\omega_4 = \Delta\omega_3 + \frac{1}{6}(\ell_{13} + 2\ell_{23} + 2\ell_{33} + \ell_{43})$$

$$= 2.1136 + \frac{1}{6}(0.1528 + 2 \times 0.0897 + 2 \times 0.0874 + 0.0265) = 2.2025 \text{ rad/sec}$$

We can write the obtained results in a tabular form:

Time (in sec)	*δ (in radian)*	*δ (in degree)*
0	0.4363	25
0.05	0.4546	26.0467
0.1	0.5093	29.1817
0.15	0.5990	34.3189
0.2	0.7089	40.6170

The MATLAB code for Example 13.11 is as follows:

```
%Prog to solve Example 13.11
Clc ; clear all
f=50;  h=8;  pm=1;pem_pre=2.5;  pem_f=0.6;  pem_post=1.5;  delta=0.4363;
delta=delta';
swcf=pi*f*pm/h; swvf=pi*f*pem_f/h; swcp=pi*f*pm/h; swvp=pi*f*pem_post/h;
tclr=0.2; t(1)=0;  dt=0.05; tstep=tclr/dt; dw(1)=0;
%Solution of swing equation by Runge-Kutta method
for k=1:tstep
 t(k+1)=t(k)+dt;
  if(t(k+1)<tclr)
 k1=dw(k)*dt;  l1=(swcf-swvf*sin(delta(k)))*dt ;
 k2=(dw(k)+l1/2)*dt;
 l2=(swcf-swvf*sin((delta(k)+k1/2)))*dt;
 k3=(dw(k)+l2/2)*dt;
 l3=(swcf-swvf*sin((delta(k)+k2/2)))*dt;
 k4=(dw(k)+l3)*dt;
 l4=(swcf-swvf*sin((delta(k)+k3)))*dt;
delta(k+1)=delta(k)+(k1+2*k2+2*k3+k4)/6;
dw(k+1)=dw(k)+(l1+2*l2+2*l3+l4)/6;
  else
k1=dw(k)*dt;  l1=(swcp-swvp*sin(delta(k)))*dt ;
k2=(dw(k)+l1/2)*dt;
l2=(swcp-swvp*sin((delta(k)+k1/2)))*dt;
k3=(dw(k)+l2/2)*dt;
l3=(swcp-swvp*sin((delta(k)+k2/2)))*dt;
k4=(dw(k)+l3)*dt;
l4=(swcp-swvp*sin((delta(k)+k3)))*dt;
delta(k+1)=delta(k)+(k1+2*k2+2*k3+k4)/6;
dw(k+1)=dw(k)+(l1+2*l2+2*l3+l4)/6    ;
  end
end
```

```
disp('Solution of swing equation by Runge-Kutta method.')
disp(sprintf('time interval=%4.3fs, total time step=%d, fault clearance
time=%4.3fs',dt,tstep,tclr))
 disp(sprintf('t(in sec)   delta(in radian)    delta(in degree)'))
 for k=1:tstep+1
 deltad(k)=delta(k)*180/pi;
 disp(sprintf('%4.3f     %10.6f          %10.6f   ',t(k),delta(k),deltad(k)))
 end
```

Execution of the computer program to solve swing equation by Runge–Kutta method for Example 13.11:

```
Solution of swing equation by Runge-Kutta method.
time interval=0.05s, total time step=4, fault clearance time=0.2s
t(in sec)   delta(in radian)    delta(in degree)
0.000            0.436300            24.998149
0.050            0.454580            26.045509
0.100            0.508937            29.159919
0.150            0.597962            34.260679
0.200            0.706389            40.473133
```

13.10 SWING EQUATION FOR A MULTI-MACHINE SYSTEM

In a multi-machine system, let

G_m = base rating of the machine,

G_{sys} = system base,

H_{sys} = inertia constant in system base = $H_m \times \dfrac{G_m}{G_{sys}}$

where, H_m = inertia constant at base rating

Thus, swing equation can be written as

$$(P_m - P_e)\frac{G_m}{G_{sys}} = \frac{H_m}{\pi f}\frac{d^2\delta}{dt^2}\frac{G_m}{G_{sys}}$$

$$\therefore \quad (P_m - P_e) = \frac{H_{sys}}{\pi f}\frac{d^2\delta}{dt^2} \tag{13.54}$$

EXAMPLE 13.12 An alternator is connected to an infinite bus and delivering a complex power of (1 + *j*0.6) p.u to the infinite bus through a short line having reactance of 0.1 p.u. (Figure E13.5(a)). The voltage of the infinite bus is 1 p.u. while the transient reactance of the machine is 0.2 p.u. and the inertia constant is 5 p.u. The base MVA is 100. Form the swing equation for the generator during fault and post-fault period if the generator is subjected to a three phase ground fault at its terminal and the fault is cleared at 0.06 sec. What will happen to the transient stability for this fault ? What will happen to the transient stability if the fault is cleared at 0.3 sec? Use modified Euler's method to find and plot the swing curve for a time increment of 0.02 sec upto 0.5 sec. The system frequency is 50 Hz. Assume that during fault the power output of the alternator drops to zero while the power output returns to prefault state once the fault is cleared. Write a MATLAB code to solve this problem.

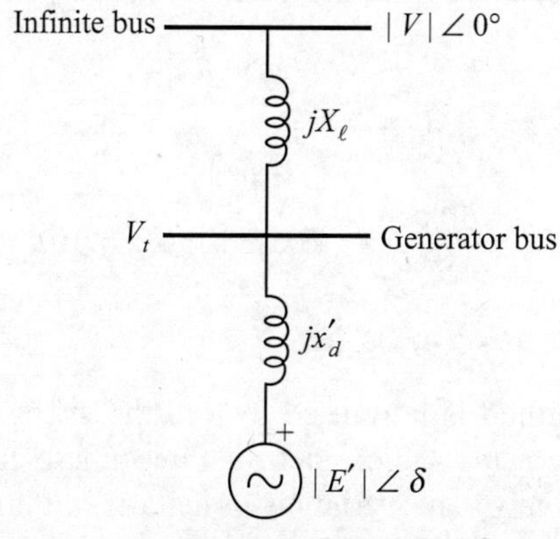

Figure E13.5(a)

Solution: Let us assume that the infinite bus and the generator bus are numbered as 1 and 2, respectively. From load flow study, voltage V_t at generator bus is 1.0525 p.u. having angle 0.0952 rad, i.e. (1.0477 + *j*0.1) p.u.

The current injected by the generator bus

$$I_g = \frac{S_g^*}{V_t^*} = \frac{(1 - j0.6)}{(1.0477 - j0.1)} = (1 - j0.4772) \text{ p.u.}$$

∴ Transient internal voltage (E') of the generator can be obtained as

$$(E') = V_t + I_g \times jx'_d = 1.0477 + j0.1 + (1 - j0.4772) \times j0.2$$

$$= 1.1431 + j0.3 \text{ p.u.} = 1.1818 \text{ p.u.} \angle 0.2566 \text{ rad}$$

Magnitude of this transient voltage will remain the same because of the constant flux linkage concept. The initial load angle is 0.2566 rad (δ_0).

∴ Series reactance between the generator internal voltage (E') and infinite bus voltage (V) is given by

$$x = x_\ell + x'_d = 0.1 + 0.2 = 0.3 \text{ p.u.}$$

The power equation in prefault condition is then given by

$$P_e = \left|\frac{VE'}{x}\right| \sin\delta = \frac{1.0 \times 1.1818}{0.3} \sin\delta = 3.9394 \sin\delta \text{ p.u.}$$

The three phase fault occurs at generator bus. Therefore, during fault P_e is zero and the swing equation is obtained as

$$\frac{d^2\delta}{dt^2} = \frac{\pi f}{H}(P_m - P_e) = \frac{50\pi}{5}(1 - 0) = 10\pi = 31.4159 \text{ rad/m}^2$$

After clearance of the fault, the system will return to its electrical state similar to pre-fault condition.

Therefore, the power equation during pre-fault clearance period is

$$P_e = 3.9394 \sin \delta \text{ p.u.}$$

The swing equation in the post-fault condition is

$$\frac{d^2\delta}{dt^2} = \frac{\pi f}{H}(P_m - P_e) = \frac{50\pi}{5} (1 - 3.673 \sin \delta) = 10\pi (1 - 3.673 \sin \delta)$$

$$= (31.4159 - 123.7599 \sin \delta)$$

The modified Euler's method is now used to find the swing curve. [Figure E13.5(b) and Figure E13.5(c)] for the alternator subjected to a three phase fault, subsequently cleared at 0.06 sec and 0.3 sec respectively. The variation in delta is within 90 degree for clearing time of 0.06 sec as shown in Figure E13.5(b). Hence transient stability is maintained in this case. But in Figure E13.5(c) delta swings beyond 90 degree for clearing time of 0.3 sec hence it is not possible to maintain transient stability if clearing time is 0.3 sec.

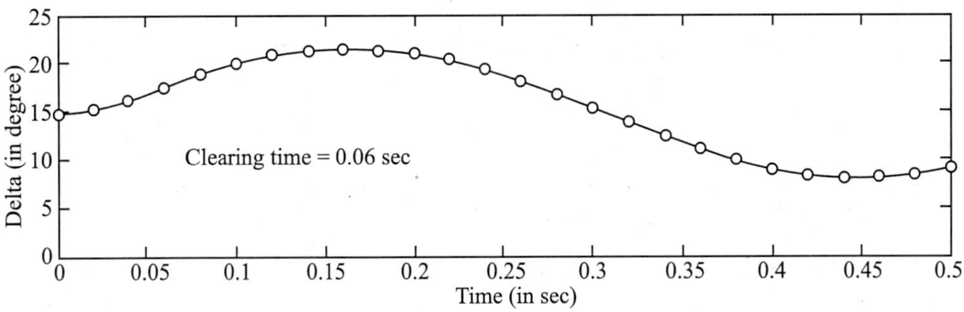

Figure E13.5(b) Variation of delta with time for for clearing time of 0.06 s.

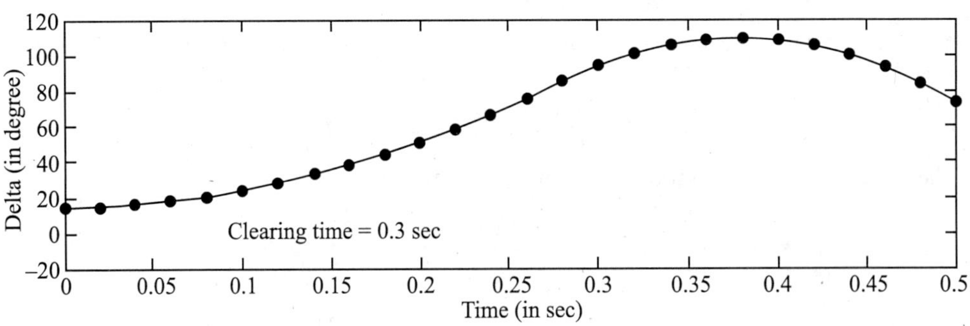

Figure E13.5(c) Variation of delta with time for clearing time of 0.3 s.

The MATLAB code for Example 13.12 is as follows:

```
%Prog to solve Example 13.12
clc ; clear all
f=50; Pg=1; Qg=0.6; Sg=Pg+i*Qg; V=1; Vt=1.0525*exp(i*0.0952);h=5;
Xline=0.1; Xd_dash=0.2; X=Xline+Xd_dash; Pm=Pg;
Ig=conj(Sg/Vt); Edash=Vt+Ig*i*Xd_dash; delta(1)=angle(Edash);
swcf=pi*f*Pm/h;  swcp=pi*f*Pm/h;  Pep=abs(Edash*V/X); swvp=pi*f*Pep/h;
t(1)=0; dt=0.02; tmax=0.5; tstep=tmax/dt; dw(1)=0;
tclr=input('Enter claring time:')
%Solution of swing equation by Modifier Euler method.
for k=1:tstep
t(k+1)=t(k)+dt;
 if(t(k+1)<tclr)
  sw(k)=swcf;
 else
  sw(k)=swcp-swvp*sin(delta(k));
 end
 alpha(k)=sw(k); ddw(k)=alpha(k)*dt; ddelta(k)=dw(k)*dt;
 deltap(k+1)=delta(k)+ddelta(k); dwp(k+1)=dw(k)+ddw(k);
 if(t(k+1)<tclr)
   sw(k+1)=swcf;
 else
   sw(k+1)=swcp-swvp*sin(deltap(k+1));
 end
 alpha(k+1)=sw(k+1);  ddw(k+1)=alpha(k+1)*dt;
 ddelta(k+1)=dwp(k+1)*dt;  dw(k+1)=dw(k)+(ddw(k)+ddw(k+1))/2;
 delta(k+1)=delta(k)+(ddelta(k)+ddelta(k+1))/2;
end
disp('Solution of swing equation by Modifier Euler method.')
disp(sprintf('time interval=%4.3fs, total time step=%d, fault clearance
time=%4.3fs',dt,tstep,tclr))
 disp(sprintf('t(in sec)   delta(in radian)   delta(in degree)'))
for k=1:tstep+1
 deltad(k)=delta(k)*180/pi;
 disp(sprintf('%4.3f     %10.6f       %10.6f  ',t(k),delta(k),deltad(k)))
end
dplot=plot(t,deltad,'o');
xlabel('Time (in sec)'),ylabel('Delta (in degree)')
title ('FigureE13.5(b): Variation of delta with time.')
set(dplot,'LineWidth',1.5,{'LineStyle'},{'-'})
set(dplot,{'Color'},{'k'})
text(0.1,10,['Clearing  time  =',  num2str(tclr),'sec'],'FontSize',10,
'HorizontalAlignment','left')
```

Execution of the computer program for Example 13.12:

```
Part(a): Solution of swing equation by Modifier Euler method.
time interval=0.02s, total time step=25, fault clearance time=0.06s
```

```
t(in sec)   delta(in radian)   delta(in degree)
0.000          0.256680           14.706704
0.020          0.262964           15.066704
0.040          0.281813           16.146704
0.060          0.306345           17.552287
0.080          0.328499           18.821626
0.100          0.347220           19.894247
0.120          0.361621           20.719367
0.140          0.371024           21.258107
0.160          0.374986           21.485136
0.180          0.373321           21.389706
0.200          0.366102           20.976090
0.220          0.353663           20.263423
0.240          0.336586           19.284930
0.260          0.315670           18.086561
0.280          0.291907           16.725030
0.300          0.266430           15.265299
0.320          0.240464           13.777591
0.340          0.215269           12.334029
0.360          0.192075           11.005090
0.380          0.172021            9.856053
0.400          0.156096            8.943654
0.420          0.145092            8.313145
0.440          0.139555            7.995938
0.460          0.139765            8.007959
0.480          0.145714            8.348810
0.500          0.157111            9.001777
Part(b): Solution of swing equation by Modifier Euler method.
time interval=0.02s, total time step=25, fault clearance time=0.3s
t(in sec)   delta(in radian)   delta(in degree)
0.000          0.256680           14.706704
0.020          0.262964           15.066704
0.040          0.281813           16.146704
0.060          0.313229           17.946704
0.080          0.357211           20.466704
0.100          0.413760           23.706704
0.120          0.482875           27.666704
0.140          0.564556           32.346704
0.160          0.658804           37.746704
0.180          0.765618           43.866704
0.200          0.884999           50.706704
0.220          1.016946           58.266704
0.240          1.161459           66.546704
0.260          1.318539           75.546704
0.280          1.488185           85.266704
0.300          1.645728           94.293272
0.320          1.766508          101.213455
0.340          1.851385          106.076562
0.360          1.901385          108.941351
```

```
0.380      1.917270      109.851461
0.400      1.899303      108.822038
0.420      1.847182      105.835733
0.440      1.760119      100.847380
0.460      1.637085       93.798067
0.480      1.477253       84.640371
0.500      1.280659       73.376340
```

13.11 STEADY STATE STABILITY

Analytically, it is more convenient to derive the criterion of steady state stability from the swing equation. Rewriting equation (13.11) we get,

$$M\frac{d^2\delta}{dt^2} + D\frac{d\delta}{dt} = P_m - P_{\max}\sin\delta$$

where $P_{\max} = \dfrac{|E'||V|}{x'_d + x_\ell}$; x_ℓ being the external reactance through which the machine delivers power to the infinite bus in the SMIB system. $|E'|$ is the induced voltage in the machine leading the infinite bus voltage $|V|$ by $\angle\,\delta$. The infinite bus voltage angle is zero and the transient reactance of the machine is x'_d. D represents the damping coefficient; its value is small but positive. For convenience, we often neglected D where we analysed transient stability for a very short period, say up to 2 seconds. For steady state stability analysis, it is advisable to consider effect of D as the system needs to be analysed for longer periods.

We apply the concept of linearization for the swing equation for obtaining the criterion of steady state stability. The linearized swing equation has already been expressed in equation (13.12), whose roots are obtained in equation (13.13) and is reproduced here.

$$s_1, s_2 = \frac{-D}{2M} \pm \sqrt{\frac{D^2}{4M^2} - \frac{K}{M}},$$

where, $K = P_S$, the synchronous power coefficient $= P_{\max}\cos\delta$.

If K is positive then both eigenvalues would have negative real parts. If K is negative, one of the eigenvalues is positive real. With smaller values of D and $K > 0$, eigenvalues are complex and may be represented as

$$\lambda_1, \lambda_2 = -\sigma \pm j\omega, \text{ where } \sigma = \frac{D}{2M};\ \omega = \sqrt{\frac{K}{M} - \frac{D^2}{4M^2}}$$

Hence, for steady state stability the necessary condition is $K > 0$, i.e. $\dfrac{dP_e(\delta)}{d\delta} > 0$.

Let us now have a relook at P-δ curve for an alternator (Figure 13.10). We find there could be two stable operating points for the machine for constant P_m. The first operating point is δ_0 while the second is at δ_m. For $\delta = \delta_0$, $K > 0$ while for $\delta = \delta_m$, $K < 0$. The two equilibrium points approach closer as P_m is increased. The maximum power supplied is $P_{\max}$ and is the steady state stability limit (at $\delta = 90°$ or $\pi/2$ rad.). Since $K > 0$ gives the feasible region of steady state stability limit, we can say steady state stability is achieved till $\delta < \pi/2$ (or 90°) and at $\delta = \pi/2$, the machine operates at its theoretical steady state stability limit. It may also be noted that with P_m increasing, δ_0 approaches $\pi/2$ and with $P_m > P_{max}$, there is no stable operating point.

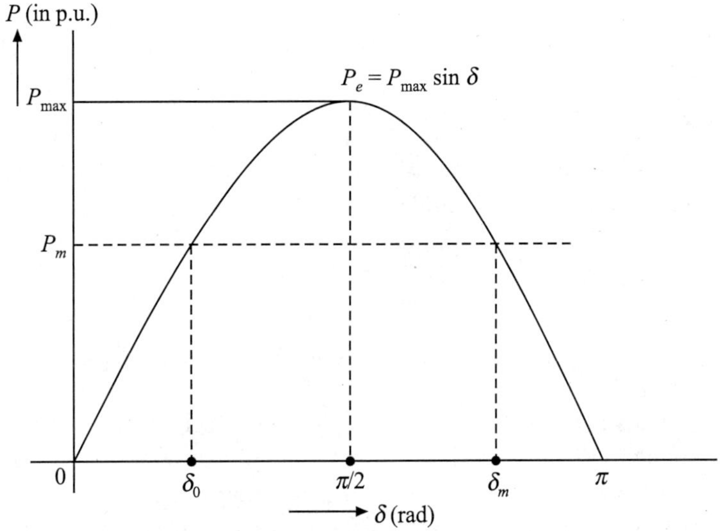

Figure 13.10 P_e vs δ curve.

It is interesting to note that the limit of steady state stability is also equal to the maximum power transferred in the network neglecting the losses. Hence, the steady state stability limit is also the network power limit. The steady state stability margin is defined as:

$$\text{Steady state stability margin} = \frac{P_{max} - P_0}{P_{max}} \times 100 \text{ (in \%)}$$

$$= \frac{\delta_m - \delta_0}{\delta_m} \times 100 = \left(1 - \frac{\delta_0}{\delta_m}\right) \times 100$$

Also, $$P_{max} = \frac{|E'||V|}{x'_d + x_\ell};\ P_0 = \frac{|E'||V|}{x'_d + x_\ell} \sin \delta_0$$

∴ We can also express this margin in terms of power angle as,

Steady state stability margin = $(1 - \sin \delta_0) \times 100$

EXAMPLE 13.13 A power transmission network has been shown in Figure E13.6. The following data represents the parameters of the system.

$x_S = 0.85$ p.u., $x_{T1} = x_{T2} = 0.157$ p.u., $x_{l1} = x_{l2} = 0.35$ p.u., $|E| = 1.5$ p.u., $|V_L| = 1.0$ p.u.

Find the minimum value of E and V_L at which the generator with 1.0 power output operates stably. Also, calculate the maximum power limit and steady state stability margin with indicated value of E and V_L.

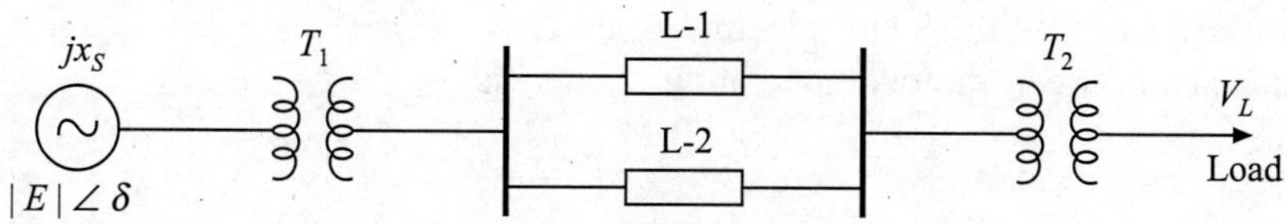

Figure E13.6

Solution: Here,

$$\Sigma x = x_S + x_{T1} + x_{T2} + \frac{x_\ell}{2}$$

$$\Sigma x = j0.85 + j0.157 + j0.157 + \frac{j0.35}{2} = j1.339 \text{ p.u.}$$

$$\therefore \qquad P_{max} = \frac{|EV_L|}{|\Sigma x|} = \frac{1.5 \times 1.0}{1.339} = 1.12 \text{ p.u.}$$

Now, steady state stability margin is obtained as

$$M = \frac{P_{max} - P_0}{P_{max}} = \frac{1.12 - 1.0}{1.12} = 0.107 = 10.7\% \quad [\text{Given, } P_0 = 1.0 \text{ p.u}]$$

$$V_{L_{min}} = P_0 \frac{|\Sigma x|}{|E|} = 1.0 \times \frac{1.339}{1.5} = 0.893 \text{ p.u.}$$

$$E_{min} = P_0 \frac{|\Sigma x|}{|V_L|} = 1.0 \times \frac{1.339}{1.0} = 1.339 \text{ p.u.}$$

EXAMPLE 13.14 A power system has been shown in Figure E13.7. Compare the magnitude of maximum power transfer if (a) a shunt inductor is connected at bus-2 and (b) a shunt capacitor is connected at bus 2.

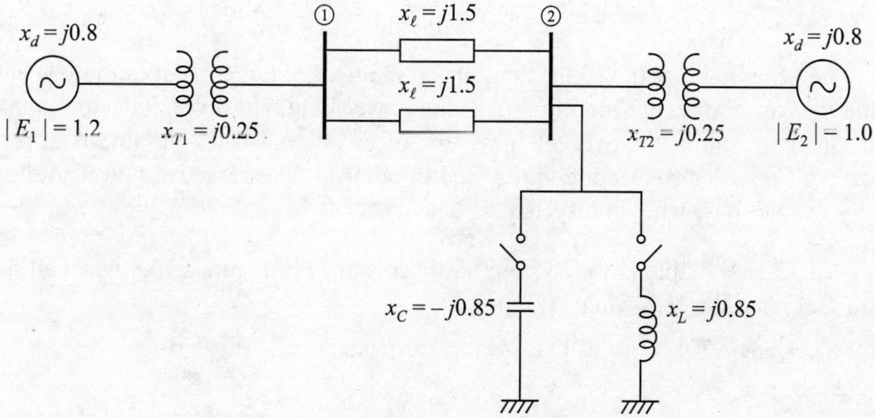

Figure E13.7

Solution: *Case (a):* With shunt inductor at bus-2,
transfer reactance between the two generating sources is

$$x = x_1 + x_2 + \frac{x_1 x_2}{x_3}$$

where $$x_1 = x_d + x_{T1} + \frac{x_\ell}{2} = j0.8 + j0.25 + \frac{j1.5}{2} = 1.8 \text{ p.u.}$$

$$x_2 = x_{T2} + x_d = j0.25 + j0.8 = j1.05 \text{ p.u.}$$

$$x_3 = x_L = j0.85 \text{ p.u.}$$

Therefore, $$x = j1.8 + j1.05 + \frac{j1.8 \times j1.05}{j0.85} = j5.07 \text{ p.u.}$$

∴ Maximum power transfer under this condition is

$$P_{\max_1} = \left| \frac{E_1 E_2}{x} \right| = \frac{1.2 \times 1.0}{5.07} = 0.237 \text{ p.u.}$$

Case (b): With shunt capacitor at bus 2,
transfer reactance between the two generating sources becomes

$$x = x_1 + x_2 + \frac{x_1 x_2}{x_3}$$

where x_1 = 1.8 p.u. and $x_2 = j1.05$ p.u., but $x_3 = x_C = -j0.85$ p.u.

Therefore, $$x = j1.8 + j1.05 + \frac{j1.8 \times j1.05}{-j0.85} = j0.626 \text{ p.u.}$$

∴ Maximum power transfer under this condition is

$$P_{\max_2} = \left| \frac{E_1 E_2}{x} \right| = \frac{1.2 \times 1.0}{0.626} = 1.915 \text{ p.u.}$$

It may be observed that the connection of the inductor at bus-2 reduces the power transfer limit while the capacitor connection improves the power limit. However, inductor connection is needed in order to control the overvoltage at the receiving end when the line is at no load or lightly loaded. Capacitor connection is required in order to inject reactive power to the system, particularly at load bus, when the system is heavily loaded.

EXAMPLE 13.15 A simple power system has been shown in Figure E13.8 when all quantities are in p.u. Compute the following:

(i) Power angle (δ) of V_2 so that $|V_2|$ = 1.0 p.u.

(ii) Q_{g_2} in p.u.

(iii) What will be the value of δ if Q_{g_2} is made zero?

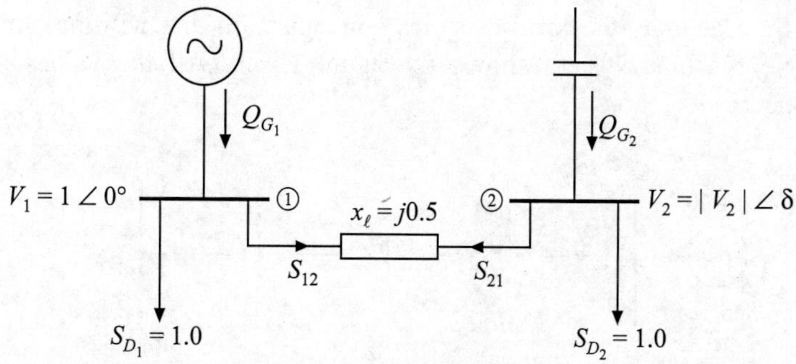

Figure E13.8

Solution: We know that $P_{12} = \left|\dfrac{V_1V_2}{x}\right| \sin\delta$ (1)

Here S_{D_2} (= 1.0) = (1 + *j*0) p.u. and Q_{G_2} is purely imaginary. (Since Q_{G_2} is the injected reactive power from a capacitor)

This gives, $$P_{12} = -P_{21} = 1 \text{ p.u.}$$

Hence, from equation (1) we have, $1 = \dfrac{1\times 1}{0.5} \sin\delta$

i.e., $$\delta = 30°$$

This also means that angle of V_2 is 30° [it should be noted that V_2 lags V_1]. (Ans. (i))

Now, $$Q_{G_2} = \frac{|V_2|^2}{|x|} - \left|\frac{V_2V_1}{x}\right| \cos\delta \quad (2)$$

$$= \frac{1.0^2}{0.5} - \frac{1.0\times 1.0}{0.5} \cos 30° = 0.268 \text{ p.u.} \quad \text{(Ans. (ii))}$$

[It is consistent as $Q_{G_2} > 0$ and it means that it is a capacitive reactive power source.]

With $Q_{G_2} = 0$, equation (2) reduces to

$$\cos\delta = V_2 \quad (3)$$

and equation (1) reduces to

$$\sin\delta = \frac{1}{2V_2} \quad (4)$$

∴ From (3) and (4), we can write,

$$\cos^2\delta + \sin^2\delta = V_2^2 + \frac{1}{4V_2^2}$$

or, $$4V_2^4 - 4V_2^2 + 1 = 0 \quad (5)$$

∴ Equation (5) has a real solution of $V_2^2 = \dfrac{1}{2}$ i.e., $V_2 = \dfrac{1}{\sqrt{2}}$.

Therefore, from equation (3), $$\delta = 45° \quad \text{(Ans. (iii))}$$

∴ The amount of power was being transmitted at $V_2 = 0.707\angle 45°$ (with $Q_{g_2} = 0$).

EXAMPLE 13.16 Calculate the level of series compensation that will maximize the received power in a single machine system shown in Figure E13.9. All the values are in p.u. The generator reactance is $j0.15$ p.u.

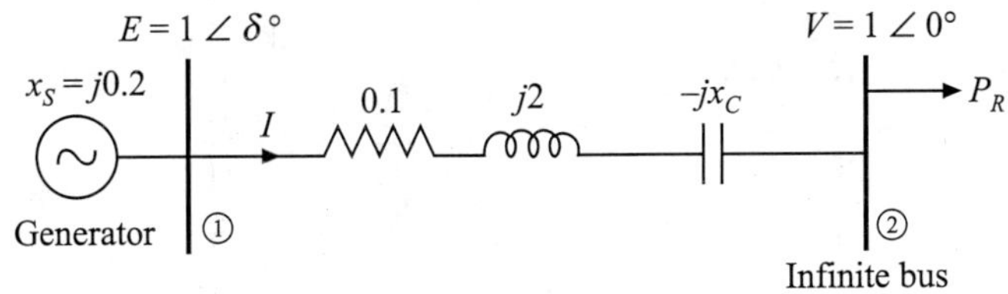

Figure E13.9

Solution: Here, generator reactance $(x_S) = j0.2$ p.u.

Line impedance $(z_\ell) = r_\ell + jx_\ell = 0.1 + j2$ p.u.

Capacitive reactance for series compensation $= -jx_C$

$$\therefore \quad \text{Total impedance } (Z) = x_S + z_\ell - jx_C = j0.2 + 0.1 + j2 - jx_C$$

$$= 0.1 + j(2.2 - x_C) = R + jX = |Z| \angle \phi$$

Assuming infinite bus voltage as reference phasor, the sending end bus voltage is expressed as

$$E = |E| \angle \delta$$

Current through the line,

$$I = \frac{E - V}{Z} = \frac{|E|\angle\delta - |V|\angle 0^\circ}{|Z|\angle\phi} = \frac{|E|}{|Z|}\angle(\delta - \phi) - \frac{|V|}{|Z|}\angle -\phi$$

$$\therefore \quad \text{i.e.,} \quad I^* = \frac{|E|}{|Z|}\angle(\phi - \delta) - \frac{|V|}{|Z|}\angle\phi$$

The receiving end complex power is obtained as

$$S_R = P_R + jQ_R = VI^* = |V| \angle 0^\circ \left[\frac{|E|}{|Z|}\angle(\phi - \delta) - \frac{|V|}{|Z|}\angle\phi\right]$$

$$= \frac{|EV|}{|Z|}\angle(\phi - \delta) - \frac{|V^2|}{|Z|}\angle\phi$$

$$\therefore \quad P_R = \frac{|EV|}{|Z|}\cos(\phi - \delta) - \frac{|V^2|}{|Z|}\cos\phi$$

From the above expression, it is clear that P_R will be maximum when $\delta = \phi$.

With $\delta = \phi$, the receiving end power is given by

$$P_R' = \frac{|EV|}{|Z|} - \frac{|V|^2}{|Z|}\cos\phi = \frac{|EV|}{|Z|} - \frac{|V|^2 R}{|Z|^2} = \frac{1}{R}(\cos\phi - \cos^2\phi)$$

where $E = V = 1$ p.u. and $|Z| = \dfrac{R}{\cos\phi}$

Hence, P'_R is a function of ϕ which is variable as x_C varies. P'_R is maximized if

$$\frac{dP'_R}{d\phi} = 0$$

or, $(\sin \phi_m - 2 \sin \phi_m \cos \phi_m) = 0$ [ϕ_m is the value of ϕ for maximum value of P'_R]

or, $$\cos \phi_m = \frac{1}{2}$$

∴ $$\phi_m = 60°$$

∴ Maximum received power is $P'_{R_{\max}} = \frac{1}{0.1}\left(\frac{1}{2} - \frac{1}{4}\right) = 2.5$ p.u.

This corresponds to $\delta = 60°$. The optimum value of x_C can then be obtained.

For the system, $\tan \phi_m = \frac{X}{R}$ i.e., $X = R \tan \phi_m$

However, $|X| = |x_S + x_\ell + x_C|$

For $\phi_m = 60°$, we now can write, $R \tan 60° = x_S + x_\ell + x_C$

∴ $x_C = R \tan 60° - x_S + x_\ell = 0.2 + 0.2 - 0.1 \times \sqrt{3} = 0.2268$ p.u. (Ans.)

EXAMPLE E 13.17 Deduce an expression for the maximum steady state power which can be transmitted over a power transmission line (neglecting shunt capacitance of the line) if the voltage at each end is kept constant. Show that if the reactance X of the line be varied, the resistance R remaining constant, the maximum steady state power that could be transmitted over the line would be greatest when $X = R$.

Solution: In Figure E13.10, let V_R and V_S be the receiving and sending end voltages, respectively. Since capacitance is neglected in the line current, sending and receiving end currents are the same, namely I, lagging by an angle θ_R with respect to V_R and by θ_S with respect to V_S. IR and IX are the components of voltage drops due to the impedance Z of the line, when a current I flows through it.

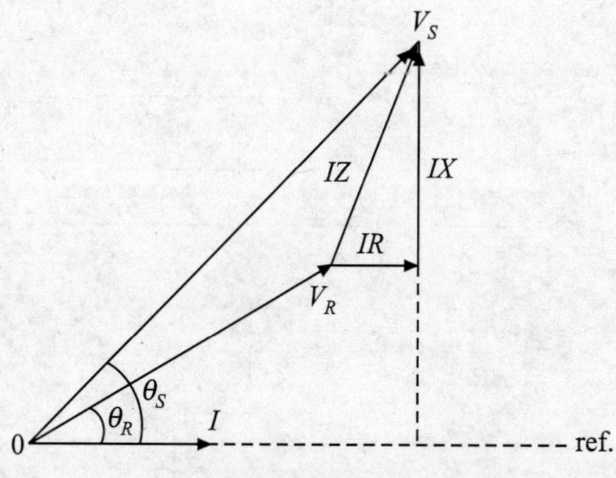

Figure E13.10

From the phasor diagram shown in Figure E13.10, we can write

$$V_S \cos\theta_S = V_R \cos\theta_R + IR$$

and $$V_S \sin\theta_S = V_R \sin\theta_R + IX$$

Squaring and adding,

$$V_S^2 = V_R^2 + 2V_R I(R\cos\theta_R + X\sin\theta_R) + I^2(R^2 + X^2)$$

However, $$V_R I\cos\theta_R = P \text{ (active power received)}$$

and $$V_R I\sin\theta_R = Q \text{ (reactive volt-amperes)}$$

∴ We can write $$V_S^2 = V_R^2 + 2PR + 2QX + \frac{(P^2+Q^2)(R^2+X^2)}{V_R^2}$$

i.e., $$-V_S^2 + V_R^2 + 2PR + 2QX + \frac{(P^2+Q^2)(R^2+X^2)}{V_R^2} = 0$$

In the above expression, V_S, V_R, R and X are constants, P and Q being the only variables. We want to determine the maximum value of P when Q is varied. Differentiating this expression with respect to Q, we get

$$2R\frac{dP}{dQ} + 2X + \left(\frac{R^2+X^2}{V_R^2}\right)\left(2P\frac{dP}{dQ} + 2Q\right) = 0$$

Equating dP/dQ to zero for maximum value of P we find,

$$2X + 2Q\left(\frac{R^2+X^2}{V_R^2}\right) = 0 \text{ or } Q = -\frac{V_R^2 X}{R^2+X^2}$$

Substituting this value of Q in equation (1)

$$-V_S^2 + V_R^2 + 2PR - \frac{2V_R^2X^2}{R^2+X^2} + \frac{R^2+X^2}{V_R^2}\left[P^2 + \frac{V_R^4X^2}{(R^2+X^2)^2}\right] = 0$$

This is a quadratic equation in P. Multiplying throughout by $\frac{V_R^2}{R^2+X^2}$ and substituting $R^2 + X^2 = Z^2$,

we get $$P^2 + 2PR\frac{V_R^2}{Z^2} + (V_R^2 - V_S^2)\frac{V_R^2}{Z^2} - \frac{V_R^4X^2}{Z^4} = 0$$

∴ $$P = -\frac{RV_R^2}{Z^2} \pm \sqrt{\frac{R^2V_R^4}{Z^4} + \frac{V_S^2V_R^2}{Z^2} - \frac{V_R^4R^2}{Z^4}}$$

$$= -\frac{RV_R^2}{Z^2} \pm \frac{V_SV_R}{Z} = \frac{V_R^2}{Z^2}\left[-R \pm \frac{V_S}{V_R}Z\right]$$

So that $$P_{\max} = \frac{V_R^2}{Z^2}\left[\frac{V_S}{V_R}Z - R\right]$$ (Ans.)

For the second part of the question, with $V_S = V_R$, the expression of P_{max} in this case becomes

$$P_{max} = \frac{V_R^2}{Z^2}(Z - R)$$

The next task is keeping R fixed and varying X such that P_{max} is greatest.

Obviously, $\frac{dP_{max}}{dX}$ should be zero for this condition so that

$$\frac{dP_{max}}{dX} = \frac{d}{dX}\left[\frac{V_R^2}{Z^2}(Z - R)\right] = 0$$

or,
$$\frac{d}{dX}\left[\frac{V_R^2}{R^2 + X^2}\left(\sqrt{R^2 + X^2} - R\right)\right] = 0$$

or,
$$\frac{d}{dX}\left[\frac{V_R^2}{(R^2 + X^2)^{1/2}} - \frac{RV_R^2}{(R^2 + X^2)}\right] = 0$$

or,
$$-\frac{V_R^2 \cdot 2X}{2(R^2 + X^2)^{3/2}} + \frac{RV_R^2 \cdot 2X}{(R^2 + X^2)^2} = 0$$

or,
$$\frac{R}{\sqrt{R^2 + X^2}} = \frac{1}{2}$$

$\therefore$
$$X = \sqrt{3}R \qquad \text{(Ans.)}$$

EXAMPLE 13.18 Consider the system shown in Figure E13.11(a). A FC-TCR (Fixed Capacitor–Thyristor Controlled Reactor) SVC (Static Var Compensator) is connected at bus no. 3. Assume a simple model of SVC where the SVC can be represented by a voltage source V_C in series with a reactance $x_C = 0.15$ p.u. Here both sending end and receiving end have same magnitude of voltage as E, i.e. 1 p.u.

(i) Obtain expression for $|V|$ and P_e as function of E, x_C and δ.
(ii) Find the stability limit for P_e and the corresponding δ.
(iii) Repeat for the case when $x_C = 0$.

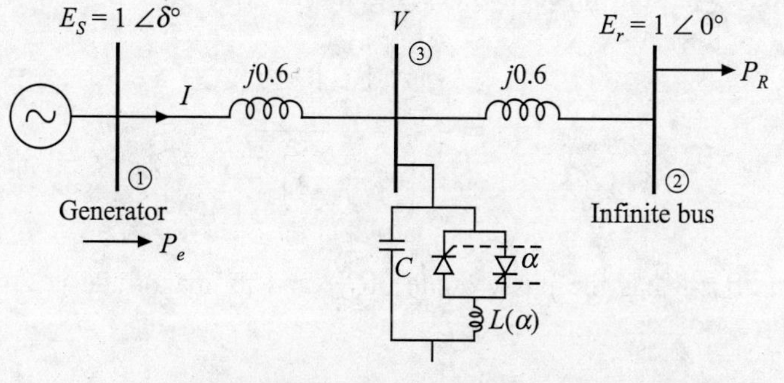

Figure E13.11(a) [α = firing angle]

Solution: At the SVC bus (i.e. at bus 3), the Thevenin's equivalent circuit has been obtained as shown in Figure E13.11(b) and the combined equivalent circuit, including the SVC, is shown in Figure E13.11(c).

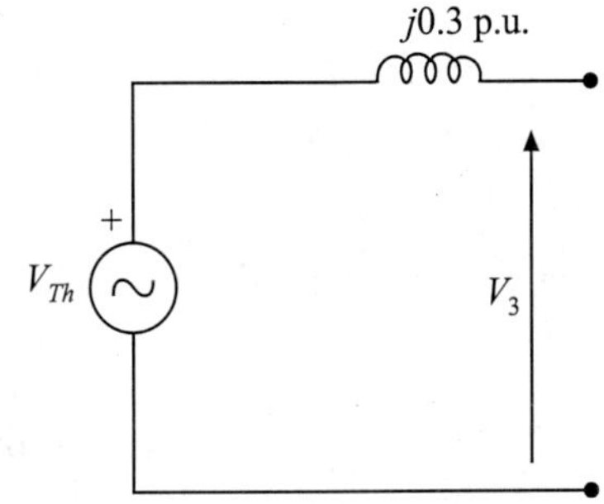

Figure E13.11(b) Thevenin's equivalent circuit across bus-3.

The current flowing into SVC is I_C (say) and this current is purely reactive (there are no losses). Therefore, the phase angle of V_c is same as that of V_{Th}. From phasor diagram, as shown in Figure E13.11(d), we can write

$$V_{Th} = \left| E\cos\frac{\delta}{2} \right| \angle\frac{\delta}{2} \; ; [E_S = E_r = E]$$

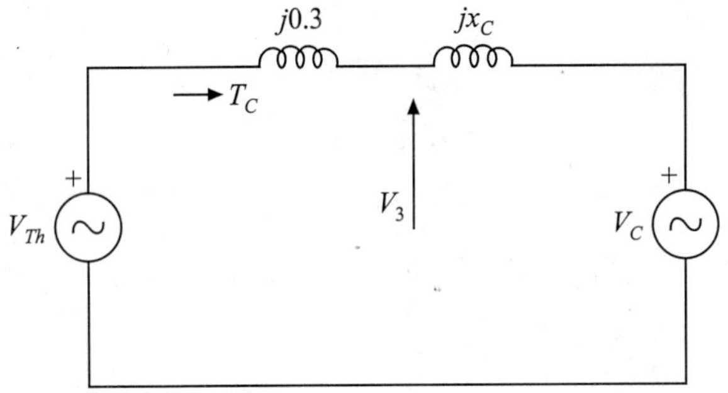

Figure E13.11(c) Complete equivalent system including SVC.

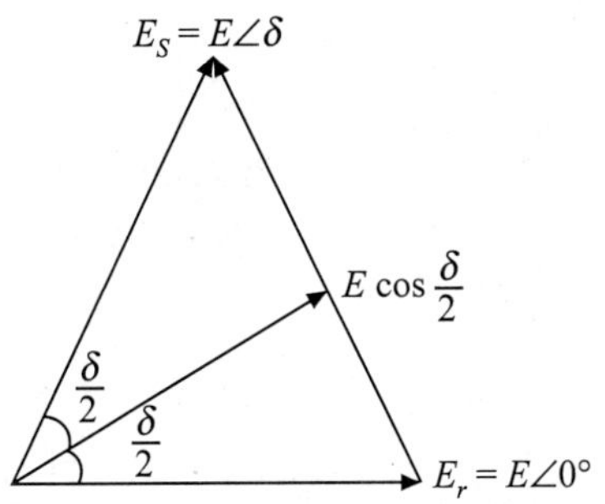

Figure E13.11(d) Phasor diagram.

The magnitude of current through SVC is given by

$$|I_C| = \frac{E\cos\left(\frac{\delta}{2}\right) - V_C}{\left(x_C + \frac{x_\ell}{2}\right)}$$

The voltage at bus-3 has the phase angle of $\frac{\delta}{2}$ and its magnitude is given by

$$V_3 = V_C + |I_C|\, x_C = V_C + \frac{E\cos\left(\frac{\delta}{2}\right) - V_C}{\left(x_C + \frac{x_\ell}{2}\right)} \times x_C$$

$\therefore$
$$V_3 = \frac{\frac{V_C x_\ell}{2}}{\left(x_C + \frac{x_\ell}{2}\right)} + \frac{E x_C \cos\left(\frac{\delta}{2}\right)}{\left(x_C + \frac{x_\ell}{2}\right)}$$

$\therefore$
$$P_e = \frac{E V_3 \sin\left(\frac{\delta}{2}\right)}{x_\ell} = \frac{E V_C \sin\left(\frac{\delta}{2}\right)}{2\left(x_C + \frac{x_\ell}{2}\right)} + \frac{E x_C \sin\left(\frac{\delta}{2}\right)\cos\left(\frac{\delta}{2}\right)}{x_\ell\left(x_C + \frac{x_\ell}{2}\right)}$$

i.e.,
$$P_e = \frac{E V_C \sin\left(\frac{\delta}{2}\right)}{(2x_C + x_\ell)} + \frac{E^2 x_C \sin\delta}{x_\ell(2x_C + x_\ell)} \qquad \text{(Ans. of (i))}$$

Let
$$P_e = A_1 \sin\left(\frac{\delta}{2}\right) + A_2 \sin\delta$$

where
$$A_1 = \frac{E V_C}{(2x_C + x_\ell)} = \frac{1 \times 1}{(2 \times 0.15 + 0.6)} = 1.1111$$

and
$$A_2 = \frac{E^2 x_C}{x_\ell(2x_C + x_\ell)} = \frac{1.0^2 \times 0.15}{0.6(2 \times 0.15 + 0.6)} = 0.2778$$

The power P_e will be maximum when $\frac{dP_e}{d\delta} = 0$

i.e.,
$$\frac{1}{2} A_1 \cos\left(\frac{\delta}{2}\right) + A_2 \cos\delta = 0$$

or,
$$\frac{1}{2} A_1 \cos\left(\frac{\delta}{2}\right) + A_2\left[2\cos^2\left(\frac{\delta}{2}\right) - 1\right] = 0$$

or,
$$2A_2 \cos^2\left(\frac{\delta}{2}\right) + \frac{1}{2} A_1 \cos\left(\frac{\delta}{2}\right) - A_2 = 0$$

$\therefore$
$$\cos\left(\frac{\delta}{2}\right) = \frac{-\frac{A_1}{2} \pm \sqrt{\frac{A_1^2}{4} + 8A_2^2}}{4A_2}$$

Here, only the positive solution for $\cos\left(\frac{\delta}{2}\right)$ is to be considered so that $\left(\frac{\delta}{2}\right)$ remains less than 90°. Substituting the respective values, we get

$$\cos\left(\frac{\delta}{2}\right) = \frac{-\frac{1.1111}{2} + \sqrt{\frac{1.1111^2}{4} + 8 \times 0.2778^2}}{4 \times 1.1111} = 0.0915$$

$\therefore$ $$\delta = \cos^{-1}(0.0915) = 84.75° \qquad \text{(Ans. of (ii))}$$

and $$P_{e_{max}} = 1.1111 \sin\left(\frac{84.75°}{2}\right) + 0.2778 \sin 84.75° = 1.025 \text{ p.u.} \qquad \text{(Ans. of (ii))}$$

When $x_C = 0$, $$P_e = \frac{EV_C \sin\left(\frac{\delta}{2}\right)}{x_\ell}$$

In this case, the power P_e will be maximum when $\delta = 180°$.

$$P_{e_{max}} = \frac{1 \times 1 \sin\left(\frac{180}{2}\right)}{0.6} = 1.6667 \text{ p.u.} \qquad \text{(Ans. of (iii))}$$

EXERCISES

1. How can you broadly define and classify power system stability? Establish the concept of power system stability analytically.
2. How do you define the transient stability of a power system? How can you model the transmission line, loads and alternators in transient stability study?
3. What are the assumptions in classical study of transient stability? Derive the swing equation. Why the swing equation is so important in the study of transient stability?
4. Derive the swing equation for a machine connected to an infinite bus in a power network. Extend it for a two machine system.
5. Linearise the swing expression including damping effect and discuss the stability aspect of the system.
6. Derive the equal area criterion. Interpret equal area criterion graphically for a generator supplying load at its bus.
7. What is critical clearing angle? What is its mathematical expression? What do you mean by critical clearing time? How can you find the critical clearing time for a generator subjected to a transient fault?
8. Show that it is possible to apply the equal area criterion to the transient stability of a synchronous motor.
9. What are the methods of improving transient stability?

10. Can you apply the equal area criterion to show improvement of transient stability for autoreclosure? Can you describe any other application of such type?
11. Briefly describe the step by step method and Runge–Kutta method to find solution of swing equation.
12. State the expression of swing equation for a multi-machine system.
13. Plot swing curves for Example 13.12 using step by step method.
14. Plot swing curves for Example 13.12 using Runge–Kutta method.
15. Develop an expression for steady state stability limit. Explain the concept of steady state stability margin.

CHAPTER 14

Power System Faults

14.1 INTRODUCTION

The operation of a power system alters following the occurrence of a fault. Faults give rise of abnormal operating conditions leading excessive high currents and abnormal voltages. Three phase faults are categorized as 'balanced faults' or 'symmetrical faults', whereas the faults in one or two phases, either involving ground or not, are called 'unbalanced faults'. Usually the fault current (kA) is multiplied by pre-fault line voltage (kV) and the product is referred to as 'fault level'. It is expressed in MVA.

14.1.1 Calculation of p.u. Quantities for Single Phase System

If VA_{base} and V_{base} are the selected base quantities for power and voltage respectively; then,

base current, $$I_{base} = \frac{V_{base} I_{base}}{V_{base}} = \frac{VA_{base}}{V_{base}} \text{ A} \tag{14.1}$$

base impedance, $$Z_{base} = \frac{V_{base}}{I_{base}} = \frac{V_{base}^2}{VA_{base}} \ \Omega \tag{14.2}$$

Also, base current, $$I_{base} = \frac{MVA_{base}}{kV_{base}} \text{ kA} \tag{14.3}$$

base impedance, $$Z_{base} = \frac{kV_{base}^2}{MVA_{base}} \ \Omega \tag{14.4}$$

In these expressions, all the quantities are single phase quantities.

14.1.2 Calculation of p.u. Quantities for Three Phase System

In the three phase representations, if $VA_{3\phi\ base}$ and $V_{LL\ base}$ be the base three phase power and line to line voltage respectively; then,

base current, $$I_{base} = \frac{VA_{base}}{V_{base}} = \frac{3VA_{base}}{3V_{base}} = \frac{VA_{3\phi\ base}}{\sqrt{3}V_{LL\ base}}\ \text{A} \tag{14.5}$$

base impedance, $$Z_{base} = \frac{V_{base}^2}{VA_{base}} = \frac{\left(\sqrt{3}\right)^2 V_{base}^2}{3VA_{base}} = \frac{V_{LL\ base}^2}{VA_{3\phi\ base}}\ \Omega \tag{14.6}$$

Also, base current, $$I_{base} = \frac{MVA_{3\phi\ base}}{\sqrt{3}\ kV_{LL\ base}}\ \text{kA} \tag{14.7}$$

base impedance, $$Z_{base} = \frac{V_{LL\ base}/\sqrt{3}}{I_{base}} = \frac{kV_{LL\ base}^2}{MVA_{3\phi\ base}}\ \Omega \tag{14.8}$$

Hence in three phase systems, the per unit (p.u.) quantities become

$$S_{p.u.} = \frac{S_{actual}(\text{MVA})}{MVA_{3\phi\ base}};\ V_{p.u.} = \frac{V_{actual}(\text{kV})}{kV_{LL\ base}}$$

$$I_{p.u.} = I_{actual}(\text{kA}) \times \frac{\sqrt{3}\ kV_{LL\ base}}{MVA_{3\phi\ base}} \tag{14.9}$$

$$Z_{p.u.} = Z_{actual}(\Omega) \times \frac{MVA_{3\phi\ base}}{kV_{LL\ base}^2} \tag{14.10}$$

$$\%\ Z = Z_{actual}(\Omega) \times \frac{kVA_{base}}{10\ kV_{base}^2} \tag{14.10a}$$

The power factor remains unchanged in p.u. system.

14.1.3 Conversion from One Base to Another Base

As different components of a power network can have different ratings, it is necessary to convert all quantities to a common base for the ease of computation. This can be done for three phase systems as follows;

$$S_{p.u.\ (new)} = S_{p.u.\ (old)} \times \frac{MVA_{3\phi\ base\ (old)}}{MVA_{3\phi\ base\ (new)}} \tag{14.11}$$

$$V_{p.u.\ (new)} = V_{p.u.\ (old)} \times \frac{kV_{LL\ base\ (old)}}{kV_{LL\ base\ (new)}} \tag{14.12}$$

$$Z_{p.u.\ (new)} = Z_{p.u.\ (old)} \times \frac{MVA_{3\phi\ base\ (new)}}{MVA_{3\phi\ base\ (old)}} \times \frac{kV_{LL\ base\ (old)}^2}{kV_{LL\ base\ (new)}^2} \tag{14.13}$$

EXAMPLE 14.1 A star-connected power system is rated for 100 MVA and 220 kV. Express 75 MVA of apparent power (3-phase) as per unit values referred to

(a) three phase system MVA as base and
(b) per phase system MVA as base.

Solution:

(a) When expressed in terms of three phase system MVA,

$$\text{p.u. MVA} = \frac{75}{100} = 0.75 \text{ p.u.} \qquad \text{(Ans.)}$$

(b) When expressed per phase basis, $\text{Base MVA} = \frac{1}{3} \times 100 = 33.33 \text{ MVA} = 1 \text{ p.u.}$

$$\text{and Base kV} = \frac{220}{\sqrt{3}} \cong 127 \text{ kV} = 1 \text{ p.u.}$$

$$\therefore \text{ p.u. MVA} = \frac{75/3}{33.33} = 0.75 \text{ p.u.} \qquad \text{(Ans.)}$$

It may be observed that p.u. value remain same whether the quantity is expressed in terms of three phase or single phase basis.

EXAMPLE 14.2 A star-connected, three phase 40 MVA, 15 kV, 50 Hz alternator has a reactance 5Ω/phase. Obtain the p.u. reactance using the rated MVA and voltage as base values then refer this obtained p.u. value of reactance to 11 kV, 50 MVA base.

Solution: Let us take the base value at first as 40 MVA, 15 kV (i.e., 40 MVA = 1 p.u. power and 15 kV = 1 p.u. voltage)

$$\text{Base current} = \frac{40 \times 10^6}{\sqrt{3} \times 15 \times 10^3} = 1539.6007 \text{ A}$$

and

$$\text{Base reactance} = \frac{\text{Base kV} \times 10^3/\sqrt{3}}{\text{Base current}} = \frac{15 \times 10^3}{\sqrt{3} \times 1539.6007} = 5.6250 \ \Omega$$

$$\therefore \quad \text{p.u. reactance} = \frac{5}{\text{Base reactance}} = \frac{5}{5.625} = 0.8889 \text{ p.u.} \qquad \text{(Ans.)}$$

This p.u. reactance is to be referred now with respect to 11kV, 50 MVA base.

$$\therefore \quad \text{New p.u. reactance} = X_{\text{p.u. (old)}} \times \frac{\text{MVA}_{3\phi \text{ base (new)}}}{\text{MVA}_{3\phi \text{ base (old)}}} \times \frac{\text{kV}^2_{\text{LL base (old)}}}{\text{kV}^2_{\text{LL base (new)}}}$$

$$= 0.8889 \times \frac{50}{40} \times \left(\frac{15}{11}\right)^2 = 2.0661 \text{ p.u.} \qquad \text{(Ans.)}$$

EXAMPLE 14.3 A three phase, 11 kV transmission line is delivering 10 MVA load to a load centre. The p.u. impedance of the line is (0.05 + j 0.1) p.u. at the base values of 10 MVA and 11 kV. Find the voltage drop in line.

Solution:

$$\text{Base current } (I_{\text{base}}) = \frac{\text{MVA}_{3\phi \text{ base}}}{\sqrt{3}\ \text{kV}_{\text{LL base}}} \text{ kA} = \frac{10\times10^6}{\sqrt{3}\times11\times10^3} = 524.8639 \text{ A } (=1 \text{ p.u.})$$

$$\text{Base impedance } (Z_{\text{base}}) = \frac{\text{V}_{\text{LL base}}/\sqrt{3}}{I_{\text{base}}} = \frac{11\times10^3/\sqrt{3}}{524.8639} = 12.1\ \Omega$$

$$\left[\text{Also, } Z_{\text{base}} = \frac{\text{kV}^2_{\text{LL base}}}{\text{MVA}_{3\phi \text{ base}}} = \frac{11^2}{10} = 12.1\ \Omega\right]$$

∴ Actual impedance of the line, Z = p.u. impedance of the line × Z_{base} = (0.05 + j0.1) × 12.1

$$= (0.605 + j1.21)\ \Omega/\text{ph}$$

Current through the line, $I_{\text{line}} = I_{\text{base}}$

∴ Voltage drop in line $= I_{\text{line}} \times Z = 524.8639 \times (0.605 + j1.21) = (317.5427 + j635.0853)$

$$= 710.047\angle 63.43° \text{ volt/ph}$$ (Ans.)

[Alternatively, voltage drop in each phase of the line

= p.u. current in the line × p.u. impedance of the line

$= 1\times(0.05 + j0.1) = (0.05 + j0.1) \text{ p.u.} = (0.05 + j0.1) \times \text{V}_{\text{LL base}}/\sqrt{3}$

$= (0.05 + j0.1)\times 11\times10^3/\sqrt{3} = 710.047\angle 63.43° \text{ volt/ph}$] (Ans.)

EXAMPLE 14.4 Single line diagram of a single phase power system is shown in Figure E14.1. Find p.u. current through the generator.

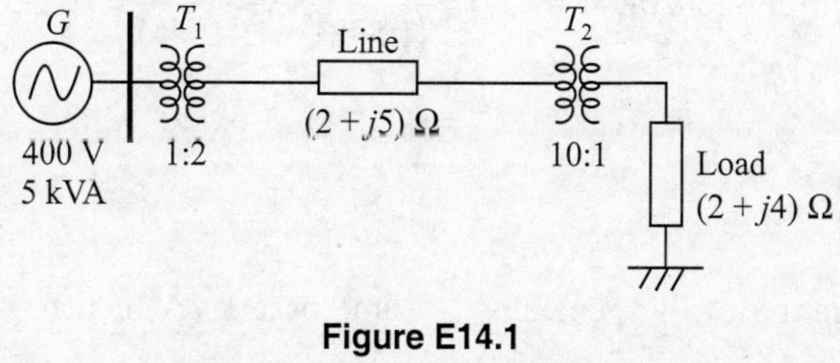

Figure E14.1

Solution: Let us assume kVA rating of the generator is same as base kVA of the system.

∴ $$\text{kVA}_{\text{base}} = \text{kVA}_{\text{base(gen)}} = 5 \text{ kVA } (=1 \text{ p.u.})$$

Along the line side, $V_{\text{base(line)}} = (400\times2)\text{ V} = 800 \text{ V} = 1 \text{ p.u.}$ [∵ turns ratio of transformer T_1 is 1:2]

$$\therefore \quad \text{kVA}_{\text{base}} = \text{kVA}_{\text{base(gen)}} = 5 \text{ kVA } (=1 \text{ p.u.})$$

$$I_{\text{base(line)}} = \frac{\text{kVA}_{\text{base}} \times 10^3}{V_{\text{base (line)}}} = \frac{5 \times 10^3}{800} = 6.25 \text{ A } (=1 \text{ p.u.})$$

$$\therefore \quad Z_{\text{base (line)}} = \frac{V_{\text{base (line)}}}{I_{\text{base (line)}}} = \frac{800}{6.25} = 128 \ \Omega \ (=1 \text{ p.u.})$$

$$\therefore \quad Z_{\text{line}} = \frac{(2 + j5)}{128} = (0.0156 + j0.0391) \text{ p.u.}$$

At load side, $V_{\text{base (load)}} = \dfrac{V_{\text{base (line)}}}{10} = \dfrac{800}{10} \text{ V} = 80 \text{ V} = 1 \text{ p.u.}$ [$\because$ turns ratio of transformer T_2 is 10:1]

$$\therefore \quad \text{kVA}_{\text{base}} = 5 \text{ kVA } (=1 \text{ p.u.})$$

$$I_{\text{base (load)}} = \frac{\text{kVA}_{\text{base}} \times 10^3}{V_{\text{base (load)}}} = \frac{5 \times 10^3}{80} = 62.5 \text{ A } (=1 \text{ p.u.})$$

$$Z_{\text{base (load)}} = \frac{V_{\text{base (load)}}}{I_{\text{base (load)}}} = \frac{80}{62.5} = 1.28 \ \Omega \ (=1 \text{ p.u.})$$

$$\therefore \quad Z_{\text{load}} = \frac{(2 + j4)}{1.28} = (1.5625 + j3.1250) \text{ p.u.}$$

Assuming transformers to be ideal, the impedances of transformers are neglected.

$$\therefore \quad Z_{\text{total}} = Z_{\text{line}} + Z_{\text{load}} = (0.0156 + j0.0391) + (1.5625 + j3.125) = (1.5781 + j3.1641) \text{p.u.}$$

$\therefore$ p.u. current through the generator $(= I_g) = \dfrac{1}{(1.5781 + j3.1641)} = (0.1262 - j0.2531)$

$$= 0.2828\angle -63.50° \text{ p.u.} \qquad \text{(Ans.)}$$

Alternate method

Let us assume kVA rating of the generator as same base kVA of the system.

$$\therefore \text{kVA}_{\text{base}} = \text{kVA}_{\text{base(gen)}} = 5 \text{ kVA } (=1 \text{ p.u.})$$

Along the line side, $V_{\text{base (line)}} = 400 \times 2 \text{ V} = 800 \text{ V} = 1 \text{ p.u.}$ [$\because$ turns ratio of transformer T_1 is 1:2]

$$Z_{\text{base (line)}} = \frac{V^2_{\text{base (line)}}}{\text{kVA}_{\text{base}} \times 10^3} = \frac{800^2}{5 \times 10^3} = 128 \ \Omega \ (=1 \text{ p.u.})$$

$$\therefore \qquad Z_{\text{line}} = \frac{(2+j5)}{128} = (0.0156 + j0.0391) \text{ p.u.}$$

At load side, $V_{\text{base (load)}} = \dfrac{800}{10}$ V = 80 V = 1 p.u. [$\because$ turns ratio of transformer T_2 is 10:1]

$$Z_{\text{base (load)}} = \frac{V^2_{\text{base (load)}}}{\text{kVA}_{\text{base}} \times 10^3} = \frac{80^2}{5 \times 10^3} = 1.28\ \Omega\ (=1 \text{ p.u.})$$

$$\therefore \qquad Z_{\text{load}} = \frac{(2+j4)}{1.28} = (1.5625 + j3.1250) \text{ p.u.}$$

Assuming the generator and transformers to be ideal, the impedances of these equipment are neglected.

$$\therefore \quad Z_{\text{total}} = Z_{\text{line}} + Z_{\text{load}} = (0.0156 + j0.0391) + (1.5625 + j3.125) = (1.5781 + j3.1641) \text{p.u.}$$

$\therefore$ p.u. current through the generator $(= I_g) = \dfrac{1}{(1.5781 + j3.1641)} = (0.1262 - j0.2531)$

$$= 0.2828\angle -63.50° \text{ p.u.} \qquad \text{(Ans.)}$$

EXAMPLE 14.5 Two generators G_1 and G_2 deliver power from bus-1 to a load centre (bus-2) through a HV transmission line (Figure E.14.2). All the parameters of the components are shown in the figure. Express all the parameters in terms of 50 MVA base.

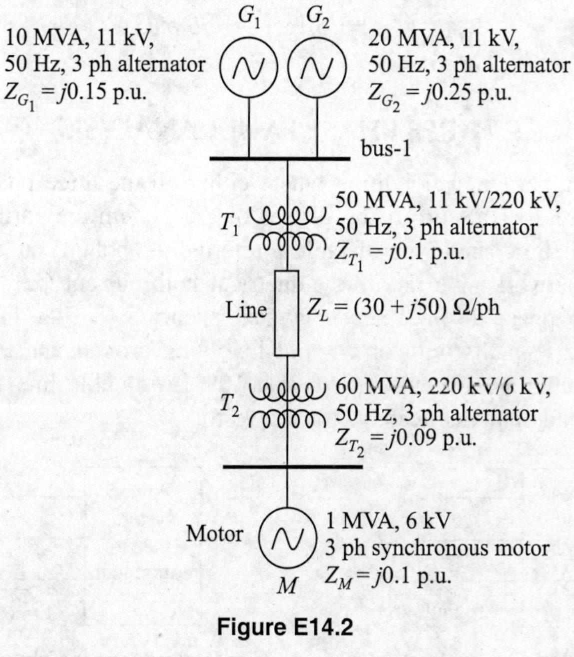

Figure E14.2

Solution: All the p.u. values of the parameters of the components are given based on their own rating.

$$\text{MVA}_{3\phi\text{ base(new)}} = 50$$

$$\therefore \quad Z_{\text{G}_1\text{ p.u. (new)}} = Z_{\text{G}_1\text{ p.u. (old)}} \times \frac{\text{MVA}_{3\phi\text{ base(new)}}}{\text{MVA}_{3\phi\text{ base (old)}}} = j0.15 \times \frac{50}{10} = j0.75\text{ p.u.} \qquad \text{(Ans.)}$$

Similarly,

$$Z_{\text{G}_2\text{ p.u. (new)}} = j0.25 \times \frac{50}{20} = j0.625\text{ p.u.} \qquad \text{(Ans.)}$$

$$Z_{\text{T}_1\text{ p.u. (new)}} = j0.1 \times \frac{50}{50} = j0.1\text{ p.u.} \qquad \text{(Ans.)}$$

Now,

$$Z_{\text{line (p.u.)}} = \frac{Z_{\text{line}}}{Z_{\text{base(line)}}} = (30 + j150) \times \frac{50}{220^2} = (0.031 + j0.155)\text{ p.u.} \qquad \text{(Ans.)}$$

$$\left[\because \; Z_{\text{line (p.u.)}} = \frac{Z_{\text{line}}}{Z_{\text{base(line)}}} = Z_{\text{line}} \times \frac{\text{MVA}_{\text{base}}}{\text{kV}^2_{\text{LL base}}}\right]$$

Again,

$$Z_{\text{T}_2\text{ p.u. (new)}} = j0.09 \times \frac{50}{60} = j0.075\text{ p.u.} \qquad \text{(Ans.)}$$

$$Z_{M\text{ p.u. (new)}} = j0.1 \times \frac{50}{1} = j5\text{ p.u.} \qquad \text{(Ans.)}$$

14.2 SYMMETRICAL THREE PHASE FAULT ANALYSIS

Three phase fault is a particular condition where either all the three phases of the system are short circuited (Figure 14.1(a)) or all the phases of the system are earthed due to earth fault (Figure 14.1(b)). Sometime such type of fault is referred as 'bolted' or 'solid' fault. Such type of fault in a power network give rise to symmetrical fault current (i.e., equal fault current in each phase with 120° phase displacement). Typically, only 5% of the initial faults in a power network are three phase faults with or without involving ground. On the other side, 80% of the unsymmetrical faults are line to earth faults, 10% are double line faults with or without ground. Broken conductor faults account for the rest.

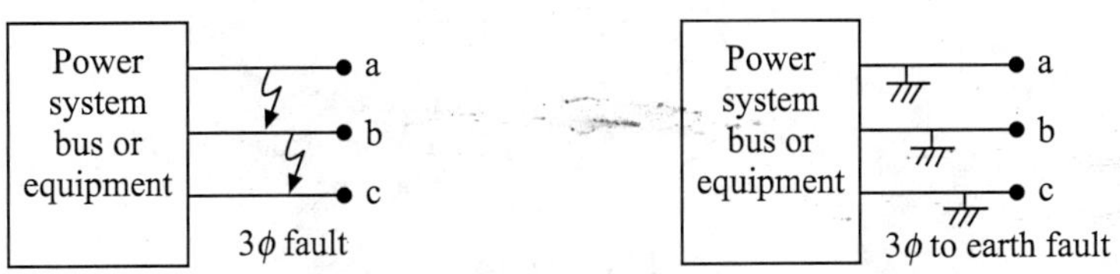

(a) Balanced three phase fault

(b) Balanced three phase to earth fault

Figure 14.1 Symmetrical three phase faults.

14.2.1 Limiting Factors of Fault Current

When a short circuit occurs at any point 'F' in a power network, the short circuit current is limited by the impedance of the system.

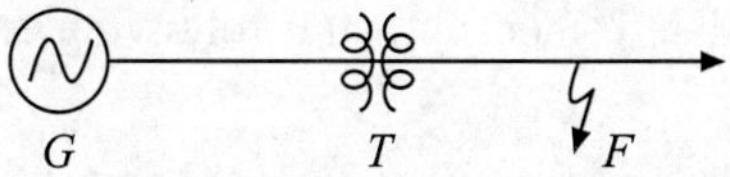

Figure 14.2 Single line diagram of a three phase fault.

Let us consider a fault occurring at point F. The short circuit current from generating station will be limited by relevant impedance of the generator (G), transformer (T) and impedance of the line between the generator and the point of fault. Generally, the fault currents are largely reactive in nature since in practical system total reactance exceeds three times the resistance. Thus the resistance can be neglected during calculation of fault current. The error for this assumption normally does not exceed 5%.

14.2.2 Fault Level Calculations

In a power network, fault level and the maximum fault current (or fault MVA) that can flow into a zero impedance fault and is necessary to be known for planning and selection of switchgear installation. The fault level usually defines the value of fault current for symmetrical fault condition. The fault level is usually expressed in MVA (or corresponding per unit value), with the maximum fault current value being considered at nominal voltage rating.

$$\text{Nominal MVA, } \text{MVA}_{\text{base}} = \sqrt{3} \times \text{Nominal voltage (kV)} \times I_{\text{base}}\text{(kA)} \tag{14.14}$$

$$\text{Fault MVA, } \text{MVA}_{\text{fault}} = \sqrt{3} \times \text{Nominal voltage (kV)} \times I_{sc}\text{(kA)} \tag{14.14a}$$

where, I_{base} = Rated current

I_{sc} = Short circuit current during fault.

Thus p.u. value of 'Fault level' can be expressed as

$$\text{Fault level}_{\text{p.u.}} = \frac{\text{Fault MVA}}{\text{Nominal MVA}} = \frac{\sqrt{3}}{\sqrt{3}} \times \frac{I_{sc}}{I_{\text{base}}} = I_{sc\ \text{p.u.}} = \frac{V_{\text{nominal p.u.}}}{Z_{\text{p.u.}}} \tag{14.15}$$

The per unit voltage for nominal value is unity. Thus we have,

$$\text{Fault level}_{\text{p.u.}} = \frac{1}{Z_{\text{p.u.}}}$$

$$\text{Fault level} = \text{Fault level}_{\text{p.u.}} \times \text{MVA}_{\text{base}} = \frac{\text{MVA}_{\text{base}}}{Z_{\text{p.u.}}} \tag{14.15a}$$

$$\therefore \quad \text{Fault level} = \frac{\text{MVA}_{\text{base}}}{\%Z} \times 100 \tag{14.16}$$

Magnitude of short circuit current is time dependent. It is initially at its high value during inception of the fault and starts decreasing to steady state value with time. This higher fault current affects existing circuit breaker adversely; hence current limiting reactors are sometimes used. The short circuit MVA is considered as a better indicator than short circuit current as breaker has to withstand the recovery voltage too across the breaker following arc interruption.

Instead of percentage impedance (or reactance), if total reactance upto fault point from generating station can be calculated, then by using following formula short circuit current can be determined.

$$\text{Short circuit current} = \frac{\text{Base MVA}}{\sqrt{3} \times \text{kV} \times \text{Total reactance upto fault point}} \times 100 \qquad (14.17)$$

14.3 THREE PHASE SUDDEN SHORT CIRCUIT OF AN UNLOADED ALTERNATOR

Whenever a three-phase sudden short circuit occurs at the terminals of an unloaded alternator, the armature current abruptly increases to a large value and voltage across the terminals drops.

At the instant of short circuit, there is mutual coupling between the stator winding, rotor winding and damper winding. The equivalent circuit is represented in Figure 14.3(a).

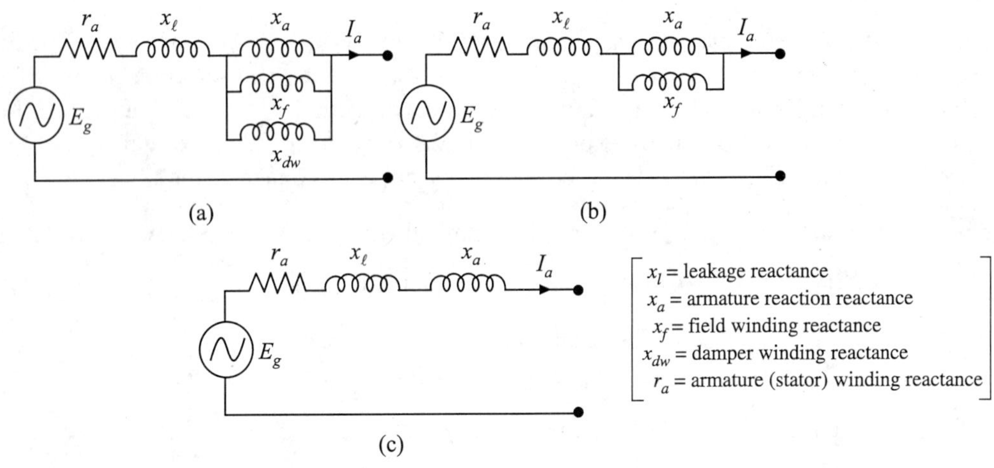

Figure 14.3 Equivalent circuit of an unloaded alternator under (a) sub-transient (b) transient and (c) steady state period.

The short circuit current can be divided into following three periods:

(a) sub-transient period
(b) transient period
(c) steady state period

In sub-transient condition, the equivalent reactance is sub-transient reactance (x_d'') and is given by,

$$x_d'' = x_\ell + \frac{1}{\dfrac{1}{x_a} + \dfrac{1}{x_f} + \dfrac{1}{x_{dw}}}$$

where,

r_a = armature winding resistance

x_ℓ = leakage reactance

x_a = reactance representing armature reaction

x_f = field winding reactance

x_{dw} = damper winding reactance

In transient condition, the equivalent reactance (transient reactance) is given by

$$x'_d = x_\ell + \frac{1}{\dfrac{1}{x_a} + \dfrac{1}{x_f}}$$

Finally, the reactance in steady state condition (synchronous reactance) is given by

$$x_d = x_\ell + x_a$$

(All reactances are expressed in ohm or in p.u.)

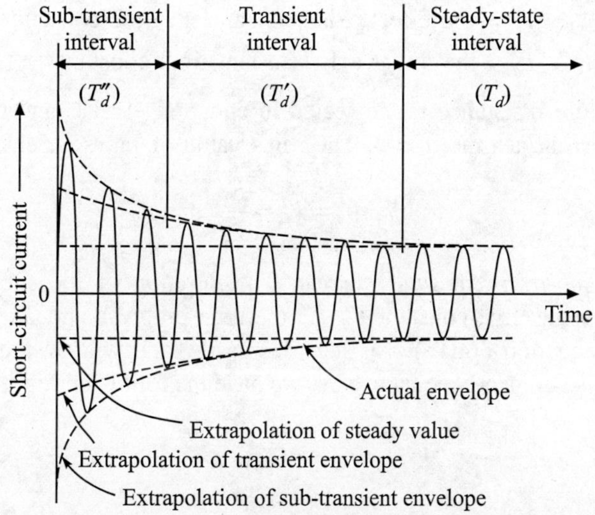

Figure 14.4 Symmetrical short circuit current oscillogram of an unloaded alternator.

The three periods can be viewed from the oscillogram of the short circuit current and shown in Figure 14.4. It has been assumed here that there is no DC offset in the armature current in the pre-fault condition. The magnitude of the current decreases exponentially from a high initial value following the fault. The instantaneous expression for the fault current is given by

$$i_f(t) = \sqrt{2}E_t\left[\left(\frac{1}{x''_d} - \frac{1}{x'_d}\right)e^{-t/T''_d} + \left(\frac{1}{x'_d} - \frac{1}{x_d}\right)e^{-t/T'_d} + \frac{1}{x_d}\right]\sin(\omega t + \alpha - \pi/2) \quad (14.18)$$

or,

$$i_f(t) = \sqrt{2}\left[\left(I''_f - I'_f\right)e^{-t/T''_d} + \left(I'_f - I_f\right)e^{-t/T'_d} + I_f\right]\sin(\omega t + \alpha - \pi/2) \quad (14.18a)$$

where

E_t = the r.m.s value of open circuit terminal voltage and α is its phase angle
x''_d = the direct axis sub-transient reactance
x'_d = the direct axis transient reactance
x_d = the direct axis synchronous reactance

with
$$x''_d < x'_d < x_d$$

$$I''_f = \text{sub-transient fault current} = \frac{\sqrt{2}E_t}{x''_d}$$

$$I'_f = \text{transient fault current} = \frac{\sqrt{2}E_t}{x'_d}$$

$$I_f = \text{steady state fault current} = \frac{\sqrt{2}E_t}{x_d}$$

The time constants are,

T''_d = the direct axis sub-transient time constant
T'_d = the direct axis transient time constant

If the effect of armature resistance be neglected in equation (14.18) and then $\alpha = \pi/2$. Let the fault be occurring at time, $t = 0$. The r.m.s value of fault current is thus given by

$$I_f(0) = I''_f = \frac{E_t}{x''_d} \tag{14.19a}$$

I''_f is called the *sub-transient fault current*. The duration of the sub-transient current is dictated by the decreasing time constant T''_d. As the time progress and $T''_d < T'_d < T_d$. The first exponential term of equation (14.18) will start decreasing and will eventually vanish.

However, since t is still nearly equal to zero, we have the following r.m.s value of fault current

$$I'_f = \frac{E_t}{x'_d} \tag{14.19b}$$

where, I'_f is called the *transient fault current*.

As the time progresses further and second exponential term also decays, we find the following r.m.s value of current which is called *steady state or sustained fault current*.

$$I_f = \frac{E_t}{x_d} \tag{14.19c}$$

14.4 SYMMETRICAL COMPONENTS

A three-phase is said to be *symmetrical* when the system has all the three phases identical to each other. Thus, in a three-phase symmetrical system the self-impedances of all the three phases are equal and mutual impedances, if any, between the three phase are also same with each other. Three-phase voltages (or currents) are said to be balanced, if the magnitude of the three voltages (or currents) are equal and they are separated from each other 120° (electrical) in phase.

When the system has unbalance, the voltages, currents and the phase impedances are in general unequal. Such a system can be resolved by a technique known as the *method of symmetrical components.*

Any three-phase system, having *unbalanced* phasor quantities can be represented in terms of three-phase balanced phasor components as follows:

(a) **Positive phase sequence components:** These components have three-phasors equal in magnitude and displaced by 120°. They have same phase sequence as original phasors. It represents a set of balanced phasors V_{r_1}, V_{y_1} and V_{b_1}.

(b) **Negative phase sequence components:** These components have three-phasors equal in magnitude and displaced by 120° but having phase sequence *opposite* to original phasors. It represents a set of balanced phasors V_{r_2}, V_{y_2} and V_{b_2}.

(c) **Zero phase sequence components:** These phasors are equal in magnitude and having zero phase displacement (set of V_{r_0}, V_{y_0} and V_{b_0}).

Since, each of the original unbalanced phasor is the sum of its components hence,

$$V_r = V_{r_0} + V_{r_1} + V_{r_2} \tag{14.20a}$$

$$V_y = V_{y_0} + V_{y_1} + V_{y_2} \tag{14.20b}$$

$$V_b = V_{b_0} + V_{b_1} + V_{b_2} \tag{14.20c}$$

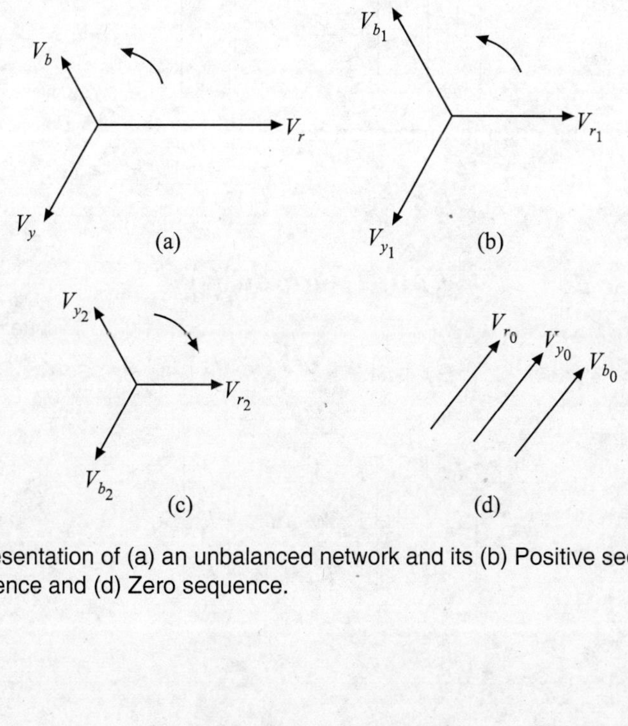

Figure 14.5 Representation of (a) an unbalanced network and its (b) Positive sequence, (c) Negative sequence and (d) Zero sequence.

$$V_r = V_{r_0} + V_{r_1} + V_{r_2} \tag{14.21a}$$

$$V_y = V_{r_0} + \alpha^2 V_{r_1} + \alpha V_{r_2} \tag{14.21b}$$

$$V_b = V_{r_0} + \alpha V_{r_1} + \alpha^2 V_{r_2} \tag{14.21c}$$

Equations (14.21a) to (14.21c) can be written as*

$$\begin{bmatrix} V_r \\ V_y \\ V_b \end{bmatrix} = \begin{bmatrix} 1 & 1 & 1 \\ 1 & \alpha^2 & \alpha \\ 1 & \alpha & \alpha^2 \end{bmatrix} \begin{bmatrix} V_{r_0} \\ V_{r_1} \\ V_{r_2} \end{bmatrix} \tag{14.21d}$$

Solving equations (14.21a) to (14.21c), we get

$$V_{r_0} = \frac{1}{3}\left(V_r + V_y + V_b\right) \tag{14.22a}$$

$$V_{r_1} = \frac{1}{3}\left(V_r + \alpha V_y + \alpha^2 V_b\right) \tag{14.22b}$$

$$V_{r_2} = \frac{1}{3}\left(V_r + \alpha^2 V_y + \alpha V_b\right) \tag{14.22c}$$

In matrix form, the three components of voltage phasors can be represented as

$$\begin{bmatrix} V_{r_0} \\ V_{r_1} \\ V_{r_2} \end{bmatrix} = \frac{1}{3} \begin{bmatrix} 1 & 1 & 1 \\ 1 & \alpha & \alpha^2 \\ 1 & \alpha^2 & \alpha \end{bmatrix} \begin{bmatrix} V_r \\ V_y \\ V_b \end{bmatrix}$$

Similarly, for current phasors, the three components are

$$I_{r_0} = \frac{1}{3}\left(I_r + I_y + I_b\right) \tag{14.23a}$$

$$I_{r_1} = \frac{1}{3}\left(I_r + \alpha I_y + \alpha^2 I_b\right) \tag{14.23b}$$

$$I_{r_2} = \frac{1}{3}\left(I_r + \alpha^2 I_y + \alpha I_b\right) \tag{14.23c}$$

In matrix form,

$$\begin{bmatrix} I_{r_0} \\ I_{r_1} \\ I_{r_2} \end{bmatrix} = \frac{1}{3} \begin{bmatrix} 1 & 1 & 1 \\ 1 & \alpha & \alpha^2 \\ 1 & \alpha^2 & \alpha \end{bmatrix} \begin{bmatrix} I_r \\ I_y \\ I_b \end{bmatrix}$$

---

* $\alpha = e^{j120°}$; $\alpha^2 = e^{j240°} = \alpha^*$
$(\alpha^2)^* = \alpha$; $\alpha^3 = 1$; $1 + \alpha + \alpha^2 = 0$

14.5 THREE PHASE POWER IN TERMS OF SYMMETRICAL COMPONENTS

The power consumed in a three phase circuit can be computed directly from the symmetrical components, provided the current and voltages are known.

Let S be the total complex power flowing into a three phase circuit through its phases R, Y, B while P and Q are the real and reactive powers in three phases. V_r, V_y and V_b indicate phase voltages at terminals and I_r, I_y and I_b the currents flowing into the circuit.

Here, $$S = P + jQ = V_r I_r^* + V_y I_y^* + V_b I_b^* \quad (14.24a)$$

In matrix notation, $$S = [V_r \; V_y \; V_b] \begin{bmatrix} I_r \\ I_y \\ I_b \end{bmatrix}^* = \begin{bmatrix} V_r \\ V_y \\ V_b \end{bmatrix}^T \begin{bmatrix} I_r \\ I_y \\ I_b \end{bmatrix}^* \quad (14.24b)$$

However, $$\begin{bmatrix} V_r \\ V_y \\ V_b \end{bmatrix} = [A] \begin{bmatrix} V_{r_0} \\ V_{r_1} \\ V_{r_2} \end{bmatrix} \quad (14.25a)$$

where, $$[A] = \begin{bmatrix} 1 & 1 & 1 \\ 1 & \alpha^2 & \alpha \\ 1 & \alpha & \alpha^2 \end{bmatrix}$$

and $$\begin{bmatrix} I_r \\ I_y \\ I_b \end{bmatrix} = [A] \begin{bmatrix} I_{r_0} \\ I_{r_1} \\ I_{r_2} \end{bmatrix} \quad (14.25b)$$

Utilising (14.25a) and (14.25b) in matrix equation (14.24b)

$$S = [AV]^T \, [AI]^* \quad (14.26a)$$

where, $$[V] = \begin{bmatrix} V_{r_0} \\ V_{r_1} \\ V_{r_2} \end{bmatrix} \text{ and } [I] = \begin{bmatrix} I_{r_0} \\ I_{r_1} \\ I_{r_2} \end{bmatrix}$$

But, in matrix algebra, the transpose of the product of the two matrices is equal to the product if the transposes of the matrices are in reverse order.

i.e. $$[AV]^T = V^T A^T$$

$\therefore$ Here, $$S = V^T \, A^T \, [AI]^* = V^T A^T \, A^* I^* \quad (14.26b)$$

Noting that $A^T = A$ and that α and α^2 are conjugates, we obtain,

$$S = \begin{bmatrix} V_{r_0} & V_{r_1} & V_{r_2} \end{bmatrix} \begin{bmatrix} 1 & 1 & 1 \\ 1 & \alpha^2 & \alpha \\ 1 & \alpha & \alpha^2 \end{bmatrix} \begin{bmatrix} 1 & 1 & 1 \\ 1 & \alpha & \alpha^2 \\ 1 & \alpha^2 & \alpha \end{bmatrix} \begin{bmatrix} I_{r_0} \\ I_{r_1} \\ I_{r_2} \end{bmatrix}^* \tag{14.27}$$

However, $A^T A^* = 3u$, u being the unity matrix.

$$[u] = \begin{bmatrix} 1 & 0 & 0 \\ 0 & 1 & 0 \\ 0 & 0 & 1 \end{bmatrix} \tag{14.28}$$

$$S = 3 \begin{bmatrix} V_{r_0} & V_{r_1} & V_{r_2} \end{bmatrix} \begin{bmatrix} I_{r_0} \\ I_{r_1} \\ I_{r_2} \end{bmatrix}^* \tag{14.29}$$

$$= 3[V_{r_0} I_{r_0}^* + V_{r_1} I_{r_1}^* + V_{r_2} I_{r_2}^*]$$

i.e. the sequence power is one-third the power in terms of phase quantities.

14.6 SEQUENCE MODEL OF AN UNLOADED ALTERNATOR

A three phase alternator with its neutral grounded through an impedance (Z_n) can be represented as shown in Figure 14.6.

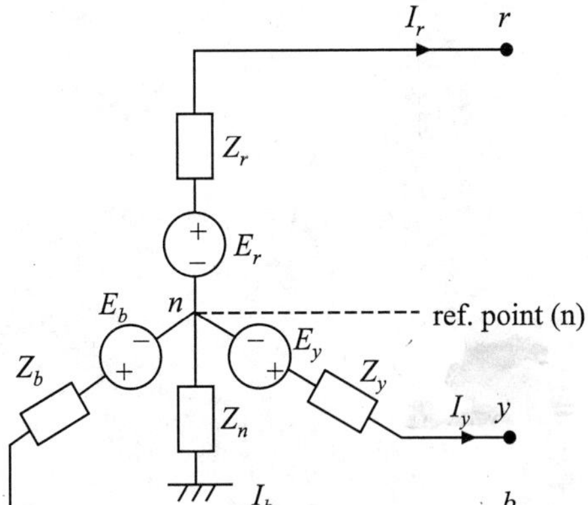

Figure 14.6 Diagrammatic representation of a three phase synchronous generator with all phase impedance same.

(a) ***Positive sequence network*** The positive sequence voltage of terminal r with respect to the reference bus (Figure 14.7(a)) is represented as

$$V_{r_1} = E_r - I_{r_1} Z_1$$

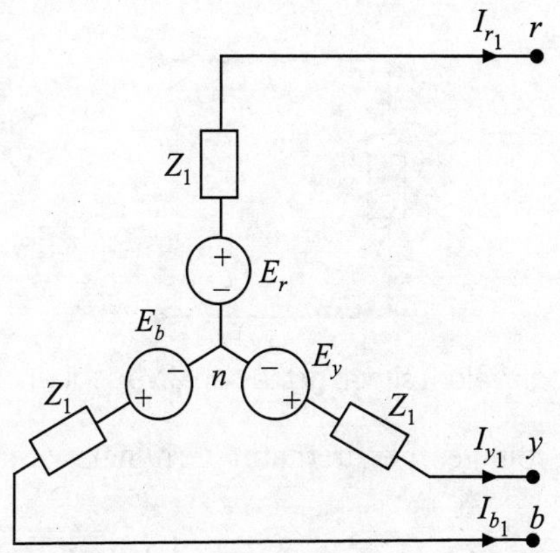

Figure 14.7(a) Three-phase positive sequence model of an alternator.

where suffix 1 is included to identify the positive sequence quantities. Figure 14.7(b) represents the positive sequence network of such generator.

Also, $$V_{y_1} = E_y - I_{y_1} Z_1 \text{ and } V_{b_1} = E_b - I_{b_1} Z_1$$

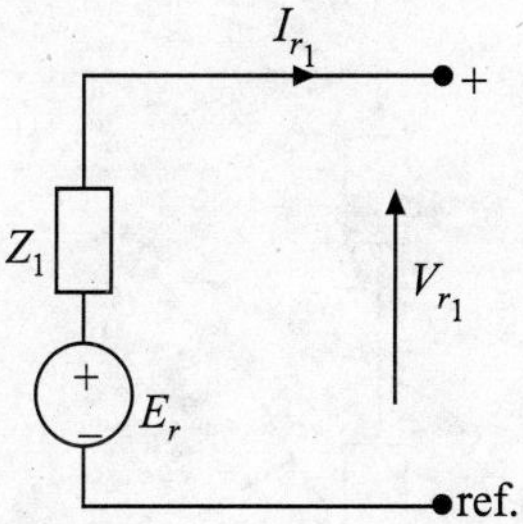

Figure 14.7(b) Single phase equivalent circuit of the positive sequence network model of an alternator.

(b) ***Negative sequence network*** Since there is no negative sequence induced emf, hence for alternator, the negative sequence network will have no emf at its phase and the network model and its single phase equivalent circuit is shown in Figures 14.8(a) and (b).

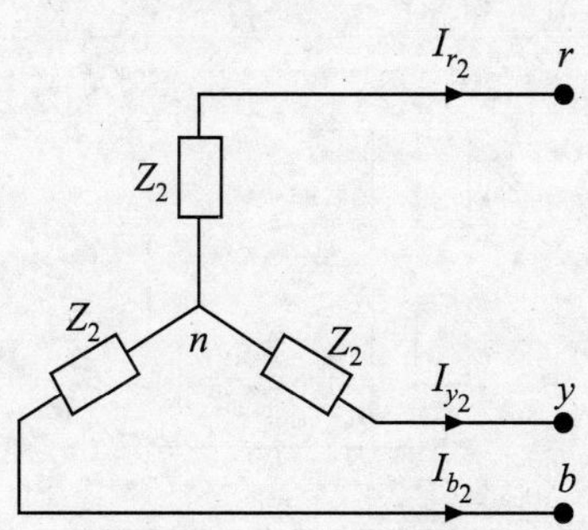

Figure 14.8(a) Three-phase negative sequence model of an alternator.

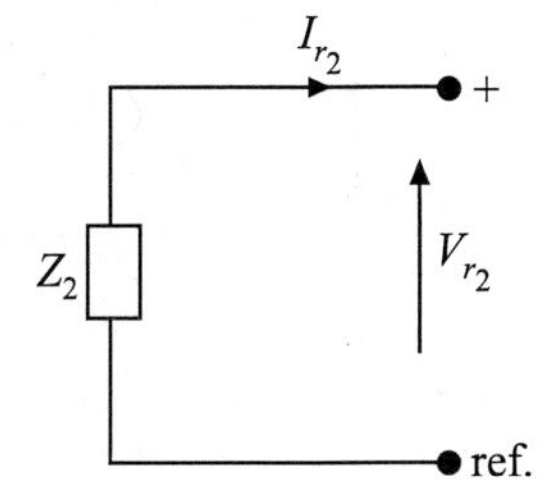

Figure 14.8(b) Single phase equivalent circuit of the negative sequence network model of an alternator.

The negative sequence voltages at alternator terminal with respect to reference bus are given by,

$$V_{r_2} = -Z_2 I_{r_2}\,;\; V_{y_2} = -Z_2 I_{y_2}\,;\; V_{b_2} = -Z_2 I_{b_2}$$

(c) ***Zero sequence network*** Let Z_{0g} be zero sequence impedance in each phase of an alternator in its three-phase zero sequence model. Since the currents are additive for a three-phase zero sequence network, hence the current flowing through Z_n is given by,

$$I_n = I_{r_0} + I_{y_0} + I_{b_0} = 3I_{r_0} \quad [\text{with } I_{r0} = I_{y0} = I_{b0}]$$

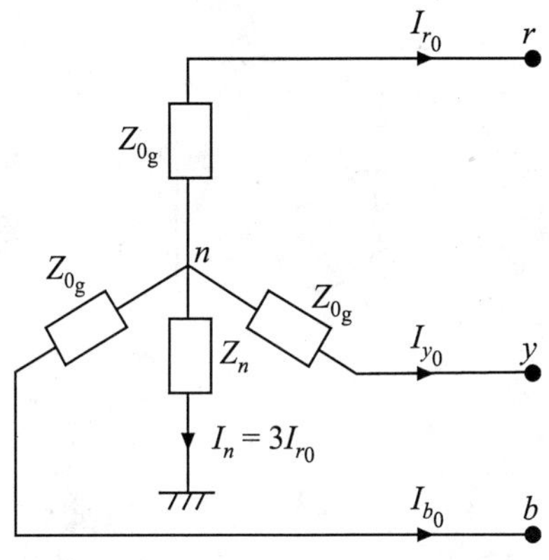

Figure 14.9(a) Three-phase zero sequence model of an alternator.

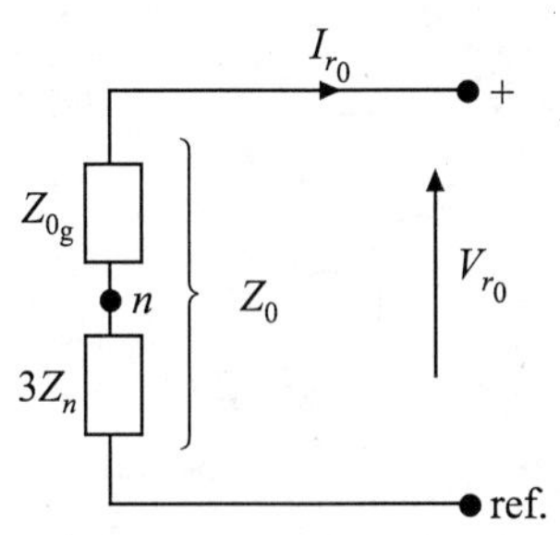

Figure 14.9(b) Single phase equivalent circuit of the zero sequence network model of an alternator.

The total zero sequence impedance is

$$Z_0 = Z_{0g} + 3Z_n$$

Applying KVL in the zero sequence equivalent network we have,

$$V_{r_0} + \left(Z_{0g} + 3Z_n\right) I_{r_0} = 0$$

or,

$$V_{r_0} + I_{r_0} Z_0 = 0$$

i.e.,

$$V_{r_0} = -Z_0 I_{r_0}$$

Therefore, the zero sequence voltages at alternator terminal with respect to reference bus are given by

$$V_{r_0} = -Z_0 I_{r_0}\ ;\ V_{y_0} = -Z_0 I_{y_0}\ ;\ V_{b_0} = -Z_0 I_{b_0}$$

In case the neutral is solidly grounded, $Z_n = 0$ and hence $Z_0 = Z_{0g}$

14.7 SEQUENCE NETWORKS OF A TRANSFORMER

For short circuit studies, the shunt magnetising branch of transformer can be neglected, as the current through this magnetising branch is negligible as compared to short circuit current. Hence the transformer can be modelled with an equivalent series impedance. Again transformer being a static device, the series leakage impedances remains unaltered even if the phase sequence of the supply voltage is reversed. *Thus, for a transformer the positive and negative sequence impedances are equal.* The zero sequence current will flow through a transformer if path for it exits on primary and secondary sides. If there is no ground return, zero sequence current cannot flow and corresponding zero sequence network may be represented by an open circuit. The positive and negative sequence networks of transformer are identical and as shown in following Figure 14.10.

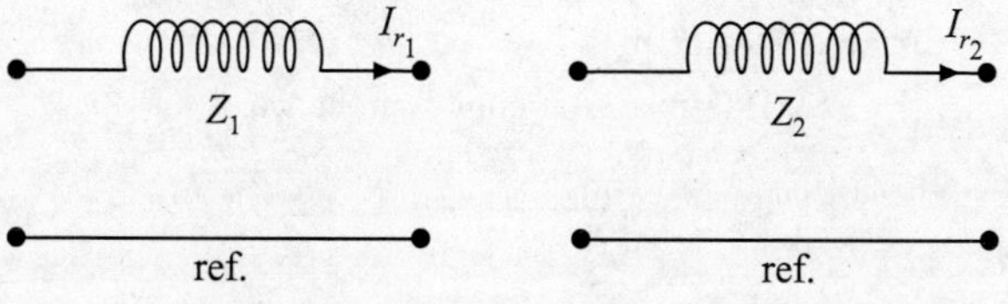

(a) Positive sequence (b) Negative sequence

Figure 14.10 The positive and negative sequence networks of transformer.

The zero sequence network of a transformer depends on the primary and secondary winding connection of the transformer as well as the grounding connection of the windings. The different three-phase transformer connections and their equivalent zero sequence networks are discussed in following section.

14.7.1 Star-star Connection with Both Neutral Grounded

Since both the neutrals are grounded, the phasor sum of the three unbalanced phase currents is equal to three times of zero sequence current I_{r_0}. The zero sequence currents can flow through the primary and secondary winding. The equivalent circuit is shown in Figure 14.11. The transformer can then be represented by equivalent zero sequence leakage impedance.

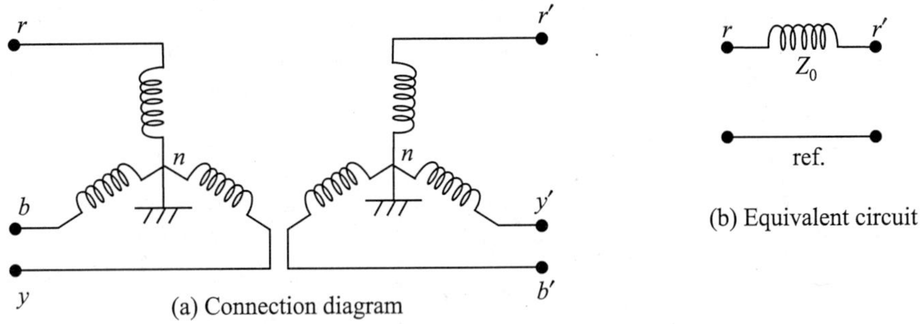

Figure 14.11 The zero sequence equivalent circuit of a star-star transformer with both the neutral grounded.

14.7.2 Star-star Connection with Only One Neutral Grounded

When the one neutral of any one winding of the transformer is grounded, the zero sequence current cannot exist. In this case, the transformer is represented as an open circuit between primary and secondary winding.

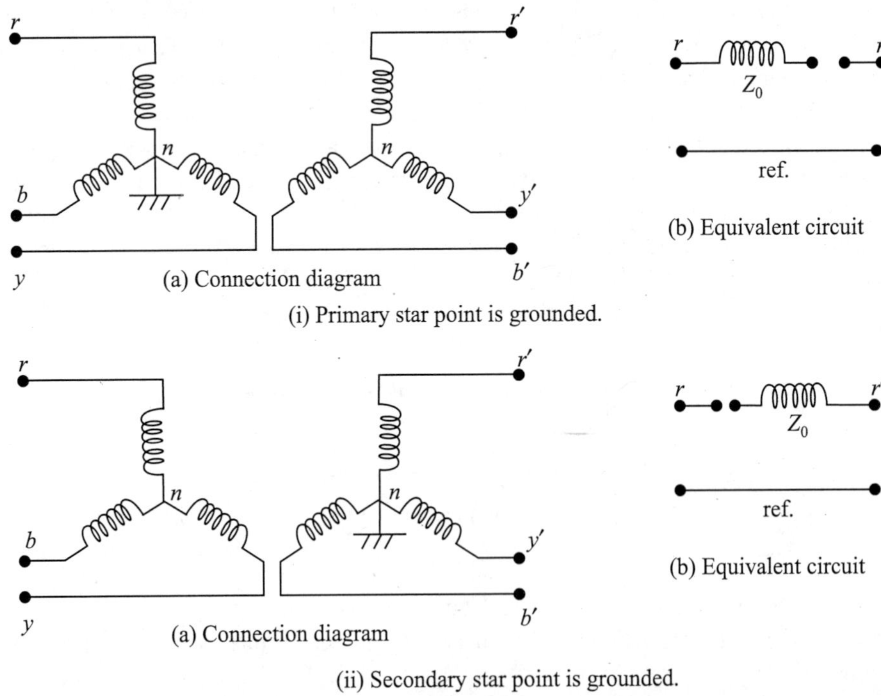

Figure 14.12 The zero sequence equivalent circuit of a star-star transformer with only one neutral grounded.

14.7.3 Star-star Connection with Ungrounded Neutral

In this case, the phasor sum of the currents of both windings is zero. Hence the zero sequence current cannot exist. The zero sequence impedance is represented as an open circuit as shown in Figure 14.13.

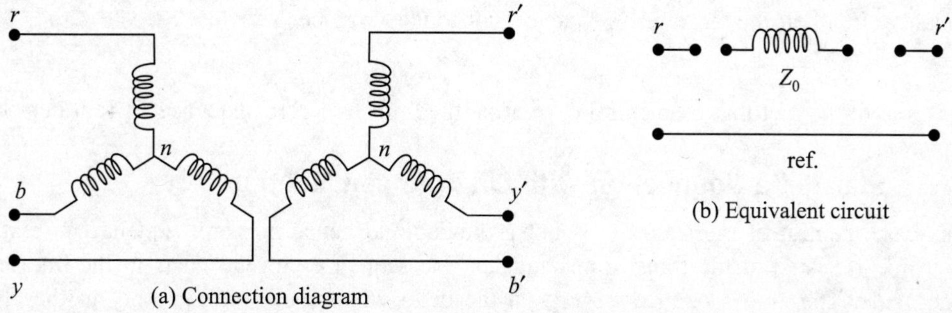

Figure 14.13 The zero sequence equivalent circuit of a star-star transformer with no neutral grounded.

14.7.4 Star-delta Connection with Neutral Grounded

A zero sequence current on the grounded star winding will cause a circulating zero sequence current in the closed delta winding. However, the zero sequence current on the delta winding

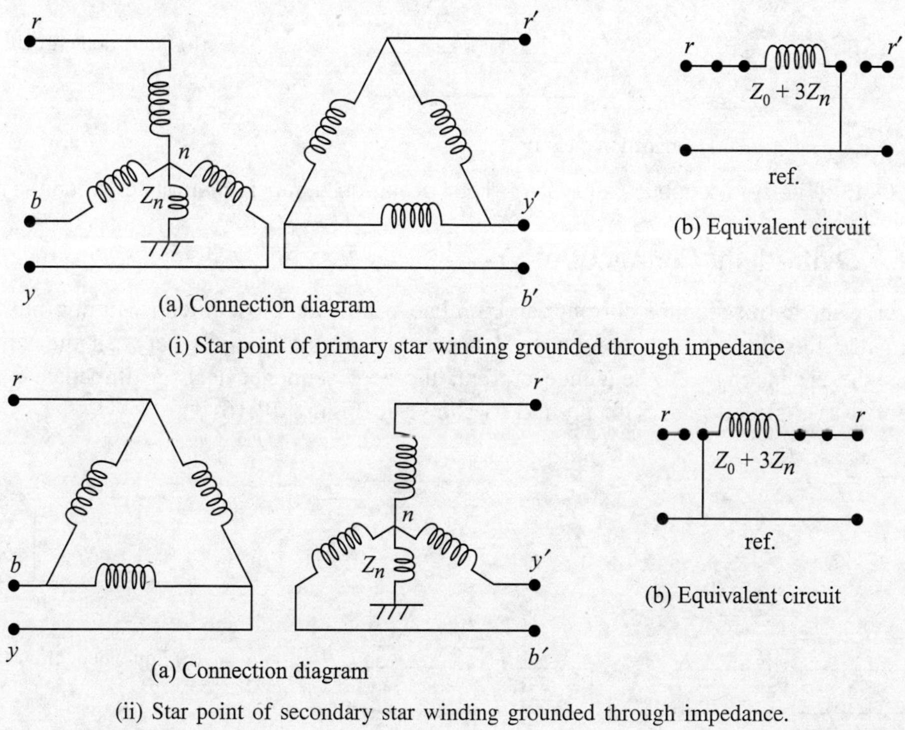

Figure 14.14 The zero sequence equivalent circuit of a star-delta transformer with star side grounded through grounding impedance.

is confined only to the closed delta winding. Thus an open circuit exist between star and delta side. As the zero sequence current exists only on the line side of the grounded star winding, the zero sequence leakage impedance of the transformer will not be connected to the ground on the star side. An open circuit will exist between two windings.

If the neutral of the star side winding of the transformer is grounded through a grounding impedance, Z_n, then the total zero sequence impedance will be

$$Z_{0\ \text{total}} = Z_0 + 3Z_n$$

This is due to the fact that the neutral current is three times of zero sequence current per phase.

14.7.5 Star-delta Connection with Ungrounded Neutral

In this case, neutral of the star side is not grounded and hence no zero sequence current can exist in the star side of the transformer and no zero sequence current exists in the line of the of delta side. The zero sequence current on the delta winding is confined only to the closed delta winding.

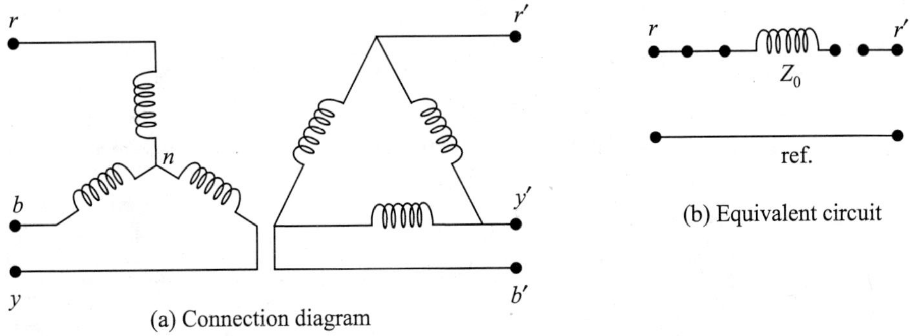

Figure 14.15 The zero sequence equivalent circuit of star-delta transformer with ungrounded neutral.

14.7.6 Delta-delta Connection

In this case the zero sequence current can circulate within the closed delta winding but cannot exist on the line side of both windings. Hence an open circuit exists between the windings. To flow the circulating zero sequence current, the zero sequence leakage impedance can be represented as a closed path with ground as shown in Figure 14.16.

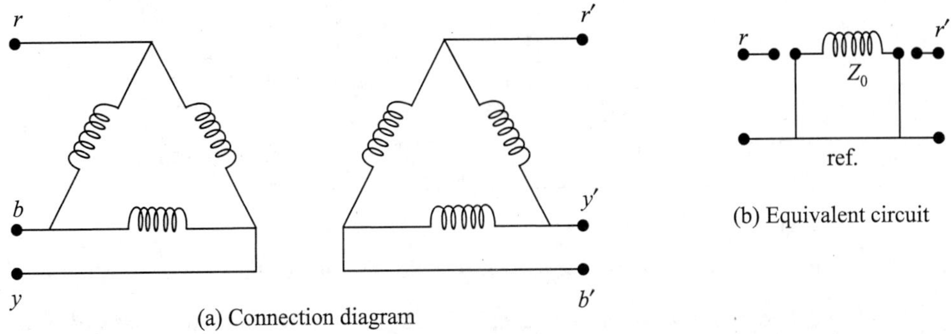

Figure 14.16 The zero sequence equivalent circuit of a delta-delta transformer.

14.8 SEQUENCE NETWORK OF TRANSMISSION LINES

Figure 14.17 shows a transmission line which is carrying unbalanced currents. Let X_s and X_m be the self reactance of each line and mutual reactance of any pair of transmission lines respectively.

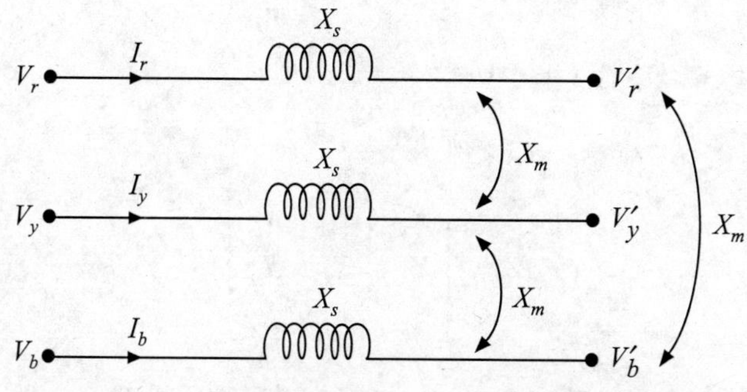

Figure 14.17 The three phases of a transmission line.

For this given network, the following equations can be written as

$$\left.\begin{aligned} V_r &= V_r' + jX_sI_r + jX_mI_y + jX_mI_b \\ V_y &= V_y' + jX_mI_r + jX_sI_y + jX_mI_b \\ V_b &= V_b' + jX_mI_r + jX_mI_y + jX_sI_b \end{aligned}\right\} \tag{14.30}$$

In matrix form, it can be written as

$$\begin{bmatrix} V_r \\ V_y \\ V_b \end{bmatrix} - \begin{bmatrix} V_r' \\ V_y' \\ V_b' \end{bmatrix} = j\begin{bmatrix} X_s & X_m & X_m \\ X_m & X_s & X_m \\ X_m & X_m & X_s \end{bmatrix}\begin{bmatrix} I_r \\ I_y \\ I_b \end{bmatrix} \tag{14.31}$$

Let, $V_{ph} = \begin{bmatrix} V_r \\ V_y \\ V_b \end{bmatrix}$ = Vector of original phasors and similarly, $V'_{ph} = \begin{bmatrix} V_r' \\ V_y' \\ V_b' \end{bmatrix}$ and $I_{ph} = \begin{bmatrix} I_r \\ I_y \\ I_b \end{bmatrix}$

And, $V_{\text{comp}} = \begin{bmatrix} V_{r_0} \\ V_{r_1} \\ V_{r_2} \end{bmatrix}$ = Vector of symmetrical components

and similarly, $V'_{\text{comp}} = \begin{bmatrix} V'_{r_0} \\ V'_{r_1} \\ V'_{r_2} \end{bmatrix}$ and $I_{\text{comp}} = \begin{bmatrix} I_{r_0} \\ I_{r_1} \\ I_{r_2} \end{bmatrix}$

The relation between V_{ph} and V_{comp} (equation (14.21d)) can be written as $V_{ph} = AV_{comp}$

where
$$A = \begin{bmatrix} 1 & 1 & 1 \\ 1 & \alpha^2 & \alpha \\ 1 & \alpha & \alpha^2 \end{bmatrix} \tag{14.32}$$

The matrix equation (14.31) can be written as:

$$V_{ph} - V'_{ph} = ZI_{ph}$$

where
$$Z = j\begin{bmatrix} X_s & X_m & X_m \\ X_m & X_s & X_m \\ X_m & X_m & X_s \end{bmatrix}$$

The equation (14.32) in terms of sequence components, can be written as:

$$A(V_{comp} - V'_{comp}) = ZAI_{comp}$$

or,
$$(V_{comp} - V'_{comp}) = A^{-1}ZAI_{comp} \tag{14.34}$$

Now,
$$A^{-1}ZA = \begin{bmatrix} 1 & 1 & 1 \\ 1 & \alpha^2 & \alpha \\ 1 & \alpha & \alpha^2 \end{bmatrix}^{-1} j\begin{bmatrix} X_s & X_m & X_m \\ X_m & X_s & X_m \\ X_m & X_m & X_s \end{bmatrix}\begin{bmatrix} 1 & 1 & 1 \\ 1 & \alpha^2 & \alpha \\ 1 & \alpha & \alpha^2 \end{bmatrix}$$

$$= \frac{1}{3}j\begin{bmatrix} 1 & 1 & 1 \\ 1 & \alpha & \alpha^2 \\ 1 & \alpha^2 & \alpha \end{bmatrix}\begin{bmatrix} (X_s + 2X_m) & (X_s - X_m) & (X_s - X_m) \\ (X_s + 2X_m) & \alpha^2(X_s - X_m) & \alpha(X_s - X_m) \\ (X_s + 2X_m) & \alpha(X_s - X_m) & \alpha^2(X_s - X_m) \end{bmatrix}$$

$$= j\begin{bmatrix} (X_s + 2X_m) & 0 & 0 \\ 0 & (X_s - X_m) & 0 \\ 0 & 0 & (X_s - X_m) \end{bmatrix} = \begin{bmatrix} Z_0 & 0 & 0 \\ 0 & Z_1 & 0 \\ 0 & 0 & Z_2 \end{bmatrix}$$

Thus, equation (14.31) can be written as:

$$\begin{bmatrix} V_r \\ V_y \\ V_b \end{bmatrix} - \begin{bmatrix} V'_r \\ V'_y \\ V'_b \end{bmatrix} = j\begin{bmatrix} Z_0 & 0 & 0 \\ 0 & Z_1 & 0 \\ 0 & 0 & Z_2 \end{bmatrix}\begin{bmatrix} I_{r_0} \\ I_{r_1} \\ I_{r_2} \end{bmatrix} \tag{14.35}$$

where,

Z_0 = Zero sequence impedance = $j(X_s + 2X_m)$
Z_1 = Positive sequence impedance = $j(X_s - X_m)$
Z_2 = Negative sequence impedance = $j(X_s - X_m)$

Thus positive sequence, negative sequence and zero sequence networks of transmission line are presented in Figure 14.18.

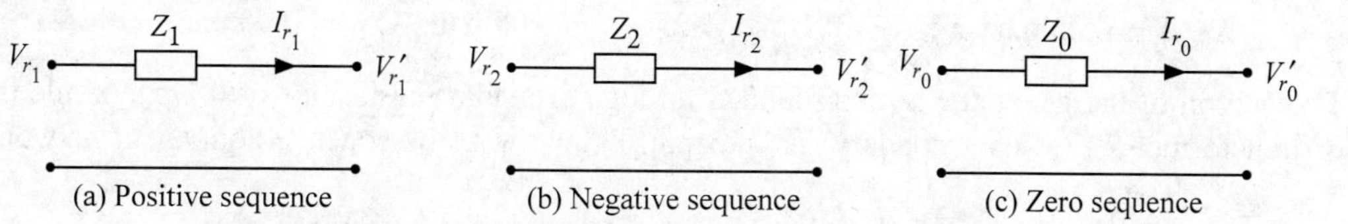

Figure 14.18 The per phase sequence networks of a transmission line.

It is to be noted that for static and balanced power system components such as transformer and power lines, positive and negative sequence impedances are same whereas they are different for rotating machines.

EXAMPLE 14.6 A three bus network is shown in Figure E14.3. The zero sequence reactance of the components in the given power system has been indicated in the figure on a common base. What is the zero sequence driving point impedance at bus-3?

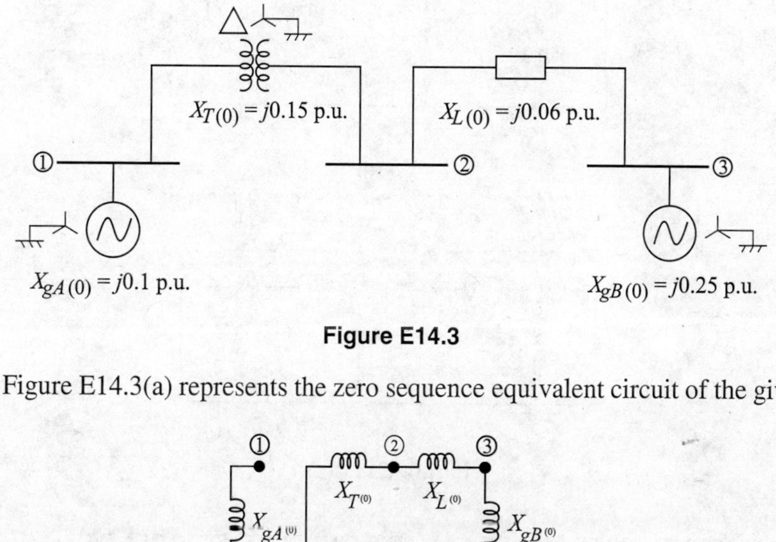

Figure E14.3

Solution: Figure E14.3(a) represents the zero sequence equivalent circuit of the given network.

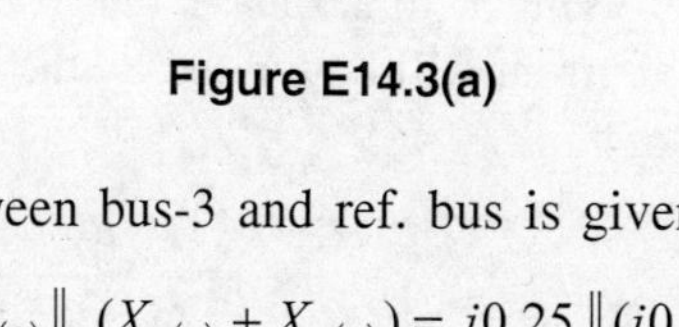

Figure E14.3(a)

Here the equivalent circuit between bus-3 and ref. bus is given by

$$X_{eqv(3)} = X_{gB(0)} \| (X_{T(0)} + X_{L(0)}) = j0.25 \| (j0.15 + j0.05)$$

$$= j0.25 \| j0.2 = \frac{j0.25 \times j0.2}{j0.25 + j0.2} = j0.11 \text{ p.u.} \quad \text{(Ans.)}$$

EXAMPLE 14.7 In the figure of simple power system model (Figure E14.4), the zero sequence reactance of each component on a common base are given as

$$X_{gA(0)} = 0.05 \text{ p.u.}, X_{TA(0)} = 0.1 \text{ p.u.} = X_{TB2(0)} = 0.05 \text{ p.u.}, X_{L(0)} = 0.06 \text{ p.u.}$$

The neutral of the generator G_A is grounded through grounding resistance of 0.2 p.u. while that of transformer T_B (at its secondary) is 0.25 p.u. Draw the zero sequence equivalent network.

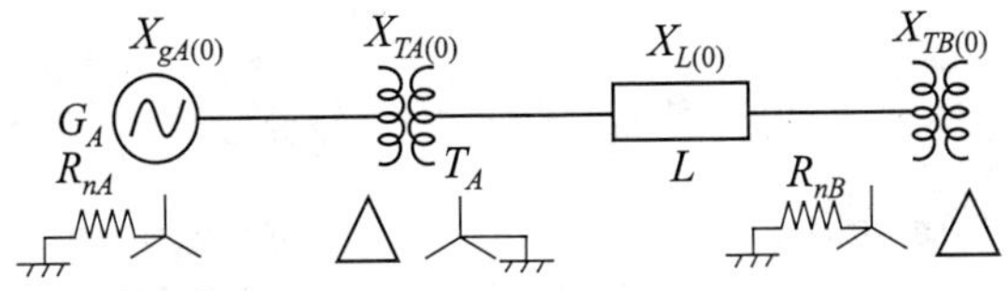

Figure E14.4

Solution: The required zero sequence equivalent network of the given power system is shown below (Figure E14.4(a)).

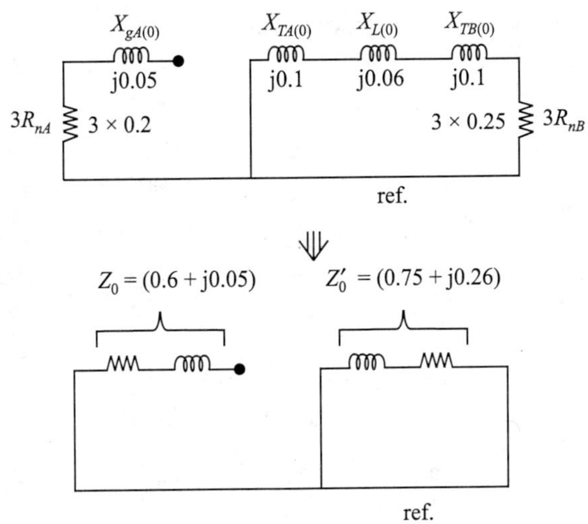

Figure E14.4(a)

14.9 REACTOR CONTROL OF SHORT CIRCUIT CURRENTS

When a short circuit current involving a three-phase fault occurs at any point F in a power network (Figure 14.19), the fault current will be limited by the impedance of the power network from the source upto the point of fault.

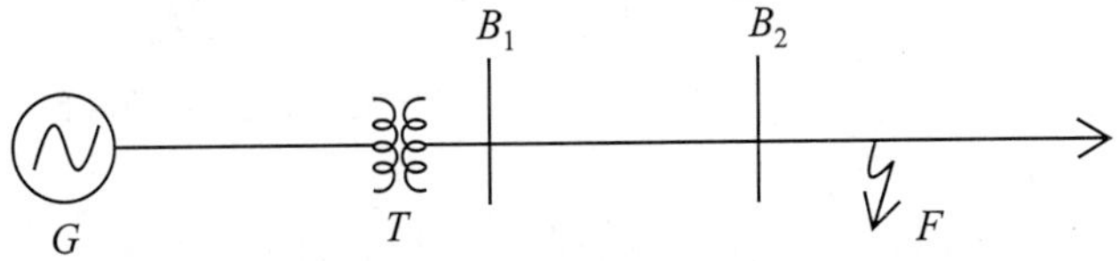

Figure 14.19 A simple power network.

For the power network shown in Figure 14.19 for a fault at point F, the fault current will be limited by the impedance of generator, transformer and transmission line. It is important to mention that the impedance of generators, transformers and other power equipments are mostly inductive, thus limiting impedance of fault current can be termed as '*limiting reactance*'. Thus to limit the fault current, within a specified value, power engineers may connect additional reactor with known reactance value in series with the power network. Since the coils of reactors have very small resistance compared with its reactor value, it hardly affects the power network efficiency. Depending on placement of current limiting reactor in a power network, the reactors have different nomenclature in power engineering. All these type of reactors have their own advantages and disadvantages.

14.9.1 Generator Reactors

In this type of reactor arrangement additional reactors are connected in series with the generators present in the power network. Thus it increases the equivalent reactance of the generator which in turn, limits short circuit currents. But since it remains present in the power network, even in normal operating condition, it becomes the most responsible factor for extra power loss and voltage drop.

14.9.2 Feeder Reactors

In this arrangement, reactors are connected in series with the feeder. Thus it increases the limiting reactance to limit fault limits short circuit currents. But its presence in power network even in normal operating condition is responsible for constant power loss and voltage drop.

14.9.3 Bus Bar Reactors

Constant voltage drop and power loss due to presence of either generator reactor or feeder reactor can be avoided with bus bar reactor. There are two type of bus bar reactor: one for ring system and another for tie bar system. Figure 14.20 shows the location of reactors for ring main system and tie bar system.

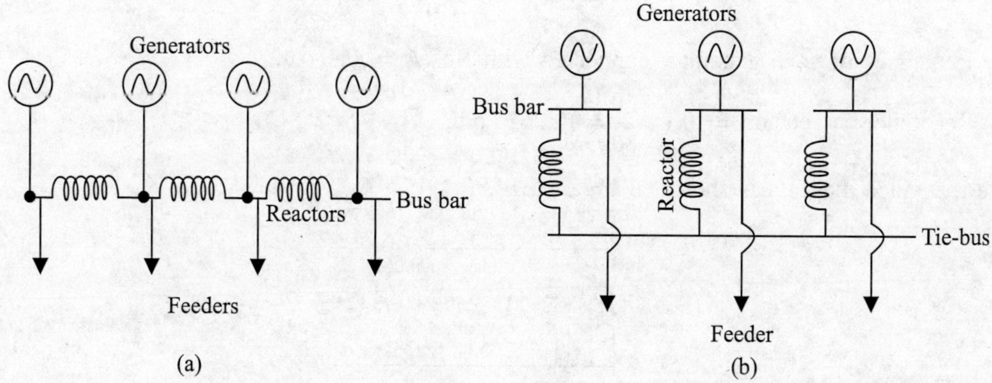

Figure 14.20 (a) Ring main system; (b) Tie bar system.

In the ring main system, the arrangement is such that, one generator is feeding only one feeder. Thus during normal operating condition, very low power loss and voltage drop will take place in the reactors. In this system the bus bar is divided into subsections which are connected through reactors. Hence during a fault in one feeder, only one generator will be affected by fault current

In the tie bar system, there are effectively two reactors in series between sections, hence same value of reactors in ring main system can provide two times protection in feeding the current.

EXAMPLE 14.8 Calculate the current feed into the fault in power system shown in Figure E14.5.

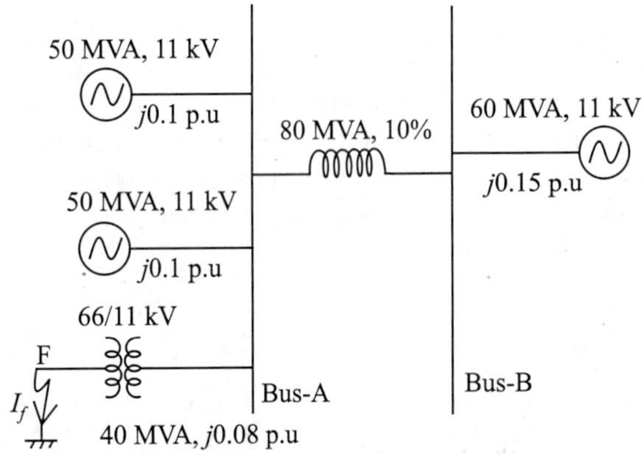

Figure E14.5 Single line diagram of power system.

Solution: Let us consider base MVA = 50. We convert each reactance to p.u. reactance with common base.

$$\text{Reactance of generator connected with Bus-B } = j0.15 \times \frac{50}{60} = j0.125 \text{ p.u.}$$

$$\text{Reactance of transformer connected } = j0.08 \times \frac{50}{40} = j0.1 \text{ p.u.}$$

Reactance of each generator connected with Bus-A = $j0.1$ p.u.

$$\text{Reactance of reactor } = j0.1 \times \frac{50}{80} = j0.0625 \text{ p.u.}$$

The reactance diagram is shown in Figure E14.5(a).

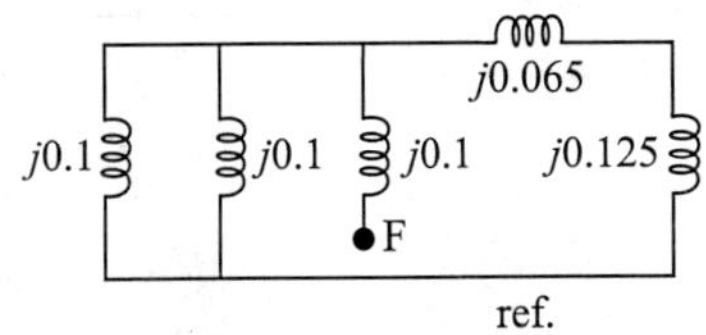

Figure E14.5(a) Reactance diagram.

Equivalent reactance across point F and ref. bus $= \left[(j0.1 \| j0.1) \| (j0.0625 + j0.125) \right] + j0.1$

$$= \left[j0.05 \| j0.1875 \right] + j0.1 = j0.1395 \text{ p.u.}$$

Short circuit current fed into the fault $= \dfrac{1}{j0.1395}$ p.u.

i.e., $$|I_f| = \frac{1}{0.1395} \times \frac{50 \times 10^6}{\sqrt{3} \times 66 \times 10^3} \text{A} = 3.1354 \text{ kA (Ans.)}$$

EXAMPLE 14.9 Figure E14.6 shows a five bus power network. Find the fault MVA and fault current for a three phase fault at bus 5. L_1 has an inductance of 1.2 mH/phase/km and length is 100 km. L_2 and L_3 have an inductance of 1 mH/phase/km and length is 50 km each. All other required data are indicated in the figure.

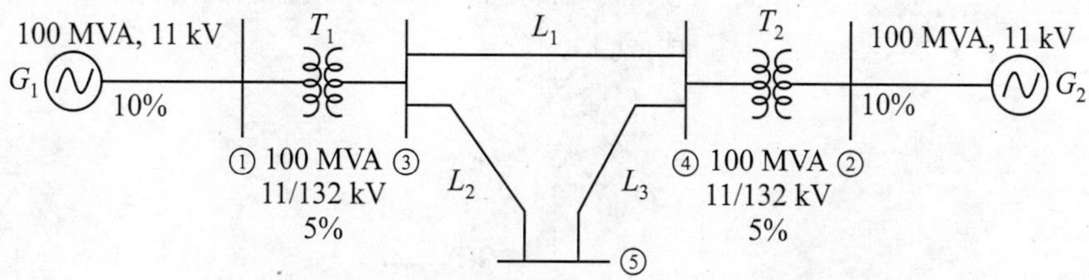

Figure E14.6 Single line diagram of power network.

Solution: Let us consider base MVA = 100, base kV as 11 kV for LV side of transformer and base kV as 132 kV for HV side of transformer.

Base impedance for HV side of transformer $= \dfrac{\text{kV}_{\text{base}}^2}{\text{MVA}_{\text{base}}} = \dfrac{132^2}{100} = 174.24\ \Omega$

Reactance of $L_1 = 2\pi \times 50 \times 1.2 \times 10^{-3} \times 100 = 37.7\ \Omega = 0.216$ p.u.

Reactance of L_2 and $L_3 = 2\pi \times 50 \times 1 \times 10^{-3} \times 50 = 15.71\ \Omega = 0.09$ p.u

The reactance diagram is shown in Figure E14.6(a).

Thus fault MVA $= \dfrac{1}{j0.121} \cong 8.26 \text{ p.u.} = 826 \text{ MVA}$

Fault current in p.u. $= \dfrac{1}{j0.121} \cong 8.26$ p.u.

However, base current $= \dfrac{100 \times 10^6}{\sqrt{3} \times 132 \times 10^3} \cong 437.39$ A

Thus fault current = 8.26 × 437.39 = 3.61 kA (Ans.)

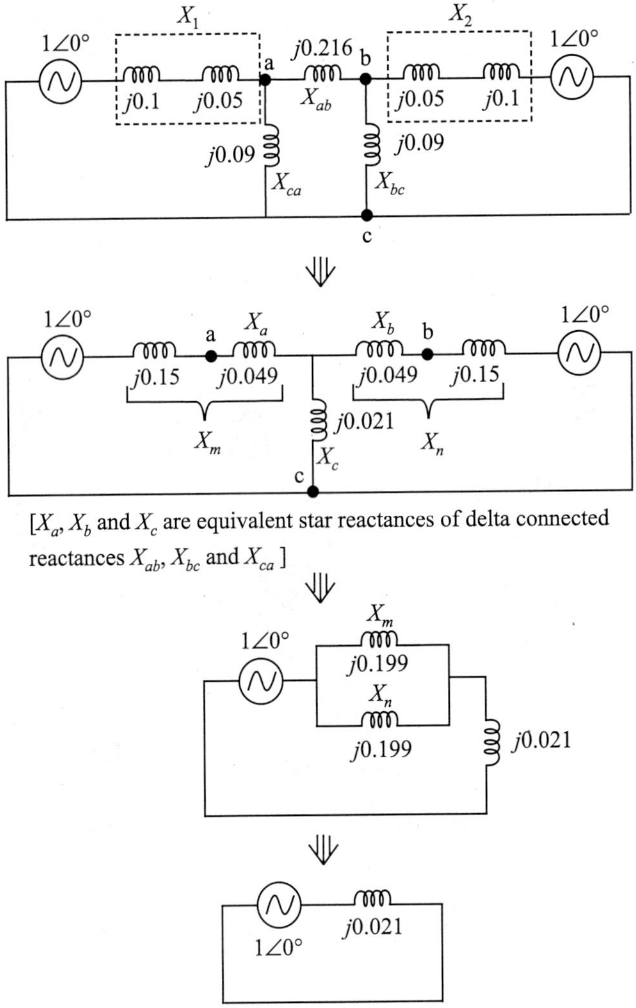

Figure E14.6(a) Reactance diagram.

EXAMPLE 14.10 Figure E14.7 shows a system having three alternators each rated 11 kV, 50 MVA, 15% sub-transient reactance. Find

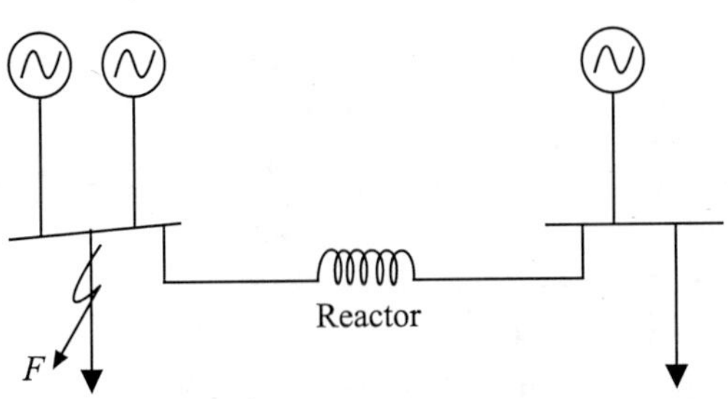

Figure E14.7 Single line diagram of power network.

(a) the fault level for a fault on one feeder with zero value of reactance.
(b) the reactance value for current limiting reactor X to limit fault level to 800 MVA.

Solution: Let us consider, base MVA = 50 and base kV as 11 kV.
The reactance diagram is shown in Figure E14.7(a).

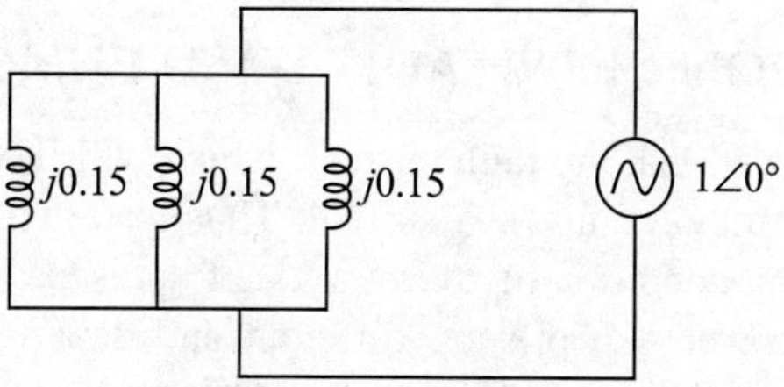

Figure E14.7(a) Reactance diagram.

Equivalent reactance, $X_{eqv} = j0.15 \| j0.15 \| j0.15 = j0.05$ p.u.

$$\text{Fault level } = \frac{1\angle 0^0}{0.05} = 20 \text{ p.u.} = 20 \times 50 = 1000 \text{ MVA}$$

Now let us consider X (p.u.) be the reactance value of the current limiting reactor to limit the fault MVA at 800 MVA.

The modified reactance diagram is shown in Figure E14.7(b).

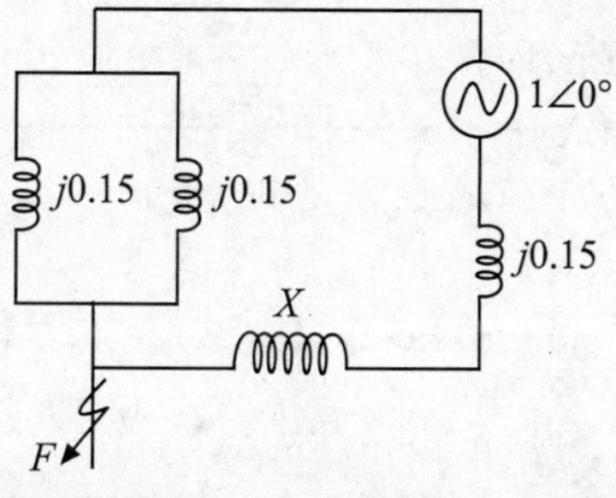

Figure E14.7(b) Reactance diagram

Equivalent reactance becomes, $X'_{eqv} = (j0.15 \| j0.15) \| (j0.15 + jX) = j\dfrac{0.075(0.15+X)}{0.225+X}$

$$\therefore \qquad \frac{1\angle 0^0}{X'_{eqv}} = \frac{800}{50}$$

$$\text{or,} \qquad \frac{0.225+X}{0.075(0.15+X)} = 16$$

$$\text{or,} \qquad 0.225 + X = 0.18 + 1.2X$$

$$\text{or,} \qquad 0.2X = 0.045$$

$$\therefore \qquad X = 0.225 \text{ p.u.}$$

$$\text{Base reactance} = \frac{\text{kV}_{\text{base}}^2}{\text{MVA}_{\text{base}}} = \frac{11^2}{100} = 2.42\ \Omega$$

Therefore reactance to be connected, $X = 0.225$ p.u. $= 0.225 \times 2.42 = 0.5445\ \Omega$ (Ans.)

14.10 CONSIDERATION OF PREFAULT LOAD CURRENT

Generally, it has been observed that the fault current, which will flow through the circuit during a fault, is much higher than the prefault load current. Thus load current can be neglected during fault analysis. During healthy condition of a circuit, due to presence of load current, the voltage at different points in the power network are different and depart from 1 p.u.. Hence, at the time of calculation of fault current, prefault voltage at the point of fault is necessary to know. *The total current during the fault will be the phasor sum of load current and fault current.*

Following numerical example will explain this with more clarity.

EXAMPLE 14.11 An 11 kV, 100 MVA alternator with 0.25 p.u. reactance is supplying a motor through a transmission line with line reactance of 0.025 p.u. at 100 MVA base. A three-phase fault occurs at alternator terminal when the motor is drawing 20 MW at 0.8 leading p.f. with terminal voltage of 11 kV. Calculate total fault current in alternator and motor. The motor rating is 50 MVA with 0.2 p.u. sub-transient reactance.

Solution: The single line diagram of the power system is shown in Figure E14.8(a).

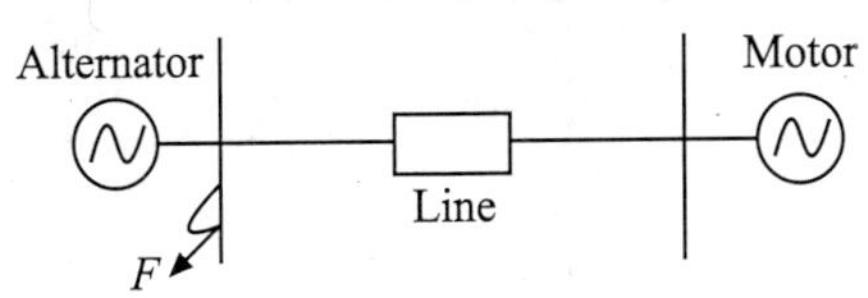

Figure E14.8(a) Single line diagram.

Let us consider, base MVA = 100 and base kV = 11.

$$\text{Base current} = \frac{100 \times 10^6}{\sqrt{3} \times 11 \times 10^3} = 5248.64\ \text{A}$$

$$\text{Magnitude of motor load current, } |I_m| = \frac{20 \times 10^6}{\sqrt{3} \times 11 \times 10^3 \times 0.8} = 1312.16\ \text{A} = \frac{1312.16}{5248.64} = 0.25\ \text{p.u.}$$

By considering the motor terminal voltage as reference phasor, the complex load current is $I_m = |I_m| \angle \cos^{-1}(0.8) = (0.2 + j0.15)$ p.u. [$\because$ motor is drawing power at 0.8 leading p.f.]

$$\text{Motor terminal voltage, } V_m = \frac{11}{11} = 1\ \text{p.u.}$$

Prefault voltage at the alternator terminal

$$V = \text{Motor terminal voltage} + \text{line drop}$$

$$V_m + I_m Z_{\text{line}} = 1 + (0.2 + j0.15) \times j0.025 = (0.996 + j0.005) \text{ p.u.}$$

Sub-transient reactance of motor, $x'_m = j0.2 \times \dfrac{100}{50} = j0.4$ p.u.

The circuit diagram under faulted condition will be as follows:

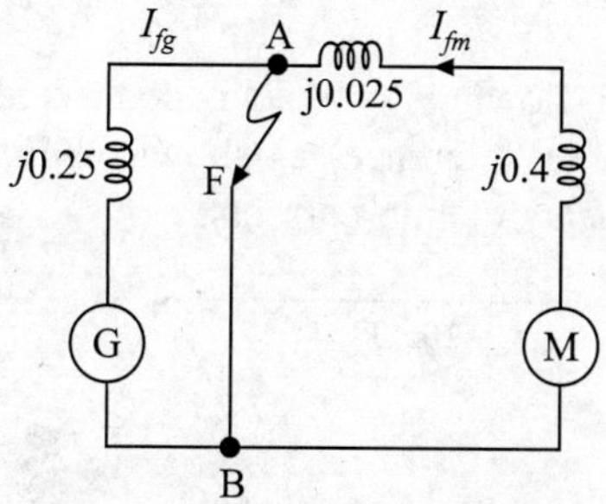

Figure E14.8(b) Reactance diagram.

The Thevenin's equivalent reactance looking from the fault terminal AB,

$$X_{eqv} = j0.25 \| j0.425 = 0.157 \text{ p.u.}$$

$\therefore$ Fault current $I_f = \dfrac{V}{X_{eqv}} = \dfrac{(0.996 + j0.005)}{j0.157} = (0.0318 - j6.3439)$ p.u.

Fault current supplied by generator

$$I_{fg} = (0.0318 - j6.3439) \times \frac{j0.425}{j0.25 + j0.425} = (0.02 - j3.9943) \text{ p.u.}$$

Fault current supplied by motor

$$I_{fm} = I_f - I_{fg} = (0.0318 + j6.3439) - (0.02 - j3.9943) = (0.0118 - j2.3496) \text{ p.u.} \quad \text{(Ans.)}$$

EXAMPLE 14.12 A 50 Hz alternator is rated for 500 MVA, 20 kV with x_d = 1 p.u. and $x''_d = j0.2$ p.u. It supplies a purely resistive load of 400 MW at 20 kV. The load is connected directly across the generator terminal. If a symmetrical fault occurs at the load terminals, what will be the value of initial r.m.s current in the generator in p.u.? **[Gate 2001]**

Solution: Let us consider, base MVA = 500; connected load, P = 400 MW at unity power factor

$$\therefore \quad P = \frac{400}{500} = 0.8 \text{ p.u.}$$

$\because$ Load is connected at the alternator terminals, load voltage = Alternator terminal voltage

$$\therefore \quad \text{Current, } I = \frac{P}{V} = \frac{0.8}{1} = 0.8 \text{ p.u.}$$

Alternator induced e.m.f. is

$$E''_d = V + Ix''_d = 1 + 0.8 \times j0.2 = (1 + j0.16) = 1.01\angle 9.09° \text{ p.u.}$$

$\therefore$ Initial r.m.s current in the generator $I'' = \dfrac{E''_d}{x''_d} = \dfrac{1.01\angle 9.09°}{j0.2} = 5.05\angle -80.91° \text{ p.u.}$ (Ans.)

EXAMPLE 14.13 The current flowing to a delta connected load through line-X is 10 A (current in line X is considered as reference) as shown in Figure E14.9. Assuming line Z is open, find symmetrical components of line currents.

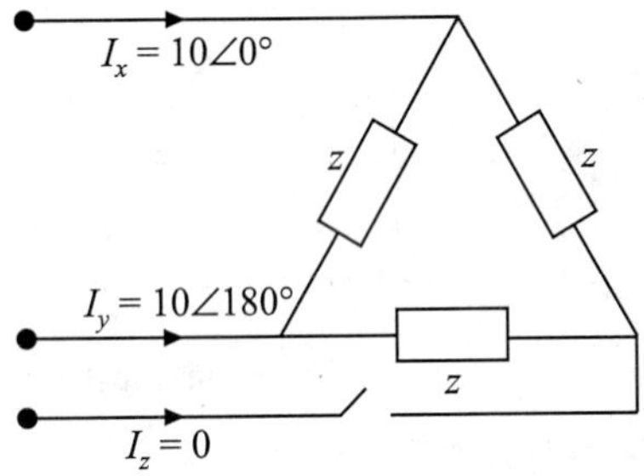

Figure E14.9

Solution: Given, $I_X = 10\angle 0° = 10$ A, $I_Y = 10\angle 180° = -10$ A and $I_Z = 0$

The zero sequence component of current in line-X will be as follows

$$I_{X_0} = \frac{1}{3}(I_X + I_Y + I_Z) = \frac{1}{3}(10 - 10 + 0) = 0 \text{ A} \quad \text{(Ans.)}$$

The positive sequence component of current in line-X will be as follows

$$I_{X_1} = \frac{1}{3}\left(I_X + \alpha I_Y + \alpha^2 I_Z\right) = \frac{1}{3}(10 + 10\angle 300°) = \frac{1}{3} \times 10(1 + 0.5 - j0.866)$$

$$= \frac{1}{3} \times 10(1.5 - j0.866) = (5 - j2.89)\text{A} = 5.7751\angle -30.03° \text{ A} \quad \text{(Ans.)}$$

The negative sequence component of current in line-X will be as follows

$$I_{X_2} = \frac{1}{3}\left(I_X + \alpha^2 I_Y + \alpha I_Z\right) = \frac{1}{3}(10 + 10\angle 420°) = \frac{1}{3} \times 10(1 + 0.5 + j0.866)$$

$$= \frac{1}{3} \times 10(1.5 + j0.866) = (5 + j2.89)\text{A} = 5.7751\angle 30.03° \text{ A} \quad \text{(Ans.)}$$

Hence, the zero, positive and negative sequence component of current in line-Y will be as follows:

$I_{Y_0} = 0$ A (Ans.)

$I_{Y_1} = I_{X_1}\angle -120° = (5 - j2.89)(-0.5 - j0.866) = (-5.002 - j2.885)\text{A} = 5.7750\angle -150.03°$ A (Ans.)

$I_{Y_2} = I_{X_2}\angle 120° = (5 + j2.89) \times (-0.5 + j0.866) = (-5.0027 + j2.885)\text{A} = 5.7750\angle 150.03°\text{A}$

(Ans.)

Similarly, for line-Z, $I_{Z_0} = 0$ A

$I_{Z_1} = I_{X_1}\angle 120° = (5 - j2.89)(-0.5 + j0.866) = (0.0027 + j5.775)\text{A} = 5.7750\angle 89.9732°\text{ A}$ (Ans.)

$I_{Y_2} = I_{X_2}\angle -120° = (5 + j2.89) \times (-0.5 - j0.866) = (0.0027 - j5.775)\text{A} = 5.7750\angle -89.97°\text{A}$

(Ans.)

EXAMPLE 14.14 A three-phase balanced star connected voltage source with frequency ω rad/sec is connected to a star connected balanced load which is purely inductive. The instantaneous line currents and phase to neutral voltages are denoted by I_a, I_b and I_c and V_{an}, V_{bn} and V_{cn} respectively. Their r.m.s valus are denoted by V and I. Find the magnitude of R, where

$$R = \begin{bmatrix} V_{an} & V_{bn} & V_{cn} \end{bmatrix} \begin{bmatrix} 0 & 1/\sqrt{3} & -1/\sqrt{3} \\ -1/\sqrt{3} & 0 & 1/\sqrt{3} \\ 1/\sqrt{3} & -1/\sqrt{3} & 0 \end{bmatrix} \begin{bmatrix} I_a \\ I_b \\ I_c \end{bmatrix}$$ [Gate 2007]

Solution:

$$R = \begin{bmatrix} V_{an} & V_{bn} & V_{cn} \end{bmatrix} \begin{bmatrix} 0 & 1/\sqrt{3} & -1/\sqrt{3} \\ -1/\sqrt{3} & 0 & 1/\sqrt{3} \\ 1/\sqrt{3} & -1/\sqrt{3} & 0 \end{bmatrix} \begin{bmatrix} I_a \\ I_b \\ I_c \end{bmatrix}$$

$$= (1/\sqrt{3})\begin{bmatrix} V_{an} & V_{bn} & V_{cn} \end{bmatrix} \begin{bmatrix} (I_b - I_c) \\ (I_c - I_a) \\ (I_a - I_b) \end{bmatrix} = \frac{V_{an}}{\sqrt{3}}(I_b - I_c) + \frac{V_{bn}}{\sqrt{3}}(I_c - I_a) + \frac{V_{cn}}{\sqrt{3}}(I_a - I_b)$$

Since $I_a = I_b = I_c$ and $V_{an} = V_{bn} = V_{cn}$

$$\therefore \quad R = \frac{V_{an}}{\sqrt{3}} \times 0 + \frac{V_{bn}}{\sqrt{3}} \times 0 + \frac{V_{cn}}{\sqrt{3}} \times 0 = 0 \quad \text{(Ans.)}$$

EXAMPLE 14.15 The magnitude of three-phase fault currents at buses A and B of a power system are 10 p.u. and 8 p.u. respectively. Neglect all resistances and consider the prefault system to be unloaded. The prefault voltage at all buses in the system is 1 p.u. The voltage magnitude at bus B during a three-phase fault at bus A is 0.8 p.u. What is the voltage magnitude at bus A during a three-phase fault at bus B in p.u.? [Gate 2016]

Solution: When a three-phase fault occurs at bus A, the voltage at bus B

$V_B = V_{B(\text{prefault})} - I_{f(\text{at bus A})} \times Z_{AB}$ (where Z_{AB} = Impedance between bus A and B)

or, $0.8 = 1 - 10 \times Z_{AB}$

$\therefore$ $Z_{AB} = 0.2$ p.u.

Also, when a three-phase fault occurs at bus B, the voltage at bus A

$$V_A = V_{A\,(\text{prefault})} - I_{f(\text{at bus B})} \times Z_{AB} = 1 - 8 \times 0.02 = 0.84 \text{ p.u.} \quad (\text{Ans.})$$

14.11 UNSYMMETRICAL FAULTS

Unsymmetrical faults are classified as:

1. Shunt faults involving short circuits
   (a) Single line to ground (SLG) faults
   (b) Line to line (LL) faults
   (c) Double line to ground (LLG) faults
2. Series faults involving open circuit faults of conductors
   (a) One conductor open faults
   (b) Two conductors open faults

   Usually any short circuit or short fault is characterized by increase in line current(s) and drop in voltage and even drop in frequency may occur. Series faults are characterized by increase in voltage and may involve increase in frequency. Obviously, the current in the faulted phase of such fault decreases.

14.11.1 Single Line to Ground (SLG) Fault

Let us assume the single line to ground fault occurs at the terminal of phase r (Figure 14.21) and $Z_r = Z_y = Z_b = Z$.

The boundary conditions are, $I_y = 0$; $I_b = 0$; $V_r = 0$

where, I is the respective phase (line) current and V is the terminal voltage.

The sequence network equations are:

$$\left.\begin{aligned} V_{r_1} &= E_r - I_{r_1} Z_1 \\ V_{r_2} &= -I_{r_2} Z_2 \\ V_{r_0} &= -I_{r_0} Z_0 \end{aligned}\right\} \quad (14.36)$$

However,

$$\left.\begin{aligned} I_{r_1} &= \frac{1}{3}\left(I_r + \alpha I_y + \alpha^2 I_b\right) = \frac{I_r}{3} \\ I_{r_2} &= \frac{1}{3}\left(I_r + \alpha^2 I_y + \alpha I_b\right) = \frac{I_r}{3} \\ I_{r_0} &= \frac{1}{3}\left(I_r + I_y + I_b\right) = \frac{I_r}{3} \end{aligned}\right\} \quad (14.37)$$

$\therefore$

$$I_{r_1} = I_{r_2} = I_{r_0} = \frac{I_r}{3} \quad (14.38)$$

and

$$V_r = V_{r_1} + V_{r_2} + V_{r_0} = 0 \quad (14.39)$$

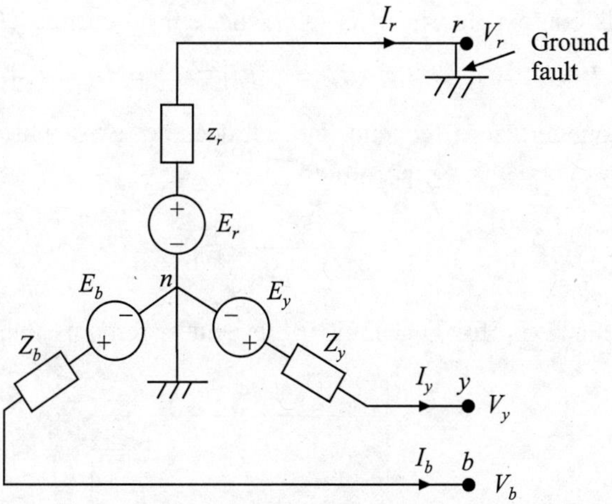

Figure 14.21 A L-G fault at the terminal (r) or an unloaded alternator.

Using equation (14.36) in equation (14.39),

$$E_r - I_{r_1}Z_1 - I_{r_2}Z_2 - I_{r_0}Z_0 = 0$$

Using equation (14.38) we have

$$E_r - I_{r_1}Z_1 - I_{r_2}Z_2 - I_{r_1}Z_0 = 0$$

i.e.,

$$I_{r_1} = \frac{E_r}{Z_1 + Z_2 + Z_0} = \frac{I_r}{3} \tag{14.40}$$

∴

$$I_r = \frac{3E_r}{Z_1 + Z_2 + Z_0} \tag{14.41}$$

Figure 14.22 represents the equivalent network of SLG fault.

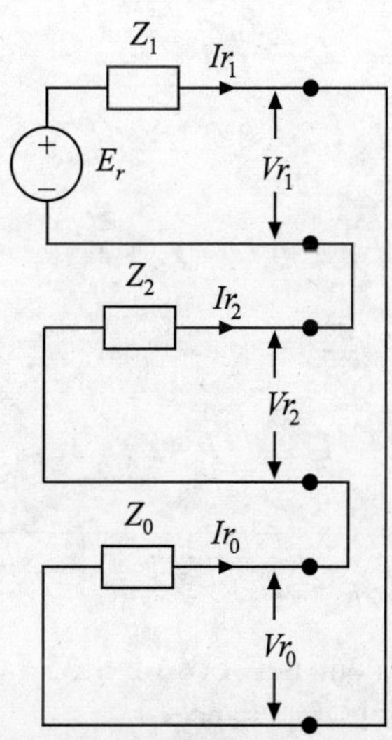

Figure 14.22 Interconnection of sequence networks for SLG fault.

It may be noted in this context that in SLG fault, the neutral current (I_n) is given by

$$I_n = I_r = I_{r_1} + I_{r_2} + I_{r_3} = 3I_{r_0} \; [\because I_{r_1} = I_{r_2} = I_{r_0}]$$

In case neutral is ungrounded, zero sequence current does not exist which in other sense means that the zero sequence impedance Z_0 is infinite.

$$\therefore \qquad I_{r_1} = \frac{E_r}{Z_1 + Z_2 + \infty} = 0$$

No sequence components exist for L-G fault if the neutral remains ungrounded.

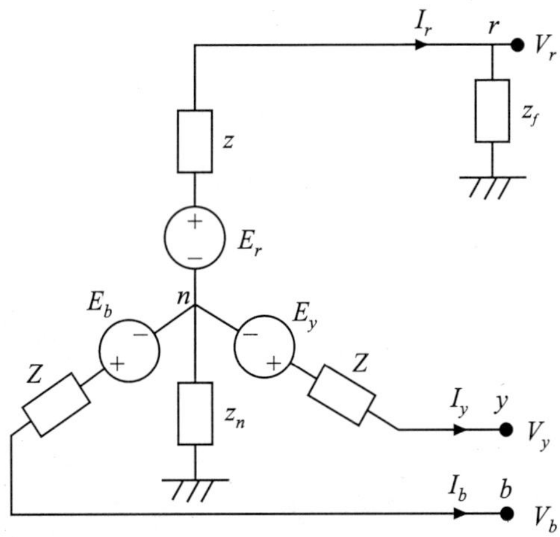

Figure 14.23 A L-G fault at the terminal (*r*) of an unloaded alternator grounded through an impedance.

In case the neutral is grounded through an impedance Z_n and there is fault impedance Z_f (Figure 14.23), the boundary conditions are:

$$I_y = I_b = 0 \,;\, V_r = I_r Z_f \,;\, V_{r_1} = E_r - I_{r_1} Z_1 \,;\, V_{r_2} = -I_{r_2} Z_2$$

and $$V_{r_0} = -I_{r_0}(Z_0 + 3Z_n) = -I_{r_0} Z_0 - 3I_{r_0} Z_n$$

Here, $$I_{r_1} = I_{r_2} = I_{r_0} = \frac{I_r}{3} \qquad (14.42)$$

and $$V_r = V_{r_1} + V_{r_2} + V_{r_0} = 3I_{r_1} Z_f \qquad (14.43)$$

or, $$E_r - I_{r_1} Z_1 - I_{r_2} Z_2 - I_{r_0}(Z_0 + 3Z_n) = 3I_{r_1} Z_f$$

$$\therefore \qquad I_{r_1} = \frac{E_r}{Z_1 + Z_2 + (Z_0 + 3Z_n) + 3Z_f} \quad \left[\text{As } I_{r_1} = I_{r_2} = I_{r_0}\right] \qquad (14.44)$$

Figure 14.24 represents the equivalent network of SLG fault with grounded neutral through impedance and the fault having fault impedance.

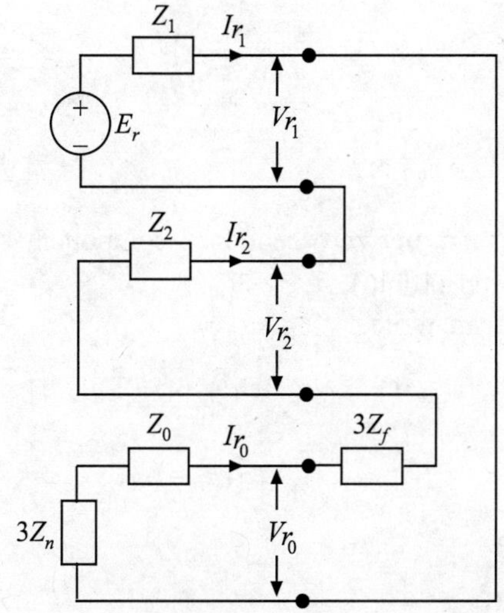

Figure 14.24 Interconnection of sequence networks for SLG fault through fault impedance.

14.11.2 Line to Line (LL) Fault

Let us assume that the line to line (LL) fault occurs involving phases y and b as shown in Figure 14.25.

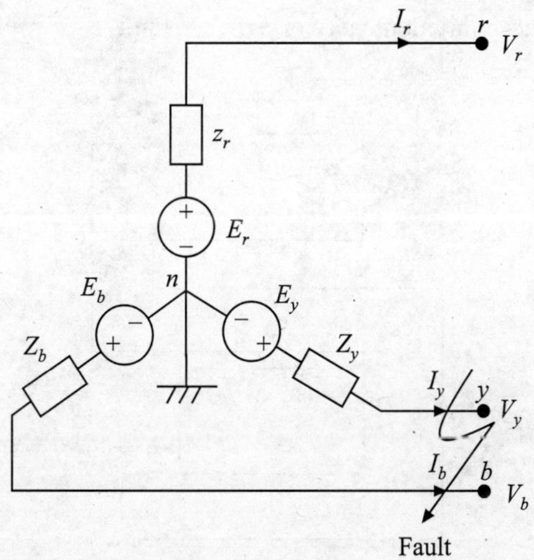

Figure 14.25 A L-L fault at the terminals of an unloaded alternator.

The boundary conditions are:

$$V_y = V_b;\ I_r = 0;\ I_y + I_b = 0,\ \text{i.e.,}\ \ I_y = -I_b$$

$$\therefore \quad I_{r_1} = \frac{1}{3}\left(0 + \alpha I_y + \alpha^2 I_b\right) = \frac{1}{3}\left(0 + \alpha I_y - \alpha^2 I_y\right) = \frac{1}{3}\left(\alpha - \alpha^2\right) I_y \tag{14.45}$$

$$I_{r_2} = \frac{1}{3}\left(0 + \alpha^2 I_y + \alpha I_b\right) = \frac{1}{3}\left(\alpha^2 - \alpha\right) I_y \tag{14.46}$$

$$I_{r_0} = \frac{1}{3}\left(0 + I_y + I_b\right) = 0 \tag{14.47}$$

Note that LL fault does not have any zero sequence component of current.

From equations (14.45) and (14.46), $I_{r_1} = -I_{r_2}$.

However, since $V_y = V_b$, we can write,

$$V_{r_0} + \alpha^2 V_{r_1} + \alpha V_{r_2} = V_{r_0} + \alpha V_{r_1} + \alpha^2 V_{r_2}$$

i.e.,
$$\left(\alpha^2 - \alpha\right) V_{r_1} = \left(\alpha^2 - \alpha\right) V_{r_2}$$

or,
$$V_{r_1} = V_{r_2} \tag{14.48}$$

or,
$$E_r - I_{r_1} Z_1 = -I_{r_2} Z_2$$

or,
$$E_r - I_{r_1} Z_1 = I_{r_1} Z_2 \quad \left[\because I_{r_1} = -I_{r_2}\right]$$

$\therefore$
$$I_{r_1} = \frac{E_r}{Z_1 + Z_2} = -I_{r_2} \tag{14.49}$$

Figure 14.26 represents the equivalent network of LL fault.

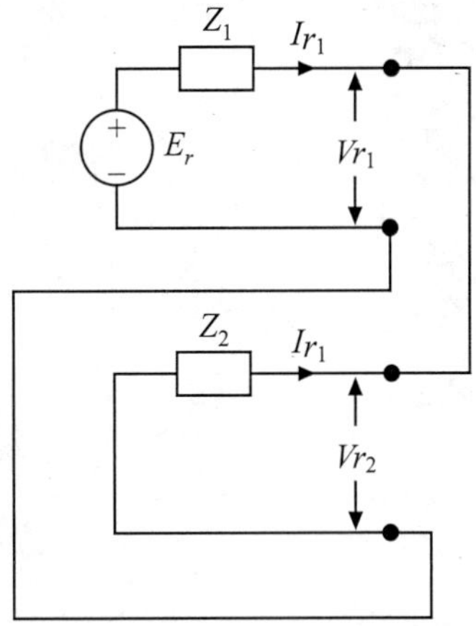

Figure 14.26 Interconnection of sequence networks for LL fault.

Figure 14.27 represents (LL) fault at phases y and b through a fault impedance Z_f while the neutral of the alternator is grounded through an impedance Z_n.

Boundary conditions are:

$$V_y = V_b + I_y Z_f;\ I_r = 0;\ I_y + I_b = 0,\ \text{i.e. } I_y = -I_b$$

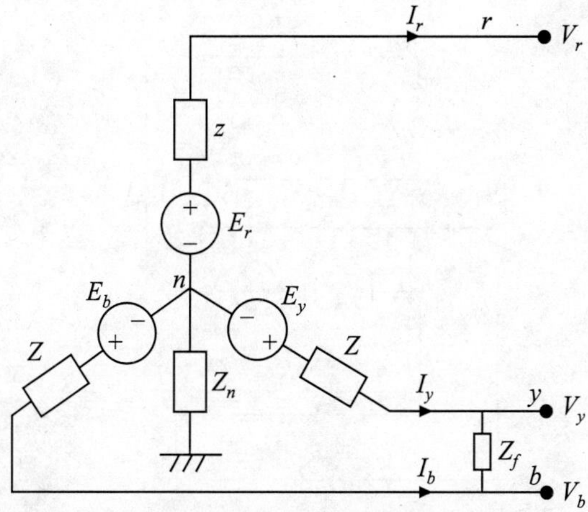

Figure 14.27 A LL fault through fault impedance at the terminals of an unloaded alternator grounded through an impedance.

The sequence network equations can be written as:

$$V_{r_1} = E_r - I_{r_1} Z_1;\ V_{r_2} = -I_{r_2} Z_2;\ V_{r_0} = -I_{r_0} Z_0$$

Since, $V_y = V_b + I_y Z_f$

$$V_{r_0} + \alpha^2 V_{r_1} + \alpha V_{r_2} = V_{r_0} + \alpha V_{r_1} + \alpha^2 V_{r_2} + \left(I_{r_0} + \alpha^2 I_{r_1} + \alpha I_{r_2}\right) Z_f$$

or, $$\left(\alpha^2 - \alpha\right) V_{r_1} = \left(\alpha^2 - \alpha\right) V_{r_2} + \left(0 + \alpha^2 I_{r_1} - \alpha I_{r_1}\right) Z_f \quad \left[\because I_{r_0} = 0 \text{ and } I_{r_1} = -I_{r_2}\right]$$

or, $$V_{r_1} = V_{r_2} + I_{r_1} Z_f$$

or, $$E_r - I_{r_1} Z_1 = -I_{r_2} Z_2 + I_{r_1} Z_f$$

or, $$E_r - I_{r_1} Z_1 = I_{r_1} Z_2 + I_{r_1} Z_f \quad \left[\because I_{r_1} = -I_{r_2}\right]$$

$\therefore$ $$I_{r_1} = \frac{E_r}{Z_1 + Z_2 + Z_f} \tag{14.50}$$

Note that for the LL fault, the neutral impedance has no role in limiting the fault current. Figure 14.28 represents the sequence equivalent circuit of LL fault with fault impedance Z_f.

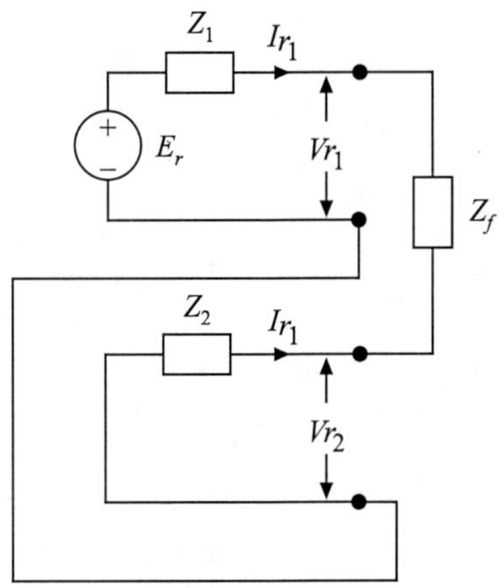

Figure 14.28 Interconnection of sequence networks for LL fault through a fault impedance.

14.11.3 Double Line to Ground (LLG) Fault

Let us assume that double line to ground (LLG) fault takes place involving phases y and b with ground as shown in Figure 14.29.

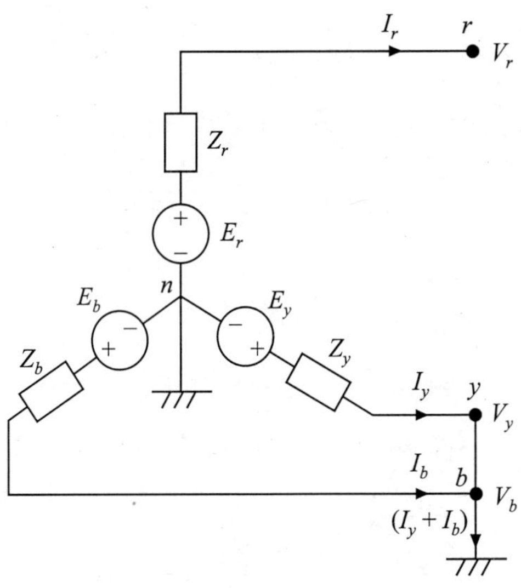

Figure 14.29 A LLG fault at the terminals of a solidly grounded, unloaded alternator.

Boundary conditions are: $V_y = 0$; $V_b = 0$; $I_r = 0$

Here,
$$V_{r_1} = \frac{1}{3}\left(V_r + \alpha V_y + \alpha^2 V_b\right) = \frac{1}{3}\left(V_r + 0 + 0\right) = \frac{V_r}{3}$$

$$V_{r_2} = \frac{1}{3}\left(V_r + \alpha^2 V_y + \alpha V_b\right) = \frac{V_r}{3}$$

$$V_{r_0} = \frac{1}{3}\left(V_r + V_y + V_b\right) = \frac{V_r}{3}$$

$\therefore$

$$V_{r_1} = V_{r_2} = V_{r_0} = \frac{V_r}{3} \tag{14.51}$$

Since $V_{r_1} = V_{r_0}$, we can write

$$E_r - I_{r_1} Z_1 = -I_{r_0} Z_0$$

$\therefore$

$$I_{r_0} = -\left(\frac{E_r - I_{r_1} Z_1}{Z_0}\right) \tag{14.52}$$

and since $V_{r_1} = V_{r_2}$, we have

$$E_r - I_{r_1} Z_1 = -I_{r_2} Z_2$$

i.e.,

$$I_{r_2} = -\left(\frac{E_r - I_{r_1} Z_1}{Z_2}\right) \tag{14.53}$$

since $I_r = I_{r_1} + I_{r_2} + I_{r_0} = 0$, hence from equations (14.52) and (14.53) we can write

$$I_{r_1} + \left[-\frac{E_r - I_{r_1} Z_1}{Z_2}\right] + \left[-\frac{E_r - I_{r_1} Z_1}{Z_0}\right] = 0$$

$\therefore$

$$I_{r_1} = \frac{E_r}{Z_1 + \dfrac{Z_0 Z_2}{Z_0 + Z_2}} = -\left(I_{r_2} + I_{r_0}\right) \tag{14.54}$$

The interconnection of the sequence network is shown in Figure 14.30.

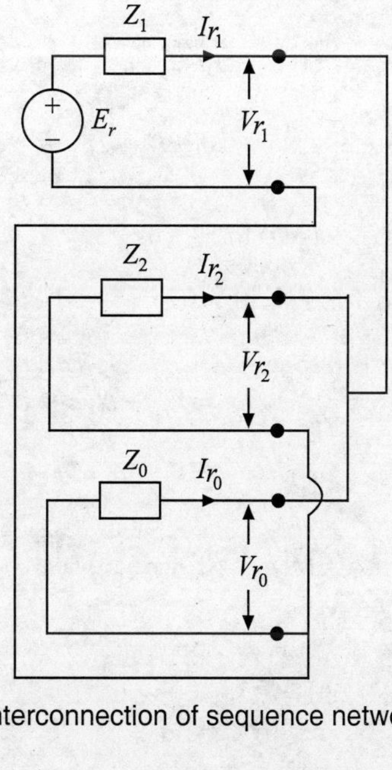

Figure 14.30 Interconnection of sequence networks for LLG fault.

In case LLG fault involves a fault impedance Z_f while the neutral of the alternator is grounded through an impedance Z_n, as shown in Figure 14.31, the boundary conditions are:

$$V_y = V_b = \left(I_y + I_b\right) Z_f;\ I_r = 0$$

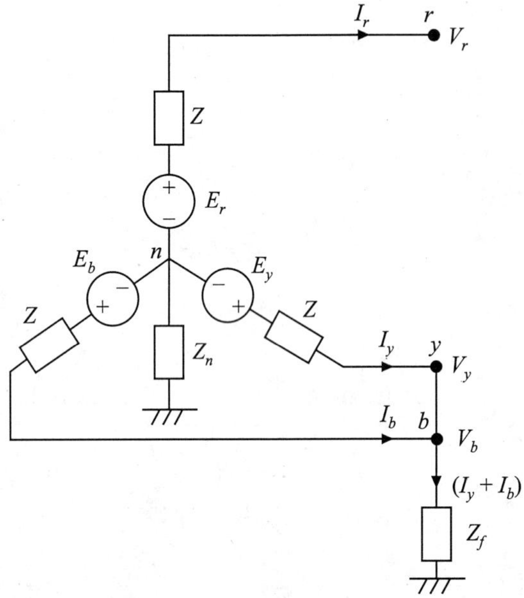

Figure 14.31 A LLG fault through fault impedance at the terminals of an unloaded alternator grounded through an impedance.

Sequence network equations are:

$$V_{r_1} = E_r - I_{r_1} Z_1;\ V_{r_2} = -I_{r_2} Z_2;\ V_{r_0} = -I_{r_0}\left(Z_0 + 3Z_n\right)$$

$$I_{r_0} = \frac{1}{3}\left(I_r + I_y + I_b\right) = \frac{1}{3}\left(0 + I_y + I_b\right)$$

$\therefore$
$$\left(I_y + I_b\right) = 3I_{r_0} \tag{14.55}$$

Note that above equation is valid for $Z_f = 0$ also.

Since, $V_y = V_b$

$$V_{r_0} + \alpha^2 V_{r_1} + \alpha V_{r_2} = V_{r_0} + \alpha V_{r_1} + \alpha^2 V_{r_2}$$

i.e.,
$$V_{r_1} = V_{r_2}$$

Also,
$$V_y = \left(I_y + I_b\right) Z_f = 3I_{r_0} Z_f \quad \left[\because I_y + I_b = 3I_{r_0}\right]$$

We can now write,

$$V_{r_1} = \frac{1}{3}\left(V_r + \alpha^2 V_y + \alpha V_b\right)$$

$$\therefore \qquad V_{r_1} = \frac{1}{3}\left[V_r + \left(\alpha^2 + \alpha\right)V_y\right] \left[\because V_y = V_b\right] \tag{14.56}$$

$$V_{r_0} = \frac{1}{3}\left(V_r + V_y + V_b\right) = \frac{1}{3}\left[V_r + 2V_y\right] \tag{14.57}$$

Subtracting equation (14.56) from equation (14.57),

$$V_{r_0} - V_{r_1} = \frac{1}{3}\left(2 - \alpha^2 - \alpha\right)V_y = V_y = 3I_{r_0} Z_f$$

or,

$$V_{r_1} = V_{r_0} - 3I_{r_0} Z_f$$

or,

$$E_r - I_{r_1} Z_1 = -I_{r_0}\left(Z_0 + 3Z_n\right) - 3I_{r_0} Z_f \quad \left[\because V_{r_0} = I_{r_0}\left(Z_0 + 3Z_n\right)\right]$$

$$\therefore \qquad I_{r_0} = -\left(\frac{E_r - I_{r_1} Z_1}{Z_0 + 3Z_n + 3Z_f}\right) \tag{14.58}$$

Similarly, $I_{r_2} = -\left(\dfrac{E_r - I_{r_1} Z_1}{Z_2}\right)$ (14.59)

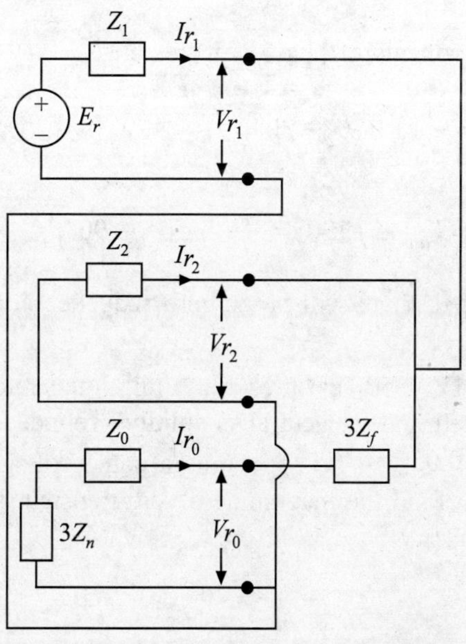

Figure 14.32 Interconnection of sequence networks for LLG fault through a fault impedance.

Since $\therefore\ I_r = I_{r_1} + I_{r_2} + I_{r_0} = 0$, we can write from above

$$I_{r_1} - \left(\frac{E_r - I_{r_1}Z_1}{Z_2}\right) - \left(\frac{E_r - I_{r_1}Z_1}{Z_0 + 3Z_n + 3Z_f}\right) = 0$$

$$\therefore \qquad I_{r_1} = \frac{E_r}{Z_1 + \dfrac{Z_2\left(Z_0 + 3Z_n + 3Z_f\right)}{Z_2 + Z_0 + 3Z_n + 3Z_f}} \tag{14.60}$$

The interconnection of the sequence networks is shown in Figure 14.32.

EXAMPLE 14.16 Three identical star connected resistors of 1.0 p.u. are connected to an unbalanced three-phase supply. The load neutral is isolated. The symmetrical components of line voltage in p.u. are $V_{ab_1} = X\angle\theta_1$, $V_{ab_2} = Y\angle\theta_2$. If all the p.u. calculations are with respective base values, find the phase to neutral sequence voltages.

Solution: Let us assume the phase sequence of the supply voltage is abc.
The positive sequence components of phase voltages are:

$$V_{an_1} = \frac{1}{\sqrt{3}}V_{ab_1}\angle -30° = \frac{1}{\sqrt{3}}X\angle\left(\theta_1 - 30°\right),\ V_{bn_1} = V_{bn_1}\angle -120° = \frac{1}{\sqrt{3}}X\angle\left(\theta_1 - 150°\right)$$

and

$$V_{cn_1} = V_{an_1}\angle 120° = \frac{1}{\sqrt{3}}X\angle\left(\theta_1 + 90°\right) \quad \text{(Ans.)}$$

The negative sequence components of phase voltages are:

$$V_{an_2} = \frac{1}{\sqrt{3}}V_{ab_2}\angle 30° = \frac{1}{\sqrt{3}}Y\angle\left(\theta_2 + 30°\right),\ V_{bn_2} = V_{bn_2}\angle 120° = \frac{1}{\sqrt{3}}Y\angle\left(\theta_2 + 150°\right)$$

and

$$V_{cn_2} = V_{an_2}\angle -120° = \frac{1}{\sqrt{3}}Y\angle\left(\theta_2 - 90°\right) \quad \text{(Ans.)}$$

As the load neutral is isolated, there will be no zero sequence components of phase voltages.

EXAMPLE 14.17 A 500 MVA, 50 Hz, three-phase turbo generator produces power at 22 kV. The generator is Y connected and its neutral is solidly grounded. Its sequence reactance are $x_1 = x_2 = 0.15$ p.u. and $x_0 = 0.05$ p.u. It is operating at rated voltage and disconnected from the rest of the system (no load). Find the magnitude of sub-transient line current for SLG fault at the generator terminal in p.u. **[Gate 2004]**

Solution: For SLG fault, $I_f = 3I_{r_1}$

It is obvious that, $I_{r_1} = I_{r_2} = I_{r_0} = \dfrac{E_r}{Z_1 + Z_2 + Z_0} = \dfrac{1\angle 0^\circ}{j0.15 + j0.15 + j0.05} = 2.857\angle -90^\circ$ p.u.

$$\therefore \qquad |I_f| = 3|I_{r_1}| = 3 \times 2.857 = 8.57 \text{ p.u.} \qquad \text{(Ans.)}$$

EXAMPLE 14.18 A two bus power network is shown in Figure E14.10, in which a SLG fault occurs. All specifications are indicated in the figure. If the generator neutral is solidly grounded and fault impedance is zero then find the value of fault current.

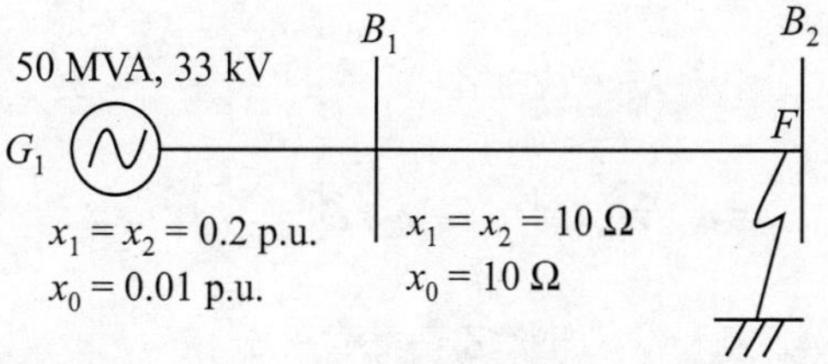

Figure E14.10 Single line diagram of power network.

Solution: Let us consider base MVA = 50, base kV = 33.

Base impedance, $Z_{\text{base}} = \dfrac{\text{kV}^2_{\text{LL base}}}{\text{MVA}_{3\phi \text{ base}}} = \dfrac{33^2}{50} = 21.78\ \Omega$

Base current, $= \dfrac{\text{VA}_{3\phi \text{ base}}}{\sqrt{3}\text{V}_{\text{LL base}}} = \dfrac{50 \times 10^6}{\sqrt{3} \times 33 \times 10^3} = 874.77$ A

Line sequence reactances in p.u. are:

$$X_{L1} = X_{L2} = \frac{10}{21.78} = 0.46 \text{ p.u. and } X_{L0} = \frac{20}{21.78} = 0.92 \text{ p.u.}$$

Generator sequence reactances in p.u. are:

$$X_{G1} = X_{G2} = 0.2 \text{ p.u. and } X_{G0} = 0.01 \text{ p.u.}$$

Let us assume, the pre fault bus voltage is 1 p.u.

$\therefore$ Fault current for SLG fault,

$$I_f = \frac{3E_g}{X_{1_{eqv}} + X_{2_{eqv}} + X_{0_{eqv}}} = \frac{1\angle 0^\circ}{j0.66 + j0.66 + j0.93} = 1.33\angle -90^\circ \text{ p.u.}$$

$$= 1.33\angle -90^\circ \times 874.77 \text{ A} = 1.16\angle -90^\circ \text{ kA} \qquad \text{(Ans.)}$$

EXAMPLE 14.19 A generator with grounded neutral has sequence impedance of z_1, z_2 and z_0. Generated voltage by this generator is E. Find V_y and V_b for a SLG fault occuring at the terminal of phase r.

Solution: Let us assume fault impedance is Z_f.

The sequence current expression will be as follows:

$$I_{r_0} = I_{r_2} = \frac{E}{Z_1 + Z_2 + Z_0 + 3Z_f}$$

Sequence components of phase voltages are:

$$V_{r_1} = E - I_{r_1} Z_1 \; ; \; V_{r_2} = -I_{r_2} Z_2 \; ; \; V_{r_0} = -I_{r_0} Z_0$$

$$V_y = V_{r_0} + \alpha^2 V_{r_1} + \alpha V_{r_2} = -I_{r_0} Z_0 + \alpha^2 \left(E - I_{r_1} Z_1\right) + \alpha\left(-I_{r_2} Z_2\right)$$

$$= \alpha^2 E - I_{r_1}\left(\alpha^2 Z_1 + \alpha Z_2 + Z_0\right) \quad \left[\because I_{r_1} = I_{r_2} = I_{r_0}\right]$$

$$= \alpha^2 E - \frac{E\left(\alpha^2 Z_1 + \alpha Z_2 + Z_0\right)}{Z_1 + Z_2 + Z_0 + 3Z_f}$$

Similarly, $V_b = V_{r_0} + \alpha V_{r_1} + \alpha^2 V_{r_2} = -I_{r_0} Z_0 + \alpha\left(E - I_{r_1} Z_1\right) + \alpha^2\left(-I_{r_2} Z_2\right)$

$$= \alpha E - I_{r_1}\left(\alpha Z_1 + \alpha^2 Z_2 + Z_0\right) = \alpha E - \frac{E\left(\alpha Z_1 + \alpha^2 Z_2 + Z_0\right)}{Z_1 + Z_2 + Z_0 + 3Z_f}$$

EXAMPLE 14.20 Three alternators of ratings 20 MVA, 40 MVA and 60 MVA are connected at a 15 kV bus bar. The transient d-axis reactance of each of them is 0.1 p.u. A three phase symmetrical fault occurs on the bus bars. With base voltage of 15 kV and base power of 60 MVA, find the fault MVA and the fault current. Neglect all resistances.

Solution: Converting the transient reactances of each of the alternators to a common base we have,

$$x_1 = 0.1 \times \frac{60}{20} = 0.3 \text{ p.u.}, \; x_2 = 0.1 \times \frac{60}{40} = 0.15 \text{ p.u. and } x_3 = 0.1 \text{ p.u.}$$

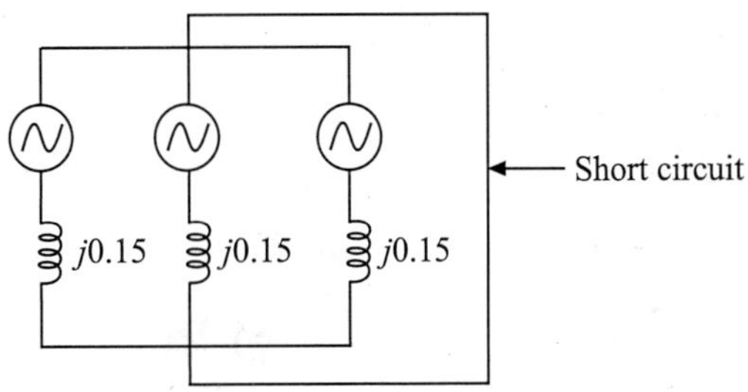

Figure E14.11 Equivalent circuit of the system.

The equivalent circuit is shown in Figure E14.11.

$$\text{Equivalent reactance, } X_{eqv} = \frac{1}{\frac{1}{j0.3}+\frac{1}{j0.15}+\frac{1}{j0.1}} = j0.05 \text{ p.u.}$$

$$\text{Fault MVA} = \frac{60}{0.05} = 1200 \quad \text{(Ans.)}$$

$$\text{Fault current} = \frac{1200\times10^6}{\sqrt{3}\times15\times10^3} = 46.19 \text{ kA} \quad \text{(Ans.)}$$

EXAMPLE 14.21 A 20 MVA, 6.6 kV, three-phase alternator is connected to a three-phase transmission line. The per unit positive, negative and zero sequence impedances of the alternator are *j*0.1, *j*0.1 and *j*0.04 p.u. respectively. The neutral of the alternator is connected to ground through an inductive reactor of *j*0.05 p.u. The per unit positive, negative and zero sequence impedances of the transmission line are *j*0.1, *j*0.1 and *j*0.3 p.u. respectively. All per unit values are based on machine rating. A solid ground fault occurs at one phase of the far end of the transmission line. The voltage of the alternator neutral with respect to ground during fault is: (a) 513.8 V; (b) 889.9 V; (c) 1112.0 V; (d) 642.2 V. **[Gate 2004]**

Solution: Here, zero sequence impedance $(=Z_0) = Z_{L_0} + Z_{G_0} = j0.3 + j0.04 = j0.34$ p.u.

where 'L' stands for transmission line and 'G' stands for the generator , and suffix '0' stands for zero sequence.

Similarly, positive sequence impedance $(=Z_1) = Z_{L_1} + Z_{G_1} = j0.1 + j0.1 = j0.2$ p.u.

Also, negative sequence impedance $(=Z_2) = Z_{L_2} + Z_{G_2} = j0.2$ p.u.

Given that, grounding impedance of alternator, $(=Z_n) = j0.5$ p.u.

Induced e.m.f of generator, $E_g = 1\angle 0°$ p.u.

$$\text{For SLG fault, } I_{r_1} = \frac{E_g}{Z_1+Z_2+Z_0+3Z_n} = \frac{1}{j0.2+j0.2+j0.34+3\times j0.05} = -j1.124 \text{ p.u.}$$

$$\text{Base current} = \frac{\text{VA}_{3\phi\text{ base}}}{\sqrt{3}\text{V}_{\text{LL base}}} = \frac{20\times10^6}{\sqrt{3}\times6.6\times10^3} = 1750 \text{ A}$$

∴ Fault current for SLG fault,

$$I_f = 3\times I_{r_1} = 3\times(-j1.124) = -j3.372 \text{ p.u.} = -j3.372\times1750 = -j5901 \text{ A}$$

∴ Magnitude of ground to neutral voltage of alternator,

$$|V_n| = |I_f Z_n| = |-j3.372\times j0.05| = 0.1686 \text{ p.u.} = 0.1686\times6.6\times10^3 = 642.45 \text{ V}$$

This answer is close to that given in option (d). Therefore option (d) is the right answer.

EXAMPLE 14.22 A 25 MVA, 11 kV, three-phase alternator has positive, negative and zero sequence impedances of $j0.25$, $j0.1$ and $j0.1$ p.u. respectively. The generator neutral is solidly grounded and the machine runs on no load. Determine the fault current at its terminals for the following types of faults:

(i) single line to ground fault
(ii) line to line fault
(iii) double line to ground fault

Solution: For single line to ground fault,

$$I_{r_1} = \frac{E_g}{Z_1 + Z_2 + Z_0} = \frac{1\angle 0^0}{j0.25 + j0.25 + j0.1} = -j1.67 \text{ p.u.}$$

[Here it is assumed that generator phase voltage is $1\angle 0°$ p.u. and r is the faulted phase.]

$$I_r = 3I_{r_1} = 3 \times (-j1.67) = -j5 \text{ p.u.}$$

$$\text{Base current} = \frac{\text{VA}_{3\phi\text{ base}}}{\sqrt{3}\text{V}_{\text{LL base}}} = \frac{50 \times 10^6}{\sqrt{3} \times 11 \times 10^3} = 2624.32 \text{ A}$$

$\therefore$ Fault current, $|I_r| = 5 \times 2624.32 \text{ A} = 13.122 \text{ kA}$ (Ans. of (i))

For line to line fault, $I_{r_0} = 0$ (see Figure 14.26).

and
$$I_{r_1} = -I_{r_2} = \frac{E_g}{Z_1 + Z_2} = \frac{1\angle 0^0}{j0.25 + j0.25} = -j2 \text{ p.u.}$$

[Here it is assumed that LL fault occurs between phases y and b.]

$\therefore$ Fault current, $I_y = I_{r_0} + \alpha^2 I_{r_1} + \alpha I_{r_2} = \alpha^2 I_{r_1} - \alpha I_{r_1} = (\alpha^2 - \alpha)I_{r_1}$

$$= \{(-0.5 + j0.866) - (-0.5 - j0.866)\} \times (-j2) = -j1.732 \times (-j2) = -3.464 \text{ p.u.}$$

$$= -3.464 \times 2624.32 = -9090.64 \text{ A} \quad \text{(Ans. of (ii))}$$

and $I_b = -I_y = -9090.64$ A (Ans. of (ii))

For double line to ground fault,

$$I_{r_1} = \frac{E_g}{Z_1 + \dfrac{Z_0 Z_2}{Z_0 + Z_2}} = \frac{1\angle 0^0}{j0.25 + \dfrac{j0.1 \times j0.25}{j0.1 + j0.25}} = -j3.115 \text{ p.u.}$$

From equation (14.51),
$$V_{r_1} = V_{r_2} = V_{r_3} = \frac{V_r}{3}$$

Now,
$$V_{r_1} = E_g - I_{r_1} Z_1 = 1 - (-j3.115) \times j0.25 = 0.221 \text{ p.u.}$$

$$\therefore \quad I_{r_0} = -\frac{V_{r_0}}{Z_0} = -\frac{0.221}{j0.1} = -2.21 \text{ p.u.} \left[\because V_{r_0} = -I_{r_0} Z_0\right]$$

Here it is assumed that LLG fault occurs between phases y and b and ground.

Therefore, fault current, $I_f = I_y + I_b$ (see Figure 14.29)

From equation (14.54) $I_f = I_y + I_b = 3I_{r_0} = 3 \times (-2.21) = -6.63$ p.u.

$$\therefore \quad |I_f| = 6.63 \text{ p.u.} = 6.63 \times 2624.32 \text{ A} = 17.4 \text{ kA} \quad \text{(Ans. of (iii))}$$

14.12 EFFECT OF NEUTRAL GROUNDING ON FAULT CURRENT

For a solid LG fault, Z_f becomes zero. Hence the fault current at phase R can be obtained as

$$I_{f_{SLG}} = \frac{3E_g}{Z_1 + Z_2 + Z_0}$$

Assuming the resistance part of the impedance component to be negligible and assuming the positive and negative sequence reactances equal, the above equation can be further modified as

$$I_{f_{SLG}} = \frac{3E_g}{2X_1 + X_0} \tag{14.61}$$

On the other hand, for a solid three-phase fault, $I_{f_{3ph}} = \dfrac{E_g}{Z_1}$

Considering the reactance part only,

$$I_{f_{3ph}} = \frac{E_g}{X_1} \tag{14.62}$$

Comparing equations (14.61) and (14.62), it can be concluded that the single line to ground fault is more severe than the three phase fault, as the zero sequence reactance is of low value, for a generator with equal positive and negative sequence reactances and neutral solidly grounded. However, if X_0 is much higher than X_1, there is a possibility that the three phase fault current is higher in magnitude than the LG fault current. In practical context, if fault occurs at the generator terminals, due to lower value of X_0, $I_{f_{SLG}} > I_{f_{3ph}}$ while for a fault in the transmission line $I_{f_{3ph}} > I_{f_{SLG}}$.

To make the fault current due to SLG fault less than three phase fault current at generator terminals we use a grounding reactance (X_n) at the neutral. This gives

$$\frac{3E_g}{2X_1 + X_0 + 3X_n} < \frac{E_g}{X_1}$$

or,

$$\frac{3E_g}{2X_1 + X_0 + 3X_n} < \frac{E_g}{X_1}$$

or, $$2X_1 + X_0 + 3X_n > 3X_1$$

$\therefore$ $$X_n > \frac{1}{3}(X_1 - X_0) \tag{14.63}$$

Thus, it is possible to select the grounding reactance (X_n) to limit the SLG fault current lower than the three phase fault current fault current.

EXAMPLE 14.23 A three phase synchronous generator has positive and negative sequence reactances as $j0.15$ p.u. each while zero sequence reactance is 0.045 p.u. The generator neutral point is grounded through an inductive reactance (X_n). What should be the value of X_n so that the severity of the single line to ground fault and three phase fault at the generator terminals of the unloaded generator will be same?

Solution: Fault current for SLG fault, $I_{f_{SLG}} = \dfrac{3E_g}{Z_1 + Z_2 + Z_0 + 3X_n}$

where X_n is the grounding reactance.

While for three-phase fault,

$$\text{Fault current } I_{f_{3ph}} = \frac{E_g}{Z_1}$$

When the fault currents for SLG fault and for three-phase fault are same

$$\frac{3E_g}{Z_1 + Z_2 + Z_0 + 3X_n} = \frac{E_g}{Z_1}$$

or, $$Z_1 + Z_2 + Z_0 + 3X_n = 3Z_1$$

or, $$X_n = \frac{2Z_1 - Z_2 - Z_0}{3}$$

or, $$X_n = \frac{2 \times j0.15 - j0.15 - j0.045}{3}$$

$\therefore$ $$X_n = j0.035 \text{ p.u.}$$

Hence, the required inductive reactance at the generator neutral should be $j0.035$ p.u. (Ans.)

14.13 OPEN CONDUCTOR FAULTS

The objective of open conductor faults analysis is to find the line currents and series voltages between broken ends of the conductors. Let us assume a three phase, three wire line as shown in Figure 14.33(a). Let an 'open' conductor fault take place at location F so that open circuit voltages at the respective phases between broken ends of the conductors are $V_{rr'}$, $V_{yy'}$ and $V_{bb'}$. Let the series currents be represented by I_r, I_y and I_b. The line currents and series voltages are then represented as

$$I = \begin{bmatrix} I_r \\ I_y \\ I_b \end{bmatrix}; \quad V = \begin{bmatrix} V_{rr'} \\ V_{yy'} \\ V_{bb'} \end{bmatrix}$$

Assuming the symmetrical currents and voltages I_{r_1}, I_{r_2}, I_{r0} and $V_{rr_1'}$, $V_{rr_2'}$ $V_{rr_0'}$ respectively, the sequence currents and voltages are shown in Figure 14.33(b). They are suitably connected to represent the complete sequence network depending on the type of fault, i.e. whether a single conductor fault or two conductor fault.

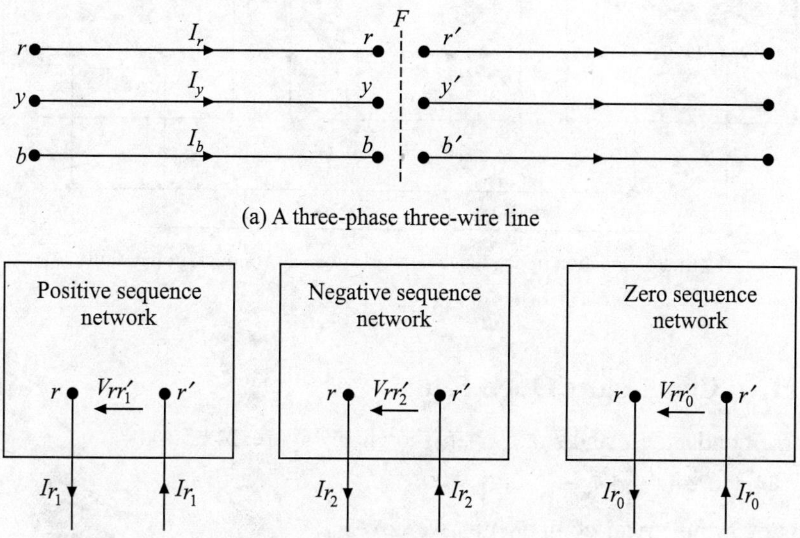

(a) A three-phase three-wire line

(b) Sequence networks for open conductor faults

Figure 14.33

14.13.1 Single Conductor Open Fault

For one conductor (phase r) open as shown in Figure 14.34(a),

$$V_{yy'} = V_{bb'} = 0 \text{ and } I_r = 0$$

In terms of symmetrical components, we have

$$V_{rr_1'} = V_{rr_2'} = V_{rr_0'} = \frac{1}{3} V_{rr'} \tag{14.64a}$$

and

$$I_{r_1} + I_{r_2} + I_{r_0} = 0 \tag{14.64b}$$

Thus, the sequence network is clearly a parallelly connected network as shown in Figure 14.34(b).

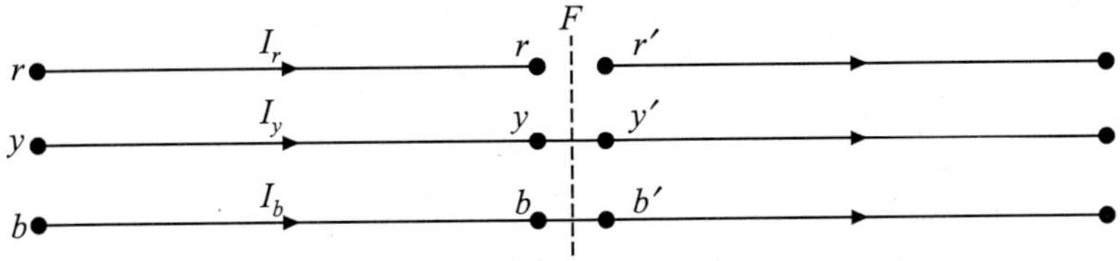

(a) A three-phase three-wire line with one conductor open

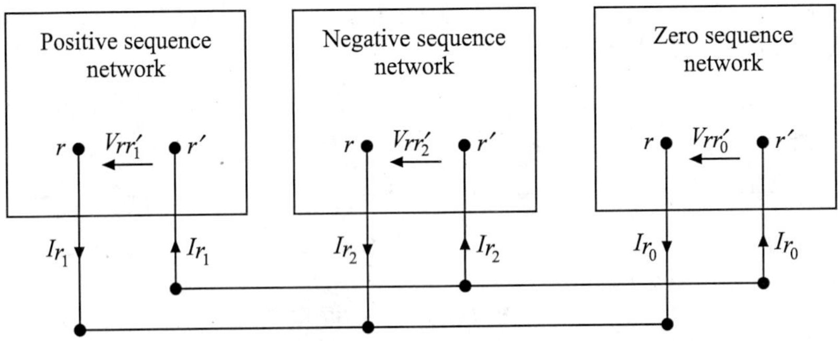

(b) Interconnection of sequence networks for one conductor open fault

Figure 14.34

14.13.2 Two Conductors Open Fault

Let us assume conductor y and b are open shown in Figure 14.35(a),

Obviously, $V_{rr'} = 0$ and $I_y = I_b = 0$

In terms of symmetrical components, we have

$$V_{rr'_1} + V_{rr'_2} + V_{rr'_0} = 0 \tag{14.65a}$$

and

$$I_{r_1} = I_{r_2} = I_{r_0} = \frac{1}{3} I_r \tag{14.65b}$$

These equations suggest that the connection diagram is a series connected sequence network as shown in Figure 14.35(b).

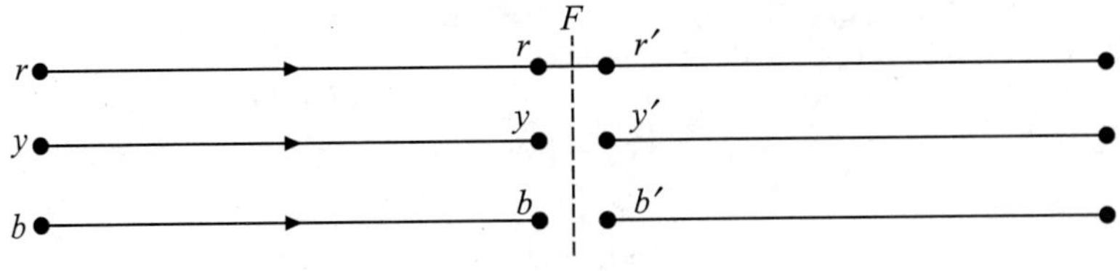

(a) A three-phase three-wire line with two conductor open

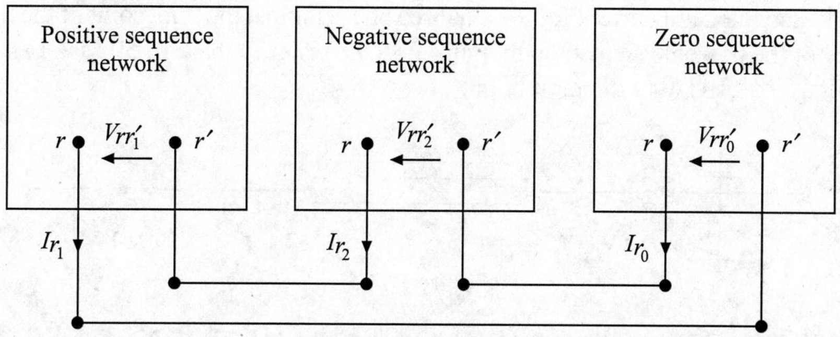

(b) Interconnection of sequence networks for two conductor open fault

Figure 14.35

EXERCISES

1. What is fault level?
2. What do understand by symmetrical components?
3. Derive the expression for three phase power using symmetrical components.
4. What are sub-transient, transient and steady state condition of a short circuit fault?
5. Derive necessary equations and draw the interconnected circuit of the three sequence networks for (a) three phase fault, (b) single line to ground fault, (c) line to line fault and (b) double line to ground fault at the terminal of an unloaded generator. Assuming the short circuit occurs through a fault resistance R_f in each type of fault.
6. Discuss about sequence network of three phase transformer.
7. Draw zero sequence network for the following types winding connections of three phase transformers (a) star-star connection with only one neutral grounded through an impedance Z_f, (b) star-star connection and both the neutral grounded through an impedance Z_f.
8. Show that the positive and negative sequence impedance of a transmission line are equal and its zero sequence impedance is more than its positive or negative sequence impedance.
9. What is open conductor fault? Derive necessary equations and draw the interconnected circuit of the three sequence networks when (a) one conductor is open, (b) two conductors are open
10. Derive equation for modification of the per unit impedance due to change in MVA and kV base values.
11. A 5 kVA, 1.1 kV/220 V single phase transformer, has impedances $z_1 = (1 + j10)\ \Omega$ and $z_2 = (0.1 + j0.5)\ \Omega$ respectively for the primary and secondary windings. Determine the per unit impedance of the transformer based on its own rating. (**Ans.** $(0.015 + j0.09)$ p.u.)

12. Calculate the fault current for a solid three phase fault at the mid-point of the transmission line of the power system shown Figure P 14.1. Assume a base of 50 MVA and a pre fault voltage of 210 kV at the fault point. (**Ans.** 526.57A)

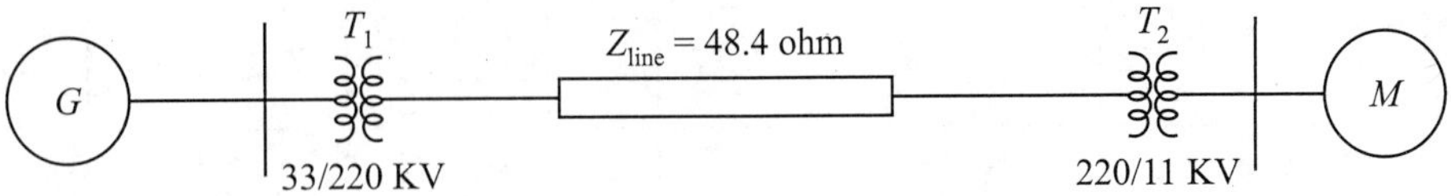

Figure P14.1

G: 50 MVA, 35 kV, X = 20%, M: 40 MVA, 11 kV, X = 30%

T_1 = 50 MVA, X = 15%, T_2 = 40 MVA, X = 15%

Line = 50 MVA, 220 kV, X = 48.4 ohm.

13. Calculate the reactance of the current limiting reactor, to be inserted between the generator and transformer T_1, so that fault current of Exercise 11 be limited to 500 A. (**Ans.** 35.62 ohm)

14. A station operating at 33 kV is divided into two sections A and B. Section A consists of three generators 20 MVA each having a reactance of 25% and section B is fed from grid through a 60 MVA transformer having 10% reactance as shown in Figure P14.2. The outgoing feeders connected to section A have a rupturing capacity of 600 MVA each. Determine the reactance of the reactor to be connected between sections A and B to prevent the breakers being overloaded if a symmetrical short circuit occurs on any of the outgoing feeders. (**Ans.** 1.21 ohm)

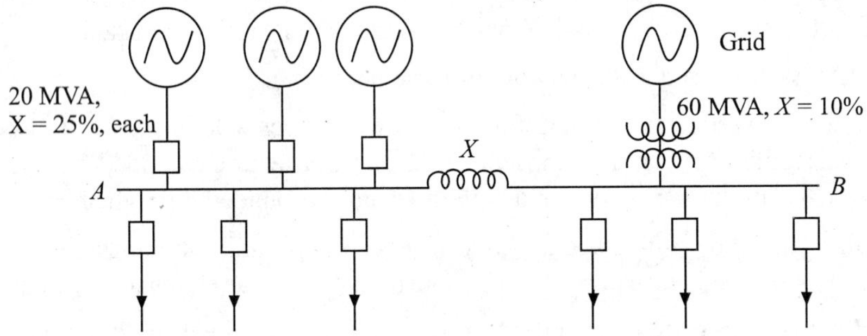

Figure P14.2

15. The alternator supplies two motors over a transmission line having transformers as shown in Figure P14.3. Line reactance is 150 Ω and zero sequence reactance of line is 300 Ω. Calculate fault current if there is a single line to ground fault at HV terminal of generator transformer (T_1). (**Ans.** 8.2 kA)

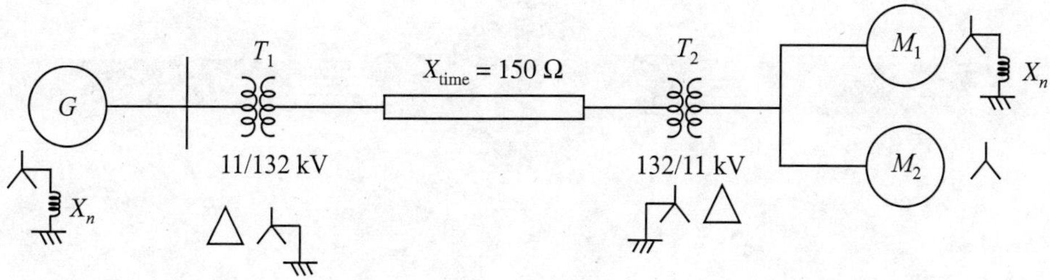

Figure P14.3

G: 20 MVA, 11 kV, $X_1 = 20\%$, $X_2 = 15\%$, $X_0 = 5\%$

$T_1 = T_2 = 25$ MVA, $X = 10\%$

M_1: 10 MVA, 11 kV, $X_1 = 25\%$, $X_2 = 20\%$, $X_0 = 5\%$, $X_n = 2.5$ ohm (in each case)

M_2: 15 MVA, 11 kV, $X_1 = 25\%$, $X_2 = 20\%$, $X_0 = 5\%$

16. Calculate fault current if there is a line to line fault at HV terminal of motor transformer (T_2) in the power system shown in Figure P14.3. (**Ans.** 5.63 kA)
17. Calculate fault current if there is a double line to ground fault at the terminal of motor M_2 in the power system shown in Figure P14.3. Assume fault reactance of j_1 ohm. (**Ans.** 6.57 kA)

CHAPTER 15

Power System Passive Compensation

15.1 INTRODUCTION

With the ongoing expansion and growth of the electric utility industry, including deregulation in many countries, numerous changes are continuously being introduced to once a predictable business. Although electricity is a highly engineered product, it is increasingly being considered and handled as a commodity. Thus, transmission systems are being pushed closer to their stability and thermal limits while the focus on the quality of power delivered is greater than ever.

In the evolving utility environment, financial and market forces continue and will continue, to demand a more optimal and profitable operation of the power system with respect to generation, transmission and distribution. Now, more than ever, advanced technologies are paramount for reliable and secure operation of power systems. To achieve both operational reliability and financial profitability, it has become clear that more efficient utilization and control of the existing transmission system infrastructure is required.

The chief objective of the power system operation being the supply of electric power from source end to load end stably and economically, it is imperative that the frequency and voltage at any point (node) of the network remains constant while the voltage waveform at any bus is free form harmonics, and the phases are balanced. It is preferred in practice that the power factor is either unity or very close to unity and bus voltage, even if it varies, remains within specified limits and the load performs its function in an optimal way. The maintenance of constant frequency in a power system needs an exact balance between the generated power and the load plus loss (i.e. power generated = power consumed at load + power loss). Also, voltage plays an important role in maintaining stable operation of the system. In the concept of voltage control and stability, it is very much evident that the receiving end voltage in a power system network is extremely sensitive to any change in reactive power status at the receiving end bus. In order to achieve optimum performance of the power system, it is required to control the reactive power flow in the network and this is possible by applying reactive power management at the different nodes in the power system. However, for reactive power management, the focus may be either towards the performance of load or line performance. Traditional solutions for upgrading the electrical transmission system infrastructure have been primarily in the form of new transmission lines, substations and associated equipment. However, as experiences have proven over the past decade or more, the process to permit site and construct new transmission lines has become extremely difficult, expensive, time-consuming and controversial. The reactive power management of load, to achieve its optimum performance, is sometimes more preferred. This is called *load compensation* while the reactive power compensation of the transmission network is called *line compensation*. In this chapter we will discuss about both types of compensation methods and will briefly describe the passive reactive compensators used for these purposes with their modelling aspects.

15.2 OBJECTIVES OF LOAD COMPENSATION

The management of reactive power at the load bus, for optimum performance of power system loads, being the prime objective of load compensation, we can subdivide the scope of load compensation into the following three components:

1. Power factor correction; 2. Improving voltage regulation; 3. Balancing of loads.

15.2.1 Power Factor Correction

Most of the real-life loads operate in lagging power factor, i.e. they absorb reactive power. Since the reactive power generation sources are limited (only the alternator at its over-excitation mode and shunt charging by cables and lines), there is a deficit of reactive power when the power system is heavily loaded. This leads to voltage depression (with higher regulation) at the load bus, resulting in higher current intake by constant power loads (e.g. induction motors). Load power factor being low, the power system efficiency decreases and the control of voltage throughout the network becomes more difficult. Supply companies penalize low power factor loads of industrial customers. All these factors encourage use of power factor correcting equipment. These pieces of equipments are basically reactive power compensation devices (i.e. they supply capacitive reactive power at the load node in order to compensate for the inductive demand at the load nodes) and are usually installed at the load buses. The following treatment highlights the theory of power factor compensating equipment at the load bus.

Following Figures 15.1(a) and 15.1(b), we can write for a load $Y_\ell(= G_\ell + jB_\ell)$ being connected to a load bus (having voltage $V\angle 0°$), the load current I_ℓ is given by

$$I_\ell = V \times Y_\ell = VG_\ell + jVB_\ell = I_w + jI_\mu \tag{15.1}$$

I_w = resistive component of load current I_ℓ

I_μ = reactive component of load current I_ℓ

also, S_ℓ (the complex power at load bus) $= V_2 I_\ell^* = V_2 G_\ell - jV_2 B_\ell = (P_\ell + jQ_\ell)$ (15.2)

and
$$\cos \phi_\ell = \frac{P_\ell}{S_\ell} \tag{15.3}$$

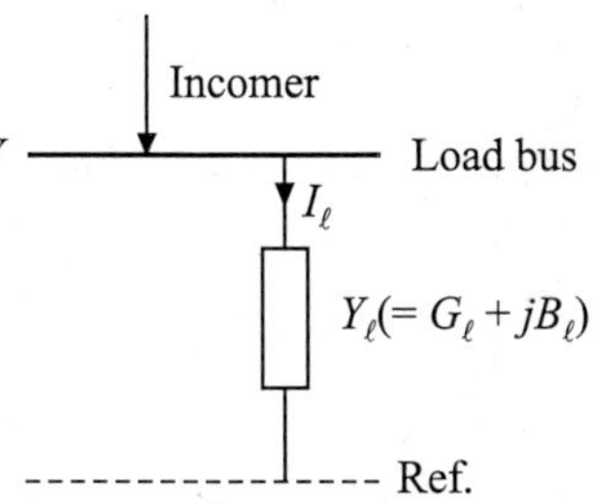

Figure 15.1(a) Load admittance (Y_ℓ) connected at load bus.

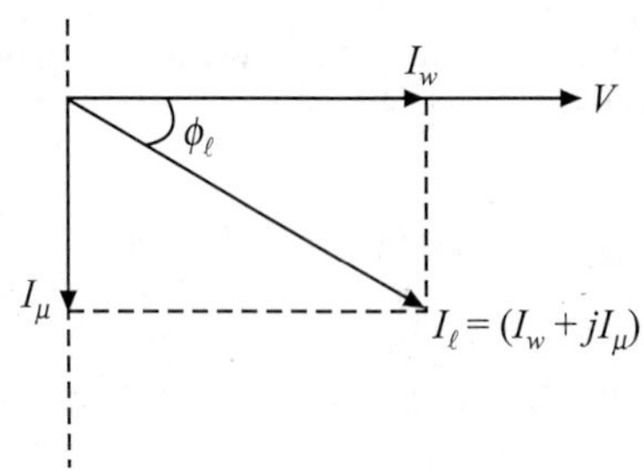

Figure 15.1 (b) Phasor diagram of Figure 15.1(a).

Let us now assume we add a compensator locally at the load bus in parallel to the load in such a way so that it has pure admittance of ($-jB_\ell$) in order to compensate for the inductive susceptance of Y_ℓ (the load). Let us assume this compensator takes current I_c. With reference to Figure 15.2(a), we can write

$$I = I_\ell + I_c = V(G_\ell + jB_\ell) - jVB_\ell = VG_\ell = I_w \quad (\text{Here, } I_c = -jVB_\ell) \tag{15.4}$$

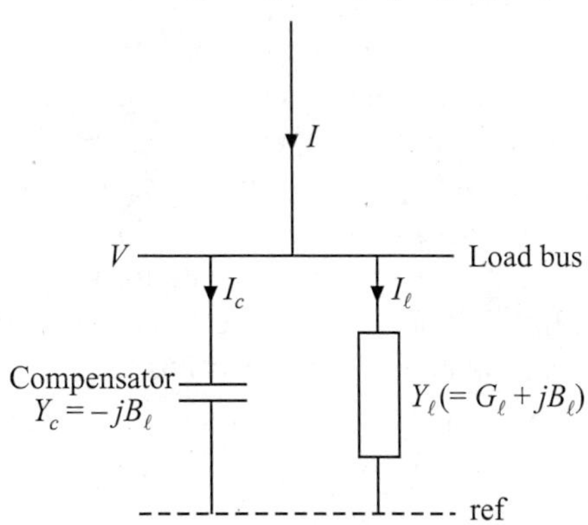

Figure 15.2 (a) Compensator added to load bus.

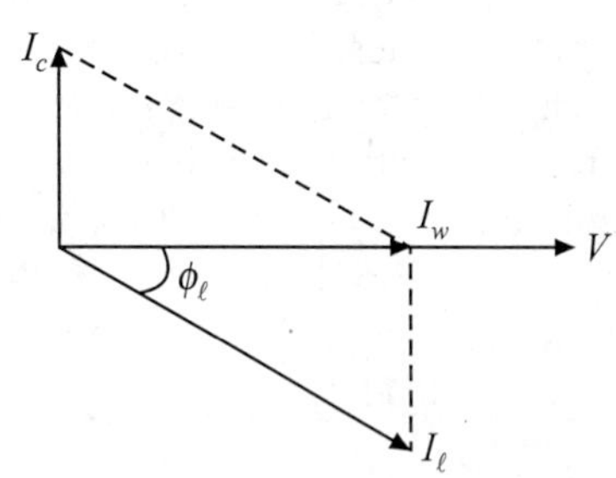

Figure 15.2 (b) Phasor diagram of Figure 15.2(a).

$$\therefore \qquad S_\ell' = VI^* = VI_w = V^2G_\ell \tag{15.5}$$

Equation (15.5) indicates that with full compensation at the receiving end bus, the power drawn (S_ℓ') by the load admittance is real power only and the power factor obviously is unity.

The compensation capacity can be obtained as follows:

$$S_{comp} = VI_c^* = jV^2B_\ell = Q_c(= -Q_\ell) \tag{15.6}$$

We thus observe that the compensator does not require any real power input and can feed the bus with capacitive reactive power. If the compensator susceptance is negative, it is a *capacitive compensator*, while for compensator susceptance of positive value, it is an *inductive compensator*.

Also, $$Q_c = S_\ell \sin \phi_\ell = S_\ell\sqrt{1-\cos^2\phi_\ell} \tag{15.7}$$

Equation (15.7) can be used to obtain the compensator rating while equation (15.6) may be used to find the compensator susceptance. It may be noted here that if $|Q_c| < |Q_\ell|$, the compensator is said to be partial and the network reactive power at the load bus is obtained as

$$|Q| = |Q_\ell - Q_c| \tag{15.8}$$

EXAMPLE 15.1 An industrial three phase load bus draws power of (100 + *j*50) KVA. If the bus voltage is 400V (L-L), find the compensator rating per phase. What is the compensator susceptance?

Solution:

$$P_\ell + jQ_\ell = (100 + j50) \text{ kVA}$$

$$\therefore \quad |S_\ell| = 111.803 \text{ kVA}$$

$$\cos\phi_\ell = \frac{100}{111.803} = 0.894$$

Compensator rating/phase is obtained from equation (15.7) as

$$\frac{Q_c}{\text{No. of phases}} = \frac{|S_\ell|}{3}\sin\phi_\ell = \frac{111.803}{3}\sqrt{1-0.894^2} = 16.67 \text{ kVAR} \quad \textbf{(Ans.)}$$

The compensator susceptance is obtained using equation (15.6),

$$B_\ell = \frac{Q_c}{jV^2} = -j\frac{16.67\times 10^3}{\left(\frac{400}{\sqrt{3}}\right)^2} = -j0.3125 \text{ mho (S) [capacitive compensator]} \quad \text{(Ans.)}$$

EXAMPLE 15.2 In Example 15.1, let us assume that we need to compensate the load bus reactive demand partially, i.e., we now desire that the existing power factor of 0.894 (lag) be improved to 0.98 (lag). What is the rating of compensator?

Solution: We have seen that the load demand in (100 + *j*50) kVA, i.e. the reactive load demand is 50 kVA, or 16.67 kVA/ph, the power factor (p.f.) being 0.894 (lag). If we are to improve the power factor to 0.98 (lag), the new reactive demand would lessen and is obtained as

$$Q_{new} = P_\ell \tan\phi_{new} \; [\because Q_\ell = P_\ell \tan\phi_\lambda] = 100\times\tan(\cos^{-1} 0.98) = 20.31 \text{ kVAR, i.e. } 6.77 \text{ kVAR/ph}$$

Therefore, we need to compensate (16.67 – 6.77) kVAR or 9.9 kVAR/ph to improve the p.f. from 0.894 to 0.98 (lag). Thus, the rating of the compensator should be 9.9 kVAR/ph.

The compensator susceptance (capacitive) would be

$$B_{\ell_{new}} = \frac{Q_c}{jV^2} = -j\frac{9.9 \times 10^3}{\left(\frac{400}{\sqrt{3}}\right)^2} = -j0.186 \text{ mho} \quad \textbf{(Ans.)}$$

15.2.2 Improving Voltage Regulation

Voltage regulation for a line may be defined as the per unit change in the sending end voltage magnitude for a specific variation in the receiving end voltage from no load to full load, and is caused by the drop in voltage due to passage of load current through the supply impedance. Thus the voltage regulation in p.u. for a simple transmission system (Figure 15.3) is given by

$$\Delta V = \frac{|E| - |V|}{|V|} \tag{15.9}$$

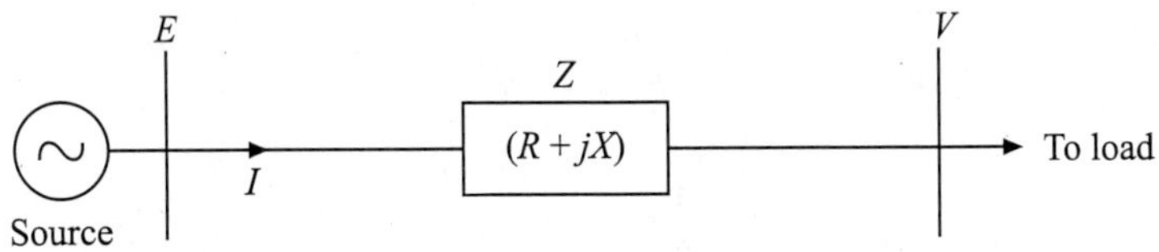

Figure 15.3 A simple transmission line model.

From the phasor diagram (Figure 15.4) corresponding to the system shown in Figure 15.3.

$$\Delta V = E - V = IZ \tag{15.10}$$

where I is the load current through the line having the line impedance Z (= $R + jX$).

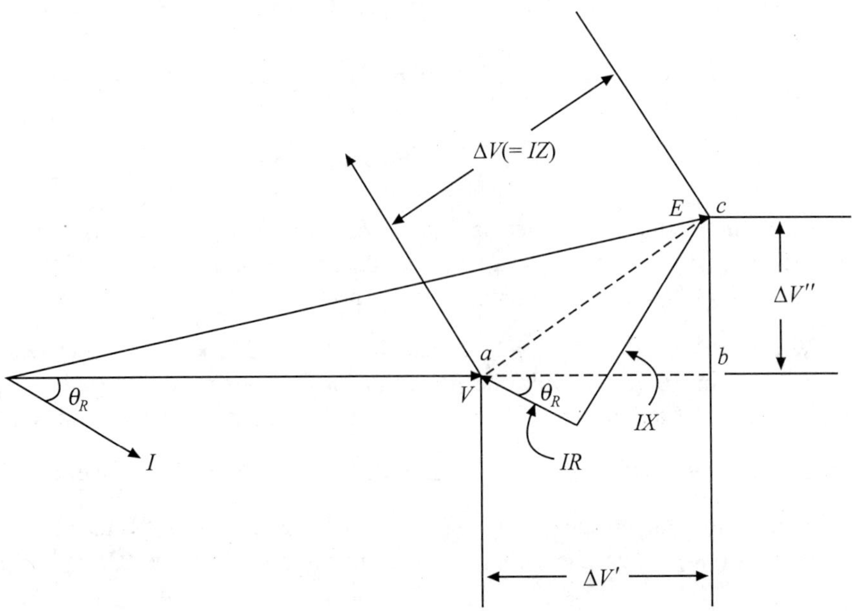

Figure 15.4 Phasor diagram of the power system shown in Figure 15.3.

Let us assume V is the reference phasor. Therefore,

$$I = \left(\frac{S}{V}\right)^* = \frac{P - jQ}{V} \quad \left[\because\ S = VI^*, \text{assuming } V \text{ to be reference phasor}\right] \tag{15.11}$$

Substituting the value of I from equation (15.11) as well as that of Z (= $R + jX$) in equation (15.10),

$$\Delta V = (R + jX)\frac{P_\ell - jQ_\ell}{V}$$

$$= \left(\frac{RP_\ell + XQ_\ell}{V}\right) + j\left(\frac{XP_\ell - RQ_\ell}{V}\right) \tag{15.12}$$

$$= \Delta V' + j\Delta V'' \tag{15.13}$$

where, $\Delta V' = (RP_\ell + XQ_\ell)/V$ and $\Delta V'' = (XP_\ell - RQ_\ell)/V$.

We may term $\Delta V'$ as the *inphase* drop while $\Delta V''$ the *quadrature* drop. Thus, voltage regulation is a function of both of these drops and subsequently depends on magnitude and phase of the bus voltage (V) and the load current. In other words, the voltage change depends upon both the real and reactive power demands of the load, precisely more on reactive power demand of the load.

Typically ±5% voltage regulation is allowed in power transmission utilities. Due to heavy loading condition, the system voltage remains depressed frequently and affects power system performance. The most ideal way to improve load bus voltage and thereby reducing the voltage regulation within limit is to increase the size and number of generating units as well as by increasing the interconnections. However, this method has got inherent limitations as it is costly and increases the fault level. On the other hand, it is better and economical to opt for reactive power management at load bus to apply compensation.

By installing a reactive compensator in parallel to the load, it is possible to make $|E| = |V|$ (i.e. voltage regulation becomes zero). This is a purely reactive compensator and Q_ℓ in the equation of voltage regulation being replaced by Q, where $Q = Q_\ell + Q_c$. For heavy inductive loading Q_ℓ is positive and the compensation is negative (i.e. $-Q_c$), the compensator being a pure capacitance. On the other hand, in order to compensate for the Ferranti effect where $|\Delta V|$ ($= |E| - |V|$) is negative, we need to have inductive compensator and Q_c is positive (Q_c is the reactive-power capability of the compensator).

Once the compensator is installed at the load bus, we can write (following the phasor diagram shown in Figure 15.4)

$$|E|^2 = \left[V + \frac{RP_\ell + XQ}{V}\right]^2 + \left[\frac{XP_\ell - RQ}{V}\right]^2 \tag{15.14a}$$

Q_ℓ being replaced by Q, where $Q = Q_\ell + Q_c$.

Rearranging equation (15.14a), we get

$$AQ^2 + BQ + C = 0 \tag{15.14b}$$

where $A = R^2 + X^2$; $B = 2V^2X$; $C = (V^2 + RP_\ell)^2 + X^2P_\ell^2 - E^2V^2$

Solution of equation (15.14b) yields

$$Q = \frac{-B \pm \sqrt{B^2 - 4AC}}{2A} = Q_\ell + Q_c \tag{15.15}$$

Here Q is obtained and Q_ℓ is given, Q_c can also be obtained from equation (15.15); the reactive load demand being compensated, the voltage regulation obtained from equation (15.12) is given by

$$\Delta V = \left(\frac{RP_\ell + jXP_\ell}{V}\right) = (R + jX)\frac{P_\ell}{V} \tag{15.16}$$

Expression (15.16) indicates that even with a purely reactive compensator (with $Q_\ell = 0$), there will be a voltage regulation being governed by line impedance and real power demand.

Algorithm to determine capacity of a reactive compensator to improve voltage regulation

Step 1: Store data of load bus voltage (V), line impedance (z) [$= r + jx$] and load demand $S_\ell = P_\ell + jQ_\ell$.

Step 2: Calculate the constants A, B and C in equation (15.14b).

Step 3: Obtain Q using equation (15.15).

Step 4: Find Q_c using equation [$Q_c = Q - Q_\lambda$].

Step 5: Calculate ΔV using equation (15.16).

Step 6: Obtain Q using equation (15.15).

15.2.3 Balancing of Load

AC power systems are usually 3-phase and are most efficient for operation with balanced loads. Unbalanced operations give rise to components of current in the negative or zero phase sequence components. Such components have the effect of additional losses in lines and rotating machinery, higher ripple in rectifiers, higher neutral current, higher triplen harmonics and generation of oscillating torques in rotating machinery and measuring instruments. Electric furnace, hoist, particular types of power electronic devices, etc. are all the loads that may create severe unbalancing. Compensators should be designed accordingly so that they have dynamic capability to compensate for these unbalances. *Thyristor controlled compensators* are used in the distribution side to compensate for these load unbalancing.

Let us assume a delta load is connected with a load bus and is represented using admittance representation (Figure 15.5).

Here, $Y_{br} \neq Y_{ry} \neq Y_{yb}$;

$$\left.\begin{aligned} Y_{br} &= G_{br} + jB_{br} \\ Y_{ry} &= G_{ry} + jB_{ry} \\ Y_{yb} &= G_{yb} + jB_{yb} \end{aligned}\right\} \tag{15.17}$$

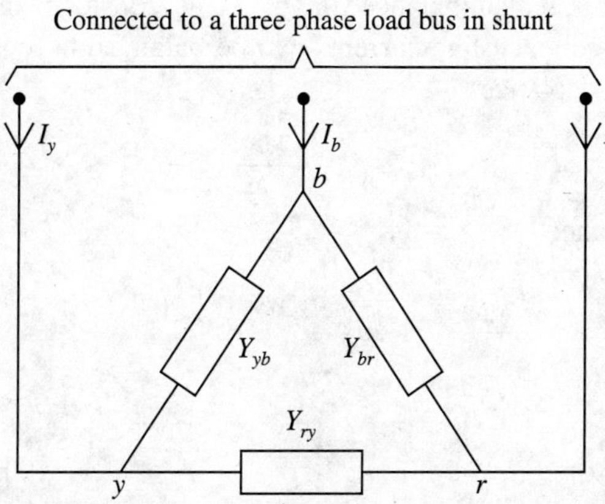

Figure 15.5 Unbalanced delta load on a three phase load bus.

The power factor of each phase can be corrected to unity by connecting compensating admittances in parallel to load admittances such that $(jB)_{br(c)} = (-jB)_{br}$; $(jB)_{ry(c)} = (-jB)_{ry}$; $(jB)_{yb(c)} = (-jB)_{yb}$. This will make the load impedance purely real as we have applied parallel compensating susceptance equal to negative of load susceptance. This leads to conversion of delta load to purely real load of unity power factor (but still unbalanced) as shown in Figure 15.6.

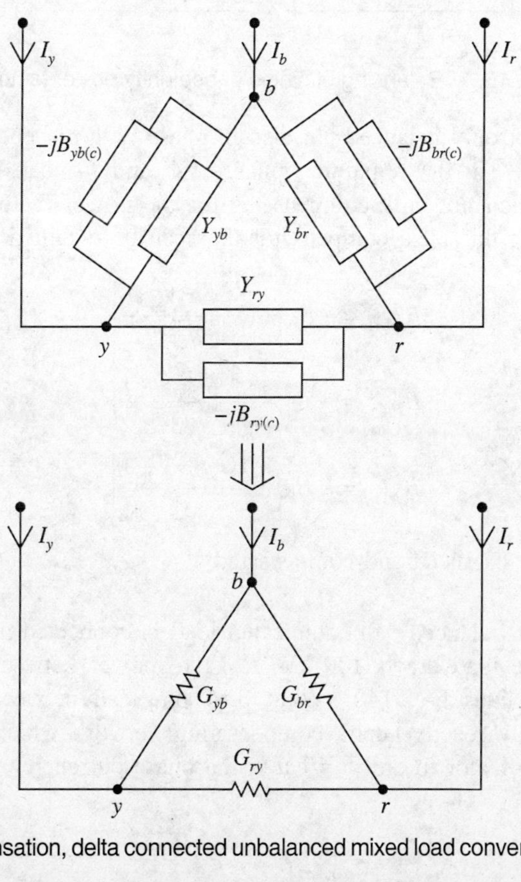

Figure 15.6 With compensation, delta connected unbalanced mixed load converts to real unbalanced load.

In order to balance this real unbalanced load, let us consider single phase real load G_{ry}. The three phase positive sequence line currents can be balanced by connecting between phases y and b the capacitive susceptance

$$B_{yb(c)} = \frac{G_{ry}}{\sqrt{3}} \tag{15.18}$$

and the inductive susceptance

$$B_{br(c)} = -\frac{G_{ry}}{\sqrt{3}} \text{ between } b \text{ and } r. \tag{15.19}$$

This is illustrated in Figure 15.7.

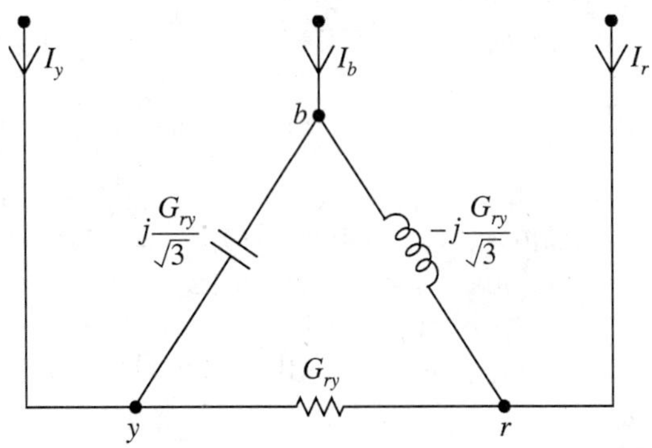

Figure 15.7 Balancing of single phase unity power factor load.

Now I_r, I_y and I_b are not only balanced but also in phase with respect to their phase voltages.

The real admittances in the remaining phases y-b and b-r can be balanced in turn by the same procedure which not only compensates for the power factor but also reduces load unbalance. The rating of this ideal compensator thus can be obtained as

$$\left.\begin{aligned} B_{ry(c)} &= -B_{ry} + \frac{(G_{br} - G_{yb})}{\sqrt{3}} \\ B_{yb(c)} &= -B_{yb} + \frac{(G_{ry} - G_{br})}{\sqrt{3}} \\ B_{br(c)} &= -B_{br} + \frac{(G_{yb} - G_{ry})}{\sqrt{3}} \end{aligned}\right\} \tag{15.20}$$

The load is now purely balanced and compensated.

EXAMPLE 15.3 An unbalanced delta connected load is connected across a 50 Hz, 3 phase, 400 V AC supply. The load draws 100 kW resistive power between lines a-b, 140 kW at 0.85 p.f. lag between lines b-c, 140 kW at 0.85 p.f. lead between lines c-a. Obtain the susceptances of a purely reactive delta compensating network which will balance this load and improve the power factor to unity. What is the phase current?

Solution: Since load is $(P + jQ)$, i.e. VI^*,

$\therefore$ We can write, $P + jQ = VI^* = V^2Y^*$

or, $Y = \dfrac{P - jQ}{V^2}$, i.e. it is now possible to find equivalent susceptance of the connected load.

For phase *a-b*, $$Y_{ab} = \frac{(100 - j0)\times 10^3}{400^2} \quad [\because \text{ load is } \Delta \text{ connected, } V_{ph} = V_{line}]$$

i.e. $$Y_{ab} = (0.625 + j0)\text{S} = G_{ab} + jB_{ab}.$$

For phase *b-c*, $$Y_{bc} = \frac{(140 - j86.74)\times 10^3}{400^2}$$

[$\because$ 140 kW at 0.85 p.f. lag means 140/0.85 sin 31.78, i.e. 86.74 kVAR].

$\therefore$ $$Y_{bc} = (0.875 - j0.542)\text{S} = G_{bc} + jB_{bc}$$

For phase *c-a*, $$Y_{ca} = \frac{(140 + j86.74)\times 10^3}{400^2} = (0.875 + j0.542)\text{S} = G_{ca} + jB_{ca}$$

[Here, the load is leading].

Using equation (15.20), we can now write

$$B_{ab(c)} = -B_{ab} + \frac{(G_{ca} - G_{bc})}{\sqrt{3}} = -\,0 + (0.875 - 0.875) = 0 \text{ S}$$

$$B_{bc(c)} = -B_{bc} + \frac{(G_{ab} - G_{ca})}{\sqrt{3}} = -\,(-0.542) + (0.625 - 0.875) = 0.292 \text{ S}$$

$$B_{ca(c)} = -B_{ca} + \frac{(G_{bc} - G_{ab})}{\sqrt{3}} = -\,0.542 + (0.875 - 0.625) = -0.292 \text{ S}$$

Here suffix (c) indicates the compensator susceptances.

Resultant $G = G_{ab} + G_{bc} + G_{ca} = 0.625 + 0.875 + 0.875 = 2.375$ S (Ans.)

$\therefore$ $$I_{ph} = \left(\frac{400}{\sqrt{3}}\right) \times 2.735 = \left(\frac{I_L}{\sqrt{3}}\right) = 548.498 \text{ A} \quad \text{(Ans.)}$$

15.3 TRANSMISSION LINE COMPENSATION

The rapid growth of electrical energy consumption, combined with the demand for cost-effective energy and widespread transmission-distribution network, has led to the development of generation sites remotely located from load centres. The generation of power at remote site necessitates the use of transmission lines to connect generation units to load centres. Also, to enhance system reliability, lines are erected to connect several load centres with generating sources, and to have enough redundancy, several points of interconnections have been established within the power network. This has led to the evolution of a complex power system network with interconnected long distance transmission lines. Transmission voltages have also gradually increased, and to keep pace with economics of power transmission, lines are needed to be controlled for power and voltage at different points/nodes. Until recently, active and reactive power control in ac transmission lines was exercised by carefully adjusting transmission line impedances, as well as regulating voltages by generator excitation adjustment and by using tap changing transformers. At times, series and shunt capacitor or reactors were employed to effectively change line impedance. Bulk ac power transmission has two fundamental requirements synchronism and voltage profile.

Synchronism

The power transmitted between two synchronous machines can be slowly changed upto a certain level known as *steady state stability limit*. Beyond this level, electrical power transfer is not possible between these two machines as they will lose synchronism. The steady state stability limit can be modified to some extent by excitation control, by adjusting the line impedance between the machines, and by controlling the reactive power flow in the line. In power system operation, there may be specific minor disturbances during power transfer and this is referred to as *dynamic stability* (or small disturbance stability). Transient stability is also responsible for maintaining the system synchronism and is concerned with the ability to recover normal operation following specified major disturbances (e.g. faults, sudden load rejection, etc.). Researches reveal that reactive power control is important for maintaining dynamic stability. Transient stability governs the appropriate margin that would be required such that synchronism is maintained following disturbances.

Voltage profile

It is evident that a correct level of voltage is to be maintained within narrow limits at all nodes in the network. Undervoltage degrades the performance of loads and decreases the efficiency of transmission. It also causes flow of more current through lines when the load bus has predominantly constant impedance loads. Overvoltage is dangerous from the point of view of insulation failure of different components within the power system network. It may also cause saturation to reactors and transformers and may lead to flashover on insulators. Usually, voltage changes are associated with or initiated by load changes, and particularly by the change in reactive components of currents flowing. When the line length is high, the voltage control becomes harder and may require reactive power compensation. As the line length increases, at light load periods the line and receiving end bus face overvoltages due to increased capacitive charging by line charging capacitances, while during heavy loading condition the line and receiving end bus face undue voltage depressions due to reactive power deficit.

Different techniques are used for controlling the voltage in the line. Generator excitation control as well as switching of capacitors and/or inductors are enough for cyclic load variation. However, to control dynamic variation of voltages in the line as well as in the load bus due to variety of reasons as well as to control the system stability in a better way, controlled reactive compensating equipment operating over time scales are needed. Some of the compensator devices may serve as a reactive power compensators while a few may act as both real and reactive power compensators. Broadly, we may classify the compensators in the following two groups.

1. Passive compensators
2. Active compensators.

Whatever be the type of compensator, an ideal reactive power compensator should have the following characteristics:

(i) It should produce a substantially flat voltage profile at any point of power transmission line and bus.
(ii) It should be capable of increasing steady state stability limit and dynamic stability.
(iii) It should improve voltage stability and voltage security.
(iv) It must be capable to supply the reactive power requirements of the system in an economical way.

15.4 PASSIVE COMPENSATORS

Shunt inductors and *capacitors* as well as *series capacitors* are *passive compensators*. These elements either absorb or generate reactive power and they constitute passive control. Passive compensators basically modify inductance and capacitance of the line and apart from switching, they are uncontrolled and incapable of continuous variation. Shunt capacitors generate capacitive reactive power and boosts receiving end voltage, thereby reducing the reactive demand of the bus on the line. They also improve the power transfer limit of the line. On the other hand, shunt inductors control the steady state overvoltage in the line during light load operation by absorbing reactive power at the node, where it is connected. In some cases these reactors are sectionalized and connected to the line at different points. Series capacitors reduce the line inductive reactance and thereby reduce the reactive burden of the line. It also enhances the real power transfer through the line and increases the steady state stability limit. During light load operation shunt capacitors need to be switched off and shunt inductors are needed to be switched on, while during heavy loading shunt capacitors are switched on and shunt reactors are switched off. The application of shunt capacitors needs to be done carefully as it may lead to overvoltage across the breaker contacts following its switching off and capacitors will retain voltages for significant period of time, and they need to be discharged through proper discharging circuit. Also, it may lead to high frequency resonance in the system and to harmonic overvoltage in the bus. Switching of shunt inductors may also create current chopping in breakers and hence transient overvoltage may appear across the breaker contacts.

Series compensation also has a few major drawbacks. It may lead to series resonance in the system causing transient overvoltages. Also, series compensation is vulnerable when a fault occurs in the line after the capacitors and heavy fault current flow through the series capacitor. It may lead to high transient voltages generated across the capacitors causing insulation failure. *Subsynchronous resonance* (SSR) of series capacitor compensated line is another aspect that needs to be carefully studied while applying series compensation in transmission lines.

15.4.1 Static Shunt Reactor

Static shunt reactors are usually used to compensate the voltage rise in the line as well as in the receiving end when a transmission line (medium or long) is lightly loaded. This phenomenon of steady state overvoltage at the receiving end of a highly loaded line is known as *Ferranti effect.* The shunt reactors absorb reactive power from the line and if properly designed, can keep a flat voltage profile in the line even when the line is unloaded. These reactors are usually placed at intervals of 60–100 km in long lines in intermediate substations (Figure 15.8).

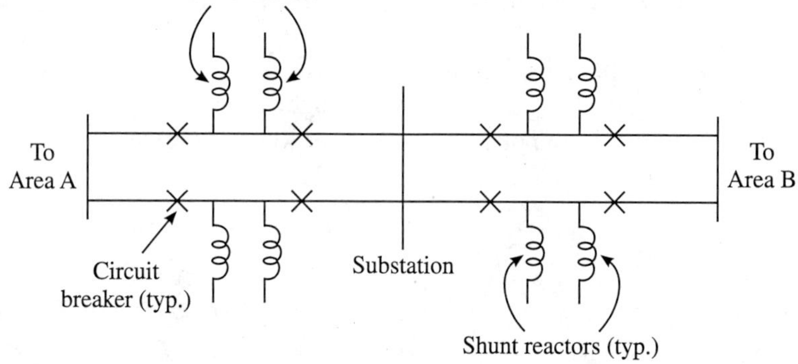

Figure 15.8 Distributed shunt reactors in long ac line.

Shunt reactors are commonly employed for EHV lines more than 150 km long. A shorter line may need a shunt reactor if supplied from a weak capacity power system, i.e. from a source of low short circuit capacity. A shunt reactor of sufficient size must be permanently connected to the line to compensate for the temporary overvoltages due to switching transients. Additional shunt reactors are required to control the steady state overvoltages in the line during light loading period.

Let us now explore analytically the effects of placement of shunt reactors in the line. First, we draw the simplest equivalent circuit of the shunt reactor compensated unload line (Figure 15.9).

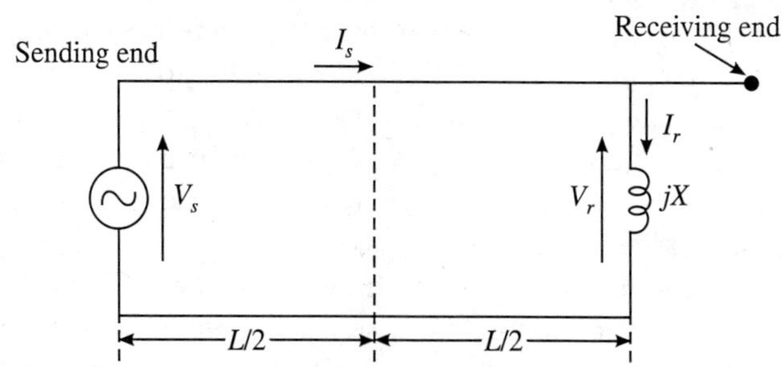

jX = Shunt reactor reactance at receiving end.

Figure 15.9 Equivalent circuit of a shunt reactor compensated line at no load.

At the receiving end,

$$V_r = jXI_r \tag{15.21}$$

From transmission line theory, for a lossless line we have

$$\text{Surge impedance } (Z_0) = \sqrt{\frac{\ell}{c}} \text{ ohm}$$

Electrical line length (θ) = βD radian, D is line length in km.
where ℓ represents line series inductance/length and c the distributed capacitance/length. D is line length.

We have propagation constant (γ) = ($\alpha + j\beta$);

[α being the attenuation constant in neper and β the phase constant in radian]

and *surge impedance loading* (or *natural loading*, P_0) = $\left(\frac{V^2}{Z_0}\right)$ MVA, when V is expressed in kV and Z_0 in ohm.

Also, $$Z_0 = \sqrt{x_\ell x_c} \quad [x_\ell = \omega\ell,\ x_c = 1/\omega c]$$

and $\gamma(=\beta) = \sqrt{\frac{x_\ell}{x_c}}$ [$\because \alpha = 0$ for lossless line and from $\gamma = \alpha + j\beta$, $\beta = j\omega\sqrt{\ell c}$, we have $\beta = \sqrt{\frac{x_\ell}{x_c}}$].

The sending end voltage is given by the following equation (V_r is reference phasor)

$$V_s = V_r \cos \beta D + jZ_0 I_r \sin \beta D \tag{15.22}$$

Using equation (15.21) in equation (15.22), we get

$$V_s = V_r \cos \beta D + \frac{Z_0}{jX} jV_r \sin \beta D = V_r\left(\cos\beta D + \frac{Z_0}{X}\sin\beta D\right) \tag{15.23a}$$

i.e. $$\frac{V_r}{V_s} = \frac{1}{\cos\theta + \frac{Z_0}{X}\sin\theta} \quad [\theta = \beta D,\ \theta \text{ being the line length in radian}] \tag{15.23b}$$

In order to have $|V_r| = |V_s|$, we can write from equation (15.23b)

$$\cos\theta + \frac{Z_0}{X}\sin\theta = 1$$

or, $$X = \frac{Z_0}{1-\cos\theta}\sin\theta \tag{15.24}$$

Equation (15.24) gives the value of shunt reactor reactance required to be connected at the receiving end such that $|V_r| = |V_s|$ even during no load. Since the receiving end is completely inductive, no real power will be transferred from sending end to receiving end.

The reactive power equations can also be obtained as follows:

From equation (15.21), $I_r = \frac{V_r\angle 0}{jX} = \frac{V_r}{j\frac{Z_0}{1-\cos\theta}\sin\theta} = -j\frac{V_r(1-\cos\theta)}{Z_0\sin\theta}$

Reactive power at receiving end, $Q_r = \text{Im}(V_r I_r^*) = \text{Im}\left[V_r \times j\dfrac{V_r(1-\cos\theta)}{Z_0 \sin\theta}\right] = \dfrac{V_r^2(1-\cos\theta)}{Z_0 \sin\theta}$.

With $|Q_s| = |Q_r|$, we have $|V_r| = |V_s|$, and no active power is being transferred. Hence, the angle between V_r and V_s is zero ($V_r = V_s$). Thus the reactive power at sending is given by

$$Q_s = -Q_r = -\frac{V_s^2(1-\cos\theta)}{Z_0 \sin\theta} \tag{15.25}$$

The charging power is divided equally between two halves of the line (Figure 15.9) and the voltage profile is flat at the mid-point as well as symmetrical about the mid-point. The charging power is negative at the left and positive at the right of the mid-point (Figure 15.10).

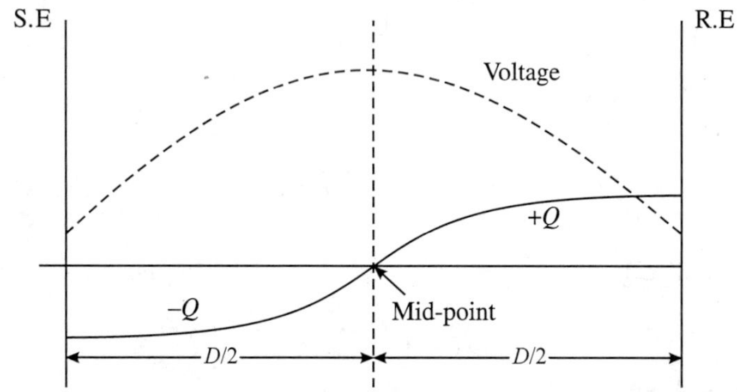

Figure 15.10 Voltage and reactive power profiles for the long line.

The maximum voltage occurs at the mid-point of uncompensated line and is given by the equation

$$V_m = V_r\left[\cos\frac{\theta}{2} + Z_0 \sin\frac{\theta}{2}\right] = \frac{V_s}{\cos\dfrac{\theta}{2}} \tag{15.26}$$

The above analysis can be generalized for a line having n sections and $(n - 1)$ reactors for compensating Ferranti effect. The reactors are assumed to be at equal intervals. To have a flat voltage profile, equation (15.24) can be generalized as

$$X = \frac{Z_0}{2}\left[\frac{\sin\left(\dfrac{\theta}{n}\right)}{1-\cos\left(\dfrac{\theta}{n}\right)}\right] \tag{15.27}$$

(The ideal condition is achieved by connecting shunt reactors of half the reactance given by equation (15.24) at every junction between two sections as shown in Figure 15.11. The shunt reactors at the ends of the line are each of twice reactance of the intermediate ones. The shunt reactors may be directly connected to the bus or may be connected through the tertiary winding of transformer at the two ends of the line).

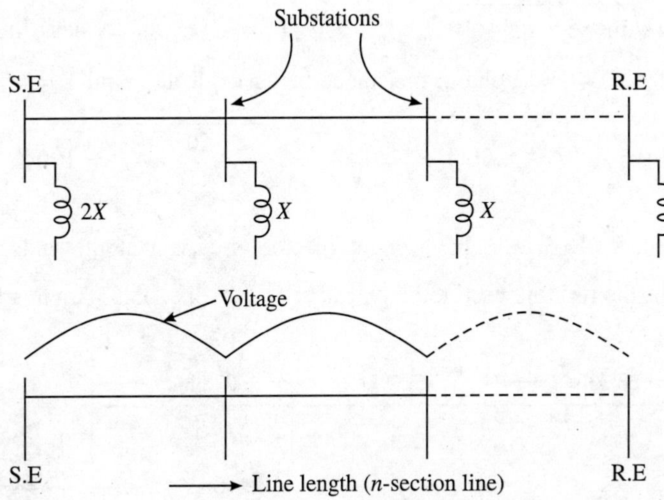

Figure 15.11 Reactor placement scheme and voltage profile in *n*-section transmission line.

Shunt reactors are similar in construction to the transformer but have only one winding per phase on the core with air gaps and are oil immersed. They may be either single phase or three phase type in construction.

EXAMPLE 15.4 Four inductive reactances are placed at equal intervals in a three phase, 400 kV, 100 km symmetrical power transmission line operating at 50 Hz to control Ferranti effect such that the no load voltage profile is flat. The line interconnects two grids A (sending end) and B (receiving end) at both sides of the line and out of four reactors, one is at the sending end while the other is at the receiving end. Assuming the line to be lossless, and the grids do not contribute reactive power to the line, find the values of the inductive reactances of the reactors. Assume the series inductive reactance of the line to be 50 Ω/phase while shunt capacitive susceptance is 5×10^3 μS/phase. What is the maximum voltage in the compensated line at no load and where does it occur?

Solution: Using equation (15.27), for $|V_r| = |V_s|$

$$X_{Sh} = \frac{Z_0}{2}\left[\frac{\sin\left(\dfrac{\theta}{n}\right)}{1-\cos\left(\dfrac{\theta}{n}\right)}\right], \text{ } (X_{Sh} \text{ being the shunt inductor reactance}).$$

Static shunt capacitors are used for compensating the drop in voltage at the load bus due to heavy loading condition. These capacitors are connected in shunt with the load bus and supply capacitive reactive power (Q_c) so that the load inductive power (Q_L) is compensated. The reactive power (Q) at the load bus is then given by

$$Q = Q_L - Q_C$$

Load bus p.f. is also improved with shunt capacitor at load bus.

Here $\theta = \sqrt{ZY} = \sqrt{50 \times 5 \times 10^3 \times 10^{-6}}$ [$\because$ $\theta = \gamma D = \sqrt{yz}$; for lossless line, $\theta = \beta D$]

[$\because$ Z (= line impedance) = 50 Ω/phase, resistance being neglected, and $Y = 5 \times 10^3$ mS/phase]

i.e., $$\theta = 0.5 \text{ rad } (= 28.66°),\ Z_0 = \sqrt{\frac{Z}{Y}} = \sqrt{\frac{50}{5 \times 10^3 \times 10^{-6}}} = 100\ \Omega.$$

As explained in Section 15.4.1, to have shunt reactor of ($2X_{Sh}$) at the ends and (X_{Sh} at the intermediate sections, the reactors need to be placed at $\left(\frac{100}{3}\right)$, i.e. 33.33 km apart (Figure E15.1).

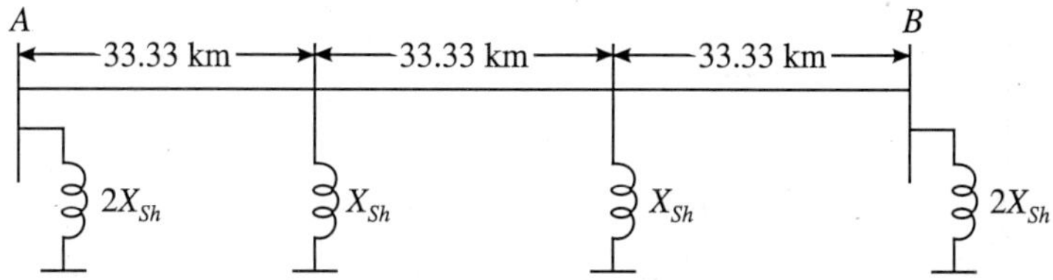

Figure E15.1 Four shunt inductors placed on a 100 km line.

$\therefore$ From equation (15.27), $$X = \frac{100}{2}\left[\frac{\sin\left(\frac{28.66°}{3}\right)}{1-\cos\left(\frac{28.66°}{3}\right)}\right] \cong 598.36\ \Omega \quad \text{(Ans.)}$$

Thus, the intermediate reactors will be 598.36 Ω each while the end reactors are (2 × 598.36), i.e. 1196.72 Ω each.

Maximum voltage is at the mid-point of the line, i.e. at 50 km length from each end. Using equation (15.26), this mid-point voltage is

$$V_m = \frac{V_s}{\cos\left(\frac{\theta}{2n}\right)}, \text{ for } n \text{ sections with } n = 3,$$

i.e. $$V_m = \frac{400}{\cos\left(\frac{28.66°}{2 \times 3}\right)} = 401.4 \text{ kV (L-L), i.e. 1.0035 p.u.} \quad \text{(Ans.)}$$

EXAMPLE 15.5 In the problem of Example 15.4, what is the network reactive power (Q_{Sh}) of the four inductive reactors used to control the no load voltage?

Solution: System voltage = 400 kV (L-L)

Inductive reactance (X_{Sh}) = 598.36 Ω/ph

Since all these reactances are in parallel in the line, being connected at intermediate sections and at ends,

$$\frac{1}{X_{Sh_{eqv}}} = \frac{1}{2X_{Sh}} + \frac{1}{X_{Sh}} + \frac{1}{X_{Sh}} + \frac{1}{2X_{Sh}} = \frac{3}{X_{Sh}}$$

$\therefore$ $$X_{Sh_{eqv}} = \frac{X_{Sh}}{3} = \frac{598.36}{3} = 198.79\ \Omega$$

Here $$Q_{Sh} = \frac{V^2}{X_{Sh_{eqv}}} = \frac{3 \times \left(\frac{400}{\sqrt{3}}\right)^2}{198.76} = 80.50 \text{ MVAR}$$

15.4.2 Uniformly Distributed Shunt Compensation

For an uncompensated lossless line we can write for surge impedance (Z_0),

$$Z_0 = \sqrt{\frac{Z}{Y}} = \sqrt{\frac{X_L}{\frac{1}{X_C}}} = \sqrt{\frac{X_L}{B_c}}$$

$\theta = \beta D$, attenuation constant α being zero, $\gamma = \alpha + j\beta = j\beta$ and θ = electrical line length ($=\gamma D$) becomes βD. D is the line length in km while θ is the line length in radian or in degree.

From transmission line theory,

$$\beta = \omega\sqrt{\ell c} = \frac{\sqrt{X_L B_c}}{D}, \text{ for entire line.}$$

Let us assume that the line has uniformly distributed shunt capacitive susceptance or shunt inductive susceptance for compensation and this is expressed as b_{Sh} per unit length. If the susceptance is b_c, we can have for the compensated line the network shunt susceptance as

$$b_{c(comp)} = b_c - b_{Sh} = b_c(1 - k_{Sh}) = b'_c \text{ (say)} \tag{15.28}$$

where k_{Sh} is the *degree of shunt compensation* and is expressed as $k_{Sh} = \left(\frac{b_{Sh}}{b_c}\right)$.

[k_{Sh} *is positive if the shunt compensation is inductive and is negative if the shunt compensation is capacitive.*]

$\therefore$ $Z_{0(comp)}$, the surge impedance of the uniformly distributed uncompensated line, is given by

$$Z_{0(comp)} = \sqrt{\frac{\frac{X_L}{D}}{\frac{B'_c}{D}}} = \sqrt{\frac{x_L}{b'_c}} = \frac{Z_0}{\sqrt{1-k_{Sh}}} \left[\because Z_0 = \sqrt{\frac{x_\ell}{b_c}}\right] \tag{15.29a}$$

and $$\beta_{(comp)} = \beta\sqrt{1-k_{Sh}} \tag{15.29b}$$

Also we have

$$\theta_{(comp)} = \theta\sqrt{1-k_{Sh}} \tag{15.29c}$$

and

$$P_{(comp)} = \frac{V_s V_r}{Z_{0(comp)} \sin\theta_{(comp)}} \sin\delta \tag{15.29d}$$

Hence for shunt compensation,

$$\frac{Z_{0(comp)}}{Z_0} = \sqrt{\frac{1}{1-k_{Sh}}} \tag{15.30a}$$

$$\frac{\beta_{(comp)}}{\beta} = \sqrt{1-k_{Sh}} \tag{15.30b}$$

$$\frac{\theta_{(comp)}}{\theta} = \sqrt{1-k_{Sh}} \tag{15.30c}$$

$$\frac{P_{(comp)}}{P_0} = \frac{\dfrac{V_s V_r}{Z_{0(comp)} \sin\theta_{(comp)}}\sin\delta}{\dfrac{V_s V_r}{Z_0 \sin\theta}\sin\delta} = \frac{Z_0 \sin\theta}{Z_{0(comp)} \sin\theta_{(comp)}} \tag{15.30d}$$

and

$$\frac{P_{0(comp)}}{P_0} = \sqrt{1-k_{Sh}} \tag{15.30e}$$

Thus, we observe that shunt inductive compensation increases Z_0 and reduces power transfer capacity of the line, virtual length of the line, surge impedance loading and β. On the other hand, shunt capacitive compensation decreases Z_0 and increases power transfer capacity (assuming reduction of Z_0 due to shunt compensation is more pronounced than increase in sin θ due to capacitive compensation), virtual length of the line, surge impedance loading and β.

EXAMPLE 15.6 A 300 km lossless three phase, 50 Hz, 500 kV line transfers power of 1000 MW from side A to side B in a grid while the voltages at both ends are maintained at 1 p.u. Assuming first the line is uncompensated, find the SIL, power angle and mid-point voltage. How these values are changed due to distributed shunt compensation (capacitive) such that the voltage profiles remain flat at both ends, as well as at the centre of the line even if the power demand increases by 25% of the SIL value? Assume that the line has series inductive reactance of 0.35 Ω/km and shunt capacitive susceptance of 5 microsiemens/km.

Solution: From the given data for the line,

$$Z_0 = \sqrt{\frac{X_L}{B_c}} = \sqrt{\frac{0.35 \times 300}{5 \times 10^{-6} \times 300}} = 264.57\ \Omega$$

$$\beta = \omega\sqrt{\ell c} = \frac{\sqrt{X_L B_c}}{D} = \frac{\sqrt{0.35 \times 300 \times 5 \times 10^{-6} \times 300}}{300} = 1.32 \times 10^{-3} \text{ rad/km}$$

$$V_s = V_r = 1.0 = 500 \text{ kV (L-L)}$$

$$\theta = \beta D = 1.32 \times 10^{-3} \times 300 = 0.397 \text{ rad} = 22.76°.$$

$$\therefore \qquad \text{SIL} = \frac{V^2}{Z_0} = \frac{500^2}{264.57} = 945 \text{ MW}.$$

Also,

$$\frac{P}{SIL} = \frac{\sin\delta}{\sin\theta}$$

$$\left[\because\ P = \frac{V_s V_r \sin\delta}{Z_0 \sin\theta};\ V_s = V_r = V,\ SIL = \frac{V^2}{Z_0} \text{ and } X_L = Z_0 \sin\theta\right]$$

or,

$$\frac{1000}{945} = \frac{\sin\delta}{\sin 22.76°}$$

$$\therefore \qquad \sin\delta = 0.41 \text{ giving } \delta = 24.20°.$$

Considering one half the symmetrical line,

$$P = \frac{V_s V_m}{Z_0 \sin\frac{\theta}{2}} \sin\frac{\delta}{2}; \text{ for the half line.}$$

In p.u., $V_s = V_r = 1.0$ and $P_0 = \frac{1}{Z_0}$; thus in p.u. terms we express the above power equation of the half line as

$$P(\text{p.u.}) = \frac{1}{Z_0} \frac{V_m \sin\frac{\delta}{2}}{\sin\frac{\theta}{2}}; \ V_m \text{ and } Z_0 \text{ both are expressed in p.u.}$$

or,

$$P = P_0 \frac{V_m \sin\frac{\delta}{2}}{\sin\frac{\theta}{2}}$$

$$\therefore \qquad V_m = \left(\frac{P}{P_0}\right)\left(\frac{\sin\frac{\theta}{2}}{\sin\frac{\delta}{2}}\right)$$

In our case, $$V_m(\text{p.u.}) = 1.058\ \frac{\sin\dfrac{22.76°}{2}}{\sin\dfrac{24.2°}{2}} = 0.995 \quad [\because \text{ here, } P/P_0 = 1.058]$$

i.e. $$V_m = 497.95 \text{ kV}.$$

Thus, it may be observed that for the given uncompensated line even if the terminal voltages are held at 1.0 p.u., the mid-point voltage sags to 0.995 p.u.

Next, we study the case of mid-point shunt capacitive compensation so that V_m is also 1.0 p.u. alongwith V_s and V_r.

With compensation, we have seen

$$\frac{P_{0(comp)}}{P} = \sqrt{1-k_{Sh}}$$

As per given data, with compensation the power flow increases by 25%.

$$\therefore \qquad \frac{P_{0(comp)}}{P} = 1.25$$

Hence, we can find the value of k_{Sh} so that the voltage profile is flat although using the relation,

$$\sqrt{1-k_{Sh}} = 1.25$$

$$\therefore \qquad k_{Sh} = -0.5625.$$

Next, we find Z_0 and θ for this mid-point compensated line.

$$Z_{0(comp)} = \frac{Z_0}{\sqrt{1-k_{Sh}}} = \frac{264.57}{\sqrt{1+0.5625}} = 211.66\ \Omega.$$

$$\theta_{(comp)} = \theta\sqrt{1-k_{Sh}} = 0.397\sqrt{1+0.5625} = 0.496 = 28.45°$$

With compensation, using the relation $\dfrac{P}{SIL} = \dfrac{Z_0 \sin\delta}{Z_{0(comp)} \sin\theta_{(comp)}}$; we get

$$\sin\delta = \frac{PZ_{0(comp)}\sin\theta_{(comp)}}{Z_0 \times SIL} = \frac{1181.25 \times 211.66 \times \sin 28.45°}{264.57 \times 945} = 0.476$$

[$\because$ 25% increase of power flow from P_0 amounts to 1.25 P_0 = 1181.25 MW]

$$\therefore \qquad \delta = 28.45°.$$

The new *SIL* value is $P_{0(comp)} = P_0\sqrt{1-k_{Sh}} = 945\sqrt{1+0.5625} = 1181.25$ MW

The mid-point voltage is given by

$$V_m = \frac{P}{P_0} \times \frac{Z_{0(comp)}}{Z_0} \times \frac{\sin\frac{\theta_{(comp)}}{2}}{\sin\frac{\delta_{(new)}}{2}} = 1.25 \times \frac{211.66}{264.57} \times \frac{\sin\frac{28.45°}{2}}{\sin\frac{28.45°}{2}}$$

$= 1.00002$ p.u. (as expected).

Thus, without compensation, $SIL = 945$ MW, $\delta = 24.2°$ and $V_m = 0.995$ p.u. With mid-point shunt compensation, $SIL = 1181.25$ MW, $\delta = 28.44°$ and $V_m = 1.00002$ p.u. **(Ans.)**

15.4.3 Shunt Compensation at Middle of the Line Using Dynamic Compensator

We have seen earlier that voltage at the mid-point of a power transmission line needs attention. In this section, an analytical treatment is presented to model the concept of mid-point compensation. We assume a lossless line having π equivalent sections connected to two subgrids at both sides of the line (referred to as sending end and receiving end) and compensated by a passive shunt compensator at the mid-point of the line. The series reactance of the line is expressed by X_L while the shunt susceptance by B_c. The mid-point shunt compensator has susceptance B (Figure 15.12).

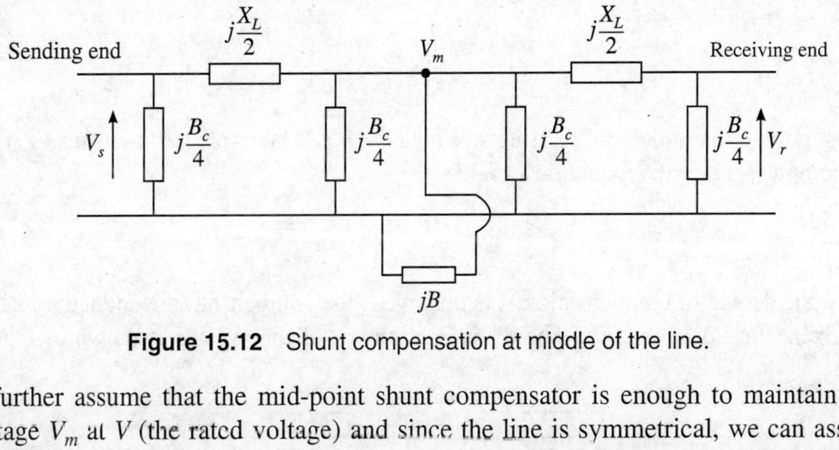

Figure 15.12 Shunt compensation at middle of the line.

We further assume that the mid-point shunt compensator is enough to maintain the mid-point voltage V_m at V (the rated voltage) and since the line is symmetrical, we can assume that in steady state the line can be sectioned into two independent halves.

The power transmission is given by

$$P = \frac{V^2}{\left(\frac{X_L}{2}\right)} \sin\frac{\delta}{2} = \frac{2V^2}{X_L} \sin\frac{\delta}{2} \tag{15.31}$$

The steady state power limit is $\left(\frac{2V^2}{X_L}\right)$, i.e. twice that of uncompensated line. It is interesting to note here that $P = P_{max}$ when $\frac{\delta}{2} = \frac{\pi}{2}$, i.e. $\delta = \pi$, i.e. power transfer is maximum at steady state when the total power angle is 180° across the whole line (or 90° across each half of the line). Figure 15.13 represents these power-angle characteristics.

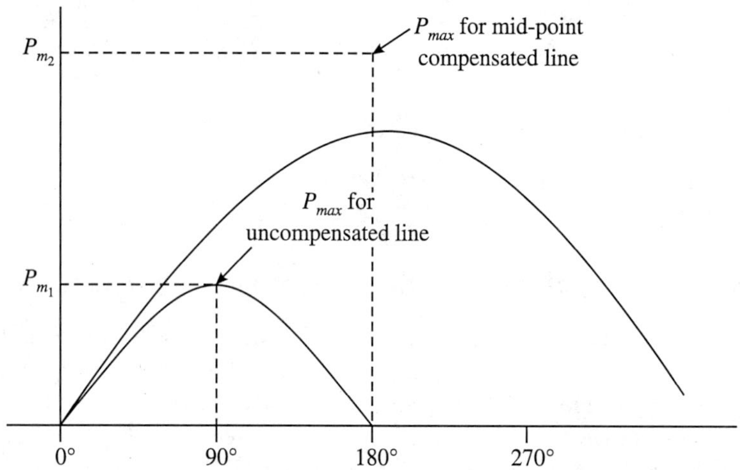

Figure 15.13 Power-angle characteristics of uncompensated and compensated line.

The reactive power requirement Q of the mid-point compensator is given by

$$Q = \frac{4V^2}{X_L}\left(1 - \cos\frac{\delta}{2}\right) \tag{15.32}$$

Let us assume that the compensating susceptance B can be expressed in terms of susceptance of the line and is given by equation (15.33)

$$jB = -k\frac{jB_c}{2} \tag{15.33}$$

where, k is the degree of compensation; k is positive if the compensation is inductive, k is negative if the compensation is capacitive. The equivalent circuit of the system is shown in Figure 15.14.

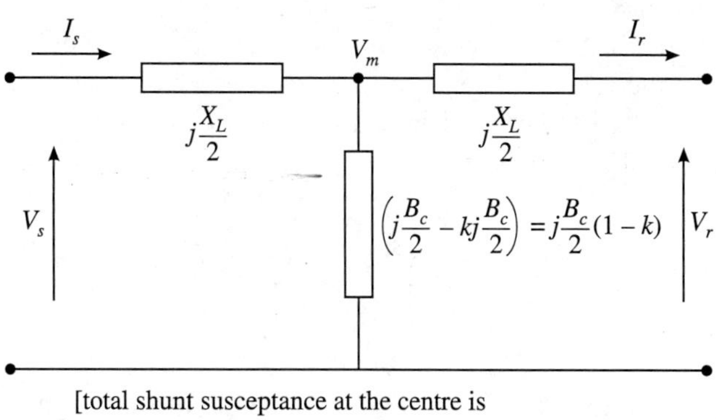

[total shunt susceptance at the centre is

$\left(2\times j\frac{B_c}{4} + jB\right) = 2\times j\frac{B_c}{4} - k\times j\frac{B_c}{2} = j\frac{B_c}{2}(1-k)$.]

Figure 15.14 Equivalent circuit of the line with mid-point compensation.

Let us now segregate the line into two equal halves as shown in Figure 15.15. The central shunt susceptance is split into two equal parallel susceptances.

In Figure 15.14, we have

$$I_s = \frac{V_s}{j\frac{X_L}{2} + \frac{1}{j\frac{B_c}{4}(1-k)}}$$

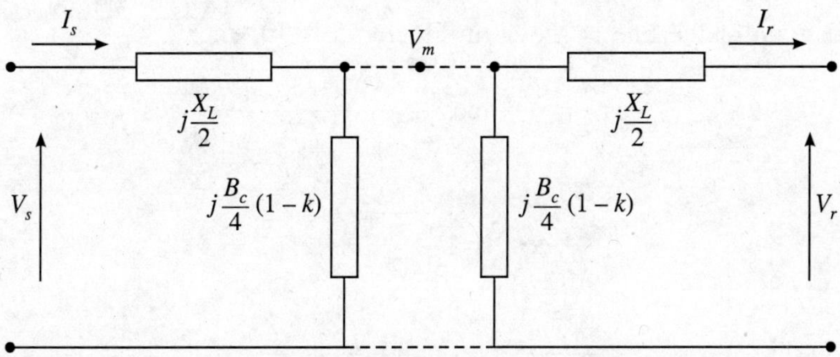

Figure 15.15 Segregation of the equivalent circuit shown in Figure 15.14.

The Thevenin's equivalent voltage at the sending end is given by

$$v_s = I_s \times \frac{1}{j\frac{B_c}{4}(1-k)} = \frac{V_s}{j\frac{X_L}{2} + \frac{1}{j\frac{B_c}{4}(1-k)}}\left(\frac{1}{j\frac{B_c}{4}(1-k)}\right)$$

$$= \frac{\frac{1}{j\frac{B_c}{4}(1-k)}}{j\frac{X_L}{2} + \frac{1}{j\frac{B_c}{4}(1-k)}} V_s = \frac{\frac{1}{j\frac{B_c}{4}(1-k)}}{\frac{-\frac{X_L}{2}\times\frac{B_c}{4}(1-k)+1}{j\frac{B_c}{4}(1-k)}} V_s$$

$$= \frac{V_s}{1-\frac{X_L}{2}\times\frac{B_c}{4}(1-k)} = \frac{V_s}{1-\alpha} \quad \text{[see Figure 15.16(a)]}$$

where

$$\alpha = \frac{X_L}{2}\times\frac{B_c}{4}(1-k) \tag{15.34}$$

Thevenin's equivalent impedance (Z or Z_{Th}) is given by

$$Z = \frac{j\frac{X_L}{2} \times \frac{1}{j\frac{B_c}{4}(1-k)}}{j\frac{X_L}{2} + \frac{1}{j\frac{B_c}{4}(1-k)}} = \frac{j\frac{X_L}{2}}{1-\alpha} \ (= Z_{Th}) \tag{15.35}$$

The phasor diagram of the line is shown in Figure 15.16(b) where,

$$V_m = v_s \cos\frac{\delta}{2} = \frac{V_s}{1-\alpha}\cos\frac{\delta}{2} = \frac{V_s}{1-\frac{X_L}{2}\frac{B_c}{4}(1-k)}\cos\frac{\delta}{2}$$

or,
$$\frac{V_s}{V_m}\cos\frac{\delta}{2} = 1 - \frac{X_L}{2}\frac{B_c}{4}(1-k)$$

or,
$$\frac{X_L}{2}\frac{B_c}{4}(1-k) = 1 - \frac{V_s}{V_m}\cos\frac{\delta}{2}$$

or,
$$\frac{B_c}{2}(1-k) = \frac{4}{X_L}\left(1 - \frac{V_s}{V_m}\cos\frac{\delta}{2}\right)$$

or,
$$-k\frac{B_c}{2} = \frac{4}{X_L}\left(1 - \frac{V_s}{V_m}\cos\frac{\delta}{2}\right) - \frac{B_c}{2}$$

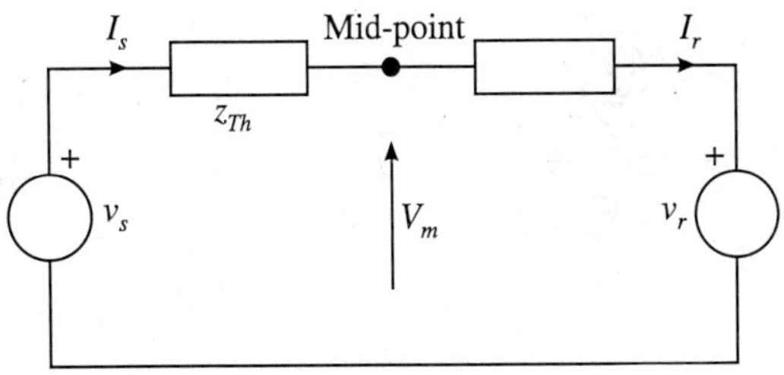

Figure 15.16(a) Thevenins equivalent circuit of mid-point compensated line.

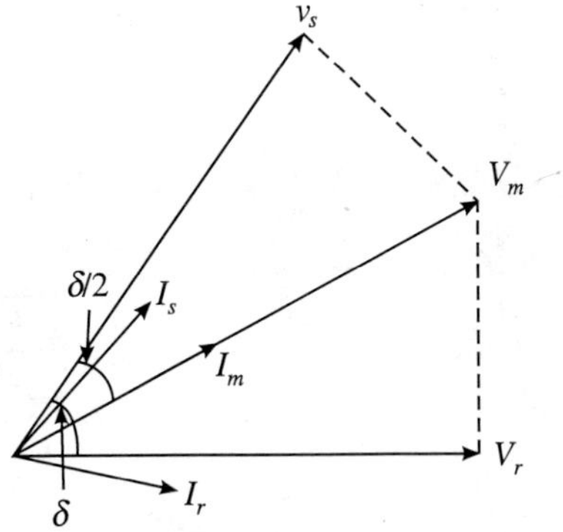

Figure 15.16(b) Phasor diagram of the compensated line.

∴
$$B = -\frac{4}{X_L}\left(1 - \frac{V_s}{V_m}\cos\frac{\delta}{2}\right) + \frac{B_c}{2}\text{; for capacitive compensator} \tag{15.36a}$$

and
$$B = \frac{4}{X_L}\left(1 - \frac{V_s}{V_m}\cos\frac{\delta}{2}\right) - \frac{B_c}{2}\text{; for inductive compensator.} \tag{15.36b}$$

It is evident from expressions (15.36a) and (15.36b) that B must be varied with power angle δ so that (V_m/V_s) is constant. This has been termed as *dynamic compensation* at mid-point of the line.

In the above derivation, the mid-point voltage (V_m) is obtained as

$$V_m = \frac{V_s}{1-\alpha}\cos\frac{\delta}{2}\left[\text{where, } \alpha = \frac{X_L}{2}\frac{B_c}{4}(1-k)\right] \tag{15.37}$$

If $\alpha < 1$, the mid-point voltage would increase by a factor $\left(\frac{1}{1-\alpha}\right)$. This factor of increase would offset the voltage drop in the line reactance and hence can compensate the voltage drop due to load. If $\left[\frac{\cos(\delta/2)}{1-\alpha}\right]$ is greater than 1, the mid-point voltage would exceed the terminal voltages even when the line is heavily loaded.

If $k = 1$ in the expression of V_m, α becomes zero; the shunt capacitance of the line is cancelled along the line. The line is then practically reduced to its series impedance. If the terminal voltages are kept at 1 p.u., then we have,

$$V_s = V_r = V\sqrt{1+\left(\frac{X_L}{2}\right)^2\left(\frac{P}{P_0}\right)^2} \tag{15.38}$$

where, P_0 (= *SIL* of the line) and P are the amount of power transfer; $V = V_m$.

EXAMPLE 15.7 A long EHV (500 kV, L-L) transmission line is heavily loaded at its receiving end, the load being of variable magnitude. The line series inductive reactance is 50 Ω while the shunt capacitive susceptance is 5×10^3 μ*S* on per phase basis. The line length is 180 km. It is compensated at the midpoint by a dynamic capacitive compensator such that the terminal voltage at the receiving end is 1.0 p.u. for all loading conditions. Obtain the value of the reactance of this mid-point compensator when the load is 1000 MW.

Solution: Using equation (15.31), we can write

$$P = \frac{2V^2}{X_L}\sin\frac{\delta}{2}$$

or,
$$1000 = \frac{2\times 500^2}{50}\sin\frac{\delta}{2}$$

or,
$$\sin\frac{\delta}{2} = 0.1 \quad \therefore\ \delta = 11.48°$$

The susceptance of the mid-point compensator can be obtained for this δ angle using equation

$$B = -\frac{4}{X_L}\left(1-\frac{V_s}{V_m}\cos\frac{\delta}{2}\right)+\frac{B_c}{2},$$ derived in the text [equation (15.36a)].

Here, $$B = -\frac{4}{50}\left(1 - \cos\frac{11.48°}{2}\right) + \frac{5 \times 10^3 \times 10^{-6}}{2}, \text{ for } V_s = V_m$$

i.e. $$B = -\,0.000401 + 0.0025 \cong 0.0021 \text{ S.}$$

∴ Capacitive reactance of the compensator (X_{comp}) is obtained as

$$X_{comp} = \frac{1}{0.0021} = 476.44\ \Omega/\text{ph.} \qquad \text{(Ans.)}$$

15.4.4 Series Capacitor Compensation

A *series capacitor compensator* not only reduces the line reactance but also increases the power transmission remarkably as well as the steady state stability limit. It virtually increases the surge impedance loading. However, shunt reactors may also be needed to control voltage during operation of the line at higher load. In long distance power transmission, series capacitors are beneficial provided proper care is taken to eliminate SSR (*sub synchronous resonance*) problem and to protect the series capacitors against puncture when encountering through-faults. Usually, the degree of series compensation is 40–80% as over compensation may give rise to high steady state voltages and very low damping in the power system. Series capacitors are placed in a lumped form in all the three phases at mid-point of the line.

In order to model the series compensated line, we assume lossless symmetrical line with a midpoint series compensator, with equal shunt reactors connected at both the sides of the line. The mid-point capacitor is equally divided to maintain two equal halves of the line (Figure 15.17).

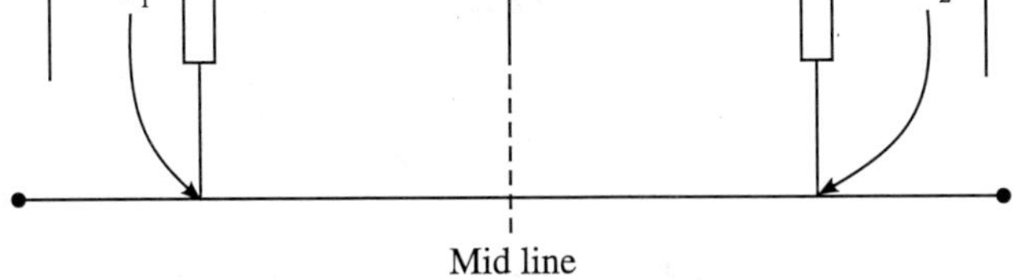

Figure 15.17 Series capacitor compensated symmetrical line.

The voltage and current relations at the sending end half of the line shown in Figure 15.17, are given by

$$V_1(=V_s) = V_m \cos\frac{\theta}{2} + jZ_0 I_m \sin\frac{\theta}{2} \tag{15.39a}$$

$$I_s = j\frac{V_m}{Z_0}\sin\frac{\theta}{2} + I_m \cos\frac{\theta}{2} \tag{15.39b}$$

Also we have, $V_c(\text{drop in series capacitor}) = V_1 - V_2 = -jI_mX_c$ (15.40)

$$V_m = V_1 - \frac{V_c}{2} = V_2 + \frac{V_c}{2} \tag{15.41}$$

and

$$\left.\begin{aligned} I_s &= I_m + \frac{V_1}{jX_{Sh}} = I_m - j\frac{V_1}{X_{Sh}} \\ I_m &= \frac{V_2}{jX_{Sh}} + I_r = I_r - j\frac{V_2}{X_{Sh}} \end{aligned}\right\} \tag{15.42}$$

If we assume X_c to be enough to keep the mid-point voltage of the line at nominal value so that $V_s(=V_1) = V_m = V_r(=V_2) = V$, we can write

$$P = \frac{V^2}{(X_L - X_c)} \sin\delta \text{ (for the whole line)}$$

$$= \frac{V^2}{X_L\left(1 - \dfrac{X_c}{X_L}\right)} \sin\delta = \frac{V^2}{X_L(1 - k_{Se})} \sin\delta \tag{15.43}$$

where k_{Se} represents the *degree of series compensation* and is expressed as

$$k_{Se} = \frac{X_c}{X_L} = \frac{X_c}{\omega \times \text{line length} \times \text{inductance of line per unit length}}.$$

At no load, the shunt reactors would compensate Ferranti effect. As derived earlier (equation 15.24), the shunt reactance connected at each side of the line would be

$$X_{Sh} = Z_0 \frac{\sin\left(\dfrac{\theta}{2}\right)}{1 - \cos\left(\dfrac{\theta}{2}\right)}$$

where,

$$Z_0 = \sqrt{\frac{X_L - X_c}{X_{c_0}}} \tag{15.44}$$

Here, X_L = inductive reactance of the line

X_c = series capacitive reactance of the series capacitor

X_{c_b} = lumped shunt capacitance of the line.

For series capacitor compensation, we can also write

$$Z_{0(comp)} = \sqrt{\frac{X_L'}{B_c}} = Z_0\sqrt{1 - k_{Se}}$$

With series capacitive compensation, both Z_0 and β decrease.

15.4.5 Comparison between Series and Shunt Compensation

Recollect the concept of applying a shunt capacitor at the mid-point of a line. Let us assume, due to this the shunt susceptance is incrementally increased by ΔB_c (Figure 15.18). The power transfer, in terms of mid-point voltage on the line, can be expressed as

$$P = \frac{V_s V_m}{\left(\frac{X_L}{2}\right)} \sin\frac{\delta}{2} = \frac{2V_s V_m}{X_L} \sin\frac{\delta}{2}$$

$$\therefore \qquad \Delta P = \frac{2V_s}{X_L} \sin\frac{\delta}{2} \Delta V_m \qquad (15.45)$$

Since the current I_c taken by the mid-point compensator is given by $I_c = V_m B_c$, we can write

$$\Delta I_c = V_m \Delta B_c \qquad (15.46)$$

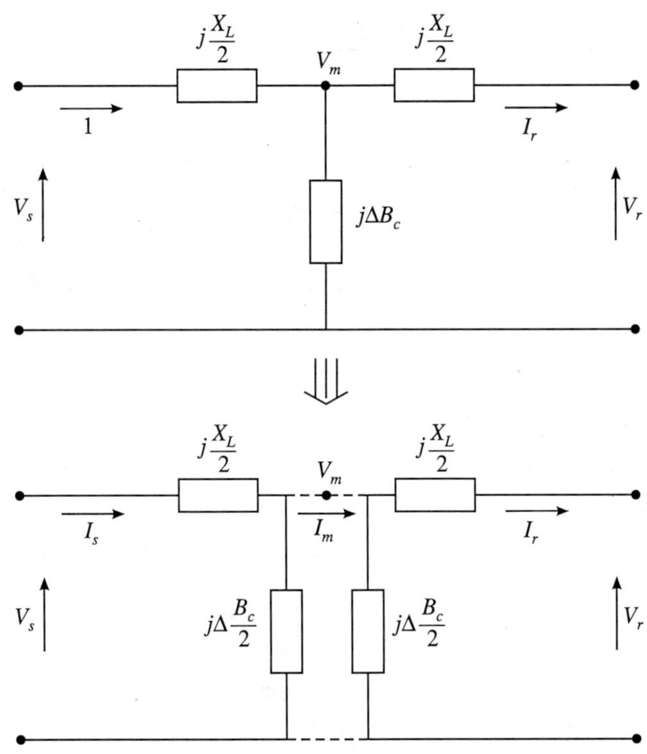

Figure 15.18 Equivalent circuit of mid-point shunt compensation.

Thus, the receiving end current is obtained as

$$I_r = I_m - V_m\left(j\frac{\Delta B_c}{2}\right) \qquad (15.47)$$

The mid-point voltage V_m can be expressed as

$$V_m = V_r + jI_r \frac{X_L}{2} = V_r + j\frac{X_L}{2}\left(I_m - jV_m \frac{\Delta B_c}{2}\right)$$

$$= V_r + j\frac{X_L}{2} I_m + \frac{V_m \Delta B_c}{2}\left(\frac{X_L}{2}\right) \tag{15.48}$$

From equation (15.48), we can write

$$\Delta V_m = V_m \frac{\Delta B_c}{2}\left(\frac{X_L}{2}\right) \tag{15.49a}$$

$$= \frac{\Delta I_c}{2} \cdot \frac{X_L}{2} \text{ [Using equation (15.46)] } = \frac{\Delta I_c X_L}{4} \tag{15.49b}$$

Substituting the value of ΔV_m from equation (15.49a) in equation (15.45), we get

$$\Delta P = \frac{V_s V_m}{2} \sin\frac{\delta}{2} \Delta B_c \tag{15.50}$$

However, the reactive power charged by the mid-point capacitor is given as

$$\Delta Q_{Sh} = V_m^2 \Delta B_c \tag{15.51}$$

From equations (15.50) and (15.51), we can write

$$\frac{\Delta P}{\Delta Q_{Sh}} = \frac{\frac{V_s V_m}{2} \sin\frac{\delta}{2} \Delta B_c}{V_m^2 \Delta B_c} = \frac{V_s}{2} \sin\frac{\delta}{2} \cdot \frac{1}{V_m}$$

$$= \frac{V_s}{2} \sin\frac{\delta}{2} \cdot \frac{1}{V_s \cos\frac{\delta}{2}} \quad \left[\text{Assuming } V_m \cong V_s \cos\frac{\delta}{2}\right]$$

$$= \frac{1}{2} \tan\frac{\delta}{2} \tag{15.52}$$

Next, we consider series compensation. The line is series capacitor compensated at the midpoint having the line reactance changed by ΔX_L, causing the line current to change by ΔI_L.
In a compensated line, the power transfer is given by

$$P = \frac{V^2}{X_L^2} \sin\delta = \frac{V^2}{X_L^2} 2\sin\frac{\delta}{2}\cos\frac{\delta}{2}$$

i.e.

$$\Delta P = -\frac{V^2}{X_L} 2\sin\frac{\delta}{2}\cos\frac{\delta}{2} \Delta X_L \tag{15.53}$$

The line being symmetrical, for the half line we can write

$$P' = \frac{V^2}{X_L/2}\sin\frac{\delta}{2} \quad [\text{Assuming } V_m = V_r = V]$$

$$= \frac{2V^2}{X_L}\sin\frac{\delta}{2} = V\cdot\frac{2V}{X_L}\sin\frac{\delta}{2} = V.I$$

$$\therefore \qquad I = \frac{2V}{X_L}\sin\frac{\delta}{2} \qquad (15.54)$$

From equation (15.53), we can write

$$\Delta P = -\left(\frac{4V^2}{X_L^2}\sin^2\frac{\delta}{2}\right)\cos\frac{\delta}{2}\cdot\frac{\Delta X_L}{2\sin\frac{\delta}{2}} = \frac{1}{2\tan\frac{\delta}{2}}(-\Delta X_L I^2) \qquad (15.55)$$

In series capacitor compensated line $Q_{Se} = -I^2\Delta X_L$, where $(-\Delta X_L)$ is the incremental change in the line reactance due to capacitive effect. Thus, from equation (15.55), we have

$$\frac{\Delta P}{\Delta Q_{Se}} = \frac{1}{2\tan\frac{\delta}{2}} \qquad (15.56)$$

Comparing equation (15.52) with equation (15.56), we have

$$\frac{\left(\frac{\Delta P}{\Delta Q_{Sh}}\right)}{\left(\frac{\Delta P}{\Delta Q_{Se}}\right)} = \frac{\left(\frac{1}{2}\tan\frac{\delta}{2}\right)}{\left(\frac{1}{2\tan\frac{\delta}{2}}\right)},$$

for the same change in real power transfer for both shunt and series compensated lines.

$$\therefore \qquad \frac{\Delta Q_{Sh}}{\Delta Q_{Se}} = \tan^2\frac{\delta}{2} \qquad (15.57)$$

Let us assume $\delta = 15°$. This gives, $\frac{\Delta Q_{Sh}}{\Delta Q_{Se}} = 0.0173$, i.e., VAR injection of series compensator is 1.73% of that required of a shunt compensator for the same change in power transfer. Thus, we can conclude that for the same change in real power, the series compensator needs much lesser rating than the shunt compensator in identical situations. Also, the series compensation adjusts the network line reactance and suitably compensates for the entire range of line loading.

Let us now consider a case when both series capacitor compensation and shunt compensation are present. The combined effects are as follows:

$$Z_{0(new)} = Z_0\sqrt{\frac{1-k_{se}}{1-k_{Sh}}} \qquad (15.58)$$

$$\beta_{(new)} = \beta\sqrt{(1-k_{se})(1-k_{Sh})} \tag{15.59}$$

$$\theta_{(new)} = \theta\sqrt{(1-k_{se})(1-k_{Sh})} \tag{15.60}$$

$$P_{0(new)} = P_0\sqrt{\frac{1-k_{Sh}}{1-k_{se}}} \tag{15.61}$$

(If the shunt compensation is inductive, k_{Sh} is positive).

Under the light load condition, a flat voltage profile can be obtained by inductive shunt compensation, with k_{Sh} = 1 (100% inductive compensation), the values of P_0 and θ approach zero and $Z_{0(comp)}$ tends to infinity, resulting in a flat voltage profile even at no load. Under the heavy loading condition, a flat voltage profile can be obtained by adding shunt capacitive compensation or a series compensation. The steady state stability limit also increases with shunt or series capacitive compensation.

Again, it may be noted that the characteristic impedance can be reduced with capacitive shunt compensation, but it is associated with increase of line length (θ). On the other hand, for inductive shunt compensation though θ reduces, it results in increase of Z_0. Only the series capacitor compensator contributes to the reduction of both Z_0 and θ. The power expression in shunt compensation is given by

$$\frac{P_{comp}}{P} = \frac{Z_0 \sin\theta}{Z_{0(comp)} \sin\theta_{(comp)}} \tag{15.62}$$

One cannot predict whether $\left(\frac{P_{comp}}{P}\right)$ would increase with shunt capacitive compensation, as with this though Z_0 reduces, θ increases. Thus, the ratio $\left(\frac{P_{comp}}{P}\right)$ is governed by the product of $Z_{0(comp)}$ and sin $\theta_{(comp)}$. On the other hand, for series capacitor compensation P increases on compensation as both Z_0 and θ reduce with series capacitor compensation and the power is given by

$$P = \frac{V_s V_r}{Z_{0(comp)} \sin\theta_{(comp)}} \sin\delta \tag{15.63}$$

EXAMPLE 15.8 A 400 kV, 3 phase, 50 Hz, power line has natural impedance of 300 Ω while the electrical length is 0.5 radian. Obtain expression for $\left(\frac{P}{P_0}\right)$ and V_m (in p.u.) in terms of power angle when the line is series capacitor compensated, the degree of compensation being 40%.

Solution: $k_{Se} = 0.4$

$\therefore$

$$Z_{0(comp)} = Z_0\sqrt{(1-k_{se})} = 300\sqrt{(1-0.4)} = 232.38\ \Omega$$

$$\theta_{(comp)} = \theta\sqrt{(1-k_{se})} = 0.5\sqrt{(1-0.4)} = 0.387\text{ rad} = 22.2°$$

Also,

$$\frac{P}{P_0} = \frac{Z_0 \sin\delta}{Z_{0(comp)}\sin\theta_{(comp)}} = \frac{300\sin\delta}{232.38\sin 22.20°} = 3.416\sin\delta$$

Thus,

$$\frac{P}{P_0} = 3.416\sin\delta \qquad \text{(Ans.)}$$

and

$$V_{m(comp)} = \frac{P}{P_0}\times\frac{Z_{0(comp)}}{Z_0}\times\frac{\sin\frac{\theta_{(comp)}}{2}}{\sin\frac{\delta}{2}} = \frac{P}{P_0}\times\frac{232.38}{300}\times\frac{\sin\frac{22.2°}{2}}{\sin\frac{\delta}{2}}$$

$$= 0.149\left(\frac{P}{P_0}\right)\times\frac{1}{\sin\frac{\delta}{2}} \qquad \text{(Ans.)}$$

EXERCISES

1. What are the components of load compensation? What is the principle of power factor correction? Find the static shunt compensator susceptance per phase if it is connected to a load bus which is drawing a power of (80 + *j*30) kVA. The compensator is required to improve load bus power factor to 0.98 (lag). Assume the load bus voltage to be 400 V (L-L).

   [*Hints:* See Examples 15.1 and 15.2. Ans. 0.086 mho (capacitive).]

2. How would you reduce voltage regulation of a line by capacitive power compensation at the receiving end bus? Derive an expression for the required capacitive or compensating power.

3. What is load balancing? Derive the theory of compensation when delta connected mixed load is connected to real unbalanced load. Also illustrate how you will convert this unbalanced load to real balanced load.

4. What do you mean by transmission line compensation? What are active and passive compensators?

5. Assume a shunt reactor placed in the receiving end of a line. To have $|V_r| = |V_s|$, what should be the value of X_L of this reactor? What is the expression for mid-point voltage? Is it possible to generalize this concept of reactor control of line voltage for a long line in a sectionalized manner?

6. In a 3 phase 220 kV, 50 Hz overhead line three inductors are placed at equal intervals to control the Ferranti effect. Assume one reactor is placed at the sending end, another at the middle, while the rest is at the receiving end. The line being lossless and no load compensated voltage being at nominal value, find the value of these inductances used as shunt reactors. What is the maximum voltage at no load before compensation and where does it occur? Assume the series line reactance to be 50 ohms/phase while the shunt capacitive susceptance is 2×10^3 μS/ph.

   [*Hints:* See Example 15.4.] (**Ans.** 220.69 kV or 1.0031 p.u.)

7. Analytically establish that shunt inductive compensation increases the surge impedance of the line and reduces the power transfer capacity of the line, virtual line length and natural loading of the line.

8. A 250 km, lossless, 3 phase, 50 Hz, 400 kV power line transfers power of 650 MW from sending end to receiving end in a grid while the voltages at both the ends are maintained at 1.0 p.u. Assuming the line to be uncompensated, find the natural loading, power angle and mid-point voltage. Next, assume a distributed shunt capacitive compensation such that the voltage profiles remain flat at both ends as well as at the centre of the line, even if the real power demand is enhanced by 30% of the SIL value. Find the new values of the natural loading and the mid-point voltage. Assume that the line has series inductive reactance of 0.25 Ω/km and shunt capacitive susceptance of 2 microsiemens/km.

   [*Hints:* See Example 15.6. **Ans.** Without compensation, SIL = 452.55 MW, $\delta = 14.63°$ and $V_m = 0.998$ p.u. With mid-point shunt compensation, SIL = 588.32 MW, $\delta = 20.18°$ and $V_m = 1$ p.u.]

9. Develop the concept of dynamic compensation at the middle of the transmission line.

10. What is series capacitor compensation? What are the relative advantages and disadvantages? Analytically compare between the series and shunt compensation of transmission line.

11. Show that for a line where both series compensation and shunt compensation are present will have the following expression for the SIL value:

$$\text{SIL} = \text{SIL without compensation} \times \sqrt{\frac{1 - k_{Sh}}{1 - k_{se}}}$$

where k_{Sh} and k_{se} are the respective degree of shunt and series compensation.

Appendix
(Test System)

A.1 IEEE 14 BUS TEST SYSTEM

A.1.1 Line Data of IEEE 14 Bus Test System

Line no.	*From bus–To bus*	*Line impedance* (p.u.)	*Line charging susceptance* (p.u.)	*Off-nominal tap ratio of transformer*
1	1–2	0.01938 + j0.05917	0.0264	
2	2–3	0.04699 + j0.19797	0.0219	
3	2–4	0.05811 + j0.17632	0.0187	
4	1–5	0.05403 + j0.22304	0.0246	
5	2–5	0.05695 + j0.17388	0.0170	
6	3–4	0.06701 + j0.17103	0.0173	
7	4–5	0.01335 + j0.04211	0.0064	
8	7–8	0.0 + j0.17615	0.0	
9	7–9	0.0 + j0.11001	0.0	
10	9–10	0.03181 + j0.08450	0.0	
11	6–11	0.09498 + j0.19890	0.0	
12	6–12	0.22291 + j0.25581	0.0	
13	6–13	0.06615 + j0.13027	0.0	
14	9–14	0.12711 + j0.57038	0.0	
15	10–11	0.05205 + j0.19207	0.0	
16	12–13	0.022092 + j0.19988	0.0	
17	13–14	0.17093 + j0.34802	0.0	
18	4–7	0.0 + j0.20912		0.978
19	4–9	0.0 + j0.55618		0.969
20	5–6	0.0 + j0.25202		0.932

A.1.2 Bus Data of IEEE 14 Bus Test System

Bus no.	*Bus type*	*Voltage magnitude* (p.u.)	*Phase angle* (rad)	*Generation* (p.u.)		*Load* (p.u.)		*Reactive power limit* (p.u.)		*Shunt capacitor data* (p.u.)
				Active	*Reactive*	*Active*	*Reactive*	Q_{min}	Q_{max}	
1	Slack	1.06	0	1.1417	– 0.169	0	0	– 0.4	0.6	0
2	PV	1.045	?	0.4	0	0.217	0.127	– 0.4	0.6	0
3	PV	1.01	?	0	0	0.942	0.19	– 0.5	0.4	0
4	PQ	?	?	–	–	0.478	– 0.039	–	–	0
5	PQ	?	?	–	–	0.076	0.016	–	–	0
6	PV	1.07	?	0	0	0.112	0.075	– 0.06	0.3	0
7	PQ	?	?	–	–	0	0	–	–	0
8	PV	1.09	?	0	0	0	0	– 0.06	0.3	0
9	PQ	?	?	–	–	0.295	0.166	–	–	0.19
10	PQ	?	?	–	–	0.09	0.058	–	–	0
11	PQ	?	?	–	–	0.035	0.018	–	–	0
12	PQ	?	?	–	–	0.061	0.016	–	–	0
13	PQ	?	?	–	–	0.135	0.058	–	–	0
14	PQ	?	?	–	–	0.149	0.05	–	–	0

A.2 IEEE 30 BUS TEST SYSTEM

A.2.1 Line Data of IEEE 30 Bus Test System

Line no.	*From bus– To bus*	*Line impedance* (p.u.)	*Line charging susceptance* (p.u.)	*Off-nominal tap ratio of transformer*
1	1–2	0.0192 + j0.0575	0.0264	
2	1–3	0.0452 + j0.1852	0.0204	
3	2–4	0.057 + j0.1737	0.0184	
4	3–4	0.0132 + j0.0379	0.0042	
5	2–5	0.0472 + j0.1983	0.0209	
6	2–6	0.0581 + j0.1763	0.0187	
7	4–6	0.0119 + j0.0414	0.0045	
8	5–7	0.0460 + j0.1160	0.0102	
9	6–7	0.0267 + j0.082	0.0085	
10	6–8	0.012 + j0.042	0.0045	

(Contd)

Line no.	*From bus–To bus*	*Line impedance* (p.u.)	*Line charging susceptance* (p.u.)	*Off-nominal tap ratio of transformer*
11	9–11	j0.208	0	
12	9–10	j0.11	0	
13	12–13	j0.14	0	
14	12–14	0.1231 + j0.2559	0	
15	12–15	0.0662 + j0.1304	0	
16	12–16	0.0945 + j0.1987	0	
17	14–15	0.2210 + j0.1997	0	
18	16–17	0.0824 + j0.1932	0	
19	15–18	0.1070 + j0.2185	0	
20	18–19	0.0639 + j0.1292	0	
21	19–20	0.0340 + j0.068	0	
22	10–20	0.0936 + j0.209	0	
23	10–17	0.0324 + j0.0845	0	
24	10–21	0.0348 + j0.0749	0	
25	10–22	0.0727 + j0.1499	0	
26	21–22	0.0116 + j0.0236	0	
27	15–23	0.1 + j0.202	0	
28	22–24	0.115 + j0.179	0	
29	23–24	0.132 + j0.27	0	
30	24–25	0.1885 + j0.3292	0	
31	25–26	0.2544 + j0.38	0	
32	25–27	0.1093 + j0.2087	0	
33	27–29	0.2198 + j0.4153	0	
34	27–30	0.3202 + j0.6027	0	
35	29–30	0.2399 + j0.4533	0	
36	8–28	0.0636 + j0.2	0.0214	
37	6–28	0.0169 + j0.0599	0.0065	
38	6–9	j0.208		1.0155
39	6–10	j0.556		0.9629
40	4–12	j0.256		1.0129
41	28–27	j0.3960		0.9581

A.2.2 Bus Data of IEEE 30 Bus Test System

Bus no.	*Bus type*	*Voltage magnitude* (p.u.)	*Phase angle* (rad)	*Generation* (p.u.)		*Load* (p.u.)		*Reactive power limit* (p.u.)		*Shunt capacitor data* (p.u.)
				Active	*Reactive*	*Active*	*Reactive*	Q_{min}	Q_{max}	
1	Slack	1.06	0	1.8722	0	0	0	–1.50	1.5	0
2	PV	1.04	?	0.5378	0	0.217	0.127	–0.17	0.5	0
3	PQ	?	?	0	0	0.024	0.012	–	–	0
4	PQ	?	?	0	0	0.076	0.016	–	–	0
5	PV	1.0100	?	0.1696	0	0.942	0.19	–0.08	0.25	0
6	PQ	?	?	0	0	0	0	–	–	0
7	PQ	?	?	0	0	0.228	0.109	–	–	0
8	PV	1.0100	?	0.1129	0	0.3	0.30	–1.4	2.0	0
9	PQ	?	?	0	0	0	0	–0.03	0.09	0
10	PQ	?	?	0	0	0.058	0.02	–	–	0.19
11	PV	1.0820	?	0.1129	0	0	0	–	–	0
12	PQ	?	?	0	0	0.112	0.075	–0.5	1.55	0
13	PV	1.0710	?	0.1336	0	0	0	–	0	0
14	PQ	?	?	0	0	0.062	0.016	–	0	0
15	PQ	?	?	0	0	0.082	0.025	–	0	0
16	PQ	?	?	0	0	0.035	0.018	–	0	0
17	PQ	?	?	0	0	0.09	0.058	–	0	0
18	PQ	?	?	0	0	0.032	0.009	–	0	0
19	PQ	?	?	0	0	0.095	0.034	–	0	0
20	PQ	?	?	0	0	0.022	0.007	–	0	0
21	PQ	?	?	0	0	0.175	0.112	–	0	0
22	PQ	?	?	0	0	0	0	–	0	0
23	PQ	?	?	0	0	0.032	0.016	–	0	0
24	PQ	?	?	0	0	0.087	0.067	–	0	0.043
25	PQ	?	?	0	0	0		–	0	0
26	PQ	?	?	0	0	0.035	0.023	–	0	0
27	PQ	?	?	0	0	0	0	–	0	0
28	PQ	?	?	0	0	0	0	–	0	0
29	PQ	?	?	0	0	0.024	0.009	–	0	0
30	PQ	?	?	0	0	0.106	0.019	–	0	0

A.3 IEEE 57 BUS TEST SYSTEM

A.3.1 Line Data of IEEE 57 Bus Test System

Line no.	*From bus– To bus*	*Line impedance* (p.u.)	*Line charging susceptance* (p.u.)	*Off-nominal tap ratio of transformer*
1	1–2	0.0083 + j0.0280	0.0645	1.0
2	2–3	0.0298 + j0.0850	0.0409	1.0
3	3–4	0.0112 + j0.0366	0.0190	1.0
4	4–5	0.0625 + j0.1320	0.0129	1.0
5	4–6	0.0430 + j0.1480	0.0174	1.0
6	6–7	0.0200 + j0.1020	0.0138	1.0
7	6–8	0.0339 + j0.1730	0.0235	1.0
8	8–9	0.0099 + j0.0505	0.0274	1.0
9	9–10	0.0369 + j0.1679	0.0220	1.0
10	9–11	0.0258 + j0.0848	0.0109	1.0
11	9–12	0.0648 + j0.2950	0.0386	1.0
12	9–13	0.0481 + j0.1580	0.0203	1.0
13	13–14	0.0132 + j0.0434	0.0055	1.0
14	13–15	0.0269 + j0.0869	0.0115	1.0
15	1–15	0.0178 + j0.0910	0.0494	1.0
16	1–16	0.0456 + j0.2060	0.0273	1.0
17	1–17	0.0238 + j0.1080	0.0143	1.0
18	3–15	0.0162 + j0.0530	0.0272	1.0
19	4–18	j0.5550	0	0.97
20	4–18	j0.4300	0	0.978
21	5–6	0.0302 + j0.0641	0.0062	1.0
22	7–8	0.0139 + j0.0712	0.0097	1.0
23	10–12	0.0277 + j0.1262	0.0164	1.0
24	11–13	0.0223 + j0.0732	0.0094	1.0
25	12–13	0.0178 + j0.0580	0.0302	1.0
26	12–16	0.0180 + j0.0813	0.0108	1.0
27	12–17	0.0397 + j0.1790	0.0238	1.0
28	14–15	0.0171 + j0.0547	0.0074	1.0
29	18–19	0.4610 + j0.6850	0	1.0
30	19–20	0.2830 + j0.4340	0	1.0
31	21–20	j0.7767	0	1.043

Line no.	*From bus– To bus*	*Line impedance* (p.u.)	*Line charging susceptance* (p.u.)	*Off-nominal tap ratio of transformer*
32	21–22	0.0736 + j0.1170	0	1.0
33	22–23	0.0099 + j0.0152	0	1.0
34	23–24	0.1660 + j0.2560	0.0042	1.0
35	24–25	j1.0820	0.0	1.0
36	24–25	j1.2300	0.0	1.0
37	24–26	j0.0473	0.0	1.043
38	26–27	0.1650 + j0.2540	0.0	1.0
39	27–28	0.0618 + j0.0954	0.0	1.0
40	28–29	0.0418 + j0.0587	0.0	1.0
41	7–29	j0.0648	0.0	0.967
42	25–30	0.135 + j0.2020	0.0	1.0
43	30–31	0.3260 + j0.4970	0.0	1.0
44	31–32	0.5070 + j0.7550	0.0	1.0
45	32–33	0.0392 + j0.0360	0.0	1.0
46	34–32	j0.9530	0.0	0.975
47	34–35	0.0520 + j0.0780	0.0016	1.0
48	35–36	0.0430 + j0.0637	0.0008	1.0
49	36–37	0.0290 + j0.0366	0.0	1.0
50	37–38	0.0651 + j0.1009	0.0010	1.0
51	37–39	0.0239 + j0.0379	0.0	1.0
52	36–40	0.0300 + j0.0466	0.0	1.0
53	22–38	0.0192 + j0.0295	0.0	1.0
54	11–41	j0.7490	0.0	0.955
55	41–42	0.2070 + j0.3520	0.0	1.0
56	41–43	j0.4120	0.0	1.0
57	38–44	0.0289 + j0.0585	0.0010	1.0
58	15–45	j0.1042	0.0	0.955
59	14–46	j0.0735	0.0	0.90
60	46–47	0.0230 + j0.0680	0.0016	1.0
61	47–48	0.0162 + j0.0233	0.0	1.0
62	48–49	0.0834 + j0.1290	0.0024	1.0
63	49–50	0.0801 + j0.1280	0.0	1.0
64	50–51	0.1386 + j0.2200	0.0	1.0
65	10–51	j0.0712	0.0	0.93

Line no.	From bus–To bus	Line impedance (p.u.)	Line charging susceptance (p.u.)	Off-nominal tap ratio of transformer
66	13–49	j0.1910	0.0	0.895
67	29–52	0.1442 + j0.1870	0.0	1.0
68	52–53	0.0762 + j0.0984	0.0	1.0
69	53–54	0.1878 + j0.2320	0.0	1.0
70	54–55	0.1732 + j0.2265	0.0	1.0
71	11–43	j0.1530	0.0	0.958
72	44–45	0.0624 + j0.1242	0.002	1.0
73	40–56	j1.1950	0.0	0.958
74	56–41	0.5530 + j0.5490	0.0	1.0
75	56–42	0.2125 + j0.3540	0.0	1.0
76	39–57	0.0 + j1.3550	0.0	0.98
77	57–56	0.1740 + j0.2600	0.0	1.0
78	38–49	0.1150 + j0.1770	0.003	1.0
79	38–48	0.0312 + j0.0482	0.0	1.0
80	9–55	j0.1205	0.0	0.94

A.3.2 Bus Data of IEEE 57 Bus Test System

Bus no.	Bus type	Voltage magnitude (p.u.)	Phase angle (rad)	Generation (p.u.)		Load (p.u.)		Reactive power limit (p.u.)		Shunt capacitor data (p.u.)
				Active	Reactive	Active	Reactive	Q_{min}	Q_{max}	
1	Slack	1.04	0	?	?	0.55	0.17	–1.5	1.5	0
2	PV	1.01	?	0	?	0.03	0.88	–0.17	0.5	0
3	PV	0.985	?	0.4	?	0.41	0.21	–0.1	0.6	0
4	PQ	?	?	0	0	0	0	–	–	0
5	PQ	?	?	0	0	0.13	0.04	–	–	0
6	PV	0.98	?	0	0	0.75	0.02	–0.08	0.25	0
7	PQ	?	?	0	0	0	0	–	–	0
8	PV	1.005	?	4.5	?	1.5	0.22	–1.4	2	0
9	PV	0.98	?	0	?	1.21	0.26	–0.03	0.09	0
10	PQ	?	?	0	0	0.05	0.02	–	–	0
11	PQ	?	?	0	0	0	0	–	–	0
12	PV	1.015	?	3.1000	?	3.77	0.24	–0.5	1.55	0
13	PQ	?	?	0	0	0.18	0.023	–	–	0
14	PQ	?	?	0	0	0.105	0.053	–	–	0
15	PQ	?	?	0	0	0.22	0.05	–	–	0

(*Contd*)

Bus no.	*Bus type*	*Voltage magnitude* (p.u.)	*Phase angle* (rad)	*Generation* (p.u.)		*Load* (p.u.)		*Reactive power limit* (p.u.)		*Shunt capacitor data* (p.u.)
				Active	*Reactive*	*Active*	*Reactive*	Q_{min}	Q_{max}	
16	PQ	?	?	0	0	0.43	0.03	–	–	0
17	PQ	?	?	0	0	0.42	0.08	–	–	0
18	PQ	?	?	0	0	0.272	0.098	–	–	0.1
19	PQ	?	?	0	0	0.033	0.006	–	–	0
20	PQ	?	?	0	0	0.023	0.01	–	–	0
21	PQ	?	?	0	0	0	0	–	–	0
22	PQ	?	?	0	0	0	0	–	–	0
23	PQ	?	?	0	0	0.063	0.021	–	–	0
24	PQ	?	?	0	0	0	0	–	–	0
25	PQ	?	?	0	0	0.063	0.032	–	–	0.059
26	PQ	?	?	0	0	0	0	–	–	0
27	PQ	?	?	0	0	0.093	0.005	–	–	0
28	PQ	?	?	0	0	0.046	0.023	–	–	0
29	PQ	?	?	0	0	0.17	0.026	–	–	0
30	PQ	?	?	0	0	0.036	0.018	–	–	0
31	PQ	?	?	0	0	0.058	0.029	–	–	0
32	PQ	?	?	0	0	0.016	0.008	–	–	0
33	PQ	?	?	0	0	0.038	0.019	–	–	0
34	PQ	?	?	0	0	0	0	–	–	0
35	PQ	?	?	0	0	0.06	0.03	–	–	0
36	PQ	?	?	0	0	0	0	–	–	0
37	PQ	?	?	0	0	0	0	–	–	0
38	PQ	?	?	0	0	0.14	0.07	–	–	0
39	PQ	?	?	0	0	0	0	–	–	0
40	PQ	?		0	0	0	0	–	–	0
41	PQ	?	?	0	0	0.063	0.03	–	–	0
42	PQ	?	?	0	0	0.071	0.04	–	–	0
43	PQ	?	?	0	0	0.02	0.01	–	–	0
44	PQ	?	?	0	0	0.12	0.018	–	–	0
45	PQ	?	?	0	0	0	0	–	–	0
46	PQ	?	?	0	0	0	0	–	–	0
47	PQ	?	?	0	0	0.297	0.116	–	–	0
48	PQ	?	?	0	0	0	0	–	–	0

Bus no.	*Bus type*	*Voltage magnitude* (p.u.)	*Phase angle* (rad)	*Generation* (p.u.)		*Load* (p.u.)		*Reactive power limit* (p.u.)		*Shunt capacitor data* (p.u.)
				Active	*Reactive*	*Active*	*Reactive*	Q_{min}	Q_{max}	
49	PQ	?	?	0	0	0.18	0.085	–	–	0
50	PQ	?	?	0	0	0.21	0.105	–	–	0
51	PQ	?	?	0	0	0.18	0.053	–	–	0
52	PQ	?	?	0	0	0.049	0.022	–	–	0
53	PQ	?	?	0	0	0.2	0.1	–	–	0.063
54	PQ	?	?	0	0	0.041	0.014	–	–	0
55	PQ	?	?	0	0	0.068	0.034	–	–	0
56	PQ	?	?	0	0	0.076	0.022	–	–	0
57	PQ	?	?	0	0	0.067	0.02	–	–	0

Bibliography

Acha, E; Fuerte-Esquivel, C.R.; Ambriz-Pérez, H. and Angeles-Camacho, César, FACTS Modelling and Simulation in Power Networks, 1st Ed., John Wiley & Sons, Ltd. UK, 2004.

Arrillaga, J., C.P. Arnold and B.H. Harker, *Computer Modelling of Electric Power Systems*, John Wiley & Sons, Inc. (USA), 1983.

Barret, J.P., P. Bornard and B. Meyer, *Power System Simulation*, Chapman & Hall, Great Britain, 1997.

Chakrabarti, A., D.P. Kothari and A.K. Mukhopadhyay, *Performance Operation and Control of EHV Power Transmission Systems*, Wheeler Publication, India, 1995.

Cutsem, T. Van and C. Vournas, *Voltage Stability of Electric Power Systems*, Second printing, Kluwer Academic Publishers, Boston/London/Dordrecht, 2001.

Elegard, O.I., *Electric Energy Systems Theory: An Introduction*, 2nd ed., Tata McGraw-Hill, New Delhi, 1983.

Grainger, J.H. and W.D. Stevenson Jr., *Power System Analysis*, International edition, McGraw-Hill International Editions (USA), 1994.

Hingorani, N.G. and L. Gyugi, *Understanding FACTS* (*concepts and technology of flexible ac transmission* systems), Wiley Interscience (USA), 2000.

Kothari, D.P. and I.J. Nagrath, *Modern Power System Analysis*, 3rd ed., Tata McGraw-Hill, New Delhi, 2003.

Kothari, D.P. and J.S. Dhillon, *Power System Optimization*, Prentice-Hall of India, New Delhi, 2004.

Kusic, G.L., *Computer Aided Power System Analysis*, Prentice-Hall of India, New Delhi, 1989.

Mahalanabis, A.K., D.P. Kothari and S.I. Ahson, *Computer Aided Power System Analysis and Control*, Tata McGraw-Hill, New Delhi, 1988.

Mariani, E. and S.S. Murthy, *Advanced Load Dispatch for Power Systems—Principles, Practices and Economics*, Springer-Verlag, London, 1997.

Mariani, E. and S.S. Murthy, *Control of Modern Integrated Power Systems*, Springer-Verlag, London, 1997.

Mathur, Mohan R. and Rajiv K. Varma, *Thyristor-based FACTS Controllers for Electrical Transmission Systems*, IEEE Press, Wiley Interscience (USA), 2002.

Nagrath, I.J. and D.P. Kothari, *Power System Engineering*, Tata McGraw-Hill, New Delhi, 1994.

Padiyar, K.R., *Power System Dynamics: Stability and Control*, 2nd ed., BS Publications, Hyderabad, 2002.

Sauer, P.W. and M.A. Pai, *Power System Dynamics and Stability*, 1st Indian reprint, Pearson Education, Asia, 2002.

Singh, L.P., *Advanced Power System Analysis and Dynamics*, 3rd ed., Wiley Eastern, India, 1992.

Taylor, C.W., *Power System Voltage Stability*, (EPSR), McGraw-Hill Inc. Edition (USA), 1994.

Wood, A.J. and B.F. Woolenberg, *Power Generation Operation and Control*, 2nd ed., John Wiley & Sons, Inc. (USA).

Index